Lecture Notes in
Compute

Lecture Notes in Computer Science

Vol. 1: GI-Gesellschaft für Informatik e.V. 3. Jahrestagung, Hamburg, 8.–10. Oktober 1973. Herausgegeben im Auftrag der Gesellschaft für Informatik von W. Brauer. XI, 508 Seiten. 1973.

Vol. 2: GI-Gesellschaft für Informatik e.V. 1. Fachtagung über Automatentheorie und Formale Sprachen, Bonn, 9.–12. Juli 1973. Herausgegeben im Auftrag der Gesellschaft für Informatik von K.-H. Böhling und K. Indermark. VII, 322 Seiten. 1973.

Vol. 3: 5th Conference on Optimization Techniques, Part I. (Series: I.F.I.P. TC7 Optimization Conferences.) Edited by R. Conti and A. Ruberti. XIII, 565 pages. 1973.

Vol. 4: 5th Conference on Optimization Techniques, Part II. (Series: I.F.I.P. TC7 Optimization Conferences.) Edited by R. Conti and A. Ruberti. XIII, 389 pages. 1973.

Vol. 5: International Symposium on Theoretical Programming. Edited by A. Ershov and V. A. Nepomniaschy. VI, 407 pages. 1974.

Vol. 6: B. T. Smith, J. M. Boyle, J. J. Dongarra, B. S. Garbow, Y. Ikebe, V. C. Klema, and C. B. Moler, Matrix Eigensystem Routines – EISPACK Guide. XI, 551 pages. 2nd Edition 1974. 1976.

Vol. 7: 3. Fachtagung über Programmiersprachen, Kiel, 5.–7. März 1974. Herausgegeben von B. Schlender und W. Frielinghaus. VI, 225 Seiten. 1974.

Vol. 8: GI-NTG Fachtagung über Struktur und Betrieb von Rechensystemen, Braunschweig, 20.–22. März 1974. Herausgegeben im Auftrag der GI und der NTG von H.-O. Leilich. VI, 340 Seiten. 1974.

Vol. 9: GI-BIFOA Internationale Fachtagung: Informationszentren in Wirtschaft und Verwaltung. Köln, 17./18. Sept. 1973. Herausgegeben im Auftrag der GI und dem BIFOA von P. Schmitz. VI, 259 Seiten. 1974.

Vol. 10: Computing Methods in Applied Sciences and Engineering, Part 1. International Symposium, Versailles, December 17–21, 1973. Edited by R. Glowinski and J. L. Lions. X, 497 pages. 1974.

Vol. 11: Computing Methods in Applied Sciences and Engineering, Part 2. International Symposium, Versailles, December 17–21, 1973. Edited by R. Glowinski and J. L. Lions. X, 434 pages. 1974.

Vol. 12: GFK-GI-GMR Fachtagung Prozessrechner 1974. Karlsruhe, 10.–11. Juni 1974. Herausgegeben von G. Krüger und R. Friehmelt. XI, 620 Seiten. 1974.

Vol. 13: Rechnerstrukturen und Betriebsprogrammierung, Erlangen, 1970. (GI-Gesellschaft für Informatik e.V.) Herausgegeben von W. Händler und P. P. Spies. VII, 333 Seiten. 1974.

Vol. 14: Automata, Languages and Programming – 2nd Colloquium, University of Saarbrücken, July 29–August 2, 1974. Edited by J. Loeckx. VIII, 611 pages. 1974.

Vol. 15: L Systems. Edited by A. Salomaa and G. Rozenberg. VI, 338 pages. 1974.

Vol. 16: Operating Systems, International Symposium, Rocquencourt 1974. Edited by E. Gelenbe and C. Kaiser. VIII, 310 pages. 1974.

Vol. 17: Rechner-Gestützter Unterricht RGU '74, Fachtagung, Hamburg, 12.–14. August 1974, ACU-Arbeitskreis Computer-Unterstützter Unterricht. Herausgegeben im Auftrag der GI von K. Brunnstein, K. Haefner und W. Händler. X, 417 Seiten. 1974.

Vol. 18: K. Jensen and N. E. Wirth, PASCAL – User Manual and Report. VII, 170 pages. Corrected Reprint of the 2nd Edition 1976.

Vol. 19: Programming Symposium. Proceedings 1974. V, 425 pages. 1974.

Vol. 20: J. Engelfriet, Simple Program Schemes and Formal Languages. VII, 254 pages. 1974.

Vol. 21: Compiler Construction, An Advanced Course. Edited by F. L. Bauer and J. Eickel. XIV. 621 pages. 1974.

Vol. 22: Formal Aspects of Cognitive Processes. Proceedings 1972. Edited by T. Storer and D. Winter. V, 214 pages. 1975.

Vol. 23: Programming Methodology. 4th Informatik Symposiu IBM Germany Wildbad, September 25–27, 1974. Edited by C Hackl. VI, 501 pages. 1975.

Vol. 24: Parallel Processing. Proceedings 1974. Edited by T. F VI, 433 pages. 1975.

Vol. 25: Category Theory Applied to Computation and Control. ceedings 1974. Edited by E. G. Manes. X, 245 pages. 1975.

Vol. 26: GI-4. Jahrestagung, Berlin, 9.–12. Oktober 1974. H ausgegeben im Auftrag der GI von D. Siefkes. IX, 748 Seit 1975.

Vol. 27: Optimization Techniques. IFIP Technical Conferen Novosibirsk, July 1–7, 1974. (Series: I.F.I.P. TC7 Optimizat Conferences.) Edited by G. I. Marchuk. VIII, 507 pages. 19

Vol. 28: Mathematical Foundations of Computer Science. Symposium at Jadwisin near Warsaw, June 17–22, 1974. Edi by A. Blikle. VII, 484 pages. 1975.

Vol. 29: Interval Mathematics. Procedings 1975. Edited by K. Ni VI, 331 pages. 1975.

Vol. 30: Software Engineering. An Advanced Course. Edited F. L. Bauer. (Formerly published 1973 as Lecture Notes in E nomics and Mathematical Systems, Vol. 81) XII, 545 pages. 19

Vol. 31: S. H. Fuller, Analysis of Drum and Disk Storage Units. 283 pages. 1975.

Vol. 32: Mathematical Foundations of Computer Science 19 Proceedings 1975. Edited by J. Bečvář. X, 476 pages. 1975.

Vol. 33: Automata Theory and Formal Languages, Kaiserslaut May 20–23, 1975. Edited by H. Brakhage on behalf of GI. 292 Seiten. 1975.

Vol. 34: GI – 5. Jahrestagung, Dortmund 8.–10. Oktober 19 Herausgegeben im Auftrag der GI von J. Mühlbacher. X, 755 Sei 1975.

Vol. 35: W. Everling, Exercises in Computer Systems Analy (Formerly published 1972 as Lecture Notes in Economics Mathematical Systems, Vol. 65) VIII, 184 pages. 1975.

Vol. 36: S. A. Greibach, Theory of Program Structures: Schem Semantics, Verification. XV, 364 pages. 1975.

Vol. 37: C. Böhm, λ-Calculus and Computer Science Theory. ceedings 1975. XII, 370 pages. 1975.

Vol. 38: P. Branquart, J.-P. Cardinael, J. Lewi, J.-P. Delesca M. Vanbegin. An Optimized Translation Process and Its Applicat to ALGOL 68. IX, 334 pages. 1976.

Vol. 39: Data Base Systems. Proceedings 1975. Edited by H. Hass meier and W. Spruth. VI, 386 pages. 1976.

Vol. 40: Optimization Techniques. Modeling and Optimization in Service of Man. Part 1. Proceedings 1975. Edited by J. Cea. 854 pages. 1976.

Vol. 41: Optimization Techniques. Modeling and Optimization in Service of Man. Part 2. Proceedings 1975. Edited by J. Cea. 852 pages. 1976.

Vol. 42: James E. Donahue, Complementary Definitions of P gramming Language Semantics. VII, 172 pages. 1976.

Vol. 43: E. Specker und V. Strassen, Komplexität von Entscheidun problemen. Ein Seminar. V, 217 Seiten. 1976.

Vol. 44: ECI Conference 1976. Proceedings 1976. Edited b Samelson. VIII, 322 pages. 1976.

Vol. 45: Mathematical Foundations of Computer Science 19 Proceedings 1976. Edited by A. Mazurkiewicz. XI, 601 pages. 19

Vol. 46: Language Hierarchies and Interfaces. Edited by F. L. Ba and K. Samelson. X, 428 pages. 1976.

Vol. 47: Methods of Algorithmic Language Implementation. Ed by A. Ershov and C. H. A. Koster. VIII, 351 pages. 1977.

Vol. 48: Theoretical Computer Science, Darmstadt, March 19 Edited by H. Tzschach, H. Waldschmidt and H. K.-G. Walter behalf of GI. VIII, 418 pages. 1977.

Lecture Notes in Computer Science

Edited by G. Goos and J. Hartmanis

74

Mathematical Foundations of Computer Science 1979

Proceedings, 8th Symposium,
Olomouc, Czechoslovakia, September 3–7, 1979

Edited by J. Bečvář

Springer-Verlag
Berlin Heidelberg New Yor[redacted] 19[redacted]9

Editor
Jiří Bečvář
Mathematical Institute
Czechoslovak Academy of Sciences
Žitná 25
115 67 Prague 1/Czechoslovakia

AMS Subject Classifications (1970): 02E10, 02E15, 02F10, 02F15, 68A05, 68A20, 68A25, 68A30, 68A45, 68A50
CR Subject Classifications (1974):

ISBN 3-540-09526-8 Springer-Verlag Berlin Heidelberg New York
ISBN 0-387-09526-8 Springer-Verlag New York Heidelberg Berlin

Printed in Germany

Printing and binding: Beltz Offsetdruck, Hemsbach/Bergstr.
2145/3140-543210

MFCS '79

FOREWORD

This volume contains the papers which were selected for presentation at the symposium on Mathematical Foundations of Computer Science - MFCS´79, held in Olomouc, Czechoslovakia, September 3 - 7, 1979.

The symposium is the eighth in a series of annual international meetings which take place alternately in Czechoslovakia and Poland. It has been organized by the Mathematical Institute of the Czechoslovak Academy of Sciences, Prague, the Faculty of Mathematics and Physics of Charles University, Prague, and the Faculty of Natural Sciences of Palacký University, Olomouc, in co-operation with the Federal Ministry for Technical and Investment Development, the Technical University, Prague, the Computing Research Centre, Bratislava, the Faculty of Natural Sciences of Komenský University, Bratislava, and the Faculty of Natural Sciences of Šafárik University, Košice.

The articles in these Proceedings include invited papers and short communications. The latter were selected from among 95 extended abstracts submitted in response to the call for papers. Selection was made on the basis of originality and relevance to theoretical computer science by the following Program Committee: J. Bečvář /Chairman/, J. Gruska, P. Hájek, M. Chytil, J. Král, M. Novotný, B. Rovan. A number of referees helped the Program Committee in the evaluation of the abstracts.

The papers included in these Proceedings were not formally refereed. It is anticipated that most of them will appear in a polished and completed from in scientific journals.

The organizers of the symposium are much indebted to all those who contributed to the program, in particular to the authors of the papers. Special thanks are due to the referees of the abstracts. Thanks are also due to all the above mentioned co-operating institutions for their valuable assistance and support, and to all the persons who helped in organizing the symposium. The Organizing Committee consisted of J. Bečvář, J. Gregor, J. Gruska, P. Hájek, I. Havel, Š. Hudák, M. Chytil, J. Král, F. Krutský, B. Miniberger, M. Novotný, A. Rázek, Z. Renc, B. Rovan, and M. Vlach /Chairman/. The Program Chairman acknowledges with gratitude the extensive assistance of I.M. Havel, P. Pudlák, and S. Žák in editing this volume.

The organizers of the symposium wish to express their thanks to the representatives of the Palacký University in Olomouc for their support and interest in the symposium.

Finally, the help of the Springer-Verlag in the timely publication of this volume is highly appreciated.

Prague, May 1979 Jiří Bečvář

CONTENTS

Invited lectures

Communications

Appendix *

* Manuscript received too late to be placed correctly in the alphabetic listing.

A SOUND AND COMPLETE PROOF SYSTEM FOR
PARTIAL PROGRAM CORRECTNESS

J.W. de Bakker
Mathematical Centre
2e Boerhaavestraat 49, Amsterdam

1. *Introduction*

We investigate soundness and completeness of a proof system dealing with partial correctness of programs in a language with assignment, composition, conditionals, block structure, subscripted variables and (possibly recursive) procedures with the parameter mechanisms of call-by-value and call-by-address (call-by-variable in PASCAL, call-by-reference in FORTRAN). The paper is a continuation of Apt & de Bakker [3] presented at MFCS '76, and of its successor Apt & de Bakker [4]. In the meantime various problems not yet well-understood at that time have been pursued further and, we hope, solved.

Section 2 presents syntax and (denotational) semantics of our language; in section 3 we are confronted with an unpleasant consequence of our way of defining the semantics of a block b new x; S e, and propose as solution to restrict our correctness considerations to programs obeying the restriction that all such local variables x be *initialized*. Section 4 introduces the proof system; in the course of trying to prove its soundness we were somewhat shocked by the discovery that essential rules such as, for example, the familiar composition rule turned out to be invalid with respect to the natural validity definition, requiring a complicated refinement of that definition to remedy the situation. Section 5, finally, discusses the completeness of the system.

All proofs are omitted in this paper; they are scheduled to appear in a forthcoming publication.

Our paper could not have been written without Apt [1]. Though the technical results differ (e.g., [1] does not treat parameter mechanisms, nor does it impose the initialization requirement), there are many similarities, in particular concerning the validity definition and soundness proof. Also, the problem signalled at the beginning of section 3 was found by K.R. Apt. Various other approaches to the topic of our paper have appeared in the literature (Cartwright & Oppen [6], Clarke [7], Cook [8], Gorelick [9], to mention only a small selection; for a detailed survey see Apt [2]). However, out treatment of both soundness and completeness of the proposed proof system differs substantially from the techniques used elsewhere; in particular, we have not encountered any analogue of our validity definition in its refined form.

Acknowledgement. As should be clear from the above, our paper owes a lot to the work of K.R. Apt. J.I. Zucker contributed many helpful comments on a previous version.

2. *Syntax and semantics*

Convention. By writing "Let $(\alpha\in)V$ be the set such that ... "we introduce a set V, with variable α ranging over V, such that ... .

2.1. *Syntax.* "≡" denotes identity between two syntactic objects. Let $(n,m,\in)$*Icon* be the set of integer constants, let $(x,y,z,u\in)$*Svar*, $(a\in)$*Avar*, $(P,Q\in)$*Pvar* be the (infinite, well-ordered) sets of *simple variables* (s.v.), *array variables* and *procedure variables*.

Let $(v,w\in)$*Ivar* be the set of *integer variables* defined by $v ::= x \mid a[s]$

... $(s,t\in)$*Iexp* ...	*integer expressions* ...	$s ::= n \mid v \mid s_1+s_2 \mid$ if b then s_1 else s_2 fi
$(b\in)$*Bexp*	*boolean expressions*	$b ::=$ true | false | $s_1=s_2 \mid \neg b \mid b_1 \supset b_2$
$(S\in)$*Stat*	*statements*	$S ::= v:=s \mid S_1;S_2 \mid$ if b then S_1 else S_2 fi | b new x; S e | P(t,v)
$(D\in)$*Decl*	*declarations*	$D ::= P_1 \Leftarrow B_1, \ldots, P_n \Leftarrow B_n$, $n \geq 0$, and, for $1 \leq i, j \leq n$, $(P_i \equiv P_j) \Rightarrow (i=j)$
$(B\in)$*Pbod*	*procedure bodies*	$B ::=$ <val x, add y:S>, $x \not\equiv y$
$(R\in)$*Prog*	*programs*	$R ::= \langle D:S \rangle$
$(p,q,r\in)$*Assn*	*assertions*	$p ::=$ true | false | $s_1=s_2 \mid \neg p \mid p_1 \supset p_2 \mid \exists x[p]$
$(f\in)$*Form*	*correctness formulae* (c.f.)	$f ::= p \mid \{p\}S\{q\} \mid f_1 \wedge f_2$
$(g\in)$*Gfor*	*generalized c.f.*	$g ::= \langle D: f_1 \Rightarrow f_2 \rangle$

Remarks.

1. We write $\langle D|S \rangle$ instead of $\langle D:S \rangle$, and similarly in B and g. If $R \equiv \langle D|S \rangle$, with $D \equiv \langle P_i \Leftarrow B_i \rangle_{i=1}^{n}$, and all procedure variables occurring in S or any of the B_i, $i = 1,\ldots,n$, are included in $\{P_1,\ldots,P_n\}$, we call R *closed*.
2. Our statements have local s.v. declarations, but, for simplicity's sake, no local array or procedure declarations, nor array *bounds*.
3. In $B \equiv$ <val x, add y|S>, x is the formal value parameter and y the formal address parameter, and, in P(t,v), t is the actual value par. and y the actual address par. (cf. also the definition of syntactic application in 2.2).
4. $\langle D|f \rangle$ is short for $\langle D|$true$\Rightarrow f\rangle$.
5. For the intended meaning of "⇒" in a formula g, cf. the remark on the validity of $\langle D|f_1 \Rightarrow f_2 \rangle$ versus the soundness of $\frac{\langle D|f_1\rangle}{\langle D|f_2\rangle}$ below.

2.2. *Substitution and syntactic application*

Substitution is denoted by [···/...], e.g. we use notations such as s[t/x], p[t/x], S[v/x], S[a'/a], S[Q/P], etc. In case a construct contains a variable binding operator (in $\exists x[p]$ and b new x; S e, occurrences of x in p and S are bound) the usual precautions preventing clashes between free and bound variables apply. A notation such as

$s[y_i/x_i]_{i=1}^n$ implies that, for $1 \le i, j \le n$, $(x_i \equiv x_j) \Rightarrow (i=j)$. Substitution in a procedure call only affects its parameters (i.e., $P(t,v)[w/x] \equiv P(t[w/x],v[w/x])$), but not the procedure body (possibly) associated with P in the accompanying declaration D. *svar*(s), *svar*(p), *avar*(S), *pvar*(f), etc., denote the set of all *free* simple variables of s, of p, all array variables of S, all procedure variables of f, etc. Note that *svar*(<<u>val</u> x, <u>add</u> y|S>) = *svar*(S)\{x,y}. Notations such as *svar*(D,p,S,q) should be clear. We also employ the substitution s[w/v] etc., for the definition of which we refer to de Bakker [5]. Constructs which differ at most in their bound variables are called *congruent*. The congruence relation is denoted by "$\cong$".

Syntactic application is a technique of associating with a procedure body B and two actual parameters t, v, a piece of program text B(t,v) such that, for B the body of procedure P, B(t,v) embodies the meaning of P(t,v) according to the customary semantics of the parameter mechanism of call-by-value and call-by-address: let B ≡ <<u>val</u> x, <u>add</u> y|S>.

(i) v ≡ z. B(t,z) ≡ <u>b</u> <u>new</u> u; u:=t; S[u/x][y/z]<u>e</u>, where u is the first s.v. not in *svar*(x,y,z,t,S)

(ii) v ≡ a[s]. B(t,a[s]) ≡ <u>b</u> <u>new</u> u_1,u_2; u_1:=t; u_2:=s; S[u_1/x][a[u_2]/y]<u>e</u>, where u_1(u_2) is the first (second) s.v. not in *svar*(x,y,s,t,S)

2.3. *Domains*

A *cpo*$(x\in)C$ is a partially ordered set with least element $\perp_C$ such that each (ascending) chain $\langle x_i\rangle_{i=0}^\infty$ has a lub $\bigsqcup_i x_i$. Let C_1, C_2 be cpo's. If f_1,f_2: $C_1 \to C_2$, we put $f_1 \sqsubseteq f_2$ iff $f_1(x) \sqsubseteq f_2(x)$, all $x \in C_1$. For f: $C_1 \to C_2$, we call f *monotonic* if $x_1 \sqsubseteq x_2 \Rightarrow f(x_1) \sqsubseteq f(x_2)$. A monotonic function is called *continuous* if, for each chain $\langle x_i\rangle_i$, $f(\bigsqcup_i x_i) = \bigsqcup_i f(x_i)$. Each continuous function f: C → C has a *least fixed point* $\mu f = \bigsqcup_i f^i(\perp_C)$.

Let V_0 be the set of integers, W_0 = {tt,ff} the set of truth-values, and E_0 the (infinite, well-ordered) set of addresses. Let $(\alpha\in)V = V_0 \cup \{\perp_V\}$, $(\beta\in)W = W_0 \cup \{\perp_W\}$, $(e\in)E = E_0 \cup \{\perp_E\}$, with $\alpha_1 \sqsubseteq \alpha_2$ iff $\alpha_1 = \perp_V$ or $\alpha_1 = \alpha_2$, and similarly for β, e. Let $(\xi\in)Intv = Svar \cup (Avar \times V_0)$ be the set of *intermediate variables*, and let *intv*(s) = *svar*(s) ∪ {<a,α> | a ∈ *avar*(s), α ∈ V_0}, etc.. Let $(\varepsilon\in)Env = Intv \to E$ be the set of *environments* which are required to satisfy: let dom(ε) = {ξ ∈ *Intv* | ε(ξ) ≠ $\perp_E$}, range(ε) = {e ∈ E | ε(ξ) = e for some ξ ∈ dom(ε)}. Then (i) ε is 1-1 on its domain (ii) ε(a,α) ≠ $\perp_E$ for some α ∈ V_0 ⇔ ε(a,α) ≠ $\perp_E$ for all α ∈ V_0 (iii) *Intv*\dom(ε) ≠ ∅, E_0\range(ε) ≠ ∅. Let, for y ∉ dom(ε), e ∈ E_0\range(ε), ε' $\stackrel{\text{df.}}{=}$ ε ∪ <y,e> denote the extension of ε such that ε'(y) = e, and similarly for $\varepsilon \cup \langle y_i,e_i\rangle_{i=1}^n$, $\varepsilon \cup \langle\langle a,\alpha\rangle, e_\alpha\rangle_{\alpha\in V_0}$, etc. Let $(\sigma\in)\Sigma = (E \to V) \cup \{\perp_\Sigma\}$ be the set of *states*, where, for σ ∈ E → V, $\sigma(e) = \perp_V$ iff $e = \perp_E$, and $\perp_\Sigma$ (⊥, for short) $\stackrel{\text{df.}}{=}$ $\lambda e\cdot\perp_V$. Let σ{α/$\bar{e}$} $\stackrel{\text{df.}}{=}$ ⊥, if σ = ⊥, and λe. <u>if</u> $\bar{e}$ = e <u>then</u> α <u>else</u> σ($\bar{e}$) <u>fi</u>, otherwise. Let $(\phi\in)M \stackrel{\text{df.}}{=} Env \to (\Sigma\to\Sigma)$, $(\eta\in)H \stackrel{\text{df.}}{=} Iexp \times Ivar \to M$, $(\gamma\in)\Gamma \stackrel{\text{df.}}{=} Pvar \to H$, and let $\gamma\{\eta_i/P_i\}_i$ be defined similarly to σ{α/e}. Let, finally $(\Phi\in)H^n \to H$.

2.4. *Semantics*

The functions R: *Iexp* $\to (\Sigma\to V)$, L: *Ivar* $\to (\Sigma\to E)$, W: *Bexp* $\to (\Sigma\to W)$, N: *Stat* $\to (\Sigma\to M)$, M: *Prog* $\to (\Sigma\to M)$, T: *Assn* $\to (Env\to(\Sigma\to\{tt,ff\}))$, F: *Form* $\to (\Gamma\to(Env\to(\Sigma\to\{tt,ff\})))$, and G: *Gfor* $\to (\Gamma\to\{tt,ff\})$ are defined by

a. If $intv(s) \not\subseteq dom(\varepsilon)$ or $\sigma = \bot$ then $R(s)(\varepsilon)(\sigma) = \bot_V$. Otherwise, $R(v)(\varepsilon)(\sigma) = \sigma(L(v)(\varepsilon)(\sigma))$, $R(m)(\varepsilon)(\sigma) = \alpha$, where α is the integer denoted by the integer constant m, $R(s_1+s_2)(\varepsilon)(\sigma) = R(s_1)(\varepsilon)(\sigma) + R(s_2)(\varepsilon)(\sigma)$, $R(\underline{if}\ b\ \underline{then}\ s_1\ \underline{else}\ s_2\ \underline{fi})(\varepsilon)(\sigma) = \underline{if}\ W(b)(\varepsilon)(\sigma)\ \underline{then}\ R(s_1)(\varepsilon)(\sigma)\ \underline{else}\ R(s_2)(\varepsilon)(\sigma)\ \underline{fi}$.

b. If $intv(v) \not\subseteq dom(\varepsilon)$ or $\sigma = \bot$ then $L(v)(\varepsilon)(\sigma) = \bot_E$. Otherwise, $L(x)(\varepsilon)(\sigma) = \varepsilon(x)$, $L(a[s])(\varepsilon)(\sigma) = \varepsilon(a,R(s)(\varepsilon)(\sigma))$.

c. $W(b)$. Omitted.

d. If $intv(S) \not\subseteq dom(\varepsilon)$ or $\sigma = \bot$ then $N(S)(\gamma)(\varepsilon)(\sigma) = \bot$. Otherwise, $N(v{:=}s)(\gamma)(\varepsilon)(\sigma) = \sigma\{R(s)(\varepsilon)(\sigma)/L(v)(\varepsilon)(\sigma)\}$, $N(S_1;S_2)(\gamma)(\varepsilon)(\sigma) = N(S_2)(\gamma)(\varepsilon)(N(S_1)(\gamma)(\varepsilon)(\sigma))$, $N(\underline{if}\ \ldots\ \underline{fi}) = \ldots$, $N(\underline{b}\ \underline{new}\ x;\ S\underline{e})(\gamma)(\varepsilon)(\sigma) = N(S[y/x])(\gamma)(\varepsilon\cup\langle y,e\rangle)(\sigma)$, where y is the first s.v. not in $dom(\varepsilon)$ and e is the first address in $E_0\backslash range(\varepsilon)$. (*Remark*. The use of a new s.v. y ensures that we obtain the *static scope* rule for procedures.) $N(P(t,v))(\gamma)(\varepsilon)(\sigma) = \gamma(P)(t,v)(\varepsilon)(\sigma)$.

e. If $intv(R) \not\subseteq dom(\varepsilon)$ or $\sigma = \bot$ then $M(R)(\gamma)(\varepsilon)(\sigma) = \bot$. Otherwise, let $R \equiv \langle D|S\rangle$, $D \equiv \langle P_i \Leftarrow B_i\rangle_{i=1}^n$. $M(R)(\gamma)(\varepsilon)(\sigma) = N(S)(\gamma\{\eta_i/P_i\}_{i=1}^n)(\varepsilon)(\sigma)$, where $\langle\eta_1,\ldots,\eta_n\rangle = \mu[\Phi_1,\ldots,\Phi_n]$, and, for $j = 1,\ldots,n$, $\Phi_j = \lambda\eta_1'\cdot\cdots\cdot\lambda\eta_n'\cdot\lambda t\cdot\lambda v,\ N(B_j(t,v))(\gamma\{\eta_i'/P_i\}_i)$.

f. If $intv(p) \not\subseteq dom(\varepsilon)$ or $\sigma = \bot$ then $T(p)(\varepsilon)(\sigma) = ff$. Otherwise, $T(\underline{true})(\varepsilon)(\sigma) = tt,\ldots$, $T(\exists x[p])(\varepsilon)(\sigma) = \exists\alpha[T(p[y/x])(\varepsilon\cup\langle y,e\rangle)(\sigma\{\alpha/e\})]$, with $\langle y,e\rangle$ as in part d.

g. If $intv(f) \not\subseteq dom(\varepsilon)$ or $\sigma = \bot$ then $F(f)(\gamma)(\varepsilon)(\sigma) = ff$. Otherwise, $F(p)(\gamma)(\varepsilon)(\sigma) = T(p)(\varepsilon)(\sigma)$, $F(\{p\}S\{q\})(\gamma)(\varepsilon)(\sigma) = \forall\sigma'[T(p)(\varepsilon)(\sigma)\wedge\sigma' = N(S)(\gamma)(\varepsilon)(\sigma)\wedge\sigma'\neq\bot \Rightarrow T(q)(\varepsilon)(\sigma')]$, $F(f_1\wedge f_2)(\gamma)(\varepsilon)(\sigma) = F(f_1)(\gamma)(\varepsilon)(\sigma)\wedge F(f_2)(\gamma)(\varepsilon)(\sigma)$.

h. Let $g \equiv \langle D|f_1 \Rightarrow f_2\rangle$, with $D \equiv \langle P_i \Leftarrow B_i\rangle_{i=1}^n$. Let $\bar\gamma = \gamma\{\eta_i/P_i\}_{i=1}^n$, with η_i as in part e. $G(g)(\gamma) = [\forall\varepsilon$ such that $intv(D,f_1) \subseteq dom(\varepsilon)$, $\sigma\neq\bot\ [F(f_1)(\bar\gamma)(\varepsilon)(\sigma)] \Rightarrow \forall\varepsilon$ such that $intv(D,f_2) \subseteq dom(\varepsilon)$, $\sigma\neq\bot\ [F(f_2)(\bar\gamma)(\varepsilon)(\sigma)]]$.

Validity and soundness (first definition, to be modified below).

a. $\models g$ (g is valid) iff $G(g)(\gamma) = tt$ for all $\gamma\in\Gamma$

b. An *inference* $\frac{g_1,\ldots,g_n}{g}$ is called *sound* whenever $\models g_1,\ldots,\models g_n$ implies $\models g$.

Remark. Observe the difference between the validity of $\langle D|f_1\Rightarrow f_2\rangle$ and soundness of $\frac{\langle D|f_1\rangle}{\langle D|f_2\rangle}$. Putting (i) $\overset{df.}{=\!=}$ $\forall\varepsilon$ such that $intv(D,f_i) \subseteq dom(\varepsilon)$, all $\sigma\neq\bot\ [F(f_i)(\gamma)(\varepsilon)(\sigma)]$, $i = 1,2$, we have that the former corresponds to $\forall\gamma[(1)\Rightarrow(2)]$, whereas the latter corresponds to the *weaker* fact that $\forall\gamma[(1)] \Rightarrow \forall\gamma[(2)]$.

2.5. *Lemmas*. A number of lemmas stating properties of our various constructs will be used below. First we have a lemma relating substitution to state modification.

<u>LEMMA 2.1.</u>

a. If $intv(s,t,v) \subseteq dom(\varepsilon)$ then $R(s[t/v])(\varepsilon)(\sigma) = R(s)(\varepsilon)(\sigma\{R(t)(\varepsilon)(\sigma)/L(v)(\varepsilon)(\sigma)\})$

b. Similarly for $b \in Bexp$, $p \in Assn$.

END 2.1.

Next, we have a useful property of *closed* programs, asserting that such programs only affect the values of variables occurring in them:

<u>LEMMA 2.2.</u> If $<D|S>$ is closed, $\gamma \in \Gamma$, $\bar{\gamma}$ as usual (cf. 2.4h) then, if (i) $intv(D,S) \subseteq dom(\varepsilon)$, (ii) $\xi \in dom(\varepsilon)\backslash intv(D,S)$, (iii) $\sigma' = N(S)(\bar{\gamma})(\varepsilon)(\sigma)$, $\sigma' \neq \bot$, then, (iv) $\sigma'(\varepsilon(\xi)) = \sigma(\varepsilon(\xi))$

END 2.2.

The last lemma is rather technical, and foreshadows a property of statements to be discussed in section 3. *Notation.* For $\delta \subseteq Intv$, $(\sigma\circ\varepsilon)|\delta$ denotes the function composed of σ and ε restricted to δ.

<u>LEMMA 2.3.</u>

a. Let $m,n \geq 0$. If (i) $intv(s)\backslash\{x_i\}_{i=1}^{n}\backslash\{<a_j,\alpha>_{\alpha\in V_0}\}_{j=1}^{m} \subseteq \delta \subseteq dom(\varepsilon) \cap dom(\bar{\varepsilon})$, (ii) $(\sigma\circ\varepsilon)|\delta = (\bar{\sigma}\circ\bar{\varepsilon})|\delta$, (iii) For $i = 1,\ldots,n$, either $\sigma(e_i) = \bar{\sigma}(\bar{e}_i)$, or $x_i \notin svar(s)$, and, for $j = 1,\ldots,m$, either, for all $\alpha \in V_0$, $\sigma(e_{\alpha,j}) = \bar{\sigma}(\bar{e}_{\alpha,j})$, or $a_j \notin avar(s)$, then, (iv) $R(s[y_i/x_i]_i[a'_j/a_j]_j)(\varepsilon \cup <y_i,e_i>_i \cup <<<a'_j,\alpha>,e_{\alpha,j}>_\alpha>_j)(\sigma) = R(s[z_i/x_i]_i[a''_j/a_j]_j)$ $(\bar{\varepsilon} \cup <z_i,\bar{e}_i>_i \cup <<<a''_j,\alpha>,\bar{e}_{\alpha,j}>_\alpha>_j)(\bar{\sigma})$

b. Similarly for $b \in Bexp$ and $p \in Assn$.

END 2.3.

3. *Initialization*

The validity definition as given in 2.4 is, though rather natural, not satisfactory for our purposes. First, it implies the validity of formulae such as (*): <|{<u>true</u>} <u>b</u> <u>new</u> x;x:=0 <u>e</u>; <u>b</u> <u>new</u> x;y:=x <u>e</u>{y=0}>, or (**): <|{<u>true</u>} <u>b</u> <u>new</u> x;y:=x <u>e</u>; <u>b</u> <u>new</u> x;z:=x <u>e</u>{y=z}>. The source of this problem is that our semantics is overspecified in that, when declaring a new local s.v., we want its initial value to be some *arbitrary* integer. Now in def. 2.4d, we take for this the value stored at the first free address and, in a situation such as (*), upon entry of the second block we find, as aftereffect of the first block, 0 stored at this address. ((**) can be explained similarly.) A solution to this problem is *either* to change the semantics (ensuring by some flag-mechanism that no address is ever used twice as first free address), which we do not adopt mainly because of severe technical complications, *or* to restrict our correctness considerations to programs in which all local s.v. are initialized. The second solution is the one elaborated below (also motivated by the idea that the correctness of programs containing uninitialized local s.v. is probably not very interesting anyway). A second problem with the validity definition is the following: For reasons to be explained below we have to consider in a formal correctness proof also *non-closed* programs in which case counter examples can be found to the validity of quite natural c.f. such as $<D|\{p\}S_1\{q\} \wedge \{q\}S_2\{r\} \Rightarrow \{p\}S_1;S_2\{r\}>$. The second problem is dealt with in section 4; we now define the notion of initialization and state the main theorem concerning it.

DEFINITION 3.1. (initialized s.v.)

a. The set $\mathit{init}(R)$ of all s.v. initialized in R is the smallest subset of Svar satisfying

(i) If $x \notin \mathit{svar}(s)$ then $x \in \mathit{init}(<D|x:=s>)$.

(ii) If $x \in \mathit{init}(<D|S_1>)$, or $x \notin \mathit{svar}(S_1)$ and $x \in \mathit{init}(<D|S_2>)$ then $x \in \mathit{init}(<D|S_1;S_2>)$.

(iii) If $x \notin \mathit{svar}(b)$, $x \in \mathit{init}(<D|S_i>)$, $i = 1,2$, then $x \in \mathit{init}(<D|\underline{\text{if}}\ b\ \underline{\text{then}}\ S_1\ \underline{\text{else}}\ S_2\ \underline{\text{fi}}>)$.

(iv) If $x \not\equiv y$, $x \in \mathit{init}(<D|S>)$ then $x \in \mathit{init}(<D|\underline{\text{b}}\ \underline{\text{new}}\ y; S\ \underline{\text{e}}>)$.

(v) If $D \equiv <P_i \Leftarrow B_i>_{i=1}^n$ then, for $i = 1,\ldots,n$, if $B_i \equiv <\underline{\text{val}}\ x_i,\ \underline{\text{add}}\ y_i | S_i>$, $x \notin \mathit{svar}(t)$, $x \equiv v$, and $y_i \in \mathit{init}(<D|S_i>)$, then $x \in \mathit{init}(<D|P_i(t,v)>)$.

b. All local s.v. in a program $<D|S>$, with $D \equiv <P_i \Leftarrow B_i>_{i=1}^n$, $B_i \equiv <\underline{\text{val}}\ x_i,\ \underline{\text{add}}\ y_i|S_i>$, are initialized whenever for each statement $\underline{\text{b}}\ \underline{\text{new}}\ x;\ S_0\ \underline{\text{e}}$ occurring as substatement of S or any of the S_i, $1 \le i \le n$, we have that $x \in \mathit{init}(<D|S_0>)$.

END 3.1.

For an initialized local s.v., the value associated with it through def. 2.4d is irrelevant. This is one of the (somewhat hidden) messages of

THEOREM 3.2. Let $<D|S>$ be a closed program in which all local s.v. are initialized. Let $n,m \ge 0$, and let γ, $\bar{\gamma}$ be as usual. If

(i) $\mathit{intv}(D) \cup (\mathit{intv}(S)\backslash\{x_i\}_i\backslash\{<a_j,\alpha>_\alpha\}_j) \subseteq \delta \subseteq \mathrm{dom}(\varepsilon) \cap \mathrm{dom}(\bar{\varepsilon})$

(ii) $(\sigma\circ\varepsilon)|\delta = (\bar{\sigma}\circ\bar{\varepsilon})|\delta$

(iii) For $i = 1,\ldots,n$, either $\sigma(e_i) = \bar{\sigma}(\bar{e}_i)$, or $x_i \notin \mathit{svar}(S)$, or $x_i \in \mathit{init}(<D|S>)$.
For $j = 1,\ldots,m$, either, for all $\alpha \in V_0$, $\sigma(e_{\alpha,j}) = \bar{\sigma}(\bar{e}_{\alpha,j})$, or $a_j \notin \mathit{avar}(S)$.

(iv) $N(S[y_i/x_i]_i[a'_j/a_j]_j)(\bar{\gamma})(\varepsilon \cup <y_i,e_i>_i \cup <<<a'_j,\alpha>,e_{\alpha,j}>_\alpha>_j)(\sigma) = \sigma'$
$N(S[z_i/x_i]_i[a''_j/a_j]_j)(\bar{\gamma})(\bar{\varepsilon} \cup <z_i,\bar{e}_i>_i \cup <<<a''_j,\alpha>,\bar{e}_{\alpha,j}>_\alpha>_j)(\bar{\sigma}) = \bar{\sigma}'$

then

(v) $(\sigma'\circ\varepsilon)|\delta = (\bar{\sigma}'\circ\bar{\varepsilon})|\delta$

(vi) For $i = 1,\ldots,n$, either $\sigma'(e_i) = \bar{\sigma}'(\bar{e}_i)$, or $x_i \notin \mathit{svar}(S)$.
For $j = 1,\ldots,m$, either, for all $\alpha \in V_0$, $\sigma'(e_{\alpha,j}) = \bar{\sigma}'(\bar{e}_{\alpha,j})$, or $a_j \notin \mathit{avar}(S)$.

END 3.2.

In section 4 two special cases of this theorem are of interest, mentioned in

COROLLARY 3.3. Let $<D|S>$, $m,n;\gamma,\bar{\gamma}\ldots$ be as in theorem 3.2.

a. If (i) $\mathit{intv}(D,S) \subseteq \delta \subseteq \mathrm{dom}(\varepsilon) \cap \mathrm{dom}(\bar{\varepsilon})$, (ii) $(\sigma\circ\varepsilon)|\delta = (\bar{\sigma}\circ\bar{\varepsilon})|\delta$, (iii) $N(S)(\bar{\gamma})(\varepsilon)(\sigma) = \sigma'$, $N(S)(\bar{\gamma})(\bar{\varepsilon})(\bar{\sigma}) = \bar{\sigma}'$, then (iv) $(\sigma'\circ\varepsilon)|\delta = (\bar{\sigma}'\circ\bar{\varepsilon})|\delta$.

b. If (i) $\mathit{intv}(D) \cup (\mathit{intv}(S)\backslash\{x_i\}_i\backslash\{<a_j,\alpha>_\alpha\}_j) \subseteq \mathrm{dom}(\varepsilon)$, (ii) $(S[y_i/x_i]_i[a'_j/a_j]_j)(\bar{\gamma})(\varepsilon \cup <y_i,e_i>_i \cup <<<a'_j,\alpha>,e_{\alpha,j}>_\alpha>_j)(\sigma) = \sigma'$, $N(S[z_i/x_i]_i[a''_j/a_j]_j)(\gamma)(\varepsilon \cup <z_i,e_i>_i \cup <<<a''_j,\alpha>,e_{\alpha,j}>_\alpha>_j)(\sigma) = \sigma''$, then (iii) $(\sigma'\circ\varepsilon)|\mathrm{dom}(\varepsilon) = (\sigma''\circ\varepsilon)|\mathrm{dom}(\varepsilon)$.

END 3.3.

Remark. Let us call a pair $\langle\sigma,\varepsilon\rangle$, $\langle\bar{\sigma},\bar{\varepsilon}\rangle$ *matching* with respect to δ if it satisfies condition (ii) of part a. We see that a program satisfying the indicated requirements preserves the property of matching. Cor. 3.3b tells us that substituting different fresh s.v. y, z (since $y,z \notin dom(\varepsilon)$, $y,z \notin intv(D) \cup (intv(S)\backslash\ldots)$) for some x makes no (essential) difference in the outcome, provided that they are associated with the same address.

4. *A sound proof system*

The following proof system will be considered:

A. Rules about "⇒".

1. $\langle D \mid f \Rightarrow \underline{true}\rangle$ (strengthening)

2. $\dfrac{\langle D \mid f_1 \Rightarrow f_2\rangle}{\langle D \mid f_1 \wedge f_3 \Rightarrow f_2\rangle}$ (weakening)

3. $\dfrac{\langle D \mid f_1 \Rightarrow f_2\rangle, \langle D \mid f_2 \Rightarrow f_3\rangle}{\langle D \mid f_1 \Rightarrow f_3\rangle}$ (transitivity)

4. $\dfrac{\langle D \mid f \Rightarrow f_1\rangle, \langle D \mid f \Rightarrow f_2\rangle}{\langle D \mid f \Rightarrow f_2 \wedge f_2\rangle}$ (collection)

5. $\langle D \mid f_1 \wedge \ldots \wedge f_n \Rightarrow f_i\rangle$, $n \geq 1$, $1 \leq i \leq n$ (selection)

B. Rules about programming concepts.

6. $\langle D \mid \{p[t/v]\} v{:=}t \{p\}\rangle$ (assignment)

7. $\langle D \mid \{p\}S_1\{q\} \wedge \{q\}S_2\{r\} \Rightarrow \{p\}S_1;S_2\{r\}\rangle$ (composition)

8. $\langle D \mid \{p \wedge b\}S_1\{q\} \wedge \{p \wedge \neg b\}S_2\{q\} \Rightarrow \{p\}\ \underline{if}\ \ldots\ \underline{fi}\ \{q\}\rangle$ (conditional)

9. $\langle D \mid \{p\}S[y/x]\{q\} \Rightarrow \{p\}\underline{b}\ \underline{new}\ x;\ S\ \underline{e}\{q\}\rangle$ (s.v. declaration)
   provided that $y \notin svar(D,p,S,q)$

10. Let Ω be a procedure constant such that $N(\Omega) = \lambda\gamma\cdot\lambda t\cdot\lambda v\cdot\lambda\varepsilon\cdot\lambda\sigma\cdot\perp\cdot$

$$\frac{\langle\langle P_i \Leftarrow B_i\rangle_i \mid f[\Omega/Q_i]_i\rangle, \langle\langle P_i \Leftarrow B_i\rangle_i \mid f \Rightarrow f[B_i'/Q_i]_i\rangle}{\langle\langle P_i \Leftarrow B_i\rangle_i \mid f[P_i/Q_i]_i\rangle} \qquad \text{(induction)}$$

   where $Q_i \notin pvar(P_1,\ldots,P_n,B_1,\ldots,B_n)$, and $B_i' \equiv B_i[Q_j/P_j]_j$, $i = 1,\ldots,n$.

C. Auxiliary rules

11. $\langle D \mid (p \supset p_1) \wedge \{p_1\}S\{q_1\} \wedge (q_1 \supset q) \Rightarrow \{p\}S\{q\}\rangle$ (consequence)

12. $<D|\{p\}S\{q_1\} \wedge \{p\}S\{q_2\} \Rightarrow \{p\}S\{q_1 \wedge q_2\}>$ (conjunction)

13. $$\frac{<<P_i \Leftarrow B_i>_i|(f_1 \Rightarrow f_2)[Q_i/P_i]_i>}{<<P_i \Leftarrow B_i>_i|f_1 \Rightarrow f_2>}$$ (instantiation)

where $Q_i \notin$ *pvar*$(P_1,\ldots,P_n,f_1,f_2)$

14. $<D|\{p\}S\{p\}>$ (invariance)
provided that *intv*(p) ∩ *intv*(D,S) = ∅

15. $<D|\{p\}S\{q\} \Rightarrow \{p[y/x][a'/a]\}S\{q\}>$ (substitution, I)
provided that $x \notin$ *svar*(D,S,q), $a \notin$ *avar*(D,S,q)

16. $<D|\{p\}S\{q\} \Rightarrow \{p[y/x][a'/a]\}S[y/x][a'/a]\{q[y/x][a'/a]\}>$
provided that $x \notin$ *svar*(D), $a \notin$ *avar*(D), $y \notin$ *svar*(D,S,q), (substitution, II)
$a' \notin$ *avar*(D,S,q)

17. $<D|\{p\}S\{q\} \Rightarrow \{p\}\tilde{S}\{q\}>$ (congruence)
provided that $S \cong \tilde{S}$

Let $\vdash<D|\{p\}S\{q\}>$ denote that $<D|\{p\}S\{q\}>$ is formally provable in the proof system consisting of

1. As axioms: Rules 1, 5, 6, 7, 8, 9, 11, 12, 14, 15, 16, 17, *together* with all c.f. $<D|p>$ such that $\models<D|p>$
2. As proof rules: Rules 2, 3, 4, 10, 13.

We are interested in showing *soundness* (if $\vdash<D|\{p\}S\{q\}>$ then $\models<D|\{p\}S\{q\}>$) and *completeness* (if $\models<D|\{p\}S\{q\}>$ then $\vdash<D|\{p\}S\{q\}>$) of the above system in case $<D|S>$ is closed and contains only initialized local s.v. In the present section we discuss soundness, in the next one completeness.

Due to the presence of the induction rule (#10), even when we start the formal proof of $<D|\{p\}S\{q\}>$ with $<D|S>$ closed, we may encounter at intermediate stages *non-closed* programs (note that the Q_i in the premise of rule 10 are not declared), and we therefore cannot restrict our attention to closed programs only. However, for non-closed programs various of the c.f. in the system are *invalid*. (E.g., take rule 14, with D empty, $p \equiv q \equiv (x=0)$, $S \equiv P(0,y)$ with $y \not\equiv x$, and let $\gamma(P) \stackrel{\text{df.}}{=} \lambda t\cdot\lambda v\cdot\lambda\varepsilon\cdot\lambda\sigma\cdot\sigma\{1/\varepsilon(x)\}$. In fact, rules 5, 7, 8, 9, 11, 14, 15, 16 are all invalid.) Hence, we have to *refine* the notion of validity. Instead of putting: $\models g$ iff $\mathcal{G}(g)(\gamma)$ holds for *all* $\gamma \in \Gamma$ (cf. 2.4), we introduce a subset $\Gamma^D \subsetneq \Gamma$, such that, for all P, t, v, and $\gamma \in \Gamma^D$, $\gamma(P)(t,v)$ satisfies the properties as mentioned in lemma 2.2 and theorem 3.2 for closed programs. More specifically, we have

<u>DEFINITION 4.1</u>. Let $D \in \mathcal{D}ecl$. Γ^D is the subset of Γ consisting of those γ which satisfy, for all P, t, v:

a. If (i) $intv(D,t,v) \subseteq dom(\varepsilon)$, (ii) $\xi \in dom(\varepsilon) \backslash intv(D,t,v)$, (iii) $\sigma' = \gamma(P)(t,v)(\varepsilon)(\sigma)$, $\sigma' \neq \bot$ then, (iv) $\sigma'(\varepsilon(\xi)) = \sigma(\varepsilon(\xi))$

b. If $n,m \geq 0$ and

(i) $intv(D) \cup (intv(t,v) \backslash \{x_i\}_i \backslash \{<a_j,\alpha>_\alpha\}_j) \subseteq \delta \subseteq dom(\varepsilon) \cap dom(\bar{\varepsilon})$.

(ii) $(\sigma \circ \varepsilon)|\delta = (\bar{\sigma} \circ \bar{\varepsilon})|\delta$.

(iii) For $i = 1,\ldots,n$, either $\sigma(e_i) = \bar{\sigma}(\bar{e}_i)$, or $x_i \notin svar(P(t,v))$.
For $j = 1,\ldots,m$, either, for all $\alpha \in V_0$, $\sigma(e_{\alpha,j}) = \bar{\sigma}(\bar{e}_{\alpha,j})$, or $a_j \notin avar(P(t,v))$.

(iv) $\gamma(P)((t,v)[y_i/x_i]_i[a'_j/a_j]_j)(\varepsilon \cup <y_i,e_i>_i \cup <<<a'_j,\alpha>,e_{\alpha,j}>_\alpha>_j)(\sigma) = \sigma'$
$\gamma(P)((t,v)[z_i/x_i]_i[a''_j/a_j]_j)(\bar{\varepsilon} \cup <z_i,\bar{e}_i>_i \cup <<<a''_j,\alpha>,\bar{e}_{\alpha,j}>_\alpha>_j)(\bar{\sigma}) = \bar{\sigma}'$,

then

(v) $(\sigma' \circ \varepsilon)|\delta = (\bar{\sigma}' \circ \bar{\varepsilon})|\delta$.

(vi) For $i = 1,\ldots,n$, either $\sigma'(e_i) = \bar{\sigma}'(e_i)$, or $x_i \notin svar(P(t,v))$.
For $j = 1,\ldots,m$, either, for all $\alpha \in V_0$, $\sigma'(e_{\alpha,j}) = \bar{\sigma}'(\bar{e}_{\alpha,j})$, or $a_j \notin avar(P(t,v))$

END 4.1.

Combining the postulated properties of Γ^D with lemma 2.2, thm. 3.2, cor. 3.3, we obtain

<u>THEOREM 4.2</u>. As lemma 2.2, thm. 3.2 and cor. 3.3, but now for arbitrary programs <D|S> (i.e., closed or non-closed), provided (all local s.v. are initialized and) $\gamma \in \Gamma^D$

END 4.2.

Using thm. 4.2 it is not difficult to prove

<u>LEMMA 4.3.</u> If $\gamma \in \Gamma^D$, $\bar{\gamma}$ as usual, then, if (i) $intv(D,f) \subseteq \delta \subseteq dom(\varepsilon) \cap dom(\bar{\varepsilon})$, (ii) $(\sigma \circ \varepsilon)|\delta = (\bar{\sigma} \circ \bar{\varepsilon})|\delta$ then, (iii) $F(f)(\bar{\gamma})(\varepsilon)(\sigma) = F(f)(\bar{\gamma})(\bar{\varepsilon})(\bar{\sigma})$.

END 4.3.

The refined validity definition takes the form

<u>DEFINITION 4.4</u>. Let $g \equiv <D|f_1 \Rightarrow f_2>$. We call g valid ($\models g$) if, for all $\gamma \in \Gamma^D$, $G(g)(\gamma)$.

END 4.4.

Remark. The definition of soundness of an inference remains unchanged (but now refers to the new notion of validity).

At last, our efforts are rewarded by

<u>THEOREM 4.5.</u> (soundness theorem). All c.f. of the above proof system are valid, and all its inferences are sound.

The proof uses more or less traditional means for rules such as the assignment rule, the induction rule (employing the continuity of $\lambda\eta'_1 \cdot \ldots \cdot \lambda\eta'_n \cdot \lambda t \cdot \lambda v \cdot N(B_j(t,v))(\gamma\{\eta'_i/P_i\}_i))$ and a number of further rules. For the rules of selection, composition and consequence it moreover uses lemma 4.3. Finally, the proofs of validity of the

s.v. declaration rule, the invariance rule and the two substitution rules rely heavily on theorem 4.2 and lemma 4.3 (and lemma 2.1, 2.3).
END 4.5.

5. *Completeness*

We now prove the completeness of the system of rules 1 to 17 (extended with all $\langle D|p\rangle$ such that $\models\langle D|p\rangle$; this follows an idea of Cook [8], and leads to what may be called *relative* completeness). Let us recall that we only consider programs in which all local s.v. are initialied. For simplicity's sake, from now on we restrict ourselves to programs with only one procedure declaration, say $D \equiv P \Leftarrow B$, with $B \equiv \langle \underline{\text{val}}\ x, \underline{\text{add}}\ y|S\rangle$. We can then specialize the induction rule to

$$10'.\quad \frac{\langle P\Leftarrow B|\langle\{p_i\}Q(t_i,v_i)\{q_i\}\rangle_{i=1}^{m} \Rightarrow \langle\{p_i\}B'(t_i,v_i)\{q_i\}\rangle_{i=1}^{m}\rangle}{\langle P\Leftarrow B|\langle\{p_i\}P(t_i,v_i)\{q_i\}\rangle_{i=1}^{m}\rangle} \qquad \text{(induction')}$$

where $m \geq 1$, $B' \equiv B[Q/P]$, $Q \notin pvar(P,B)$.

For the completeness proof we need some auxiliary results. First, we introduce the notion of *strongest postcondition* of a program $\langle D|S\rangle$ with respect to an assertion p. Let r be an assertion satisfying

1. $\models\langle D|\{p\}S\{r\}\rangle$
2. For all $q \in \mathit{Assn}$, if $\models\langle D|\{p\}S\{q\}\rangle$, then $\models\langle D|r\supset q\rangle$.

We then say that r is a strongest postcondition of $\langle D|S\rangle$ with respect to p, and denote it (with slight abuse of language) by $sp(p,\langle D|S\rangle)$.

LEMMA 5.1.

a. For each *closed* $\langle D|S\rangle$ and $p \in \mathit{Assn}$ there exists $r \in \mathit{Assn}$ such that $r \equiv sp(p,\langle D|S\rangle)$ and $intv(r) \subseteq intv(p,D,S)$.
b. For p, D, S, r as in part a, and ε such that $intv(p,D,S) \subseteq dom(\varepsilon)$, $\mathcal{T}(r)(\varepsilon)(\sigma) = (\sigma\neq\bot) \wedge \exists\sigma'[\mathcal{T}(p)(\varepsilon)(\sigma') \wedge (\sigma\circ\varepsilon)|dom(\varepsilon) = (\mathcal{N}(S)(\bar{\gamma})(\varepsilon)(\sigma')\circ\varepsilon)|dom(\varepsilon)]$ (here $\bar{\gamma} = \gamma\{\eta/P\}$ (η as usual) which, since $\langle D|S\rangle$ is closed, is independent of γ).
c. $\models sp(p,\langle D|S\rangle)[y/x] = sp(p[y/x],\langle D|S[y/x]\rangle)$, provided $x \notin svar(D)$, $y \notin svar(p,D,S)$.

END 5.1.

Remark. The proof of part a of this lemma is non-trivial and uses tools from recursive function theory.

Next, we need

LEMMA 5.2.

a. Let $\langle D|S_1;S_2\rangle$ be a closed program. Then $\models\langle D|\{p\}S_1;S_2\{q\} \Rightarrow \{p\}S_1\{r\} \wedge \{r\}S_2\{q\}\rangle$, where $r \equiv sp(p,\langle D|S_1\rangle)$
b. $\models\langle D|\{p\}S[y/x]\{q\} \Rightarrow \{p\}\underline{\text{b}}\ \underline{\text{new}}\ x;S\ \underline{\text{e}}\{q\}\rangle$, provided $y \notin svar(D,p,S,q)$

END 5.2.

The proof of the completeness theorem uses a key lemma (following Gorelick [9]):

<u>LEMMA 5.3</u>. Let $p,q \in \mathit{Assn}$, $\langle D|S\rangle$ a closed program, and let $\tilde{S} \cong S$ be such that no bound s.v. of $\tilde{S}$ occurs free in D. Let $P(t_1,v_1),\ldots,P(t_m,v_m)$ be the occurrences of procedure calls (of P, by closedness no other calls are possible) in $\tilde{S}$. For each $i = 1,\ldots,m$ let $p(t_i,v_i) \overset{\text{df.}}{\equiv} (x_1^{(i)} = z_1^{(i)}) \wedge \ldots \wedge (x_{k_i}^{(i)} = z_{k_i}^{(i)}) \wedge (a_1^{(i)} = \bar{a}_1^{(i)}) \wedge \ldots \wedge (a_{\ell_i}^{(i)} = \bar{a}_{\ell_i}^{(i)})$, where $\{x_1^{(i)},\ldots,x_{k_i}^{(i)}\} = \mathit{svar}(D,t_i,v_i)$, $\{a_1^{(i)},\ldots,a_{\ell_i}^{(i)}\} = \mathit{avar}(D,t_i,v_i)$, $z_1^{(i)},\ldots,z_{\ell_i}^{(i)}, \bar{a}_1^{(i)},\ldots,\bar{a}_{\ell_i}^{(i)}$ are completely fresh, and $a' = a''$ abbreviates $\forall x[a'[x] = a''[x]]$. Let moreover, $r(t_i,v_i) \overset{\text{df.}}{\equiv} sp(p(t_i,v_i),\langle D|P(t_i,v_i)\rangle)$, $i = 1,\ldots,m$. We have: If $\models\langle D|\{p\}S\{q\}\rangle$ then $\vdash\langle D|\langle\{p(t_i,v_i)\}Q(t_i,v_i)\{r(t_i,v_i)\}\rangle_{i=1}^{m} \Rightarrow \{p\}\tilde{S}[Q/P]\{q\}\rangle$.

PROOF (sketch). The proof uses induction on the complexity of S. If $\tilde{S} \equiv v{:=}t$ or $\tilde{S} \equiv$ <u>if</u> ... <u>fi</u>, the result is clear by the rules for assignment and conditionals. If $\tilde{S} \equiv S_1;S_2$, apply lemma 5.2a. If $\tilde{S} \equiv$ <u>b</u> <u>new</u> $x;S_1$ <u>e</u> (note that $x \notin \mathit{svar}(D)$ by assumption) use lemma 5.1c, 5.2b, and the rules of s.v. declaration and substitution II. If $\tilde{S} \equiv P(t_i,v_i)$ for some i, $1 \le i \le m$, we need a more elaborate argument involving the rules of invariance and conjunction, and both substitution rules.
END 5.3.

From now on we assume that S_0 (the statement of the procedure body B) contains precisely one recursive call of P, say $S_0 \equiv \ldots P(t_0,v_0)\ldots$. (We expect that, at the cost of a rather tedious elaboration of the argument to follow, the general case can also be taken care of.) Let t,v be arbitrary parameters. We show that, using the notation of lemma 5.3, (*): $\vdash\langle D|\{p(t,v)\}P(t,v)\{q(t,v)\}\rangle$. First, we have the following corollary of lemma 5.3.

<u>COROLLARY 5.4</u>. $\vdash\langle D|\langle\{p(t_i,v_i)\}Q(t_i,v_i)\{r(t_i,v_i)\}\rangle_{i=1}^{4} \Rightarrow \langle\{p(t_i,v_i)\}B'(t_i,v_i)^{\sim}\{r(t_i,v_i)\}\rangle_{i=1}^{3}\rangle$, where $t_1 \equiv t$, $v_1 \equiv v$, and, for $i = 2,3,4$, $Q(t_i,v_i)$ is the procedure call occurring in $B'(t_{i-1},v_{i-1})^{\sim}$ (and $B' \equiv B[Q/P]$).
END 5.4.

Next, we use

<u>LEMMA 5.5</u>. (Notation as in cor. 5.4). $\vdash\langle D|\{p(t_3,v_3)\}Q(t_3,v_3)\{r(t_3,v_3)\} \Rightarrow \{p(t_4,v_4)\} Q(t_4,v_4)\{r(t_4,t_4)\}\rangle$.

PROOF (sketch). We distinguish eight cases, depending on the form of v_0 and v_1 ($v_0 \equiv x$, $v_0 \equiv y$, $v_0 \equiv z_0 (\not\equiv x,y)$, $v_0 \equiv a[s_0]$, $v_1 \equiv z_1$, $v_1 \equiv a[s_1]$), apply the definition of syntactic application repeatedly in order to determine t_i, v_i, $i = 2,3,4$, and, finally, use substitution rule II twice.
END 5.5.

By corollary 5.4, lemma 5.5 and the congruence rule we obtain $\vdash\langle D|\langle\{p(t_i,v_i)\}Q(t_i,v_i)\{r(t_i,v_i)\}\rangle_{i=1}^{3} \Rightarrow \langle\{p(t_i,v_i)\}B'(t_i,v_i)\{r(t_i,v_i)\}\rangle_{i=1}^{3}\rangle$, so by the induction' rule (#10') $\vdash\langle D|\langle\{p(t_i,v_i)\}P(t_i,v_i)\{r(t_i,v_i)\}\rangle_{i=1}^{3}\rangle$, implying that (*): $\vdash\langle D|\{p(t,v)\}P(t,v)\{r(t,v)\}\rangle$ is indeed satisfied. We are now sufficiently prepared for the proof of

THEOREM 5.6. (completeness theorem). Let p,q $\in$ *Assn*, <D|S> a closed program, and assume $\models$<D|{p}S{q}>. Then $\vdash$<D|{p}S{q}>.

PROOF (sketch). Let $\tilde{S}$ be as in lemma 5.3., and let $P(t_1,v_1),\ldots,P(t_m,v_m)$ be the occurrences of a procedure call in $\tilde{S}$. By lemma 5.3, $\vdash\langle D|\langle\{p(t_i,v_i)\}Q(t_i,v_i)\{r(t_i,v_i)\}\rangle_{i=1}^m \Rightarrow \{p\}\tilde{S}[Q/P]\{q\}\rangle$. Thus, by the instantiation rule, $\vdash\langle D|\langle\{p(t_i,v_i)\}P(t_i,v_i)\{r(t_i,v_i)\}\rangle_{i=1}^m \Rightarrow \{p\}\tilde{S}\{q\}\rangle$. By (*) (and the collection rule) $\vdash\langle D|\langle\{p(t_i,v_i)\}P(t_i,v_i)\{r(t_i,v_i)\}\rangle_{i=1}^m\rangle$, and the desired result follows by the (transivity and) congruence rule.

END 5.6.

Altogether, we have achieved the goal of our paper: We have presented a proof system for partial correctness, and proved is soundness (theorem 4.5) and completeness (theorem 5.6).

References

1. Apt, K.R., A sound and complete Hoare-like system for a fragment of PASCAL. Report IW 97/78, Mathematisch Centrum (1978).
2. Apt, K.R., Ten years of Hoare's logic, a survey. Proc. 5th Scandinavian Logic Symposium, Aalborg University Press, to appear.
3. Apt, K.R. & J.W. de Bakker, Exercises in denotational semantics. Proc 5th Symp. Math. Foundations of Computer Science (A. Mazurkiewicz, ed.), pp. 1-11, Lecture Notes in Computer Science 45, Springer (1976).
4. Apt, K.R. & J.W. de Bakker, Semantics and proof theory of PASCAL procedures. Proc. 4th Coll. Automata, Languages and Programming (A. Salomaa & M. Steinby, eds.), pp. 30-44, Lecture Notes in Computer Science 52, Springer (1977).
5. De Bakker, J.W., Correctness proofs for assignment statements. Report IW 55/76, Mathematisch Centrum (1976).
6. Cartwright, R & D. Oppen, Unrestricted procedure calls in Hoare's logic.Proc. 5th Symp. Principles of Programming Languages, pp. 131-140 (1978).
7. Clarke, E.M., Programming language constructs for which it is impossible to obtain good Hoare-like axioms. Journal ACM, vol. 26, pp. 129-147 (1979).
8. Cook, S.A., Soundness and completeness of an axiom system for program verification. SIAM J. on Computing, vol. 7, pp. 70-90 (1978).
9. Gorelick, G.A., A complete axiomatic system for proving assertions about recursive and non-recursive programs. Technical Report no. 75, Department of Computer Science, University of Toronto (1975).

THE PROBLEM OF REACHABILITY AND VERIFICATION OF PROGRAMS

J.M.Barzdin
Computing Centre
Latvian State University
Riga, USSR

ABSTRACT

A new method for verification of programs is proposed. The main idea is to reduce assertions about programs to the problem of reachability. It is shown in [1] that the reachability problem is algorithmically solvable for a wide class of programs.

1. INTRODUCTION

The following problem of reachability of instructions in programs is considered: given a program P and an instruction K in the program P, the question is whether the instruction K is reachable, i.e. whether there is an input data sample S such that while running the program P on the sample S, the instruction K is executed. It was shown in [1] that this problem is algorithmically solvable for a sufficiently wide class of data processing programs. An experimental system was built for reachability testing of programs in a COBOL-like language [1,2]. The results obtained make us believe in the possibility of implementing a system which solves the reachability problem for most of real data processing programs.

There arises a question: is it possible to utilize the solvability of the reachability problem for proving assertions about programs. In other words, we ask what advantage can be got by introduction of the following type of derivation rules:

$$\frac{\text{Program } P, \text{ instruction } K, \ K \text{ in } P}{\begin{array}{l} R(P, K), \ \text{ if } K \text{ is reachable in } P, \\ \neg R(P, K), \ \text{ if } K \text{ is not reachable in } P \end{array}}$$

The aim of this paper is to show that a wide class of assertions about programs is reducible to the problem of reachability. The idea of this reducibility is as follows. We add new blocks to the program P, and these blocks check the validity of the given assertion about the input-output relation in the program P. This, in a sense, is a further development of ideas by Panzl [3] for a wider class of assertions.

2. THE MAIN IDEAS AND EXAMPLES

In order to expose the main ideas we consider a very simple programming language L_0 introduced in [1]. This language could be used for processing of sequential files. We consider file as a variable whose values are finite sequences $(X_1, X_2, \ldots, X_p)$ of integers, and X_i is the i-th record of the file. Each program has a finite number of input files and a finite number of output files. There are internal variables in the program. These variables take integer values. The initial value of each internal variable is 0.

Let A be an input file and Z be an output file. Let t and u be the internal variables and c be a constant, i.e. a fixed integer. Instructions of the following type are allowed.

1°. $A \rightarrow t$. The current record of the file A is assigned to the variable t. Thus, if $A = (A_1, A_2, \ldots, A_p)$, then the first occurence of the instruction $A \rightarrow t$ assigns record A_1 to the variable t, the second assigns record A_2 to the same variable t etc. The instruction has two exits, namely, the exit " + " for the case if the current record exists, and the exit " - " for the case, if the current record does not exist. In the latter case the value of the variable t is not changed.

2°. $t \rightarrow Z$. The value of the variable t is assigned to the current record of the file Z.

3°. $u \rightarrow t$ (respectively, $c \rightarrow t$). The value of the variable u (the value of the constant c) is assigned to the variable t.

4°. $u \odot t$ (respectively, $c \odot t$), where $\odot \in \{<, >, \leq, \geq, =, \neq\}$. The instruction has two exits, namely, " + " for the case, if the condition holds, and " - ", otherwise.

5°. STOP

Let L_o be the language generated by the instructions 1°-5°, where the programs are given as flowcharts over this instruction set. Fig. 1 gives an example of a program, which creates a new sorted file Z by merging sorted files A and B.

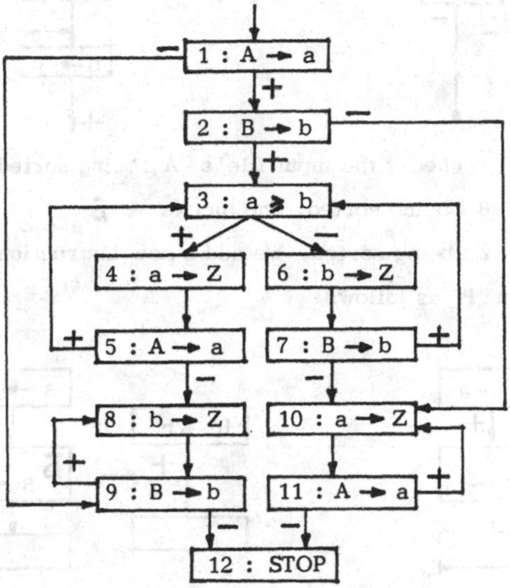

Fig. 1.

Now we consider the following assertion $\mathfrak{A}$ about programs with input files A and B, and output file Z: if the input files A and B are sorted, then the output file Z also is sorted. Now we describe how to reduce checking of the assertion $\mathfrak{A}$ to the problem of reachability. Let P be a program. We try to check the assertion $\mathfrak{A}$ for it. Let the program P contain instructions of the following types: $A \to a$, $B \to b$, $a \geq b$, $a \to Z$, $b \to Z$, STOP. We use no more information about the program P, while describing the reduction. We use the following blocks for the reduction:

$Б_{A \to a} =$

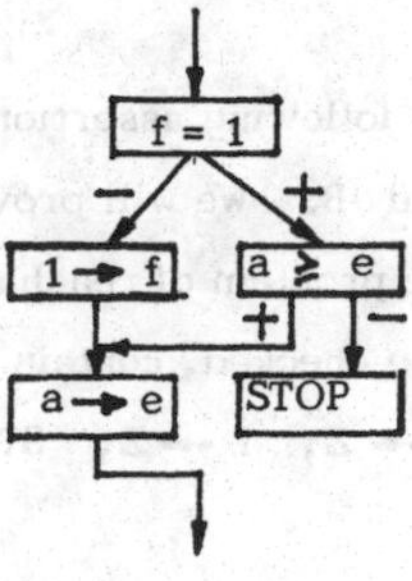

$Б_{B \to b} =$

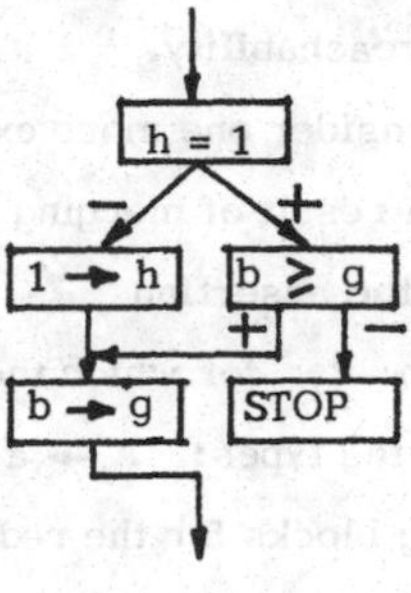

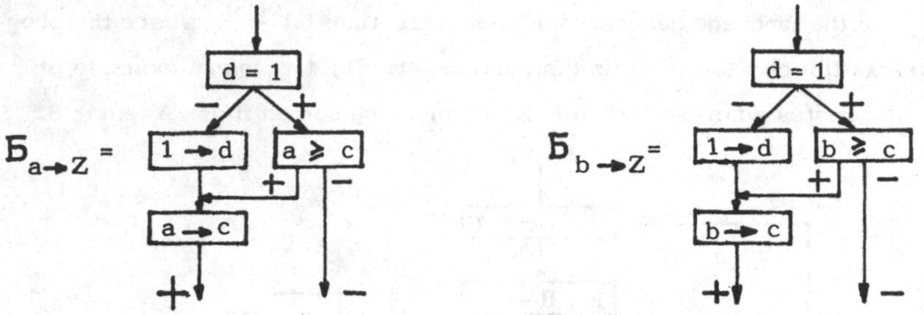

The block $Б_{A \to a}$ checks the input file's A being sorted, the block $Б_{B \to b}$ checks the input file's B being sorted. The blocks $Б_{a \to Z}$ and $Б_{b \to Z}$ check the output file's Z being sorted. We add a new instruction K and the Б-blocks into the program P as follows:

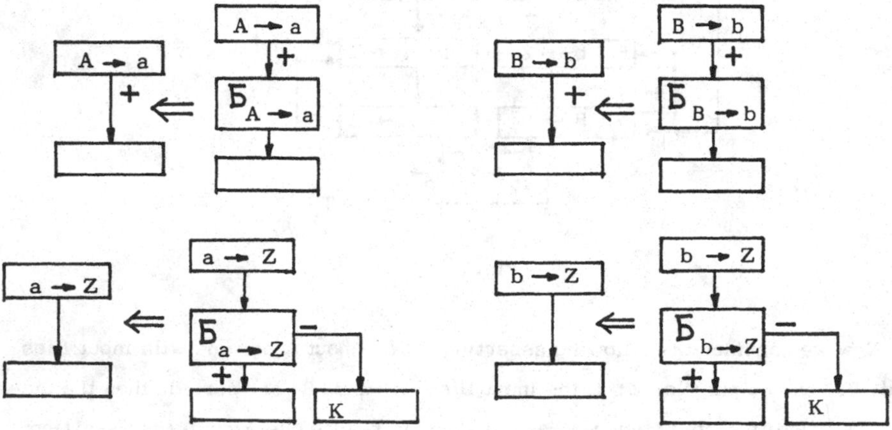

The result of this substitution is a new program P'. Fig.2 shows the result of the substitution for the program P shown in Fig.1. It is obvious that the instruction K is reachable in the program P' if and only if the assertion $\mathcal{A}$ does not hold for the program P. Hence, the checking of the assertion $\mathcal{A}$ for P is reduced to the problem of reachability.

We consider one more example. Let $\mathcal{B}$ be the following assertion: the output file Z is a result of merging of the input files A and B. We will prove that the checking of the assertion $\mathcal{B}$ is also reducible to the problem of reachability. Let the program P, for which the assertion $\mathcal{B}$ is to be checked, contain instructions of the following types: $A \to a$, $B \to b$, $a \geq b$, $a \to Z$, $b \to Z$, STOP. We use the following blocks for the reduction:

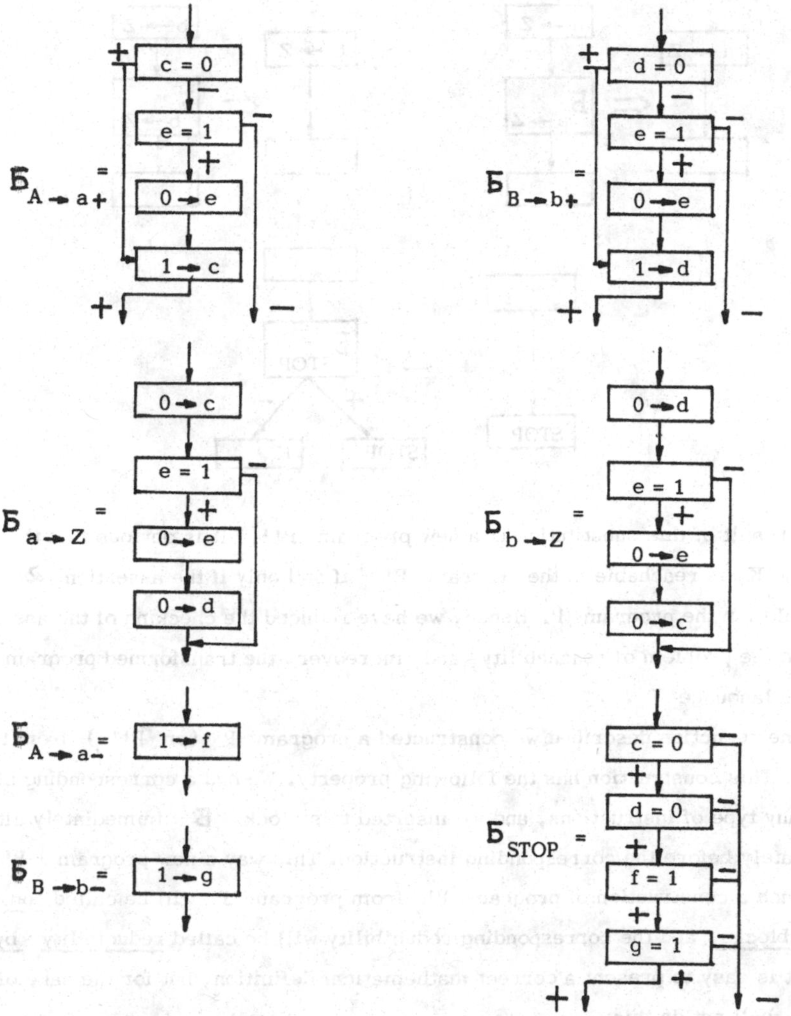

We add a new instruction K and perform the following substitutions of Б -blocks into the program P:

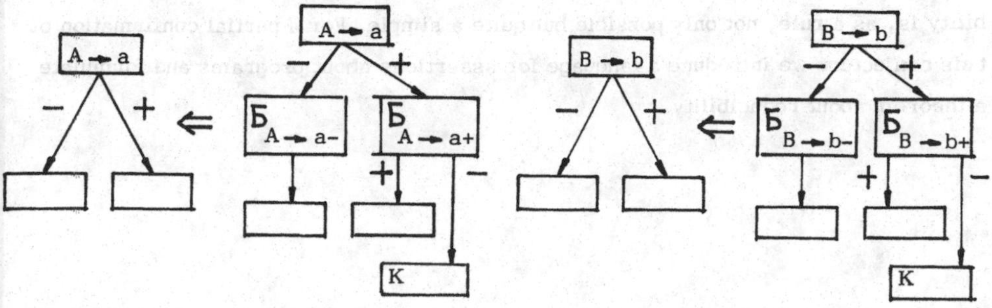

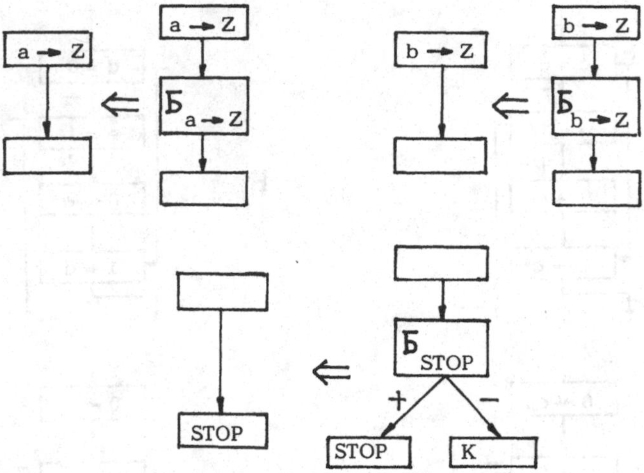

The result of this substitution is a new program P''. It is obvious that the instruction K is reachable in the program P'' if and only if the assertion $\mathcal{L}$ does not hold for the program P. Hence, we have reduced the checking of the assertion $\mathcal{L}$ to the problem of reachability, and, moreover, the transformed program is in the same language L_o.

In the reduction described we constructed a program P' (or P'') from the program P. This construction has the following property. We had a corresponding block Б for any type of instructions, and we inserted this block Б immediately after or immediately before the corresponding instruction. This way a new program P' was got. Such a construction of program P' from program P will be called *construction by blocks*, and the corresponding reducibility will be called *reducibility by blocks*. It is easy to present a correct mathematical definition, but for the sake of brevity we shall not do this.

A question arises: are the other natural assertions about programs reducible to the problem of reachability as well ? Our main conjecture asserts that such a reducibility is, as a rule, not only possible but quite a simple. For a partial confirmation of this conjecture we introduce a language for assertions about programs and formulate a theorem about reducibility.

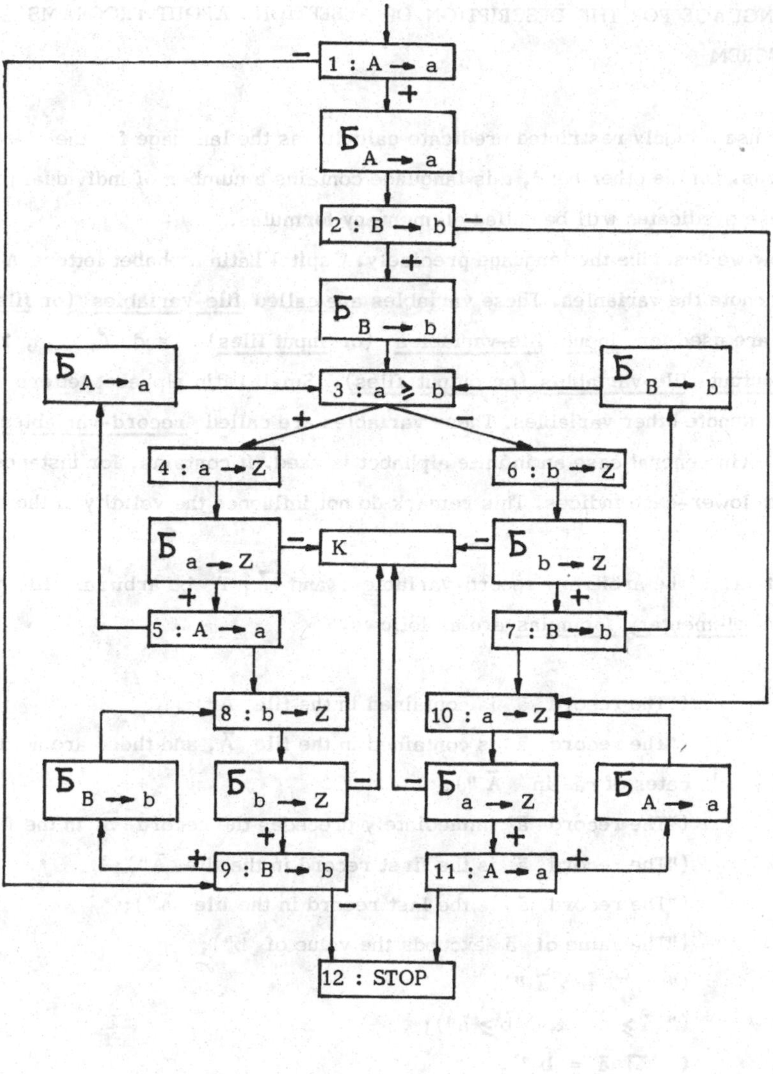
1 : A → a
Б A → a
2 : B → b
Б B → b
3 : a ≥ b
Б A → a
Б B → b
4 : a → Z
6 : b → Z
Б a → Z
K
Б b → Z
5 : A → a
7 : B → b
8 : b → Z
10 : a → Z
Б B → b
Б b → Z
Б a → Z
Б A → a
9 : B → b
11 : A → a
12 : STOP
+
−

Fig.2.

3. A LANGUAGE FOR THE DESCRIPTION OF ASSERTIONS ABOUT PROGRAMS AND THE MAIN THEOREM

We use a highly restricted predicate calculus as the language for the description of assertions. On the other hand, this language contains a number of individual predicates. These predicates will be called elementary formulas.

Now we describe the language precisely. Capital Latin alphabet letters A, B, ... , Z denote the variables. These variables are called *file-variables* (or *files*). A, ... S are used for *input file-variables* (or *input files*), and T, ... , Z are used for *output file-variables* (or *output files*). Small Latin alphabet letters a, b, ... , z denote other variables. These variables are called *record-variables* (or *records*). (In general case an infinite alphabet is used. It contains, for instance, letters with lower-case indices. This remark do not influence the validity of the result below).

Let $\bar{a}$, $\bar{b}$ be arbitrary record-variables, and $\bar{A}$, $\bar{T}$ be arbitrary file-variables. The *elementary formulas* are as follows.

$\bar{a} \in \bar{A}$	("The record $\bar{a}$ is contained in the file $\bar{A}$");
$\bar{a} \in\cdot \bar{A}$	("The record $\bar{a}$ is contained in the file $\bar{A}$, and there are no duplicates of $\bar{a}$ in $\bar{A}$");
$\bar{a} \succ \bar{b} / \bar{A}$	("The record $\bar{a}$ immediately precedes the record $\bar{b}$ in the file $\bar{A}$");
$\bar{a} / \bar{A}$	("The record $\bar{a}$ is the first record in the file $\bar{A}$");
$\bar{a} // \bar{A}$	("The record $\bar{a}$ is the last record in the file $\bar{A}$");
$\bar{a} > \bar{b}$	("The value of $\bar{a}$ exceeds the value of $\bar{b}$");
$\bar{a} \geq \bar{b}$	(" $\neg\, \bar{b} > \bar{a}$ ");
$\bar{a} = \bar{b}$	(" $\bar{a} \geq \bar{b}$ & $\bar{b} \geq \bar{a}$");
$\bar{a} \neq \bar{b}$	(" $\neg\, \bar{a} = \bar{b}$ ").

Formulas are the formulas of first-order predicate calculus built from elementary formulas using &, $\vee$, $\supset$, $\neg$ and quantifiers. There is a severe restriction, namely, only quantifiers over free record-variables are allowed. Hence all the file-variables in our formulas are free. For example, in the formula

$$\forall a \; (a \succ b / A \supset a > b)$$

the free variables are b and A.

<u>Simple formulas</u> are the quantifier-free formulas, i.e. the formulas built from elementary formulas by use of $\&$, $\vee$, $\supset$, $\neg$.

Let $\mathcal{A}(a, b, \ldots)$ be a formula. We consider the formula

$$\forall a \quad \forall b \ \ldots \quad \mathcal{A}(a, b, \ldots),$$

where the universality quantifiers are taken over all the record-variables in $\mathcal{A}$. We denote such a formula by $\{\mathcal{A}\}$. It is easy to see that only file-variables are free in $\{\mathcal{A}\}$.

<u>Local assertion about the file</u> $\bar{A}$ is a formula $\{\mathcal{A}\}$, where $\mathcal{A}$ is a simple formula, and its only file-variable is $\bar{A}$.

<u>Assertion about the file</u> $\bar{T}$ <u>with respect to files</u> $\bar{A}, \bar{B}, \ldots$ is a formula $\{\mathcal{B} \supset \mathcal{C}\}$, where $\mathcal{B}$ is a simple formula containing no other file-variables than $\bar{A}, \bar{B}, \ldots$, and $\mathcal{C}$ is a simple formula containing only one file-variable $\bar{T}$.

<u>Assertion about programs with input files</u> A, B, ... <u>and output files</u> T, U, ... is a formula

$$(\{\mathcal{A}_1\} \& \{\mathcal{A}_2\} \& \cdots) \supset$$
$$\supset (\{\mathcal{B}_1 \supset \mathcal{C}_1\} \& \{\mathcal{B}_2 \supset \mathcal{C}_2\} \& \cdots \{\mathcal{D}_1\} \& \{\mathcal{D}_2\} \& \cdots)$$

where

every $\{\mathcal{A}_i\}$ is a local assertion about an input file $\bar{A} \in \{A, B, \ldots\}$,

every $\{\mathcal{B}_i \supset \mathcal{C}_i\}$ is an assertion about an output file $\bar{T} \in \{T, U, \ldots\}$ with respect to input files A, B, ... ,

every $\{\mathcal{D}_i\}$ is a local assertion about an output file $\bar{T} \in \{T, U, \ldots\}$.

It is easy to see that the free variables in assertions of this type are the file-variables A, B, ... , T, U, ...

We denote the language for the description of assertions about programs by $\mathfrak{a}$.

Two examples of assertions about programs with input files A, B and an output file Z are considered below.

<u>Example 1</u>

$$(\{(x \in A \ \& \ y \in A \ \& \ x \succ y / A) \supset x \leq y\} \ \&$$
$$\& \{(x \in B \ \& \ y \in B \ \& \ x \succ y / B) \supset x \leq y\}) \supset$$
$$\supset (\{x \in Z \ \& \ y \in Z \ \& \ x \succ y / Z) \supset x \leq y\})$$

<u>Example 2</u>

$$\{(x \in A \vee x \in B) \supset x \in Z\}$$

It will follow from the semantics described below that Example 1 is in a sense equivalent to the assertion $\mathfrak{A}$ of Section 2, and Example 2 is equivalent to the assertion $\mathfrak{B}$.

Now we describe the semantics of the language $\mathfrak{a}$. To do this we slightly change the interpretation of the record and of the file given in Section 2. Now we interpret a record as a pair (I, n), where the identifier I of the record is an arbitrary word (I may be the empty word as well) consisting of Latin alphabet letters and digits, and the numerical value n of the record is an arbitrary integer. We interpret a file as a finite sequence of records

$$((I_1, n_1) , (I_2, n_2) , \ldots , (I_k, n_k)).$$

An example of the value of a file:

((DE 1, 7) , (E, -2), (A 1, 5) , (E, -2) , (A 1, 7)).

This example contains two occurrences of the record (E, -2) in the value of the file. By the way, it is easy to express in the language $\mathfrak{a}$ an assertion about a file A prohibiting multiple occurrences of records:

$$\{a \in A \supset a \in\cdot A\}.$$

Now let records $\bar{a}$, $\bar{b}$ and a file $\bar{A}$ take certain values:

$\bar{a} = \bar{a}^o = (I^o_{\mathfrak{a}} , n^o_{\mathfrak{a}})$, $\bar{b} = \bar{b}^o = (I^o_{\mathfrak{b}} , n^o_{\mathfrak{b}})$,

$\bar{A} = \bar{A}^o = ((I^o_1 , n^o_1) , \ldots , (I^o_k , n^o_k)).$

Then the values of the elementary formulas are defined as follows.

$\bar{a} \in \bar{A}$ is true, if $\bar{a}^o$ is contained in $\bar{A}^o$, and false otherwise;

$\bar{a} \in\cdot \bar{A}$ is true, if $\bar{a}^o$ is contained in $\bar{A}^o$ and there are no multiple occurrences of $\bar{a}^o$ in $\bar{A}^o$, and false otherwise;

$\bar{a} \succ \bar{b}/\bar{A}$ is true, if there are occurrences of $\bar{a}^o$ and $\bar{b}^o$ in $\bar{A}^o$ such that $\bar{a}^o$ precedes $\bar{b}^o$, and false otherwise; e.g. for $\bar{a}$ = (A, 5), $\bar{b}$ = (B, 7), $\bar{c}$ = (C, 2) and $\bar{A}$ = ((C, 2), (A, 5), (B, 7), (A, 5)), $\bar{a} \succ \bar{b}/\bar{A}$ is true, $\bar{b} \succ \bar{a}/\bar{A}$ is true, $\bar{a} \succ \bar{c}/\bar{A}$ is false;

$\bar{a} / \bar{A}$ is true, if the first record in $\bar{A}^o$ is $\bar{a}^o$, and false otherwise;

$\bar{a} // \bar{A}$ is true, if the last record in $\bar{A}^o$ is $\bar{a}^o$, and false otherwise;

$\bar{a} > \bar{b}$ is true, if $n^o_{\mathfrak{a}} > n^o_{\mathfrak{b}}$, and false otherwise;

$\bar{a} \geq \bar{b}$ is true, if $n_{\varepsilon}^{o} \geq n_{\delta}^{o}$, and false otherwise;

$\bar{a} = \bar{b}$ is true, if $n_{\varepsilon}^{o} = n_{\delta}^{o}$, and false otherwise;

$\bar{a} \neq \bar{b}$ is true, if $n_{\varepsilon}^{o} \neq n_{\delta}^{o}$, and false otherwise.

The values of simple formulas and assertions about programs for given values of the free variables are defined in the natural way.

Since we have slightly changed the interpretation of files and records, we are to change the interpretation of instructions of the language L_0 as well. Now the values of the records, i.e. arbitrary pairs of (I, n) type, will be considered as the values of the inner variables of the program. The constants of the program will be denoted as (λ, c) , where λ is the empty word, and c is an integer, namely, the numerical value of the given constant. The instructions $A \rightarrow t$, $t \rightarrow u$, $t \rightarrow Z$ now mean the assignment of the values of the records, i.e. pairs of the mentioned type. On the other hand, instructions of $t < u$ type compare only numerical values of the records. Hence the performance of the program in fact do not depend on the identifiers of the records used. The identifiers of the records are introduced only for the interpretation of assertions about programs.

Let F be an assertion about programs with input files $A, B, \ldots$ and output files $T, U, \ldots$ Let a program P use the same files. We say that the <u>assertion F is true for the program P</u> if for all possible values of input files $A = A^{o}$, $B = B^{o}, \ldots$ such that the program P terminates and for the obtained values $T = T^{o}$, $U = U^{o}, \ldots$ of the output files the assertion F is true.

When the language for the description of assertions about programs is fixed, the following principal question arises: given an arbitrary assertion F in the language and given an arbitrary program P , is it possible to find out effectively whether the assertion F is true for the program P. The following theorem gives an affirmative answer in the case of the language α .

<u>THEOREM.</u> There is an algorithm which for an arbitrary assertion F in the language α and for arbitrary program P in the language L_0 constructs by blocks a program P' in the same language L_0 and an instruction $K \in P'$ such that the instruction K is reachable in the program P' if and only if the assertion F is false for the program P.

The proof of the theorem is a bit lengthy and, therefore, we are not able to include it in the paper.

4. CONCLUSION

From the theoretical point of view our research resembles the problem of equivalence of programs in the language L_0. In the general case this problem is apparently extremely hard problem. A highly particular case of this problem, namely the equivalence problem for two-tape finite automata, was solved by Bird [4]. From this point of view our Theorem can be interpreted as the solution of the equivalence problem for programs in the language L_0 under the restriction providing the description of the performance of the given program in the language α. (It is evident that such a description is possible not for every program in the language L_0).

It seems that our Theorem can be generalized by adding new elementary predicates to the language α. However, when extending the language α, it should be remembered that algorithmic unsolvability can appear easily. For example, let a language α' contain α and at the same time let it contain means to express identity of two input files. Then it is easy to prove (by usage of Post's combinatorial problem) that there is no algorithm to recognize whether an assertion in the language α' about programs in the language L_0 is true.

ACKNOWLEDGMENTS

A considerable part of this research was done when the author was a visiting scientist in Cornell University. I wish to express gratitude to Professors J.Hartmanis, R.Constable and D.Gries for stimulating discussions.

REFERENCES

1. Barzdin J., Bicevskis J. and Kalninsh A., Automatic construction of complete sample system for program testing. In: Information Processing 77, (B.Gilchrist, Ed.), pp.57-62, North-Holland Publishing Co., 1977.

2. Bicevskis J., Borzovs J., Straujums U., Zarins A. and Miller E.F., SMOTL - A system to construct samples for data processing program debugging, IEEE Trans. Software Engineering, vol.SE-5, No.1, 1979, pp.60-66.

3. Panzl D.J., Automatic software test drivers, Computer, vol.11, No.4, 1978, pp.44-50.

4. Bird M.R., The equivalence problem for deterministic two-tape automata, J. Comput. and Syst. Scien., vol.7, 1973, pp.218-236.

ASSERTION PROGRAMMING

Andrzej J. Blikle
Institute of Computer Science, Polish Academy of Sciences, PKiN, P.O. Box 22, 00-901 Warsaw, Poland

ABSTRACT. The paper describes a method of the systematic development of programs supplemented by specifications. A program in our sense consists of an instruction (the virtual program) plus the specification which in turn consists of a precondition, a postcondition and a set of assertions. Such a program is called correct if the instruction is totally correct wrt the pre- and postcondition, if it does not abort and if the set of assertions is adequate for the corresponding proof of correctness. In the presented method correct programs are developed by sound transformations. The method is illustrated by the derivation of a bubblesort program.

1. INTRODUCTION

The motivation for structured programming (Dijkstra [14]) was to help the programmer in developing, understanding, documenting and possibly also proving correct programs. The latter goal, however important and worth of effort (Dijkstra [16]), may be quite cumbersome if we first develop the complete program and only then try to prove it correct. It is much easier to develop and prove programs simultaneously since in such a case in structuring a program we may also structure the proof. This idea stimulated many authors to formalize the process of programming by describing its steps as more or less formal transformations (Dijkstra [15], Darlington [11,12], Spitzen, Levitt and Lawrence [30], Wegbreit [31], Bär [2], Burstal and Darlington [10], Bjørner [3]). In such a case every step of the proof of correctness has the same scheme: we prove the correctness of the current version of the program on the strength of the correctness of all former versions. This raises immediately a new idea. Instead of checking each time that our transformation does not violate the correctness, we can prove once and for all that in a certain class of programs this transformation always preserves the correctness (Dershowitz and Manna [13], van Emden [17], Irlik [24], Blikle [4,5,7], Back [1]). In this way the correctness proofs of programs are replaced by the soundness proofs of transformation rules.

In developing programs by sound transformations we avoid the necessity of proving programs correct but at the same time we lose a substantial source of information

An early version of this paper was presented at the international conference Formale Methoden und Mathematische Hilfsmittel für die Softwarekonstruktion, Oberwolfach, January 1979. The present version was prepared when the author was a visiting Mackay professor in the Dept. of Electrical Engineering and Computer Science of the UC Berkeley and was also partly supported by NSF Grant MCS 77-09906.

about the program which is contained in the proof of correctness (cf. Dijkstra [16]). Even if this information may be implicitly accessible to the programmer through the way he has developed the program, it certainly will not be seen by the user. In order to maintain all the advantages of programming by sound transformations without loosing the advantages of having the proof of correctness in an explicit form we propose in this paper to enrich the input-output specification of programs by the specification of the proof of correctness. Technically, the proof of correctness is specified by a set of assertions nested in the appropriate places between the instructions of the program. Program correctness is understood as partial correctness plus non-looping and non-abortion. Such correctness is stronger than so called total correctness which usually means (cf. Manna and Pnueli [28] and Manna [27]) partial correctness plus non-looping. The fact that we deal with the abortion problem makes the classical logic inadequate for the treatment of conditions and assertions in our method. We are using therefore McCarthy's [29] partial logic which perfectly fits to that goal.

Another method of developing programs with assertions has been described by Lee, de Roever and Gerhard [25]. In that method, however, the underlined concept of program correctness is the partial correctness and the assertions coincide with Floyd's invariants.

The fact that we extend the I/O specification of programs (the pre- and postconditions) by adding assertions seems to have the following advantages: First of all, assertions describe local properties of programs which may be helpful not only in program understanding but also in its maintenance and testing. Secondly, knowing assertions we can recheck program correctness in a nearly mechanical way. This option may be of interest whenever we need an extra high reliability of programs; i.e., in microprogramming. Finally, since our assertions satisfy the requirements of proofs of termination, they adequately describe the time complexity of all loops.

The paper is organized as follows: Sec. 2 contains the description of an abstract programming language. Sec. 3 is devoted to the particular logical framework which is needed in order to handle the problem of abortions. Sec. 4 introduces the concept of the assertion-correctness of programs. Program development rules and the problem of their soundness are discussed in Sec. 5. The last Sec. 6 illustrates the method by an example of the bubblesort program. Due to space limitations the proofs of theorems, many comments and also some facts have been omitted in this paper. For more details see Blikle [9].

2. THE LANGUAGE OF PROGRAMS' DEVELOPMENT AND SPECIFICATION

It is not the aim of this paper to concentrate on the technical details of a programming language suitable for our method of program derivation. All we want to convey to the reader is the general idea of such a language along with some technical suggestions about the transformation rules and the programming techniques. The language which is described below should be considered as only an experimental version. Since it represents a certain method of programming and since it is the first approximation of what we may expect to have in the future, we shall call it PROMETH-1 (programming method, 1).

PROMETH-1 is an abstract programming language which allows abstract programming constructions and abstract data types. These abstract elements are used only in program development and documentation and therefore need not be necessarily implementable. Data types in PROMETH-1 are user definable and may be developed together with the development of programs. Each time we derive a program we fix our initial version of data type thus establishing the initial primitives of a problem-oriented version of the language. PROMETH-1 constitutes a mathematical environment for the development and the documentation of programs, rather than a particular programming language. It is not intended for implementation. Instead, we may establish links between PROMETH-1 and any implementable language (like FORTRAN or ALGOL) and define sound transformations from programs in PROMETH into programs in that language.

In this paper by an <u>abstract data type</u> (cf. Guttag [23], Liskov and Zilles [26] ADJ [22]) we mean a many sorted relational system of the form $DT=(D,f_1,\ldots,f_n,q_1\ldots,q_m)$ where D is a nonempty many-sorted carrier and $f_i \in [D^{a_i} \to D]$ and $g_j \in [D^{b_j} \to \{\text{true},\text{false}\}$ are partial functions and partial predicates respectively. The partiality of functions and predicates is an essential point in our approach since it allows us to introduce and investigate the problem of abortion (Sec. 3). The problem of data-type specification is skipped. For the sake of this paper we simply assume that our data type is always somehow defined (e.g., in the set theory). A very elegant formalism for data-type specification is provided by the initial-algebra approach (see ADJ [22], Goguen [21], Erig, Kreowski, Padawitz [19] and the papers referenced there).

Given DT we first establish the primitive syntactical components of PROMETH-1. With each f_i and q_j we associate the symbols F_i and Q_j respectively. For simplicity "=" denotes both, the identity relation in D and the corresponding predicate symbol. For each sort in D we have in the set of q_j's the corresponding <u>sort predicate</u>. This is a unary total predicate which gives the value true for arguments of the given sort and gives false for all other arguments. Typical sort predicates are <u>integer</u> n, <u>array</u> a, etc. We also have constant predicates <u>true</u> and <u>false</u> defined in an obvious way. Having introduced the data-type oriented syntax we establish the infinite set of <u>identifiers</u> (individual variables) IDE and we are ready to define the class EXP of <u>expressions</u> and the class CON of <u>conditions</u>. These classes are mutually recursive:

EXP is the least syntactical class with the following properties:

1) $IDE \subseteq EXP$

2) $F_i(E_1,\ldots,E_{a_i}) \in EXP$ for any $i \leq n$ and any $E_1,\ldots,E_{a_i} \in EXP$

3) <u>if</u> c <u>then</u> E_1 <u>else</u> E_2 <u>fi</u> $\in$ EXP for any $c \in CON$ and any $E_1,E_2 \in EXP$. □

CON is the least syntactical class with the following properties:

1) $Q_j(E_1,\ldots,E_{b_j}) \in CON$ for any $j \leq m$ and any $E_1,\ldots,E_{b_j} \in EXP$

2) $c_1 \to c_2,c_3 \in CON$ for any $c_1,c_2,c_3 \in CON$

3) $(\forall x)c \in CON$ and $(\exists x)c \in CON$ for any $x \in IDE$ and $c \in CON$ □

<u>Remark</u>: In the applications we identify F_i with f_i and Q_j with q_j and allow the infix notation. Typical elementary expressions are therefore $x+\sqrt{y}$, $(x+y)-z$, $\underline{max}\{k|k<2^n\}$, etc. and typical elementary conditions are of the form $z \leq y$, a <u>is sorted</u> $i \leq$ <u>length</u> a, etc. □

Having defined IDE, EXP and CON we can define subsequent syntactical classes: ASR - of <u>assertions</u>, TES - of <u>tests</u>, ASG - of <u>assignments</u>, EIN - of <u>elementary instructions</u>, and INS - of <u>instructions</u>. We use the BNF formalism for this purpose:

ASR::=<u>as</u> CON <u>sa</u>
TES::=<u>if</u> CON <u>fi</u>
ASG::=IDE:=EXP
EIN::=<u>skip</u>|<u>abort</u>|TES|ASG|EIN;EIN
INS::=EIN|INS <u>as</u> CON <u>sa</u> INS|<u>if</u> CON <u>then</u> INS <u>else</u> INS <u>fi</u>|
 <u>while</u> CON <u>do</u> INS <u>as</u> CON <u>sa</u> EXP <u>od</u>|<u>inv</u> CON;INS <u>vni</u>

So far we have defined rather usual programming concepts, although with somewhat extravagant syntax. The latter is the consequence of the assumption that our instructions (programs) are enriched by assertions. In the semantics of instructions these assertions play the role of comments and are simply skipped in the execution. Their role becomes essential in the semantics of <u>assertion specified programs</u> (abbreviated <u>a.s. programs</u>). This class is denoted by ASP and is defined by the equation:

ASP::=<u>pre</u> CON; INS <u>post</u> CON

In a.s. programs the conditions following <u>pre</u> and <u>post</u> are called the <u>precondition</u> and the <u>postcondition</u> respectively. In contrast to instructions, which describe algorithms the a.s. programs are statements about algorithms. This is formalized in Sec.4.

In the description of the semantics of PROMETH-1 we use the algebra of binary relations. For any sets A_1,A_2,A_3 any relations $R_1 \subseteq A_1 \times A_2$, $R_2 \subseteq A_2 \times A_3$ and any subset $C \subseteq A_2$ we define the <u>composition of relations</u> by $R_1R_2=\{(a,b)|(\exists c)(aR_1c \& cR_2b)\}$, the <u>coimage</u> of C w.r.t. R_1 by $R_1C=\{a|(\exists c)(aR_1c \& c \in C)\}$. For any A, by I_A - or simply by I if A is understood - we denote the <u>identity relation</u> in A, i.e., $I=\{(a,a)|a \in A\}$.

By Φ we denote the empty relation and the empty set. If $R \subseteq A \times A$, then $R^0=I$ and $R^{i+1}=RR$ for any integer $i \geq 0$. By the iteration of R we mean $R^*=\cup_{i=0}^{\infty} R^i$. For more details see Blikle [6,8]. By $[A_1 \to A_2]$ and $[A_1 \to A_2]_t$ we denote the set of all partial resp. total functions from A_1 into A_2. By $S=[IDE \to D]_t$ we denote the set of <u>states</u>. Consequently states are valuations of identifiers in the carrier of our data type DT.

In this paper semantics is understood as a function (strictly speaking a many-sorted homomorphism) (ADJ [22]) which assigns meanings to all the investigated syntactical entities. This function is denoted by [ ] hence [X] denotes the meaning of X, where X may be an expression, a condition, an instruction, etc. Of course, depending on the class where X belongs, [X] is of appropriate type:

1) $[\ ]: EXP \to [S \to D]$
2) $[\ ]: CON \to [S \to \{true, false\}]$
3) $[\ ]: INS \to [S \to S]$
4) $[\ ]: ASP \to \{true, false\}$

The semantics of the class EXP is defined by the following recursive (schemes of) equations:

1) $[x](s)=s(x)$
2) $[F_i(E_1,\dots,E_{a_i})](s)=f_i([E_1](s),\dots,[E_{a_i}](s))$
3) $$\underline{\text{if}}\ c\ \underline{\text{then}}\ E_1\ \underline{\text{else}}\ E_2\ \underline{\text{fi}}\ (s) = \begin{cases} [E_1](s) & \text{if } [c](s)=\text{true} \\ [E_2](s) & \text{if } [c](s)=\text{false} \\ \text{undefined} & \text{if } [c](s) \text{ undefined} \end{cases}$$

This coincides with the usual understanding of expressions both in programming languages and in mathematical logic. The semantics of CON is postponed to Sec. 3. Below we define the semantics of INS using standard denotational equations. Let x,E,c and IN, possibly with indices, denote identifiers, expressions, conditions and instructions respectively.

1) $[\underline{\text{as}}\ c\ \underline{\text{sa}}]=[\underline{\text{skip}}]=I$
2) $[\underline{\text{if}}\ c\ \underline{\text{fi}}]=\{(s,s) \mid [c](s)=\text{true}\}$
3) $[x:=E]=\{(s_1,s_2) \mid s_2(x)=[E](s_1) \text{ and } s_2(y)=s_1(y) \text{ for all } y \in IDE-\{x\}\}$
4) $[\underline{\text{abort}}]=\Phi$
5) $[IN_1;IN_2]=[IN_1][IN_2]$
6) $[\underline{\text{if}}\ c\ \underline{\text{then}}\ IN_1\ \underline{\text{else}}\ IN_2\ \underline{\text{fi}}]=[\underline{\text{if}}\ c\ \underline{\text{fi}}][IN_1] \cup [\underline{\text{if}}\ {\sim}c\ \underline{\text{fi}}][IN_2]$
7) $[\underline{\text{while}}\ c\ \underline{\text{do}}\ IN\ \underline{\text{as}}\ c_a\ \underline{\text{sa}}\ E\ \underline{\text{od}}]=([\underline{\text{if}}\ c\ \underline{\text{fi}}][IN])^*[\underline{\text{if}}\ {\sim}c\ \underline{\text{fi}}]$
8) $[IN_1\ \underline{\text{as}}\ c\ \underline{\text{sa}}\ IN_2]=[IN_1][IN_2]$
9) $[\underline{\text{inv}}\ c;\ IN\ \underline{\text{vni}}]=[IN]$

The semantics of the class ASP is described in Sec.4.

3. ON THE PARTIALITY OF CONDITIONS AND THE UNDERLINED LOGIC

In the majority of approaches to the problem of program correctness expressions and conditions represent total functions. This assumption simplifies the mathematical model but is hardly acceptable from the practical point of view. Every programmer knows that both expressions and conditions may lead to abortion if evaluated in an improper environment. For instance we cannot evaluate the condition $a(i)\leq a(j)$ whenever either i or j is outside of the scope of a. Despite this fairly clear motivation the partiality of conditions has not been widely recognized in the theory of programs, although this problem was pointed by McCarthy [29] as early as in 1961. We adopt McCarthy's model as the base for our semantics of conditions.

Similarly as for case of EXP the semantics of CON is defined by a set of (schemes of) recursive equations

1) $[Q_j(E_1,\ldots E_{b_j})](s)=q_j([E_1](s),\ldots,[E_{b_j}](s))$.

2) $$[c_1\rightarrow c_2,c_3](s)=\begin{cases}[c_2](s) & \text{if } [c_1](s)=\text{true}\\ [c_3](s) & \text{if } [c_1](s)=\text{false}\\ \text{undefined} & \text{if } [c_1](s) \text{ undefined}\end{cases}$$

3) $$[(\forall x)c](s)=\begin{cases}\text{true if for any state } s_1 \text{ which differs}\\ \quad\text{from } s \text{ at most in } x, [c](s)=\text{true}\\ \text{false if there exists a state } s_1 \text{ which differs from } s\\ \quad\text{at most in } x, \text{ such that } [c](s_1)=\text{false}\\ \text{undefined in all other cases}\end{cases}$$

The existential quantification is defined in a similar way. For better explanation consider the following example. Let the carrier D contain just two sorts - integers and integer arrays. Let $s\in S$, $y\in IDE$ with $[\underline{\text{integer}}\ y](s)=\text{true}$. Then

1) $[(\forall x)(x+y)^2>0](s)=\text{false}$
2) $[(\forall x)(x+y)^2\geq 0](s)$ is undefined
3) $[(\forall x)(\underline{\text{integer}}\ x\rightarrow(x+y)^2\geq 0,\ \underline{\text{true}})](s)=\text{true}$
4) $[(\exists x)(x+y)^2\leq 0](s)=\text{true}$
5) $[(\exists x)(x+y)^2<0](s)$ is undefined
6) $[(\exists x)(\underline{\text{integer}}\ x\rightarrow(x+y)^2<0,\ \underline{\text{false}})](s)=\text{false}$

Examples 3) and 6) show that it may be worth to extend the syntax of CON by allowing the conditions of the form $(\forall\ \underline{\text{sort}}\ x)c$ and $(\exists\ \underline{\text{sort}}\ x)c$, with the following semantics:

$[(\forall\ \underline{\text{sort}}\ x)c]=[(\forall x)(\underline{\text{sort}}\ x\rightarrow c,\ \underline{\text{true}})]$
$[(\exists\ \underline{\text{sort}}\ x)c]=[(\exists x)(\underline{\text{sort}}\ x\rightarrow c,\ \underline{\text{false}})]$

Now, 3) and 6) can be written in a more readable way: $[(\forall\ \underline{\text{integer}}\ x)(x+y)^2\geq 0](s)=$

true and $[(\exists\ \underline{\text{integer}}\ x)(x+y)^2 < 0](s)=$false. For further convenience we extend CON again by introducing the usual connectives such as $\vee$, &, and $\supset$. We define their semantics after McCarthy [29]:

$$
\begin{aligned}
&1)\quad [c_1 \vee c_2] = [c_1 \to \underline{\text{true}},\ c_2]\\
&2)\quad [c_1 \& c_2] = [c_1 \to c_2,\ \underline{\text{false}}]\\
&3)\quad [\sim c_1] \ = [c_1 \to \underline{\text{false}},\ \underline{\text{true}}]\\
&4)\quad [c_1 \supset c_2] = [c_1 \to c_2,\ \underline{\text{true}}]
\end{aligned}
\tag{3.1}
$$

These connectives constitute a natural generalization of the classical case. In the present case, however, $\vee$ and & are not commutative.

In our approach we frequently have to describe certain metarelationships which may hold between conditions. In order to define them we introduce an auxiliary notation. Let for any c $\{c\}=\{s \mid [c](s)=\text{true}\}$. As is easy to prove, for any c_1 and c_2

$$\{c_1 \& c_2\} = \{c_1\} \cap \{c_2\}$$

$$\{c_1 \vee c_2\} \subseteq \{c_1\} \cup \{c_2\}$$

Now, we define four relations in the set CON:

$c_1 \simeq c_2$ if $[c_1]=[c_2]$ read: c_1 is <u>strongly equivalent</u> to c_2

$c_1 \sqsubseteq c_2$ if $[c_1] \subseteq [c_2]$ read: c_1 is <u>less defined than</u> c_2

$c_1 \Leftrightarrow c_2$ if $\{c_1\}=\{c_2\}$ read: c_1 is <u>equivalent to</u> c_2

$c_1 \Rightarrow c_2$ if $\{c_1\} \subseteq \{c_2\}$ read: c_1 implies c_2

Our strong equivalence coincides with the McCarthy's strong equivalence but our equivalence is not his weak equivalence. The set CON may be regarded as a relational system with the operations $\simeq, \sqsubseteq, \Leftrightarrow, \Rightarrow$. Below we sketch some properties of this system. Here and in the sequel we adopt the convention of using the words <u>equivalent</u> and <u>implies</u> homonymously: in the sense attached to $\Leftrightarrow$ and $\Rightarrow$ and in a colloquial sense, e.g., in saying that $c_1 \sqsubseteq c_2$ implies $c_1 \Rightarrow c_2$. The appropriate meaning is always defined by the context.

<u>THEOREM</u> 3.1. The relations $\simeq$ and $\Leftrightarrow$ are equivalence relations in CON. Moreover $\simeq$ is a congruence, but $\Leftrightarrow$ is not. □

The relation $\Leftrightarrow$ is not a congruence since $c_1 \Leftrightarrow c_2$ does not imply $\sim c_1 \Leftrightarrow \sim c_2$.

<u>THEOREM</u> 3.2. The relations $\sqsubseteq$ and $\Rightarrow$ are partial orderings in $\text{CON}/_{\simeq}$ and $\text{CON}/_{\Leftrightarrow}$ respectively. The operations $\vee$ and & are monotone wrt both these orderings and the remaining operations are monotone only wrt $\sqsubseteq$. □

THEOREM 3.3. The equivalence $\approx$ is strictly stronger than $\Leftrightarrow$, i.e., $c_1 \approx c_2$ implies $c_1 \Leftrightarrow c_2$ but not vice versa. Also the ordering $\sqsubseteq$ is strictly stronger than $\Rightarrow$, i.e., $c_1 \sqsubseteq c_2$ implies $c_1 \Rightarrow c_2$ but not vice versa. □

McCarthy's propositional calculus with the strong equivalence is quite similar to the classical calculus (see Blikle [9]). We have all the associativities and distributivities for v and & as well as de Morgan's laws. We do not have the commutativity of v and & and we have $c_1 v{\sim}c_1 \sqsubseteq$ true and $c_1\&{\sim}c_1 \sqsubseteq$ false instead of the usual tautologies. There are also some laws which hold for $\Leftrightarrow$ and $\Rightarrow$ but does not hold for $\approx$ and $\sqsubseteq$:

$$c_1 \Rightarrow c_1 v c_2$$

$$c_1 \& c_2 \Rightarrow c_1$$

$$c_1 \& c_2 \Leftrightarrow c_2 \& c_1$$

Observe that in $CON/_{\Leftrightarrow}$, & is commutative but v is not. In particular $c_1 \Rightarrow c_2 v c_1$ does not hold!

4. THE CORRECTNESS AND THE ASSERTION CORRECTNESS OF PROGRAMS

As was already mentioned in Sec.2 assertion specified programs are statements about programs and therefore their semantical meanings are truth values. Accordingly to the traditional wording of the field we shall say, however, that an a.s. program is correct rather than true. Below we define two concepts of correctness. An assertion specified program pre c_{pr};IN post c_{po} is called correct if

$$\{c_{pr}\} \subseteq [IN]\{c_{po}\} \tag{4.1}$$

This means that for any state s which satisfies c_{pr} the execution of IN terminates successfully; i.e., neither aborts nor runs indefinitely, and the output state satisfies c_{po}. Observe that in the usual understanding of total correctness (Manna and Pnueli [28], Manna [27]) the problem of abortion is neglected: successful termination simply means no indefinite execution. Consequently, the correctness defined by (4.1) is stronger than the total correctness.

The above defined concept of correctness is restricted to global properties of programs. Below we define the assertion correctness which refers not only to the pre- and postcondition but also to the assertions of the program. Intuitively pre c_{pr}; IN post c_{po} is assertion correct if it is correct and if the assertions which occur in IN may be used in the proof of (4.1). The formal definition is inductive w.r.t. the syntax of INS:

(A) For any elementary instruction IN the a.s. program pre c_{pr}; IN post c_{po} is assertion correct if it is correct. Notice that elementary instructions contain no assertions.

(B) The a.s. program <u>pre</u> c_{pr}; <u>if</u> c <u>then</u> IN_1 <u>else</u> IN_2 <u>fi</u> <u>post</u> c_{po} is <u>assertion correct</u> if

(B1) $c_{pr} \Rightarrow c \vee \sim c$

(B2) <u>pre</u> c_{pr}&c; IN_1 <u>post</u> c_{po} is assertion correct

(B3) <u>pre</u> c_{pr}&~c; IN_2 <u>post</u> c_{po} is assertion correct

(C) The a.s. program <u>pre</u> c_{pr}; <u>while</u> c <u>do</u> IN <u>as</u> c_a <u>sa</u> E <u>od</u> <u>post</u> c_{po} is <u>assertion correct</u> if

(C1) $c_{pr} \Rightarrow c_a \& E \geq 0$

(C2) <u>pre</u> $c_a \& E \geq 1$; <u>if</u> c <u>fi</u>; IN <u>post</u> $c_a \& E \geq 0$ is a.c.

(C3) <u>pre</u> $c_a \& E < 1$; <u>if</u> ~c <u>fi</u> <u>post</u> c_{po} is a.c.

(C4) [<u>if</u> $c_a \& E \geq 1$ <u>fi</u>][IN][E] $\subseteq$ [E-1]

Here E is the <u>loop counter</u>, i.e., a real expression whose integer value gives the number of cycles through IN which must be performed in order to exit from the loop. This concept may be easily generalized using well founded sets (Floyd [20]). The condition c_a is called the <u>loop assertion</u> and loosely speaking describes the global effect of IN.

(D) The a.s. program <u>pre</u> c_{pr}; IN_1 <u>as</u> c_a <u>sa</u> IN_2 <u>post</u> c_{po} is <u>assertion correct</u> if

(D1) <u>pre</u> c_{pr}; IN_1 <u>post</u> c_a is assertion correct

(D2) <u>pre</u> c_a; IN_2 <u>post</u> c_{po} is assertion correct

(E) The a.s. program <u>pre</u> c_{pr}; <u>inv</u> c_i; IN <u>vni</u> <u>post</u> c_{po} is <u>assertion correct</u> if the a.s. program <u>pre</u> c_{pr}; IN_1 <u>post</u> c_{po} where IN_1 results in from IN by the substitution for each assertion <u>as</u> c_a <u>sa</u> in IN the assertion <u>as</u> $c_a \& c_i$ <u>sa</u>, is <u>assertion correct</u>.

The condition c_i in (E) is called the <u>permanent invariant</u> and <u>inv</u> c_i is called its <u>declaration</u>. The mirror key-word <u>vni</u> defines the <u>scope</u> of this declaration. Permanent invariants are used to "factorize" conditions which are permanently satisfied in a segment of a program. Typical factorizable conditions are these which describe the unchangable properties of the environment, e.g., the sort of identifiers.

<u>THEOREM</u> 4.1. Every a.s. program which is assertion correct is correct. □

5. THE RULES OF THE COMPOSITION AND THE TRANSFORMATION OF PROGRAMS

Due to space limitations we show only a few rules, mainly these which we need in our example of Sec.6. First of all we have five composition rules which are implicit in the definition (A)-(E) of assertion correctness (Sec.4). Below we give an example of a transformation which introduces an invariant into a program. Generally speaking, given an assertion correct program pre c_{pr}; IN post c_{po} and a condition c we say that we are introducing the invariant c into this program if we transform IN onto a certain IN_1 such that pre $c_{pr}\&c_1$; IN_1 post $c_{po}\&c_1$ is assertion correct. In the description of our transformation we use the inclusion of the form $[IN][E]\subseteq[E]$, where $IN\in INS$ and $E\in EXP$. This inclusion is equivalent to the implication $(\forall s_1,s_2)(s_1[IN]s_2 \Rightarrow [E](s_1)=[E](s_2))$, which says that the value of E before and after the execution of IN is the same. E.g. $[x:=x/z][y+z]\subseteq[y+z]$.

THEOREM 5.1. If pre c_{pr}; while c do IN as c_a sa E od post c_{po} is assertion correct, then for any $c_1,c_1'\in CON$ and any $IN_1\in INS$ if

1) pre $c_a\&c_1\&E\geq 1$; if c fi; IN post $c_a\&c_1'$ is assertion correct

2) pre $c_a\&c_1'$; IN_1 post $c_a\&E\geq 0\&c_1$ is assertion correct

3) [if $c_a\&c_1'\&E\geq 0$ fi]$[IN_1][E]\subseteq[E]$

then

pre $c_{pr}\&c_1$; while c do IN as $c_a\&c_1'$ sa IN_1 as $c_a\&c_1$ sa E od post $c_{po}\&c_1$

is assertion correct. □

COMMENT. Since IN violates the required invariant c_1 (assumption 1)), we have to supply the loop body with a recovery instruction IN_1 leading back to c_1 (assumption 2)). To make it sure that the alteration of the loop does not violate the termination property, we assume that IN_1 preserves the value of the loop counter E (assumption 3)). □

A large group of programming rules consists of transformations which modify only the condition in a program. Two examples of such rules are given below.

THEOREM 5.2. If the a.s. program pre c_{pr}; IN post c_{po} is assertion correct and $c'_{pr} \Rightarrow c_{pr}$ and $c_{po} \Rightarrow c'_{po}$, then pre c'_{pr}; IN post c'_{po} is assertion correct. □

THEOREM 5.3. If in an arbitrary assertion correct program we replace:

1) any while-do or if-then-else condition c by c_1 such that $c\approx c_1$,

2) any precondition, postcondition or assertion c by c_1 such that $c \Leftrightarrow c_1$,

then the resulting program is assertion correct. □

The essential point in the last theorem is that we cannot replace $\approx$ by $\Leftrightarrow$ in 1).

An appropriate example is given in Sec.6. We shall also see in that section that many conditions, appearing in a.s. programs are of the form $c_1\&...\&c_n$. Since & commutes in CON/$_{\Leftrightarrow}$ but does not commute in CON/$_{\approx}$ (Sec.3), our theorem indicates that the ordering of c_i's in $c_1\&...\&c_n$ is irrelevant in preconditions, postconditions and assertions but is relevant in while-do and if-then-else conditions.

6. AN EXAMPLE OF PROGRAM DERIVATION; BUBBLESORT

To get started we recall the intuitive idea of bubblesort. Suppose that we are given a vertical column of bubbles, each bubble having a certain weight. Suppose that our bubbles are immersed in an environment which satisfies the following Archimedes' principle: each bubble which is lighter than its upper neighbor tends to swap with this neighbor in moving up. At some initial moment all the bubbles are glued together. In the first step of bubblesort we free the first bubble from the top. Of course nothing happens since this bubble has no upper neighbor. Next we free the second bubble. This time a swap may occur if the second bubble is lighter than the first one. In each successive step of our procedure we free the successive bubble and this bubble immediately starts to move up in searching for such a position in the column which does not violate the Archimedes' principle.

The systematic development of the bubblesort program requires, first of all, the establishment of an appropriate data type. This data type will be developed in a stepwise manner along with the development of the program. Since in this paper we skip the problem of the formal specification of data type, we are using below a mixture of formal and intuitive mathematics. We start by the first approximation of our data type and program.

SORTS: Int - integers; Arr - arrays; each array is a total function, $a:\{0,...,n\} \to$ Int, where $n \geq 0$; Bol - {true, false}.

FUNCTIONS: +, -, 0, 1 - the arithmetical functions and constants
length: Arr $\to$ Int - the length of an array
component: Arr x Int $\to$ Int - the i-th component of an array; according to the common style we shall write a(i) rather than component(a,i)
seg: Arr x Int $\to$ Arr - the initial segment;
seg $(a,j)=(a(0),...,a(j))$ for $0 \leq j \leq$ length a

PREDICATES:
integer, array - the sort predicates (Sec.2)
$\leq$, < - the usual arithmetical inequalities
is sorted: Arr $\to$ Bol; a is sorted: $\approx$ ($\forall$integer i)($0 \leq i <$ length a $\supset a(i) \leq a(i+1)$)
perm: Arr x Arr $\to$ Bol; a_1 perm a_2: $\approx a_1$ is a permutation of a_2

Now, we may establish the first approximation of our program which we informally call the propulsion loop. Here and in the sequel the operational part of the program will be framed in order to distinguish it visually from the specification part.

pre array A&a=A&j=0&k=length A

inv k=length a&a perm A&0≤j≤k

| while j<k do j:=j+1 | (P$_1$)

as true sa k-j od

vni

post j=k

This program only defines the framework of further approximations and is, obviously, assertion correct. Into this program we introduce the invariant seg (a,j) is sorted using Theorem 5.1. Let

c_1:≈seg (a,j) is sorted

c_1':≈seg (a,j-1) is sorted & j≥1

c:≈k=length a&a perm A&0≤j≤k

Of course, c is the permanent invariant declared in P$_1$. Now, according to Theorem 5.1 we have to check that the program

pre c&c_1&k-j≥1; | if j<k fi; j:=j+1 | post c&c_1'

is assertion correct and we have to construct an instruction IN$_1$ such that the following two conditions are satisfied:

pre c&c_1'; IN$_1$ post c&c_1&k-j≥0   is assertion correct   (6.1)

[if c&c_1'&k-j≥0 fi][IN$_1$][k-j]⊆[k-j]   (6.2)

The first requirement is, of course, satisfied. Therefore, on the strength of Theorem 5.1, for any IN$_1$ which satisfies (6.1) and (6.2) the subsequent program is assertion correct. We write it already in a simplified form removing c_1 from the precondition - since for j=0 it is always true - and replacing j=k&c_1 in the postcondition by j=k&a is sorted, since j=k&k=length a & seg(a,j) is sorted implies j=k&a is sorted. Formally we apply here the Theorems 5.2 and 5.3.

pre array A&a=A&j=0&k=length A

inv c

```
while j<k do
   j:=j+1
```
(P_2)

as seg (a,j-1) is sorted & j≥1 sa

```
      IN1
```

as seg (a,j) is sorted sa k-j od

vni

post j=k&a is sorted

Since there are many IN_1 which satisfy the conditions (6.1) and (6.2), our P_2 represents a class of sorting procedures organized according to the following iterative scheme: given an array a where seg(a,j) has already been sorted, increase j by 1 and permute a in such a way that the new seg(a,j) is sorted again. Our prospective bubblesort belongs to this class. In order to describe it we extend our data type by two new sorts, four new functions and one new predicate.

SORTS

Vec - vectors; each vector is a total function $v:N \to Int$ where N is an arbitrary finite set of integers

Set - finite subsets of Int

FUNCTIONS

swap: Arr x Int x Int → Arr; swap (a,i,j) is, for 0≤i,j<length a, the result of swapping the i-th with the j-th element in a

but: Arr x Int → Vec; a but i is, for 0≤i<length a, the restriction of array a to the domain {0,..., length a} - {i}

max: Set → Int; max B is the maximal element of the set B

bd: Arr x Int → Int; read: bubbledepth

bd(a,i)=if i≤0 ∨ a(i)≥a(i-1) then 0 else max {d|a(i)<a(i-d)}

PREDICATES

First we extend the formerly defined predicate is sorted to the sort of vectors. We also assume that the empty vector satisfies this predicate. Now, we define the new predicate.

bubbles in seg(,): IntxIntxArr → Bol

i bubbles in seg(a,j):≡0≤i≤j≤length a & seg(a,j) but i is sorted & i<j⊃a(i+1)>a(i)

The following may be proved easily:

bd(a,i)≥1≡i>0&a(i)<a(i-1) (6.3)

i=j&j≥1&i bubbles in seg(a,j) ⇔ i=j&j≥1& seg(a,j-1) is sorted (6.4)

bd(a,i)=0&i bubbles in seg(a,j) ⇒ seg(a,j) is sorted (6.5)

Using the predicate i bubbles in seg(a,j) we may construct the assertion-specified program which describes the bubbling process:

```
pre c&i=j&j≥1&i bubbles in seg(a,j)
inv c
   while bd(a,i)≥1 do
      a:=swap(a,i-1,i)
      as i-1 bubbles in seg(a,j) sa                    (P3)
      i:=i-1
      as i bubbles in seg(a,j) sa bd(a,i) od
vni
post c&bd(a,i)=0&i bubbles in seg(a,j)
```

The assertion correctness of this program may be proved directly from the definitions (C) and (D) of Sec.4. Now, we modify (P_3) into the form required by the conditions (6.1) and (6.2). This is done in the following steps.

(1) The pre- and postcondition are modified on the strength of (6.4) and (6.5); cf. Theorem 5.2.

(2) The while condition bd(a,i)≥1, which is inacceptable from the practical viewpoint, is replaced by i>0&a(i)<a(i-1); cf. (6.3) and Theorem 5.3.

(3) The program which results in from (1) and (2) is combined sequentially (rule (D) of Sec.4) with the program

```
pre c&j≥1&seg(a,j-1) is sorted
   i:=j
post c&i=j&j≥1&seg(a,j-1) is sorted
```

We get

```
pre c&j≥1&seg(a,j-1) is sorted
inv c
 ┌─────┐
 │ i:=j │
 └─────┘
  as i=j&j≥1&seg(a,j-1) is sorted sa
 ┌──────────────────────────────┐
 │ while i>0&a(i)<a(i-1) do     │
 │   a:=swap(a,i-1,i)           │                                (P4)
 └──────────────────────────────┘
    as i-1 bubbles in seg(a,j) sa
   ┌────────┐
   │ i:=i-1 │
   └────────┘
    as i bubbles in seg(a,j)sa bd(a,i) od
vni
post c&seg(a,j) is sorted
```

Since c ⇒ k-j≥0, the latter condition may be added to the postcondition of (P_4). Therefore, the instruction of (P_4) satisfies (6.1). It also satisfies (6.2) since neither j nor k is modified in (P_4). Consequently, the instruction of (P_4) has all the properties required for IN_1 of (P_2) and may be substituted there. In this way we get the final version of our program:

```
pre array A&a=A&j=0&k=length A
inv k=length A&a perm A&0≤j≤k
 ┌─────────────┐
 │ while j<k do │
 │   j:=j+1     │
 └─────────────┘
    as j≥1&seg(a,j-1) is sorted sa
   ┌──────┐
   │ i:=j │
   └──────┘
    as i=j&j≥1&seg(a,j-1) is sorted sa
   ┌──────────────────────────────┐
   │ while i>0&a(i)<a(i-1) do     │
   │   a:=swap(a,i-1,i)           │
   └──────────────────────────────┘
      as i-1 bubbles in seg(a,j) sa
     ┌────────┐
     │ i:=i-1 │
     └────────┘
      as i bubbles in seg(a,j) sa bd(a,i) od
    as seg(a,j) is sorted sa k-j od
vni
post j=k&a is sorted
```

Observe that if in the inner loop we replace the while condition i>0&a(i)<a(i-1) by the condition a(i)<a(i-1)&i>0 which is equivalent - but not strongly equivalent - to the former, then we get a program which is no longer correct. That new program aborts whenever the value of i reaches 0 since in that case a(i)<a(i-1) cannot be evaluated.

ACKNOWLEDGEMENTS. I wish to express my thanks to Hans Eckart Sengler who discovered a mistake in the early version of my example and to Antoni Mazurkiewicz who suggested that I use McCarthy's logic as a partial logic in my approach. I also wish to express my gratitude to Krzysztoft Apt with whom I have discussed some general logical problems of my paper. Finally my thanks are addressed to Andrzej Tarlecki who conveyed to me many interesting remarks about the earlier versions of this paper.

REFERENCES

1. Back, R.J.,On the correctness of refinement steps in program development, Dept. of Computer Science, University of Helsinki, Report A-1978-4.

2. Bär, D.,A Methodology for simultaneously developing and verifying PASCAL programs, in: Constructing Quality Software (Proc. IFIP TC-2 Working Conf., May 1977, Novosibirsk), North Holland, Amsterdam 1978.

3. Bjørner, D., The Vienna development method (VDM): Software specification & program synthesis, in: Mathematical Studies of Information Processing (Proc. Int. Conf. Kyoto, August 1978), 307-340,to appear in LNCS by Springer Verlag.

4. Blikle, A., A mathematical approach to the derivation of correct programs, in: Semantics of Programming Languages (Proc. Int. Workshop, Bad Honnef, March 1977), Abteilung Informatik, Universität Dortmund, Bericht Nr. 4,1 (1977),25-29.

5. Blikle, A., Towards mathematical structured programming, in: Formal Description of Programming Concepts (Proc. IFIP Working Conf. St. Andrews, N.B. Canada, August 1-5, 1977, E.J. Neuhold, ed.),183-202, North Holland, Amsterdam 1978.

6. Blikle, A., A comparative review of some program-verification methods. Mathematical Foundations of Computer Science 1977 (Proc. 6th Symp. Tatranska Lomnica, 1977),Lecture Notes in Computer Science 53,Springer Verlag, Heidelberg 1977, 17-33.

7. Blikle, A., Specified programming, in: Mathematical Studies of Information Processing (Proc. Int. Conf. Kyoto, August 1978).

8. Blikle, A., A survey of input-output semantics and program verification. ICS PAS Reports 344 (1979).

9. Blikle, A., On the development of correct programs with the documentation, College of Eng., University of California, Berkeley, Memo UCB/ERL, M7925, April 1979.

10. Burstall, R.M. and Darlington, J., A transformation system for developing recursive programs, J. ACM, 24 (1977), 44-67.

11. Darlington, J., Applications of program transformation to program synthesis, Proc. Symp. of Proving and Improving Programs, Arc-et-Senans 1975, pp. 133-144.

12. Darlington, J., Transforming specifications into efficient programs. New Directions in Algorithmic Languages 1976 (ed. S.A. Schuman), IRIA Rocquencourt 1976.

13. Dershowitz, N. and Manna, Z., Inference rules for program annotation, Report No. STAN-CS-77-631 (1977).

14. Dijkstra, E.W., A constructive approach to the problem of program correctness, BIT 8 (1968), 174-186.

15. Dijkstra, E.W., Guarded commands, non-determinism and a calculus for the derivation of programs, Proc. 1975 Int. Conf. Reliable Software, pp. 2.0-2.13. Also in: Comm. ACM, 18 (1975), 453-457.

16. Dijkstra, E.W., Formal techniques and sizable programs, Proc. 1st Conf. Eur. Coop. Inf. Amsterdam 1976, Lecture Notes Comput. Sci. 44, 225-235 (1976).

17. Emden van, M.H., Verification conditions as representations for programs, manuscript (1975).

18. Emden van, M.H., Unstructured systematic programming, Dept. CS, Univ. Waterloo, CS-76-09 (1976).

19. Erig, H., Kreowski, H.J., Padawitz, P., Stepwise specification and implementation of abstract data types, in: Automata Languages and Programming (Proc. Fifth Coll. Udine, July 1978), Springer-Verlag LNCS 62, New York 1978, 205-226.

20. Floyd, R.W., Assigning meanings to programs, Proc. Sym. in Applied Math., 19 (1967), 19-32.

21. Goguen, J., Some ideas in algebraic semantics, (manuscript) presented at IBM Conference in Kyoto (Japan), 1978.

22. Goguen, J.A., Thatcher, J.W., Wagner, E.G., Wright, J.B., Abstract data types as initial algebras and correctness of data representations, Proc. Conf. on Comp. Graphics, Pattern Recognition and Data Structure, May 1975, 89-93.

23. Guttag, J., Abstract data types and the development of data structures, Comm. ACM, 20(1977), 396-404.

24. Irlik, J., Constructing iterative version of a system of recursive procedures, In: MFCS (Proc. Int. Symp. MFCS'76) Lecture Notes in CS, Springer-Verlag, Heidelberg 1976, vol. 45.

25. Lee, S., de Roever, W.R., Gerhart, S.L., The evolution of list-copying algorithms, 6th ACM Symposium on Principles of Programming Languages, January 1979.

26. Liskov, B.H., Zilles, S.N., Specification techniques for data abstraction, IEEE Trans. on SE Vol.Se-1,No.1 (1975), 7-19.

27. Manna, Z., Mathematical Theory of Computation, McGraw-Hill, New York 1974.

28. Manna, Z., Pnueli, A., Axiomatic approach to total correctness of programs, Acta Informatica (1974).

29. McCarthy, J., A basis for a mathematical theory of computation, in: Computer Programming and Formal Systems, R. Braffort and D. Hirschberg eds., North-Holland,Amsterdam 1976, pp. 33-70.

30. Spitzen, J.M., Levitt, K.N., Lawrence, R., An example of hierarchial design and proof. New Directions in Algorithmic Languages 1976 (ed. S.A. Schuman), IRIA, Rocquencourt 1976.

31. Wegbreit, B., Goal-directed program transformations, IEEE TSE. Vol. SE-2, No. 2, (1976), 69-79.

COMPLEXITY CLASSES OF FORMAL LANGUAGES†

(Preliminary Report)

Ronald V. Book

Department of Mathematics

University of California at Santa Barbara

Santa Barbara, California 93106, U.S.A.

Introduction

How do time and space relate? How is the structure of a complexity class related to the structure of its various relativized versions? What classes of sets can serve as bases or sub-bases for generating the most common complexity classes of formal languages?

In this paper these questions are investigated by studying complexity classes of formal languages and considering their positive closure properties. Recently, a number of new characterization theorems for complexity classes have been developed [3-5,10] and these characterization theorems are discussed and interpreted here in the context of the questions above. While each of the characterization theorems in [3-5,10] may be considered to be a technical result of limited interest, the usefulness of these results becomes clear when the entire collection is viewed as a whole (and this could not be done until the results of [5] were discovered). This paper is a summary of the complete development [6].

In studying computational complexity one must specify a model for computation and then specify the measure of difficulty to be used in studying this model. When studying dynamic measures, the most common examples are the running time (the number of steps in the computation) and the amount of space (memory cells) used in a computation. One wishes to know the inherent space (time) complexity of a problem. Also, if the

---

†The research reported here was supported in part by the National Science Foundation under Grant MCS77-11360.

space (time) complexity of a problem or a class of problems is known, then one wishes to bound the time (space) needed in terms of space (time) used. Here we study classes of formal languages specified by nondeterministic oracle machines with time bounds or space bounds taken from certain sets of bounding functions, e.g., for any language A, the class NP(A) of languages accepted in polynomial time by nondeterministic oracle machines using the language A as oracle set. Characterization theorems for classes specified by time-bounded (space-bounded) oracle machines are developed. Using the class of regular sets as the sub-basis, closure under a small set of operations characterizes the class specified by the time-bounded machines, and the class specified by the space-bounded machines is obtained by adding one more closure operation, an operation that captures the notion of the transitive closure of a length-preserving binary operation.

We will say that a class of languages is "weakly transitively closed" if a language encoding the transitive closure of length-preserving binary relation is in the class of languages whenever a language encoding the relation itself is in that class. (This notion of being "weakly transitively closed" is defined in Section 3.) <u>For a suitably nice set of bounding functions and any oracle set, the class specified by time-bounded machines is equal to the class specified by space-bounded machines if and only if the class specified by time-bounded machines is weakly transitively closed</u>. For example, for any language A, if PSPACE(A) is the class of languages accepted by machines using polynomial space and oracle set A, then NP(A) = PSPACE(A) if and only if NP(A) is weakly transitively closed. Thus we have an algebraic closure operation that characterizes the difference between time and space, and the definition of this operation does not depend on the set of bounds used.

The notion of relativizing the computations of an algorithm or of a Turing machine plays an important role in recursive function theory. In complexity theory there are several reasons for introducing the notion of relativization: to see if the methods that have been successful in recursive function theory can be successful in complexity theory; to understand the structure of certain classes and the structural differences between different types of classes; to better understand the ideas of complexity-bounded reducibilities and the appropriate notions of complete sets; to develop a theory showing how knowledge of the computations of one algorithm can help in computing another algorithm. Research on complexity-bounded reducibilities has led to the study of complexity classes of formal languages specified by various types of oracle machines. In this paper the classes NTIME($\mathcal{B}$,A) and NSPACE($\mathcal{B}$,A) are characterized in terms of closure operations and the oracle set A, where $\mathcal{B}$ is a suitable set of bounding functions. This algebraic characterization of the relativized version of a class in terms of the oracle set and the operations used to characterize the class itself does not require additional operations which depend on the measure or on the set of bounds used. _To go from the class_ NSPACE($\mathcal{B}$) _to its relativized version_ NSPACE($\mathcal{B}$,A) _we require exactly the same information as used to characterize_ NTIME($\mathcal{B}$,A) _in terms of_ NTIME($\mathcal{B}$), _and this information is uniform in the oracle set_ A. Further, for certain classes of languages encoding relations such as the rudimentary relations, this method can be used to characterize the relativized version of the class in terms of the operations used to characterize the class itself and the oracle set.

Section 1. Preliminaries

It is assumed that the reader is familiar with the basic notions from the theories of automata, formal languages, and automata-based complexity. Here we establish notation.

For a string w, the length of w is denoted by $|w|$, the reversal of w by w^R, and $w^1 = w$.

A homomorphism $h : \Sigma^* \to \Delta^*$ is _nonerasing_ if for all $w \in \Sigma^*$, $h(w) = e$ implies $w = e$. If f is a function, $h : \Sigma^* \to \Delta^*$ a homomorphism, and $L \subseteq \Sigma^*$ a language, then h is _f-erasing on_ L if there is some $k > 0$ such that for all $w \in L$ with $|w| > k$, $|w| \leq kf(|h(w)|)$. If $\mathcal{B}$ is a set of functions, $h : \Sigma^* \to \Delta^*$ a homomorphism, and $L \subseteq \Sigma^*$ a language, then h is _$\mathcal{B}$-erasing on_ L if there is some $f \in \mathcal{B}$ such that h is f-erasing on L.

A function f _majorizes_ a function g if for all $n \geq 0$, $f(n) \geq g(n)$.

Section 2. Time Classes

In this section we define "oracle machines" and state the characterizations of the classes specified by time-bounded machines.

An _oracle machine_ is a multitape Turing machine M with a distinguished work tape, the _oracle_ tape, and three distinguished states, QUERY, YES, and NO. At some step of a computation on an input string, M may transfer into state QUERY. In state QUERY, M transfers into state YES if the string currently appearing on the oracle tape is in some _oracle set_ A; otherwise, M transfers into state NO; in either case, the oracle tape is instantly erased. The set of strings _accepted by_ M relative to the oracle set A is L(M,A) = {w | there is an accepting computation of M on input w when the oracle set is A}.

When we consider an oracle machine operating within some time bound, then the oracle calls are also counted within that time bound.

Let A be any language and let T be a function (of the length of the input string) that serves as a time bound. Let DTIME(T,A) = {L(M,A) | M is a deterministic oracle machine that operates in time bound T} and let NTIME(T,A) = {L(M,A) | M is a nondeterministic oracle machine that operates in time bound T}.

Let $\mathcal{B}$ be a set of functions such that for all $f \in \mathcal{B}$ and all $m, n \geq 0$, $f(m) + f(n) \leq f(m+n)$ and $f(n) \geq n$. For any language A, the class of languages that are nondeterministic (deterministic) $\mathcal{B}$-time in A is NTIME($\mathcal{B}$,A) = $\cup$\{NTIME(f,A) | $f \in \mathcal{B}$\} (DTIME($\mathcal{B}$,A) = $\cup$\{DTIME(f,A) | $f \in \mathcal{B}$\}).

If a machine's instructions do not allow any oracle calls, then the oracle tape may as well not be present. Another way of interpreting this is to say that the oracle set is empty. When A is the empty set, we write NTIME($\mathcal{B}$) (DTIME($\mathcal{B}$)) for NTIME($\mathcal{B}$,ϕ) (resp., DTIME($\mathcal{B}$,ϕ)).

For any oracle set A, there is one class of particular interest here. Let $NP(A) = \bigcup_{k\geq 1} NTIME(n^k,A)$ so that NP(A) is the class of languages that are nondeterministic polynomial time in A. Also, NP is the class of languages accepted in polynomial time by nondeterministic machines.

The classes to be characterized in this section are those of the form NTIME($\mathcal{B}$,A) and NTIME($\mathcal{B}$). We assume that each set $\mathcal{B}$ of time bounds is closed under composition.

To obtain our characterizations we need an operation on languages that allows us to perform "nondeterministic copying."

Let n be a positive integer and let ρ be a function from {1,...,n} to {1,R}. Let L be a language and let $h_1,\ldots,h_n$ be a sequence of n homomorphisms. The language $\langle \rho; h_1,\ldots,h_n \rangle (L) = \{h_1(w)^{\rho(1)} \ldots h_n(w)^{\rho(n)} \mid w \in L\}$ is a homomorphic replication of L. Let $\mathcal{L}$ be a class of languages and let $\mathcal{B}$ be a class of functions. If for every n > 0, every function $\rho : \{1,\ldots,n\} \to \{1,R\}$, every language $L \in \mathcal{L}$, and every sequence $h_1,\ldots,h_n$ of n homomorphisms, each of which is nonerasing ($\mathcal{B}$-erasing on L), the language $\langle \rho; h_1,\ldots,h_n \rangle (L)$ is in $\mathcal{L}$, then $\mathcal{L}$ is closed under nonerasing ($\mathcal{B}$-erasing) homomorphic replication.

Clearly a class of languages closed under nonerasing ($\mathcal{B}$-erasing) homomorphic replication is closed under nonerasing ($\mathcal{B}$-erasing) homomorphism.

If for each i, $\rho(i) = 1$, then we have a homomorphic duplication.

Notation. For a language A, let Δ be the smallest alphabet such that $A \subseteq \Delta^*$, define $\bar{A} = \Delta^* - A$ and let $A \oplus \bar{A} = \{c\}A \cup \{d\}\bar{A}$ for any two symbols, c, d $\notin \Delta$.

Now we can state the characterization theorem.

Theorem 1. Let $\mathcal{B}$ be a set of time bounds containing some function that majorizes the function $f(n) = n^2$.

(a) For any language A, the class NTIME($\mathcal{B}$,A) of languages that are nondeterministic $\mathcal{B}$-time in A is the smallest class containing all regular sets and the language $(A \oplus \bar{A})^*$, and closed under intersection, $\mathcal{B}$-erasing homomorphic replication (duplication), and inverse homomorphism.

(b) The class NTIME($\mathcal{B}$) of languages accepted in time $\mathcal{B}$ by nondeterministic machines is the smallest class containing all regular sets and closed under intersection and $\mathcal{B}$-erasing homomorphic replication (duplication).

To what classes does Theorem 1 apply? Since it is assumed that each set $\mathcal{B}$ of time bounds is closed under composition, the requirement that $f(n) = n^2$ be majorized in $\mathcal{B}$ implies that for every polynomial p there is a function $q \in \mathcal{B}$ such that for all $n \geq 0$, $p(n) \leq q(n)$. Thus for each set $\mathcal{B}$ and every language A, NP $\subseteq$ NTIME($\mathcal{B}$) and NP(A) $\subseteq$ NTIME($\mathcal{B}$,A). If $\mathcal{B}$ is the class of functions obtained from $f(n) = n^2$ by composition, then for any language A, NTIME($\mathcal{B}$,A) = NP(A). If $\mathcal{B}$ is the class of recursive (primitive recursive, partial recursive) functions, then NTIME($\mathcal{B}$,A) is the class of languages that are recursive (resp., primitive recursive, partial recursive) in A. Similarly, we obtain for each $k \geq 3$, the class $\mathcal{E}^k_*(A)$ of languages that are $\mathcal{E}^k$ in A where $\mathcal{E}^k$ is the Grzegorczyk class [12]. Other classes to which these characterizations apply include each of the Σ-classes in the arithmetic hierarchy and in the arithmetic hierarchy relativized to language A [15,17], and in the polynomial-time hierarchy and in the polynomial-time hierarchy relativized to A [20-22]. Also, these characterizations can be applied to the linear-time hierarchy, the linear-time hierarchy relativized to A, and the appropriate Σ-classes even though linear functions do not majorize $f(n) = n^2$ [22-24].

Suppose that in Theorem 1 we require closure under complementation, hence, under the Boolean operations. In three cases this characterizes a class of languages that is of interest here. If $\mathcal{B}$ is the class of partial recursive functions, then for any language A the smallest class containing all regular sets and $((A \oplus \bar{A})^*)$ and closed under the Boolean operations, $\mathcal{B}$-erasing homomorphic replication, and inverse homomorphism is the class of sets that are arithmetic in A [10]. If $\mathcal{B}$ is the class of polynomials, then for any language A the smallest class containing all regular sets and $(A \oplus \bar{A})^*$ and closed under the Boolean operations, $\mathcal{B}$-erasing homomorphic replication, and inverse homomorphism is the class of sets that are extended rudimentary in A [10]. If $\mathcal{B}$ is the class of linear functions (so that $f(n) = n^2$ is not in $\mathcal{B}$), then for any language A the smallest class containing all regular sets and $(A \oplus \bar{A})^*$, and closed under the Boolean operations, $\mathcal{B}$-erasing homomorphic replication (duplication), and inverse homomorphism is the class of sets that are rudimentary in A [10].

In Theorem 1 the sets of time bounds are required to contain $f(n) = n^2$. Let us consider the case when the set $\mathcal{B}$ is the set of linear functions. It is not known whether every language in NTIME($\mathcal{B}$) can be obtained from the class of regular sets using the operations of intersection and homomorphic replication; in fact, it is conjectured that the Dyck sets cannot be so obtained [8].

Consider the behavior of a machine to be restricted so that its read-write heads can make only a bounded number of changes of direction. Such a machine is called <u>reversal-bounded</u>. Let $\mathcal{L}_{BNP}$ be the class of languages accepted in linear time by nondeterministic reversal-bounded machines [9]. If an oracle machine has its work tape heads reversal-bounded but has no restriction on the oracle tape head, then we say that the oracle machine is reversal-bounded. For any language A, let $\mathcal{L}_{BNP}(A) = \{L(M,A) \mid M$ is a reversal-bounded nondeterministic oracle machine that operates in linear time$\}$.

A <u>reset tape</u> is a one-way infinite tape with one read-write head which moves only left-to-right and can be reset once to the left end of the tape. A nondeterministic multiple-reset machine is an acceptor with a one-way input tape, finite-state control, and some finite number of reset tapes as work tapes. Let MULTI-RESET denote the class of languages accepted in linear time by nondeterministic multiple-reset machines [7].

Just as in the case of reversal-bounded machines, we can consider oracle machines that have reset tapes as work tapes and an unrestricted oracle tape. For any language A, let MULTI-RESET(A) = $\{L(M,A) \mid M$ is a nondeterministic oracle machine that has reset tapes as work tapes and that runs in linear time$\}$.

The classes $\mathcal{L}_{BNP}(A)$ and MULTI-RESET(A) can be characterized in a manner similar to that of Theorem 1.

<u>Theorem 2</u>

(i) For any language A the class $\mathcal{L}_{BNP}(A)$ (MULTI-RESET(A)) is the smallest class containing all of the regular sets and the language $(A \oplus \bar{A}$

and closed under intersection, linear-erasing homomorphic replication (duplication), and inverse homomorphism.

(ii) The class $\mathcal{L}_{BNP}$ (MULTI-RESET) is the smallest class containing all of the regular sets and closed under intersection and linear-erasing homomorphic replication (duplication).

Finally we have a characterization of the class of languages accepted in linear time. Let LIN be the set of linear functions.

Theorem 3

(i) For any language A, the class NTIME(LIN,A) of languages that are nondeterministic linear-time in A is the smallest class containing all context-free languages and $(A \oplus \bar{A})^*$ and closed under intersection, linear-erasing homomorphic replication and inverse homomorphism.

(ii) The class NTIME(LIN) is the smallest class containing all context-free languages and closed under intersection and linear-erasing homomorphic replication.

Section 3. Space Classes

In this section we state the characterizations of the classes specified by space-bounded oracle machines.

When we consider an oracle machine operating within some space bound, then the oracle tape is required to operate within that same bound.

Let A be any language and let S be a function (of the length of the input string) that serves as a space bound. We only consider those functions such that for all m, n $\geq$ 0, $S(m) + S(n) \leq S(m+n)$ and $S(n) \geq n$. Let DSPACE(S,A) = {L(M,A) | M is a deterministic oracle machine that operates in space bound S} and let NSPACE(S,A) = {L(M,A) | M is a nondeterministic oracle machine that operates in space bound S}.

Let $\mathcal{B}$ be a set of space bounds that is closed under composition. For any language A, the class of languages that are <u>nondeterministic</u> (<u>deterministic</u>) $\mathcal{B}$-<u>space</u> <u>in</u> A is NSPACE($\mathcal{B}$,A) = $\cup$ {NSPACE(f,A) | f $\in \mathcal{B}$} (DSPACE($\mathcal{B}$,A) = $\cup$ {DSPACE(f,A) | f $\in \mathcal{B}$}).

If a machine's instructions do not allow any oracle calls, then the oracle tape may as well not be present. Another way of interpreting this is to say that the oracle set is empty. When A is the empty set, we write NSPACE($\mathcal{B}$) (DSPACE($\mathcal{B}$)) for NSPACE($\mathcal{B}$,ϕ) (resp., DSPACE($\mathcal{B}$,ϕ)).

It is known that for every nondeterministic oracle machine M_1 operating with space bound S, there is a deterministic oracle machine M_2 operating within space bound $S^2(n) = (S(n))^2$ such that for all languages

A, $L(M_2,A) = L(M_1,A)$. Thus if $\mathcal{B}$ is a set of functions containing some function majorizing $f(n) = n^2$ and closed under composition, then for all languages A, NSPACE($\mathcal{B}$,A) = DSPACE($\mathcal{B}$,A) [18].

Of particular interest is the case where $\mathcal{B}$ is the set of polynomials. In this case we see that for every language A, DPSACE($\mathcal{B}$,A) = NSPACE($\mathcal{B}$,A) and so we write PSPACE(A) for DSPACE($\mathcal{B}$,A) and we write PSPACE for DSPACE($\mathcal{B}$,ϕ).

The classes of languages to be characterized in this section are of the form NSPACE($\mathcal{B}$,A) and NSPACE($\mathcal{B}$). We assume that each set $\mathcal{B}$ of space bounds is closed under composition.

To state the characterizations it is necessary to discuss relations on strings and their encodings as languages.

Consider n-ary relations on strings. If R is a binary relation on strings over the alphabet Σ, then the <u>transitive closure</u> of R is $R^* = \{\langle x,y\rangle \mid x, y \in \Sigma^*$ and either $x = y$ or there exist $n \geq 1$ and $z_0,\ldots,z_n \in \Sigma^*$ such that $z_0 = x$, $z_n = y$, and for each $i = 1,\ldots,n$, $R(z_{i-1},z_i)$ holds$\}$. A binary relation R is <u>length-preserving</u> if for all x, y, when R(x,y) holds, then $|x| = |y|$.

Let R be an n-ary relation on strings over the alphabet Σ. Let # be a symbol not in Σ. The language $SE_\#(R) = \{w_1\#\ldots\#w_n \mid$ for $i = 1,\ldots,n$ $w_i \in \Sigma^*$; $R(w_1,\ldots,w_n)$ holds$\}$ is the <u>sequential #-encoding of</u> R.

By using sequential encodings, relations can be interpreted as languages. For example, the concatenation relation over an alphabet Σ gives rise to the language $\{x\#y\#z \mid x, y, z \in \Sigma^*$ and $xy = z\}$, where # is a symbol not in Σ.

We are interested in interpreting a language as an encoding of a binary relation. Let L be a language and let Σ be a finite alphabet such that $L \subseteq \Sigma^*$. For any $a \in \Sigma$ the <u>binary relation a-encoded by</u> L is $R_a(L) = \{\langle x,y\rangle \mid x, y \in (\Sigma-\{a\})^*$ and $xay \in L\}$.

Notice that $SE_a(R_a(L)) = (\Sigma-\{a\})^*\{a\}(\Sigma-\{a\})^* \cap L$ and that if T is a binary relation on strings over Σ and $\# \notin \Sigma$, then $R_\#(SE_\#(T)) = T$.

To say that a relation R is transitively closed is to say that $R^* = R$. Here we develop the notion of "a class of languages being transitively closed" by considering the relations encoded by the languages in the class.

Let $\mathcal{L}$ be a class of languages. From a language L in $\mathcal{L}$, we consider the relation $R_a(L)$ a-encoded by L. Then we take the transitive closure $R_a^*(L)$ of $R_a(L)$ and consider the language $SE_a(R_a^*(L))$, that is, the sequential a-encoding of the relation $R_a^*(L)$. For our purposes it is sufficient to restrict attention to the cases where $R_a(L)$ is length-preserving and in that case to demand that the language $SE_a(R_a^*(L))$ is in $\mathcal{L}$. More formally, we have the next definition.

A class $\mathcal{L}$ of languages is <u>weakly transitively closed</u> if the following condition holds: Let L be any language in $\mathcal{L}$, let Σ be the smallest finite alphabet such that $L \subseteq \Sigma^*$, and let a be a symbol in Σ. If $R_a(L)$ is length-preserving, then $SE_a(R_a^*(L))$ is in $\mathcal{L}$.

Now we can state our characterization theorems.

<u>Theorem 4</u>. Let $\mathcal{B}$ be a set of space bounds.

(i) For any language A, the class NSPACE($\mathcal{B}$,A) of languages that are <u>nondeterministic</u> $\mathcal{B}$-<u>space</u> <u>in</u> A is the smallest class of languages that contains all regular sets and the language $(A \oplus \bar{A})^*$, is closed under intersection, inverse homomorphism, and $\mathcal{B}$-erasing homomorphic replication (duplication), and is weakly transitively closed.

(ii) The class NSPACE($\mathcal{B}$) of languages accepted by nondeterministic machines using at most $\mathcal{B}$ space is the smallest class of languages that contains all regular sets, is closed under intersection and $\mathcal{B}$-erasing homomorphic replication (duplication), and is weakly transitively closed.

Examples of classes to which Theorem 4 applies include the case where $\mathcal{B}$ is the class of linear functions so that NSPACE($\mathcal{B}$) is the class of context-sensitive languages and the case where $\mathcal{B}$ is the class of polynomials so that NSPACE($\mathcal{B}$) is PSPACE.

Theorem 4 is related to the work of Jones [13,14] and of McCloskey [16]. In both cases different sets of operations are used. Jones applies transitive closure to the class of strictly rudimentary relations and McCloskey uses AFL operations as well as others.

Section 4. Comparing Time and Space

What is the difference between time and space? Let us use Figure 1 to aid in focussing our attention on this question. Let $\mathcal{B}$ be a set of functions that can be either time bounds or space bounds and assume that $\mathcal{B}$ contains some function that majorizes $f(n) = n^2$. Let A be an arbitrary language. From Theorem 1 we see that NTIME($\mathcal{B}$,A) is the smallest class containing all regular sets and $(A \oplus \bar{A})^*$ and closed under intersection, $\mathcal{B}$-erasing homomorphic replication, and inverse homomorphism. From Theorem 4 we see that NSPACE($\mathcal{B}$,A) is characterized in precisely the same way except that it is required to be weakly transitively closed.

<u>Theorem 5</u>. Let $\mathcal{B}$ be a set of functions that can be either time bounds or space bounds. Suppose that $\mathcal{B}$ contains some function that majorizes $f(n) = n^2$.

(i) For any language A, NTIME($\mathcal{B}$,A) = NSPACE($\mathcal{B}$,A) if and only if NTIME($\mathcal{B}$,A) is weakly transitively closed.

(ii)   NTIME($\mathcal{B}$) = NSPACE($\mathcal{B}$) if and only if NTIME($\mathcal{B}$) is weakly transitively closed.

Consider the situation when the set of bounding functions is the set LIN of linear functions. See Figure 2. From theorems 2-4 we see that once again the requirement of being weakly transitively closed is crucial.

$$h^{-1}$$
$$(A \oplus \bar{A})*$$

NTIME($\mathcal{B}$) $\longrightarrow$ NTIME($\mathcal{B}$,A)

weakly transitively closed

weakly transitively closed

$$h^{-1}$$
$$(A \oplus \bar{A})*$$

NSPACE($\mathcal{B}$) $\longrightarrow$ NSPACE($\mathcal{B}$,A)

Figure 1

$$h^{-1}$$
$$(A \oplus \bar{A})*$$

MULTI-RESET $\longrightarrow$ MULTI-RESET(A)

$$h^{-1}$$
$$(A \oplus \bar{A})*$$

$\mathcal{L}_{BNP}$ $\longrightarrow$ $\mathcal{L}_{BNP}$(A)

$$h^{-1}$$
$$(A \oplus \bar{A})*$$

NTIME(LIN) $\longrightarrow$ NTIME(LIN,A)

$$h^{-1}$$
$$(A \oplus \bar{A})*$$

NSPACE(LIN) $\longrightarrow$ NSPACE(LIN,A)

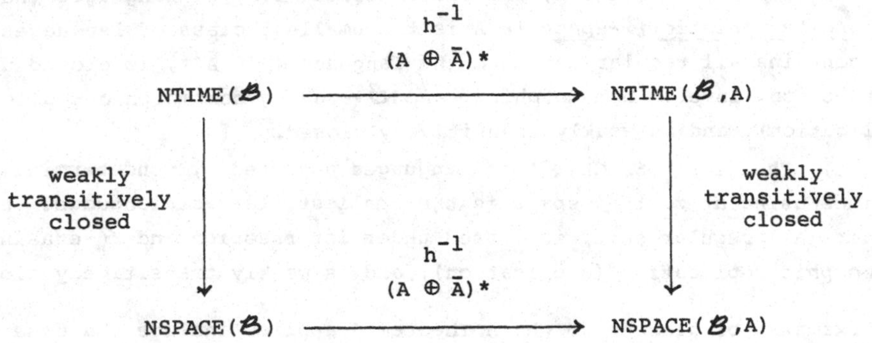

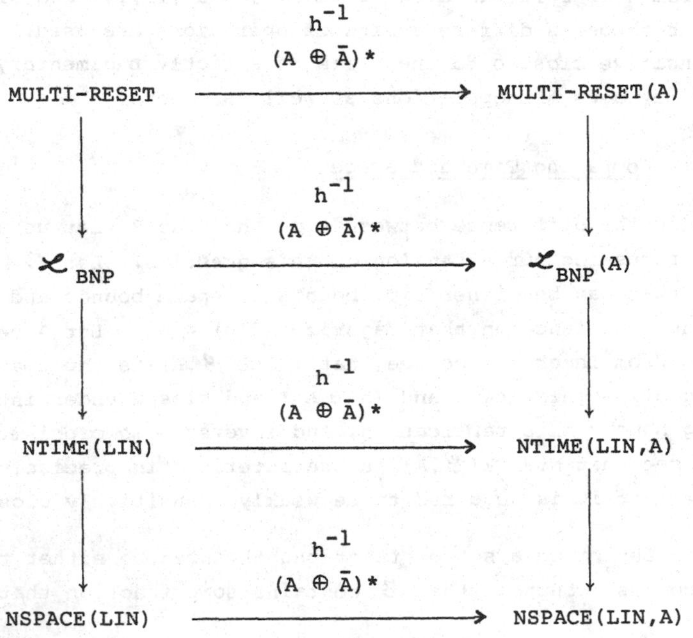

Figure 2

Theorem 6

(i) For any language A, let $\mathcal{L}$(A) be in {MULTI-RESET(A), $\mathcal{L}_{BNP}$(A), NTIME(LIN,A)}. Then $\mathcal{L}$(A) = NSPACE(LIN,A) if and only if $\mathcal{L}$(A) is weakly transitively closed.

(ii) Let $\mathcal{L}$ be in {MULTI-RESET, $\mathcal{L}_{BNP}$, NTIME(LIN)}. Then $\mathcal{L}$ = NSPACE(LIN) if and only if $\mathcal{L}$ is weakly transitively closed.

Thus we see that the difference between time and space can be characterized by the requirement of being weakly transitively closed. It is known that if a set $\mathcal{B}$ of bounding functions contains a function majorizing $f(n) = 2^n$ and $\mathcal{B}$ is closed under composition, then DTIME($\mathcal{B}$) = NTIME($\mathcal{B}$) = NSPACE($\mathcal{B}$) = DSPACE($\mathcal{B}$). This is also true when these classes are relativized in terms of an arbitrary oracle set.

An important case arises when $\mathcal{B}$ is taken to be the set of polynomials. It is not known whether NP = PSPACE.

Consider the following statements.

(a) NP = PSPACE.

(b) For every language A, NP(A) = PSPACE(A).

While it is obvious that (b) implies (a), it is not known whether (a) implies (b). It is known that there exists a language A such that NP(A) ≠ PSPACE(A) so that (b) is false [2,19].

We can prove that one statement is equivalent to (a).

Theorem 7. For any language A, let $\mathcal{L}$(A) = {L(M,A) | M is a nondeterministic oracle machine that uses at most polynomial work space and calls the oracle at most a polynomial number of times}. Then NP = PSPACE if and only if for any language A, $\mathcal{L}$(A) = NP(A).

The proofs in [2,19] showing that statement (b) is false do not appear to apply to classes such as $\mathcal{L}$(A) since the number of oracle calls is very large.

We note that one can apply the form of Theorem 5 to classes such as the class of rudimentary languages and the class of extended rudimentary languages to find that the class of rudimentary languages equals NSPACE(LIN) if and only if the class of rudimentary languages is weakly transitively closed, and the class of extended rudimentary languages equals PSPACE if and only if the class of extended rudimentary languages is weakly transitively closed. This is also true of the relativized versions.

It is important to note that the property of being weakly transitively closed is defined algebraically. It does not depend on a measure and it does not depend on the set of bounds used.

Section 5. Relativizations

Consider Figures 1 and 2.

Theorem 8. Let $\mathcal{B}$ be any set of time or space bounds. Let $\mathcal{L}$ be any of the classes NTIME($\mathcal{B}$), NSPACE($\mathcal{B}$), $\mathcal{L}_{BNP}$, MULTI-RESET, the rudimentary languages, the extended rudimentary languages, or the arithmetical sets. For any language A, let $\mathcal{L}$(A) be the relativized version of $\mathcal{L}$ with A as oracle set. Then $\mathcal{L} = \mathcal{L}$(A) if and only if every language of the form $h^{-1}((A \oplus \bar{A})^*)$, where h is a homomorphism, is in $\mathcal{L}$.

Under the assumptions about the sets of bounds considered here, each class NTIME($\mathcal{B}$) and each class NSPACE($\mathcal{B}$) is closed under union, inverse homomorphism, and Kleene *, so that every language of the form $h^{-1}((A \oplus \bar{A})^*)$ is in that class if and only if both A and $\bar{A}$ are in the class. However, the class $\mathcal{L}_{BNP}$ is not known to be closed under Kleene * so that for any oracle set A, $\mathcal{L}_{BNP}=\mathcal{L}_{BNP}(A)$ if and only if $(A \oplus \bar{A})^*$ is in $\mathcal{L}_{BNP}$.

If $\mathcal{B}$ is a set of space bounds that contains a function majorizing $f(n) = n^2$, then $\mathcal{B}$ is closed under squaring so that NSPACE($\mathcal{B}$) = DSPACE($\mathcal{B}$) and so NSPACE($\mathcal{B}$) is closed under complementation. In this case, NSPACE($\mathcal{B}$) = DSPACE($\mathcal{B}$,A) if and only if A $\in$ NSPACE($\mathcal{B}$).

If we wish to characterize the relativized version of the class, then we can do this uniformly by requiring closure under inverse homomorphism and the presence of the language $(A \oplus \bar{A})^*$. Thus the information required about the oracle set is uniform in that set and this information does not depend on the measure used to define the class nor on the set of bounds. However, using information about the measure and the set of bounds may allow one to limit the information required of the oracle set.

Section 6. The Class of Regular Sets as a Sub-Basis

Let $\mathcal{L}$ be a class of languages and let Ψ be a class of operations such that $\mathcal{L}$ is closed under each operation in Ψ. If $\mathcal{L}_0 \subset \mathcal{L}$ has the property that $\mathcal{L}$ can be characterized as the smallest class of languages containing every language in $\mathcal{L}_0$ and closed under each operation in Ψ, then we say that $\mathcal{L}_0$ is a sub-basis for $\mathcal{L}$ with respect to Ψ. When there is no chance of ambiguity, we say that $\mathcal{L}_0$ is a sub-basis for $\mathcal{L}$.

Except for the classes NTIME(LIN) and NTIME(LIN,A) characterized in Theorem 3, each of the classes considered here has the class of regular sets or the class of regular sets plus information about an oracle set as a sub-basis. Somewhat stronger results can be obtained if we consider any of the space classes and only those time classes specified by a set of bounding functions that contain a function majorizing $f(n) = n^2$. In these cases the class of all regular sets can be replaced by the class of all regular sets of star height two without adding to the set of closure operations. It is not known whether this can be done for classes such as MULTI-RESET, $\mathscr{L}_{BNP}$, or NTIME(LIN), but results in [8] suggest that this may not be the case.

Finally, let us note that if we use the class of linear context-free languages as a sub-basis, then we can replace the operation of homomorphic replication, a very powerful operation, with that of homomorphism. Thus there is a real "trade-off" between the power of the operations used and the class that forms the sub-basis.

Section 7. Remarks

What has been presented is an attempt at a unified treatment of certain types of complexity classes of formal languages in such a way that the essential differences between time and space and between a class and its relativized version are emphasized. An exposition that generalizes and strengthens the results of [3-5,10] in a unified framework is in preparation [6].

References

1. Baker, T., Gill, J. and Solovay, R., Relativization of the P = ? NP question. SIAM J. Computing, 4(1975), 431-422.

2. Baker, T. and Selman, A., A second step towards the polynomial hierarchy. Theoret. Comput. Sci., to appear.

3. Book, R., Simple representations of certain classes of languages. J. Assoc. Comput. Mach., 25(1978), 23-31.

4. Book, R., Polynomial space and transitive closure. SIAM J. Computing, 8(1979), to appear.

5. Book, R., On languages accepted by space-bounded oracle machines. Acta Informatica, to appear.

6. Book, R., Time, space, and relativizations. In preparation.

7. Book, R., Greibach, S. and Wrathall, C., Reset machines. J. Comput. System Sci., to appear.

8. Book, R. and Nivat, M., Linear languages and the intersection closures of classes of languages. SIAM J. Computing, 7(1978), 167-177.

9. Book, R., Nivat, M. and Paterson, M., Reversal-bounded acceptors and intersections of linear languages. SIAM J. Computing, 3(1974), 283-295.

10. Book, R. and Wrathall, C., On languages specified by relative acceptance. Theoret. Comput. Sci., 7(1978), 185-195.

11. Ginsburg, S., Greibach, S. and Hopcroft, J., Studies in Abstract Families of Languages, Memoir No. 87, Amer. Math. Soc., 1969.

12. Grzegorczyk, A., Some classes of recursive functions. Rozprawy Matematyczne, IV(1953), 1-46.

13. Jones, N., Formal Languages and Rudimentary Attributes, Ph.D. dissertation, University of Western Ontario, 1967.

14. Jones, N., Classes of automata and transitive closure. Info. Contr. 13(1968), 207-229.

15. Kleene, S.C., Introduction to Metamathematics. Van Nostrand, Princeton, N.J., 1952.

16. McCloskey, T., Abstract families of length-preserving processors. J. Comput. System Sci., 10(1975), 394-427.

17. Rogers, H., Theory of Recursive Functions and Effective Computability. McGraw-Hill, New York, 1967.

18. Simon, I., On Some Subrecursive Reducibilities, Ph.D. dissertation, Stanford University, 1977.

19. Simon, I. and Gill, J., Polynomial reducibilities and upwards diagonalizations. Proc. 9th ACM Symp. Theory of Computing (1977), 186-194.

20. Stockmeyer, L., The polynomial-time hierarchy. Theoret. Comput. Sci., 3(1976), 1-22.

21. Wrathall, C., Subrecursive Predicates and Automata, Ph.D. dissertation, Harvard University, 1975.

22. Wrathall, C., Complete sets and the polynomial-time hierarchy. Theoret. Comput. Sci., 3(1976), 23-33.

23. Wrathall, C., Rudimentary predicates and relative computation. SIAM J. Computing, 7(1978), 194-209.

24. Wrathall, C., The linear and polynomial-time hierarchies. In preparation.

FAST PROBABILISTIC ALGORITHMS

Rūsiņš Freivalds
Computing Centre, Latvian State University
Riga, USSR

1. INTRODUCTION

When probabilistic algorithms are considered it is usually supposed that the right result is obtained only "in most cases", i.e. for sufficiently many input data. The well-known Monte-Carlo method, moreover, produces only approximate values of the result. The following question arises naturally: are there any advantages of probabilistic algorithms over deterministic ones, when the results are of yes - no type (and therefore no approximation is possible), and we demand the result to be right with high probability for any values of the input data.

We use the term "probabilistic algorithms" for deterministic algorithms using random number generators. In order to avoid undesirable effects described in [12,17] we use only the simplest Bernoulli random number generators with two equiprobable outputs 0 and 1.

The only example of proved running time advantages of probabilistic algorithms over deterministic ones is the recognition of palindromes, as shown in [4] (see also [5]). A polynomial - time probabilistic algorithm for primality testing is found in [15]. Deterministic polynomial - time algorithms for primality testing are not known but, on the other hand, nobody has proved their non-existence.

The paper deals with fast probabilistic algorithms for testing of multiplication of integers, matrices and polynomials. The algorithm for testing of multiplication of integers is presented as one-head probabilistic Turing machine. Running time advantage over deterministic Turing machines is proved, and unimprovability of probabilistic running time is proved as well. The algorithms for testing of multiplication of matrices and polynomials are presented as computation schemes. This allows us to show that the advantages of probabilistic algorithms are not implied by specific character of one -

head Turing machines. On the other hand, this disables us to prove any nontrivial lower bounds of time complexity, just like the case of primality testing. Section 4 takes a quite opposite line and considers computation under severe restrictions, namely, it considers real - time recognition of languages by counter machines. It is proved that some languages can be recognized real - time by probabilistic multi-counter machines and cannot be recognized real - time by deterministic ones.

2. TESTING OF MULTIPLICATION OF INTEGERS

We say that a probabilistic off-line Turing machine M recognizes language L with probability p in time $t(x)$ if, when M works at arbitrary x, the probability of the following event exceeds p: the machine stops in no more than $t(x)$ steps with the result 1, if $x \in L$; and 0, if $x \in \overline{L}$.

As usual in the theory of computational complexity, we consider along with $t(x)$ as a function on strings, a more clear measure of complexity

$$t(n) = \max t(x),$$

where the maximum is taken over all strings consisting of no more than n letters.

The machine receives three numbers a, b, c (in binary representation), and produces result 1 if $a \cdot b = c$ is true and 0 otherwise. Formally, let $\overline{x}$, $\overline{y}$, $\overline{z}$, be the natural numbers whose binary representations are strings x, y, z. We consider recognition of the language A, consisting of all strings in form $x * y * z$, where x, y, z are strings in $\{0,1\}^+$ and $\overline{x} \cdot \overline{y} = \overline{z}$.

<u>THEOREM 1.</u> For every $\varepsilon > 0$ there is a one-head probabilistic Turing machine recognizing A in $\text{const} \cdot n \cdot \log n$ time with probability $1-\varepsilon$.

Theorem 1 was published in [6]. Recently the author found that a similar result for multi-tape Turing machines was proved in [18]. Let us note for sake of contrast that Barzdin [1] has proved a lower bound $\text{const} \cdot n^2$ for running time of deterministic one-head Turing machines recognizing A.

<u>THEOREM 2.</u> Let $\varepsilon < \frac{1}{2}$ and let a probabilistic Turing machine recognize the language A with probability $1-\varepsilon$ in time $t(n)$. Then there is $c > 0$ such that $t(n) > c \cdot n \cdot \log n$ for infinitely many n.

PROOF differs from the proof of Theorem 5 in [5] only in nonessential details.

3. TESTING OF MULTIPLICATION OF MATRICES AND POLYNOMIALS

Unfortunately, Turing machine is a highly restricted model of computers (in comparison with, say, multitape machines or Kolmogorov algorithms). In this section we try to explore advantages of a wider class of probabilistic algorithms, namely, we consider computation chemes.

Computation schemes are understood in the same sense as in [9]. They are ordered lists of arithmetic operations over real numbers. These operations are performed either over input data or over results of previous operations. In a sense, they are programs without loops. [9] contains an impressive collection of computation schemes, among them being schemes for computation of a fixed polynomial for various values of the argument, schemes for multiplication of $n \times n$ size matrices etc.

We consider probabilistic schemes, i.e. schemes that contain only arithmetical operations but also coin tossing, i.e. usage of Bernoulli random number generator with two equiprobable outputs 0 and 1.

It is proved in this section that for several problems there are probabilistic computation schemes containing essentially less operations than any known deterministic scheme for the same problem. The next step in this direction would be to prove that every deterministic computation scheme for this problem contains more operations. The well-known difficulties of establishing lower bounds of complexity prevent us from it.

We consider schemes for testing of multiplication of $n \times n$ size matrices and schemes for testing of multiplication of one-argument n-th degree polynomials. The input of the first scheme consists of $3n^2$ real numbers, namely, the elements of three $n \times n$ size matrices A, B, C. The result is to be 1, if $A \cdot B = C$, and 0, otherwise. The input of polynomial multiplication testing schemes consists of $4n + 3$ real numbers, namely, the coefficients of the given polynomials $P_1(x)$, $P_2(x)$, $P_3(x)$. The result is to be 1, if $P_1(x) \cdot P_2(x) = P_3(x)$, and 0, otherwise.

THEOREM 3. For every $\varepsilon > 0$ and for every natural n there is a probabilistic computation scheme with $\text{const} \cdot n^2$ operations such that it tests with probability $1 - \varepsilon$ multiplication of any two matrices of size $n \times n$.

PROOF. All the logarithms below are supposed to be binary ones. $\lfloor\delta\rfloor$ denotes the greatest integer less than or equal to δ (the "floor" of δ). $\lceil\delta\rceil$ denotes the least integer larger than or equal to δ (the "ceiling" of δ).

Let $k = \lceil \log \frac{1}{\varepsilon} + 1 \rceil$. The scheme generates k random vectors X of size $1 \times n$ with elements from $\{-1,1\}$. Each of these vectors is used to test whether $A\cdot(B\cdot X) = C\cdot X$. If at least one test fails, then $A \cdot B \neq C$. If the matrices pass all k tests, then the scheme produces result 1.

We will prove that the probability of the right result exceeds $1-\varepsilon$. We use c_{ij} for the elements of C, and d_{ij} for the elements of $A \cdot B$. Then the identity of i-th elements of matrices - columns $A\cdot B\cdot X$ and $C \cdot X$ may be expressed as follows:

$$(d_{i1} - c_{i1})\cdot x_1 + (d_{i2} - c_{i2})\cdot x_2 + \ldots + (d_{in} - c_{in})\cdot x_n = 0$$

This identity can be regarded as orthogonality of two vectors:

$F_i = \{(d_{i1} - c_{i1}), (d_{i2} - c_{i2}), \ldots, (d_{in} - c_{in})\}$,

$X = \{x_1, x_2, \ldots, x_n\}$.

It is easy to prove that for any $F_i \neq 0$ no more than 2^{n-1} out of all 2^n a priori possible vectors X are in the hyperplane orthogonal to F_i.

Therefore for any fixed $F_i \neq 0$ the probability to generate a vector X such that $F_i \cdot X \neq 0$ exceeds 1/2. When k independent vectors X are generated, the probability of inequality is $1-2^{-k} > 1-\varepsilon$. If $A \cdot B = C$, then for any i the vector F_i is zero - vector, and the scheme produces 1. Q.E.D.

Note for comparison that the best known deterministic computation scheme for this problem is based on results by Strassen [16] and Pan [14] and contains $\text{const} \cdot n^{2.795}$ operations.

THEOREM 4. For every $\varepsilon > 0$ and for every natural n there is a probabilistic computation scheme with $\text{const} \cdot n$ operations such that it tests with probability $1-\varepsilon$ multiplication of any two one - argument polynomials of n-th degree.

PROOF. Using the random number generator, the scheme produces an integer x_o, choosing it equiprobably among

$$-w+1,\ -w+2,\ \ldots,\ -1,\ 0,\ 1,\ \ldots,\ w\ ,$$

where $w = 2^{\lceil \log kn \rceil}$ and $k = \lceil \frac{1}{\varepsilon} \rceil + 1$. The values $P_1(x_o)$, $P_2(x_o)$, $P_3(x_o)$

are evaluated, and it is tested whether $P_1(x_o) \cdot P_2(x_o) = P_3(x_o)$. If <u>not</u>, then the polynomials are multiplied incorrectly; if <u>yes</u>, the scheme produces result 1.

The degree of $P_1(x) \cdot P_2(x) - P_3(x)$ does not exceed 2n. If it is not equal to zero identically, then it takes the value 0 for no more than 2n values of the argument. Therefore, the frequency of those values of x_o, for which $P_1(x_o) \cdot P_2(x_o) = P_3(x_o)$, does not exceed ε. Q.E.D.

Next we consider computation schemes where only operations with bits are allowed.

<u>THEOREM 5.</u> For every $\varepsilon > 0$ and for every natural n and r there is a probabilistic computation scheme with $\text{const}\ (n+1)^2 \cdot (r+1)^2 \cdot \log(n+r)$ bit-operations such that it tests with probability $1-\varepsilon$ multiplication of any two matrices of size $n \times n$ the elements of which are natural numbers not exceeding 2^r.

PROOF. The algorithm used is the same as in Theorem 3.

<u>THEOREM 6.</u> For every $\varepsilon > 0$ and for every natural n and r there is a probabilistic computation scheme with $\text{const} \cdot (n+1) \cdot (r+1) \cdot \log(n+r)$ bit-operations such that it test with probability $1-\varepsilon$ multiplication of any two one-argument polynomials of n-th degree, the coefficients of which are natural numbers not exceeding 2^r.

PROOF. Ideas of proofs of Theorem 1 and Theorem 4 are combined for this proof. We use $\pi(\ell)$ for the number of primes among the first ℓ natural number. Let $R_1(\ell) = \pi(2^{\lceil \log \ell \rceil})$. Let $R_2(\ell, N', N'')$ be the number of different prime divisors of $|N' - N''|$ not exceeding $2^{\lceil \log \ell \rceil}$. Let $R_3(\ell, s)$ be maximum of $R_2(\ell, N', N'')$ over all pairs (N', N'') such that $N' \leqslant 2^s$, $N'' \leqslant 2^s$, $N' \neq N''$.

Let $N' \neq N''$, $N' \leqslant 2^s$, $N'' \leqslant 2^s$, and let $p_1^{\alpha_1} \cdot p_2^{\alpha_2} \dots p_k^{\alpha_k}$ be the canonical expansion of $|N' - N''|$ into product of prime divisors. Since all $\alpha_1, \alpha_2, \dots \alpha_k$ are positive and $p_1, p_2, \dots, p_k$ are different, $p_1^{\alpha_1} p_2^{\alpha_2} \dots p_k^{\alpha_k} > k!$ Hence $2^s \geqslant k!$ and $k \leqslant \text{const} \cdot s / \log s$. Consequently, there is $c_1 > 0$ such that $R_2(\ell, N', N'') \leqslant c_1 \cdot s / \log s$. By the Čebyšev theorem on $\pi(\ell)$ (see Theorem 324 in [2]) there are constants a and b such that $0 < a < b$ and for all $\ell \geqslant 2$

$$a \cdot \ell / \log \ell < \pi(1) < b \cdot \ell / \log \ell.$$

Therefore, for every $\varepsilon > 0$ there is a natural $c(\varepsilon)$ such that

$$\lim_{s\to\infty} \frac{R_3(c(\varepsilon)\ s, s)}{R_1(c(\varepsilon)\ s)} < \frac{\varepsilon}{3} .$$

(Indeed, we can use any natural number exceeding $3c_1/a\varepsilon$ for $c(\varepsilon)$.).

Let d denote $\lceil \frac{3}{\varepsilon} \rceil$, w denote $2^{\lceil \log dn \rceil}$, u denote $r + (n+1) \log (kn)$, and v denote $\lceil \log (c(\varepsilon) \cdot u) \rceil$.

The computation scheme begins with generation of a random string consisting of v letters from $\{0,1\}$. This string is regarded as binary representation of a natural number m, and scheme tests whether this number is prime. If not, a new number m is generated and its primality tested, etc. The scheme allows to generate this number $\log (\frac{\varepsilon}{3}) / \log (1 - \frac{a}{v})$ times. If no m has turned out prime, the scheme produces result 1. If yes, a random x_o is generated, its value being equiprobably chosen of the following: $-w+1, -w+2, \ldots, -1, 0, 1, \ldots, w$. Then residues modulo m are calculated for $P_1(x_o)$, $P_2(x_o)$, $P_3(x_o)$ and the following test performed:

$$P_1(x_o)\ P_2(x_o) \equiv P_3(x_o) \pmod{m} \qquad (*)$$

If false, the polynomials have been multiplied incorrectly, and the scheme produces 0; if true, the scheme produces 1.

The number m is generated $\log (\frac{\varepsilon}{3}) / \log (1 - \frac{a}{v})$ times. For large $n+r$ this is a value of the same order of magnitude as $\log u$. One test of primality of a number, not exceeding 2^v, can be performed using $\text{const} \cdot v^2 \cdot 2^{v/2}$ operations with bits. Hence no more than $o(n+r)$ bit-operations are used for the generation of m and all the primality tests. No more than $o(n+r)$ bit-operations are used for the generation of x_o as well. Computation of $P_1(x_o) \pmod{m}$, $P_2(x_o) \pmod{m}$, $P_3(x_o) \pmod{m}$ and the test (*) takes $\text{const} \cdot (n+1) \cdot (r+1) \cdot \log (n+r)$ bit-operations.

Now we prove that for sufficiently large $(n+1) \cdot (r+1)$ the probability of the right result exceeds $1-\varepsilon$. Indeed, the degree of $P_1(x) \cdot P_2(x) - P_3(x)$ is not higher than $2n$. If this polynomial does not equal zero identically, it takes value 0 for no more than $2n$ values of x. Therefore, the probability to choose a number x_o, for which $P_1(x_o) \cdot P_2(x_o) = P_3(x_o)$, does not exceed $\varepsilon/3$.

It follows from the definition of $c(\varepsilon)$ and from the way how the value of m is chosen that, if $P_1(x_o) \cdot P_2(x_o) \neq P_3(x_o)$, then the probability of $P_1(x_o) \cdot P_2(x_o) \not\equiv P_3(x_o) \pmod{m}$ exceeds $1 - \varepsilon/3$. By the Čebyšev theorem cited above, the probability to get a prime m at the very first time is

$$\frac{a\,2^v/v}{2^v} = \frac{a}{v} \quad .$$

The number $t = \log\left(\frac{\mathcal{E}}{3}\right) / \log\left(1-\frac{a}{v}\right)$ of repetitions of generations of m is fixed to guarantee that a prime modulo m can be generated in t trials with the probability $1-\mathcal{E}/3$. Hence, for large $(n+1)\cdot(r+1)$ the probability of the right result exceeds $1-\mathcal{E}$. This implies our theorem for small n and r as well because the estimate includes an arbitrary multiplicative constant. Q. E. D.

4. REAL-TIME RECOGNITION OF LANGUAGES

Counter machines were first investigated by M.Minsky [12]. We refer [3] for a formal definition. Informally, k-counter machine is a Turing machine with k working tapes but there is a severe restriction on the usage of these tapes. A non-blank symbol is allowed only in the initial square on the tape. Therefore, the only information that matters is the position of the head on the tape. This way, a natural number is stored on this tape. At any moment this number can be changed only by adding +1, 0 or -1.

Probabilistic k-counter machine differs from deterministic one only in the possibility to use a Bernoulli random number generator with 2 equiprobable outputs 0 and 1 at any moment.

A machine is real-time if it reads one symbol from the input at every moment and produces a result concerning the string consisting of all symbols that have already entered.

We say that a probabilistic counter machine recognizes a language L with probability p, if the probability of producing a result

$$c_L(x) = \begin{cases} 1 \;, & \text{if } x \in L \\ 0 \;, & \text{if } x \in \overline{L} \end{cases}$$

exceeds p.

We consider the problem whether there are languages recognizable in real time by probabilistic counter machines and not recognizable by deterministic ones.

Let $\{L_k\}$ be the following family of languages. $L_k \subset \{a_1, b_1, a_2, b_2, \ldots, a_k, b_k\}^+$. L_k is the language of all strings x such that for all $i \in \{1,2,\ldots, k\}$ the number of symbols a_i in the string x equals to the number of symbols b_i in x. L_k can be interpreted as the language of all trajectories in k-dimensional integer space returning to the zero-point.

THEOREM 7. (Laing [11]) (1) For every $k \geq 1$ there is a deterministic k-counter machine recognizing L_k in real time. (2) For any $k \geq 2$ there is no deterministic (k-1)-counter machine recognizing L_k in real time.

We prove that probabilistic counter machines can recognize real-time more languages than deterministic ones.

THEOREM 8. For every $k \geq 1$ and every $\varepsilon > 0$ there is a probabilistic 1-counter machine recognizing L_k in real time with probability $1-\varepsilon$.

PROOF. First of all the machine chooses equiprobably an integer $r \in \{1,2,3,\ldots, 2^{\lceil \log k/\varepsilon \rceil}\}$. On every occurrence of symbols $a_1, b_2, a_2, b_2, \ldots, a_k, b_k$ the machine adds numbers $+r, -r, +r^2, -r^2, \ldots, +r^k, -r^k$, respectively, to the counter. The string is accepted if and only if the counter becomes empty. If $x \in L_k$, then the counter surely becomes empty. If the string x contains $n(a,1)$ symbols a_1, $n(b,1)$ symbols $b_1, \ldots, n(a,k)$ symbols a_k, and $n(b,k)$ symbols b_k, the counter contains the number

$$r\ (n(a,1) - n(b,1)) + r^2(n(a,2) - n(b,2)) + \ldots + r^k(n(a,k) - n(b,k)).$$

We shall prove that if $x \notin L_k$, then the probability of the counter's not becoming empty exceeds $1-\varepsilon$. Indeed, let $x \notin L_k$, i.e. there is an i such that $n(a,i)-n(b,i) \neq 0$. Then at most for $k-1$ different values of r the equality

$$1\ (n(a,1) - n(b,1)) + r^1\ (n(a,2) - n(b,2)) + \ldots + r^{k-1}(n(a,k) - n(b,k)) = 0$$

holds. Otherwise we would get a system of k linear algebraic equations (with unknowns $n(a,j) - n(b,j)$) the déterminant of which is not equal to 0 as Vandermonde determinant. Q.E.D.

We are going to prove that a probabilistic multi-counter machine can recognize in real time with probability $1-\varepsilon$ a language nonrecognizable by any deterministic multi-counter machine. For this we consider the following language which in a sense is a union of all languages L_k.

To describe strings of this language we introduce the following blocks:

$0[k]$: $\underbrace{00\ldots0}_{k \text{ times}}$

A_k : $0[k]1\,0[2^0]1\,0[2^1]1\,0[2^2]1\ldots1\,0[2^{k-1}]2\,0[2^k]$

$B_{m,i}$: $1\,0[m^1]1\,0[m^2]1\ldots1\,0[m^{i-1}]1\,2[m^i]1\,0[m^{i+1}]1\ldots1\,0[m^{k-1}]3\,0[m^k]1$

code (a_i): $4\,B_{1,i}\;3\,B_{2,i}\;3\,B_{3,i}\;3\ldots3\,B_{2^k-1,i}\;4\,B_{2^k,i}$

code (b_i): $5\,B_{1,i}\;3\,B_{2,i}\;3\,B_{3,i}\;3\ldots3\,B_{2^k-1,i}\;5\,B_{2^k,i}$

code $(x(1)x(2)\ldots x(n))$: A_k code $(x(1))$ code $(x(2))\ldots$ code $(x(n))$

We consider two languages D and E.

$D=\{y \mid y\in\{0,1,2,3,4,5\}^+\&(\exists k)(\exists x\in\{a_1,b_1,\ldots a_k,b_k\}^+)(|x|\geq k\ \&\ y=\text{code }(x))\}$,

$E=\{y \mid y\in\{0,1,2,3,4,5\}^+\&(\exists k)(\exists x\in L_k)(|x|\geq k\ \&\ y=\text{code }(x))\}$.

<u>LEMMA</u>. There is a deterministic 9-counter machine recognizing D in real time.

PROOF. The machine tests the syntactic correctness of all blocks, including the testing whether k is the same for all blocks. Two counters are used to contain the parameter k. When $0[k]$ of the block A_k is entered, the first counter counts up to k. After that whenever $+1$ (or -1, respectively) is added to one of the two counters, -1 $(+1)$ is added to the other. This way, the sum of the contents of these two counters is always equal to k. Using these two counters, the machine tests whether: 1) there are $k+1$ blocks of $0[2^i]$ type in A_k, and 2) there are k blocks of $0[m^\ell]$ and $2[m^i]$ type in every $B_{m,i}$. Two more counters are used to contain 2^k. Using these two counters, the machine tests whether: 1) every block $0[2^{i+1}]$ is twice as long as the preceding $0[2^i]$ in A_k, and 2) there are 2^k blocks of $B_{m,i}$ type in every code (a_i) and in every code (b_i). Two counters are used to contain m, and two counters are used to contain m^ℓ. Using these 4 counters, the machine tests whether: 1) the length of every $0[m^{\ell+1}]$ is equal to the product of the length of $0[m^\ell]$ and $0[m^1]$ in every $B_{m,i}$, and 2) every block $B_{m+1,i}$ in every code

(a_i) (or in code (b_i)) is preceded by a block $B_{m,i}$. The ninth counter accumulates k from $0[k]$ in A_k and uses it to test whether $|x|>k$. Q.E.D.

THEOREM 9. (1) For every $\varepsilon>0$ there is a probabilistic 12-counter machine recognizing E in real time with probability $1-\varepsilon$. (2) There is no deterministic multi-counter machine recognizing E in real time.

PROOF. The assertion (2) follows from Theorem 7. The probabilistic machine mentioned in (1) produces right results for all short strings. For long strings the probabilistic machine uses 9 counters to recognize the language D. If $x \notin D$, then $x \notin E$. Two counters are used to produce and to contain a random number r $(1 \leq r \leq 2^k)$. While A_k is entered, the random number generator is used at every moment when input symbol 1 enters, i.e. immediately before every block $0[2^0]$, $0[2^1]$, ... , $0[2^{k-1}]$. If the generator produces 0, the block $0[2^i]$ is ignored. If the generator produces 1, the length 2^i of the block $0[2^i]$ is added to the counter. The values of r are equiprobable.

The rest of the performance of the machine and of the proof resembles the proof of Theorem 8. The value $+r^i$ is added to the twelfth counter whenever code (a_i) is entered, and $-r^i$ is added whenever code (b_i) is entered. The output of the probabilistic machine is 1 if and only if the twelfth counter is empty. Q.E.D.

COROLLARY. The language E is not in Boolean closure of languages recognizable in real time by deterministic multi-counter machines.

Counter machine is a particular case of pushdown automata, stack automata, etc. It turns out that probabilistic counter machines can recognize some languages not recognizable by deterministic pushdown automata, by deterministic stack automata, and even by nondeterministic stack automata,

Stack automata were introduced by S.Ginsburg, S.Greibach and M.Harrison [7]. We consider one-way nondeterministic stack automata. Such an automaton works like an one-head on-line nondeterministic Turing machine whose entire tape to the right of the head at any moment when the head writes (i.e. changes a symbol on the tape) is filled only with empty symbols. Recall that for pushdown automata entire tape to the right of the head is empty at any moment. The difference is in the possibility of the head of a stack automaton to read symbols from inner parts of the stack.

THEOREM 10. There is a language F such that: 1) for every $\varepsilon > 0$ there is a probabilistic 1-counter machine recognizing F in real time with probability $1-\varepsilon$, and 2) there is no nondeterministic stack automaton accepting F.

PROOF. It follows from corollary 1 in [10] that the language

$$F = \left\{ 1^1 0^{n-1} 2 1^2 0^{n-2} 2 1^3 0^{n-3} 2 \dots 2 1^{n-1} 0^1 \right\}$$

cannot be recognized by a deterministic stack automaton and, moreover, cannot be accepted by a nondeterministic stack automaton.

We describe a probabilistic 1-counter machine. If the input string is not of

$$1^{i_1} 0^{j_1} 2 1^{i_2} 0^{j_2} 2 1^{i_3} 0^{j_3} 2 \dots 2 1^{i_m} 0^{j_m}$$

type, where $i_1 = 1$ and $j_m = 1$, the machine rejects it. If the input string x is of this type, the machine should check whether

$$i_2 = i_1 + 1 \quad , \quad i_3 = i_2 + 1 \ , \ \dots \ , \ i_m = i_{m-1} + 1 \ ,$$

$$j_2 = j_1 - 1 \quad , \quad j_3 = j_2 - 1 \ , \ \dots \ , \ j_m = j_{m-1} - 1 \ .$$

Since it is not possible to check all these equalities precisely, the machine chooses random coefficients $a_1, a_2, \dots, a_{m-1}, b_1, b_2, \dots, b_{m-1}$ and checks whether

$$a_1(i_2-i_1-1) + \dots + a_{m-1}(i_m-i_{m-1}-1) + b_1(j_2-j_1+1) + \dots + b_{m-1}(j_m-j_{m-1}) = 0 \ .$$

The coefficients are chosen equiprobably among 1,2,3, ... , r , where r is an integer degree of 2 exceeding $\frac{1}{\varepsilon}$. Q.E.D.

S.Greibach [8] has called a counter-machine blind if: 1) the instructions of it do not depend on counters' being empty, and 2) the string is accepted only if all the counters are empty at this moment.

THEOREM 11. If a language L is recognized in real time by a blind deterministic multi-counter machine, then for every $\varepsilon > 0$ the language L is recognized in real time by a blind probabilistic 1-counter machine with probability $1-\varepsilon$.

PROOF. The same idea as in Theorem 8 is used. Let k be the number of counters of the deterministic machine. For the beginning a random number r is chosen just like the proof of Theorem 8. When the deterministic machine adds ± 1 to the i-th counter, the probabilistic machine adds $\pm r^{i-1}$ to its only counter. If the deterministic machine accepts a string, the probabilistic machine surely accepts it as well. If the deterministic machine rejects a string, the probability of rejecting it by the probabilistic machine exceeds $1-\varepsilon$. Q.E.D.

REFERENCES

1. BARZDIN, J.M., Complexity of symmetry recognition by Turing machines. In: Problemy kibernetiki 15, pp.245-248, Nauka, Moscow, 1965 (Russian).

2. BUKHŠTAB, A.A., Number theory. Učpedgiz, Moscow, 1960 (Russian).

3. FISCHER, P.C., MEYER, A.R. and ROSENBERG A.L., Counter machines and counter languages. Mathematical Systems Theory, 2(1968), No.3, 265-283.

4. FREIVALDS, R., Fast computations by probabilistic Turing machines. In: Učenye Zapiski Latviĭskogo gos.universiteta, 233, pp.201-205, Riga, 1975 (Russian).

5. FREIVALDS, R., Probabilistic machines can use less running time. In: Information Processing 77, (B.Gilchrist, Ed.), pp.839-842, IFIP, North-Holland, 1977.

6. FREIVALDS, R., Probabilistic algorithms in proof-oriented computations. In: Proc.All-Union symposium "Artificial intelligence and automation of research in mathematics", 102-104, Institute of Cybernetics, Kiev, 1978 (Russian).

7. GINSBURG, S., GREIBACH, S.A. and HARRISON, M.A., Stack automata and compiling. Journal of the ACM, 14 (1967), 172-201.

8. GREIBACH, S.A., Remarks on blind and partially blind one-way multicounter machines. Theoretical Computer Science, 7 (1978), No.3, 311-324.

9. KNUTH, D.E., The art of computer programming. Vol.2. Addison-Wesley Publishing Co., Reading,Mass., 1969.

10. LIVŠIN, D.I., On a class of stack languages. In: Mathematical linguistics and theory of algorithms, (A.V.Gladkij, Ed.), pp.113-129, Kalinin State University, Kalinin, 1978 (Russian).

11. LAING, R., Realization and complexity of commutative events. University of Michigan, Technical Report 03105-48-T, 1967.

12. LEEUW, K. de, MOORE E.F., SHANNON, C.E. and SHAPIRO, N., Computability by probabilistic machines. Automata Studies, (C.E.Shannon and J.McCarthy, Eds.), pp.183-212, Princeton University Press, 1956.

13. MINSKY, M., Recursive unsolvability of Post's problem of Tag and other topics in the theory of Turing machines. Annals of Mathematics, 74 (1961), 437-455.

14. PAN, V., Strassen's algorithm is not optimal. Proc. 19th annual symposium on foundations of computer science, pp.166-176, IEEE, 1978.

15. SOLOVAY, R. and STRASSEN, V., A fast Monte-Carlo test for primality. SIAM Journal on Computing, 6 (1977), No.1, 84-85.

16. STRASSEN, V., Gaussian elimination is not optimal. Numerische Mathematik, 13 (1969), H.4, 354-356.

17. TRAKHTENBROT, B.A., Remarks on the computational complexity by probabilistic machines. In: Teorija algoritmov i matematičeskaja logika, pp.159-176, Comp. Ctr.Acad.Sci.USSR, Moscow, 1974 (Russian).

18. YAO, A.C., A lower bound to palindrome recognition by probabilistic Turing machines. STAN-CS-647, Stanford University, 1977.

RELATIVE SUCCINCTNESS OF REPRESENTATIONS OF LANGUAGES AND SEPARATION OF COMPLEXITY CLASSES

Juris Hartmanis*
Department of Computer Science
Cornell University
Ithaca, NY 14853

T.P. Baker
Department of Computer Science
University of Iowa
Iowa City, IA 52242

Abstract

In this paper we study the relative succinctness of different representations of deterministic polynomial time languages and investigate what can and cannot be formally verified about these representations. We also show that the relative succinctness of different representations of languages is directly related to the separation of the corresponding complexity classes; for example, PTIME ≠ NPTIME if and only if the relative succinctness of representing languages in PTIME by deterministic and nondeterministic clocked polynomial time machines is not recursively bounded, which happens if and only if the relative succinctness of these representations is not linearly bounded.

Furthermore, we discuss the problem of approximating the recognition of complete languages in NPTIME by deterministic polynomial time machines which accept finite initial segments of these languages. We conclude by discussing the relative succinctness of optimal and near-optimal programs and the nature of the families of minimal machines for different representations.

Introduction

In this paper we discuss the relative succinctness of different representations of languages and draw several consequences from these results about the problems of separating computational complexity classes. To give an easily understood interpretation of these results, which are not technically difficult, we concentrate this study on representations of the class of deterministic polynomial time languages,

* Research supported in part by National Science Foundation grant MCS 78-00418

PTIME, which is currently accepted as a good mathematical model for the feasible computations.

We define a variety of deterministic representations for the languages in PTIME and show between which pairs of these representations the relative succinctness is not recursively bounded (leaving a tantalizing open problem).

Several of these representations require that there be formal proofs that the machines involved are true descriptions of languages in PTIME, the extreme case being verified representations, which include as a part of each description a formal proof of its validity. We also consider nondeterministic polynomial time representations of languages in PTIME and show that the families of deterministic and nondeterministic polynomial time computable languages are equal, PTIME = NPTIME, if and only if the relative succinctness between the deterministic and nondeterministic representations can be linearly bounded. The size of the constants in the linear succinctness bound determine how large a deterministic polynomial time machine must be (if it exists) to recognize a complete NPTIME problem; the linear relationship could possibly be used to show that the desired deterministic polynomial time machine must be immensly large (and therefore incomprehensible) or that it does not exist.

It is interesting to note that our current understanding of these problems leaves open the possibility that PTIME = NPTIME, but not in a way that knowing this information can be of any use. Conceivably, it might be provable that there are deterministic polynomial time machines that accept complete languages for NPTIME, for example the satisfiable Boolean formulas in conjunctive normal form, CNF-SAT, but it might also be that no specific one of these machines can be proven to do so. Or, the smallest such machine might be so large in size that we are not likely to find it, rendering its existence meaningless for practical purposes. (See the later discussion why this possibility is not eliminated by the related result in [7].)

We consider also the problem of approximating the recognition of NPTIME complete sets by deterministic polynomial time machines which accept correctly all strings in the complete set up to a given size. For example, we say that a machine, M accepts an initial segment of CNF-SAT if for some n_0 an input is accepted if and only if it is a satisfiable formula no longer than n_0.

On the one hand, we show that under the assumption that PTIME ≠ NPTIME there is no recursive succinctness bound for several pairs of deterministic and nondeterministic polynomial time representations of the initial segments of CNF-SAT. On the other hand, we show that there are succinct deterministic polynomial time machines (recursively related in size to the equivalent nondeterministic machines) which recognize the initial segments of CNF-SAT. Unfortunately, our proofs also show that the polynomial degree of these machines must grow nonrecursively in their size and so must the length of proofs that they accept an initial segment of the satisfiability problem.

In conclusion, we observe that there is no recursive succinctness bound between the representations of languages by optimally fast and near-optimally fast algorithms. We discuss the possibility that the difficulty of finding optimal algorithms and proving algorithms optimal for many practical problems may be caused in some cases by the possibly immense difference in size between the known near-optimal and the desired optimal algorithms. It would be very interesting to find a natural language for which one could prove that its optimal algorithms must be immense while exhibiting a small near-oprimal algorithm.

The results reported in this paper are not technically hard, and their proofs make use of refinements of techniques which can already be found in earlier work on program size [3,8]. At the same time, they do address new questions, not answered by these previous studies, which arise naturally from more recent work on polynomial time complexity. We believe that these results gain relevance from their relation to classic open problems in theory of computation, from their possible contributions to our overall understanding of complexity of computations and from the problems they suggest for further investigation.

Preliminaries

In this section we establish our notation and define the basic concepts and representations used in this paper.

Let $M_1, M_2, \ldots$ be a standard enumeration of all deterministic multitape Turing machine (Tm's), let $L(M_i)$ denote the language accepted by M_i, let $|x|$ denote the length of x and define

$$T_i(n) = \max\{m \mid m \text{ is number of operations performed by } M_i \text{ on input } x, |x| = n\}.$$

Let $|M_i|$ be the length of the description of M_i in a fixed alphabet. Similarly, let $N_1, N_2, \ldots$ be a standard enumeration of all nondeterministic multitape Tm's.

Let the set of all polynomial time languages [1] be denoted by

$$\text{PTIME} = \{L(M_i) \mid T_i(n) \le k+n^k,\ k \ge 1\}$$

and let the corresponding set of nondeterministic polynomial time languages be denoted by NPTIME. PTIME consists of all the languages accepted by deterministic Turing machines in polynomial time. The main object of this paper is to investigate the relative succinctness of different representations of this family of languages and relate these results to the classic problem of determining whether PTIME = NPTIME.

A number of representations studied in this paper, and previously discussed in [2], consist of machines about which we can formally prove that they accept the desired type of language. To make these concepts more precise let FS be an axiomatizable, sound formal mathematical system which is sufficiently rich to formulate and prove elementary facts about Turing machines and their computations. Since FS is axiomatizable we know that the set of provable theorems is recursively enumerable and that soundness of FS assures that only true theorems can be proven. The particular choice of FS is not important for this study, so instead of specifying FS in detail we will state explicitly (later) what has to be provable in FS. (The reader can think of FS as formalized Peano Arithmetic with an agreed upon representation of Turing machines and their computations.) For a formal statement T provable in FS we write

$$FS \vdash [T].$$

We now define several well known deterministic representations of deterministic polynomial time languages:

1. Clocked polynomial time machines, <u>CPTM</u>; this is the set of Tm's

$$CPTM = \{M_{\sigma(i,k)} \mid i,k = 1,2,\dots\},$$

with standard, easily recognizable polynomial clocks, which shut off the computation of M_i after $k+n^k$ steps. The set CPTM is seen to be recursive.

2. Verified polynomial time machines, <u>VPTM</u>; this is the set of Tm's for which there is a proof in FS that they run in polynomial time,

$$VPTM = \{M_i \mid FS \vdash (\exists k)\ [T_i(n) \le k+n^k]\}.$$

The set VPTM is recursively enumerable, but not recursive (because FS is assumed to be sufficiently rich to make $M_i \in ?$ VPTM an undecidable problem).

3. Verified polynomial time language machines, <u>VPLM</u>; this is the set of total Turing machines for which it can be shown in FS that they accept a language in PTIME,

$$VPLM = \{M_i \mid FS \vdash [M_i \text{ is total and } L(M_i) \in PTIME]\}.$$

Again we note that VPLM is recursively enumerable but not recursive.

4. Polynomial time machines, <u>PTM</u>; this is the set of Tm's which run in polynomial time,

$$PTM = \{M_i \mid (\exists k)\ [T_i(n) \le k+n^k]\}.$$

Clearly this set is not recursively enumerable.

5. Polynomial language machines, <u>PLM</u>; this is the set of total Tm's which accept languages in PTIME,

$$PLM = \{M_i \mid M_i \text{ is total and } L(M_i) \in PTIME\}.$$

PLM is not recursively enumerable.

6. Partial polynomial language machines, <u>PPLM</u>; this is the set of partial Tm's which accept languages in PTIME,

$$PPLM = \{M_i \mid L(M_i) \in PTIME\}.$$

This set is not recursively enumerable.

To link the succinctness results to the classic open problem about nondeterministic polynomial time computations we consider nondeterministic representations of languages in PTIME. They are defined similarly to the corresponding deterministic machines.

7. Nondeterministic clocked polynomial time machines, NCPTM; this is the set of nondeterministic machines with standard polynomial clocks which shut them off and which accept languages in PTIME,

$$NCPTM = \{N_{\delta(i,k)} \mid L(N_{\delta(i,k)}) \in PTIME\}.$$

This set is not recursively enumerable.

8. Nondeterministic verified polynomial time machines for which it can be proven in FS that they run in polynomial time and accept languages in PTIME,

$$NVPTM = \{N_i \mid FS \vdash [(\exists k)\ T_i(n) \leq k+n^k \text{ and } L(N_i) \in PTIME]\}.$$

This set of machines is recursively enumerable.

For a family of machines C, we write

$$LANG[C] = \{L(M_i) \mid M_i \in C\}.$$

Clearly,

$$LANG[CPTM] = LANG[VPTM] = \ldots = LANG[NVPTM] = PTIME.$$

We observe in passing that under very mild assumptions about FS, it can be shown that

$$FS \vdash [LANG[CPTM] = PTIME].$$

On the other hand

$$LANG[VPTM] = PTIME$$

can be shown not to be a theorem in FS, since

$$FS \vdash [LANG[VPTM] \subseteq PTIME]$$

leads to a consistancy proof of FS in FS, which is impossible by Goedel's second Incompleteness Theorem.

Let R_1 and R_2 be two different representations of a family of languages F. We will say that the relative succinctness of representation of languages in F by R_1 and R_2 is recursively bounded if there exists a recursive function f such that for any A in R_2 there is an equivalent A' in R_1 satisfying

$$f(|A|) > |A'|.$$

Therefore, we see that the succinctness gained by going from representations in R_1 to representations in R_2 is bounded by f.

If such a recursive succinctness bound does not exist then we say that the relative succinctness of the representation R_1 and R_2 of F is not recursively bounded.

In the following we will investigate the relative succinctness between representations of languages in PTIME and relate the relative succinctness between deterministic and nondeterministic representations of languages in PTIME to the classic PTIME = ? NPTIME problem. Furthermore, we will investigate what happens if we attach formal proofs to these representations that they accept the desired languages. We will also investigate the possibility of the existence of succinct deterministic polynomial time machines which accept correctly all strings from an NPTIME complete problem up to a given length. We will conclude with some observations about the relative succinctness of representations of optimal and near-optimal algorithms and a look at the nature of the sets of minimal machines for the different representations discussed in this paper.

Succinctness Results

The relative succinctness of different representations of languages has been investigated extensively before and succinctness results related to computer science problems can be found, for example, in [3,5,8,9,10,11]. Blum's pioneering paper [3] contains the result that any infinite r.e. sequence of machines contains some machines whose size can be reduced by an arbitrary recursive amount. Meyer [8] presents a stronger theorem, to the effect that given any recursive enumeration of total functions that includes the zero-one functions of finite support, another similar enumeration can be constructed so that, for the zero-one functions of finite support, the gain in succinctness going from the first representation to the second is not recursively bounded. While this theorem does not apply directly, we make use of an important technique found in its proof, refining it to handle the more stringent requirements of our present application, where we must deal with verified and non-r.e. representations.

In Fig. 1 we have summarized schematically the information about some of the pairs of representations of languages in PTIME for which we will prove that there do not exist recursive succinctness bounds; for example the arrow from VPTM to CPTM indicates that we have nonrecursive succinctness loss going from verified polynomial time representations to clocked polynomial time representations.

We will start these proofs by explaining a very simple proof technique which has been used before and showing why it is not sufficiently powerful for our purposes (since it does not remain valid under the addition of proofs of correctness to these representations).

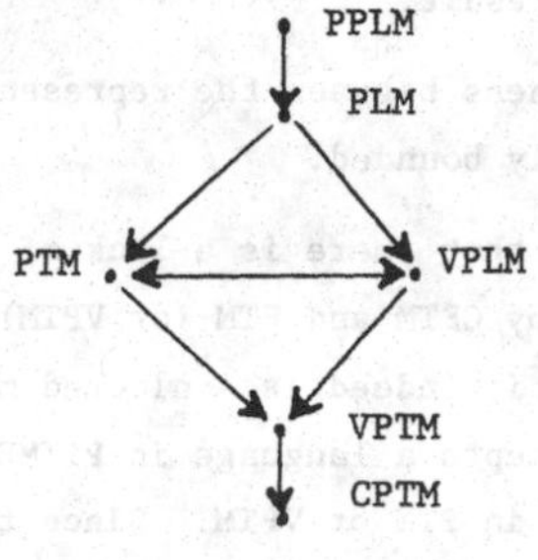

Fig. 1

Fact 1: The relative succinctness of representing languages in PTIME by machines from CPTM and PTM is not recursively bounded.

Proof: For a Tm, M_i, which halts on blank tape let m_i be the number of steps performed by M_i before halting. Clearly, m_i is not recursively bounded in the size of M_i, or the halting problem would be solvable. To show that the relative succinctness between CPTM and PTM cannot be recursively bounded we will construct a Tm $M_{d(i)}$ which runs in polynomial time and accepts a finite set not accepted by any clocked machine, $M_{\sigma(j,k)}$ such that

$$|M_{\sigma(j,k)}| \leq m_i.$$

Construction of $M_{d(i)}$: for input x $M_{d(i)}$ computes m; and enumerates all clocked machines $M_{\sigma(j,k)}$ such that $|M_{\sigma(j,k)}| \leq m_i$, say $M_{i_1}, M_{i_2}, \ldots, M_{i_p}$. The input x is reject if $|x| > p$, else

$$M_{d(i)}(x) = 1 \dot{-} M_{i_{|x|}}(x).$$

It is seen that if M_i halts on blank tape then $M_{d(i)}$ runs in polynomial time and accepts a finite set not accepted by any $M_{\sigma(j,k)}$ of size less than m_i. Since $|M_{d(i)}|$ is recursively related to $|M_i|$ and $|M_i|$ is not recursively related to m_i, we see that there is no recursive succinctness bound between the representations in CPTM and PTM. □

If the formal system FS is sufficiently rich then for any machine M_i which halts on blank tape, we can prove this fact in FS and therefore get a proof in FS that $M_{d(i)}$ is in PTM. We make this assumption explicit.

Assumption about FS: We assume that for any Tm, M_i which halts on x we can prove this fact in FS, i.e.

$$FS \vdash [M_i(x) \text{ halts}].$$

Furthermore, we assume that (for a simple and uniform construction d(i)) we can prove in FS that $M_{d(i)}$ runs in polynomial time (and that this proof in FS implies that $M_i(-)$ halts).

It should be observed that in any reasonably rich logic designed to reason about computations and machines, we should be able to prove the above stated theorems.

Under the above assumptions about FS we know that $M_{d(i)}$ is in VPTM iff $M_i(-)$ halts and therefore we get the next result.

Fact 2: The relative succinctness between the representations of PTIME languages by CPTM and VPTM is not recursively bounded.

It is interesting to note that there is a lack of symmetry between the representations of languages in PTIME by CPTM and PTM (or VPTM). For any machine $M_{\sigma(i,k)}$ in CPTM we can easily verify that it indeed is a clocked machine and that it therefore runs in polynomial time and accepts a language in PTIME. On the other hand, no such certainty exists for machines in PTM or VPTM. Since the set PTM is not recursively enumerable we have no way of verifying that the machines in PTM run in polynomial time. For machines in VPTM there are proofs that they run in polynomial time but no decision procedure. These considerations suggest that to make the representations symmetric we should consider "certified" representations by including with the machine a proof that the machine accepts a language of the desired type. Such representations are discussed in [5].

In view of these considerations let

$$\mathrm{VPTMp} = \{(M_i, p_i) \mid p_i \text{ is a proof in FS that } M_i \text{ runs in polynomial time}\}$$

and let

$$|(M_i, p_i)| = |M_i| + |p_i|.$$

Let VPLMp be defined similarly.

It should be pointed out that the previously constructed proof of Fact 2 does not prove that there is no recursive succinctness bound between CPTM and VPTMp, as shown by the next observation.

Fact 3: For any axiomatizable, sound formal system FS there exists a recursive function H such that

$$H[|\text{proof in FS that } M_i(-) \text{ halts}|] \geq m_i.$$

Proof: The recursive function H is given by

$$H(n) = \max[\{0\} \cup \{m_i \mid |\text{proof in FS that } M_i(-) \text{ halts}| \leq n\}]. \quad \square$$

Let us now consider representations of PTIME by machines from VPTM with added proofs that they run in polynomial time, VPTMp. A proof in FS that $M_{d(i)}$, used in the proof of Fact 1, runs in polynomial time implies a proof in FS that $M_i(-)$ halts. Therefore, by Fact 3 we must conclude that m_i is recursively bounded with respect to

$$|M_{d(i)}| + |\text{proof in FS that } M_{d(i)} \text{ runs in polynomial time}|$$

and therefore the relative succinctness between machines in CPTM and the $M_{d(i)}$'s with attached proofs that they run polynomial time is also recursively bounded.

It has been pointed out in [5] that, though many of the previous succinctness constructions lose their succinctness when we attach proofs of their validity as a part of the representation, there are other constructions for these succinctness results which have short proofs of validity and therefore extend these results to representations with attached proofs. (We will see later in this paper that the representations of approximations to NPTIME complete problems behave quite differently and that these representations lose their relative succinctness if we demand that they contain proofs of validity.)

In the following we describe a construction which will be used in our proofs and state what we assume is provable in FS about this construction. For a related but differently formulated construction used in succinctness results see [8].

Let $M_{b(r)}$ be a Tm which:

a) Enumerates all Tm's up to size r, i.e. $M_{i_1}, M_{i_2}, \ldots, M_{i_t}$ such that $|M_{i_j}| \le r$, $1 \le j \le t$,

b) simulates in a dove-tail manner these machines on blank tape and prints (between special markers) the maximal running time so far achieved by these machines before halting.

It is seen that $M_{b(r)}$ computes in the limit the nonrecursive and not recursively bounded function

$$B(r) = \max\{m_i \mid |M_i| \le r\}.$$

That is, eventually $B(r)$ will be printed on the tape of $M_{b(r)}$ and no larger value will ever be printed, (but as increasing values are printed, we have no effective way of determining whether $B(r)$ has been reached).

Assumption about FS: We assume that we can prove in FS, with proofs of length recursively bounded in r, that $M_{b(r)}$ prints (between special markers) only finitely many outputs.

We are now ready to prove a number of succinctness results about the representation of PTIME. The notation VPTM(p) in the result indicates that the result holds for the verified representations as well as for the verified representations with the proof as part of the representation.

Theorem 4: The relative succinctness between the following pairs of representations of languages in PTIME is not recursively bounded:

CPTM and VPTM(p),		VPLM(p) and PTM,	
VPTM(p) and PTM,		PTM and PLM,	
VPTM(p) and VPLM(p),		VPLM(p) and PLM,	
PTM and VPLM(p),		PLM and PPLM.	

Proof: We give only the proofs of the more interesting cases, the other cases follow by very similar methods.

CPTM and VPTMp: Below we construct a Tm $M_{d(r)}$ which can easily be proved in FS to run in polynomial time and accepts a finite set not accepted by any clocked polynomial time Tm, $M_{\sigma(i,k)}$, with

$$|M_{\sigma(i,k)}| \le B(r).$$

Since the size of $M_{d(r)}$ will be seen from the construction to be recursively related to r (and the length of the proof that $M_{d(r)}$ runs in polynomial time also recursively bounded in r) we conclude that the relative succinctness between the representations CPTM and VPTM as well as CPTM and VPTMp is not recursively bounded.

In the construction of $M_{d(r)}$ we use an auxiliary list, L_n, of machines in CPTM which have been found not to be equivalent to $M_{d(r)}$ by processing inputs up to length n-1.

Description of $M_{d(r)}$: Let $L_1 = \emptyset$, reject the null string and all inputs not in 0*. For $w = 0^n$, $n \geq 1$, lay off [logn] tape squares and in this amount of tape simulate $M_{b(r)}$ on blank tape. If $M_{b(r)}$ has no output reject, else let N be maximal output of $M_{b(r)}$.

Reconstruct the list L_n, and try to find the smallest $M_{\sigma(i,k)}$ in CPTM such that

$$M_{\sigma(i,k)} \notin L_n \text{ and } |M_{\sigma(i,k)}| \leq N.$$

If no such $M_{\sigma(i,k)}$ can be found on the available tape then reject input and set $L_{n+1} = L_n$. If $M_{\sigma(i,k)}$ is found then simulate $M_{\sigma(i,k)}$ on 0^n an do the opposite and set

$$L_{n+1} = L_n \cup \{M_{\sigma(i,k)}\}.$$

It is seen that $M_{d(r)}$ runs in polynomial time and that $|M_{d(r)}|$ is recursively bounded in r (and so is the proof in FS that $M_{d(r)}$ runs in polynomial time), but that the size of the smallest equivalent $M_{\sigma(i,k)}$ is not recursively bounded in r. This completes the proof.

VPTM(p) and PTM: This case follows by a very similar construction and the use of the fact that FS is sound to guarantee that all the machines in VPTM indeed run in polynomial time.

PTM and VPLM: Recall that PTM is not recursively enumerable and therefore this is a somewhat different case (and does not follow directly from the general formulations of succinctness results). Nevertheless, we can easily prove this result by constructing the appropriate $M_{d(r)}$. Let $M_{d(r)}$ be a simply and uniformly constructed total machine (using $M_{b(r)}$) which accepts a finite set not accepted by any M_i which converges for infinitely many inputs of length n in 2^n steps and for which $|M_i| \leq B(r)$. Since the size of $M_{d(r)}$ is recursively related to the size of r and since no machine in PTM of size less than B(r) can be equivalent to $M_{d(r)}$, we conclude directly that there is no recursive succinctness bound between

PTM and PLM.

Furthermore, since we can prove in FS (by short proofs) that $M_{d(r)}$ is total and accepts a finite set, there is no recursive succinctness bound between

PTM and VPLM(p).

By transitivity we also conclude that there does not exist a recursive succinctness bound between

VPTM and VPLM(p).

The other succinctness results follow by slight veriations on the above arguments. □

Note: It is interesting to observe that for example, the proof that there is no recursive succinctness bound between PTM and VPLM can be formulated and proven in FS, provided we can prove in FS that any machine in PTM converges for infinitely many inputs in 2^n steps. On the other hand, the proof that there is no recursive succinctness bound between VPTM and PLM cannot be formulated and proven in FS, since we assume the soundness of FS in this proof.

Nondeterministic Representations

In this section we consider the relative succinctness of representing languages in PTIME by deterministic and nondeterministic polynomial time Turing machines. These representations are particularly interesting because the relative succinctness of some of these representations is directly linked to the classic problem of determining whether the families of deterministic and nondeterministic polynomial time recognizable languages are different, PTIME = ? NPTIME. Furthermore, these results suggest how we could try to show that PTIME ≠ NPTIME or, even if PTIME = NPTIME, that the size of a deterministic polynomial time machine recognizing the satisfiability problem in NPTIME must be very large.

We first observe that without any assumptions about the relation between PTIME and NPTIME, we can derive some succinctness results which show that certain deterministic representations are nonrecursively more succinct than certain other nondeterministic representations.

Corollary 5: The relative succinctness between the following pairs of representations of languages in PTIME is not recursively bounded:

NCPTM and VPTM,
NVPTM and PTM,
NVPTM and VPLM.

Proof: Similar to the ones given for Theorem 4. □

Theorem 6: Assuming that PTIME ≠ NPTIME, the relative succinctness of representing PTIME by CPTM and NCPTM as well as by VPTM and NCPTM is not recursively bounded.

Proof: CPTM and NCPTM. To obtain this succinctness result we will construct a nondeterministic machine $N_{g(r)}$ which runs in n^5 time and by the delayed diagonalization method accepts a set not accepted by any $M_{\sigma(i,k)}$ in CPTM with

$$|M_{\sigma(i,k)}| \leq B(r).$$

From the construction of $N_{g(r)}$ it will be seen that $|N_{g(r)}|$ is recursively bounded in r, which will yield the desired result.

Construction of $N_{g(r)}$:

a) for input w $N_{g(r)}$ determines in $|w|^3$ steps whether w is in CNF-SAT, after that,

b) for n^3 steps $N_{g(r)}$ recomputes the previous computations (on shorter inputs) and tries to find a $M_{\sigma(i,k)}$, $|M_{\sigma(i,k)}| \le B(r)$, for which there has not yet been found a witness that $M_{\sigma(i,k)}$ is not equivalent to $N_{g(r)}$. If such a machine is not found the input is rejected. If $M_{\sigma(i,k)}$ is found then $N_{g(r)}$ accepts the input if it is a satisfiable formula and rejects it otherwise.

Since we have assumed that PTIME ≠ NPTIME, for each $M_{\sigma(i,k)}$ with $|M_{\sigma(i,k)}| \le B(r)$ the machine $N_{g(r)}$ by its delayed diagonalization will accept enough elements from CNF-SAT to find a witness that the machines are not equivalent. Because of the construction of $N_{g(r)}$ we will have the desired nonrecursive succinctness on the representations of finite sets and therefore on PTIME.

The other case follows by similar reasoning. □

From the previous results we see that if PTIME ≠ NPTIME then the use of nondeterministic algorithms to describe deterministic programs can yield recursively unbounded succinctness. Therefore, even if we know that a nondeterministic polynomial time algorithm has an equivalent deterministic polynomial time algorithm we may not be able to use the deterministic polynomial time algorithm because of its immense size. Clearly, (by contrast) an equivalent deterministic algorithm which does not have to run in polynomial time can be effectively computed from the nondeterministic algorithm and its length will be recursively related to the length of the nondeterministic algorithm.

Next we show that the nonrecursive succinctness given by Theorem 6 occurs iff PTIME ≠ NPTIME and that, furthermore, PTIME = NPTIME iff the relative succinctness can be bounded linearly.

Lemma 7: If PTIME = NPTIME then there exists a recursive mapping F, which maps every $N_{\rho(i,k)}$ in NCPTM onto an equivalent $M_{\sigma(j,t)}$ in CPTM, and there are two constants c_1 and c_2 such that for all i and k

$$c_1|N_{\rho(i,k)}| + c_2 \ge |F(N_{\rho(i,k)})| = |M_{\sigma(j,t)}|.$$

Proof: Note that the language

$$U = \{N_{\rho(i,k)} \# w(\#^{|N_{\rho(i,k)}||w|^k}) \mid N_{\rho(i,k)} \text{ accepts } w\}$$

is a complete language for NPTIME [6]. If PTIME = NPTIME then U is accepted by some $M_{\sigma(j,t)} = M_{i_0}$ and therefore for every $N_{\rho(i,k)}$ we can write down a deterministic polynomial time machine $F(N_{\rho(i,k)})$ which for any input w first writes down

$$N_{\rho(i,k)} \# w(\#^{|N_{\rho(i,k)}||w|^k})$$

and then starts M_{i_0} on this input. It is easily seen that

$$L[F(N_{\rho(i,k)})] = L(N_{\rho(i,k)})$$

and that for $c_2 = |M_{i_0}|$ and a $c_1 > 0$

$$c_1 |N_{\rho(i,k)}| + c_2 \geq |F(N_{\rho(i,k)})| = |M_{\sigma(j,t)}|,$$

as was to be shown (the constant c_1 can be computed easily if we fix a representation of Tm's). □

Combining these results we get a direct relation between the PTIME = ? NPTIME problem and the problem of succinctness of deterministic representation of sets in PTIME.

Corollary 8: PTIME ≠ NPTIME iff the relative succinctness of representing languages in PTIME by deterministic and nondeterministic clocked polynomial time machines (CPTM and NCPTM) is not recursively bounded (and this happens iff the relative succinctness is not linearly bounded).

Proof: Follows from Theorem 6 and Lemma 7. □

The above theorem could give further insights into the classic separation problems. First of all, if we could show for arbitrarily large B that the size of a deterministic equivalent of some nondeterministic machines $N_{\rho(i,k)}$ must exceed

$$c_1 |N_{\rho(i,k)}| + B$$

then we would have shown that PTIME ≠ NPTIME. Similarly, if we could show that B in the above equation must be at least of a given size then we would have a lower bound for the size of the (possibly non-existent) polynomial time machine M_{i_0} which recognizes the set U (Lemma 7); if B is very large then so must be M_{i_0}, indicating that it will be very hard to find.

Our current understanding of this problem area is still so limited that we cannot eliminate the strange possibility that PTIME = NPTIME and that the smallest deterministic polynomial time machine recognizing CNF-SAT is so large that it is of no practical importance. In the following we discuss some results related to this possibility.

First we recall L.A. Levin's observation [7] that if PTIME = NPTIME then we can exhibit a small deterministic machine which is guaranteed to find in polynomial time an assignment for any satisfiable Boolean formula in CNF-SAT. This is achieved very simply by a machine M_{j_0} which slowly enumerates (say in n^3 time) more and more deterministic polynomial time machines and runs them and checks (which can be done in polynomial time) whether their computed assignments indeed satisfy the given formula. It is easily seen that if PTIME = NPTIME then M_{j_0} will find the desired assignment in polynomial time. Unfortunately there is no known way to make M_{j_0} stop in polynomial time if the formula is not satisfiable (so that M_{j_0} is only in PPTM). Therefore, though M_{j_0} is a small machine it is not a small machine which solves the satisfiability problem in case PTIME = NPTIME.

Next we discuss an observation made in discussions with Seth Breidbart of Cornell University, between the size of machines recognizing CNF-SAT and what can be formally proven about PTIME and NPTIME. If we can show in FS for some specific machine $M_{\sigma(i,k)}$ that it accepts CNF-SAT (specified by some easily understood machine M_{SAT}),

$$FS \vdash [L(M_{\sigma(i,k)}) = L(M_{SAT})]$$

then there exists a "small" deterministic polynomial time machine accepting CNF-SAT. To see this consider the machine M_{i_0} which searches in a dove-tail manner for a proof in FS that some $M_{\sigma(i,k)}$, $i,k = 1,2,\ldots$, accepts CNF-SAT and then efficiently simulates on the original input the first $M_{\sigma(i,k)}$ for which it derives the desired proof. Clearly, the size of M_{i_0} is linearly bounded in the size of FS and it recognizes CNF-SAT in deterministic polynomial time.

The strange possibility is still open that we can prove in FS that PTIME = NPTIME and that we cannot prove for any specific $M_{\sigma(i,k)}$ that it accepts CNF-SAT, i.e., we may have

$$FS \vdash [L(M_{SAT}) \in \text{PTIME}]$$

and that for no $M_{\sigma(i,j)}$

$$L(M_{\sigma(i,j)}) = L(M_{SAT})$$

is provable in FS. In other words, we may have only nonconstructive proofs in FS of the fact that PTIME = NPTIME and therefore even a formal proof of PTIME = NPTIME would not assure the existance of succinct deterministic polynomial time recognizers for CNF-SAT.

It should furthermore be pointed out that so far we have not been able to prove that there is no recursive succinctness bound between the representation of languages in PTIME by PTM and NCPTM under the assumption that PTIME $\neq$ NPTIME. This leaves the possibility open that for all languages in PTIME we can only gain a recursively bounded succinctness by going from PTM to NCPTM representations. (Recall that going the other way, from NCPTM to PTM the succinctness gain cannot be recursively bounded.) On the other hand, we know from our results that if this is possible then the polynomial running time of the machines in PTM has to grow nonrecursively in the size of these machines (Theorem 6) and that the proofs that they accept sets in PTIME also grow nonrecursively in their size. Therefore we see that if there is recursively bounded succinctness between the representations of PTIME language by PTM and NCPTM that this is achieved at the expense of immense running times and very long proofs that the machines accept the right languages.

Still we conjecture that there exists no recursive relative succinctness bound between PTM and NCPTM for languages in PTIME.

The previous results, Lemma 7 and Corollary 8, can easily be extended to other separation problems such as deterministic and nondeterministic context-sensitive languages, PTIME and PTAPE, etc. Similarly, for example, we can prove that the relative succinctness of representing the languages accepted in time n^3 by k tape machines and by k+1 tape machines running in n^3 time, is not recursively bounded iff k+1 tape machines are faster than k tape machines. This result, stated more completely below, links the old open problem about speed of k and k+1 tape machines with the succinctness of representations.

Corollary 9: The relative succinctness of representing sets accepted by k tape machines in time n^3 by k tape and k+1 tape machines running in time n^3 is not recursively bounded iff there is a set accepted in time n^3 by a k+1 tape machine which is not accepted by any k tape machine in time n^3 (and this happens iff the relative succinctness of these representations is not linearly bounded).

Approximations of Complete NPTIME Problems

In this section we make a few observations about the problem of recognizing initial segments of complete NPTIME problems by deterministic polynomial time machines.

Let the initial part of the satisfiability problem CNF-SAT, be denoted by

$$[SAT_n] = \{x \mid |x| \leq n \text{ and } x \in \text{CNF-SAT}\}.$$

Theorem 10: If PTIME $\neq$ NPTIME then relative succinctness of the representation of $[SAT_n]$, n = 1,2,..., is not recursively bounded for machines from CPTM and NCPTM as well as for VPTM and NCPTM.

Proof: Let M_r be a Tm which halts on blank tape in m_r steps. Let $N_{d(r)}$ be a uniformly constructed nondeterministic machine which runs in n^5 time and for input w:

a) determines if w is in CNF-SAT (in $|w|^3$ time),

b) it then spends $|w|^3$ steps to enumerate as many machines $M_{i_1}, M_{i_2}, \ldots, M_{i_j}$ as possible from CPTM such that $|M_{\sigma(i,k)}| \leq m_r$,

c) another $|w|^3$ steps are spent checking the preceding inputs (in lexicographic order) for witnesses that $N_{d(r)}$ is not equivalent to $M_{i_1}, M_{i_2}, \ldots, M_{i_j}$. If it is found that $N_{d(r)}$ is not equivalent to any $M_{\sigma(i,k)}$ with $|M_{\sigma(i,k)}| \leq m_r$ then the input is rejected, otherwise it is accepted iff it is in CNF-SAT.

Since we know that CNF-SAT is not in PTIME there is a sufficiently large finite part $[SAT_n]$ of CNF-SAT which is not accepted by any $M_{\sigma(i,k)}$ such that $|M_{\sigma(i,k)}| \leq m_r$, and $N_{d(r)}$ will eventually discover witnesses for all these machines. Note that in searching for the witnesses $N_{d(r)}$ may lag behind and search through much shorter strings then the current input to be able to compute what $M_{\sigma(i,k)}$ is doing on these inputs in time $|w|^3$. Nevertheless, eventually by this delayed diagonalization $N_{d(r)}$ will find all the desired witnesses and reject all longer inputs. Thus it is seen that the size of $N_{d(r)}$ is recursively related to the size of r, but since m_r is not so related, we get our desired succinctness result.

A similar proof yields the second part of this theorem. □

It is interesting to observe that the proofs that $N_{d(r)}$ accepts $[SAT_n]$ for some n are not recursively related to its size and that in general the above succinctness results cannot be extended to representations with proofs.

Theorem 11: There is a recursive succinctness bound for the representation of $[SAT_n]$, $n = 1,2,\ldots$, by CPTM and (nondeterministic) Turing machines with attached proofs that they accept $[SAT_n]$.

Proof: The desired recursive succinctness bound F can be constructed as follows:

a) for n enumerate all machines N_i such that in FS
$|N_i| + |\text{proof that } L(N_i) = [SAT_k] \text{ for some } k| \le n$,

b) compute for each such N_i the length of the first satisfiable formula not accepted by N_i

c) F(n) is the maximum of all these values.

Because of soundness of FS we see that F is the desired recursive bound. □

It is also interesting to observe that the relative succinctness of the representations of $[SAT_n]$, $n = 1,2,\ldots$, by PTM and NCPTM is recursively bounded.

Corollary 12: The relative succinctness of representing $[SAT_n]$, $n = 1,2,\ldots$, by PTM and NCPTM is recursively bounded. On the other hand, the polynomial degree of the running time of these machines in PTM (which recognize $[SAT_n]$, $n = 1,2,\ldots$, and whose size is recursively bounded in the size of the equivalent machines in NCPTM) must grow nonrecursively in their size as must the length of the proofs that they accept $[SAT_n]$.

Proof: For any $N_{\rho(i,k)}$ such that

$$L(N_{\rho(i,k)}) = [SAT_n]$$

we can effectively compute the first place where $N_{\rho(i,k)}$ stops accepting CNF-SAT. To construct a succinct deterministic polynomial time machine which accepts $[SAT_n]$ let M_j use $N_{\rho(i,k)}$ to compute the above bound (i.e. n) and reject all inputs longer than n; the shorter inputs are accepted or rejected after an exhaustive search through all possible assignments. Clearly M_i accepts $[SAT_n]$, (trivially) runs in polynomial time, and has size recursively related to $|N_{\rho(i,k)}|$.

The last assertion follows from the previous results about CPTM and VPTM (Theorems 10 and 11). □

An Observation about Optimal Algorithms

By techniques similar to the ones used in previous sections or from results in [8] we can easily prove that for tape or time bounded computations a "slight" increase in the resource bound permits a recursively unbounded shortening of the representations. We state just one special case of this general result. Let TAPE [t(n)] denote the set of languages accepted on t(n) tape [1].

Theorem 13: The relative succinctness of representing languages in TAPE $[n^2]$ by machines using n^2-tape and machines using $n^{2+\varepsilon}$-tape, $\varepsilon > 0$, is not recursively bounded.

This result has some interesting implications for optimization of algorithms. It is well known that we can not recursively decide whether, for example, algorithms running on n^3-tape have equivalent algorithms running on n^2-tape. At the same time, there is a feeling that for any particularly important n^3-tape algorithm by hard work and cleverness we will find either a proof that a faster algorithm does not exist or find a faster algorithm. The above result shows that this may not always be the case. There are n^3-tape algorithms for which equivalent n^2-tape algorithms exist and we may even be able to prove that they exist (!), but we cannot ever obtain them because of their immense size. It is not the lack of cleverness or the weakness of our formal mathematical system (which we are willing to change) which prevents us from using the fast algorithms for these computations, it is their enormous physical size which makes them inaccessible to us.

Similarly, if PTIME $\neq$ NPTIME and we use nondeterministic machines (programs) to specify deterministic polynomial time computations we may not be able to ever write down the equivalent deterministic polynomial time algorithm because of their size.

It is not clear whether there exist any natural problems for which there are reasonable-size near-optimal algorithms, but whose optimal algorithms are so large that they are of no practical importance. On the other hand, we may be able to prove for some natural problem that it has fast algorithms but that their length must exceed a large bound, as it was done in our proofs for the "unnatural" sets constructed to show the existence of recursively unbounded relative succinctness for these representations.

It is indeed possible that our difficulties with finding optimal algorithms or proving algorithms optimal for natural problems are partially caused by the immensity of the size of the optimal algorithms. It seems well worth to investigate this possibility further.

Similarly, when we consider bounded resource algorithms we observe nonrecursive succinctness gains as we go from resource bounded algorithms to asymptotically resource bounded algorithms. We state a special case of this observation.

Corollary 14: The relative succinctness of representing languages in TAPE$[n^2]$ by Tm's which work on n^2-tape for all inputs and by Tm's which work on n^2-tape for almost all inputs is not recursively bounded.

Proof: Obvious. □

Again it would be very interesting to determine whether there are some natural problems with short algorithms running asymptotically on n^2-tape for which the corresponding algorithms running everywhere on n^2-tape must grow very large.

Sets of Minimal Machines

Since most of the results in this paper deal with relative succinctness of machines and relate the growth of minimal machines to separation of complexity classes, we conclude this study by a few observations about sets of minimal machines for different representations.

First, it is well known that minimal size Turing machines (ordered lexicographically) form random sequences, in the sense of Kolmogorov and Chaitin [4] and that any axiomatizable formal mathematical system can only prove a finite number of them minimal. In other words, the set of minimal size Tm's is an immune set.

It turns out that even for several of the restricted representations of languages in PTIME discussed in this paper, the minimal machines form immune sets, indicating that proving machines minimal in these representations is a hopeless task.

Theorem 15: The minimal size machines in

VPTM, PTM, VPLM, TPLM and PPLM

are immune sets.

Proof: We will show that for VPTM there exists a constant c_0 such that no minimal machine M_i in VPTM can be printed out by a machine M_p (starting on blank tape) if

$$|M_p| + c_0 < |M_i|.$$

This immediately implies that the minimal set of machines in VPTM is immune and that in any formal system we can only prove those machines of VPTM minimal which are "not much larger" then the size of the formal system.

To prove the above assertion assume that M_p can print out the description of M_i for an M_i in VPTM. Then we can construct a machine $M_{r(p,i)}$ which:

a) for input w, using M_p prints the description of M_i on its tape
b) simulates in polynomial time M_i on input w.

If M_i is in VPTM then $M_{r(p,i)}$ is also in VPTM since we just have to verify that M_p prints the description of M_i (which has been proven in FS to be in VPTM) and verify that the simulator works in polynomial time. All of these proofs are very simple and provable in any reasonable FS. Therefore we see that

$$|M_{r(p,i)}| \leq |M_p| + c,$$

where c is basically the size of the simulator.

If M_i is a minimal machine then

$$|M_i| \leq |M_{r(p,i)}| \leq |M_p| + c,$$

which by setting $c_0 = c$, shows that

$$|M_p| + c_0 < |M_i|$$

cannot hold, as was to be shown.

The other cases follow by similar arguments. □

On the other hand, it is easily seen that the set of minimal machines in CPTM can be recursively enumerated and therefore there are complete and sound proof rules for proving machines minimal in CPTM.

Theorem 16: The set of minimal machines in CPTM is recursive enumerable.

Proof: Standard. □

Similarly we observe that the minimal machines running on, say, n^2-tape almost everywhere is an immune set, but the set of minimal machines running everywhere on n^2-tape is recursively enumerable.

References

1. Aho, A.V., J.E. Hopcroft and J.D. Ullman, "The Design and Analysis of Computer Algorithms," Addison-Wesley, Reading, Massachusetts, 1974.

2. Baker, T.P., "On "Provable" Analogs of P and NP," Mathematical Systems Theory (to appear).

3. Blum, M., "On the Size of Machines," Information and Control, Vol. 11 (1967), 257-265.

4. Chaitin, G.J., "Information Theoretic Limitations of Formal Systems," J. ACM 21, (1974), 403-424.

5. Hartmanis, J., "On the Succinctness of Different Representations of Languages," SIAM J. Computing (to appear).

6. Hartmanis, J. and J. Simon., "On the Structure of Feasible Computations," Advances in Computers Vol. 14, Morris Rubinoff and Marshall C. Yovits, eds., 1-43, Academic Press, New York, 1976.

7. Levin, L.A., "Universal Sequential Search Problems," Problemy Peredachi Informatsii, Vol. 9 (1973), 265-266.

8. Meyer, A.R., "Program Size in Restricted Programming Languages," Information and Control, Vol. 21 (1972), 382-394.

9. Meyer, A.R. and M.J. Fischer., "Economy of Description by Automata, Grammars and Formal Systems," Conference Record IEEE 12th Annual Symposium on Switching and Automata Theory (1971), 188-190.

10. Schmidt, E.H. and T.G. Szymanski., "Succinctness of Descriptions of Unambiguous Context-Free Language," SIAM J. Computing, Vol. 6 (1977), 547-553.

11. Valiant, L.G., "A Note on the Succinctness of Description of Deterministic Languages," Information and Control, Vol. 32, (1976), 139-145.

ON TWO TYPES OF LOOPS

Ivan M. Havel
Institute of Information Theory and Automation
Czechoslovak Academy of Sciences
182 08 Prague, Czechoslovakia

1. INTRODUCTION

Whether directly or indirectly, the phenomenon of looping undoubtedly contributes to the high level of mathematical sophistication in current computer science. In a rather general setting we can approach loops by considering them as a special property of processes evolving in a suitably defined state space. A process enters a loop if from a certain state it returns, possibly through several intermediate states, again to the same state. Here the word 'same' simply means that the new state cannot be, within a certain discrimination level, distinguished from the original state. (Note that, in agreement with common programmers' jargon, we use the term 'loop' where a graph theorist would prefer 'cycle'.)

In the theory of loops to be reported here the chosen discrimination level does not take into account the internal structure of a particular state and considers only its "environment": the possible transitions to other states. This alone would hardly yield any interesting theory yet. What is novel in our treatment of loops is their combination with branching. We consider the case where the transitions from any given state are lumped together into (several, possibly overlapping) groups and the process in question always proceeds from such a state by taking <u>all</u> transitions from exactly <u>one</u> chosen group (the choice being dependent, in general, on the past history of the process). The evolving process thus have a structure resembling more a branching tree than just a single sequence.

As a mathematical tool we use the formalism of the theory of automata and formal languages. In particular, the mentioned concept of a state space takes the shape of a finite automaton with special branching relations (the finite branching automaton of [1]; alternatively we could talk about AND/OR graphs [2] with labeled arcs). Using a fixed abstract alphabet Σ as a source of names of state transitions we can conveniently represent processes by means of languages (with emphasis on their branching structure - cf. [3]) rather than by trees.

Our aim is to classify various types of loops which can occur in our automata and relate them to certain natural properties of corresponding families of languages. We believe that the research in this direction may contribute to our understanding of some combinatorial aspects of such concepts as iteration, choice, infinite processes, variability of processes etc.

(The results reported in this paper, together with some other related material, will appear with complete proofs in a forthcoming paper [4]. Here we mostly present just outlines of proofs, leaving out most technical details.)

2. AUTOMATA AND BRANCHING

Let Σ be a nonempty finite alphabet. We use the symbols Σ^*, Σ^+, Λ with their usual meaning. Moreover, $\Sigma_\Lambda := \Sigma \cup \{\Lambda\}$. By $lg(u)$ we denote the length of a string $u \in \Sigma^*$. Throughout this paper let $\mathcal{L}(\Sigma)$ be the set of all nonempty languages over Σ.

The principal notion of our theory, the finite branching automaton, was introduced in [1]. Its definition rests on the standard concept of a finite automaton. Let $\mathcal{A} = \langle Q, \delta, q_0, F\rangle$ be an ordinary deterministi finite automaton over Σ (Q is a finite set of states, $\delta: Q \times \Sigma \rightarrow Q$ is the transition function, $q_0 \in Q$ is the initial state, $F \subseteq Q$ is the set of final states). We extend δ in the usual way to $Q \times \Sigma^* \rightarrow Q$ and if there is no danger of confusion, we write qu instead of $\delta(q,u)$ for any $u \in \Sigma^*$. We denote by $|\mathcal{A}|$ the language recognized by $\mathcal{A}$, i.e.,

$$|\mathcal{A}| := \{u \in \Sigma^* \mid q_0 u \in F\} . \tag{1}$$

Let $\mathcal{A} = \langle Q, \delta, q_0\rangle$ be a finite automaton (over Σ) without final states. Let $B \subseteq Q \times 2^{\Sigma_\Lambda}$. The <u>finite branching automaton</u> (over Σ), shortly the <u>fb-automaton</u>, is the quadruple

$$\mathcal{B} = \langle Q, \delta, q_0, B\rangle .$$

We call B the <u>branching relation</u> (of $\mathcal{B}$).

To be able to describe the accepting behavior of fb-automata we need some tools for expressing the related structural properties of languages. Let $\leqslant$ be the natural prefix relation on Σ^*, viz. $u \leqslant v$ iff $v = uw$ for some $w \in \Sigma^*$. Let $L \subseteq \Sigma^*$, $u \in \Sigma^*$. We define

$$\begin{aligned} \text{pref } L &:= \{u \mid u \leqslant v \text{ for some } v \in L\} \ , \\ \partial_u L &:= \{v \mid uv \in L\} \ , \\ \Delta_L(u) &:= (\text{pref } \partial_u L \cap \Sigma) \cup (\partial_u L \cap \{\Lambda\}) \ . \end{aligned}$$

The poset $\langle \text{pref } L, \leqslant \rangle$ corresponds to the <u>prefix-tree structure</u> of L, $\partial_u L$ is the <u>derivative</u> (or <u>quotient</u>) of L with respect to u, and the function $\Delta_L : \Sigma^* \to 2^{\Sigma_\Lambda}$ expresses the <u>local branching structure</u> of L: for any $a \in \Sigma$, $a \in \Delta_L(u)$ iff $ua \in \text{pref } L$, and $\Lambda \in \Delta_L(u)$ iff $u \in L$.

A language $L \in \mathcal{L}(\Sigma)$ (note that $\emptyset \notin \mathcal{L}(\Sigma)$) is <u>accepted</u> by $\mathcal{B}$ iff

$$(q_o u, \Delta_L(u)) \in B \tag{2}$$

for all $u \in \text{pref } L$. We define

$$\|\mathcal{B}\| := \{L \mid L \text{ is accepted by } \mathcal{B}\},$$

the family <u>recognized</u> by $\mathcal{B}$. A family $X \subseteq \mathcal{L}(\Sigma)$ is <u>recognizable</u> iff it is recognized by some fb-automaton. Thus, as recognizing devices the fb-automata do not represent languages but families of languages.

For technical reasons we shall always assume that

$$(q, \emptyset) \in B \tag{3}$$

for all $q \in Q$. Since $\Delta_L(w) \neq \emptyset$ for all $w \in \text{pref } L$, this assumption is irrelevant in (2) and thus it has no influence on $\|\mathcal{B}\|$.

Let us discuss the intuitive role of branching in these devices. Given a state q in an fb-automaton $\mathcal{B}$ let us define the set

$$\mathbb{B}_q := \{\Gamma \subseteq \Sigma_\Lambda \mid (q, \Gamma) \in B\} .$$

(Note that by (3) $\mathbb{B}_q \neq \emptyset$.) The "dynamics" of the fb-automaton is best visualized by imagining it as a game of two players, One and Two. In each state q One chooses $\Gamma \in \mathbb{B}_q$ and Two chooses $a \in \Gamma$ (unless $\Gamma = \emptyset$ in which case Two wins). Now either $a \in \Sigma$ and the next state is $qa = \delta(q,a)$, or $a = \Lambda$ and One is the winner. It is not important at what time One's or Two's decisions are actually made (or at what time they are disclosed to the oponent). Now any language L accepted by $\mathcal{B}$

represents One's strategy (in the wide sense, i.e. One's decisions may depend on the history of the game: $\Gamma = \Delta_L(u)$, $q_0 u = q$). In agreement with this intuition we may call elements of $\mathbb{B}_q$ the primary branching options and, if $\Gamma \in \mathbb{B}_q$, elements of Γ the secondary branching options. If $\Lambda \in \Gamma$, Λ is the terminating option.

Formally our concept of an fb-automaton resembles the alternating finite automaton of [5] and [6], which are distinguished by the transition function of the form

$$\delta : Q \times \Sigma \longrightarrow 2^{2^Q}$$

(in our case we have $\delta : Q \times \Sigma \rightarrow Q$ but $\mathbb{B} : Q \rightarrow 2^{2^{\Sigma_\Lambda}}$). The alternating finite automata are, however, designed to recognize languages rather than families of languages.

According to our definition of an fb-automaton there are no general restrictions on the choice of an element of $\mathbb{B}_q$ (the primary branching option) in every subsequent visit of any given state q . Due to this variability the family $\|\mathcal{B}\|$ is, in general, rather rich of languages, which, in turn, may be quite arbitrary elements of $\mathcal{L}(\Sigma)$. (Note that $\mathcal{L}(\Sigma)$ itself is a recognizable family. For other examples of recognizable families cf. [3] We can, however, consider a much more restrictive mode of acceptance by allowing, for each accepting behavior, only one permanent choice of primitive branching option for every state. Let $\mathcal{B} = \langle Q, \delta, q_0, B \rangle$ be an fb-automaton. We define a branching function (for $\mathcal{B}$) as a function $\beta : Q \rightarrow 2^{\Sigma_\Lambda}$, for each $q \in Q$ satisfying

$$(q, \beta(q)) \in B \tag{4}$$

(note that due to (3) it is justifiable to assume β being a total function).

Intuitively, say in our game-playing analogy, a branching function represents a strategy in the narrow sense: One's decisions depend only on the current position, not on the previous history of the game (let us call such a strategy "simple").

For a given branching function β define a partial transition function as follows. For $a \in \Sigma$ let

$$\delta_\beta(q,a) := \delta(q,a) \text{ if } a \in \beta(q) \ , \tag{5}$$

otherwise undefined. Moreover, let

$$F_\beta := \{ q \in Q \mid \Lambda \in \beta(q) \} \ . \tag{6}$$

This yields an ordinary finite (partial) automaton over Σ,

$$\mathcal{B}/\beta := \langle Q, \delta_\beta, q_0, F_\beta \rangle ,$$

called the __β-factor__ of $\mathcal{B}$. We use the following abbreviated notation related to δ_β : for $u \in \Sigma^*$, $a \in \Sigma$,

$$(q\Lambda)_\beta := q \quad ,$$
$$(qua)_\beta := \delta_\beta((qu)_\beta, a)$$

-thus $(qua)_\beta$ is defined iff $(qu)_\beta$ is defined and $a \in \beta((qu)_\beta)$; note that, in general, $((qu)v)_\beta \neq (quv)_\beta = ((qu)_\beta v)_\beta)$; for $L \subseteq \Sigma^*$

$$(qL)_\beta := \{ (qu)_\beta \mid u \in L \} .$$

A β-factor of $\mathcal{B}$ can be viewed as a "subautomaton" of $\mathcal{B}$ with transitions restricted by β and thus with limited accessibility of states. We develop specific notation for the reduced accessibility.

Let $q \in Q$, $K \subseteq Q$. We say that K is __β-accessible__ from q, symbolically $q \xrightarrow[\beta]{} K$ (or $q \xrightarrow[\beta]{} p$ if $K = \{p\}$), iff $(qu)_\beta \in K$ for some $u \in \Sigma^*$. Trivially, $q \xrightarrow[\beta]{} K$ whenever $q \in K$. We talk about a __nontrivial__ β-accessibility, and write $q \longrightarrow\xrightarrow[\beta]{} K$, if $(qu)_\beta \in K$ for some $u \neq \Lambda$.

We say that K is __strongly β-accessible__ from q, symbolically $q \xRightarrow[\beta]{} K$, iff $q \xrightarrow[\beta]{} p$ implies $p \xrightarrow[\beta]{} K$ for each $p \in Q$, i.e., iff K is β-accessible from every state β-accessible from q. We call a branching function β for $\mathcal{B}$ __perfect__ iff $q_0 \xRightarrow[\beta]{} F_\beta$.

It can be easily shown that any perfect branching function for $\mathcal{B}$ defines an accepting behavior:

__Fact 2.1__ If β is a perfect branching function for $\mathcal{B}$ then

$$|\mathcal{B}/\beta| \in \|\mathcal{B}\| .$$

In view of the fact that $|\mathcal{B}/\beta|$ is a regular language one can show that any nonempty recognizable family of languages contains a regular language just by constructing a perfect branching function for any fb-automaton $\mathcal{B}$. (Such a construction, utilizing a "sample" language from $\|\mathcal{B}\|$, is described in [4].) In the formalism of AND-OR graphs, solving a problem essentially amounts to finding a particular branching function. A corresponding algorithm can be found in [7].

3. TAXONOMY OF LOOPS

The branching functions are a convenient tool for expressing various properties of loops. Let $\mathcal{B}$ be an fb-automaton, β a branching function for $\mathcal{B}$ and q a state in $\mathcal{B}$. We say that β <u>induces a loop</u> (through q) iff q is nontrivially β-accessible from itself, i.e. iff

$$q \longrightarrow \xrightarrow[\beta]{} q \, . \tag{7}$$

If, moreover,

$$q \Longrightarrow_{\beta} F_{\beta} \tag{8}$$

we say that β induces a <u>productive loop</u> (through q). On the other hand, if (7) is combined with

$$q \Longrightarrow_{\beta} q \tag{9}$$

and with

$$q \not\longrightarrow_{\beta} F_{\beta} \, , \tag{10}$$

(read: no state in F is β-accessible from q) we say that β induces an <u>idle loop</u> (through q).

Note that due to (8) and (10) the existence of both types of loops depends on the accepting power of the corresponding branching function. This agrees with our intention to study looping as a behavioral phenomenon rather than just a structural property of the underlying state graph.

Examples of loops of both types are in Fig. 1 - 4. In Fig. 1 the branching function β induces an idle loop through both states, in Fig. 2 β induces a productive loop through both states.

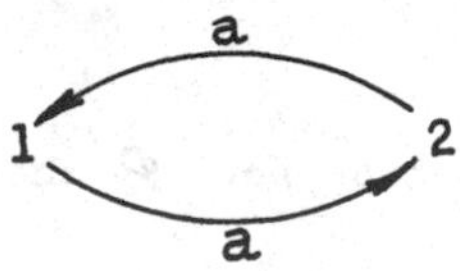

Fig. 1. $\beta(1) = \beta(2) = \{a\}$

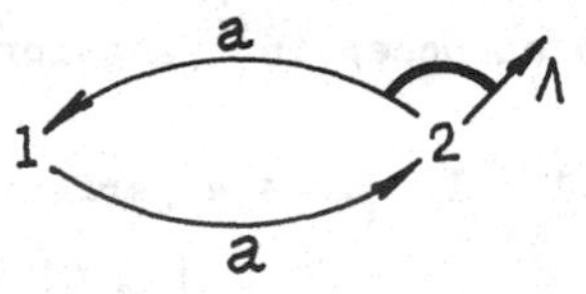

Fig. 2. $\beta(1) = a$, $\beta(2) = \{a, \Lambda\}$

In Fig. 3 there are two alternative branching functions: β induces an idle loop through both states, while β' induces a productive loop through state 1 alone. In Fig. 4 β induces an idle loop through state 2 (and neither productive nor idle loop through state 1).

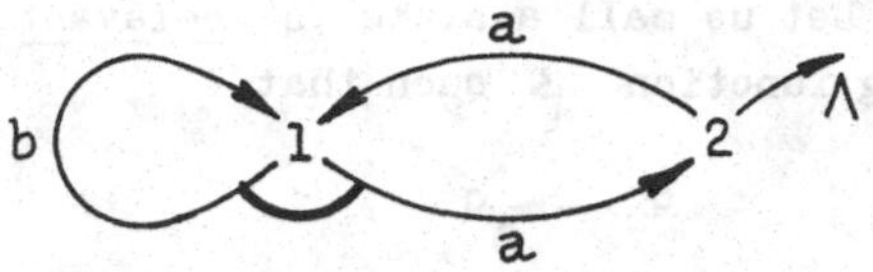

Fig. 3. $\beta(1) = \beta'(1) = \{a,b\}$, $\beta(2) = \{a\}$, $\beta'(2) = \{\Lambda\}$.

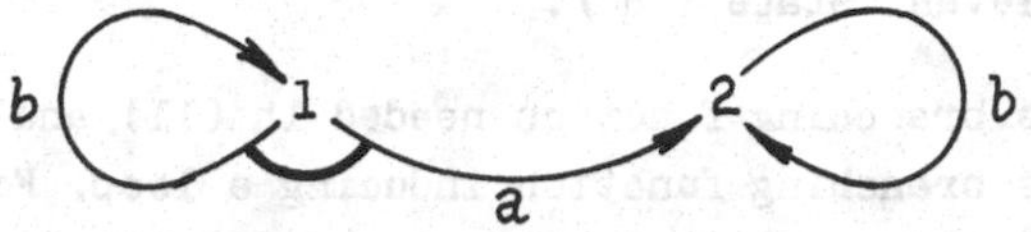

Fig. 4. $\beta(1) = \{a,b\}$, $\beta(2) = \{b\}$.

To explain the intuitive difference between idle and productive loops let us invoke our game-playing interpretation, where these two types of loops represent various ways how the power of keeping the game in a loop may be divided among the two players (One and Two). A loop is idle when One has a simple (i.e. state dependent) strategy for permanent looping so that neither One nor Two can win. On the other hand, a loop is productive when One has a simple strategy in which looping is possible but only if Two likes - otherwise I can always exit from the loop by letting One win. (Surprisingly enough, in most recreational games loops are regarded as pathological events.)

A more realistic interpretation can be given to the productive loops in action planning. Consider a physical action (like striking a nail with a hammer) for which we have reasons to believe that some (unpredictable) number of repetitions lead to a desired goal. A finite-state representation of the environment cannot incorporate this belief "locally" (in terms of a transition from one state into another). One possibility is to use a probabilistic approach (an face the problem how to convert our beliefs into real numbers). The other possibility is to represent such a situation "globally", by means of a productive loop. Obviously, the resulting plan - in its explicit form - is infinite (cf. Theorem 5.1).

Up to now we have related loops to branching functions. Since we want to attribute them directly to fb-automata it is natural to restrict ourselves only to loops passing through states which can contribute to the accepting behavior. Let us call a state q <u>relevant</u> (in $\mathcal{B}$) iff there exists a branching function β such that

$$q_0 \xrightarrow[\beta]{} q \tag{11}$$

and

$$q_0 \overset{}{\Longrightarrow}_{\beta} F_\beta \cup \{q\} \ . \tag{12}$$

We say that an fb-automaton <u>has a (productive or idle) loop</u> (through q) iff there is a branching function for $\mathcal{B}$ inducing a (productive or idle) loop through a relevant state (q).

Note that the branching function needed in (11) and (12) need not be identical with the branching function inducing a loop. For example the fb-automaton in Fig. 5 has a productive loop through state 2 induced by β, but since $1 \not\xrightarrow[\beta]{} 2$ we need for the relevancy of 2 another branching function, β' .

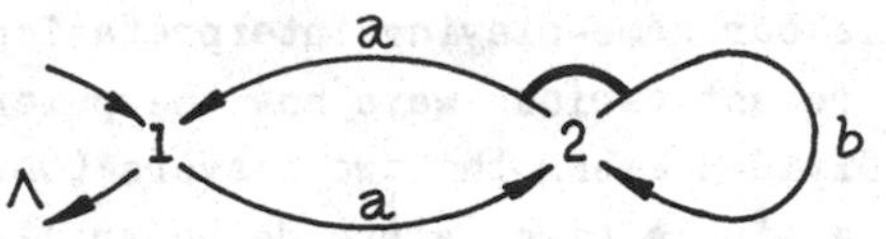

Fig. 5. $\beta(1) = \{\Lambda\}$, $\beta'(1) = \{a\}$,
$\beta(2) = \beta'(2) = \{a,b\}$, $q_0 = 1$.

It appears that idle loops, viewed from the point of view of accepting behavior of the fb-automaton, are by no means as sterile as they seem at first sight. Recall that our definition of acceptance counts with the possibility of using various branching options at different visits of the same state. Thus after a certain period of idle looping through a given state (with respect to some branching function β) a switch to another function β' may lead to final acceptance. In this case we shall talk about a "soft" idle loop.

Formally we say that $\mathcal{B}$ has a <u>soft idle loop</u> (through q) iff it has an idle loop through q and

$$q \Longrightarrow_{\beta} F_\beta \tag{13}$$

for some branching function β . (Indeed, β has to differ from the function inducing the loop in question.) For instance, in the example of Fig. 5 the fb-automaton has an idle loop through both states 1 and 2 : it is induced by β' while β satisfies (13).

4. IDLE LOOPS AND HOLES IN LANGUAGES

In this section we shall concentrate on the idle loops. First we need two auxiliary lemmas which we present here without proofs. Let $\mathcal{B}$ be an fb-automaton and let $L \in \|\mathcal{B}\|$ be any language accepted by $\mathcal{B}$.

Lemma 4.1 Let $v \in L$. Then there exists a branching function β such that

$$q_0 \longrightarrow_\beta q_0 v \tag{14}$$

and for each prefix $u \leqslant v$

$$q_0 u \Longrightarrow_\beta F_\beta \quad . \tag{15}$$

The second lemma is a consequence of the first one and of the definition of relevant states (11), (12).

Lemma 4.2 Let $v \in \text{pref}\, L$. Then $q_0 v$ is a relevant state in $\mathcal{B}$.

We have introduced soft idle loops which have a nontrivial role in the accepting behavior of fb-automata. On the other hand, idle loops which are not soft, once entered, block further acceptance and thus they have the same effect as dead states (similarly as in ordinary finite automata). Therefore the recognizing power of an fb-automaton is not altered by their removal. This suggests that if an idle loop is present in all equivalent fb-automata, it has to be a soft loop. This observation is more precisely expressed by the following theorem (which, in fact, holds in both directions, as we shall see later on).

Theorem 4.1 Let X be a recognizable family of languages with the property that every fb-automaton recognizing X has an idle loop. Then X is recognized by an fb-automaton with a soft idle loop.

Proof. It is enough to show that every idle loop which is not soft can be eliminated. Indeed, if there is a non-soft idle loop through a state p in some fb-automaton $\mathcal{B}$ recognizing X, by (13) there is no branching function β for $\mathcal{B}$ with the property that $p \Longrightarrow_\beta F_\beta$. Thus by Lemma 4.1, (15), for no $L \in X$ there is $u \in \text{pref}\, L$ such that $q_0 u = p$. Thus p can be easily "removed" from $\mathcal{B}$ without altering its accepting behavior. □

The following lemma often helps to decide whether a branching function induces an idle loop through a particular state.

Lemma 4.3 Let $\mathcal{B}$ be an fb-automaton with a state q and a branching function β such that $q \to_\beta F_\beta$ and such that $q \to_\beta p$ umplies $\beta(p) \neq \emptyset$ for every p. Then β induces an idle loop through some state which is β-accessible from q.

Proof. Assuming the hypothesis of the lemma use an induction on $\mathrm{card}(q\Sigma^*)_\beta$. Basis: If $q \Rightarrow_\beta q$ then $\beta(q) \neq \emptyset$ implies $q \to\to_\beta q$ and we are done. Induction step: The property of q in the hypothesis of the lemma is obviously shared by all $q' \in (q\Sigma^*)_\beta$. Thus, if $(q'\Sigma^*)_\beta$ is a proper subset of $(q\Sigma^*)_\beta$, the result is established by induction hypothesis. Consider, therefore, the case when

$$(q'\Sigma^*)_\beta = (q\Sigma^*)_\beta \tag{16}$$

for all $q' \in (q\Sigma^*)_\beta$. Then β induces an idle loop directly through q : By (16) for all q', $q \to_\beta q'$ implies $q' \to_\beta q$. Hence $q \Rightarrow_\beta q$. Since $\mathrm{card}(q\Sigma^*)_\beta > 1$, we have $q \to\to_\beta q$ and the result follows.□

Our next objective is to introduce a certain interesting automaton-independent property of families of languages which holds for a family recognized by an fb-automaton $\mathcal{B}$ if and only if $\mathcal{B}$ has a soft idle loop.

Consider a language L together with its prefix closure $\mathrm{pref}\, L$. Having $L \subseteq \mathrm{pref}\, L$, we may be interested in analyzing the way how elements of L are distributed in the set $\mathrm{pref}\, L$. An obvious extreme case leads to the family $\{L \mid L = \mathrm{pref}\, L\}$ of prefix-closed languages, while the other extreme are the prefix-free languages, $\{L \mid L \cap L\Sigma^+ = \emptyset\}$. Let us assume a more detailed viewpoint by asking for existence of certain specific subsets of $\mathrm{pref}\, L - L$.

Let $L \in \mathcal{L}(\Sigma)$ and let $n \geq 0$. We say that L has a hole of size at least n iff there exists $u \in \mathrm{pref}\, L$ such that

$$\partial_u L \subseteq \Sigma^n \Sigma^+ .$$

(Note that $\partial_u L \neq \emptyset$ for $u \in \mathrm{pref}\, L$.) Let $X \subseteq \mathcal{L}(\Sigma)$ be a family of languages. We say that X has the unbounded hole property iff for each $n \geq 0$, X contains a language with a hole of size at least n. Conversely, if there is a bound on the size of holes in languages belonging to X, we say that X has the bounded hole property.

In [3] we have introduced the metric space of languages $\langle \mathcal{L}(\Sigma), d\rangle$ where d is the natural distance function

$$d(L_1,L_2) = \begin{cases} 0 & \text{if } L_1 = L_2 , \\ 2^{-s(L_1,L_2)} & \text{if } L_1 \neq L_2 , \end{cases}$$

where $s(L_1,L_2)$ is the length of a shortest string $u \in \Sigma^*$ such that $\Delta_{L_1}(u) \neq \Delta_{L_2}(u)$.

We have shown (cf [3], Theorem 3a) that every recognizable family is closed in the metric space $\langle \mathcal{L}(\Sigma), d\rangle$. On the other hand, not every recognizable family is compact in this space. The following theorem establishes the connection between compactness and occurence of holes in recognizable families.

<u>Theorem 4.2</u> A recognizable family is compact iff it has the bounded hole property.

<u>Proof</u>. ($\Rightarrow$). Assume X compact but with the unbounded hole property. Then there exists an infinite sequence (L_n) of languages in X such that L_n has a hole of size at least n for $n = 0, 1, 2, \ldots$ Consider the infinite sequence

$$(M_n) := (\partial_{u_n} L_n) ,$$

where

$$\emptyset \neq \partial_{u_n} L_n \subseteq \Sigma^n \Sigma^* .$$

Since X is recognizable, by [1],Theorem 4.1, there is only a finite number of distinct families

$$X_v := \{ \partial_v L \mid L \in X \} , \quad v \in \Sigma^* .$$

Thus at least one among them, say X_w, contains an infinite subsequence $(M_{n_i})_i$ of (M_n). Let $(K_i)_i$ be the sequence of languages in X such that

$$\partial_w K_i = M_{n_i} = \partial_{u_{n_i}} L_{n_i} \subseteq \Sigma^{n_i} \Sigma^*$$

for $i = 0,1,2,\ldots$. Thus K_i has a hole of size at least n_i. By assumption, X is compact and thus (K_i) has a convergent subsequence, say $(K_{i_j})_j$. Consider the language

$$K := \lim_{j\to\infty} K_{i_j} \in X .$$

Since $w \in \text{pref } K_i$ for all i, also $w \in \text{pref } K$. There is v such that $wv \in K$. Then there is $m \geq 0$ such that for all $j > m$, $wv \in K_{i_j}$. But this is impossible since for sufficiently large j, K_{i_j} has a hole "after" w of size greater than the length of v.

($\Leftarrow$). Let $m \geq 1$ be such that no language in X has a hole of size m. The family X is precompact (cf [3], Theorem 1) and closed (since recognizable). So it is enough to show that every Cauchy sequence in X has a limit. Let (L_n) be a Cauchy sequence. There exist $n_0 \leq n_1 \leq n_2 \leq \ldots$ such that

$$d(L_{n_k}, L_n) < 2^{-k} \tag{17}$$

for all $n \geq n_k$, $k \geq 0$. Define

$$L_\infty := \bigcup_{k=0}^{\infty} L_{n_k}^{[k]}$$

where we use the notation $L^{[k]}$ for the set of all strings in L of length at most k. It can be shown (we omit the details) that for $k \geq m$

$$d(L_\infty, L_{n_k}) < 2^{-(k-m)} \quad . \tag{18}$$

By the triangular inequality we obtain from (18) and (17)

$$d(L_\infty, L_n) < 2^{-k}(2^m+1) \quad .$$

Therefore $L_\infty = \lim_{n \to \infty} L_n$. $\square$

Note that while the compactness is a general topological property of a family of languages, the presence (or absence) of holes tells something about the internal structure of particular languages. Now we relate these properties to the occurence of soft idle loops in fb-automata.

<u>Theorem 4.3</u> Let X be a recognizable family with the unbounded hole property. Then each fb-automaton recognizing X has a soft idle loop.

<u>Proof.</u> Let $X = \|\mathcal{B}\|$ for an n-state fb-automaton $\mathcal{B}$. By the hypothesis there is a language $L \in X$ with a hole of size at least $n+1$, i.e.,

$$\partial_u L \subseteq \Sigma^{n+1}\Sigma^* \tag{19}$$

for some $u \in \text{pref } L$. Consider the state $q := q_0 u$. For this state we shall define inductively an auxiliary partial function $\vartheta: Q \to \Sigma^*$ and a branching function $\beta: Q \to 2^{\Sigma_\Lambda}$. First set $\vartheta(q) := \Lambda$ and $\beta(q) := \Delta_L(u)$. For the general case assume that $\vartheta(q')$ and $\beta(q')$ has been defined for some $q' \in Q$, $\vartheta(q') = v$, but for some $p := q'a$, where

$a \in \beta(q')$, $\vartheta(p)$ has not yet been defined. Then set

$$\vartheta(p) := va \quad , \tag{20}$$

$$\beta(p) := \Delta_L(uva) \quad . \tag{21}$$

Eventually, when such q' does not exist, the construction ends and ϑ remains undefined for the remaining states, while β is extended to a total function by setting $\beta(p) = \emptyset$ (cf. (13)). We shall show that q an β meet the hypothesis of Lemma 4.3.

Let us first observe that for any $p \in Q$, if $q \xrightarrow{}_{\beta} p$ then $\vartheta(p)$ is defined. Moreover, each extension of the domain of ϑ (by executing (20)) increases just by one the maximal length of a string in the range of ϑ. Since there are only n states in Q , $\lg \vartheta(p) \leqslant n$. At the same time, $\vartheta(p) \in \text{pref } \partial_u L$. Now let $p \in Q$ be such that $q \xrightarrow{}_{\beta} p$. By (20) and (21) $\beta(p) = \Delta_L(u\vartheta(p))$ and thus $\beta(p) \neq \emptyset$. Moreover, $\Lambda \notin \beta(p)$ since otherwise $\partial_u L$ would contain $\vartheta(p)$, a string of length less than n , contrary to the assumption (19). We can now use Lemma 4.3 for q and β to conclude that β induces an idle loop through some state p where $q \xrightarrow{}_{\beta} p$.

It remains to show that p is relevant and that the loop is soft. Since $q \xrightarrow{}_{\beta} p$, we have $p = qv = q_o uv$ where $v = \vartheta(p)$. Thus $uv \in \text{pref } L$ and by Lemma 4.2 p is relevant state in $\mathcal{B}$. By Lemma 4.1 (15), $p \Longrightarrow_{\beta'} F_{\beta'}$ for some branching function β' . Thus $\mathcal{B}$ has a soft idle loop (through p). □

<u>Theorem 4.4</u> Let X be a family recognized by an fb-automaton with a soft idle loop. Then X has the unbounded hole property.

<u>Proof</u>. Let $\mathcal{B}$ be the fb-automaton assumed by the theorem and let n be an arbitrarily large number, $n \geqslant 1$. We need to construct a language $L_n \in X$ with a hole of size at least n . The property of $\mathcal{B}$ having a soft idle loop involves, in fact, three branching functions, β_1, β_2, β_3 , and a state q satisfying altogether six relations (viz. (11) and (12) for β_1 , (7), (9), and (10) for β_2 , and (13) for β_3):

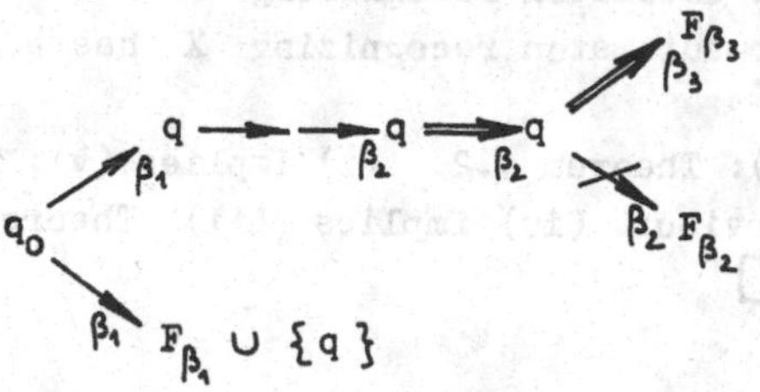

(22)

The idea is to combine β_1, β_2, and β_3 to an auxiliary function $G_n : \Sigma^* \to 2^{\Sigma_\Lambda}$ with the aim to use G_n as the function Δ_{L_n} for L_n. We define three distinguished sets of strings according to their behavior in (22):

(i) W_1 is the set of all strings $w \in \Sigma^*$ such that $(q_o w)_{\beta_1}$ is defined and w does not pass through q;

(ii) W_2 is the set of all strings $w = v_1 v_2$, where v_1 is the shortest prefix of w such that $(q_o v_1)_{\beta_1} = q$, $(qv_2)_{\beta_2}$ is defined, and w passes through q at most n times;

(iii) W_3 is the set of all strings $w = v_1 v_2 v_3$, where v_1 is the shortest prefix of w such that $(q_o v_1)_{\beta_1} = q$, $(qv_2)_{\beta_2} = q$, $(qv_3)_{\beta_3}$ is defined and $v_1 v_2$ passes through q exactly $(n+1)$-times.

Let $G_n(w) := \beta_i(q_o w)$ if $w \in W_i$ $(i=1,2,3)$; otherwise $G_n(w) := \emptyset$. Define $L_n := \{ w \mid \Lambda \in G_n(w) \}$. From the definition of W_i, $i=1,2,3$, and G_n it follows that $W_1 \cup W_2 \cup W_3 = \text{pref } L_n$ and $L_n = \| \mathcal{B} \|$.

To prove the theorem it remains to show that L_n has a hole of size at least n. Since $q_o \to_{\beta_1} q$, there exists $w_o \in \Sigma^*$ such that $(q_o w_o)_{\beta_1} = q$. Without loss of generality, assume that no proper prefix of w_o passes through q. Thus $w_o \in W_2$. We have $w_o \in \text{pref } L_n$; let us show that $\partial_{w_o} L_n \subseteq \Sigma^n \Sigma^*$. Assume the contrary: $\lg(u) < n$ for some $u \in \partial_{w_o} L_n$. Then since w_o passes through q only once, $w_o u$ passes through q at most n times. Since $w_o u \in \text{pref } L_n = W_1 \cup W_2 \cup W_3$ and $w_o \in W_2$, we have $w_o u \in W_2$. Thus $(qu)_{\beta_2}$ is defined and $G_n(w_o u) = \beta_2(q_o w_o u) = \beta_2(qu)$. Since $w_o u \in L_n$, $\Lambda \in \beta_2(qu)$. Hence $q \to_{\beta_2} F_\beta$ which contradicts our assumption that β_2 induces an idle loop. □

The following corollary summarizes our above results.

<u>Corollary 1</u> For any recognizable family X the following properties are equivalent:

(i) X has the unbounded hole property.

(ii) X is not compact.

(iii) Some fb-automaton recognizing X has a soft idle loop.

(iv) Every fb-automaton recognizing X has an idle loop.

(v) Every fb-automaton recognizing X has a soft idle loop.

<u>Proof.</u> (i) iff (ii): Theorem 4.2. (i) implies (v): Theorem 4.3. (v) implies (iv): obvious. (iv) implies (iii): Theorem 4.1. (iii) implies (i): Theorem 4.4. □

Note that properties (i) and (ii) are automaton independent while the remaining properties concern the looping behavior of fb-automata. The corollary was expressed in terms of the whole family X and the whole class of fb-automata recognizing X . If we focus on a concrete fb-automaton and a property of a particular accepted language we can state some further interesting results.

Corollary 2 An fb-automaton with n states has soft idle loop iff it accepts a language with a hole of size at least n + 1 .

Proof. ($\Rightarrow$): Immediate from Theorem 4.4. ($\Leftarrow$): The proof of Theorem 4.3 is actually a proof of this result for a concrete (but arbitrary) fb-automaton. □

Corollary 3 If an fb-automaton with n states accepts a language with a hole of size at least n + 1 then it accepts infinitely many languages.

Proof. Immediate from Corollary 2 and Theorem 4.4. □

5. PRODUCTIVE LOOPS, DEFLECTIONS, AND INFINITY

We shall now consider the case of productive loops. There are strong reasons to expect an intimate relation between existence of productive loops in fb-automata and the ability to accept infinite languages. The objective of this section is to give a detailed account of this relation.

Theorem 5.1 Any fb-automaton with a productive loop accepts an infinite language.

Proof. Let $\mathcal{B}$ be an fb-automaton with a productive loop through q . There are two branching functions β_1 and β_2 satisfying

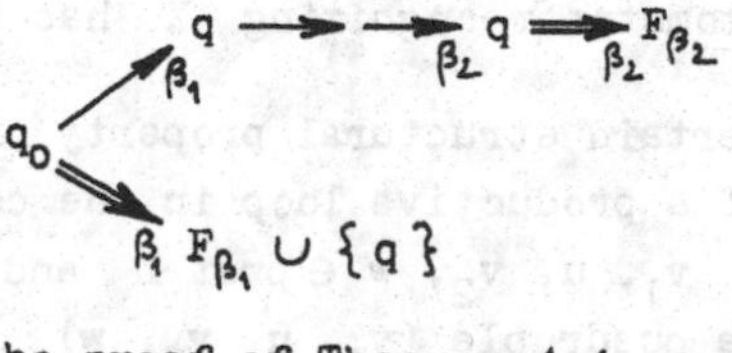

$$\begin{array}{l} q_0 \underset{\beta_1}{\longrightarrow} q \longrightarrow \underset{\beta_2}{\longrightarrow} q \underset{\beta_2}{\Longrightarrow} F_{\beta_2} \\ q_0 \underset{\beta_1}{\Longrightarrow} F_{\beta_1} \cup \{q\} \end{array} \qquad (23)$$

Analogously as in the proof of Theorem 4.4 we combine β_1 and β_2 into a function $G : \Sigma^* \longrightarrow 2^{\Sigma_\wedge}$ to be used as Δ_L for an infinite language L . Let

(i) W_1 be the set of all strings $w \in \Sigma^*$ such that $(q_o w)_{\beta_1}$ is defined and w does not pass through q ; let

(ii) W_2 be the set of all strings $w = v_1 v_2$, where v_1 is the shortest prefix of w such that $(q_o v_1)_{\beta_1} = q$, and $(qv_2)_{\beta_2}$ is defined.

Now let $G(w) := \beta_i(q_o w)$ if $w \in W_i$ $(i=1,2)$, otherwise $G(w) := \emptyset$. Define $L := \{ w \mid \Lambda \in G(w) \}$.

Now similarly as in the proof of Theorem 4.4 we can show that $W_1 \cup W_2 = \text{pref } L$ and $L \in \|\mathcal{B}\|$. Let us show that L is infinite. By (23) there are w_o , $u \in \Sigma^*$, $u \neq \Lambda$, such that $(q_o w_o)_{\beta_1} = q$ and $(qu)_{\beta_2} = q$. Assume, without loss of generality, that no proper prefix of w_o passes through q . Thus for arbitrary $n \geqslant 0$, $w_o u^n \in W_2 \subseteq$ pref L .

Is the converse also true, i.e., if $\|\mathcal{B}\|$ contains an infinite language, does it follow that $\mathcal{B}$ has a productive loop? The following example shows that this is not always the case. The fb-automaton in Fig. has a (soft) idle loop through states 1 and 2 but no productive loop. Despite of that it accepts some infinite languages, e.g. $(aa)^*b$, $(ba)^*a$

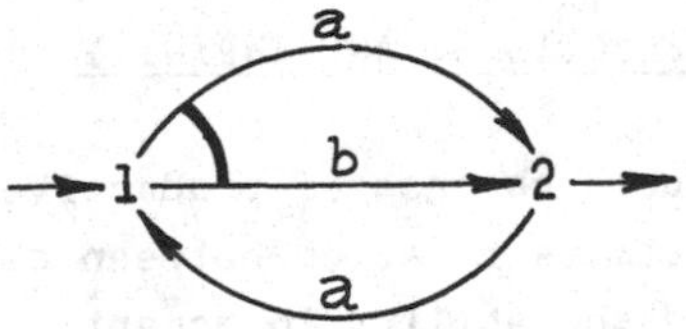

Fig. 6

Thus for a recognizable family X ,

(i) X contains an infinite language

fails to entail

(ii) Every fb-automaton recognizing X has a productive loop.

Nevertheless, we can still show (cf. Theorem 5.2 below) that (i) entails

(iii) Some fb-automaton recognizing X has a productive loop.

Let us introduce a certain structural property of languages which may reveal the presence of a productive loop in the corresponding fb-automaton. Let $L \subseteq \Sigma^*$, v_1, u, v_2, $w \in \text{pref } L$ and let $\mathcal{B}$ be an fb-automaton. We say that the quadruple (v_1, u, v_2, w) is a <u>deflection</u> in L (with resp. to $\mathcal{B}$) iff the following four conditions hold:

$$v_1 \leqslant u < v_2 \quad , \tag{24}$$

$$u \text{ is the longest common prefix of } v_2 \text{ and } w\ , \tag{25}$$

$$q_o v_1 = q_o v_2\ , \tag{26}$$

$$w \in L\ . \tag{27}$$

A useful notation for a deflection is

$$v_1 \text{ --- } u \begin{smallmatrix} \diagup w \\ \text{--- } v_2 \end{smallmatrix} \tag{28}$$

The presence of a deflection in a language accepted by $\mathcal{B}$ is, in a certain sense, a property of an intermediate character between $\mathcal{B}$ having a productive loop and $\|\mathcal{B}\|$ containing an infinite language (the letter property being automaton-independent). Note that in the definition of a deflection the dependence on a particular fb-automaton appears only in (26). The following two lemmas establish the role of deflections in languages. Let $\mathcal{B}$ be a fixed fb-automaton.

<u>Lemma 5.1</u> Any infinite language $L \in \|\mathcal{B}\|$ has a deflection with respect to $\mathcal{B}$.

<u>Proof</u>. Let $L \in \|\mathcal{B}\|$ be an infinite language. The prefix tree of L is also infinite and by Kõnig's Lemma there exists an infinite string α with all its prefixes in pref L. Thus there are infinitely many prefixes u of α such that, for some $w \in L$, u is the longest common prefix of α and w. At the same time there is a state q in $\mathcal{B}$ such that for infinitely many prefixes v of α, $q_o v = q$. Thus there necessarily exist strings $v_1 \leqslant u < v_2 < \alpha$ and $w \in L$ such that

$$v_1 \text{ --- } u \begin{smallmatrix} \diagup w \\ \text{--- } v_2 \end{smallmatrix}$$

is a deflection in L. □

We say that a deflection (28) is <u>strict</u> iff for all $x, y \in \Sigma^*$,

$$x \leqslant v_2 \text{ and } u < y \leqslant w \text{ implies } q_o x \neq q_o y\ . \tag{29}$$

<u>Lemma 5.2</u> Let $L \in \|\mathcal{B}\|$ and let (28) be a strict deflection in L with resp. to $\mathcal{B}$. Then $\mathcal{B}$ has a productive loop through $q_o u$.

We omit the proof of this lemma which depends on specific features of a standard construction of a perfect branching function from a given accepted language.

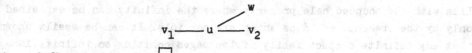

Theorem 5.2 If a recognizable family X contains an infinite language then some fb-automaton recognizing X has a productive loop.

Proof. By Lemma 5.1 an infinite language L has a deflection with resp. to any fb-automaton $\mathcal{B}$ recognizing X . We can modify $\mathcal{B}$ (by duplicating its states while preserving its recognizing power) in such a way that the deflection in question acquires the property of being strict. Now the theorem follows from Lemma 5.2.

We can summarize our results as follows:

Corollary. For any recognizable family X the following properties are equivalent:

(i) X contains an infinite language .

(ii) X contains a language with a strict deflection with respect to some fb-automaton recognizing X.

(iii) Some fb-automaton recognizing X has a productive loop.

Proof. (i) implies (ii): Lemma 5.1 and the proof of Theorem 5.2. (ii) implies (iii): Lemma 5.2. (iii) implies (i): Theorem 5.1. □

We have seen that the property of recognizable families of containing languages with arbitrarily large holes is characterized by presence of idle loops in fb-automata. Now recognizable families with the unbounded hole property are all infinite (the recognizability is needed for that the singleton family $\{\{a^{2^n} \mid n \geqslant 0\}\}$ has obviously the unbounded hole property). Thus the presence of idle loops yields an infinite variety of accepted languages. Of course, there are also infinite recognizable families with the bounded hole property where the infinity can be explained only by the presence of loops which are not idle. It can be easily shown that any infinite compact family of languages contains an infinite language. By Theorem 4.2 a recognizable family with the bounded hole property is compact and thus if it is infinite it contains an infinite language. By Theorem 5.2 it is then recognized by an fb-automaton with a productive loop.

We shall now state a much stronger result (analogous to Theorem 4.3) viz. that every fb-automaton recognizing such a family has a productive loop. The following lemma covers the essential part of the proof.

Lemma 5.3 Let $\mathcal{B}$ be an fb-automaton qnd let $X = \|\mathcal{B}\|$ be an infinite family with the bounded hole property. Then there exists $L \in X$ with a strict deflection with respect to $\mathcal{B}$.

Proof. As noted above X contains an infinite language, say L_o . By Lemma 5.1 L_o contains a deflection with resp. to $\mathcal{B}$. However, this deflection need not be strict. Since we are commited to a fixed fb-automaton $\mathcal{B}$ we cannot obtain a strict deflection just by modifying $\mathcal{B}$ (as in the proof of Theorem 5.2); the only possibility is to prove that a strict deflection exists, if not in L_o , then in another language in X . The argument can be made by contradiction. Assuming that no language in X has a strict deflection we can, starting with L_o , construct an infinite sequence of languages in X with arbitrary large holes, contrary to the hypothesis of the lemma. (The construction and its proof are relatively involved and will be therefore omitted.) □

From Lemma 5.2 and 5.3 we immediately obtain the theorem:

Theorem 5.3 Let X be an infinite fb-recognizable family with the bounded hole property. Then every fb-automaton recognizing X has a productive loop.

Let us conclude by combining Theorem 5.3 with its twin theorem for idle loops (Theorem 4.3):

Corollary. Let X be an infinite recognizable family. Then either all fb-automata recognizing X have idle loops or they all have productive loops.

REFERENCES

1. Havel, I.M., Finite branching automata. Kybernetika, 10 (1974), 281-302.
2. Nilsson, N.J., Problem Solving Methods in Artificial Intelligence. McGraw-Hill, New York, 1971.
3. Havel, I.M., On the branching structure of languages. In: Mathematical Foundations of Computer Science 1976 (A. Mazurkiewicz, Ed.), pp. 81-98, Springer-Verlag, Berlin, 1976.
4. Havel, I. M., On branching and looping. Submitted for publication.
5. Chandra, A.K. and Stockmeyer, L.J., Alternation. In: Proc. 17th Symp. Foundations of Computer Science, IEEE, 1976, pp. 98-108.
6. Kozen, D., On parallelism in Turing machines. Ibid. pp. 89-97.
7. Havel, I.M., Some remarks on searching AND-OR graphs. In: Proc. Symp. Algorithms'79, ČVTS Bratislava, 1979, pp. 45-52.

FULL ABSTRACTION FOR A SIMPLE PARALLEL PROGRAMMING LANGUAGE

M.C.B. Hennessy
G.D. Plotkin
Department of Artificial Intelligence
University of Edinburgh
Hope Park Square Meadow Lane
EDINBURGH EH8 9NW Scotland

INTRODUCTION

In [Plo1] a powerdomain was defined which was intended as a kind of analogue of the powerset construction, but for (certain kinds) of cpos. For example the powerdomain $\mathcal{P}(S_\perp)$ of the flat cpo $S_\perp$, formed from a set S, is the set $\{X \subseteq S_\perp | (X \neq \emptyset)$ and $((\perp \in X)$ or X is finite$)\}$ with the Egli-Milner ordering :

$$X \sqsubseteq_{E\text{-}M} Y \equiv (\forall x \in X. \exists y \in Y.\ x \sqsubseteq y) \wedge (\forall y \in Y. \exists x \in X.\ x \sqsubseteq y).$$

This enabled nondeterminism to be modelled by an analogue of set-theoretic union and a denotational semantics for a simple language with parallelism was given, treating parallelism in terms of non-deterministic mergeing of uninterruptible actions. Expected identities such as the associativity and commutativity of the parallel combinator were true in this semantics.

Unfortunately, other reasonable identities do not hold, in particular the distributivities :

$$P\ ;\ (Q\ \underline{or}\ R) \equiv (P\ ;Q)\ \underline{or}\ (P\ ;\ R)$$

$$(Q\ \underline{or}\ R)\ ;\ P \equiv (Q;\ P)\ \underline{or}\ (R\ ;\ P)$$

and so, with a suitable definition of behavior, the semantics will not be fully abstract [Mil 1] , [Plo 2]. Analysis of the problem leads us to the desire for a variant of the product of two powerdomains and a definition of union, $\cup$, on the new structure and a pairing function, $\otimes$, so that :

$$x \otimes (y \cup z) = (x \otimes y) \cup (x \otimes z)$$

$$(y \cup z) \otimes x = (y \otimes x) \cup (z \otimes x).$$

The ordinary product and pointwise union will not do as then the stronger equation :

$$(x \cup x') \otimes (y \cup y') = (x \otimes y) \cup (x' \otimes y')$$

holds and the corresponding equivalence for programs should be false.

In the present paper we further develop the idea of non-deterministic domains [Egl] [Hen] which are cpos with an associative, commutative, absorptive continuous binary function (called union). Their connection to cpos gives a definition of the powerdomain for all cpos, extending [Plo1][Smy1]; there is a tensor product which satisfies the above desire ; we can give a semantics using non-deterministic domains for a simple parallel programming language like that in [Plo1] which in at least one sense, is fully abstract. Interestingly most of the manipulation of sets explicit in [Plo1] disappears here as it is "built in" to the domains and their constructions.

2. THE PROGRAMMING LANGUAGE

Syntax Our language has three sets of syntactic items.

1. BExp - a given set of Boolean expressions, ranged over by the metavariable b.
2. Act - a given set of primitive actions, ranged over by a.
3. Stat - a set of statements, ranged over by s, and given by the grammar :

s::= a|(s;s)|(if b then s else s)|(while b do s)|(s or s)|(s par s)|(s co s).

It is not necessary here to assume anything about the structure of BExp or Act ; standard examples of elements would be "x ≥ y" or "x:=y+5" for an arithmetic language. The statements provide a simple imperative language with parallelism, which will be treated in terms of interleaving of atomic actions, and with a somewhat strange " coroutine" facility which gives a very strict interleaving of the atomic actions.

Operational Semantics

We will use the set T = {tt, ff} of truthvalues and a given set, S, of states, ranged over by σ. The behaviours of the Boolean expressions and the primitive actions are given, rather abstractly, by two functions :

1. $\mathcal{E}$: BExp → (S → T),
2. $\mathcal{A}$: Act → (S → T).

For the statements we axiomatise a relation → : Str×Str where Str $=_{\text{def}}$ S∪(Stat×S); the relation <s,σ> → σ' (<s,σ> → <s',σ'>) is to mean that executing the first uninterruptible step of s, starting from σ, results in σ' with the termination of s (respectively, with s' being the remainder of s) ; no other relations are possible :

I1. $\langle a,\sigma\rangle \to \mathcal{A}[\![a]\!](\sigma)$.

II1. $$\frac{\langle s_1,\sigma\rangle \to \sigma'}{\langle (s_1;s_2),\sigma\rangle \to \langle s_2,\sigma'\rangle}\ ,\qquad 2.\ \frac{\langle s_1,\sigma\rangle \to \langle s_1',\sigma'\rangle}{\langle (s_1;s_2),\sigma\rangle \to \langle (s_1';s_2),\sigma'\rangle}\ .$$

III.1 $$\frac{\langle s_1,\sigma\rangle \to str}{\langle (\underline{if}\ b\ \underline{then}\ s_1\ \underline{else}\ s_2),\sigma\rangle \to str}\qquad (\mathcal{E}[\![b]\!](\sigma) = tt\ ;\ str \in Str),$$

2. $\dfrac{\langle s_2,\sigma\rangle \to str}{\langle(\underline{if}\ b\ \underline{then}\ s_1\ \underline{else}\ s_2),\sigma\rangle \to str}$ $(\mathcal{E}[\![b]\!](\sigma) = ff\ ;\ str \in Str)$.

IV1. $\dfrac{\langle s,\sigma\rangle \to \sigma'}{\langle(\underline{while}\ b\ \underline{do}\ s),\sigma\rangle \to \langle(\underline{while}\ b\ \underline{do}\ s),\sigma'\rangle}$ $(\mathcal{E}[\![b]\!](\sigma) = tt)$,

2. $\dfrac{\langle s,\sigma\rangle \to \langle s',\sigma'\rangle}{\langle(\underline{while}\ b\ \underline{do}\ s),\sigma\rangle \to \langle(s';(\underline{while}\ b\ \underline{do}\ s)),\sigma'\rangle}$ $(\mathcal{E}[\![b]\!](\sigma) = tt)$,

3. $\langle(\underline{while}\ b\ \underline{do}\ s),\sigma\rangle \to \sigma$ $(\mathcal{E}[\![b]\!](\sigma) = ff)$.

V1. $\dfrac{\langle s_i,\sigma\rangle \to str}{\langle(s_1\ \underline{or}\ s_2),\sigma\rangle \to str}$ $(i = 1, 2\ ;\ str \in Str)$.

VI1. $\dfrac{\langle s_1,\sigma\rangle \to \sigma'}{\langle(s_1 \underline{par}\ s_2),\sigma\rangle \to \langle s_2,\sigma'\rangle}$, $\dfrac{\langle s_2,\sigma\rangle \to \sigma'}{\langle(s_1\ \underline{par}\ s_2),\sigma\rangle \to \langle s_1,\sigma'\rangle}$,

2. $\dfrac{\langle s_1,\sigma\rangle \to \langle s_1',\sigma'\rangle}{\langle(s_1\ \underline{par}\ s_2),\sigma\rangle \to \langle(s_1'\ \underline{par}\ s_2),\sigma'\rangle}$, $\dfrac{\langle s_2,\sigma\rangle \to \langle s_2',\sigma'\rangle}{\langle(s_1\ \underline{par}\ s_2),\sigma\rangle \to \langle(s_1\ \underline{par}\ s_2'),\sigma'\rangle}$.

VII1. $\dfrac{\langle s_1,\sigma\rangle \to \sigma'}{\langle(s_1 \underline{co}\ s_2),\sigma\rangle \to \sigma'}$, 2. $\dfrac{\langle s_1,\sigma\rangle \to \langle s_1',\sigma'\rangle}{\langle(s_1\ \underline{co}\ s_2),\sigma\rangle \to \langle(s_2\ \underline{co}\ s_1'),\sigma'\rangle}$

Now we can give a definition of the behavior of a statement in terms of a non-deterministic state transformation function :

3. $\mathcal{B} : Stat \to (S \to \mathcal{P}(S_\perp))$,

where : $\mathcal{B}[\![s]\!](\sigma) = \{\sigma' \in S \mid \langle s,\sigma\rangle \overset{*}{\to} \sigma'\} \cup$

$\{\perp \mid$ there is an infinite sequence $\langle s,\sigma\rangle \to .. \to \langle s_n,\sigma_n\rangle \to ..\}$.

As $\{str \mid \langle s,\sigma\rangle \to str\}$ is always finite and nonempty, Königs lemma shows $\mathcal{B}[\![s]\!](\sigma)$ is always finite or contains $\perp$ and so is in $\mathcal{P}(S_\perp)$ as required.

Note the flexibility of the method for specifying interruption points ; if we had wished conditionals to be interruptable after the test, instead of the present "test and set" capability, we would have written :

III'1. $\langle(\underline{if}\ b\ \underline{then}\ s_1\ \underline{else}\ s_2),\sigma\rangle \to \langle s_1,\sigma\rangle$ $(\mathcal{E}[\![b]\!](\sigma) = tt)$,

2. $\langle(\underline{if}\ b\ \underline{then}\ s_1\ \underline{else}\ s_2),\sigma\rangle \to \langle s_2,\sigma\rangle$ $(\mathcal{E}[\![b]\!](\sigma) = ff)$.

3. NON-DETERMINISTIC DOMAINS

We discuss the extra structure provided by the union function, the connections with powerdomains and useful constructions such as the tensor product.

Definition 3.1 : A complete partial order (cpo) is a partial order, $\langle D,\sqsubseteq\rangle$, with a least element, $\perp_D$, and lubs, $\sqcup_D x_n$, of increasing ω-chains ; a function f: D → E of partial orders is strict, monotonic or continuous according, respectively, as it preserves the least element, the order or the order and all existing lubs of increasing

ω-chains. We let CPO be the category of cpo's and continuous functions, let $CPO_\perp$ be the subcategory of strict functions, and let O be the category of partial orders with lubs of all increasing ω-chains and continuous functions.

The reason for considering the three categories, $CPO_\perp \subset CPO \subset O$ is that the main one of interest, CPO, lies between the more natural $CPO_\perp$ and O. All three have all small products given by Cartesian product ; $CPO_\perp$ and O are small complete.

Definition 3.2 : A non-deterministic partial order (nd-po) is a structure $\langle D,\sqsubseteq,\cup\rangle$ where $\langle D,\sqsubseteq\rangle$ is a po and $\cup : D^2 \rightarrow D$ is a monotonic function (called union) where :

1. Associativity For all x, y, z in D, $(x \cup y) \cup z = x \cup (y \cup z)$.
2. Commutativity For all x,y in D, $(x \cup y) = (y \cup x)$.
3. Absorption For all x in D, $(x \cup x) = x$.

A function $f: D \rightarrow E$ of nd-po's is linear if it preserves union. We let ND be the category of non-deterministic domains (nd-pos which are cpo's and have a continuous union) and continuous linear functions, let $ND_\perp$ be the subcategory of strict functions and let NO be the category of nd-po's which are O-objects and which have a continous union and continuous linear functions.

Again, $ND_\perp \subseteq ND \subseteq NO$ and all three have all small products given by Cartesian product and $ND_\perp$ and NO are small complete. Note that in any nd-po, D, we can define a "subset relation" by : $x \subseteq y$ iff $(x \cup y) = y$; this is a partial order and if D is NO then $\subseteq$ is inductive in the sense that if $\langle x_n\rangle,\langle y_n\rangle$ are two increasing ω-chains with $x_n \subseteq y_n$ then $(\bigsqcup x_n) \subseteq (\bigsqcup y_n)$.

For constructions on $ND_\perp$ we use the Freyd Adjoint Functor Theorem (FAFT-see [Mac]) in conjunction with a useful lemma.

Definition 3.3 : An O -category is one whose hom sets are equipped with a partial order so that they form an O-object and so that composition is continuous in each argument ; an O-functor $G:A \rightarrow X$ of O-categories is one which is continuous with respect to the order on the hom-sets. An NO-category is an O-category whose hom -sets are equipped with a binary function so that they form an NO-object and so that composition is linear in each argument ; an NO-functor $G:A \rightarrow X$ of NO-categories is an O-functor which is linear with respect to the union on the hom-sets.

Note that all the above categories are O-categories with respect to the natural pointwise ordering of morphisms ; further NO, ND and $ND_\perp$ are all NO-categories with respect to the natural pointwise union. Any small product of NO-categories is an NO-category and so the product functor is an NO-functor (which is also strict on the hom -sets).

Definition 3.4 : Let $G:A \rightarrow X$ be an O-functor. Then $f:x \rightarrow Ga$ is a G-orderepi iff whenever $a \xrightarrow{g\ ,\ g'} a'$ are such that $(Gg)f \sqsubseteq (Gg')f$ then $g \sqsubseteq g'$.

Lemma 3.5 : Let $G:A\rightarrow X$ be an O-functor such that every $f:x\rightarrow Ga$ factorises as

$x \xrightarrow{f'} Ga' \xrightarrow{Gg} Ga$ where f' is a G-orderepi. Then the left adjoint of G is also an $\underline{\underline{O}}$-functor and if G is an $\underline{\underline{NO}}$-functor its left adjoint is an $\underline{\underline{NO}}$-functor too.

Powerdomains : The evident forgetful functor $V_2 : \underline{\underline{NO}} \to \underline{\underline{O}}$ has a left-adjoint $\mathcal{P} : \underline{\underline{O}} \to \underline{\underline{NO}}$ which is an $\underline{\underline{O}}$-functor ; further $\mathcal{P}$ cuts down to a left-adjoint to each of the forgetful functions $V_1 : \underline{\underline{ND}} \to \underline{\underline{CPO}}$, $V_0 : \underline{\underline{ND}}_\perp \to \underline{\underline{CPO}}_\perp$. The powerdomain construction in [Plo1][Smy1] is the restriction of $\mathcal{P} : \underline{\underline{CPO}} \to \underline{\underline{ND}}$ to the ω-algebraic case and then the unit map is singleton, $\{|\cdot|\} : D \to \mathcal{P}(D)$ and the "big union" $\uplus : \mathcal{P}(\mathcal{P}(D)) \to \mathcal{P}(D)$ is the multiplication of the associated monad.

Other powerdomain constructions [Smy1][Mil1] can be treated similarly. Smyth's one can be obtained by adding the inequation :

4. $(x \cup y) \sqsubseteq x$.

If instead we add :

5. $(x \cup y) \sqsupseteq x$

we would obtain a construction involving the "other half" of the Egli-Milner ordering. Variations with an empty set [Mil1] are obtained by considering algebras, $\langle D, \sqsubseteq, \cup, \emptyset \rangle$ where $\emptyset$ is an element of D satisfying :

6. $(x \cup \emptyset) = x$.

Other possibilities are to consider a strict construction with the equation :

7. $(x \cup \perp) = \perp$

or to drop the absorption axiom to obtain a kind of multiset construction.

In all cases where we have the experience, all the neccessary auxiliary functions can be obtained from categorial considerations ; it is not clear however whether the required properties can be (conveniently) so obtained and we might need detailed constructions as in [Plo1][Smy1][Mil1], although they are not necessary in the present paper.

Sums : The category $\underline{\underline{ND}}_\perp$ has binary sums ; that is for any nd-domains, D_0, D_1 there is another (D_0+D_1) and strict continuous linear functions, $in_i : D_i \to (D_0+D_1)$ $(i=0,1)$ such that for any other nd-domain, F, and strict continuous linear functions $f_i : D_i \to F$ $(i=0,1)$ there is a unique such function $[f_0, f_1] : (D_0+D_1) \to F$ such that the following diagrams commute :

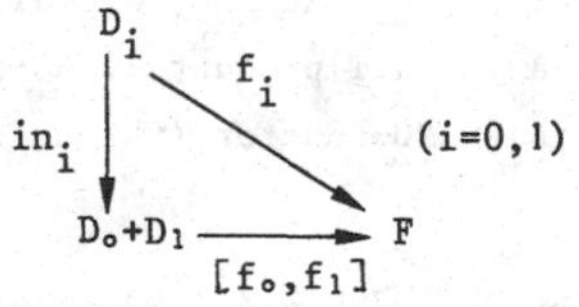

Further, [.,.] is strict continuous and linear on the hom-sets and so is $+ : \underline{\underline{ND}}^2 \rightarrow \underline{\underline{ND}}_\perp$ considered as a functor (it is an $\underline{\underline{NO}}$-functor).

<u>Lifting</u> : The forgetful functor $V_\perp : \underline{\underline{ND}}_\perp \rightarrow \underline{\underline{NO}}$ has a left adjoint $(.)_\perp : \underline{\underline{NO}} \rightarrow \underline{\underline{ND}}_\perp$; that is for any $\underline{\underline{NO}}$ object, D, there is an nd-domain, $(D)_\perp$, and a continuous linear up : $D \rightarrow (D)_\perp$ such that for any other continuous linear f: $D \rightarrow E$ there is a unique strict continuous linear lift(f) : $(D)_\perp \rightarrow E$ such that the following diagram commutes:

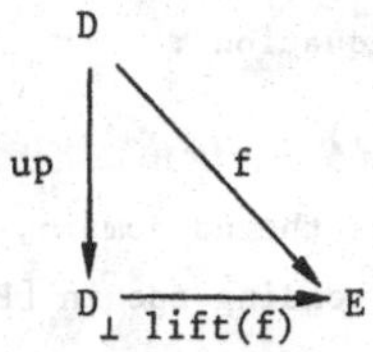

Further lift(.) is continuous and linear as is $(.)_\perp$ on the hom-sets. Finally, we note that $(.)_\perp$ cuts down to a left adjoint to the forgetful functor from $\underline{\underline{ND}}_\perp$ to $\underline{\underline{ND}}$.

<u>Tensor products</u> : The point of the tensor product is to reduce multilinear functions to linear ones.

<u>Definition 3.6</u> : Let A,B,C be nd-domains. A continuous function $f : A \times B \rightarrow C$ is <u>bistrict</u> iff for all b in B $f(\perp,b) = \perp$ and for all a in A, $f(a,\perp) = \perp$; it is bilinear iff for all a, a' in A and b in B, $f(a \cup a',b) = f(a,b) \cup f(a',b)$ and for all a in A, b, b' in B, $f(a,b \cup b') = f(a,b) \cup f(a,b')$. (<u>N.B.</u> We are <u>not</u> assuming f either strict or linear). We let Bislin (A,B;C) be the set of bistrict, bilinear continuous functions from A×B to C equipped with the pointwise order and union (and so an nd-domain).
Note the bistrict functions are strict and linear binary functions are bilinear.

Any nd-domains A,B have a tensor product $A \otimes B$; that is, there is a bistrict, bilinear continuous $\otimes : A \times B \rightarrow A \otimes B$ which is universal in the sense that for any bistrict, bilinear f: $A \times B \rightarrow C$ there is a unique strict, linear continuous slin (f): $A \otimes B \rightarrow C$ such that the following diagram commutes :

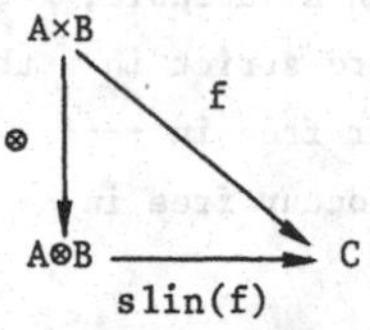

Further slin : Bislin (A,B;C) $\cong \mathrm{Hom}_{\underline{\underline{ND}}_\perp}(A \otimes B,C)$ is an isomorphism of nd-domains. We extend $\otimes$ to a functor $\otimes : \underline{\underline{ND}}_\perp \times \underline{\underline{ND}}_\perp \rightarrow \underline{\underline{ND}}_\perp$ by requiring that the following diagram always commutes :

$$\begin{array}{ccc} A\times B & \xrightarrow{f\times g} & A'\times B' \\ \big\downarrow \otimes & & \big\downarrow \otimes \\ A\otimes B & \xrightarrow{f\otimes g} & A'\otimes B' \end{array}$$

Then we find that ⊗ is continous, linear and bistrict on the hom-sets.

Domain Equations : All the theory in [Smy2] applies to $\underline{\underline{ND}}_\perp$ and as constructions we can use ×, powers, $\mathcal{P} \circ V_\circ$, +, $(.)_\perp \circ V_\perp$, ⊗ (and others not mentioned here). In the present work we only need to solve the one equation :

$$\alpha\colon R \cong (\mathcal{P}(S_\perp) + (\mathcal{P}(S_\perp) \otimes (R)_\perp))^S$$

(where we have omitted the $V_\perp$). This gives us the nd-domain, R, of *resumptions*, which have the same motivation as the corresponding cpo in [Plo1]. Below we shall treat the isomorphism as an equality, omitting to write α or α^{-1}.

4. DENOTATIONAL SEMANTICS

We present a useful abbreviation. Suppose a,b,c are different variables of types $\mathcal{P}(S_\perp)$, $\mathcal{P}(S_\perp)$, R, respectively, and ——— , --- , ... , are expressions of types $(\mathcal{P}(S_\perp) + (\mathcal{P}(S_\perp) \otimes R_\perp))$, D, D respectively where --- is strict and linear in a, and .... is strict in b and linear in b and c ; then the expression, e, where :

$$e = (\underline{\text{cases}} \text{ ——— } \underline{\text{first}}\ a : \text{----}\ \underline{\text{second}}\ b, c, : \ldots.)$$

is of type D and abbreviates :

$$[\lambda a \in \mathcal{P}(S_\perp).\ \text{----}\ ,\ \text{slin}\ (\lambda b \in \mathcal{P}(S_\perp), d \in R_\perp.\ \text{lift}(\lambda c \in R.\ \ldots.)(d))]\ (\text{———})$$

where d is a new variable of type $R_\perp$ not free in .... . There are two "evaluation" values for e :

$$(\underline{\text{cases}}\ in_0(a)\ \underline{\text{first}}\ a : \text{----}\ \underline{\text{second}}\ b,c : \ldots.) = \text{---},$$

$$(\underline{\text{cases}}\ in_1(b\otimes up(c))\ \underline{\text{first}}\ a : \text{---}\ \underline{\text{second}}\ b,c : \ldots.) = \ldots. \,.$$

From now on we will often omit in_0, in_1, up and {.} when they are clear from the context. If ——— , ----, ... are all continuous in a variable, x , so is e ; if ——— is strict in x or else if both ---- and .... are strict in x then e is strict in x ; if ——— is linear in x, but x does not occur free in ---- , ... or else if both ---- and .... are linear in x, but x does not occur free in ——— then e is linear in x.

We now consider various useful combinators.

Sequence : The sequence combinator is the least continuous $*\colon R\times R \to R$ such that :

$$r_1 * r_2 = \langle \underline{\text{cases}}\ r_1(\sigma)\ \underline{\text{first}}\ a : a\otimes r_2\ \underline{\text{second}}\ b,c : b\otimes(c* r_2)\rangle\ \sigma \in S$$

The sequence combinator is bilinear and left-strict.

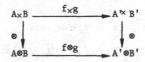

Parallelism : The parallelism combinator is the least continuous $*: R \times R \to R$ such that

$$r_1 \parallel r_2 = \langle \underline{\text{cases}}\ r_1(\sigma)\ \underline{\text{first}}\ a{:}a\otimes r_2\ \underline{\text{second}}\ b,c{:}b\otimes(c \parallel r_2)\rangle \quad \sigma \in S$$
$$\cup\langle \underline{\text{cases}}\ r_2(\sigma)\ \underline{\text{first}}\ a{:}a\otimes r_1\ \underline{\text{second}}\ b,c{:}b\otimes(r_1 \parallel c)\rangle \quad \sigma \in S.$$

It is bilinear.

Coroutine : The coroutine combinator is the least continuous co : $R\times R \to R$ such that

$$r_1 \text{ co } r_2 = \langle \underline{\text{cases}}\ r_1(\sigma)\ \underline{\text{first}}\ a{:}a\ \underline{\text{second}}\ b,c{:}b\otimes(r_2 \text{ co } c)\rangle \quad \sigma \in S$$

It is bilinear and left-strict.

The denotational semantics of our language is given by a function $\mathcal{V}$:Stat $\to$ R defined by structural induction on statements :

I. $\mathcal{V}[\![a]\!] = \langle \mathcal{A}[\![a]\!](\sigma)\rangle\ \sigma\in S$

II. $\mathcal{V}[\![s_1;s_2]\!] = \mathcal{V}[\![s_1]\!] * \mathcal{V}[\![s_2]\!]$

III. $\mathcal{V}[\![\underline{\text{if}}\ b\ \underline{\text{then}}\ s_1\ \underline{\text{else}}\ s_2]\!] = \langle \underline{\text{if}}\ \mathcal{E}[\![b]\!](\sigma)\ \underline{\text{then}}\ \mathcal{V}[\![s_1]\!]_\sigma\ \underline{\text{else}}\ \mathcal{V}[\![s_2]\!]_\sigma\rangle \quad \sigma \in S$

IV. $\mathcal{V}[\![\underline{\text{while}}\ b\ \underline{\text{do}}\ s]\!] = Y(\lambda r \in R.\ \langle \underline{\text{if}}\ \mathcal{E}[\![b]\!](\sigma)\ \underline{\text{then}}\ (\mathcal{V}[\![s]\!] * r)_\sigma\ \underline{\text{else}}\ \sigma\rangle\ \sigma \in S)$

V. $\mathcal{V}[\![s_1\ \underline{\text{or}}\ s_2]\!] = \mathcal{V}[\![s_1]\!] \cup \mathcal{V}[\![s_2]\!]$

VI. $\mathcal{V}[\![s_1\ \underline{\text{par}}\ s_2]\!] = \mathcal{V}[\![s_1]\!] \parallel \mathcal{V}[\![s_2]\!]$

VII. $\mathcal{V}[\![s_1\ \underline{\text{co}}\ s_2]\!] = \mathcal{V}[\![s_1]\!]\ \text{co}\ \mathcal{V}[\![s_2]\!]$.

Again the method is flexible for specifying interruption points ; if we had conditionals to be interruptable after the test we would have written :

III'. $\mathcal{V}[\![\underline{\text{if}}\ b\ \underline{\text{then}}\ s_1\ \underline{\text{else}}\ s_2]\!] = \langle \underline{\text{if}}\ \mathcal{E}[\![b]\!](\sigma)\ \underline{\text{then}}\ \sigma\otimes\mathcal{V}[\![s_1]\!]\ \underline{\text{else}}\ \sigma\otimes\mathcal{V}[\![s_2]\!]\rangle$
$\sigma \in S.$

5. RELATIONS BETWEEN THE TWO SEMANTICS

We begin by showing that the denotational semantics, $\mathcal{V}$, can be derived from the operational semantics as "the least model of $\to$". Specifically we can regard (Stat $\to$ R) as an nd-domain - the power R^{Stat} - and define a continuous map $\psi : R^{\text{Stat}} \to R^{\text{Stat}}$ by :

$$\psi(\xi)\ [\![s]\!]_\sigma = \cup\{\sigma' \mid \langle s,\sigma\rangle \to \sigma'\}\cup$$
$$\cup\{\sigma'\otimes \xi[\![s']\!] \mid \langle s,\sigma\rangle \to \langle s',\sigma'\rangle\}$$

The definition makes sense as, by the properties of $\to$, at least one of the sets on the right is non-empty and both are finite.

So putting $\mathcal{W}_n = \psi^n(\bot)$, the least fixed-point of ψ is $\mathcal{W} =_{\text{def}} \bigsqcup_{n\geq 0} \mathcal{W}_n$.

<u>Lemma 5.1</u> $\mathcal{V} = \mathcal{W}$.

<u>Proof</u> <u>(Outline)</u> One proves that $\mathcal{W}$ satisfies the equations defining $\mathcal{V}$.
For example to show $\mathcal{W}[\![s_1 \ \underline{\text{par}}\ s_2]\!] = \mathcal{W}[\![s_1]\!] \,\|\, \mathcal{W}[\![s_2]\!]$ one proves by induction on n that $\mathcal{W}_n [\![s_1 \ \underline{\text{par}}\ s_2]\!] \sqsubseteq \mathcal{W}[\![s_1]\!] \,\|\, \mathcal{W}[\![s_2]\!]$ and similarly that $\mathcal{W}[\![s_1]\!] \,\|_n\, \mathcal{W}[\![s_2]\!] \sqsubseteq \mathcal{W}[\![s_1 \ \underline{\text{par}}\ s_2]\!]$ where $\|_n$ is the nth approximant to $\|$. ∎

Next we recast the definition of $\mathcal{B}:\mathrm{Stat} \to (S \to \mathcal{P}(S_\perp))$ in the same style defining $\Phi : (\mathcal{P}(S_\perp)^S)^{\mathrm{Stat}} \to (\mathcal{P}(S_\perp)^S)^{\mathrm{Stat}}$ by :

$$\Phi(\mathcal{C})[\![s]\!]_\sigma = \bigcup\{\sigma' \mid <s,\sigma> \to \sigma'\} \ \cup$$
$$\bigcup\{\mathcal{C}[\![s']\!]_{\sigma'} \mid <s,\sigma> \to <s',\sigma'>\}$$

<u>Lemma 5.2</u> $\mathcal{B} = Y(\Phi)$

<u>Proof</u> Put $\mathcal{C}_n = \Phi^n(\perp)$ and then $\mathcal{C} =_{\mathrm{def}} Y(\Phi) = \bigsqcup_n \mathcal{C}_n$.
One proves by induction on the length of the derivation that $<s,\sigma> \overset{*}{\to} \sigma'$ implies $\sigma' \in \mathcal{C}[\![s]\!]_\sigma$ and by induction on n that $\sigma' \in \mathcal{C}_n[\![s]\!]_\sigma$ implies $<s,\sigma> \overset{*}{\to} \sigma'$; next one shows that if $<s,\sigma> = <s_o,\sigma_o> \to \ldots \to <s_m,\sigma_m> \to \ldots$ is an infinite derivation sequence then, by induction on n, $\perp \in \mathcal{C}_n[\![s]\!]_\sigma$ and finally one shows by induction on n that if for all s,σ we have $\perp \in \mathcal{C}_n[\![s]\!]_\sigma$ then there is a derivation sequence of length n from $<s,\sigma>$ and then applies Konig's lemma. ∎

Now let β be the least continuous function from R to $(\mathcal{P}(S_\perp))^S$ such that :

$\beta(r) = \underline{<\text{cases}}\ r(\sigma)\ \underline{\text{first}}\ a{:}a\ \underline{\text{second}}\ b,c : \mathrm{Ext}(\beta(c))(b)> \quad \sigma \in S.$

Here $\mathrm{Ext} : \mathcal{P}(S_\perp)^S \to (\mathcal{P}(S_\perp) \to \mathcal{P}(S_\perp))$ is defined so that Ext(f) is the unique strict continuous linear extension of f: $\mathrm{Ext}(f)(X) = \{\sigma : \exists x \in X.\ \sigma \in \bar{f}(x)\}$ where $\bar{f}: S_\perp \to \mathcal{P}(S_\perp)$ is the unique strict extension of f.

<u>Lemma 5.3</u> The continuous function $\beta^{\mathrm{Stat}} : R^{\mathrm{Stat}} \to (\mathcal{P}(S_\perp)^S)^{\mathrm{Stat}}$ is strict and the following diagram commutes :

$$\begin{array}{ccc}
R^{\mathrm{Stat}} & \xrightarrow{\ \Psi\ } & R^{\mathrm{Stat}} \\
\downarrow \beta^{\mathrm{Stat}} & & \downarrow \beta^{\mathrm{Stat}} \\
(\mathcal{P}(S_\perp)^S)^{\mathrm{Stat}} & \xrightarrow{\ \Phi\ } & (\mathcal{P}(S_\perp)^S)^{\mathrm{Stat}}
\end{array}$$

<u>Proof</u> Straightforward calculation. ∎

<u>Theorem 5.5</u> $\mathcal{B} = \beta \circ \mathcal{V}$

<u>Proof</u>
$$\begin{aligned}
\mathcal{B} &= Y(\Phi) && \text{(lemma 5.2)} \\
&= \beta^{\mathrm{Stat}}(Y\Psi) && \text{(lemma 5.3)} \\
&= \beta \circ \mathcal{W} && \text{(by definition)} \\
&= \beta \circ \mathcal{W} && \text{(lemma 5.1).} \quad ∎
\end{aligned}$$

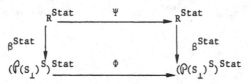

Thus the semantics, $\mathcal{V}$, determines the behavior, $\mathcal{B}$ (via β).

Full abstraction

Given a measure of behaviour, such as $\mathcal{B}$, and relations between behaviours, such as $=$, $\sqsubseteq$, $\subseteq$ on $(\mathcal{P}(S_\perp))^S$, we can define corresponding substitutive behavioural relations, $\approx$, $\precsim$, $\lesssim$. First, a context is a statement, $C[\cdot,\ldots,\cdot]$ with several "holes" which can be filled by any statements $s_1,\ldots,s_n$ to give a statement, $C[s_1,\ldots,s_n]$; a formal definition is obtained by adding the production $s ::= [\ ]$ to those given above for Stat. Now the relations $\approx$, $\precsim$, $\lesssim$ on Stat are defined by :

$$
\begin{aligned}
s_1 \approx s_2 &\quad \text{iff} \quad \forall C[.].\ \ \mathcal{B}[\![C[s_1]]\!] = \mathcal{B}[\![C[s_2]]\!] \\
s_1 \precsim s_2 &\quad \text{iff} \quad \forall C[.].\ \ \mathcal{B}[\![C[s_1]]\!] \sqsubseteq \mathcal{B}[\![C[s_2]]\!] \\
s_1 \lesssim s_2 &\quad \text{iff} \quad \forall C[.].\ \ \mathcal{B}[\![C[s_1]]\!] \subseteq \mathcal{B}[\![C[s_2]]\!]
\end{aligned}
$$

Clearly $s_1 \approx s_2$ iff $(s_1 \precsim s_2 \precsim s_1)$ iff $(s_1 \lesssim s_2 \lesssim s_1)$.

Proposition 5.5
1. $\mathcal{V}[\![s_1]\!] = \mathcal{V}[\![s_2]\!] \Rightarrow s_1 \approx s_2$
2. $\mathcal{V}[\![s_1]\!] \sqsubseteq \mathcal{V}[\![s_2]\!] \Rightarrow s_1 \precsim s_2$
3. $\mathcal{V}[\![s_1]\!] \subseteq \mathcal{V}[\![s_2]\!] \Rightarrow s_1 \lesssim s_2$

Proof 1. $\mathcal{V}[\![s_1]\!] = \mathcal{V}[\![s_2]\!] \Rightarrow \forall C[.].\ \mathcal{V}[\![C[s_1]]\!] = \mathcal{V}[\![C[s_2]]\!]$ (by definition of $\mathcal{V}$) $\Rightarrow s_1 \approx s_2$ (by theorem 5.4).

2. $\mathcal{V}[\![s_1]\!] \sqsubseteq \mathcal{V}[\![s_2]\!] \Rightarrow \forall C[.].\ \mathcal{V}[\![C[s_1]]\!] \sqsubseteq \mathcal{V}[\![C[s_2]]\!]$ (by the continuity of $*$, $\cup$, $\|$, co and the definition of $\mathcal{V}$) $\Rightarrow s_1 \precsim s_2$ (by the continuity of β and theorem 5.4).

3. As 2, but using the linearity of $\cup$, the bilinearity of $*$, $\|$, co and the easily proved monotonicity in $\subseteq$ of the conditional and while constructs and β. ∎

The rest of the section establishes, under certain reasonable assumptions, the converse of these implications, therely obtaining three full abstraction results. The assumptions are :

1. S is infinite but denumerable.

2. For each σ in S there is an element K_σ of A such that for all σ' in S, $\mathcal{A}[\![K_\sigma]\!](\sigma') = \sigma$ (a next instruction).

3. For each σ in S there is an element is_σ of BExp such that for all σ' in S, $\mathcal{E}[\![is_\sigma]\!](\sigma') = tt$ iff $(\sigma = \sigma')$.

Under these assumptions we have :

<u>Lemma 5.6</u> If $\mathcal{V}\llbracket s_1 \rrbracket \neq \mathcal{V}\llbracket s_2 \rrbracket$ then there is a context, $C[.]$ and a state σ so that $\perp \notin \mathcal{B}\llbracket C[s_1]\rrbracket(\sigma) \cup \mathcal{B}\llbracket C[s_2]\rrbracket(\sigma)$ and $\mathcal{B}\llbracket C[s_1]\rrbracket(\sigma) \neq \mathcal{B}\llbracket C[s_2]\rrbracket(\sigma)$.

which is enough to show :

<u>Theorem 5.7.</u>

$$1.\ \mathcal{V}\llbracket s_1 \rrbracket = \mathcal{V}\llbracket s_2 \rrbracket \Leftrightarrow s_1 \approx s_2 \Leftrightarrow$$
$$\mathcal{V}\llbracket s_1 \rrbracket \subseteq \mathcal{V}\llbracket s_2 \rrbracket \Leftrightarrow s_1 \lesssim s_2$$
$$2.\ \mathcal{V}\llbracket s_1 \rrbracket \subseteq \mathcal{V}\llbracket s_2 \rrbracket \Leftrightarrow s_1 \lesssim s_2$$

<u>Proof 1</u> The lemma implies that $\mathcal{V}\llbracket s_1 \rrbracket \neq \mathcal{V}\llbracket s_2 \rrbracket$ implies $s_1 \not\lesssim s_2$. So $\mathcal{V}\llbracket s_1 \rrbracket = \mathcal{V}\llbracket s_2 \rrbracket \Rightarrow \mathcal{V}\llbracket s_1 \rrbracket \subseteq \mathcal{V}\llbracket s_2 \rrbracket \Rightarrow s_1 \lesssim s_2 \Rightarrow \mathcal{V}\llbracket s_1 \rrbracket = \mathcal{V}\llbracket s_2 \rrbracket$ and the rest of 1 is immediate.

2. First $s_1 \lesssim s_2 \Rightarrow (s_1$ <u>or</u> $s_2) \approx s_2$. For let $C[.]$ be a context ; then $\mathcal{B}\llbracket C[s_1$ <u>or</u> $s_2]\rrbracket \subseteq \mathcal{B}\llbracket C[s_2$ <u>or</u> $s_2]\rrbracket$ (as $s_1 \lesssim s_2$) $= \mathcal{B}\llbracket C[s_2]\rrbracket$ (as $\mathcal{V}\llbracket s_2$ <u>or</u> $s_2\rrbracket = \mathcal{V}\llbracket s_2 \rrbracket$ shows s_2 <u>or</u> $s_2 \approx s_2$) and conversely as $\mathcal{V}\llbracket s_2 \rrbracket \subseteq \mathcal{V}\llbracket s_1$ <u>or</u> $s_2\rrbracket$, we have $\mathcal{B}\llbracket C[s_2]\rrbracket \subseteq \mathcal{B}\llbracket C[s_1$ <u>or</u> $s_2]\rrbracket$. But then by 1, $\mathcal{V}\llbracket s_1 \rrbracket \subseteq \mathcal{V}\llbracket s_1$ <u>or</u> $s_2 \rrbracket = \mathcal{V}\llbracket s_2 \rrbracket$. ∎

We now outline the proof of the lemma. First for any pair $\langle s,\sigma\rangle$ put :

$$A_{s,\sigma} = \{\sigma' \mid \langle s,\sigma\rangle \rightarrow \sigma'\}$$

$$B_{s,\sigma} = \{\sigma' \mid \exists s'. \langle s,\sigma\rangle \rightarrow \langle s',\sigma'\rangle\}$$

and for any σ' in $B_{s,\sigma}$:

$$st[s,\sigma,\sigma'] = (s'_1 \ \underline{or}\ (\ldots(s'_{n-1}\ \underline{or}\ s'_n)\ldots))$$

where $\{s'_1,\ldots,s'_n\} = \{s' \mid \langle s,\sigma\rangle \rightarrow \langle s',\sigma'\rangle\}$ and clearly $n \neq 0$.

We say $\langle s,\sigma\rangle$ is of types 1,2,3 according as $B_{s,\sigma}$, $A_{s,\sigma}$ or neither is $\emptyset$.

We have the useful formulae :

$$\mathcal{W}_{n+1}\llbracket s \rrbracket_\sigma = \cup A_{s,\sigma} \quad \cup \bigcup_{\sigma' \in B_{s,\sigma}} \sigma' \otimes \mathcal{W}_n\llbracket st[s,\sigma,\sigma'] \rrbracket$$
$$\mathcal{V}\llbracket s \rrbracket_\sigma = \cup A_{s,\sigma} \quad \cup \bigcup_{\sigma' \in B_{s,\sigma}} \sigma' \otimes \mathcal{V}\llbracket st[s,\sigma,\sigma'] \rrbracket$$
$$\mathcal{B}\llbracket s \rrbracket_\sigma = \cup A_{s,\sigma} \quad \cup \bigcup_{\sigma' \in B_{s,\sigma}} \mathcal{B}\llbracket st[s,\sigma,\sigma'] \rrbracket(\sigma')$$

Define the relation $\sim$ between such pairs by :

$$\langle s,\sigma\rangle \sim \langle s',\sigma'\rangle \text{ iff } (A_{s,\sigma} = A_{s',\sigma'}) \text{ and } (B_{s,\sigma} = B_{s',\sigma'})$$

<u>Lemma 5.8</u> If $\mathcal{W}_n\llbracket s_1 \rrbracket \neq \mathcal{W}_n\llbracket s_2 \rrbracket$ then there are statements $s_i = s_i^0,\ldots,s_i^m$ $(m \geq 0)$ states $\sigma^0,\ldots,\sigma^m$ and states $\bar{\sigma}^j$ in $B_{s_i^j,\sigma^j}$ $(j < m)$ such that $s_i^{j+1} = st[s_i^j,\sigma^j,\bar{\sigma}^j]$ $(j < m)$, $\langle s_1^j,\sigma^j\rangle \sim \langle s_2^j,\sigma^j\rangle$ $(j < m)$ but $\langle s_1^m,\sigma^m\rangle \not\sim \langle s_1^m,\sigma^m\rangle$.

This sets up the path we want to follow to extract a difference. For assuming $\mathcal{V}[\![s_1]\!] \neq \mathcal{V}[\![s_2]\!]$ we have $\mathcal{W}_n[\![s_1]\!] \neq \mathcal{W}_n[\![s_2]\!]$ for some n. And we apply the lemma to obtain $s_i^j (j < m\ ;\ i=1,2)$, $\sigma^j\ (j \leq m)$, $\bar{\sigma}^j\ (j < m)$.

Now for a state $\surd$ and a statement P^m (to be chosen later) we define statements $P^j (0 \leq j < m)$ by :

$$P^j = (\underline{\text{if}}\ \text{is}_{\bar{\sigma}^j}\ \underline{\text{then}}\ (K_{\sigma^{(j+1)}}\ ;\ P^{j+1})\ \underline{\text{else}}\ K_{\surd})$$

and for a statement $\mathcal{Q}'$ (to be chosen later) we set :

$$\mathcal{Q} = (\underline{\text{if}}\ \text{is}_{\sigma^1}\ \underline{\text{then}}\ K_{\surd}\ \underline{\text{else}}$$

$$(\underline{\text{if}}\ \text{is}_{\sigma^2}\ \underline{\text{then}}\ K_{\surd}\ \underline{\text{else}}$$

$$\cdots$$

$$(\underline{\text{if}}\ \text{is}_{\sigma^m}\ \underline{\text{then}}\ K_{\surd}\ \underline{\text{else}}\ \mathcal{Q}')\ldots))$$

and then calculate the following two formulae :

$$\mathcal{B}[\![s_i^0\ \underline{\text{co}}\ P^0]\!] (\sigma^0) = (\bigcup_{j<m} A^j) \cup C \cup \mathcal{B}[\![s_i^m\ \underline{\text{co}}\ P^m]\!] (\sigma^m)$$

(where $C \subseteq \{\surd\}$)

$$\mathcal{B}[\![(s_i^0\ ;\ \mathcal{Q})\ \underline{\text{co}}\ P^0]\!] (\sigma^0) = C \cup \bigcup\ \mathcal{B}[\![(s_i^m\ ;\ \mathcal{Q})\ \underline{\text{co}}\ P^m]\!] (\sigma^m)$$

(where $C \subseteq \{\surd\}$)

Then the proof is completed by considering various cases based on the types of $<s_1^m, \sigma^m>$, $<s_2^m, \sigma^m>$ respectively and using one of the contexts ([] $\underline{\text{co}}$ P^0) or (([] ; $\mathcal{Q}$) $\underline{\text{co}}$ P^0) and choices of $\surd$, P^m, $\mathcal{Q}'$ as appropriate for the case at hand.

6. DISCUSSION

We make a few critical remarks to obtain some perspective on the above results. First the notion of behaviour chosen is inappropriate for languages for writing continuously interacting programs expressly written not to terminate. One should study our language with the addition of some I/O instructions and a different notion of behaviour. Again the coroutine instruction is somewhat peculiar and its rôle is somewhat similar to that of the "parallel or" in [Plo2] ; without it our semantics would, we conjecture, not be fully abstract, as we would have :

$$(x:=x)\ ;\ (x:=x) \approx (x:=x)$$

$$(x:=g(f(x))) \lesssim (x:=f(x)\ :\ x:=g(x))$$

We could also study definability questions, as in [Plo2] , and look for proof rules for $\approx, \lesssim, \lesssim$ using the semantics. Most importantly, our language is hardly a good model of communicating processes, and we feel it is rather important to study many other models of parallelism ([Bri], [Hoa] [Mil] and others) before claiming that the semantics of parallelism is understood.

ACKNOWLEDGEMENTS

We thank the Science Research Council and Orsay University, Paris for their support.

REFERENCES

[Bri] Brinch Hansen, P. (1978), Distributed Processes : A Concurrent Programming Concept. Comm. ACM 21, 11, pp. 934-940.

[Egl] Egli, H. (1975) : A Mathematical Model for Nondeterministic Computation.

[Hen] Hennessy, M.C.B. and Ashcroft, E.A. (1979) : A Mathematical Semantics for a Nondeterministic Typed λ-calculus. To appear.

[Hoa] Hoare, C.A.R. (1978) : Communicating Sequential Processes. Comm. ACM 21, 8, pp. 666-677.

[Mac] Mac Lane, S. (1971) : Categories for the Working Mathematician. Berlin, Springer Verlag.

[Mil1] Milne, G.J. and Milner, R. (1977) : Concurrent processes and their syntax. To appear in the J.A.C.M.

[Mil2] Milner, R. (1977) : Fully Abstract Models of Typed λ-calculi. T.C.S.

[Plo1] Plotkin, G.D. (1976) : A Powerdomain Construction. SIAM Journal on Computing 5, 3, pp. 452-487.

[Plo2] Plotkin , G.D. (1977) : LCF Considered as a Programming Language. T.C.S. 5, pp. 223-255.

[Smy1] Smyth, M. (1978) : Powerdomains. J.C.S.S. 16, 1.

[Smy2] Smyth, M. and Plotkin, G.D. (1978) : The Category Theoretic Solution of Recursive Domain Equations. University of Edinburgh : D.A.I. Research Report N° 60.

ON SOME DEVELOPMENTS IN CRYPTOGRAPHY AND THEIR APPLICATIONS TO COMPUTER SCIENCE

Hermann A. Maurer
Institut für Informationsverarbeitung
Techn.Univ.Graz, Steyrerg.17, A-8010 Graz/Austria

SUMMARY

This paper presents a discussion and summary of results obtained jointly with K.Culik [1] and M.Nivat [2].

The idea of public key systems as applicable to information systems is reviewed. The development of information systems based on the new concept of encryption triple is outlined. Although the construction of encryption triples using number-theoretic tools is possible, systems of encryption triples obtained that way do not possess all properties desirable for designing information systems. As first step towards an alternative approach, regular bijections i.e. bijections definable by a-transducers are investigated, and a complete characterization of such bijections is given.

INTRODUCTION

The main question considered in this paper is how to assure that information stored in an information system can be accessed by authorized persons only.

To achieve such a goal, two completely different methods are currently in use. The first involves *access-path control*, the second *encryption techniques*. With the access-path control approach, unauthorized access is prevented by building into the (software) system appropriate controls (mainly based on secret passwords and the like) which presumably can be passed only by authorized users. There are a number of obvious disadvantages inherent to this technique. For example, "internal security" is comparatively small (that is, data processing personell intimately familiar with the system can circumvent access-path controls fairly easily). A still more serious drawback is the fact that it does not suffice to guard data stored internally in the computer, but it is also necessary to protect data stored externally (on punched cards, magnetic tapes etc.) from unauthorized access. Nothing short of physical protection is possible for this purpose. Because of this, information systems requiring an extreme level of security (e.g. in military environments) are often physically guarded as a whole, unauthorized usage of information being prevented by the sheer fact that only authorized personell is permitted to get near the system and its external data storages. In contrast to the access-path control approach just described, access to information is guarded by the use of encryption techniques by despositing information in encrypted form. Although the encrypted version of information may easily be obtainable by unauthorized persons it is presumably useless to anyone not knowing the decryption process. Hence, assuming that "hard-to-break" encryption methods are used and the corresponding decryption methods are known to authorized users only, any data stored (internally or externally) in encrypted form is protected against unauthorized usage.

In what follows some of the problems connected with encryption techniques, and how they can be reduced by employing public key systems are explained. Next, a more advanced set-up based on encryption triples is discussed. Finally, recent results on bijections defined by a-transducers which may serve as a general model for encryption and decryption processes are mentioned.

ENCRYPTION OF INFORMATION AND INFORMATION SYSTEMS

Encryption and decryption of information is accomplished by a *complementary pair,* i.e. a pair (E,D) of total functions where D is the inverse of E, $D = E^{-1}$. To encrypt information I, we apply the encryption algorithm E yielding E(I). The decryption algorithm D applied to encrypted information E(I) delivers the original information I because of $D(E(I)) = I$.

Complementary pairs (E,D) ordinarily used in cryptography have an additional property which is important to keep in mind: knowledge of the encryption algorithm E immediately implies knowledge of the decryption algorithm D, e.g. because both algorithms are based on a common parameter usually called *key*. Indeed one usually does not keep secret the algorithms E and D "as such", but just the key used.

Throughout the rest of this paper the following simple model of information system will be considered. An *information system* has a finite set $\mathcal{U}$ of *users* and contains m different *types* of information. The set of all users authorized to access information of type i ($1 \leq i \leq m$) is denoted by A_i and called an *authorization class* (for type i information).

$\mathcal{A} = \{A_1, A_2, \ldots, A_m\}$ denotes the set of authorization classes. Observe that $A_i \subseteq U$ for all i ($1 \leq i \leq m$) but that $A_i \cap A_j$ ($i \neq j$) may well be non-empty; a user may belong to a number of authorization classes. An information system as described is *secure* if information of type i can only be retrieved by the users in authorization class A_i.

A naive attempt to use encryption techniques for designing an information system as described is to choose a complementary pair (E_i, D_i) for each authorization class A_i, to deposit information I of type i as $E_i(I)$, secrecy being preserved (to some extent) by making the algorithm D_i known only to members of A_i. Observe that this attempt has a number of drawbacks:

(1) Each user who belongs to r authorization classes has to know r different decryption algorithms.

(2) Any user who wants to deposit information of type i has to know the encryption algorithm E_i and hence also will know D_i. Thus, a large number of users know the decryption algorithm D_i increasing the security risk. Still worse, the system is only secure if a user entitled to supply information of type i is also member of authorization class A_i. (In many situations this is not a realistic assumption: assume that information of type i is information concerning various departments within a company; A_i would then typically consist of company management but should not include the heads of departments: such heads should be able to provide information about their departments but probably should not be able to access information concerning some of their rival departments.)

(3) Information deposited in the system is not safe against fraud committed by members of an authorization class. For the sake of two examples, consider three users U_1, U_2, U_3 of some authorization class. U_1 may deposit information for U_3 pretending to be U_2. U_3 has no way of finding out (without re-checking with U_2). (Observe that in ordinary "paper-based" information systems such fraud is impossible since U_3 presumably recognizes the signature of U_2 as fake.) A con-

tract (purchase agreement, cheque, etc.) deposited by U_1 for U_3 in the information system may be fraudulently changed by U_3: in contrast to a "paper-based" information system no traces of physical tampering with the contract would be visible.

It is not at all obvious, to start with, that the difficulties (1) - (3) mentioned can be overcome by using only cryptographic methods. However, an ingenious idea due to Diffie and Hellman [3] allows to overcome difficulty (2) and (3) as explained in the next section, while difficulty (1) disappears with the use of safe systems of encryption triples of [1] as discussed in the next but one section.

DIFFIE - HELLMAN PAIRS

A *Diffie - Hellman pair, DH-pair* for short (a terminology introduced in [1]), is a complementary pair (E,D) as follows:

(i) E and D are easily computable;

(ii) The algorithm for computing E does not give a tractable way of deriving the algorithm for computing D.

The systematic construction of DH pairs based on results in number theory was first accomplished by Rivest, Shamir and Adleman in [4]. Rather then repeating their arguments, an example which shows that the existence of DH-pairs is "plausible" should suffice here. Consider a phone-directory of a large city, and an "inverted" directory (arranged in order of ascending phone numbers). To encrypt information, encrypt it letter-by-letter as follows: for a given letter, look up a random name starting with that letter in the phone directory and write down the corresponding phone number. To decrypt, use the inverted directory in the obvious way. Note that both encryption and decryption are simple processes but knowledge of the encryption process does not give an easy way for decrypting: without the "inverted" directory, decryption remains difficult.

DH-pairs can be used for information systems as follows: for each authorization class A consider a DH-pair (E_A, D_A). For each user U consider a DH-pair (E_U, D_U). The encryption algorithms E_A $(A \in \mathcal{A})$ and E_U $(U \in \mathcal{U})$ are made public. (E.g. by putting them in a universally accessible part of the information system. Hence the term "public key systems". For practical purposes, not a full algorithm but just a key as in ordinary cryptographic systems suffices). The decryption algorithms

D_A are known only to members of authorization class A, the decryption algorithm D_U are known only to user U.

Information I is deposited for authorization class A as $E_A(I)$. It can be accessed only by persons knowing D_A, i.e. by users belonging to A. Observe two important advantages over the previously discussed method: firstly, the decryption algorithms D_A are not derivable from the encryption algorithms E_A, hence the decryption algorithms are known to fewer people, decreasing the security risk; secondly, users not belonging to authorization class A may indeed supply information for authorization class A. Thus, the difficulty (2) mentioned above has disappeared. Still more surprisingly, difficulty (3) can also be overcome, as follows:

Information I is deposited by user U in a fraud-proof form for authorization class A as $E_A(D_U(I))$. By applying D_A, any member of authorization class A (and only such members) can decrypt this to $D_U(I)$. Next, the publicly known algorithm E_U is used to obtain I. Note that the intended information I has been retrieved via $D_U(I)$ which can only be computed from I using D_U, the algorithm known only to U. Thus, U is conclusively the originator of I. Also, it is impossible to tamper with the information I. For suppose a user U' belonging to A wants to (fraudulently) change I into I'. He has to replace $E_A(D_U(I))$ in the information system by $E_A(D_U(I'))$. This is clearly impossible, since U' does not know the algorithm D_U.

Thus, DH-pairs can be used to overcome the difficulties (2) and (3) mentioned in the previous section. We now consider difficulty (1) that every user has to know as many decryption algorithms as the number of authorization classes he belongs to.

ENCRYPTION TRIPLES

For functions f and g, denote by $f \circ g$ their composition defined by $f \circ g(x) = g(f(x))$.

According to [1], an *encryption triple* is a triple of functions (E,R,D) such that $(E \circ R, D)$ and $(E, R \circ D)$ are DH-pairs.

Observe that $D(R(E(I))) = D(E \circ R(I)) = I$.

Consider an information System S with users $\mathcal{U}$ and authorization classes $\mathcal{A}$. A set $F = \{(E_A, R_{A,U}, D_U) \mid A \in \mathcal{A}, U \in A\}$ is called a *safe system of encryption triples for S* if

(i) $(E_A, R_{A,U}, D_U)$ is an encryption triple for each $A \in \mathcal{A}$ and each $U \in A$ and

(ii) the knowledge of all algorithms E_A and $R_{A,U}$ for all $A \in \mathcal{A}$ and $U \in A$ and of one specific D_U provides an algorithm for easily retrieving I from $E_A(I)$ and $R_{A,U}(E_A(I))$ iff $U' = U$ and $U' \in A$.

To use a safe system of encryption triples F as described, all encryption algorithms E_A and all "recryption" algorithms $R_{A,U}$ are made public. Each user keeps a single decryption algorithm, the algorithm D_U, secret.

Information I is deposited for authorization class A as $E_A(I)$. If user U requests the information from the system, the recryption algorithm $R_{A,U}$ is applied first (possibly by the information system itself) yielding $R_{A,U}(E_A(I))$. From this, I can be retrieved using the secret decryption algorithm D_U. Observe that each user has to know only a single secret decryption algorithm, overcoming difficulty (1) mentioned earlier.

To see that only authorized users can access information, apply D_U^{-1} to both sides of the equation $D_U(R_{A,U}(E_A(I))) = I$ yielding

(*) $R_{A,U}(E_A(I)) = D_U^{-1}(I)$, i.e. $D_U^{-1} = E_A \circ R_{A,U}$.

Put differently, we have $R_{A,U}(X) = D_U^{-1}(E_A^{-1}(X))$, i.e. $R_{A,U} = E_A^{-1} \circ D_U^{-1}$. Thus, $R_{A,U}$ is the composition of a function E_A^{-1} which allows the decryption of information encrypted by E_A and of a function D_U^{-1} which encrypts information decryptable by D_U. Despite the fact D_U^{-1} is easily computable by the public because of (*), the definition of encryption triple assures that neither D_U nor $R_{A,U} \circ D_U = E_A^{-1}$ can be computed. Hence I cannot be retrieved based on E_A and $R_{A,U}$ without knowing D_U.

Observe that condition (ii) assures that the other encryption triples cannot be used for unauthorized information retrieval, either.

A user U can store fraud-proof ("signed" and "tamper proof") information I for authorization class A by depositing $E_A(D_U(I))$. A user U' belonging to authorization class A applies $R_{A,U'}$ followed by $D_{U'}$ to obtain $D_U(I)$. He now applies E_A followed by $R_{A,U}$ yielding (because of

(*)) $R_{A,U}(E_A(D_U(I))) = D_U^{-1}(D_U(I)) = I$. As in the preceding section, the pair $(D_U(I),I)$ is conclusive proof that U is originator of I, nor is it possible to tamper with the information $E_A(D_U(I))$ deposited by U.

Thus, using safe systems of encryption triples, secure information systems can be designed in which each user has to keep secret a single decryption algorithm and in which information can be deposited fraud-proof (i.e. "signed" and "tamper-proof").

Before discussing the problem of constructing safe systems of encryption triples a side-remark concerning all "public key" systems seems in order. Any public key system is safe only if the information I is taken from a very large set of possible informations. For suppose I is known to be one of a small number of possible sequences of symbols, $I \in \{I_1,I_2,\ldots,I_t\}$. Then the information $E_A(I)$ can be decrypted without knowledge of a decryption algorithm by just computing $E_A(I_1),E_A(I_2),\ldots,E_A(I_t)$ and checking for which j, $E_A(I_j) = I$ holds.

Extending methods of [4] a method for constructing encryption triples has been obtained in [1]. Unifortunately, the combination of such triples does not yield a safe system of encryption triples: as Rivest [5] has pointed out, condition (ii) of the definition of safe system of encryption triples is violated. It remains an important open problem to construct safe systems of encryption triples.

One approach in this direction is to look for alternative ways of constructing DH-pairs or, more broadly, of constructing complementary pairs. The combination of such pairs could hopefully lead to new ways of obtaining DH-pairs and encryption triples. Since encryption and decryption algorithms always appear to be special cases of mappings defined by a-transducers, a systematic investigation of bijections defined by such machines seems to be required. A first step in this direction has been made by Nivat and the author in [2]. A brief review of some of the results is given in the next section.

REGULAR BIJECTIONS

Recall that an a-transducer is a nondeterministicly working finite state machine with accepting states which is able to produce output words on both empty and nonempty inputs. More precisely, an *a-transducer* T is a six- tuple $T=(\Phi,\Sigma,\Delta,H,q_-,A)$, where Φ, Σ and Δ are finite sets of states, inputs and outputs respectively, where $q_o \in \Phi$ is the start-state, $A \subseteq \Phi$ is a set of accepting states and where H is a finite set of transition rules, $H \subseteq \Phi \times \Sigma^* \times \Phi \times \Delta^*$. For each word $y \in \Sigma^*$ one defines $T(y) = \{z \mid y = y_1 \ldots y_n,\ z = z_1 \ldots z_n,\ (q_{i-1}, y_i, q_i, z_i) \in H$ for $i = 1,2,\ldots,n$ and $q_n \in A\}$.

Two sets of words R and A are in *regular bijection*, symbolically $R \overset{f}{\longleftrightarrow} S$ or just $R \longleftrightarrow S$, if $f: R \longmapsto S$ is a bijection of R onto S such that there are a-transducers T and T^{-1} with $T(y) = \{f(y)\}$ for $y \in R$ and $T^{-1}(z) = \{f^{-1}(z)\}$ for $z \in S$. The bijection f is then called a *regular bijection.*

It is easy to see that $\longleftrightarrow$ is an equivalence relation.

Since complementary pairs (E,D) as used in cryptography can often be interpreted as pairs (f,f^{-1}), where f is a regular bijection, the study of regular bijections seems of importance. Probably the most basic question is to determine when two regular sets are in regular bijection. For instance, consider the following six pairs of regular sets:

$R_1 = a^*$, $S_1 = (a^2)^*$;
$R_2 = \{a,b\}^*$, $S_2 = \{ab,ab^2\}^*$;
$R_3 = \{x_1,x_2,\ldots,x_n\}$, $S_3 = \{y_1,y_2,\ldots,y_n\}$, x_i,y_i arbitrary words;
$R_4 = \{x_1,x_2,\ldots,x_n\}$, $S_4 = \{y_1,y_2,\ldots,y_m\}$, $n \neq m$, x_i,y_i arbitrary words;
$R_5 = a^* \setminus \{\varepsilon\}$, $S_5 = \{1\}\{0,1\}^*$;
$R_6 = \{a^2,a^3\}^*$, $S_6 = \{a,b\}^*$.
$R_1 \overset{f_1}{\longleftrightarrow} S_1$ holds by choosing $f_1(a) = a^2$;
$R_2 \overset{f_2}{\longleftrightarrow} S_2$ holds by choosing $f_2(a) = ab$, $f(b) = ab^2$;
$R_3 \overset{f_3}{\longleftrightarrow} S_3$ holds by choosing $f_3(x_i) = y_i$ for $i = 1,2,\ldots,n$.

However, R_4 and S_4 are not in regular bijection (since they are finite sets with differing number of elements). R_5 and S_5 are not in regular bijection (despite the fact $f(a^n) = b$, b the binary representation of n, is a bijection) since very long words would have to be mapped onto rather short words which is impossible for an a-transducer (to be made precise below). Finally, R_6 and S_6 are not in regular bijection for analogous reasons, R_6 being "similar" to R_5, R_6 being "similar" to S_5.

The examples R_i, S_i for $i = 4,5,6$ suggest to consider the number of words up to a certain length, leading to the following notion of *population function.*

Let R be a set of words. The *population function* of R, denoted by

p_R is defined by $p_R(n) = \text{card}\ \{x \mid x \in R,\ |x| \leq n\}$. Two languages R and S are *equally populated* if there exist numbers $c_1, c_2 > 0$ and an integer n_0 such that $p_R(c_1 n) \leq p_s(n) \leq p_R(c_2 n)$ for all $n > n_0$.

Based on this definition of population function a first necessary condition for two languages to be in regular bijection is obtained in [2]:

If $R \longleftrightarrow S$ holds, then R and S are equally populated.

Since R_5 and R_6 are equally populated and S_5 and S_6 are equally populated, but R_5 and S_5 are not equally populated, this result proves our above intuition concerning R_i, S_i (i=5,6).

It remains to consider regular bijections of sets which are equally populated. As a start, consider once more an example: $R = a^*\{b,c\}^*$, $S = \{b,c\}\{a\}^*$. First intuition might seem to indicate that R and S are not in regular bijection since all of the obvious bijections such as $f(a^n x) = xa^n$ for $n \geq 0$, $x \in \{b,c\}^*$ can be shown to be non-regular. Somewhat surprisingly, R and S are, nevertheless, in rational bijection. This follows from the main theorem of [2]:

<u>Theorem:</u>

Let R and S be regular sets. $R \longleftrightarrow S$ if and only if R and S are equally populated.

For the constructive proof, based on a sequence of about a dozen lemmas, see [2]. However, to give a feeling for the type of bijection obtained, the bijection $f\colon a^*\{b,c\}^* \longmapsto \{b,c\}\{a\}^*$ obtained by systematically following the proof of above theorem is the composition of the bijections

$g\colon a^*\{b,c\}^* \longrightarrow \{b,c\}^*$ and

$h\colon \{b,c\}^* \longrightarrow \{b,c\}^*a^*$, where

the bijection h is given by

$$h(x) = \begin{cases} (bc)^i y, & x = (bc)^{2i-1}y,\ i \geq 1,\ y \in \{bc^2, bc^3\}^* \\ a^i g(y), & x = (bc)^{2i}y,\ i \geq 1,\ y \in \{bc^2, bc^3\}^*, g(bc^2)=b, g(bc^3)=c \\ x, & x \in \{b,c\}^* \setminus (bc)^+\ \{bc^2, bc^3\}^*, \end{cases}$$

and where g is defined similarly.

Above theorem leads directly to the following characterization of the existence of regular bijections between regular sets:

Corollary:

Every regular set is in regular bijection with exactly one of the following sets:

$\{a_1,a_2,\ldots,a_n\}$ $(n \geq 1)$, or $a_1^* a_2^* \ldots a_n^*$ $(n \geq 1)$, or $\{a_1,a_2\}^*$.

Further, the construction of the theorem yields arbitrarily complex regular bijections of Σ^* onto itself for arbitrary Σ:

Choose two regular sets R and S which are equally populated. Construct regular bijections f_1: $R \longleftrightarrow S$ and f_2: $\Sigma^* \setminus R \longleftrightarrow \Sigma^* \setminus S$ and combine f_1 and f_2 into a single regular bijection f defined by:

$$f(x) = \begin{cases} f_1(x) & \text{for } x \in R \\ f_2(x) & \text{for } x \in \Sigma^* \setminus R. \end{cases}$$

It remains to be seen whether regular bijections obtained in this fashion can be used successfully for cryptographic purposes.

REFERENCES

1. Culik II, K. and Maurer, H.A., Secure information storage and retrieval using new results in cryptography. Information Processing Letters (to appear), also: Report 16, IIG Graz (1978).

2. Maurer, H.A. and Nivat, M., Rational bijections of rational sets. Report 29, IIG Graz (1979).

3. Diffie, W. and Hellman, M., New directions in cryptography. IEEE Transactions on Information Theory (1976), 644-654.

4. Rivest, R., Shamir, A. and Adelman, L., A method for obtaining digital signatures and public-key cryptosystems. C.ACM 21 (1978), 120-126.

5. Rivest, R., Private communication. (1978).

SEARCHING, SORTING AND INFORMATION THEORY

Kurt Mehlhorn
Fachbereich 10 - Angewandte
Mathematik und Informatik
Universität des Saarlandes
66oo Saarbrücken, BRD

I. Introduction

We consider the relationship between searching, sorting and information theory. In the first section algorithms for the construction of optimal and nearly optimal search trees and a-priori bounds for the cost of such search trees are presented. Many of these results become readily available if search trees are interpreted as alphabetic prefix codes. Next TRIES and dynamic search trees are briefly discussed. In the last section the results are applied to sorting. Recent developments on sorting presorted files are described.

II. Searching and Information Theory

II.1 Search Trees and Alphabetic Codes

Consider the following ternary search tree. It has 3 internal nodes

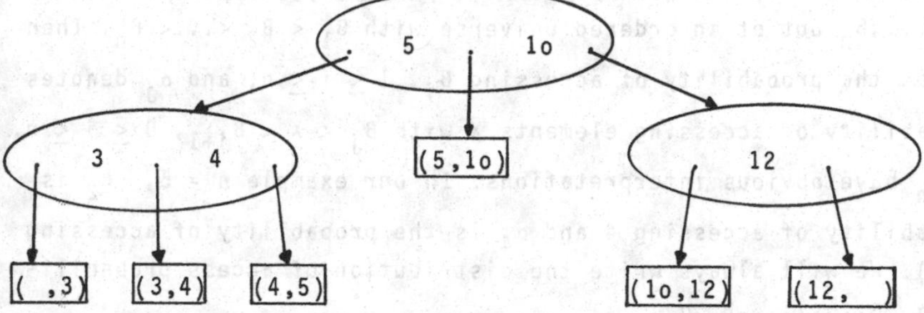

and 6 leaves. The internal nodes contain the keys {3,4,5,1o,12} in sorted order and the leaves represent the open intervals between keys.

The standard strategy to locate X in this tree is best described by the following recursive procedure SEARCH

```
proc SEARCH (int X ; node v)
if v is a leaf
then "X is not in the tree"
else begin let K1,K2 be the keys in node v;
           if X < K1 then SEARCH (X, left son of v)
           if X = K1 then exit (found);
           if K2 does not exist
           then SEARCH (X, right son of v)
           else begin if X < K2 then SEARCH (X, middle son of v);
           if X = K2 then exit (found);
           SEARCH (X, right son of v)
           end
       end
   end
```

Apparently, the search strategy is unsymmetric. It is cheaper to follow the pointer to the first subtree than to follow the pointer to the second subtree and it is cheaper to locate K_1 than to locate K_2.

We will also assume that the probability of access is given for each key and each interval between keys. More precisely, suppose we have n keys $B_1,\dots,B_n$ out of an ordered universe with $B_1 < B_2 <\dots< B_n$. Then β_i denotes the probability of accessing B_i, $1 \le i \le n$, and α_j denotes the probability of accessing elements X with $B_j < X < B_{j+1}$, $0 \le j \le n$. α_o and β_n have obvious interpretations. In our example $n = 5$, β_2 is the probability of accessing 4 and α_4 is the probability of accessing $X \in (4,5)$. We will always write the distribution of access probabilities as $\alpha_o,\beta_1,\alpha_1,\dots,\beta_n,\alpha_n$.

Ternary trees, in general (t+1)-ary trees, correspond to prefix codes in a natural way. We are given letters $a_o,a_1,a_2,\dots,a_{2t}$ of cost $c_o,c_1,c_2,\dots,c_{2t}$ respectively; $c_\ell \in \mathbb{R}_+$ for $0 \le \ell \le 2t$. Here letter $a_{2\ell}$ corresponds to following the pointer to the $(\ell+1)$-st subtree, $0 \le \ell \le t$, and letter $a_{2\ell+1}$ corresponds to a successful search terminating in the

$(\ell+1)$-st key of a node, $0 \le \ell < t$.

A search tree is a prefix code $C = \{V_o, W_1, V_1, \ldots, W_n, V_n\}$ with

1) $V_j \in \Sigma^*$, $W_i \in \Sigma^* \Sigma_{end}$

where $\Sigma = \{a_o, a_2, \ldots, a_{2t}\}$, $\Sigma_{end} = \{a_1, a_3, \ldots, a_{2t-1}\}$, $0 \le j \le n$, $1 \le i \le n$. W_i describes the search process leading to key B_i and V_j describes the search process leading to interval (B_j, B_{j+1}).

2) the ordering of the keys is reflected in the lexicographic ordering of the code words, i.e.

$$V_j \prec W_i \prec V_{j'}$$

for $j < i \le j'$ and $\prec$ denoting the lexicographic ordering of strings based on the ordering $a_o \prec a_1 \prec a_2 \prec \ldots \prec a_{2t}$ of the letters.

Conversely, every prefix code for $\alpha_o, \beta_1, \ldots, \beta_n, \alpha_n$ satisfying 1) and 2) corresponds to a search tree in a natural way. We refer to these codes as <u>alphabetic prefix codes</u>. Codes, which not necessarily satisfy 1) and 2) will be called <u>non-alphabetic prefix codes</u>, or simply prefix codes. In order to stress the distinction in the sequel, we will denote the probability distribution by $(p_1, \ldots, p_n)$, the letters by $b_1, \ldots, b_s$ and their costs by $d_1, \ldots, d_s$ in the non-alphabetic case.

The cost of word $a_{i_1} a_{i_2} \ldots a_{i_k}$ is defined as $c_{i_1} + c_{i_2} + \ldots + c_{i_k}$, i.e. as the sum of the letter costs. The cost of code C is then defined as

$$\text{Cost}(C) = \sum_{i=1}^{n} \beta_i \text{Cost}(W_i) + \sum_{j=o}^{n} \alpha_j \cdot \text{Cost}(V_j)$$

The following two problems are of immediate interest.

1) Given letters, their costs and a probability distribution, find an alphabetic code of (nearly) minimal cost.

2) Give good a-priori bounds for the cost of an optimal alphabetic code. (an alphabetic "noiseless coding theorem").

Table I gives a survey of algorithms for the construction of optimal codes. For the general problem the input is a probability distribution and letter costs. No efficient algorithm is known in the non-alphabetic case; however, it is also not known whether the corresponding recognition

	alphabetic	non-alphabetic
general problem	$O(t^2n^2)$ Itai	$\in$ NP
binary, equal cost general	$O(n^2)$ Knuth (a)	unnatural problem
leaf-oriented, $\beta_i = 0$	$O(n \log n)$ Hu/Tucker, Garsia/Wachs	$O(n \log n)$ Huffman

Table 1: Algorithms for the construction of optimal codes

problem is NP-complete. Itai's algorithm is a dynamic programming approach and so is Knuth's. In the binary equal cost case we have $t = 1$ and $c_0 = c_1 = c_2 = 1$. The input is a probability distribution $\alpha_0,\beta_1,\alpha_1,\ldots,\beta_n,\alpha_n$. In the leaf-oriented case we have in addition $\beta_i = 0$ for all i. Huffman's algorithm runs in linear time when the probabilities are sorted [van Leeuwen (b)] and the algorithms of Hu/Tucker and Garis/Wachs are isomorphic [Mehlhorn/Tsagarakis].

In summary, we can state that no efficient algorithm is known in the general case. Note that Itai's and Knuth's algorithm also need space $\Theta(n^2)$. Therefore approximation algorithms were considered early in the game [Bruno/Coffman, Gotlieb/Walker]. Most of these algorithms are based on an "alphabetic noiseless coding theorem".

A plausibility argument: Consider a prefix code over a two letter alphabet of cost d_0,d_1 respectively. In the root of the code tree the set of probabilities is split into two sets of probability p and 1-p, say. The letter of cost d_0 (d_1) is assigned to the first (second) set. Hence the average cost arising in the root of the code tree is $d_0\cdot p+(1-p)\cdot d_1$ and the average information gain is the binary entropy $H(p,1-p) = -p \log p - (1-p)\log(1-p)$. Information gain per unit cost $(H(p,1-p)/(d_0p+d_1(1-p)))$ is maximized (elementary calculus) with value d for $p = 2^{-dd_0}$ and $1-p = 2^{-dd_1}$ where $d \in \mathbb{R}$ is such that $2^{-dd_0} + 2^{-dd_1} = 1$. Since $H(p_1,\ldots,p_n) = -\Sigma p_i \log p_i$ bits have to be gained in any non-alphabetic prefix code for distribution $(p_1,p_2,\ldots,p_n)$ the cost of such a code has to be at least $H(p_1,\ldots,p_n)/d$. This is made precise in Theorem 2.1. Moreover, the plausibility argument also suggests an approximation algorithm. Try to split the distribution into two sets of probability about 2^{-dd_0} and 2^{-dd_1} respectively. Then proceed recursively on the two subsets.

Theorem 2.1:

a) [Krause, Ciszar]. Let $C = \{U_1,\dots,U_n\}$ be a prefix code for distribution $(p_1,\dots,p_n)$ over alphabet $\Sigma = \{b_1,\dots,b_s\}$ with costs $d_1,\dots,d_s$. Then

$$\mathrm{Cost}(C) = \sum_{i=1}^{n} p_i \cdot \mathrm{Cost}(U_i) \geq H(p_1,\dots,p_n)/d$$

where

$$\sum_{i=1}^{n} 2^{-d_i d} = 1 .$$

b) [Altenkamp/Mehlhorn]. Let $h \in \mathbb{R}$, $h \geq o$ and

$$L_h = \{i;\ d\cdot \mathrm{Cost}(U_i) \leq -\log p_i - h\}$$

Then

$$\sum_{i\in L_h} p_i \leq 2^{-h}$$ □

Remark: Part b) of Theorem 1 shows that the inequality of a) is almost true for corresponding terms of the two sums. Part a) is a noiseless coding theorem for arbitrary letter costs.

Of course, Theorem 2.1 also gives a lower bound in the alphabetic case. However, a better bound can be proved in that case.

Theorem 2.2 [Altenkamp/Mehlhorn]. Let $C = \{V_o,W_1,\dots,W_n,V_n\}$ be an alphabetic prefix code for distribution $(\alpha_o,\beta_1,\alpha_1,\dots\ )$ over letters $a_o,a_1,\dots,a_{2t}$ with cost $c_o,\dots,c_{2t}$. Then

$$\mathrm{Cost}(C) \geq \max\{H(\alpha_o,\beta_1,\dots)/c(x) - (x-1)\cdot(\Sigma\beta_i)\cdot \max_{k \text{ odd}} c_k;\ 1 \leq x \leq \infty\}$$

where $c(x)$ is such that $\sum_{k=o}^{t} 2^{-c(x)c_{2k}} + \sum_{k=o}^{t-1} 2^{-c(x)\cdot x\cdot c_{2k+1}} = 1.$

Proof (sketch): The idea of the proof is to approximate the combinatorial restriction that letters in Σ_{end} can be used only at the end of code words by an artificial increase in the cost of those letters. So define new costs by $\tilde{c}_k = c_k$ for k even and $\tilde{c}_k = x\cdot c_k$ for k odd and $1 \leq x \leq \infty$. Then relate the cost of code C under the new and the old cost function and apply Theorem 2.1 for the new cost function. □

Corollary 2.3 [Altenkamp/Mehlhorn]: Let C be an alphabetic prefix code for distribution $(\alpha_0,\beta_1,\ldots)$ with respect to costs $c_0,c_1,\ldots,c_{2t}$. Let c,d be such that

$$\sum_{k=o}^{2t} 2^{-cc_k} = 1 \quad \text{and} \quad \sum_{k=o}^{t} 2^{-dc_{2k}} = 1$$

Then there are constants u,v (depending on the c's but not on Cost(C) and the α's and β's) such that

$$H(\alpha_0,\beta_1,\ldots) \leq d\cdot\mathrm{Cost}(C) + \frac{c\cdot\Sigma\beta_i}{u}\cdot\max_{i\ \mathrm{odd}} c_i[1+\ln(u\cdot v\cdot\mathrm{Cost}(C))] + \frac{1}{2u}$$

Remark: Corollary 2.3 shows that the lower bound for the alphabetic code is essentially the lower bound (d·Cost(C)) for the non-alphabetic code over Σ plus a small correction of order $(\Sigma\beta_i\cdot\ln \mathrm{Cost}(C))$ which reflects the restricted usage of letters in Σ_{end}.

Proof (sketch): Let $c(x) = d+\delta(x)$ where c(x) is as in Theorem 2.2 and c,d are as in Theorem 2.3. Then $0 \leq \delta(x) \leq c-d$ and $\delta(x) \leq v\cdot e^{-u(x-1)}$ for some constants u and v. This can be verified by substituting into the definitions of c(x),c,d and δ(x). Next substitute the upper bound for $c(x) = d+\delta(x)$ into Theorem 2.2 and apply differential calculus to find x which gives the best bound.

Table 2 summarizes the lower bounds for the costs of alphabetic and non-alphabetic prefix codes. Bayer's lower bound is a special case of Theorem 2.2.

	alphabetic	non-alphabetic
general problem	2.2 and 2.3	2.1
binary, equal cost general	$\max\{(H-d\Sigma\beta_i)/\log(2+2^{-d});d \in \mathbb{R}\}$ Bayer	
leaf-oriented		

Table 2: Lower bound for the costs of prefix codes

Next we turn to the construction of nearly optimal alphabetic prefix codes. The clue is the plausibility argument preceeding Thereorem 2.1. We illustrate the methods by way of example.

Example: Let $c_o = c_1 = c_2 = 1$ and $(\alpha_o,\beta_1,\alpha_1,\ldots,\beta_4,\alpha_4) = (1/6,1/24,0,1/8,0,1/8,1/8,0,5/12)$. Then $c = 1$ and $d = \log 3$ as defined in Corollary 2.3.

Figure 2.1 shows the probability distribution drawn on the unit line. The plausibility argument suggests that we should split the distribution into two subsets of probability about 2^{-cc_1} and 2^{-cc_2} respectively,

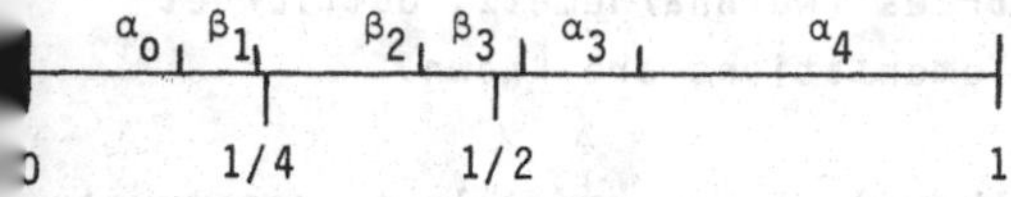

Figure 2.1: The distribution (1/6, 1/24, 0, 1/8, 0, 1/8, 1/8, 0, 5/12).

here 1/2 and 1/2. In our example 1/2 lies in the left half of α_3. Hence we should assign letter a_o to $\alpha_o,\beta_1,\alpha_1,\beta_2,\alpha_2$, letter a_1 to β_3 and letter a_2 to $\alpha_3,\beta_4,\alpha_4$. Next we work on the subproblem $\alpha_o,\beta_1,\alpha_1,\beta_2,\alpha_2$.

Method 1: Apply the same strategy recursively, i.e. try to split the distribution $(\alpha_o,\beta_1,\alpha_1,\beta_2,\alpha_2)$ in the relation 1/2 : 1/2. This would assign letter a_o to α_o, letter a_1 to β_1 and letter a_2 to $\alpha_1,\beta_2,\alpha_2$.

Theorem 2.4 [Mehlhorn (d)]. Method 1 constructs an alphabetic prefix code with $\mathrm{Cost}(C) \le H(\alpha_o,\beta_1,\ldots\;)/d + (\Sigma\alpha_j)[1/d + \max_{k\ \mathrm{even}} c_k] + \Sigma\beta_i[\max_{k\ \mathrm{odd}} c_k]$

where d is defined as in Corollary 2.3. □

In the binary case, a slightly better bound is due to Horibe.

Method 2: In order to solve the subproblem $(\alpha_o,\ldots,\alpha_2)$ method 1 splits this distribution in the relation $2^{-cc_1}:2^{-cc_2}$. Method 2 splits the interval $[0:2^{-cc_1}]$ in the relation $2^{-cc_1}:2^{-cc_2}$; this yields 1/4 in our case. Hence letter a_o is assigned to $\alpha_o,\beta_1,\alpha_1$, letter a_1 to β_2 and letter a_2 to α_2. Note that the two methods are identical if exact splits are always possible.

Theorem 2.5 [Csizar, Krause, Altenkamp/Mehlhorn]:

a) Method 2 constructs a prefix code C with

$$Cost(W_i) \leq [-\log \beta_i]/d + \max_{k \text{ odd}} c_k$$

$$Cost(V_i) \leq [-\log \alpha_j + 1]/d + \max_{k \text{ even}} c_k$$

b) Theorem 2.4 with Method 1 replaced by Method 2. □

In the binary, equal cost case results of type a) of 2.5 are also known for method 1 and for optimal trees [Katona/Nemetz, Güttler et al]. For both methods efficient implementations are known.

Theorem 2.6 [Fredman, Altenkamp/Mehlhorn]. It is possible to construct an alphabetic prefix code according to methods 1 and 2 in time $O(t \cdot n)$.

Method 3: There is a third way to interpret the plausibility argument. Try to maximize the information gain per unit cost at every step. A partial analysis of method 3 is available in the binary, equal cost case [Horibe/Nemetz, Güttler et al]. Recent results of Horibe/Nemetz suggest that a linear average time implementation is possible.

Finally we want to mention that yet another approach was followed by Cot. His method is based on Kraft's inequality.

II.2 Tries

In this section we briefly consider the case that the keys are tuples over some set Δ and that only comparisons between components of the keys are possible. One popular search strategy is to iteratively determine the components of the keys (TRIE).

Let $S \subseteq \Delta^m$ be a set of m-tuples over Δ. For $w \in \Delta^*$ let $p_w = |\{x \in S; w \text{ is a prefix of } x\}|$. Now, let $y = (y_1, y_2, \ldots, y_m)$ be an arbitrary element of S. In order to identify y we proceed as follows.

$i \leftarrow 1$; $w \leftarrow \varepsilon$ (the empty string)

while $i \leq m$ do

<u>begin</u> identify y_i in a binary search tree for Δ and probability distribution $(p_{wa}/p_w; a \in \Delta)$;

$i \leftarrow i+1$; $w \leftarrow wy_i$

<u>end</u>

If the search trees are constructed according to method 2 above then $-\log(p_{y_1 \ldots y_i}/p_{y_1 \ldots y_{i-1}}) + 2$ comparisons suffice to identify y_i according to theorem 2.5 a. Hence $2m + \log p_\varepsilon/p_y = 2m + \log |S|$ comparisons suffice to identify y.

<u>Theorem 2.7</u> [Fredman (a),(b), v.Leeuwen (a), Güttler et al.]
Let $S \subseteq \Delta^m$. Then searching in S requires no more than $2m + \log |S|$ comparisons between elements of Δ.

II.3 Dynamic Search Trees

In many applications the probability distribution varies over time. Recently, some methods were proposed to keep a search tree nearly optimal as probabilities change [Allen/Munro, Unterauer, Baer, Mehlhorn (a)].

III. Sorting and Information Theory

In this section we want to apply the knowledge of the previous section to sorting. We will only consider sorting algorithms which sort by means of comparisons with binary outcome. Such algorithms correspond to prefix codes, usually called decision trees in this context, in a natural way.

Let Γ be a subset of the set of permutations of n elements. For instance, Γ could be all permutations with $\leq$ F inversions (for a definition see below). We want to consider algorithms which sort under the assumption that only permutations in Γ are legitimate input sequences.

Theorem 3.1: Let A be a sorting algorithm for sequences in Γ. Then A uses at least $\log|\Gamma|$ comparisons on the average.

Proof: A is a prefix code for probability distribution $(p_1,\ldots,p_{|\Gamma|})$ with $p_i = 1/|\Gamma|$ for all i. □

Theorem 3.1 is usually referred to as the information theoretic lower bound in sorting. Fredman has shown that this lower bound is sharp up to a linear factor of O(n). This is most easily seen be describing a permutation by its inversion table.

Let $x_1x_2\ldots x_n$ be a sequence of distinct elements from an ordered universe. For $1 \le i \le n$ let $f_i = |\{j; j > i \text{ and } x_j < x_i\}|$ be the number of elements to the right of i yet smaller than x_i. The tuple $(f_1,\ldots,f_n)$ is called the inversion table and Σf_i is called the total number of inversions of the sequence. Knowledge of the inversion table permits a simple insertion sort of the sequence: Start with an empty sequence and then iteratively insert x_i at the f_i-th position of the sorted version of $x_{i+1},\ldots,x_n$. Hence determination of the inversion table is tantamount to sorting.

Theorem 3.2 [Fredman (a),(b)]. There is a sorting algorithm A_Γ for Γ which never uses more than $\log|\Gamma| + 2n$ comparisons.

Proof: Let $\tilde{\Gamma}$ be the set of inversion tables corresponding to permutations in Γ. By theorem 2.7 searching in $\tilde{\Gamma}$ can be done with at most $\log|\tilde{\Gamma}| + 2n$ comparisons. □

Remark: An explicite construction of algorithm A_Γ requires complete knowledge of $\tilde{\Gamma}$ which one does not have in general.

Examples:

a) Sorting X+Y. Let X and Y be sets of m distinct reals each. Then $X+Y = \{x+y;\ x \in X,\ y \in Y\}$ has m^2 elements. Only $2^{O(m \log m)}$ out of $(m^2)!$ permutations are possible for a set of form X+Y [Harper et al] and hence X+Y may be sorted with $O(m^2)$ comparisons [Fredman]. Note that $n = m^2$ in theorem 3.2.

b) Let Γ be the set of permutations of n elements with $\le F$ inversions. Then $\log|\Gamma| = O(n \log \frac{F}{n})$ and hence sequences with $\le F$ inversions can be sorted with $O(n \log \frac{F}{n})$ comparisons. In other words, presorted files $(\log F/n = o(\log n))$ can be sorted with strictly less than n log n comparisons. Note however, that Fredman's result does not say anything about the possible running time of an algorithm for sorting presorted files. Recently, algorithms with run time $O(n(1+\log F/n))$ were described by Guibas et al, Brown/Tarjan and Mehlhorn. Mehlhorn's solution is the most efficient. We describe a blend of Mehlhorn's and Brown/Tarjan's algorithm here.

Definition [Adelson-Velskii/Landis]. A binary tree T (every node of T has either 2 or no sons) is an AVL-tree if for every node v of T the difference between the heights of the left and right subtree of v is at most one. The difference is denoted hb(v). An ordered set S is stored in an AVL-tree by labelling the nodes of the tree from left to right with the elements of S. (cf. figure 3.1).

Suppose now that we want to use an AVL-tree for an insertion sort. Say $x_{i+1},\ldots,x_n$ are stored in sorted order in AVL-tree T and x_i is to be inserted next. By the definition of f_i, x_i will be inserted at the f_i-th position of the sorted version of $x_{i+1},\ldots,x_n$. If the file is presorted, i.e. Σf_i is small, the elements x_i will tend to be inserted near the beginning of the sorted sequence. Hence the standard insertion algorithm will on the average deviate from the left spine (= the path from the root to the left-most leaf) near the left-most leaf. Thus it is much cheaper to look for that point by starting at the left-most leaf y_1 and walking towards the root until a node y_{k_i} with $x_i < y_{k_i}$ is found. (cf. figure 3.2). Next x_i is inserted into the right subtree of y_{k-1} as usual. Since the height of node y_{k_i} is at most $2k_i$ it takes $O(k_i)$ units of time to find the position where x_i is to be inserted. After the correct position is found and a new leaf with label x_i is inserted, the AVL-tree needs to be rebalanced. Rebalancing is restricted to the path from the new leaf to the first node z with balance $hb(z)\neq 0$ on the path from the new leaf to the root. [see Knuth b, pp. 451-463 for details]. Let ℓ_i be the length of that path. Then rebalancing after the insertion of x_i takes time $O(\ell_i)$.

In summary, it takes $O(k_i+\ell_i)$ units of time to process x_i and hence the complete sort takes time $\sum_{i=1}^{n} O(k_i+\ell_i)$.

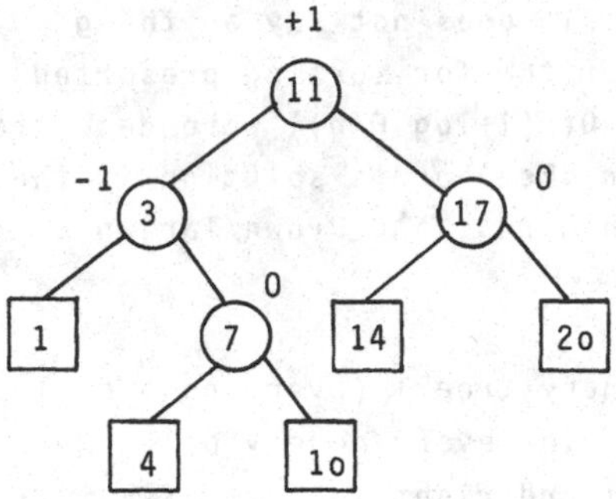

Figure 3.1: An AVL-tree for S = {1,3,4,7,1o,11,14,17,2o}. Height-balances are shown above nodes.

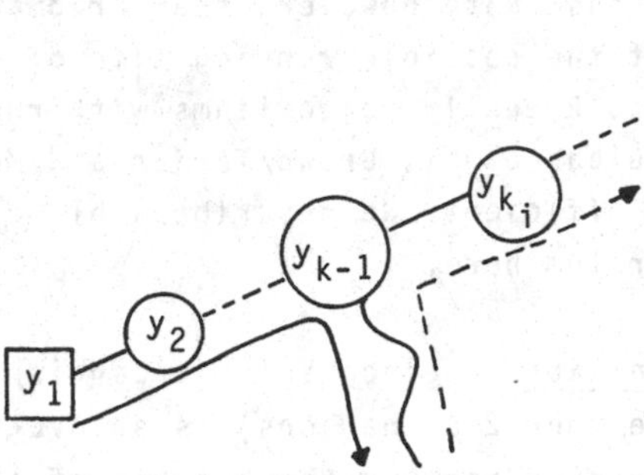

Figure 3.2: The insertion (———) and rebalancing (------) paths.

Lemma 3.3:

a) $k_i = O(\log f_i)$

b) $\Sigma k_i = O(n(1+\log \frac{\Sigma f_i}{n}))$

c) $\Sigma \ell_i = 5n$

Remark: Part c) of Lemma 3.3 is interesting in its own right. It shows that the total number of rebalancing operations (= balance changes + rotations and double rotations) is linear in the number of insertions. A more careful analysis shows $\Sigma \ell_i \leq 4n$.

Proof:

a) x_i is larger than all elements in the subtree with root y_{k_i-2}. Since the height of y_{k_i-2} is at least k_i, this subtree contains at least Fib(k_i-2+2)-1 nodes [see Knuth, page 453], where Fib(m) is the m-th Fibonacci number. This shows a).

b) From a) we conclude

$$\Sigma k_i = \Sigma O(\log f_i) = O(n + \log \Pi f_i)$$

$$= O(n + \log(\frac{\Sigma f_i}{n})^n) = O(n(1+\frac{\Sigma f_i}{n}))$$

The third inequality follows from the fact that Πf_i is maximal for

$f_i = f_j$ for all i,j.

c) Rebalancing after the insertion of x_i is tantamount to changing the balance of all nodes between the new leaf and node z from 0 to ± 1 followed by either a change of balance at node z (from ± 1 to 0) or a rotation or double rotation about z. A rotation (double rotation) changes the balance of at most 3 nodes. In summary, rebalancing after the insertion of x_i generates at most 4 nodes with balance 0 (the father of the new leaf and the 3 nodes affected by the rotation) and changes the balance of at least $\ell_i - 1$ nodes from 0 to ± 1. Hence $\Sigma(\ell_i - 1) \leq 4n$ since only zeroes which were created before can be changed. □

As an immediate consequence of Lemma 3.3 we obtain

Theorem 3.4: It is possible to sort sequences with $\leq$ F inversions in time $O(n(1+\log F/n))$ where n is the length of the sequence.

The algorithm behind 3.4 may be dubbed an adaptive sorting method. Hard sorting problems ($F \approx n^2/2$) are solved in time $O(n \log n)$ and simple sorting problems ($F \ll n^2$) are solved in time strictly less than $O(n \log n)$. In [Mehlhorn] the algorithm is compared with Quicksort and it is shown to be superior for $F \leq 0.25\, n^{1.8}$.

Bibliography:

Adel'son-Velskii, Landis : An algorithm for the organization of information, Soviet Math. Dokl, 3, 1259-1262.

Allan, B., Munro, I. : Self-Organizing Binary Search Trees, JACM, Vol. 25 (1978), pp. 526-535.

Altenkamp, D., Mehlhorn, K. : Codes : Unequal Letter Costs, Unequal Probabilities, Technical Report A 78/18, FB lo, Universität des Saarlandes (preliminary version: 5th Colloquium on Automata, Languages and Programming, Udine, 1978, Lecture Notes in Computer Science 62, pp. 15-25.

Baer, J. L. : Weight-balanced trees, Proc. AFIPS, Vol. 44, 1975, pp. 467-472.

Bayer, P. J. : Improved Lower Bounds on the Cost of Optimal and Balanced Binary Search Trees, Technical Report, Dept. of Computer Science, MIT.

Brown, M.R., Tarjan, R.E. : A Representation for Linear Lists with Movable Fingers, 1oth ACM Symposium on Theory of Comouting, pp. 19-29, 1978.

Bruno, J., Coffman, E.G.: Nearly Optimal Binary Search Trees (IFIP 1971, North Holland 1972, pp. 99-1o3).

Cot, H. : Characterization and Design of Optimal Prefix Codes, Ph. D. Thesis, Stanford University, June 1977.

Csiszar : Simple Proofs of some theorems on noiseless channels, Inf. and Control 14, pp. 285-298, 1969.

Fredman, M.L. (a) : Two Applications of a Probabilistic Search Technique : Sorting X+Y and Building Balanced Search Trees, 7th ACM Symposium on Theory of Computing, 1975, pp. 24o-244.

Fredman, M.L. (b) : How good is the information theory bound in sorting, TCS 1, (4), pp. 355-362, 1976.

Garsia, A.M., Wachs, M.L. : A new algorithm for minimum cost binary trees, SICOMP, No 4, 1977, pp. 622-642.

Gotlieb, Walker : A Top-Down Algorithm for Constructing Nearly Optimal Lexicographical Trees, Graph Theory and Computing, Academic Press, 1972.

Guibas, L.J., McCreight, E.M., Plass, M.F., Roberts, J.R.: A new representation for linear lists, 9th ACM Symposium on Theory of Computing, 1977, pp. 49-6o.

Güttler, R., Mehlhorn, K., Schneider, W.: Binary Search Trees : Average and Worst Case Behavior, GI-Jahrestagung 1976, Informatik Fachberichte 5, pp. 3o1-317.

Harper, Pague, Savage, Straus : Sorting X+Y, CACM 18, 6 (1975), pp. 347-35o.

Horibe, Y. : An improved bound for weight balanced trees, Inf. and Control 34, 1977.

Horibe, Y., Nemetz, T. : On the Max-Entropy Rule for a Binary Search Tree, Technical Report, Mathematical Institute, Hungarian Academy of Sciences.

Hu, Tucker : Optimum Computer Search Trees, SIAM J. of Applied Math. 21, 1971, pp. 514-532.

Huffman : A method for the Construction of Minimum-Redundancy Codes, Proc. IRE 4o, 1o98-11o1, 1952.

Itai, A. : Optimal Alphabetic Trees, SIAM J. on Computing, Vol. 5, No. 1, March 1976, pp. 9-18.

Katona, G., Nemetz, T. : Huffman Codes and Self Information, IEEE Transactions on Information Theory, May 1976, pp. 337-34o.

Knuth, D.E. (a) : Optimum Binary Search Trees, Acta Informatica 1, 1971, pp. 14-25.

Knuth, D.E. : The Art of Computer Programming, Vol 3 : Sorting and Searching, Addison Wesley, 1973.

Krause : Channels which transmit letters of unequal duration, Inf. and Control 5, pp. 13-24, 1962.

Mehlhorn, K. (a) : Dynamic Binary Search, 4th Colloquium on Automata, Languages and Programming, Turku, 1977, Springer Lecture Notes 52, pp. 323-336 (to appear SIAM J. on Computing).

Mehlhorn, K. (b) : Sorting Presorted Files, 4th GI-Conference on Theoretical Computer Science, Aachen, 1979.

Mehlhorn, K. (c) : Effiziente Algorithmen, Teubner Verlag, Studienbücher Informatik, 1977.

Mehlhorn, K. (d) : An Efficient Algorithm for the Construction of Nearly Optimal Prefix Codes, Technical Report A 78/13, FB 1o, Universität des Saarlandes, submitted for publication.

Mehlhorn, K., Tsagarakis, M. : On the isomorphism of two algorithms : Hu/Tucker and Garsia/Wachs, 4ième Colloque de Lille "Les Ärbes en Algebre et en Programmation", Lille 1979.

van Leeuwen, J. (a): The complexity of data organization, 1976, Mathematical Centre Tract 81, pp. 37-147.

van Leeuwen, J. (b) : On the construction of Huffmann Trees, 3rd ICALP (1976), pp. 382-41o, Ed. S. Michaelson and R. Milner, Edinburgh University Press.

Unterauer, K. : Optimierung gewichteter Binärbäume zur Organisation geordneter dynamischer Dateien, Doktorarbeit, TU München, 1977.

LCF: A WAY OF DOING PROOFS WITH A MACHINE

Robin Milner
Department of Computer Science,
University of Edinburgh,
King's Buildings,
Mayfield Road,
Edinburgh EH9 3JZ.
Scotland.

1. Introduction

The connection between programs and logic is now recognized as a leading topic of research in the theory of computing. It has many different aspects. For example logical formulae may be used to characterize the set of states attainable at a given program point [1,6], or the state transformation represented by a program segment [12]; predicate calculus formulae may be regarded as programs themselves [19] ; in special logics, programs and declarative phrases may be intermixed with considerable freedom, in making deductions [4,15] ; programs may be modal operators in a modal logic [14]; this by no means exhausts the possibilities under study. Distinct from the above examples is the use of programs to discover [16] or to check [5] proofs in logical calculi which are not in themselves particularly concerned with computation.

The LCF system originated at Stanford University as an implementation of a formal calculus due to D. Scott, which in turn arose from his work with C. Strachey [18] on the denotational semantics of programming languages; Scott's fundamental contribution was the theory of continuous function spaces [17], which provides models for functiona calculi. LCF (Logic for Computable Functions), as a logic in which to discuss programming languages as well as programs, plays a different rôle from the program-logics cited above. Moreover, its implementation embodied yet another connection between programming and logic; the idea was not to prove theorems automatically, nor to check proofs, but to use a metalanguage of commands to generate proofs step by step. But even though the system possessed features to allow many steps to be generated by a single command, and to attack goals (intended theorems) by resolving them into subgoals, our interest was mainly in the subject matter of the proofs: the meanings of programming languages. An example was the proof of correctness of a compiling algorithm [13] .

This work has continued at Edinburgh [10] with a shift of emphasis. The metalanguage of commands has been developed into a full programming language, called ML [9], in which we are experimenting with various styles of generating or performing proofs interactively. Since these styles are somewhat independent of the formal calculus itself, the acronyn LCF is now a little misleading, though it does indicate that our object-language is still a Logic for Computable Functions.

For this reason, we have chosen here to illustrate our methodology with a simple example not primarily concerned with programming - but which can be treated in a suitable restriction of our calculus. The example will show how the rich type structure of ML allows one to define procedures to represent not only derived inference rules of the calculus (given the basic influence rules as built-in procedures) but also tactics or strategies for finding proofs.

It is worth exhibiting one simple change among those which led from the Stanford LCF metalanguage of commands to the metalanguage ML. In the command language, a proof consisted of a sequence of steps (theorems), indexed by the natural numbers, each following from previous steps by inference. For example, if 50 steps have been generated, and the 39th step is

$$39 \quad \vdash \forall x.F$$

(for some logical formula F) then the command

SPEC "a + 1" 39

will generate, by specialization, the step

$$51 \quad \vdash F[a+1/x] .$$

The simple change is not to index the proof by natural numbers, but instead to bind theorems to metavariables of metatype <u>thm</u>. (Other metatypes are <u>term</u> and <u>form</u>(ula) - e.g. "a+1 is a term, and "∀X.X+0 = = X" is a form). Thus, if a metavariable th is currently bound to the theorem ⊢ ∀x.F, the specialization command can be replaced by the ML phrase

<u>let</u> th' = SPEC"a + 1" th ;;

which binds the new step (or theorem) to th'. This change is not profound, but very influential. For now the identifier SPEC stands for a ML procedure (representing a basic inference rule) whose metatype is

<u>term</u> → <u>thm</u> → <u>thm</u>

and it is a simple matter to define derived inference rules by ordinary programming. There is nothing new in representing inference rules as meta-procedures (see [11] for example); what is perhaps new - and very effective - is that the metatype discipline distinguishes the metatype <u>thm</u> from the metatypes <u>term</u> and <u>form</u>, so that - whatever complex procedures are defined - all 'values' of metatype <u>thm</u> must be theorems, since only inferences can compute such values. (We shall continue to use the word 'metatype' for types in ML, since we are later concerned also with types in the object language.)

This robustness released one from the need to preserve the whole proof as a sequence (though does not deny this possibility); a user of LCF should consider that he is performing a proof, or guiding its performance, not generating it. As we shall see, he can program his guidance in different styles - one of which is guidance towards an explicitly stated goal.

Our example concerns Boolean Algebra. Apart from illustrating proof guidance, it also illustrates how an applied theory is set up and used in LCF. Space does not allow us to explain how theories may be combined and extended to form higher theories; a good example may be found in [3], where it is shown how the proof of compiler-correctness can be structured by using several theories (for example, separate theories for syntax and semantics, and a theory for the compiler itself). An obvious use for the theory-methodology is in proving properties of programs with abstract data types; the recent work of Burstall and Goguen [2] is very relevant here, and indeed suggests ways in which LCF treatment of theories should be further developed.

2. <u>Setting up the Theory of Boolean Algebra</u>

We wish to establish the operators and axioms [8] of an arbitrary Boolean Algebra (with carrier set A)

$$< A \; ; \; +, *, \neg, 0, 1 >$$

and then to do a little work in the theory.

The normal sequence in setting up a theory falls into four parts:

(i) Mention the <u>parent</u> theories, of which the new theory will be an extension.

(ii) Introduce <u>type</u> symbols for the sets or domains of individuals of the theory (these are object-language types).

(iii) Introduce symbols for the individual or function <u>constants</u> of the theory, giving their types.

(iv) Introduce the <u>axioms</u> of the theory, with names for referring to them.

Here there are no parent theories. We present the necessary sequence of ML phrases, interspersed with a few comments. (We shall use the above symbols for the five operators, though for unimportant reasons our present implementation will not allow all of them).

```
newtype 0 `A` ;;
```

The zero here indicates that A is to be a type operator with arity zero, i.e. a type constant. <u>Tokens</u> , like \`A\` or \`+\` , are used to build all new syntactic constructs of the object-language.

```
newconstant( `O` , ":A") ;;
newconstant( `1` , ":A") ;;
newconstant( `¬` , ":A→A") ;;
newolinfix( `+` , ":A×A→A") ;;
newolinfix( `*` , ":A×A→A") ;;
```

All object-language constructs appear in double quotation marks; in the case of type expressions a colon precedes them within the quotation marks. Infixed binary function constants are introduced by newolinfix, not by newconstant.

Each of the axioms which follow is introduced with its reference token. To avoid eight separate calls of the form

newaxiom (token, formula)

we use the ML map function, which maps a function over a list (represented by [-;...;-]).

```
map newaxiom
   [ `orcomm` , "X + Y = = Y + X" ;
     `andcomm`, "X * Y = = Y * X" ;
     `ordist`,  "X + (Y * Z) = = (X + Y) * (X + Z)" ;
     `anddist`, "X * (Y + Z) = = (X * Y) + (X * Z)" ;
     `oride` ,  "X + O = = X" ;
     `andide`,  "X * 1 = = X" ;
     `orinv` ;  "X + (¬X) = = 1" ;
     `andinv`,  "X * (¬X) = = O" ] ;;
```

Each axiom is placed on a file in closed form, by prefixing universal quantifiers. Variables of the object-language need no separate introduction. The theory is now determined, and we deny the introduction of any further symbols or axioms by the phrase

```
maketheory `BA` ;;
```

which also names the theory. It may now serve as a parent for further theories, and we may work in the theory to prove and preserve theorems, which will also be accessible when working in any daughter (or descendant) theories.

Let us illustrate a short proof; the idempotency of the join operation + . The normal deduction is

$$\begin{aligned} X &= X + 0 && \text{(oride)} \\ &= X + (X * (\neg X)) && \text{(andinv)} \\ &= (X + X) * (X + (\neg X)) && \text{(ordist)} \\ &= (X + X) * 1 && \text{(orinv)} \\ &= X + X && \text{(andide)} \end{aligned}$$

Now the term on the second line can be reduced - or simplified - to X by using the axioms \`andinv\` and \`oride\` as left-to-right rewriting rules, and to (X + X) by similarly using \`ordist\` , \`orinv\` and \`andide\` . Each half of the proof can then be done in two stages : (i) Build a simplification set from an appropriate list of equational theorems, using a function

ss : <u>thm</u> <u>list</u> → <u>simpset</u>

which can easily be defined from the basic functions provided in ML ; (ii) Use the ML derived inference rule

simpterm : <u>simpset</u> → <u>term</u> → <u>term</u> × <u>thm</u>

which is such that simpterm s t yields the pair

$t' \;,\quad \vdash t == t'$

where t´ is the term gained by applying the theorems in s (as left-to-right rewriting rules) as often as possible to t , and the second component is the theorem which asserts the validity of the reduction. Such simplifications not only perform term-rewriting, but also must perform the validating inference, since a theorem is to be computed. This inference may be quite long, involving many uses of transitivity and substitutivity of = = ; our aim of complete robustness forced us to impose this task upon the simplification procedure, rather than giving it special status as a very complex built-in inference rule.

We now give an entire ML sequence for our proof, but omitting the machine's response to each phrase (a few explanatory remarks follow):

```
let  t = "X + (X * ( ¬ X))"  ;;
let  s1 = ss(map (AXIOM `BA`)[`andinv` ; `oride`] );;
let  t1,th1 = simpterm  s1  t ;;
let  s2 = ss (map(AXIOM `BA`) [`ordist` ; `orinv`; `andide`] );;
let  t2,th2 = simpterm  s2  t ;;
let  th = TRANS(SYM th1, th2) ;;
```

The function AXIOM \`BA\` : <u>token</u> → <u>thm</u> fetches an axiom (which is, of course, a theorem). The metavariables th1,th2 are given as values the theorems

$\vdash t == X \;, \qquad \vdash t == X + X$

and the last line uses symmetry and transitivity to give the metavariable th the value

$$\vdash X == X + X$$

All that remains is to name and store away this theorem for later use:

```
newfact `oridemp` th ;;
```

It may later be found (in quantified form) by

```
FACT `BA` `oridemp` ;;
```

The ML sequence above was not the only possible, or the most efficient. We wished to illustrate the simplification mechanism, and also that in LCF proofs need not be provided line-by-line for checking (a laborious task) ; rather theorems are computed by applying basic or derived inference rules as functions. In the next section we illustrate how a proof procedure which may take a very large number of primitive inference steps - a number dependent upon the size and nature of the theorem required - may be invoked by a short, fixed, incantation. It is in fact a complete proof procedure for a class of problems.

3. Computing Normal Forms in Boolean Algebra

To simplify our example, we shall assume in this paper that a Disjunctive Normal Form (DNF) is a disjunction of conjunctions of literals, and a Conjunctive Normal Form (CNF) is a conjunction of disjunctions of literals, where a literal is a possibly negated atom (0,1 or a variable). We therefore ignore the removal of occurrences of 0 and 1, and of duplicate or complementary members of conjunctions or disjunctions.

Now it is an easy matter to write, in ML, any one of a variety of procedures of metatype <u>term</u> $\rightarrow$ <u>term</u> to convert an arbitrary Boolean expression to DNF or to CNF. However, to have applied this procedure to a term and obtained a term as result is no guarantee that the resulting term is indeed equivalent, by the axioms, to the original term.

A more reliable approach is to write a procedure

DNF : <u>term</u> $\rightarrow$ <u>thm</u>

such that DNF(t) yields a theorem $\vdash t == dn$, where dn is in DNF. Of course our ML programming may give us the <u>wrong</u> theorem, but in our experience the confidence that the result <u>is</u> a theorem is of much greater importance.

Again, there is no difficulty in writing a general procedure DNF to the above specification; it is just a little harder than a procedure which produces only the term dn as result, since to produce a theorem it must appeal explicitly to the axioms and inference rules.

But we ask the reader to forget his sound knowledge of Boolean algebra, and to imagine that all he can see at first is (i) how to reduce the goal of converting t to a Normal Form to one or more subgoals of the same kind, and (ii) how to justify his reduction by axioms and inference. We argue that this is more typical of what happens in trying to prove interesting theorems.

More particularly, suppose that the reader can see that to convert t to DNF dn

(i) If t is a disjunction $(t_1 + t_2)$, then it will be enough to obtain DNF's dn_1, dn_2 for t_1, t_2 and disjoin them;

(ii) If t is a conjunction $(t_1 * t_2)$, then it will be enough to obtain DNF's dn_1, dn_2 for t_1, t_2 and to obtain dn by repeated $*$- distribution;

(iii) If t is a negation $(\neg t_1)$, then it will be enough to obtain a CNF cn_1 for t, and obtain dn by repeatedly applying deMorgan's laws and the cancellation of double negations;

(iv) If t is atomic then dn = t.

There is an exactly dual subgoaling process for conversion to CNF.

We may now see the process as a potentially interactive one. We consider that the reader is interested in goals of the form

$$(t\ ,\ b)$$

where t is an arbitrary term, and b is a truth value <u>true</u> (resp. <u>false</u>) , and in which t is to be converted to DNF (resp. CNF). So he would naturally define the metatype <u>goal</u> = <u>term</u> × <u>bool</u> , and he will say that a goal (t,b) is <u>achieved</u> only by a theorem of the form

$$\vdash t == n$$

where n is a DNF (resp. CNF) if b is <u>true</u> (resp. <u>false</u>).

We can now view a uniform, or complete, strategy for achieving goals as a procedure of metatype <u>goal</u> → <u>theorem</u> whose result always achieves its argument. But since we are supposing that the reader cannot immediately intuit this strategy, we wish to represent the above subgoaling process as a collection of <u>tactics</u>, which he can invoke repeatedly. He may see later how to build the collection into a complete strategy.

We now digress to discuss tactics in general.

4. Goals, events, and tactics

Let us now suppose that <u>goal</u> is a type of object whose achievement is desired, and that <u>event</u> is a type of object which may achieve a goal. (In this section we do not use the word 'metatype', since we are not here considering ML as a metalanguage.) We ask what the type <u>tactic</u> should be if a tactic is considered as a procedure for producing subgoals from a goal, with some validation of its action. To say that it produces a list of subgoals from a goal - i.e. to define the type

$$\underline{tactic} = \underline{goal} \to \underline{goal}\ \underline{list}$$

is not enough; there is no validation. We therefore adopt the definition

$$\underline{tactic} = \underline{goal} \to (\underline{goal}\ \underline{list} \times \underline{validation})$$
$$\underline{validation} = \underline{event}\ \underline{list} \to \underline{event}$$

with the following intention : if the tactic T produces, for a goal g ,

$$T(g) = [g_1 ;\ldots ; g_n], v$$

then the validation v should be such that for any events $e_1,\ldots,e_n$ which achieve $g_1,\ldots,g_n$ respectively, the event $v[e_1 ;\ldots; e_n]$ should achieve g .

This raises the immediate question: what is achievement? Clearly achievement must be a binary relation between events and goals. Moreover, in the particular case that events are theorems of some theory (as in our example of Boolean Algebra), achievement is a meta-theoretic notion which cannot in general be formulated within the theory. It will be possible therefore in ML, our programming meta-language, to define objects of type <u>tactic</u> which do not preserve achievement in the sense defined above; such tactics will yield spurious validations. Those which yield correct validations we shall call <u>valid</u> tactics, and to demonstrate their validity requires a meta-theoretic argument. However, it must be noted that merely the type-discipline of our meta-language ensures that the value produced by a validation <u>is</u> an event (i.e. is a theorem, in our example).

Even a valid tactic can be useless. It may be that a tactic T produces, for any goal g , a non-empty sequence $g_1,\ldots,g_n$ of subgoals which are unachievable, and then T is vacuously valid! We say therefore that T is <u>strongly valid</u> if it is valid and, whenever g is achievable (i.e. there exists an event which achieves g) and

$$T(g) = [g_1 ;\ldots; g_n], v$$

then g1,...,gn are also achievable.

A certain robustness of tactic-programming can be ensured if we can define operations for combining tactics to form more complex tactics in such a way that both validity and strong validity are preserved. We then have an algebra of tactics which, it turns out, not only gives considerable power in building sophisticated tactics (some of which are even complete proof procedures) but is also a natural and appealing form of expression. Such a set of tactic-operations (which we may call <u>tacticals</u>, by analogy with functionals) is as follows:

(i) The nullary tactical ID : tactic , given by

```
let ID (g) = [g] , λ[e].e ;;
```

Its use is in tactic-combinations; it just passes the goal on unchanged, and provides the identity validation (note that the ML abstraction λ[e].e will fail when applied to an event list of length $\neq 1$).

(ii) The binary (infixed) tactical THEN : $tactic^2 \rightarrow tactic$.
(T_1 THEN T_2)g applies T_2 to all subgoals produced by $T_1(g)$, and the resulting subgoal-lists are concatenated. We omit the definition of THEN ; its only slight complication is that the validation produced by T_1 THEN T_2 must be built from those produced by T_1 and T_2 .

(iii) The binary (infixed) tactical ORELSE : $tactic^2 \rightarrow tactic$.
(T_1 ORELSE T_2)g first applies T_1 to g , and only if this fails will it apply T_2 to g . It is given in ML by

```
let (T1 ORELSE T2)g = T1(g) ? T2(g)
```

(In ML, expression evaluation may fail; the value of the expression $e_1?e_2$ is e_1's value if e_1 does not fail, otherwise it is e_2's value).

(iv) The unary tactical REPEAT : tactic → tactic ;
(REPEAT T)g applies T to g , then again to all subgoals produced, and so on until T fails. It can be declared in ML by

```
letrec(REPEAT T)g = ((T THEN (REPEAT T)) ORELSE ID)g ;;
```

These tacticals provide a pleasantly simple, though not always sufficient, algebra over tactics. (In passing, compare it with an algebra of regular sets of strings).

One final point should be mentioned before returning to our example. If T is valid, and T(g) = [],v - i.e. the subgoal list is empty - then the event v[] achieves g . We may thus call T a complete strategy if T(g) has this form whenever g is achievable.

5. Tactics for Normal Form conversion

Let us now write some simple tactics for Normal Form conversion, assuming the metatypes goal = term × bool and event = thm . We will follow the simple sub-goaling process outlined in section 3. We assume that we have first defined some simple decomposition functions over terms:

orparts : term → term × term

orparts(t) = (t_1,t_2) if t is a disjunction (t_1 + t_2),
otherwise it fails

```
andparts : term → term × term  }
notpart  : term → term         }  similar
```

We also assume that we have derived from the logical axioms (which characterize '= =' as a congruence) three inference rules

ORCONG : thm list → thm
If $th_i = \vdash t_i = = u_i \quad (1 \le i \le n)$ then
$ORCONG[th_1;\ldots;th_n] = \vdash (t_1 + \ldots + t_n) = = (u_1 + \ldots + u_n)$

```
ANDCONG : thm list → thm  }
NOTCONG : thm → thm       }  similar
```

We shall only use ORCONG and ANDCONG on theorem lists of length 2.

Our first tactic, ORTACTIC, works on a goal whose term is a disjunction (it fails on other goals, since 'orparts' fails) :

```
let ORTACTIC (t,b) =
    let t1,t2 = orparts(t) in
        ([ (t1,b) ; (t2,b)] , v
          where v = if b then ORCONG else (ORDIST ∘ ORCONG)) ;;
```

To understand the validation v , we must explain the derived rule ORDIST : thm → thm . In the case b = false , v must convert the theorem list $[\vdash t_1 = = cn \; ; \vdash t_2 = = cn_2]$ into a theorem $\vdash t = = cn$ where cn_1, cn_2 and cn are CNFs . Now ORCONG will yield $\vdash t = = (cn_1 + cn_2)$ from the theorem list ; all that ORDIST needs to do to its theorem argument is to apply the axiom \`ordist\` (as a rewriting rule) repeatedly to its right hand side. ORDIST can be derived easily from the function simpterm, using a simpset containing only the axiom \`ordist\` .

The tactic ANDTACTIC , for conjunctive goals, is of course dual.

The tactic NOTTACTIC, for negated goals, is defined

```
let NOTTACTIC (t,b) =
      let t1 = notpart(t) in
          ( [ (t1, b ) ] , v
            where v[th] = NOTDIST (NOTCONG(th)) ) ;;
```

The derived rule NOTDIST : thm → thm must distribute negations in the right hand side of its argument (yielding a CNF from a DNF and conversely). It therefore differs from ORDIST only in the simplification set which it uses; these are of course deMorgan's laws and the cancellation of double negation, and we imagine that these have been proved and stored on the \`BA\` theory file, where they will appear in the form `deMorgan`   "∀X.∀Y. ¬ (X + Y) = = (¬ X) * (¬ Y)"

`deMorgan`   "∀X.∀Y. ¬ (X * Y) = = (¬ X) + (¬ Y)"

`doubleneg`   "∀X.   ¬ (¬ X) = = X "

Now from these tactics we can define

RESOLVETACTIC = REPEAT(ORTACTIC ORELSE ANDTACTIC ORELSE NOTTACTIC)

whose effect will be to reduce any goal to a list of subgoals with atomic terms. But atoms are both CNF and DNF, and if we define

```
let ATOMTACTIC(t,b) = ([ ], v
                        where v[ ] = REFL t ) ;;
```

(REFL is the reflexivity axiom schema) then the tactic

RESOLVETACTIC THEN ATOMTACTIC

is a complete strategy for normal form reduction. Applied to any goal, whose term is a Boolean form t , it will yield an empty subgoal list, with a validation v such that

v[ ]

will evaluate to the theorem which states the equivalence between t and its Normal Form. (The discerning reader may have noticed that ATOMTACTIC is not a valid tactic; consider its application to a non-atomic goal. But RESOLVETACTIC THEN ATOMTACTIC is valid, since it will only invoke ATOMTACTIC on atomic goals. It may be preferable for ATOMTACTIC to fail on non-atomic goals; then it also is valid).

6. Other interpretations of goal and event

In most of our work with LCF we have used a different metatype goal (though with event = thm as before). This arises partly from a fact which we concealed in our example; a theorem of LCF is actually of the sequent form

$$H \vdash F$$

where F is a formula, and H is a list of formulae (the hypotheses of the theorem). The sort of goal we have used is a triple (F,s,A) where F is a formula to be proved, s is a simpset which may be used in proof, and A is a list of assumptions (formulae) which may be used as hypotheses. So we have the metatype

goal = form × simpset × form list .

The achievement relation is as follows. The theorem H ⊢ F' achieves the goal (F,s,A) iff, up to renaming of bound variables,

(i) F = F'

(ii) $H \subseteq A \cup \bigcup_i H_i$, where H_i is the hypothesis list of the ith simplification rule in s .

We have provided in the system several simple tactics which are all valid (and mostly strongly valid). Our choice of the metatype goal may be sufficiently justified by considering a very simple tactic which, from a goal

$$(F_1 \supset F_2 , s, A)$$

(where F_1 is an equation we would like to use as a simplification rule) generates the single subgoal

$$(F_2 , s', A')$$

where A' results from A by inclusion of F_1 , and s' results from s by the addition of the theorem $F_1 \vdash F_1$ (a tautology). The validation produced by the tactic is essentially the rule of Implication Introduction (i.e. discharge of assumption). Indeed, many of our tactics are in a similar sense just inverses of basic inference rules. (In some cases assumptions may be added, without inclusion in the simpset.)

To see that goals, events and achievement have wide application, consider a totally different case. Let goals be natural numbers, let events be lists of natural numbers, and let the achievement relation be :

$[m_1 ;\ldots; m_k]$ achieves n iff the m_i are all prime, and their product is n .

The reader may like to experiment with various tactics, or complete strategies, for prime factorization, and to ask whether they are valid or strongly valid.

It is easy to think of applications in Artificial Intelligence, though space prevents us from giving examples. It may be that goals, events, achievement, tactics and validity provide a simple but useful tool to assist understanding and programming in such applications, but we have not pursued the question.

7. Discussion

In much of our current research we seek tactics which abbreviate the labour of proof for particular problems, and ask how widely applicable they are over problems in the same area. Two case studies are reported [3,7] , and others are forthcoming. The first of these is a final year undergraduate project, and illustrates how general strategies may be found for proofs about data structures. It concerns proofs of properties of linear lists; the outcome is that once a rule (and corresponding tactic) for structural induction has been derived, a standard combination of this tactic with tactics for case analysis and simplification give a strategy for proving a fairly wide class of theorems. The second of the case studies, which we have mentioned already, concerns the compiler-correctness problem; though this is a larger single problem, it was found that many parts of the proof yield to rather general tactics. We believe that much more can be done along these lines in constructing tactic-kits for proof in widely different problem areas, and we hope by our example to encourage others to apply our methods to different proof calculi.

The question of the meta-theoretic justification (proof of validity) of tactics and of derived inference rules is an important one. It is well known that such justification reduces the work of proof (even if only the work done by the machine), since the application of a complex inference rule which has been so justified need no longer involve a long sequence of primitive inferences. We hope that the work of Weyhrauch [20] on reflexion principles may be applied to this problem.

8. Acknowledgements

I am deeply grateful to my colleagues Avra Cohn, Mike Gordon, Lockwood Morris, Malcolm Newey, Chris Wadsworth and Richard Weyhrauch. The development of the LCF project over the past eight years has been in close and rewarding cooperation with them.

References

1. Blikle, A., Specified Programming, ICS PAS Report 333, Inst. of Computer Science Polish Academy of Sciences, Warsaw, 1978.
2. Burstall, R.M., and Goguen, J.A., Putting theories together to make specification Proc. 5th International Joint Conference on Artificial Intelligence, Cambridge Mass., published by Dept. Comp. Sci., Carnegie Mellon, 1045-1058, 1977.
3. Cohn, A., High level proof in LCF, Proc. 4th Workshop on Automated Deduction, Austin, Texas, 73-80, 1979.
4. Constable, R.L., and O'Donnell, M.J. , A Programming Logic, Winthrop, 1978.
5. deBruijn, N.G., The mathematical language AUTOMATH, Symposium in Automatic Demonstration, Lecture Notes in Math, Vol. 125, Springer-Verlag, New York, 29-61, 1970.
6. Floyd, R.W., Assigning meanings to programs, Proc. Symposia in Applied Mathematics, Vol. XIX, Amer. Math. Soc., Providence, 19-32, 1967.
7. Giles, D.A., The theory of LISTS in LCF, Report CSR-31-78, Computer Science Dept. Edinburgh University, 1978.
8. Goodstein, R.L., Boolean Algebra, Pergamon Press, Macmillan Company, New York, 196
9. Gordon, M., Milner, R., Morris, L., Newey, M. and Wadsworth, C., A metalanguage for interactive proof in LCF, Proc. 5th ACM SIGACT-SIGPLAN Conference on Principle of Programming Languages, Tucson, Arizona, USA, 1978.
10. Gordon, M., Milner, R. and Wadsworth, C., Edinburgh LCF, Report CSR-11-77, Computer Science Dept., Edinburgh University, 1977.
11. Hewitt, C., PLANNER: a language for manipulating models and proving theorems in a robot, AI Memo 168, Project MAC, MIT, Cambridge, Mass., 1970.
12. Hoare, C.A.R., An axiomatic basis for computer programming, Comm. ACM, 12, 576-581, 1969.
13. Milner, R. and Weyhrauch, R.W., Proving compiler correctness in a mechanized logi Machine Intelligence 7, ed. B. Meltzer & D. Michie, Edinburgh University Press, 51-72, 1972.

14. Pratt, V.R., Semantical considerations on Floyd-Hoare logic, Proc. 17th Annual IEEE Symposium on Foundations of Computer Science, 109-121, 1976.

15. Rasiowa, H., Algorithmic Logic, Inst. of Computer Science, Polish Academy of Sciences, Warsaw, 1977.

16. Robinson, J.A., A machine-oriented logic based on the resolution principle, JACM, 12, 23-41, 1965.

17. Scott, D., Data types as lattices, SIAM J. Computing, 5, 1976, 522-587.

18. Scott, D. and Strachey, C., Towards a mathematical semantics for computer languages, Proc. Symposium on Computers and Automata, Vol.21, Microwave Research Inst. Symposia Series, Polytech. Inst. of Brooklyn, New York, 19-46, 1971.

19. Van Emden, M.H. and Kowalski, R.A., The semantics of predicate logic as a programming language, J.ACM, 23, 733-742, 1976.

20. Weyhrauch, R.W., Prolegomena to a theory of formal reasoning, Memo AIM-315, Computer Science Dept., Stanford University, 1978.

AXIOMS OR ALGORITHMS

Vaughan R. Pratt
Massachusetts Institute of Technology
Cambridge, MA 02193, USA

ABSTRACT

Traditional formal proof systems have been found unusable by those working on such applications of logic as program verification. They demand too much from the proof generator and too little from the proof checker. The notion of proof sketch or informal proof is an unsatisfactory substitute both because it is imprecise and because it treats the symptom rather than the disease. We propose to replace axiomatic proof systems by algorithmic proof systems, which explicitly incorporate a quantitative notion of computational complexity. This proposal depends on the existence of tractable decision procedures for many substantial fragments of logic, the "easy fragments."

INTRODUCTION

Logic supplies the mechanics of reasoning. With the advent of powerful computers logicians have hoped to be able to implement logical machinery and so automate mathematics.

The initial attempts dealt with automatic theorem proving. Based on progress over the past two decades, mathematicians need not suffer automation anxiety; their jobs will remain secure for the remainder of this century.

But mathematicians do not only prove theorems, they also check the proofs of other mathematicians. In fact if we assume that most published theorems are new and are checked by at least two readers (e.g. the referees), the average mathematician does more checking than proving. Theorems that survive the transition to the pedagogical arena and are taught to students undergo an even greater ratio of checking to proving.

The essence of proving is creativity, that of checking, accuracy. While creativity does not come easily to computers, accuracy is their forte. Thus for the immediate future it would seem more appropriate to automate the checking process and leave the proving to humans.

With this division of labor comes a division of responsibility: the human bears the responsibility for quantity, the computer for quality. The human no longer need entertain thoughts of abandoning his career because a substantial fraction of his output is flawed; in fact he may produce more incorrect proofs than correct ones with impunity. The computer as quality controller will filter out the former. The only proviso is that his output remain at a useful level as measured at the output of the filter.

Unlike humans, computers call for extreme formality in the specification of their tasks, especially when they are expected to perform those tasks reliably. Fortunately this need has been anticipated by the development of formal logic, which spells out mechanical rules of reasoning with the necessary precision.

The problem with this picture of human-machine symbiosis is that the extant formal criteria for correct proofs are not well matched to the abilities of either humans or machines. These criteria call for a level of detail in proofs not exceeded by the most pedantic of human proof checkers. Thus a human may spend a week, at a cost of $1000 in salary and support, producing a proof that is checked by a computer in five seconds at a cost of $0.50.

Not only are formal proofs long, they are difficult to produce. Mathematicians who have no difficulty proving informally that the volume of a sphere is $4\pi r^3/3$ may experience considerable difficulty proving $P \supset P$ in some formal systems of logic. The problem is that they do not necessarily know how to translate their way of seeing the validity of $P \supset P$ into a form acceptable in some formal system in which they lack experience.

Not surprisingly therefore, formal proofs have not been the medium of exchange between humans and computers. Instead attempts have been made to find some sort of middle position between automatic theorem proving and formal proof checking. However no middle position has been identified to date that lessens the complexity of traditional proofs while meeting the following criteria.

(i) There should be a formal specification of which user-supplied inputs will be accepted as sound arguments.

(ii) The set of such acceptable arguments should be recursive (and preferably checkable at a cost comparable to that of generating them manually).

Without criterion (i) the user is left to guess at what the system will accept. The unfortunate thing is that system designers generally have difficulty appreciating the need for such a specification, since they are sufficiently familiar with the features and limitations of their system themselves that it seems to them obvious what the system will handle. They seem to feel that this obviousness will become apparent as soon as the user has seen a few examples of the use of the system.

Without criterion (ii) there is of course no algorithm to back up the definition of acceptable input given in (i). This is the situation confronted by those in the automatic theorem proving business. There is a perfectly good definition of theoremhood, what is lacking is an algorithm to back up the definition.

We propose a new definition of proof that meets these criteria while addressing the problem of eliminating some of the detail from formal proofs.

THE NATURE OF PROOF

Proofs are generally thought of as syntactic in nature, dealing with formulas rather than with their meanings. The notion of proof we wish to advocate cannot conveniently be thought of in syntactic terms alone, having a strong semantic component. Thus it is tempting to say that we are replacing syntactic proofs with semantic ones.

The difficulty here is that there really is a semantic element even in traditional proofs, without which proofs would amount to meaningless and hence worthless computations. We shall begin therefore by considering traditional proofs from a viewpoint that emphasizes their semantic nature. By so doing we hope to focus attention on the truly novel aspects of our proposed notion of proof, which have more to do with computational complexity than with the distinction between syntax and semantics.

SYNTAX OF PROOFS

There is a tradition of viewing a proof as a series of lines that is gradually giving way to viewing it as a dag (directed acyclic graph) with one output vertex (the root). (Viewing proofs as trees, while not uncommon, seems to have little intrinsic merit beyond permitting proofs to be written as expressions.) Such a dag may be viewed as a circuit through which flow formulas. The inputs are axioms, the formulas are acted on by gates (functional or relational elements) drawn from a supply of inference rules, and the output of the circuit is the theorem of which the computation as a whole is the proof.

A simple example of a proof system is given by the axioms K: $p \supset q \supset p$ (we rely on spaces to indicate which way $\supset$ associates) and S: $p \supset (q \supset r) \supset (p \supset q \supset p \supset r)$ together with the single inference rule $MP(p \supset q, p) = q$. Then $MP(S,K) = (p \supset q) \supset (p \supset p)$. Since MP is the only rule, we can abbreviate $MP(p,q)$ to pq, so we have $SK = (p \supset q) \supset (p \supset p)$. Similarly (SK)K (or SKK, letting MP associate to the left) proves $p \supset p$, and S(KS)K proves $(q \supset r) \supset (p \supset q) \supset (p \supset r)$.

SEMANTICS OF PROOFS

It is as reasonable to consider proofs as circuits working with formulas as it is to consider arithmetic circuits (with gates + - * / etc) working with bit vectors. In the case of arithmetic one prefers to think of the arithmetic functions as acting on abstract numbers, only contemplating the bit vector representation when considering implementation issues. By the same token it would seem reasonable to consider inference rules to act on the abstract predicates denoted by the formulas rather than on the formulas themselves.

We can imagine stepping back far enough from the proof that we can only resolve what each formula denotes and not be able to distinguish representational details. From this distance the action of each gate in a proof is to output 1 (the unit of whatever Boolean algebra the predicates form, corresponding the the notion of validity) whenever all its inputs are 1. What it does with other than 1 is not specified; in fact, it may not even be determinate. For example, 0⊃q is 1 independent of q, yet MP(0,0⊃q) = q, demonstrating that the output of MP may vary as q varies even though the input appears from a distance to be constant. Looking more closely of course reveals that the variation is attributable to variations in the representation of 1, namely as 0⊃q. This nondeterminacy of MP is of no concern as it only appears outside the intended domain of operation of MP (both inputs equal to 1).

We thus have two views of a proof, a distant view and a close-up. Both views perceive the proof as a circuit with data flowing through it. When the proof is sound the distant view reveals only 1's everywhere in the circuit. The close-up view reveals the various representations of 1, along with the rules used in performing each inference at each gate. The close-up view is necessary both to support the distant view (by calculation) and to extract the theorem that has been proved.

We shall refer to an uninterpreted gate (one not distinguished as to whether it employs Modus Ponens or some other rule) together with the formulas at its inputs and output as an inference.

ADDING NEW RULES

We diagnose our complaint with the traditional approach to formal proofs in terms of the supply of rules, which are simply inadequate for constructing reasonably succinct proofs. How does one go about enlarging the set of rules?

Ignoring for the moment the question of effectiveness, the only property we need for correctness of proofs is that the output of each inference in the proof be valid. Thus a rule should preserve validity. In this way, if the inputs to the proof are axioms, and hence valid, all formulas in the proof will be valid.

However, a rule should also be effectively checkable. This brings us to the main contribution of this paper.

SOUNDNESS AND TRACTABILITY

The central idea of our approach is to make use of two orthogonal predicates on inferences, one natural, one artificial. The natural one is that of soundness: if the premises of the inference are valid, so is the conclusion. The artificial one is that of tractability. A set of tractable inferences has two properties. Membership in the set is easily determined by inspection, and soundness of inferences in the set is easily determined by computer.

A proof is then a circuit whose inputs are axioms and whose gates are sound tractable inferences.

The most obvious source of tractable sets of inferences is the large collection of decision methods available for various fragments of logic. A fragment is usually defined in terms of the permissible constructs of the fragment. Such a characterization of a fragment makes it possible to tell by inspection whether an inference belongs to a given fragment. The complexity of these decision methods ranges from polynomial time (e.g for the conjunctive quantifier-free theories of equality, or of successor and inequality) through non-elementary recursive time (e.g. Rabin's algorithm for weak second order theory of n successors) to remarkably large bounds for problems involving polymorphic functional languages.

The tractable part of such a recursive set depends on its complexity. If it is polynomial, say $20n^2$ microseconds for a given implementation where n is some measure of the size of the input, the tractable part would extend to n = 1000 if the user were willing to spend 20 seconds of computer time for testing soundness. If it were 20.2^n microseconds, the tractable part would only extend to n = 20 for the same outlay of computer time. For 20.2^{2^n}, the bound becomes n = 4, with n = 5 being totally inaccessible.

Those methods permitting n > 5 or so represent a substantial improvement over offering a fixed set of inference rules for testing soundness of an inference. There are roughly K^n inferences of length n where K is the alphabet size (some allowance should be made for syntactically ill-formed inferences). A fixed fraction of these must be sound, as can be seen by taking the shortest sound inference in the language, say of size a, and padding it with any of roughly K^{n-a} irrelevancies to yield a sound inference of size n. (Almost any language will permit such soundness-preserving padding.) The fraction of sound inferences of length n is then at least roughly K^{-a}. Hence the availability of a reasonably fast decision method, even one taking time 2^n, offers the user a flexibility comparable to having a large collection of sound inference rules.

While short inferences padded with irrelevancies may seem uninteresting objects to count, they nevertheless are the bane of automatic theorem provers. A nice, albeit extreme, example of how irrelevancies ("dead rats," as they are sometimes called) can slow things down is provided by the problem of testing existence of feasible solutions in linear programming, i.e. testing rational satisfiability of conjunctions of linear inequations. A "minimally infeasible" system (one in which the removal of any inequality would lead to existence of a solution, corresponding to the negation of a theorem containing no dead rats) can be tested for infeasibility by treating the inequations as equations and solving by Gaussian elimination to demonstrate infeasibility in polynomial time. A non-minimal system cannot be treated in this way because the spurious inequations interfere with the elimination process; indeed, no polynomial-time decision method is known for the non-minimal case. The simplex method for this problem can be viewed as eliminating irrelevant inequations as it moves from vertex to vertex.

In fact, without an automatic decision method, the elimination of dead rats becomes one of the duties of the writer of the proof. Even so, provision of decision methods only lessens this obligation.

It is important to realize that, while soundness inheres in an inference by definition, tractability does not, being an artifact. There is no universal set of tractable inferences. Rather, tractability is determined by the current availability of fast decision methods for language fragments, availability of fast computers to run those methods on, and the user's patience with the computer.

USER'S VIEW OF THE SYSTEM

The user views a system of the kind we have described via the user manual. This manual would enumerate language fragments for which decision methods were offered by the system, and for each would supply a method of estimating the running time of the method.

The user would use this information as follows. First he would prepare his proof as though explaining it to a human. Next he would take each inference of this proof and refine it as necessary to a series of inferences each of which are handled by one of the decision methods of the system and each of which are of a length guaranteeing efficient solution by the implementation. Then he would submit his proof for machine verification of soundness of its inferences. Presumably he would then enter a "debug cycle" in which inferences found to be unsound were repaired and resubmitted.

FEATURES OF ALGORITHMIC PROOF SYSTEMS

One appealing feature of this approach to proof systems is that it makes effective use of many of the decision methods that have been developed in the past half-century. There is a tendency to think of these methods as being of academic interest, both because most of them are of exponential complexity or more and because none of them by themselves offer general-purpose problem solving ability. However, in a verification context exponential complexity is not as damaging as in autonomous processing, as the user has control over the length of each inference. Moreover, each inference need not invoke a general problem solver, merely a specialist for the particular inference. Again the user has control over the scope of the specialization of the inference; if a single inference deals with too broad a topic it to be broken down into inferences of narrower scope.

Another appealing feature is that essentially all the technology to build a proof system is already developed. This is because there is little more to the system than a collection of independent decision methods. About the only component needed for the system would be an algorithm for detecting cycles in proofs, which as we have defined things are clearly undesirable. This degree of independence of the system components represents an extreme in system modularity that permits fast implementation (each decision method can be implemented by a small team of programmers, independently of other methods which can be being implemented at the same time) and straightforward extensibility (each newly discovered decision method can be implemented by itself and then added to the system with no additional need to interface it to anything).

Recent work by Nelson and Oppen [6] shows how to combine quantifier-free decision methods for independent fragments to yield a decision method of comparable complexity to the slowest method (assuming that the slowest method requires exponential time) for the combination of the languages. (The language of addition and that of multiplication would not be independent because of the distribution of over +. However the languages of addition and of pairing functions are independent.) This work promises to increase considerably the variety of possible inferences that systems will be able to check efficiently. Their result is particularly significant as concerns modularity, as their procedure for combining decision methods is automatic. Thus it will not be necessary to implement separate methods for the individual fragments and for their various combinations; this will be taken care of by the system.

A brief survey of extant decision methods is in order here. We cover only those with at most exponential time. The methods we survey presently fall into three classes: those dealing with fragments of predicate calculus, for which H. Lewis [4] has developed provably optimal exponential time decision methods; those dealing with quantifier-free theories of various data types (numbers, lists, arrays, in combination with equality and functional application), for which Nelson-Oppen [6], Shostak [9], Downey-Sethi [3] and others have developed exponential time or better decision methods; and those dealing with iterative program constructs and notions of partial correctness, termination, and equivalence, for which the author has developed provably optimal exponential time decision methods [8].

A question arises as to whether there are situations where the user loses something by giving up traditional formal proof rules. There is an easy answer to this: any single proof rule can be considered to constitute an "easy" (indeed trivial) fragment of logic, and as such could have an associated decision method. In fact one can expect that all rules the user may have ever had access to will have been replaced by considerably more general decision methods.

Consider for example first order logic. The propositional axioms and rules of any known finite axiomatization of this language would be subsumed by a deterministic exponential time decision method for propositional calculus. An axiom such as $\forall x(p \supset q) \supset (\forall xp \supset \forall xq)$ or a rule such as $p/\forall xp$ (Generalization) would be subsumed by a deterministic exponential time decision for modal logic, both of these really being modal properties if we treat $\forall x$ as a modality. And calculations involving instantiation of a universally quantified variable, or elimination of a quantifier that binds no free occurrences of its variable, can be provided by a modified version of a deterministic exponential time method for one of the decidable fragments of first order logic such as Hilbert-Ackermann's or Goedel's, cf. [4].

Our approach to defining proofs might be considered equivalent to an approach where one retained the traditional notion of formal proof but offering a facility for automatically deriving new inference rules as needed. From this point of view derivability of an inference rule is defined in terms of the basic axiom system.

This does not exactly characterize our approach, as it does not eliminate the basic axiom system, the disease of which long proofs are only a symptom. In the derived-rule account there are two things being defined: derivability of a rule in terms of the axiom sytem, and soundness of the axiom system in terms of the semantics. The disease remains, and the derived rules treat only the symptoms. Our approach sidesteps the detour through axiom systems by eliminating them altogether and so curing the disease.

Our approach formalizes an informal trend in modern verification systems towards the incorporation of powerful decision methods. The systems of Oppen, Constable [3], and the author [5] exhibit this trend. None of these systems have yet made the complete break with traditional axiom systems that we have advocated in this paper. We believe that such a break with tradition would be of substantial benefit to those implementing verifiers.

PERSONAL NOTE

My five-year-old daughter Jennifer asked me what I was doing. I said I was writing a paper on proving things. She asked me what it meant to prove something, so I said it meant to argue with somebody to make them believe something. I asked her how she would prove to me that our pond was full of water and she said she would take me to see it. I asked her how she would prove that 2 and 2 made 5, and she said it wasn't even true. I asked her how she would prove that 2 and 2 was 4, and to my surprise she said to use the calculator.

I infer that as calculating tools become more readily available to people who need to prove things to others, this answer will be heard more and more often, and traditional axiom systems will go the way of tables of logarithms as a usable but inefficient relic of the pre-computer era.

ACKNOWLEDGMENTS

I am indebted to Derek Oppen, Rob Shostak, and Bob Constable for providing the moral support necessary to sustain my belief in the views presented here. I hope I have done no more than formalize their intuititions.

BIBLIOGRAPHY

1. Constable, R.L., On the Theory of Programming Logics, Proc. 9th Ann. ACM Symp. on Theory of Computing, 269-285, Boulder, Col., May 1977.

2. Constable, R.L. and M.J. O'Donnell, A Programming Logic, Winthrop Publishers, Inc., 17 Dunster St., Cambridge, Mass., 1978.

3. Downey, P., H. Samet, and R. Sethi, Off-line and On-line Algorithms for Deducing Equalities, Conference Record of the Fifth Annual ACM Symposium on Principles of Programming Languages, 158-170, Tucson, Arizona, Jan. 1978.

4. Lewis, H., Complexity of Solvable Cases of the Decision Problem for the Predicate Calculus, 19th Annual Symposium on Foundations of Computer Science, Ann Arbor, Michigan, Oct., 1978.

5. Litvintchouk, S.D. and V.R. Pratt., A Proof-checker for Dynamic Logic, Proc. 5th Int. Joint Conf. on AI, 552-558, Boston, Aug. 1977.

6. Nelson, G. and D.C. Oppen., A Simplifier Based on Efficient Decision Algorithms, Proceedings of the Fifth Annual ACM Symposium on Principles of Programming Languages, 141-150, Tucson, Arizona, Jan. 1978.

7. Oppen, D.C., Complexity of Combinations of Quantifier-Free Theories, Proceedings of the Fourth Workshop on Automated Deduction, 67-72, Austin, Texas, Feb. 1979.

8. Pratt, V.R., A Near Optimal Method for Reasoning About Action, MIT/LCS/TM-113, M.I.T., Sept. 1978.

9. Shostak, R., Deciding Linear Inequalities by Computing Loop Residues, Proceedings of the Fourth Workshop on Automated Deduction, 81-89, Austin, Texas, Feb. 1979.

POWER FROM POWER SERIES

Arto Salomaa
Mathematics Department
University of Turku, Finland

1. Introduction. The purpose of this paper is to discuss some recent results in the area of formal power series in noncommuting variables. As with most mathematical formalisms, the formalism of power series is capable of unifying and generalizing known results. However, it is also capable of establishing specific results which are difficult if not impossible to establish by other means. Examples of such specific results are given also below.

For the sake of completeness, we give here the most important definitions. The reader is referred to [12] for a more detailed discussion. For unexplained notions in language theory, we refer to [11].

Consider a monoid M and a semiring A. Mappings r of M into A are called formal power series. The values of r are denoted by (r,w), where $w \in M$, and r itself is written as a formal sum

$$r = \sum_{w \in M} (r,w)w .$$

The values (r,w) are also referred to as the coefficients of the series. We shall be interested in this paper in the case where M is the free monoid X^* generated by an alphabet X. Then we also say that r is a series with (noncommuting) variables in X. The identity of X^* (i.e., the empty word) is denoted by λ.

The collection of all formal power series in this set-up is denoted by $A \ll M \gg$ or $A \ll X^* \gg$. Given r, the subset of M defined by

$$\text{supp}(r) = \{w \mid (r,w) \neq 0\}$$

is termed the support of r. The subset of $A \ll M \gg$ consisting of all series with a finite support is denoted by $A < M >$. Elements of $A < M >$ are referred to as polynomials.

In most applications to automata and formal languages, the semiring A will be either N (the semiring of nonnegative integers) or Z (the semiring, in fact a ring, of all integers). For simplicity, we shall assume below that $M = X^*$, although some of the results are valid also in the more general case. However, observe that for instance neither the definition of product given below nor Schützenberger's Representation Theorem is valid for arbitrary monoids.

The sum and product of two series r and s are defined by

$$r + s = \sum_{w \in X^*} ((r,w) + (s,w))\, w$$

and

$$rs = \sum_{w \in X^*} \Big(\sum_{w_1 w_2 = w} (r,w_1)(s,w_2)\Big)\, w \; .$$

Sum, product and the quasi-inverse defined below are referred to as rational operations.

Quasi-inverse r^+ is defined for quasi-regular series, i.e. series r satisfying $(r,\lambda) = 0$, by

$$r^+ = \lim_{k \to \infty} \sum_{i=1}^{k} r^i \; .$$

(We say that a sequence of series $r_1, r_2, \ldots$ converges to a limit r, in symbols $\lim_{j \to \infty} r_j = r$. if for all n there exists an m such that the conditions $lg(w) \leq n$ and $j > m$ imply the condition $(r_j, w) = (r, w)$.)

The family of A-rational series, in symbols $A^{rat} \ll X^* \gg$, is the collection of series obtained from polynomials by (finitely many applications of) rational operations.

For a semiring S, we denote by $S^{m\times m}$ the collection (in fact, a semiring) of $m \times m$ matrices with entries in S. A series r of $A \ll X^* \gg$ is termed A-recognizable (in symbols $r \in A^{rec} \ll X^* \gg$) if

$$r = (r,\lambda)\lambda + \sum_{w \neq \lambda} (\pi h(w)\eta)\, w \ ,$$

where $h : X^* \to A^{m\times m}$, $m \geq 1$, is a homomorphism and π (resp. η) is an m-dimensioned row (resp. column) vector. (Instead of π and η, more general projections can be considered.)

By Schützenberger's Representation Theorem, the families $A^{rat} \ll X^* \gg$ and $A^{rec} \ll X^* \gg$ coincide. For a proof, we refer to [12].

Consider an alphabet $Z = \{Z_1, \ldots, Z_n\}$ disjoint with X. A proper algebraic system (with respect to the pair (A,X) and with variables in Z) is a set of equations of the form

$$(1) \qquad Z_i = p_i \ , \quad i = 1, \ldots, n,$$

where each p_i is in $A \langle (X \cup Z)^* \rangle$ and, for each i and j,

$$(p_i, \lambda) = (p_i, Z_j) = 0 \ .$$

An n-tuple $(\sigma_1, \ldots, \sigma_n)$ of quasiregular series in $A \ll X^* \gg$ is a solution of (1) if (1) is satisfied when Z_i is replaced by σ_i, for $i = 1, \ldots, n$.

It is shown in [12] that every proper algebraic system (1) possesses a unique solution. A quasiregular series in $A \ll X^* \gg$ is termed A-algebraic if it appears as a component of the solution of a proper algebraic system. The family of A-algebraic series is denoted by $A^{alg} \ll X^* \gg$.

Shamir's Theorem (cf. [12]) gives a representation result for algebraic series, analogous to Schützenberger's Representation Theorem.

There is a natural correspondence between context-free grammars (having no chain rules and no λ-rules) and proper algebraic systems of equations : the variables Z_i correspond to nonterminals. In this correspondence, only the supports of the polynomials p_i are significant.

(The coefficients correspond to weights in a weighted context-free grammar.) If we begin with a context-free grammar G, write the productions as a proper algebraic system of equations (with coefficients equal to 1), and consider the series r in the solution corresponding to the initial letter of G, then $\mathrm{supp}(r) = L(G)$ and, moreover, the coefficient of an arbitrary word w in r equals the ambiguity of r according to G. (It is assumed that the basic semiring is N.)

The family of supports of N-rational (resp. N-algebraic) series equals the family of regular (resp. λ-free context-free) languages. The families of supports of Z-rational and Z-algebraic series (referred to as families of Z-rational and Z-algebraic languages) are strictly larger. Indeed, there are non-context-free Z-rational languages, for instance the language

$$\{a^m b^n \mid m, n \geq 1 \text{ and } n \neq m^2\} .$$

We would like to emphasize that very little is known about Z-rational and Z-algebraic languages although , at least from the mathematical point of view, these families are very natural in formal language theory.

<u>2. On algebraic series.</u> We shall establish first a "super normal form" result for the generation of algebraic series, analogous to a result concerning context-free grammars, [6].

<u>Definition.</u> Assume that $u \geq 3$ and $t_1, \ldots, t_u$ are nonnegative integers. We say that a proper algebraic system (1) is in the <u>$(t_1, \ldots, t_u)$ normal form</u> if

$$\mathrm{supp}(p_i) \subseteq X^+ \cup X^{t_1} Z X^{t_2} Z \ldots X^{t_{u-1}} Z X^{t_u} ,$$

for every $i = 1, \ldots, n$. (Cf. the equation (1) in the Introduction. The notations p_i, X, Z are the same as there.)

Thus, $(t_1, \ldots, t_u)$ normal form means that the right sides of the equations contain only (i) words over the "terminal" alphabet, and (ii) words involving exactly $u-1$ variables, separated by terminal words of fixed lengths determined by the tuple $(t_1, \ldots, t_u)$.

Theorem 2.1. For every $u \geq 3$ and nonnegative integers $t_1, \ldots, t_u$, every A-algebraic series can be generated by a proper algebraic system in the $(t_1, \ldots, t_u)$ normal form.

Proof. Let r be generated by the system (1), i.e., r is a component in the solution, say, the component corresponding to Z_1. To get the required normal form, we make a sequence of transformations to (1) preserving r (i.e., r will always equal the first component in the solution). Such a transformation may alter the set of variables Z: in general, the new set of variables Z' is bigger than the original one. For simplicity, we will denote the set of variables always by Z.

It is shown in [12] that r is generated by a proper algebraic system in which the supports of the right sides are included in the set

(2) $$X \cup X^2 \cup XZX \cup XZXZX .$$

Our first transformation consists in bringing the given system into this form.

Thus, we may assume that the supports of the right sides are included in the set (2). Our next step is to show that, given an integer $i \geq 0$, we may assume that the supports of the right sides are included in the set

(3) $$X^+ \cup X^i Z X^i \cup X^i Z X^i Z X^i .$$

(Thus, there may be longer terminal words in the two last terms of the union but then also $X \cup X^2$ might not be sufficient.) The step from (2) to (3) is accomplished exactly as the corresponding step for grammars in [6]. (In [6], this is called "reduction of subgoal 2 to subgoal 1".) The only additional observations needed are the following. (These observations are needed also in the remaining two reduction steps.)
(i) Whenever a substitution preserving a language is performed, we have to make sure that the same substitution preserves also the coeffi-

cient of each word in the language. But this is easily taken care of by preserving the original coefficient at the starting stage and making the coefficients all equal to 1 at other stages. (As customary when dealing with albegraic series, we assume that A is a semiring with identity.) (ii) Taking the union of languages corresponds to summing up the series.

In the last two reduction steps we first go from (3) to the set

$$(4) \qquad X^+ \cup X^j Z X^k Z X^l ,$$

for an arbitrary triple (j, k, l) of nonnegative integers and, finally, from (4) to the set

$$X^+ \cup X^{t_1} Z \ldots Z X^{t_u} ,$$

determined by the given tuple $(t_1, \ldots, t_u)$. These two reduction steps are established exactly as "reduction of subgoal 3 to subgoal 2" and "reduction of subgoal 4 to subgoal 3" in [6]. This completes the proof of Theorem 2.1.

The set X^+ in the statement of Theorem 2.1 can be replaced by a finite set

$$X \cup X^2 \cup \ldots \cup X^i ,$$

where i depends on the tuple $(t_1, \ldots, t_u)$. An explicit upper bound for i, in terms of the tuple, can be given. The whole research area concerning trade-offs between the number of variables in the systems and the numbers $u, t_1, \ldots, t_u$ is open.

As the second topic in this section, we consider a typical result justifying the title of this paper. The result is originally due to Semenov, [13], and a proof is given also in [12]. This proof uses facts concerning sums of infinite series, viewed as Taylor expansions, which is a rather unusual approach in the theory of formal power series. We give below a different proof.

The main result we are aiming at is that it is decidable whether a given unambiguous context-free language L and a given regular language R are equal. Several other results similar to this one can also be formulated. Observe the following two facts. (i) When a context-free grammar for L is given, an oracle has to tell it is unambiguous because this property is undecidable for context-free grammars. (ii) The result is a considerable generalization of the easily established fact that equality between deterministic and regular languages is decidable. The proof method of this fact is not applicable for the generalization because unambiguous context-free languages are not closed under complementation.

The following lemma is our crucial tool.

Lemma 2.2. It is decidable whether or not a given series $r \in Z^{alg} \ll \{x\}^* \gg$ is identically 0.

Proof. Observe first that, from the system of equations defining r, we may compute arbitrarily many first terms in r. The question is: how long must we continue if we encounter only 0's ?

Using elimination techniques (cf. [3]), we can effectively construct an equation

(5) $$a_0(x)z^m + a_1(x)z^{m-1} + \dots + a_m(x) = 0, \quad a_0(x) \neq 0,$$

where each $a_i(x)$ is a polynomial in $Z < \{x\}^* >$, such that r satisfies (5). Essential for this transition from many variables Z_i to one variable z is that the alphabet X consists of one letter x only. (5) can have solutions other than r, but r is certainly among the solutions.

Let now K be the maximum among the degrees of $a_i(x)$, for $i = 0, \dots, m$. We claim that if r is not identically 0 then

(6) $$(r, x^j) \neq 0 \quad , \text{ for some } j \leq K .$$

To prove this claim, assume first that $a_m(x)$ is not identically 0. Then if r satisfies (5), we must have $(r, x^j) \neq 0$, where j is less than or equal to the degree of $a_m(x)$. (Otherwise, the necessary cancellation cannot take place. Recall also that r is quasiregular.)

Consequently, (6) holds.

If $a_m(x) = 0$, we divide (5) by z and proceed as above, provided $a_{m-1}(x) \neq 0$. We verify that (6) is also now satisfied. If $a_{m-1}(x) = 0$, we divide (5) again by z. Continuing in this way, we see that (6) is always satisfied.

Lemma 2.2 now follows because we have seen that it suffices to test $K+1$ first coefficients of r.

Since Z-algebraic series are closed under difference, the following result is an immediate consequence of Lemma 2.2.

Lemma 2.3. It is decidable whether or not two given series r_1 and r_2 in $Z^{alg} < \{x\}^* >$ are identical.

Theorem 2.4. It is decidable whether or not a given unambiguous context-free language L and a given regular language R coincide.

Proof. We construct an unambiguous grammar for the language $L_1 = L \cap R$ and decide whether or not

(7) $$L_1 = L \quad \text{and} \quad L_1 = R .$$

Let $r(L_1)$, $r(L)$, $r(R)$ be the series obtained from the characteristic series of L_1, L, R by identifying all variables : $x_1 = x_2 = \ldots = x_k = x$. Then (7) holds if and only if

(8) $$r(L_1) = r(L) \quad \text{and} \quad r(L_1) = r(R) .$$

But the decisions (8) can be made by Lemma 2.3. This completes the proof.

The reader is referred to [12] for several other related decidability results for context-free and weighted context-free grammars. In all cases some oracular information is needed about the unambiguity or the degree of ambiguity.

3. An important open problem.

We have already encountered in the previous section series over an alphabet $\{x\}$ with only one letter. The coefficients of such a series constitute a sequence in the natural way.

Thus, we can speak of Z-rational and N-algebraic sequences.

A very intriguing, still open, decision problem is the following. Is it decidable whether or not the number 0 appears in a given Z-rational sequence? By Schützenberger's Representation Theorem, this problem can also be formulated in the following very simple way. Consider square matrices M with integral entries. Is it decidable whether or not, given such an M, the number 0 appears in the upper right-hand corner of some power M^i, for $i = 1, 2, \ldots$?

Still an equivalent version of this problem is the following. Is it decidable whether or not a given DOL length sequence contains two consecutive equal numbers? (For this and other related equivalent versions of the problem, the reader is referred to [12].)

Further importance to the problem is brought about by the fact that many other, even purely language-theoretic, problems have been reduced to it. Typical examples are given in [10]. It is shown that the decidability of the problem we are considering implies the decidability of many language-theoretic problems (whose decidability status is unknown at present). Morewer, it would give new proofs for the decidability of some celebrated problems such as the DOL sequence equivalence problem.

Intuitively, the problem we are considering seems to be "very decidable": only one variable is involved and the process is very deterministic. It seems that the known undecidability tools, such as the Post Correspondence Problem, are just impossible to encode into this problem. We would like to conjecture that even the following much more general problem is decidable: Does the number 0 occur in a given Z-algebraic sequence?

4. Recent results on rational series. In this final section of the paper we give a brief survey of some recent results concerning rational series. For further such results the reader is referred to [1], [4],

[5], [7] - [9].

The DTOL transformation introduced by Reutenauer, [7], combines ideas of L systems with ideas of power series, and the approach seems to be very promising in many respects.

By definition, a DTOL system is a tuple $G = (X, h_1, \dots, h_m, w_0)$, where X is an alphabet, $m \geq 1$ and $h_i : X^* \to X^*$ are homomorphisms for $i = 1, \dots, m$, and $w_0 \in X^*$.

Consider also the alphabet $Y = \{1, \dots, m\}$. Then G transforms a series r in $A \langle\langle X^* \rangle\rangle$ into the series $G(r) = s$ in $A \langle\langle Y^* \rangle\rangle$, defined as follows. Let $i_1 \dots i_t$ be an arbitrary nonempty word over Y (each i_j is a letter). Then

$$(s, i_1 \dots i_t) = (r, (w_0) h_{i_1} \dots h_{i_t}) .$$

Moreover, $(s, \lambda) = (r, w_0)$.

A series r in $Z \langle\langle X^* \rangle\rangle$ is polynomially bounded (resp. linearly bounded) if there is a polynomial (resp. linear polynomial) P such that

$$|(r, w)| \leq P(\lg(w)) .$$

It is shown in [7] that if r is a polynomially bounded N-rational series and G is a DTOL system (with the same alphabet) then $G(r)$ is also N-rational (but not necessarily polynomially bounded). The same holds true with "N-rational" replaced by "Z-rational".

This result shows that the equation $G(r) = G'(r')$ is decidable for arbitrary given polynomially bounded Z-rational series r and r', and DTOL systems G and G'. In particular, if $G = (X, h, w_0)$ is a DOL system (i.e., $m = 1$) and $u \in X^*$, then the series

$$\sum_n \binom{h^n(w_0)}{u} x^n \quad , \tag{9}$$

where $\binom{w}{u}$ is the Eilenberg binomial coefficient, cf. [2], is N-rational. Consequently, the equality of two series (9), possibly coming from different DOL systems, is decidable.

Every linearly bounded Z-rational series can be expressed as a polynomial of degree at most 2 in terms of characteristic series of regular languages. It is shown in [8] that, for two given linearly bounded Z-rational series r and r' expressed in this way, it is decidable whether or not (i) all coefficients of r are nonnegative, (ii) $\mathrm{supp}(r) = \mathrm{supp}(r')$. Problem (i) is undecidable for arbitrary Z-rational series, cf. [12]. Further results established in [8] concerning linearly bounded Z-rational series are the following. (iii) A linearly bounded Z-rational series with nonnegative coefficients in N-rational. (This result is not valid any more if the growth order of the Z-rational series is quadratic.) (iv) The support of a linearly bounded Z-rational series is an unambiguous context-free language.

The result (iv) gives some information concerning the characterization problem of Z-rational languages, mentioned in the Introduction. Again, supports of quadratic Z-rational series can be non-context-free.

Let us call an <u>iteration</u> every language of the form $u\,v^*w$, where u, v, w are words and v is not empty. A language over X is <u>dense</u> if it intersects every iteration over X. Let L be dense over X and $r \in Z^{rat} \ll X^* \gg$. Then the set of (distinct) coefficients of r is finite if the set of coefficients of the form (r, w), where w is in L, is finite. Morewer, the set of coefficients of r is finite if and only if, for every iteration L_1, the set of coefficients (r, w) with w in L_1 is finite. For these results, cf. [9].

Finally, [5] contains interesting applications of rational series to the computation of the height and average height of certain derivation trees.

References

1. J. BERSTEL (ed.), Séries formelles en variables non commutatives et applications. Lab. d'Informatique Théorique et Programmation, 2, place Jussieu, Paris, 1978.

2. S. EILENBERG, Automata, Languages and Machines, Vol. B. Academic Press 1976.

3. N. JACOBSON, Basic Algebra I. W.H. Freeman 1974.

4. J. KARHUMÄKI, Remarks on commutative N-rational series. Theoretical Computer Science 5 (1977), 211-217.

5. W. KUICH, Quantitative Aspekte bei Ableitungsbäumen. Institut für Mathematische Logik und Formale Sprachen, TU Wien, Technical Report,1979.

6. H. MAURER, A. SALOMAA and D. WOOD, On generators and generative capacity of EOL forms. Institut für Informationsverarbeitung, TU Graz, Bericht 5, 1978.

7. C. REUTENAUER, Sur les séries associées à certains systèmes de Lindenmayer. Theoretical Computer Science, to appear.

8. C. REUTENAUER, Sur les séries rationnelles en variables non commutatives. Springer Lecture Notes in Computer Science, Vol. 62 (1978), 372-381.

9. C. REUTENAUER, Propriétés arithmétiques de séries rationnelles et ensembles denses. Acta Arithmetica, to appear.

10. K. RUOHONEN, Zeroes of Z-rational functions and DOL equivalence. Theoretical Computer Science 3 (1976), 283-292.

11. A. SALOMAA, Formal Languages. Academic Press 1973.

12. A. SALOMAA and M. SOITTOLA, Automata-Theoretic Aspects of Formal Power Series. Springer-Verlag 1978.

13. A.L. SEMENOV, Algoritmicneskie problemy dlja stepennykh rjadov i kontekstnosvobodnykh grammatik. Dokl. Akad. Nauk SSSR, Vol. 212 (1973), 50-52.

COMPUTATIONAL COMPLEXITY OF STRING AND GRAPH IDENTIFICATION

A.O.Slisenko

LOMI, Fontanka 27

Leningrad, 191011, USSR

The main aim of my report is to draw attention to some ideas which were successfully exploited for constructing fast algorithms for string-matching and related problems, and to discuss ways of making use of these ideas in more complicated identification problems. The ideas to be discussed are the idea of identifier and the idea of substructure automorphism. They may prove to lead to essential decreasing of upper bounds on the computational complexity of the problems. Lower bounds remain outside the discussion. The recent progress in investigating lower bounds on the complexity of concrete problems is at best implicit. My exposition pursues purely theoretical objectives and the choice of problems for discussion is entailed by theoretical considerations and not by applications.

By the complexity I mean the (worst-case) time complexity for address machines introduced by Slisenko [1]-[2] (in Angluin and Valiant [3]a similar model is named RAC), i.e. random access machines with relatively bounded registers - the length of computer words is not greater than the binary logarithm of the time (plus some function of lower order, if necessary). More or less detailed description of address machines can be found in Slisenko [2]. The choice of computational model is essential for "low-level" complexity. If we wish to stay on the firm ground of practicalness then this area of computational complexity becomes one of the most important. And the address machine model seems to be the most adequate model (among those used in computational complexity) for computations by one processor in homogeneous random access memory. The place of address machines among such models as Turing machines and Kolmogorov algorithms (or their slight generalization named "storage modification machines" by

Schönhage [4]) is as follows. Let $AM(\varphi)$ be the class of functions (or predicates) computable by address machines with the time complexity not greater than φ , $TM_n(\varphi)$ - the similar class for Turing machines with n-dimensional tapes (one can take either the class of one-head machines or of multi-head ones), $KA(\varphi)$ - the similar class for Kolmogorov algorithms or storage modification machines. Within these notations: $TM_n(\varphi) \subseteq KA(C_1\cdot\varphi) \subseteq AM(C_2\cdot\varphi) \subseteq TM_1(C_3\cdot\varphi^2 \log^3\varphi)$; $TM_n(\varphi) \subseteq AM(C_4\cdot\varphi\cdot(\log\varphi)^{-1/n})$, where C_i's are constants (the first inclusion is due to Schönhage [4] , the last is virtually some generalization of a theorem from Hopcroft, Paul and Valiant [5]).

As a starting point for the discussion I take the following "embedding framework" for representation of the problems to be considered. This framework is wider than I really need but it can play a certain organizing role. Let $G_i = (N_i, S_i)$, $i=1,2$, be two "graph-like" structures, where N_i is a finite set of nodes, and S_i is a finite set of links (i.e. partial mappings $S_i \to S_i$) and functions (i.e. partial mappings $S_i \to$ some set of words, e.g. integers). "Straight" embedding problem looks as follows. There are given mappings (or embeddings) $f_k: N_2 \to N_1$, $1 \le k \le m$, which are injective as a rule. And we have some "global" criterion Ψ of the quality of a set of embeddings. The problem consists in computing $\Psi(G_1, G_2, \{f_k\}_1^m)$. Surely, the particular Ψ's below will be rather easy to compute. Several "inverse" problems correspond to a given "straight" problem, e.g. for a given G_1 and a value of Ψ (or some its property, such as "to be maximal") we are to find G_2 and f_k of a certain type.

One of the simpliest embedding problems, namely, string-matching, will illustrate the main ideas discussed in this report. We consider strings in some alphabet $A=\{a_1,\dots,a_\nu\}$ and are interested in recognizing the set

$$\{u*v:\ v \text{ is a substring of } u\} \tag{1}$$

In an input $u*v$, $*\notin A$, we discern the text u and the pattern v. Now we represent the problem within the embedding framework. The length of w is denoted by $|w|$ and the i-th letter (character) in w by $w(i)$. Let $N_1=\{1,2,\dots,|u|\}$, $N_2=\{1,2,\dots,|v|\}$ and S_1, S_2 describe the string structure on N_1 and N_2 (in fact, S_1 and S_2 will be incorporated in f_k), and $f_k(j)=k+j-1$. Then

$$\Psi(u,v,\{f_k\}) = \max_k\{\Phi(u,v,f_k)\}, \quad \text{where}$$

$$\Phi(u,v,f_k) = \max\{i \le |v|:\ \&_{1\le j\le i}\ v(j)=u(f_k(j))\}.$$

(Generally speaking, to find $\Psi(u, v, \{f_k\})$ is a more difficult problem than to recognize the set (1), but this will prove to be unessential.) A linear-time algorithm for computing Ψ was devised by several authors about 10 years ago (for references see Slisenko [1], [6]; formally the shortest description is due to Slisenko [2]). But this algorithm neither is real-time nor solves interesting inverse problems, e.g. finding the longest repetitions in a string. Nevertheless, the algorithm can be regarded as based on a particular case of the identifier approach. In itself the notion of identifier is very simple. A segment $[i,j]$ is a repetition in w if $w[i,j] = w(i)w(i+1)\dots$ $\dots w(j)$ has two different occurrences in $w[1,j]$. A segment $[i,j]$ is an identifier in w if it is not a repetition in w, and $[i, j-1]$ is a repetition unextendible to the left, i.e. $[i-1, j-1]$ is not a repetition in w. A compact representation of identifiers as a tree was firstly devised by Weiner [7] ; more simple forms are due to Pratt [8] and Slisenko [6]. These methods gave linear-time (even on-line) procedures for string-matching, for finding the longest repetitions and several other problems. However, the algorithms were not real-time. The first real-time algorithms for these problems sketched by Slisenko [6] exploited essentially a new idea - they used some succinct representation of the periodical substructure of an input text-string. Moreover, the problem of finding all the periodicities also proved to be solvable in real-time — see Slisenko [6]. I accentuate the latter result because of the following reasons. It is of a theoretical interest by itself. Though in the case of the particular problem, i.e. string-matching, making use of periodicities now seems to be almost redundant, in some other situations (see Slisenko [9]) it remains a necessary instrument of fast algorithms. The periodicities as a gear for speeding up computations and "physical" periodicities have different origins. In the case of exact string-matching they coincide. Later we shall see (e.g. in the case of approximate string-matching) that they can be diverse. In computations periodicities appear as a method for regulating overlappings of equal strings. We can look at a periodicity as at an appropriate substructure of a string with nontrivial automorphism group which is convenient for fast computations. In the case of strings such substructures and their automorphism groups are rather poor. Surely, both ideas - that of identifier and of substructure automorphisms, are rather simple when taken for themselves, and, moreover, are often used, for example, in pattern classification and recognition. So while speaking about any of them I shall tacitly assume that they are sustained by appropriate algorithmic constructions or presumably can be sustained. I conclude this part of the exposition by an example of a data structure which can be used for realization of the identifier approach; an input text-string is 0101102101.

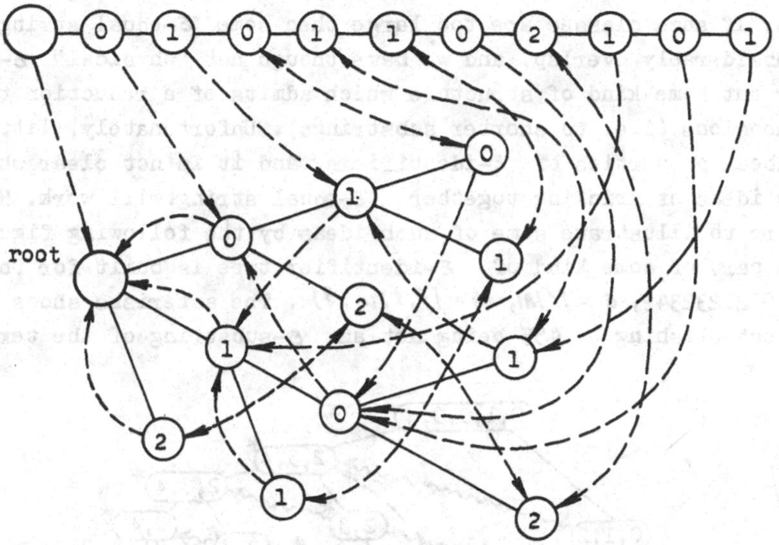

For the further discussion I make explicit 3 features of the problem considered above (i.e. string-matching): (a) the input text-structure is 1-dimensional; (b) patterns standing for identification are connected; (c) identification (matching) is exact (not approximate). I shall try to describe difficulties that arise in attempts to apply the ideas mentioned above to more complicated structures. Preserving (a) in full and (b) as much as possible, I loosen (c) and consider one kind of approximate identification in a metric space. Given $\varepsilon \geq 1/|A|$ and two string u and v, find k such that $1 \leq k \leq |u|-|v|+1$ and

$$\max_{1 \leq i \leq |v|} \rho(u(k+i-1), v(i)) \leq \varepsilon , \qquad (2)$$

where $\rho(a_i, a_j) = |i-j| \cdot |A|^{-1}$. The left part of the inequality (2) defines some metric on strings of equal lengths; I shall denote it also by ρ. If we replace string equality by "ε-equality" (that means $\rho(x,y) \leq \varepsilon$ for strings x and y), the general framework of identifier approach preserves its main outward features: if we have an ε-identifier tree, it will work. But the methods of constructing identifier trees used in the case of exact string-matching do not fit here, since the ε-equality is nontransitive. Different classes of ε-equal strings can overlap. A class of ε-identifiers whose prefixes up to the last character (excluding it) are ε-equal, can be rather large. However we need not to represent all the strings in such a class. If such classes are too large then some ε-equal strings must considerably overlap, and we have though not "physical" ε-periodicity but some kind of structure which admits of a reduction to lower dimensions (i.e. to shorter substrings). Unfortunately, little is known about properties of ε-identifiers, and it is not clear whether various ideas of sticking together ε-equal strings will work. However I try to illustrate some of such ideas by the following figure, where a part of some kind of ε-identifier tree is built for the string 0121232349, $\varepsilon = 1/|A|$, $A=\{0,1,\dots,9\}$. The asterisks shows the way of establishing 455 being not an ε-substring of the text-string.

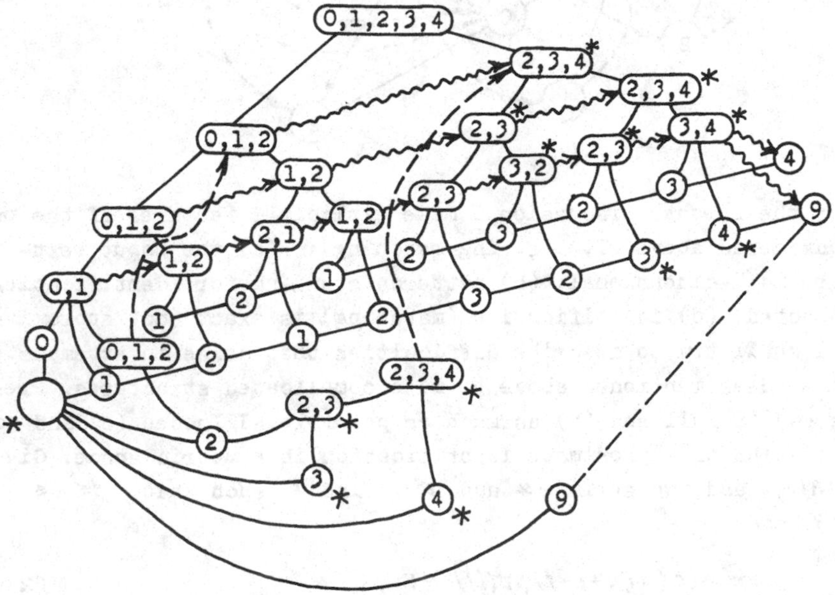

As for searching "physical" ε-periodicities, the situation is more complicated. A string w is an ε-periodicity if $w = v_1 v_2 \dots v_k$, where $k \geq 2$, $|v_i| = |v_j| > 1$, $\rho(v_i, v_j) \leq \varepsilon$ for $1 \leq i,j \leq k$. In the case of exact string-matching, periodicities can be characterized as overlapping occurrences of two equal strings. Now two overlapping ε-equal strings do not necessarily give an ε-periodicity. So we are to find some other easily tested characterization of ε-periodicities. It may prove that a fast algorithm for pure ε-string-matching will give some key for operating with ε-periodicities. Thus the main difficulty in the ε-matching described above is entailed by nontransitivity of ε-equality. If we direct our attention to ε-matching in some integral metric, e.g. that of $l^p, p\geq 1$, we stumble on one more obstacle, that is singularities in "local" behaviour of differences between two ε-equal strings. And here it is not clear whether to try to overcome these difficulties by straight-forward developing of the idea of identifier or to try methods based on other principles. There is one more way to simplify the problem. A random string (with respect to uniform distribution) has rather a poor and easily computed structure of substrings. Thus, in practical situations we are interested not in random strings. Analysing difficulties arising on the way of developing identifier approach, we can try to exclude some types of unmanageable strings by physical considerations. So, an identifier approach may be an instrument for seeking for such simplifying reformulations of identification problems.

I classify the two types of identification discussed above (i.e. exact and approximate) as combinatorial and analytical. And now I go on to arithmetical identification, which is characterized by essential loosening of (b) - we identify disconnected structures.

By arithmetical identification I mean such problems as string-matching with don't cares, convolution, integer multiplication. From now on I assume that our input alphabet A consists of integers, including zero. As it was recently shown by Schönhage [4] integer multiplication (and, apparently, integer convolution) has linear time complexity even for storage modification machines, and the more so, address machines. Thus by Fischer and Paterson [10], matching with don't cares has quasilinear complexity $O(n \log n)$. Though arithmetical identification seems to be the most universal with respect to identification of the other types, I do not know good reduction of, say, analytical problems to arithmetical ones. For further discussion I take the following example. Let $A=\{0,1\}$; for given u and v find $\max_k \zeta(k)$, where $\zeta(k)=\sum_{i\geq 1} u(k+i-1)v(i)$. Suppose we are seeking for a linear time on-line algorithm solving the problem, the input being of the form $u*v$. Is it possible to find an acceptable realization for the idea of identifier in this case? I am unable to give a definite answer to the question. To illustrate difficulties arising here let us replace u by $\bar{u}=i_0, i_1, \ldots, i_k$, where $i_0=1$ and $i_1, \ldots, i_k$ is the list of distances between consecutive units in

u . Let $\mathcal{U}$ be the set of all lists which can be derived from $\bar{u}$ with the help of the productions of the form: $i,j \vdash i+j$. We need identifiers for the whole set $\mathcal{U}$ (we treat lists as strings). Whether it is possible to compactify these identifiers or not, not speaking about ways of fast constructing them, is an open question. More attractive (at the first glance) approach may be based on the idea of using substructures with easily computed automorphisms. Grids of units are substructures of this type. (A slight analogy with the discrete Fourier transform, where some other grids are formed by powers of roots of unity.) Strictly speaking, usual grids, i.e. arithmetical progressions, do not work in the case of sparse numbers, but we put aside such numbers. One can devise a representation of u as some hierarchical, tree-like data structure of grids. But we need a representation which gives a fast procedure for comparing u and v, and this is one of the most obscure points in this approach. The problem of computing $max\,\zeta(k)$ when u and v go to the input in parallel may prove to be a more workable problem for analysis of the ideas under discussion.

The difficulties we met in treating arithmetical identification have some features common to the difficulties we meet when the dimension of patterns under consideration rises; the reason is the same — the growth of degree of freedem. Let us throw a glance on 2-dimensional (combinatorial) pattern matching. The pure idea of identifier though it works in principal, gives no advantages as one can easily see on $n \times n$-square filled with the same character — the total volume of identifiers is n^4 in order, that is equal to the complexity of the most trivial algorithm (the role of strings is played by connected configurations with fixed starting points; we are seeking for connected neighborhoods separating all the points of the square). Here the situation seems to be simpler than in the case of arithmetical identification. One of the ideas of representing a given text-structure is illustrated by the figure:

0	0	0	0	0	0	0	0	0
0	0	0	0	0	0	0	0	0
0	0	1	0	0	0	0	0	1
0	0	0	0	0	0	0	0	0
0	0	0	0	0	0	0	0	0
0	0	0	0	0	0	0	0	1
0	0	0	0	0	0	0	0	0
0	0	0	0	0	0	0	0	0
0	0	0	0	0	0	0	0	0

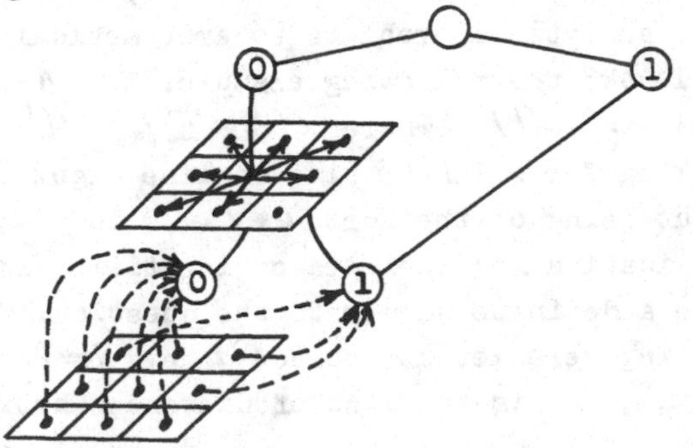

Here we construct the "identifier tree" starting from the centre.The resulting tree depends on the starting point, generally speaking.But it remains sufficiently flexible for testing matching.

The problem becomes essentially more arduous if we permit rotations (in Euclidean plane). Then we come to some sort of analytical identification (approximate matching concerns only the boundary of a pattern). New difficulties may be clarified by the following question. Let a string u be ascribed to some circle, i.e. u forms a cycle. We wish to construct an identifier tree for ordinary string-matching (a pattern is a usual string) in some more or less symmetric fashion, not breaking the circle. How to do it?

The described way of representing 2-dimensional patterns is rather inviting for studying graph isomorphism. An example of complete graph shows that the straight-forward idea of identifier does not work here. Nevertheless all known to me algorithms for graph isomorphism (many of them are surveyed by Read and Corneil [11])try to manage only with identifieres. This is done by embedding a given graph into more rich structures which are used for constructing appropriate identifiers. And if a graph is "very regular" such approaches fail, i.e. they,virtually, turn into an exhaustive search. I would like to draw attention to the fact that while treating graph automorphisms we are not to solve difficult classical problems concerning structure and properties of the automorphism groups. All we need is a representation of these groups which is convenient for fast computations.

To conclude the discussion I shall try to discribe structures which arise when attempting to apply the identifier approach to a NP-complete problem. I consider the problem of existence of a Hamiltonian cycle in a plane graph. As substructures of a given graph we can take connected subgraphs with cyclic boundaries (an analogue of spheres)

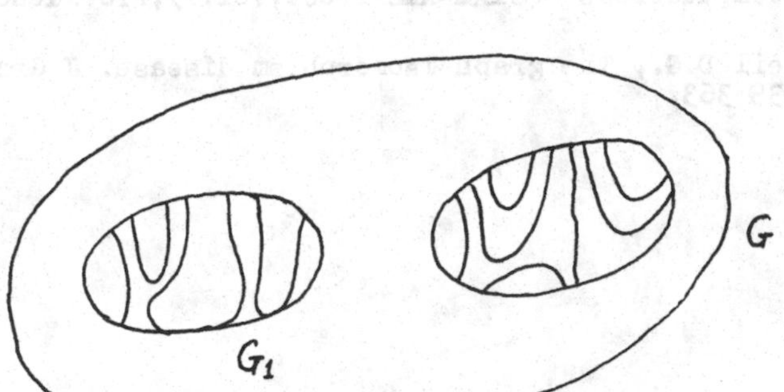

Let G be a given plane graph, G_1 - its subgraph with the boundary S .A Hamiltonian partition of G_1 is a partition of G_1 into simple paths and simple cycles $g_1, g_2, \dots, g_k$, such that $S \cap g_i \neq \emptyset, g_i \cap g_j = \emptyset$; if g_i is a path then its end points belong to S , $1 \leq i \neq j \leq k$. The problem is to represent the set of all Hamiltonian partitions of various G_1's to gain fast updating of these representations when these G_1's merge (or extend themselves).

References

1. Slisenko A.O., String-matching in real-time: some properties of the data structure. Lect.Notes Comput.Sci,1978,64,493-496.
2. Слисенко А.О., Сложностные задачи теории вычислений. Препринт. ВИНИТИ,М.,1979.
3. Angluin D., Valiant L.G., Fast probabilistic algorithms for Hamiltonian circuits and matching. CSR-17-77, Dept.Comput.Sci.,Univ.of Edinburgh,1977.
4. Schönhage A., Storage modification machines. Univ.Tübingen,Math. Inst.,1979,45 pp.
5. Hopcroft J., Paul W., Valiant L., On time versus space and related problems. IEEE 16th Annu.Symp.Found.Comput.Sci.,Berkeley,California,1975,57-64.
6. Слисенко А.О., Распознавание предиката вхождения в реальное время. Препринт ЛОМИ Р-7-77, Ленинград,1977.
7. Weiner P., Linear pattern-matching algorithm. IEEE 14th Symp. Switch. and Automata Theory,Iowa,1973,1-11.
8. Rodeh M., Pratt V., Even S., A linear algorithm for finding repetitions and its applications in data compression. TR 72,April,1976, Technion-Israel Inst. of Technology,Dept.Comput.Sci.,Haifa.
9. Слисенко А.А.,Упрощенное доказательство распознаваемости симметричности слов в реальное время. Записки научных семинаров ЛОМИ, 1977,68,123-139.
10. Fischer M.J., Paterson M.S.,String-matching and other products. "Complexity of Computations" (SIAM-AMS Proc.,Vol.7),Providence,R. I.,1974,113-125.
11. Read R.C., Corneil D.G., The graph isomorphism disease. J.Graph Theory,1977,1,339-363.

A SURVEY OF GRAMMAR AND L FORMS-1978

Derick Wood
Unit for Computer Science
McMaster University, Hamilton, Ontario, Canada, L8S 4K1

ABSTRACT

The present paper gives an overview of grammar and L form theory as of January 31, 1979. It is intended to complement Ginsburg's 1977 survey of grammar forms [G]. Hence although we present some new results on grammar forms the main thrust of the paper is to survey L forms.

The number of papers in this area is now over 65, see [W1] hence it may be observed that this is indeed a fast developing area, as is claimed in [M]. A more detailed exposition of the area can be found in [W2].

Because of this many open problems are stated in the hope that our own [G and MSW] excitement with this new area of formal language theory will be catching.

INTRODUCTION

In [CG] the notions of a grammar form and its interpretation grammars were first introduced. A grammar form can be considered to be a "master grammar" and its interpretation grammars can similarly be considered to be those grammars which "look like" the master grammar. Therefore each grammar form gives rise to a family of interpretation grammars and also to a family of languages. In a similar way a master L grammar gives rise to a family of interpretation L grammars and a family of L languages; these were first introduced in [MSW1].

In Section 2 the basic definitions and some examples are given, while Section 3 deals with grammar forms and Section 4 with L forms.

2. PRELIMINARIES

Intuitively the two notions of "looks like" can be motivated by considering the following context-free productions:

(1) $S \to aAbB$;

(we use the convention that early lower case roman letters denote terminals and early upper case nonterminals).

(2) $D \to abbEF$;

(3) S' → a'A'b'B';
(4) D' → E'E'a'; and
(5) S → aACbB.

We say (2) looks like (1) since the nonterminals D, E, F correspond to nonterminals S, A, B, respectively, while abb corresponds to a and λ (the empty word) to b. This corresponds to the notion of g-interpretation. Similarly (3) looks even more like (1), this corresponds to s-interpretation. However (4) does not look like (1) on two counts, firstly that E' corresponds to both A and B, which we do not allow, and secondly a' corresponds to λ in (1) which we also do not allow. If we do allow (4) to look like (1) then it implies that linear grammars look like regular grammars and expansive grammars look like non-expansive ones. Similarly we say (5) does not look like (1) since the number of nonterminals has increased.

We now formalize these notions.

Definition: Let $G_i = (V_i, \Sigma_i, P_i, S_i)$ i = 1, 2, be two (context-free) grammars and μ a finite substitution on V_2^*. We say G_1 is a g-interpretation of G_2 modulo μ, denoted $G_1 \triangleleft_g G_2(\mu)$ (or simply $G_1 \triangleleft_g G_2$ if μ is understood), if μ satisfies the following conditions:

(i) μ(a) is a finite subset of Σ_1^*, for all a in Σ_2,
(ii) μ(A) if a finite subset of $V_1 - \Sigma_1$, for all A in $V_2 - \Sigma_2$,
(iii) $\mu(A) \cap \mu(B) = \emptyset$, for all A, B in $V_2 - \Sigma_2$, $A \neq B$
(iv) $P_1 \subseteq \mu(P_2) = \bigcup_{A\to\alpha \text{ in } P_2} \mu(A\to\alpha)$, where $\mu(A\to\alpha) = \{A' \to \alpha' : A' \text{ in } \mu(A) \text{ and } \alpha' \text{ in } \mu(\alpha)\}$
(v) S_1 is in $\mu(S_2)$.

We say that G_1 is a (g-) interpretation grammar. The (context-free) grammar G_2 is called a (context-free) grammar form when used in this way.

We say G_1 is an s-interpretation of G_2 modulo μ, denoted $G_1 \triangleleft_s G_2(\mu)$ (or simply $G_1 \triangleleft_s G_2$ if $G_1 \triangleleft_g G_2(\mu)$ and μ also fulfills:

(i)' μ(a) is a finite subset of Σ_1, for all a in Σ_2.
(iii)' $\mu(X) \cap \mu(Y) = \emptyset$, for all X, Y in V_2, $X \neq Y$.

Note especially that in s-interpretation terminals and nonterminals are treated in a consistent manner.

In the case that G is an EOL grammar we only define s-interpretations of G. The reasons for this are simply that EOL grammars have terminal rewriting rules, which (i) under g-interpretation could give rise to productions of the form: λ → α and ab $\rightarrow \alpha$ neither of which are EOL productions, and (ii) allow derivations in the interpretation grammar which are not images of derivations in the master grammar. An EOL form is simply an EOL grammar.

For either a grammar or an EOL form G we define the s-grammar family of G by: $\mathcal{G}_s(G) = \{G' : G' \triangleleft_s G\}$, while for a grammar form G the g-grammar family is defined by $\mathcal{G}_g(G) = \{G' : G' \triangleleft_g G\}$. Similarly we can define $\mathcal{L}_s(G)$ and $\mathcal{L}_g(G)$, the s- and g-grammatical families of G by $\mathcal{L}_s(G) = \{L(G') : G' \triangleleft_s G\}$ and $\mathcal{L}_g(G) = \{L(G') : G' \triangleleft_g G\}$. Whether G in $\mathcal{L}_s(G)$ is a grammar or L form is usually clear from the context.

Consider the following examples.

(a) Let G_1 be defined by:
$S \rightarrow aSa$; $S \rightarrow a$
then $\mathcal{L}_g(G_1) = \mathcal{L}(\mathrm{LIN})$,
while $\mathcal{L}_s(G_1) \neq \mathcal{L}(\mathrm{LIN})$.

(b) Let G_2 be defined by:
$S \rightarrow S$; $S \rightarrow a$; $S \rightarrow SS$; $a \rightarrow S$
then $\mathcal{L}_s(G_2) = \mathcal{L}(\mathrm{EOL})$.

(c) Let G_4 be defined by:
$S \rightarrow a$; $S \rightarrow SS$
then $\mathcal{L}_g(G_4) = \mathcal{L}(\mathrm{CF}) = \mathcal{L}_s(G_4)$.

In the following we also mention various other rewriting systems, for example, non-context free grammars, matrix grammars, EIL grammars, OL and DOL grammars, etc. Their definitions can be found in [S], while the extension of s- and/or g-interpretations to these cases essentially follows the definitions given above.

Definition: Two languages are equal if they differ by at most λ. Two language families are equal if for every language in one there is an equal language in the other and vice versa.

We say two forms (both grammar or both EOL) G_1 and G_2 are strongly s-form equivalent if $\mathcal{G}_s(G_1) = \mathcal{G}_s(G_2)$ and s-form equivalent if $\mathcal{L}_s(G_1) = \mathcal{L}_s(G_2)$. Define (strongly) g-form equivalent similarly.

3. GRAMMAR FORMS

Since the survey in [G] treats the grammar form case more fully, we restrict attention to some problem areas. There are at least three major problem areas for grammars.

3.1 The RE-CF gap conjecture: It has been conjectured that for all non-context-free grammar froms G, $\mathcal{L}_g(G) \supsetneq \mathcal{L}(\mathrm{CF})$ implies $\mathcal{L}_g(G) = \mathcal{L}(\mathrm{RE})$.

It has been shown in [MPSW] that under s-interpretation there are non-context-free grammar forms G with the property that:

$$\mathcal{L}(CF) \subsetneq \mathcal{L}_s(G) \subsetneq \mathcal{L}(RE).$$

Also in [MPSW] it has been shown that for a non-context-free grammar form G with a single nonterminal

$$\mathcal{L}(CF) \subsetneq \mathcal{L}_g(G) \text{ implies } \mathcal{L}_g(G) = \mathcal{L}(RE).$$

Although this can be extended to the case of two nonterminals the problem is, in general, still open.

However there is a matrix form G with $\mathcal{L}_g(G)$ in the gap, in fact G given by: $[S_1 \to aS_1a;\ S_2 \to aS_2a]$, $[S_1 \to a;\ S_2 \to a]$, $[S \to S_1S_2]$, $[S \to S_3]$, $[S_3 \to a]$, $[S_3 \to S_3S_3]$, has this property.

3.2. Form equivalence: Remarkably it has recently been shown to be decidable whether $\mathcal{L}_g(G_1) = \mathcal{L}_g(G_2)$, for two arbitrary grammar forms G_1 and G_2. This result is reported in two papers [GS1] and [GS2].

However it is still not known whether $\mathcal{L}_s(G_1) = \mathcal{L}_s(G_2)$ is or is not decidable, for two grammar forms G_1 and G_2.

3.3. Completeness: Letting $\mathcal{L}$ be a family of languages, we say a grammar form G is $\mathcal{L}$-x- complete, for x = g or s, if $\mathcal{L}_x(G) = \mathcal{L}$.

It has been shown in [CG] to be decidable whether a reduced grammar G is finite-, regular- or context-free- g-complete, since in this case $\mathcal{L}$-g-completeness reduces to the question of whether G is finite, non-self-embedding or expansive, respectively.

However in the case of s-interpretations $\mathcal{L}$-s-completeness is far from trivial. In [MWS10] regular- and linear-s-completeness are characterized. Hence it is shown in [MWS10] that G_1: $S \to aSa$; $S \to a$; $S \to aa$ does not generate $\mathcal{L}(LIN)$, while G_2: $S \to aS$; $S \to Sa$; $S \to a$ does.

The major open problem here [MSW9] is the characterization of context-free-s-completeness.

4. L FORMS

We will provide in this section a rapid survey of the highlights of EOL form theory and also pose some open problems. Since only s-interpretations are considered we omit the prefix or subscript "s".

4.1 <u>Normal Forms</u>: Most of the normal form transformations which preserve the language of an EOL or ETOL grammar also carry over to the form situation. For example, we obtain:

<u>Theorem 1</u>: For each ETOL form G there exists a form equivalent ETOL form H which is reduced, binary and has only two tables.

However form equivalence is not preserved under synchronization or under λ-removal.

<u>Theorem 2</u>: Let G: $S \to a$; $a \to b$; $b \to b$ be an EOL form. Then for each synchronized EOL form H,

$$\mathcal{L}(H) \neq \mathcal{L}(G).$$

We have a second exception in the case of propagation.

<u>Theorem 3</u>: Let G: $S \to abba$; $b \to \lambda$; $a \to c$; $c \to c$. Then for each propagating EOL form H, $\mathcal{L}(H) \neq \mathcal{L}(G)$.

In [Sk] it is shown how G of Theorem 2 can be synchronized by using 2 tables, that is:

<u>Theorem 4</u>: Let G: $S \to a$; $a \to b$; $b \to b$ and

H: $\{S \to \bar{a}$; $\bar{a} \to \bar{b}$; $\bar{b} \to \bar{b}$; $a \to N$; $b \to N$; $N \to N\}$,

$\{S \to N$; $\bar{a} \to a$; $\bar{b} \to b$; $a \to N$; $b \to N$; $N \to N\}$,

where $\bar{a}$, $\bar{b}$ and N are new nonterminals.

Then $\mathcal{L}(G) = \mathcal{L}(H)$.

<u>Open problem</u>: What is the relationship of the classes of language families obtained from EOL forms, synchronized EOL forms and synchronized ETOL forms?

For EIL forms the normal form results are closest to those for DIL grammars [MSW6]. An EIL grammar has, for some fixed integers m and n, m, n ≥ 0, a set of productions, with context, of the type $(\beta, A, \gamma) \to \alpha$, where $|\beta| \leq m$, $|\gamma| \leq n$. This is read as: A is replaced by α if its left context is β and its right context is γ. The set of productions includes at least one production for all symbols in all contexts fulfilling the above conditions.

Although EOL-like reduction results can be obtained the main interest for EIL forms is the reduction of context.

<u>Theorem 5</u>: Let m, n > 0, then for each E(m,n)L form G an E(1,1)L form H can be constructed such that $\mathcal{L}(G) = \mathcal{L}(H)$.

Theorem 6: Let m, n be such that m > 0, then for each synchro- E(m,n)L form G a synchro- E(1,0) form H can be constructed such that $\mathcal{L}(G) = \mathcal{L}(H)$.
That Theorem 5 is the best possible follows from:

Theorem 7: There is an E(1,1)L form which has no form equivalent E(1,0)L or E(0,1)L form.

4.2 Completeness and vompleteness: An EOL form G is said to be $\mathcal{L}$-vomplete (very complete) if for all EOL forms F with $\mathcal{L}(F) \subseteq \mathcal{L}$, there exists G' ◅ G such that $\mathcal{L}(F) = \mathcal{L}(G')$.

In [CMO] the problem of EOL-completeness is tackled for short (that is, the right sides of productions have length at most two) {S,a}-forms (that is, EOL forms whose only nonterminal is S and only terminal is a). Even in this restrictive situation a complete classification is not forthcoming. For example, G_1: S → a; S → aS; a → a; a → S is EOL-complete via the chain-free normal form theorem of [CM], while G_2: S → a; S → aS; a → a; a → SS is not EOL-complete. It is not known whether G_3: S → a; S → S; S → Sa; S → aS; a → a; a → aS is EOL-complete or not.

Open problem: Classify all {S,a}-forms with respect to EOL-completeness.

A synchro-EOL form G is always assumed to have the production N → N for the universal blocking symbol and the productions a → N for all terminals in G.

In [MSW 11] it has been shown that it is decidable whether an {S,a}-synchro-EOL form is EOL-complete.

Open problem: If F is an arbitrary EOL-complete EOL form does this imply there is an a-restriction of F which is EOL-complete? Cf. [MSW10].

The decidability of EOL-completeness for unary synchro-EOL forms (that is, one terminal symbol) depends upon the following:

Open problem: If F is an arbitrary unary EOL grammar, is it decidable whether L(F) = a* or not?

Turning to EOL-vompleteness we have

H_1: S → a; S → SS; S → S; a → S

is not EOL-vomplete, because of Theorem 3 while H_2: S → λ; S → a; S → SS; S → S; a → S is EOL-vomplete. Recently it has been shown [AiM] that any EOL-vomplete form F = (V,Σ,P,S) must contain a production A → λ, where A is in V - Σ. Hence, in contrast to H_2,

H_3: S → a; S → S; S → SS; a → S; a → λ

is not EOL-vomplete.

4.3 <u>Generative capacity</u>: Here we are concerned with the language families generated by particular forms. Two of the most interesting results, both of which are non-trivial, are:

<u>Theorem 8</u>: There are no context-free-complete EOL forms [AM].

<u>Theorem 9</u>: There are context-free-complete EIL forms [MSW6] under CC-interpretations.

Recently in [MSW11] it has been shown that

<u>Theorem 10</u>: For all context-free grammar forms G such that $\mathcal{L}_s(G) \subseteq \mathcal{L}(LIN)$, there is an $\mathcal{L}_s(G)$-complete EOL form.

In other words, every sublinear s-grammatical family can be generated by an EOL form.

<u>Open problems</u>: (1) Is the converse true? (2) If G is a context-free grammar form such that $\mathcal{L}_s(G) \supseteq \mathcal{L}(LIN)$ then is it true that there is no $\mathcal{L}_s(G)$- complete EOL-form? [MR] have shown this to be the case for OL forms and the so-called <u>clean</u> EOL forms.

For a language family $\mathcal{L}$, we say that an EOL form G is $\mathcal{L}$-sufficient if $\mathcal{L}(G) \supseteq \mathcal{L}$. Similarly we say G is <u>$\mathcal{L}$-bounded</u> if $\mathcal{L}(G) \subseteq \mathcal{L}$. [AMR] investigate this notion for $\mathcal{L} = \mathcal{L}(CF)$ under a restricted form of interpretation, the <u>uniform interpretation</u>. They succeed in classifying {S,a}-EOL forms completely, hence for example; both

G_1: $S \to a$; $S \to SS$; $a \to a$ and G_2: $S \to a$; $S \to SS$; $a \to \lambda$

are context-free bounded, while

G_3: $S \to a$; $S \to SS$; $a \to a$; $a \to \lambda$

is not. Recently [AMO] have shown that it is decidable whether an OL form is regular-bounded.

<u>Open problems</u>: Investigate $\mathcal{L}$-sufficiency and $\mathcal{L}$-boundedness.

Let L be an EOL language and $\mathcal{L}$ a language family, $\mathcal{L} \subseteq \mathcal{L}(EOL)$. We say L is a <u>generator for $\mathcal{L}$</u> if for all synchro-EOL forms F such that L = L(F), F is $\mathcal{L}$-sufficient, i.e. $\mathcal{L} \subseteq \mathcal{L}(F)$. L is a <u>proper generator</u> if L is in $\mathcal{L}$.

It has been shown in [MSW7] that a* is a proper generator of $\mathcal{L}(REG)$ and a generator of $\mathcal{L}(FIN)$. It has also been shown that there is no generator for $\mathcal{L}(EOL)$.

<u>Open problem</u>: It is conjectured there is no generator for $\mathcal{L}(LIN)$. Prove or disprove this.

4.4 Decision problems: The major decidability result is:

Theorem 11 [CMORS]: Given two PDOL forms G_1 and G_2 it is decidable whether or not $\mathcal{L}(G_1) = \mathcal{L}(G_2)$.

The major undecidability results are:

Theorem 12 [MSW2]: Under uniform interpretations form-equivalence of EOL forms is undecidable.

Theorem 13 [MSW6]: For EIL forms, form-equivalence, CC-form-equivalence, completeness and CC-completeness are undecidable.

Open problems: (1) Is EOL-completeness for EOL forms decidable? (2) Is form-equivalence of EOL forms undecidable? (3) Is the equation L(F) = a* decidable for an arbitrary synchronized EOL grammar F? (4) Is it decidable whether an EOL form F is context-free-bounded (in the case of uniform interpretations it is)?

REFERENCES

[AiM] Ainhirn, W., and Maurer, H. A., On ε-productions for terminals in EOL forms, TU Graz (1979) IIG Report 25.

[AM] Albert J., and Maurer, H. A., The class of context-free languages is not an EOL family. Information Processing Letters 6,6(1978), 190-195.

[AMO] Albert, J., Maurer, H. A., Ottmann, Th., On sub-regular OL forms. TU Graz TR 15, 1978.

[AMR] Albert J., Maurer, H. A., and Rozenberg, G., Simple EOL forms under uniform interpretation generating CF languages. Fundamenta Informaticae(1979), to appear.

[CG] Cremers, A. B., and Ginsburg, S., Context-free grammar forms. Journal of Computer and System Sciences 11(1975), 86-116.

[CM] Culik II, K., and Maurer, H. A., Propagating chain-free normal forms for EOL systems. Information and Control 36(1978), 309-319.

[CMO] Culik II, K., Maurer, H. A., and Ottmann, Th., Two-symbol complete EOL forms. Theoretical Computer Science 6(1978), 69-92.

[CMORS] Culik II, K., Maurer, H. A., Ottmann, Th., Ruohonen, K., and Salomaa, A., Isomorphism, form equivalence and sequence equivalence of PDOL forms. Theoretical Computer Science 6(1978), 143-174.

[G] Ginsburg, S., A survey of grammar forms -- 1977. Acta Cybernetica 3, 4 (1978), 269-280.

[GS1] Ginsburg, S., and Spanier, E. H., Prime and coprime decomposition theorems for grammatical families. Manuscript (1979).

[GS2] Ginsburg, S., and Spanier, E. H., Manuscript in preparation (1979).

[L] Leipala, R., On context-free matrix forms. Dissertation, Mathematics Department, University of Turku, Finland, 1977.

[M] Maurer, H. A., New aspects of homomorphism. Invited paper, The 4th GI Conference on Theoretical Computer Science, Aachen, 1979.

[MPSW] Maurer, H. A., Pentonnen, M., Salomaa, A., and Wood, D., On non context-free grammar forms. Mathematical Systems Theory (1979), to appear.

[MR] Maurer, H. A., and Rozenberg, G., Sub-context-free L forms. IIG Report 22, TU Graz, Austria (1979).

[MSW1] Maurer, H. A., Salomaa, A., and Wood, D., EOL forms. Acta Informatica 8 (1977), 75-96.

[MSW2] Maurer, H. A., Salomaa, A., and Wood, D., Uniform interpretations of L forms. Information and Control 36(1978), 157-173.

[MSW3] Maurer, H. A., Salomaa, A., and Wood, D., ETOL forms. Journal of Computer and System Sciences 16(1978), 345-361.

[MSW4] Maurer, H. A., Salomaa, A., and Wood, D., On good EOL forms. SIAM Journal of Computing 7(1978), 158-166.

[MSW5] Maurer, H. A., Salomaa, A., and Wood, D., Relative goodness of EOL forms. Revue Francaise d'Automatique, Informatique et Recherche Operationelle 12 (1978).

[MSW6] Maurer, H. A., Salomaa, A., and Wood, D., Context-dependent L forms. Information and Control (1979), to appear.

[MSW7] Maurer, H. A., Salomaa, A., and Wood, D., On generators and generative capacity of EOL forms. Report 4, Institut für Informationsverarbeitung, TU Graz, (1978) and McMaster University Report 78-CS-1.

[MSW8] Maurer, H. A., Salomaa, A., and Wood, D., Syncrhonized EOL forms under uniform interpretations. Report 78-CS-10, McMaster University, Hamilton, Canada (1978).

[MSW9] Maurer, H. A., Salomaa, A., and Wood, D., Strict context-free grammar forms: completeness and decidability. Report 78-CS-19, McMaster University, Hamilton, Canada (1978).

[MSW10] Maurer, H. A., Salomaa, A., and Wood, D., Context-free grammar forms with strict interpretations. Journal of Computer and System Sciences (1979), to appear.

[MSW11] Maurer, H. A., Salomaa, A., and Wood, D., Synchronized EOL forms. Report 79-CS-3, McMaster University, Hamilton, Ontario, Canada (1979).

[RS] Rozenberg, G., and Salomaa, A., The Mathematical Theory of L Systems. In preparation (1979).

[S] Salomaa, A., Formal Languages. Academic Press, New York (1975).

[Sk] Skyum, S., On good ETOL forms. Theoretical Computer Science, (1978), to appear.

[W1] Wood, D., A bibliography of grammatical similarity, available from the author.

[W2] Wood, D., Grammar and L Forms. In preparation (1979).

ACKNOWLEDGEMENT:

Work supported by a Natural Sciences and Engineering Research Council of Canada Grant No. A-7700 and encouraged by Seymour Ginsburg, Hermann Maurer and Arto Salomaa.

A THEORETICAL STUDY ON THE TIME ANALYSIS OF PROGRAMS

Akeo Adachi
Department of Academic and Scientific Programs, IBM Japan
Roppongi, Minato-ku, Tokyo, 106 Japan
Takumi Kasai
IBM Thomas J. Watson Research Center, P.O. Box 218
Yorktown Heights, N.Y. 10598, U.S.A.
Etsuro Moriya
Department of Mathematics, Tokyo Woman's Christian University
Zenpukuji, Suginami-ku, Tokyo, 167 Japan

0. Introduction

Meyer and Ritchie [1][2] introduced the notion of loop programs and classified the primitive recursive functions syntactically with the help of loop programs into the hierarchy $L_0 \subset L_1 \subset L_2 \subset \ldots$ by restricting the depth of loop nesting, i.e., L_n is the class of functions computed by loop programs whose depth of loop nesting is not greater than n. They also showed that each class L_n can be characterized by computational complexity, measured by the amount of time required on loop programs to compute the functions. In particular, a function is elementary in the sense of Kalmár [3], i.e., belongs to L_2 if and only if it can be computed by a loop program whose computing time is bounded by a k-fold exponential function of its inputs for some k, while there is general agreement that those computations which are exponentially difficult in time are practically intractable. In this respect, many people feel that even the class L_2 is still too inclusive in the sense of "practical computation".

In the earlier paper [4] the authors attempted an investigation to obtain a substantial subclass of L_2 reflecting practical computation, where the notion of loop programs is extended so as to include additional types of primitive statements such as $x \leftarrow x \dot{-} 1$ and IF-THEN-ELSE, and the use of arrays is allowed as well. It is proved that if such an extended loop program satisfies a certain syntactical restriction called "simpleness", then the computing time of the program is bounded by a polynomial of its inputs whose degree can be effectively determined only by the depth of loop nesting. This is worthy of notice, since it says that we can know syntactically a *practical* estimation of the time required to execute a given "simple loop program" *before* execution.

Here in this paper, on the basis of our earlier work [4], we make a rather precise analysis of time complexity of simple loop programs. We present an algorithm which gives an accurate upper bound of the computing time of any given simple loop program; a slight modification of the algorithm gives a lower bound.

1. Simple Loop Programs

First we review a basic result from Kasai & Adachi [4], being the starting point of our investigation.

Definition. Let C, S and A be fixed mutually disjoint countable sets of symbols. An element of C, S or A is called a *control variable*, *simple variable* or *array name*, respectively. Let Var denote the set of all variables, that is,

$$Var = C \cup S \cup \{ a[i] \mid a \in A,\ i \in N \},$$

where $N = \{ 0, 1, 2, \ldots \}$. A *loop program* is a statement over Var defined recursively as follows, where

$$\bar{A} = \{ a[i] \mid a \in A,\ i \in C \cup S \},$$

$u \in C \cup S \cup \bar{A}$, $v \in S \cup \bar{A}$, $c \in N$ and $w_1, w_2 \in C \cup S \cup \bar{A} \cup N$:

<atomic statement> ::= $u \leftarrow u+1 \mid v \leftarrow v \dot{-} 1 \mid v \leftarrow u \mid u \leftarrow c$
<loop statement> ::= LOOP x DO <statement> END
<condition> ::= $w_1 = w_2 \mid w_1 \neq w_2$
<statement> ::= <atomic statement> | <loop statement> |
IF <condition> THEN <statement> ELSE <statement> END |
<statement>;<statement>

Definition. A function $d: Var \to N$ is called a *memory*. We denote the set of memories by D. Let P be a loop program, then P realizes the partial function $\bar{P}: D \to N$. The *time complexity* of P is the function $time_P: D \to N$ such that $time_P(d)$ is the number of atomic statements executed by P under an initial memory d. The definition of $\bar{P}$ and $time_P$ is straightforward so that we omit the details.

Definition. For each loop program P, we define the relation $>_P$ on C_P as follows, where C_P denotes the set of control variables appearing in P.

We write $x >_P y$ if and only if the program P includes a statement of the form LOOP x DO Q END, and Q includes $y \leftarrow y+1$.

We say that P is *simple* if there is no sequence of control variables x_1, x_2, ..., x_k, $k > 1$, such that

$$x_1 >_P x_2 >_P \cdots >_P x_k$$

and $x_1 = x_k$.

Thus, if P is simple, then the transitive closure of $>_P$ is a partial order on C_P.

Definition. Let P be a loop program. Let s be an occurrence of an atomic statement in P. Then the *depth* of s, denoted by $\delta(s)$, is the number of loop statements which include s. Let $s_1, s_2, \ldots, s_k$ be the occurrences of the atomic statements in P such that $\delta(s_i) \geq 1$, $1 \leq i \leq k$. Then the *loop complexity* of P is $\delta(s_1)\cdot\delta(s_2)\cdot\ldots\cdot\delta(s_k)$ and is denoted by $\ell c(P)$. If P contains no atomic statements whose depth ≥ 1, then $\ell c(P) = 0$.

Example 1. Consider the program P defined as follows:

```
z ← 0;
LOOP x DO
   LOOP y DO  z ← z+1  END;
   LOOP x DO
      LOOP x DO  y ← y+1  END
   END
END
```

Obviously $x >_P y >_P z$ so that P is simple, and $\ell c(P) = 2\times 3 = 6$. If d is the memory defined by $d(x) = c$ and $d(w) = 0$ for each w other than x, then

$$\bar{P}(d)(x) = c,\ \bar{P}(d)(y) = c^3,\ \bar{P}(d)(z) = c^3(c-1)/2$$

and

$$\mathrm{time}_P(d) = c^3(c+1)/2 + 1.$$

Theorem 1. (Kasai and Adachi [4]) *Let* P *be a simple loop program. Then there exist constants* c *and* c' *such that*

$$\mathrm{time}_P(d) \leq c\cdot(\max_{x\in C_P} d(x))^{\ell c(P)} + c'$$

for all d *in* D.

2. A Precise Analysis

By Theorem 1, we can get a rough upper bound of the time complexity of any given simple loop program. For example,

$$\mathrm{time}_P(d) \leq c\cdot n^6 + c'$$

for the program P in Example 1, where $d \in D$ and $n = \max_{z\in C_P} d(z)$. However, as will be shown in this section, we can construct an algorithm which gives a more precise upper bound:

$$\mathrm{time}_P(d) \leq d(x)^4 + d(x)^3 + d(x)\cdot d(y) + 1.$$

By an analogous algorithm, it is also possible to get a lower bound:

$$m(m^3 + m^2 + d(y) \dot{-} 1) + 1 \leq \mathrm{time}_P(d),$$

where $m = d(x) \dot{-} 1$.

Definition. Let P be a loop program. We define a *term* as follows, where $c \in N$ and $x \in C$:

<term> ::= c | x | <term> + <term> | <term> $\dot{-}$ <term> |
<term>·<term> | max{<term>, <term>} |
min{<term>, <term>}

For all terms t, t_1 and t_2, and each d in D, we define $d(t)$ as follows:

$$d(c) = c, \qquad c \in N,$$
$$d(t_1 + t_2) = d(t_1) + d(t_2); \quad d(t_1 \dot{-} t_2) = d(t_1) \dot{-} d(t_2),$$
$$d(\max\{t_1, t_2\}) = \max\{d(t_1), d(t_2)\},$$
$$d(\min\{t_1, t_2\}) = \min\{d(t_1), d(t_2)\}.$$

Let P be a simple loop program. Here we consider an algorithm to obtain, for each $z \in C_P$, a term t_z which satisfies

(2.1) $\quad \bar{P}(d)(z) \leq d(t_z)$ for all $d \in D$.

Definition. Let P be a simple loop program, and let s be a control variable not appearing in P. Then the *step counting version*, $\hat{P}$, of P is defined by

$$s \leftarrow 0;\ Q$$

where Q is the loop program obtained from P by inserting $s \leftarrow s+1$ immediately after each occurrence of atomic statements of P. The variable is called the *step counting variable*.

Since $\mathrm{time}_P(d) = \bar{\hat{P}}(d)(s)$, in order to obtain an upper bound of the time complexity of P, it suffices to apply the following algorithm to its step counting version $\hat{P}$ with s as the step counting variable. Note that $\hat{P}$ is simple whenever P is simple.

Algorithm 1.

Input: A simple loop program P and a control variable $z \in C_P$.

Output: A term t_z which satisfies (2.1).

Method: Let $C_P = \{x_1, x_2, \ldots, x_n\}$. For each substatement Q of P, we construct an n-tuple of terms $T(Q) = (t_1, t_2, \ldots, t_n)$ which satisfies the following three conditions:

(2.2) $\bar{Q}(d)(x_i) \leqq d(t_i)$ for all $d \in D$ and $1 \leqq i \leqq n$.

(2.3) t_i is a term of the form $\max\{u, x_i + v\}$, where u and v are terms (possibly empty) which contain no occurrence of x_i.

(2.4) If t_i contains an occurrence of x_j $(i \neq j)$, then $x_j >_P x_i$.

If t_i is of the form $\max\{w\}$, then we simply write w instead of $\max\{w\}$.

Case 1: Suppose that Q is an atomic statement. Then $T(Q) = (t_1, \ldots, t_n)$ is defined as follows:

Case 1.1: If Q is $x_\ell \leftarrow x_\ell + 1$, then let

$$t_i = \begin{cases} x_i & \text{if } i \neq \ell \\ x_\ell + 1 & \text{if } i = \ell. \end{cases}$$

Case 1.2: If Q is $x_\ell \leftarrow c$, $c \in N$, then let

$$t_i = \begin{cases} x_i & \text{if } i \neq \ell \\ c & \text{if } i = \ell. \end{cases}$$

Case 1.3: If Q is an atomic statement other than above, then let

$$t_i = x_i, \quad 1 \leqq i \leqq n.$$

Case 2: Suppose that Q is IF q THEN Q_1 ELSE Q_2 END. Assuming that $T(Q_1) = (t_{11}, \ldots, t_{1n})$ and $T(Q_2) = (t_{21}, \ldots, t_{2n})$ have already been obtained, define $T(Q) = (t_1, \ldots, t_n)$ by

$$t_i = \max\{\max\{u_{1i}, u_{2i}\}, x_i + \max\{v_{1i}, v_{2i}\}\}, \quad 1 \leqq i \leqq n,$$

where $t_{ji} = \max\{u_{ji}, x_i + v_{ji}\}$ for $j = 1, 2$.

Case 3: Suppose that Q is $Q_1;Q_2$. Assume that $T(Q_1) = (t_{11}, \ldots, t_{1n})$ and $T(Q_2) = (t_{21}, \ldots, t_{2n})$, then t_i $(1 \leqq i \leqq n)$ is obtained from t_{2i} by replacing each occurrence of x_j, $1 \leqq j \leqq n$, in t_{2i} by t_{1j}.

Case 4: Suppose that Q is LOOP x_ℓ DO Q_1 END. Let $T(Q_1) = (t'_1, \ldots, t'_n)$. We calculate t_i, $1 \leqq i \leqq n$, in the descending order with respect to $>_P$, that is, we calculate t_j before t_i if $x_j >_P x_i$. Assume that t_j has been obtained for all j such that $x_j >_P x_i$. Let

$$t'_i = \max\{u, x_i + v\}.$$

Then t_i is defined by

$$t_i = \max\{\bar{u} + x_\ell \cdot \bar{v}, x_i + x_\ell \cdot \bar{v}\},$$

where $\bar{w}$ denotes the term obtained from w by replacing the variables x_j, $1 \leqq j \leqq n$, appearing in w by t_j. (Note that if t_i' contains x_j, then $x_j >_P x_i$ from (2.4), and thus t_j must have been obtained.)

Example 2. Consider the following program P:

```
y ← 0; v ← 0; w ← 0;
LOOP x DO
   IF A[y] = 0 THEN
      w ← 100;
      LOOP y DO  v ← v+1  END
   ELSE
      LOOP v DO  w ← w+1  END
   END;
   LOOP w DO  B[w] ← w  END;
   y ← y+1; v ← v+1
END
```

Then $x >_P y >_P v >_P w$. Clearly P is simple. Insert the statement $s \leftarrow s+1$ immediately after each occurrence of the atomic statements in P to calculate $T(\hat{P})$ according to Algorithm 1. We obtain

$$T(\hat{P}) = (t_x, t_y, t_v, t_w, t_s), \text{ where}$$

$$t_x = t_y = x,\ t_v = x(x+1),\ t_w = 100+x^2(x+1),$$

and the desired term t_s is $x(x(x+1)+(100+x^2(x+1)+x(x+1))+2)+3$. Thus we have

$$\mathrm{time}_P(d) \leqq d(x^4 + 3x^3 + 2x^2 + 102x + 3).$$

The correctness of the algorithm is stated in the following theorem, the proof of which is by induction on the syntactical structure of input program P. It is enough to prove that each substatement Q of P satisfies the conditions (2.2) to (2.4).

Theorem 2. For any simple loop program P, *Algorithm 1 terminates and satisfies (2.1). Thus we can obtain an upper bound of the time complexity of* P.

To obtain a lower bound of the time complexity of P, in the description of Algorithm 1, replace every occurrence of the word "max" by "min", and, in addition, modify Case 4 as follows:

Define t_i by

$$t_i = \min\{\bar{u},\ x_i + (x_\ell \dot{-} 1)\cdot\bar{v}\}$$

Then we have *Algorithm 2* for which we have

Theorem 3. *Algorithm 2 terminates and satisfies*

$$d(t_z) \leqq \bar{P}(d)(z) \quad \text{for all} \quad d \in D \ ,$$

if P *is a simple loop program.*

3. Concluding Remarks

We described algorithms which calculate syntactically an upper and a lower bound of the time complexity of programs in a restricted class. Note that, in addition, it is possible to test if a subscripted variable is beyond the scope of an array. For example, by using $T(P) = (t_1, \ldots, t_n)$ obtained by Algorithm 1, we have

$$d'(x_i) \leqq d(t_i)$$

for any memory d' at any arbitrary instance in the computation of P under an initial memory d. In other words, the content of the variable x_i does not exceed the value of t_i during the computation.

We think it valuable to apply this theory to practical programming languages. Then, for example, it should be necessary to extend the condition for simpleness, and the syntax must be loosened enough to be of practical use.

References

1. Meyer, A.R. and D.M.Ritchie, The complexity of loop programs. Proc. 22nd ACM National Meeting (1967), 465-469.
2. Meyer, A.R. and D.M.Ritchie, Computational complexity and program structures. IBM research paper RC-1817, Watson Research Center, 1967.
3. Kalmár, L., Egyszerű példa eldönthetetlen aritmetikai problémára. Matematikai és Fizikai Lapok 50 (1943), 1-23.
4. Kasai, T. and A.Adachi, A characterization of time complexity by simple loop programs. Technical report RIMS-250, Research Institute for Math. Sci., Kyoto Univ., 1978.
5. Brainerd, W.S. and L.H.Landweber, Theory of Computation. Wiley, New York, 1974.

COMPLETENESS PROBLEMS IN VERIFICATION OF PROGRAMS AND PROGRAM SCHEMES

H. Andréka
I. Németi
I. Sain

Mathematical Institute of the
Hungarian Academy of Sciences
Budapest, Reáltanoda u. 13-15
H-1053 Hungary

Abstract: Thm 1 states a negative result about the classical semantics $\models^{\omega}$ of program schemes. Thm 2 investigates the reason for this. We conclude that Thm 2 justifies the Henkin-type semantics $\models$ for which the opposite of the present Thm 1 was proved in Andréka-Németi[1],[2],[3] and also in a different form in part III of Gergely-Ury[8]. The strongest positive result on $\models$ is Corollary 6 in Andréka-Németi[3].

Basic concepts

First we recall some basic notions and notations from textbooks on Logic Monk[10], Chang-Keisler[5] and from Program Schemes Theory, e.g. Manna[9], Andréka-Németi[1],[2],[3], Gergely-Ury[8].

ω denotes the set of natural numbers.

d denotes an arbitrary similarity type. I.e.: d correlates arities to some fixed function symbols and relation symbols. See Sacks[12], p.11.

$Y = \{ y_z : z \in \omega \}$ denotes the set of variable symbols.

F_d is the set of all classical first order formulas of type d with variables in Y . See Chang-Keisler[5], p.22.

M_d is the class of all classical first order <u>models</u> of type d . See Chang-Keisler[5] or Monk[10], Def.11.1. or Sacks[12], p.11.

$\models \subseteq F_d \times M_d$ is the usual validity relation. See Chang-Keisler[5] or Sacks[12], p.21.

τ denotes a <u>term</u> of type d in the usual sense of first order logic, see Chang-Keisler[5], p.22 or Monk[10], p.166.Def.10.8.(ii).

$\underset{\sim}{D}$ and $\underset{\sim}{E}$ denote elements of M_d , the universes of which are D and E respectively.

P_d denotes the set of <u>program schemes</u> of type d . P_d is defined as in Manna[9], Andréka-Németi[1],[2], Gergely-Ury[8], p.72. E.g., let t be the similarity type of arithmetic. Then the following sequence is in P_t , i.e. it is a program scheme of type t :

$$\langle (0: y_0 \leftarrow 0), \; (1: \text{IF } y_0=y_1 \text{ THEN } 4), \; (2: y_0 \leftarrow y_0+1), \; (3: \text{IF } y_1=y_1 \text{ THEN } 1), \; (4: \text{HALT}) \rangle$$

$P_d \times F_d$ is the set of <u>output statements</u> about programs. An output statement $(p,\psi) \in P_d \times F_d$ means intuitively that the program scheme p is partially correct w.r.t. output condition ψ .

$\underset{\sim}{D} \overset{\omega}{\models} (p,\psi)$ is meaningful if $\underset{\sim}{D} \in M_d$ and $(p,\psi) \in P_d \times F_d$. Now $\underset{\sim}{D} \overset{\omega}{\models} (p,\psi)$ holds if the program scheme p is partially correct w.r.t. ψ in the model $\underset{\sim}{D}$. I.e.: If p is started in $\underset{\sim}{D}$ with any <u>input</u> $q:\omega \longrightarrow D$ then whenever p <u>halts</u> with some output $k:\omega \longrightarrow D$, the formula ψ will be true in $\underset{\sim}{D}$ under the valuation k of its free variables, i.e. $\underset{\sim}{D} \models \psi[k]$. See Manna [9], Chapter 4 . Note that a precise definition of $\overset{\omega}{\models}$ would strongly use the structure $\langle \omega, \leq \rangle$ of <u>natural numbers</u>. See [14], Gergely-Ury[8], p.78, Andréka-Németi[1], p.116, [2], [3]. The letter ω above the sign $\overset{\omega}{\models}$ serves to remind us of this fact.

For any set $Th \subseteq F_d$ of formulas, " $Th \overset{\omega}{\models} (p,\psi)$ " is defined in the usual way:

$$\mathrm{Th} \models^{\omega} (p,\psi) \quad \text{iff} \quad (\forall D \in M_d)\left[\underset{\sim}{D} \models \mathrm{Th} \Rightarrow \underset{\sim}{D} \models^{\omega} (p,\psi) \right] \quad .$$

PROPOSITION 0 :
Let the type d contain the similarity type of successor arithmetic $\langle \omega, s, 0 \rangle$. Let $\mathrm{Th} \subseteq F_d$ be such that

$$\mathrm{Th} \supseteq \{ s^z 0 \neq s^r 0 \ : \ z < r \in \omega \} \overset{d}{=} \mathrm{Th}'$$

where $s^0 0 \overset{d}{=} 0$ and $s^{r+1} 0 \overset{d}{=} s s^r 0$ for $r \in \omega$.
Let $\underset{\sim}{E} \in M_d$ be an arbitrary but fixed model of Th' such that $\underset{\sim}{E} =$ $= \langle \omega, s, 0, \ldots \rangle$.
Suppose H is an arbitrary set such that

$$\{ (p,\psi) \ : \ \mathrm{Th} \models^{\omega} (p,\psi) \ \} \supseteq H \supseteq$$

$$\{ (p,\psi) \ : \ \mathrm{Th} \models^{\omega} (p,\psi) \text{ and } p \text{ \underline{terminates} in } \underset{\sim}{E} \text{ for every input and } \psi \text{ is quantifier-free} \} .$$

Then H is not recursively enumerable.

Proof: The present proposition is a special case of Thm 1 to be formulated later.
QED

Now we turn to relax the conditions made on d and Th in the above proposition. I.e. we are going to generalize Proposition 0 .

From now on c and τ denote arbitrary terms of type d such that c contains no variable and τ contains one variable y_0 . To make this explicit, we write $\tau(y_0)$.

Notation: $\tau^0 \overset{d}{=} c$, and $\tau^{z+1} \overset{d}{=} \tau(\tau^z)$ for every $z \in \omega$.

Note that the terms τ^0 , τ^1 , ... , τ^z , ... contain no variable.

DEFINITION 1 :

$Th \subseteq F_d$ is said to be good if there exist terms c and $\tau(y_o)$ such that

$$Th \supseteq \{ \tau^z \neq \tau^r : z < r \in \omega \} \overset{d}{=} Th' .$$

Let $\underset{\sim}{E} \in M_d$ be an arbitrary model of Th' such that

$(\forall b \in E)(\exists z \in \omega)$ $[\tau^z$ denotes b in $\underset{\sim}{E}]$. Then we define

$$May(Th) \overset{d}{=} \{ (p,\psi) \in P_d \times F_d : Th \overset{\omega}{\models} (p,\psi) \} .$$

$Must(Th) \overset{d}{=} \{ (p,\psi) \in May(Th) :$ p terminates in $\underset{\sim}{E}$ for every input; and ψ is an atomic formula or the negation of an atomic formula such that $Th \vdash \exists y_o \psi$ $\}$.

Remark: To a fixed Th , Must(Th) is not unique since it may depend on the choice of c , $\tau(y_o)$, and $\underset{\sim}{E}$. This makes the following theorem even stronger since it will hold for any choice of c , τ , and $\underset{\sim}{E}$. Observe that Must(Th) is a reasonably small set of output statements since ψ contains no quantifiers, no "$\vee$" or "$\wedge$" and at the same time p is such that it terminates in $\underset{\sim}{E}$ for every input. Thus Must(Th) contains no tricky statement about the "halting problem" (since p has to terminate) and no "strange sentence" since ψ has to be simple; moreover, $\exists y_o \psi$ is provable from Th .

THEOREM 1 :

Let d be arbitrary and let $Th \subseteq F_d$ be good (in sense of Def.1.) and consistent. Let H be an arbitrary set such that

$$May(Th) \supseteq H \supseteq Must(Th) .$$

Then H is not recursively enumerable.

Proof: We shall treat the constant-term "c" as zero and $\tau(y_o)$ as the successor function. E.g. τ^z will be considered to be the name of the natural number $z \in \omega$. By using successor τ and zero c we can write programs "add" $\in P_d$ and "mult" $\in P_d$ for addition and multiplication. By using these programs, for an arbitrary Diophantine equation $e(y_2,\dots,y_m)$ we can write a program $\bar{e} \in P_d$ such that after having executed $\bar{e}$ we shall have $y_0=y_1$ iff $e(y_2,\dots,y_m)$ was true before starting $\bar{e}$.

Let p be an m-variable version of the program scheme given as an example at the beginning of this paper. Namely, p starts with

$(0: y_{m+1} \leftarrow c)$, $(1:$ IF $y_2=y_{m+1}$ THEN $4)$, $(2: y_{m+1} \leftarrow y_{m+1})$, $(3:$ IF TRUE THEN $1)$, $(4: y_{m+1} \leftarrow c)$, $(5:$ IF $y_3=y_{m+1}$ THEN $8)$, ...

This program p terminates iff all the initial values of $y_2,\dots,y_m$ can be reached from "c" by finitely many applications of τ. Now, by writing $\bar{e}$ after p we obtain a program $p\bar{e} \in P_d$ which first checks whether $y_2,\dots,y_m$ can be reached from "c" by applications of τ and if yes then results $y_0=y_1$ if $e(y_2,\dots,y_m)$ was true, $y_0 \neq y_1$ if $e(y_2,\dots,y_m)$ was false for the initial values. Now to each Diophantin equation $e(\bar{y})$ correlate $\bar{\bar{e}} = (p\bar{e},\ y_0 \neq y_1)$.

Clearly $\bar{\bar{e}} \in P_d \times F_d$. Also $Th \overset{\omega}{\models} \bar{\bar{e}}$ iff e has no solution in the standard model $\langle \omega,+,\cdot,0,1 \rangle$ of arithmetic. If there were a recursively enumerable H as in the statement of the present theorem then

$$Eq \overset{d}{=} \{ e \in \text{"Diophantine equations"} : Th \overset{\omega}{\models} \bar{\bar{e}} \}$$

would be recursively enumerable since the construction of $\bar{\bar{e}}$ from e was "constructive". But, since Hilbert's tenth problem is unsolvable (Davis[6] or Monk[10]), this is impossible.

<u>QED</u>

The following theorem says that if one "avoids Logic" and proves properties of programs by using "Mathematics in general" then this will <u>not help</u> one to avoid the "shortcoming" formulated in Thm 1.

<u>THEOREM 2</u>:

Let the real world $\langle V,\in \rangle \models$ ZFC of Set Theory (see Devlin[7], p.3, line 4 from below or Chang-Keisler[5], p.476) be fixed. I.e.: V is the class of all sets and $\in$ is the "element of" relation between them.

Then the following is true:

There exist

- a similarity type d , and
- a model $\langle W,E\rangle \models ZFC$ of Set Theory inside of $\langle V,\in\rangle$ (i.e. $\langle W,E\rangle$ is an element of V and $\langle W,E\rangle \models ZFC$ is true inside of $\langle V,\in\rangle$, see Devlin[7], p.14, line 6)

such that (i) and (ii) below hold.

(i) There are a finite set $Th \subseteq F_d$ of axioms and an output statement (p,ψ) such that

$Th \overset{\omega}{\models} (p,\psi)$ is true, but inside of $\langle W,E\rangle$ we have
$Th \overset{\omega}{\not\models} (p,\psi)$.

More precisely:

$\langle V,\in\rangle \models$ " $Th \overset{\omega}{\models} (p,\psi)$ " but

$\langle W,E\rangle \models$ " $Th \overset{\omega}{\not\models} (p,\psi)$ " .

(Observe that " $Th \overset{\omega}{\models} (p,\psi)$ " is a statement of the language of ZFC.)

(ii) There is an output statement (p,ψ) such that

$\langle V,\in\rangle \models$ " $M_d \overset{\omega}{\models} (p,\psi)$ " while

$\langle W,E\rangle \models$ " $M_d \overset{\omega}{\not\models} (p,\psi)$ " .

As a <u>contrast</u> we note that:
For all $\varphi \in F_d$ and for every model $\langle W,E\rangle \in V$ of ZFC ,

$\langle V,\in\rangle \models$ " $M_d \models \varphi$ " implies $\langle W,E\rangle \models$ " $M_d \models \varphi$ " .

<u>Proof</u>:

(i)

Let $d \overset{d}{=} \{ \langle 0,0\rangle , \langle s,1\rangle \}$.

Let Th consist of the following two axioms:

$\forall y\,(sy \neq 0)$

$\forall y_1 \; \forall y_2 \; (sy_1 = sy_2 \rightarrow y_1 = y_2)$.

We know that Hilbert's tenth problem is unsolvable. This implies the existence of a Diophantine equation $e(\bar{y})$ such that the set theoretic formula

$$\text{"}\langle \omega, s, +, \cdot, 0, 1\rangle \models \exists \bar{y}\; e(\bar{y})\text{"}$$

is false in $\langle V,\in\rangle$ but is true in $\langle W,E\rangle$ for some model $\langle W,E\rangle \in V$ of ZFC .

Now let the output statement $\bar{\bar{e}} = (p\bar{e}\ ,\ y_0 \neq y_1)$ be the one defined in the proof of Thm 1. There it was observed that

$$Th \models^{\omega} \bar{\bar{e}} \quad \text{iff} \quad \langle \omega,+,\cdot,0,1 \rangle \not\models \exists \bar{y}\ e(\bar{y}) \quad .$$

(Note that the present Th satisfies the conditions of Thm 1 .)

Thus $\langle V,\in \rangle \models$ " $Th \models^{\omega} \bar{\bar{e}}$ " and $\langle W,E \rangle \models$ " $Th \not\models^{\omega} \bar{\bar{e}}$ " .

(ii)

The proof of (ii) is an easy modification of the proof of (i) above. Namely, let us choose the above e , $\langle W,E \rangle$, and $\bar{\bar{e}} = (p\bar{e},\ y_0 \neq y_1)$. Let φ be the conjunction of all elements of Th . (Note that Th is finite and therefore $\varphi \in F_d$.) Let $\psi \stackrel{d}{=} (\varphi \rightarrow y_0 \neq y_1)$. Now,

$\langle V,\in \rangle \models$ " $M_d \models^{\omega} (p\bar{e},\psi)$ " while

$\langle W,E \rangle \models$ " $M_d \not\models^{\omega} (p\bar{e},\psi)$ " .

<u>QED Thm 2</u> (For a more detailed proof cf. Andréka-Németi-Sain[4].)

The above Thm 2 says that something is wrong with the classical semantics (or model theory) $\models^{\omega}$ of program schemes. Namely: There exists a good program (p,ψ) which is <u>not provable</u> by mathematics, i.e. the goodness of (p,ψ) is not "a mathematical truth" i.e. it is not implied by ZFC despite of the fact that it happens to be the case that (p,ψ) <u>is</u> good. See Németi-Sain[11]Def.2 and Andréka-Németi-Sain[4] about " $Th \models^{\omega} (p,\psi)$ " -s being a formula of Set Theory. This way Thm 2 supports the Henkin-type semantics introduced in Andréka-Németi [1]-[3], the consequence concept $(Th \models (p,\psi))$ of which does <u>not</u> have the above shortcoming.

By Thm 2 above there exists an output statement (p,ψ) which is valid, i.e. " $\models^{\omega} (p,\psi)$ ", but the validity of which is <u>not</u> a mathematical truth, i.e. ZFC $\not\models$ " $\models^{\omega} (p,\psi)$ " . A semantics with this paradoxical property was called <u>instable</u> in Andréka-Németi-Sain[4]. It was proved in [4] that any "reasonable" semantics has to be stable. Indeed, the Henkin-type semantics introduced in Andréka-Németi[1]-[3] was proved to be <u>stable</u> there.

On basis of Thm 2 above an effective inference system for program correctness was given in Andréka-Németi-Sain[4] such that if (p,ψ) cannot be proved then there exists a model of ZFC Set Theory in which the program p <u>is actually not</u> correct w.r.t. ψ . Cf. Andréka-Németi [1]-[3], too.

A HENKIN TYPE SEMANTICS FOR PROGRAM SCHEMES

Now to every classical (one-sorted) similarity type d we define an associated <u>3-sorted similarity</u> type td . About many-sorted logic and its model theory see Monk[10], p.483.

As before, d is an arbitrary type. Let t denote the similarity type of <u>Peano Arithmetic</u> and let t be disjoint from d . The type td is defined as follows:

<u>There are 3 sorts of td</u> : $\bar{t}$, $\bar{d}$, $\bar{i}$ called "time , "data" , and "intensions" respectively.

<u>The operation symbols of td</u> are the following: The operation symbols of t , the operation symbols of d , and an additional operation symbol "ext" .

<u>The sorts (or "arities") of the operation symbols of td</u> : The operation symbols of t go from sort $\bar{t}$ to sort $\bar{t}$. The operation symbols of d go from sort $\bar{d}$ to sort $\bar{d}$. The operation symbol "ext" goes from sort $(\bar{i},\bar{t})$ to sort $\bar{d}$.: I.e. "ext" has two arguments, the first is of sort $\bar{i}$, the second is of sort $\bar{t}$, and the result or value of "ext" is of sort $\bar{d}$. Now the definition of the 3-sorted type td is completed.

$TL_d = \langle TF_d , TM_d , \models \rangle$ denotes the 3-sorted language <u>of type td</u> , see Monk[10], p.483. In more detail:

(i) TM_d is the class of all <u>models of type td</u> , see Monk [10], Def.29.27. I.e. a model $\mathfrak{M} \in TM_d$ <u>has</u>

1. <u>three universes</u> throughout denoted by T , D , and I , of sort $\bar{t}$, $\bar{d}$, and $\bar{i}$ respectively.
2. Operations "$T^n \to T$" originating from the type t , operations "$D^n \to D$" originating from the type d , and an operation $ext: I \times T \to D$.

Roughly speaking, we could say that $\mathfrak{M}$ consists of structures $\underset{\sim}{T} \in M_t$, $\underset{\sim}{D} \in M_d$, and an additional operation $ext: I \times T \to D$.

Therefore we shall use the sloppy notation:

$\mathfrak{M} \overset{d}{=} \langle \underset{\sim}{T}, \underset{\sim}{D}, I, ext \rangle$ for elements of TM_d .

(ii) TF_d is the set of first order (3-sorted) formulas of type td . Roughly speaking, we can say that F_t and F_d are contained in TF_d , and there are additional terms of the form "ext(y,τ)", where τ is a term of type t and y is a <u>variable of sort $\bar{i}$</u> . Further "ext(y,τ)" is defined to be a term of sort $\bar{d}$.

(iii) $\models \subseteq (TM_d \times TF_d)$ is the usual, see Monk[10], p.484.

Now we define the <u>meanings of program schemes</u> $p \in P_d$ in the 3-sorted models $\mathfrak{M} \in TM_d$. Let $p \in P_d$ be a fixed program scheme. Let $y_1, \ldots, y_m$ be the variables occurring in p . Let $\mathfrak{M} \in TM_d$ be fixed. Recall that I is the universe of sort $\bar{i}$ of $\mathfrak{M}$.

A <u>trace</u> of p in $\mathfrak{M}$ is a sequence $\langle s_0, \ldots, s_m \rangle \in {}^{(m+1)}I$ of elements of I satisfying (*) below. (I.e. a trace of p in $\mathfrak{M}$ is a sequence of i-sorted elements of $\mathfrak{M}$.) To formulate (*), observe that if $s \in I$ then "ext(s,-)" is a function $\langle ext(s,z) : z \in T \rangle$ from T into D . We shall use y_0 as <u>"the control-variable"</u> of p . I.e. $ext(s_0,z)$ is considered to be the "value of the control or execution" at time point z . Thus "$ext(s_0,z)$" is supposed to be a "<u>label</u>" in the program scheme p .

(*) The sequence $\langle ext(s_0,-), \ldots, ext(s_m,-) \rangle$ of functions should be a <u>history of an execution</u> of p in $\underset{\sim}{D}$ along the "time axis" $\underset{\sim}{T}$.

The only difference from the classical definition (cf. Manna[9], Andréka-Németi[1]-[3], Gergely[14],[8]) of a trace of p in $\underset{\sim}{D}$ is that now the "time-axis" of execution is not necessarily $\langle \omega, s, +, \cdot, 0, 1 \rangle$ but, instead, it is $\underset{\sim}{T}$.
Condition (*) above can be made precise by replacing ω with $\underset{\sim}{T}$ in the classical definition, see Andréka-Németi[1]-[3], Gergely-Ury[8].

The trace $\langle s_0, \ldots, s_m \rangle$ of p in $\mathfrak{M}$ terminates if $ext(s_0,z)$ is the label of the HALT statement, for some $z \in T$. If the trace $\langle s_0, \ldots, s_m \rangle$ terminates at time $z \in T$ then its <u>output</u> is

$\langle ext(s_1,z), \ldots, ext(s_m,z)\rangle$. Now we define for $\psi \in F_d$:

$\mathfrak{M} \models (p,\psi)$ holds iff for every <u>terminating</u> trace of p in $\mathfrak{M}$ the output satisfies ψ in $\underset{\sim}{D}$. Cf. Def.8 of Szőts-Gergely [14], Andréka-Németi[1]-[3], and def. of $\overset{\omega}{\models}$ in the present paper.

For an arbitrary theory $Th \subseteq TF_d$ the consequence relation

$Th \models (p,\psi)$

is defined in the usual way.

<u>THEOREM 3 (Completeness of Programverification)</u> :

Let $Th \subseteq TF_d$ be recursively enumerable. Then the set

$$\{ (p,\psi) \in P_d \times F_d \quad : \quad Th \models (p,\psi) \}$$

of all its consequences is also recursively enumerable.

<u>Proof</u>: The proof can be found in Andréka-Németi-Sain[4]. Moreover, a complete inference system is explicitly given there, with decidable proof concept.

<u>QED</u>

To execute programs in arbitrary elements of TM_d might look counter--intuitive. However, we may require Th to contain the Peano Axioms for the sort $\bar{t}$ and some Induction Axioms for the sort $\bar{i}$. The set of these axioms was denoted by Ax in Andréka-Németi[3]. The induction axioms for $\bar{i}$ are of the kind:

$$\forall y\left[\Big(\varphi(ext(y,0)) \wedge \forall z\left[\varphi(ext(y,z)) \rightarrow \varphi(ext(y,z+1))\right]\Big) \rightarrow \forall z\ \varphi(ext(y,z))\right],$$

for every $\varphi(x) \in F_d$. Now the models $\mathfrak{M} \in TM_d$ of Ax do satisfy all the intuitive requirements about <u>time</u> and about <u>processes</u> "happening in time" .

PROPOSITION 4 :

Let $Th \supseteq Ax$ be a subset of TF_d . Suppose that (p,ψ) is Floyd-Hoare provable from Th .
Then $Th \models (p,\psi)$.

Proof is in Andréka-Németi[3].
QED

REFERENCES

1. Andréka, H. and Németi, I., Completeness of Floyd Logic. Bulletin of Section of Logic 7(1978), 115-121, Wrocław.
2. Andréka, H. and Németi, I., A characterization of Floyd provable programs. Submitted to Proc.Coll.Logic in Programming, Salgótarján 1978. Coll.Math.Soc.J.Bolyai, North Holland.
3. Andréka, H. and Németi, I., Classical many-sorted model theory to turn negative results on program schemes to positive. Preprint 197
4. Andréka, H., Németi, I., and Sain, I., Abstract model theory, semantics, logics. Preprint 1979.
5. Chang, C.C. and Keisler, H.J., Model Theory. North Holland, 1973.
6. Davis, M., Hilbert's tenth problem is unsolvable. Amer.Math.Monthl 80(1973), 233-269.
7. Devlin, K.J., Aspects of Constructibility. Lecture Notes in Math. 354, Springer Verlag, 1973.
8. Gergely, T. and Ury, L., Mathematical Programming Theories.
9. Manna, Z., Mathematical Theory of Computation. McGraw Hill, 1974.
10. Monk, J.D., Mathematical Logic. Springer Verlag, 1976.
11. Németi, I. and Sain, I., Connections between Algebraic Logic and Initial Algebra Semantics of CF Languages. Submitted to Proc.Coll. Logic in Programming, cf.[2].
12. Sacks, G.E., Saturated Model Theory. W.A.Benjamin, Inc. Publ., Reading, Massachusetts, 1972.
13. Sain, I., On the General Theory of Semantics of Languages. Preprin 1979.
14. Szőts, M. and Gergely, T., On the incompleteness of proving partia correctness. Acta Cybernetica Tom 4, Fasc 1, Szeged 1978, pp.45-57

RELATIONSHIPS BETWEEN AFDL'S AND CYLINDERS

Jean-Michel AUTEBERT

Laboratoire associé du CNRS : Informatique Théorique et Programmation
et
Institut de Programmation, Université Paris VI

Introduction

One of the most studied problem in languages theory is the problem of generating families of languages from reduced sets of languages in the concerned families with only a few operations. We can part this problem in two : find a reasonable set of operations E for which the families of languages we want to study are closed, and for every family $\mathcal{L}$ closed under the operations of E, find a subset $\mathcal{R}$ of $\mathcal{L}$, as small as possible (best if composed of one element only), such that every language in $\mathcal{L}$ can be obtained from languages in $\mathcal{R}$ through operations of E. Call $\mathcal{R}$ an E-base of $\mathcal{L}$; if $\mathcal{R}$ can be reduced to one language only : $\mathcal{R} = \{L_0\}$, we will say that $\mathcal{L}$ is E-principal and that L_0 is an E-generator of $\mathcal{L}$.

In particular, the most classical cases are : the set E of operations contains only rational transduction, a family closed under rational transduction being a rational cone [6, 9, 10, 18] or the set E of operations of union, product, star, homomorphism, inverse homomorphism and intersection with regular sets, a family closed under these operations being an AFL [12, 13, 14, 15]. These two notions have closed connections studied in [5, 14]. But, in both cases, they do not allow to describe yet classical families of languages such as the nonambiguous languages or the C.F. deterministic languages . To study this last family, W.J. Chandler has introduced the notion of AFDL, family closed under inverse gsm mappings (abbreviated as gsm^{-1}), marked union and marked star, and removal of endmarkers. Among these operations, inverse gsm mapping is the most powerfull one, and S. Greibach has studied the $\{gsm^{-1}\}$- principality of some families of languages : She proves that Det, the family of deterministic context-free languages, is not $\{gsm^{-1}\}$-principal [17], and that Alg, the family of all context-free languages, is [16]. But, when looking at the proof of this last proposition, one can notice that all the power of inverse gsm mappings is not operating, since only inverse homomorphisms are used. This leads to introduce the notion of cylinder, family closed under inverse homomorphism and intersection with regular sets : to solve positively the question of principality of a family of languages as a cylinder (or C-principality) solves the question of principality of this family as an AFDL (or D-principality) - this was done in the case of deterministic realtime C.F. languages [4]-, and a negative answer gives a first step towards the solution of this question - L. Boasson and M. Nivat have proved that the family of linear C.F. languages is not C-principal [7], and we proved that the family of one counter languages is not either [1]. The two notions of AFDL and cylinder seem very close to each other and the question arises whether their relations are as tight as those between the notions of AFL and of rational cone.

We study here how the D-principality of a family of languages may reflect its C-principality. For that purpose, we introduce, in addition to the usual [11] notion of gsm^{-1}, the operation of $agsm^{-1}$ (inverse "gsm with accepting states" mapping, as defined in [19]), as an intermediate between the set D of operations of AFDL and the set C of operations of cylinder. Thus we have :

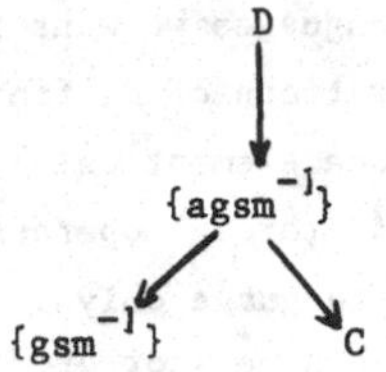

We first prove that if $\mathcal{L}$ is an AFDL which is D-principal with L as D-generator, then $\mathcal{L}$ is $\{agsm^{-1}\}$-principal and $(Ld)^*$ is an $\{agsm^{-1}\}$-generator of $\mathcal{L}$. This relation is exactly the same as the one relating the generators of an AFL and its generators as a rational cone. We then prove that a family of languages $\mathcal{L}$ closed under $agsm^{-1}$ is $\{agsm^{-1}\}$-principal if and only if it is $\{gsm^{-1}\}$- principal, and that it may be $\{agsm^{-1}\}$-principal, and not C-principal.

We give an additional condition for $\mathcal{L}$ which is sufficient to get this converse. We conjecture that this condition is necessary.

1. Preliminaries

We assume that the reader is familiar with the classical definitions of languages theory, as exposed in [11].

Definition 1 [11] :

A gsm is a 6-uple $<X,Y,Q,q_o,t,h>$, where X and Y are respectively the input and the output alphabets, Q is a finite set of states, $q_o \in Q$ is the initial state, t is a mapping from $Q \times X$ in Q called transition function and h is a mapping from $Q \times Y$ in Y^* called output function.

Definition 2 [19] :

An agsm is a 7-uple $<X,Y,Q,q_o,Q',t,h>$, where all the symbols have the same meaning as in definition 1 and Q' is a subset of Q called set of final states.

If one forgets the outputs, one has got exactly a deterministic finite automaton which will be called the automaton associated to the agsm.

t and h can be extended to $Q \times X^*$ in the usual way :

$\forall q \in Q, \forall x \in X, \forall w \in X^*$:

$t(q,\varepsilon) = q$ and $t(q,wx) = t(t(q,w),x)$,

$h(q,\varepsilon) = \varepsilon$ and $h(q,wx) = h(q,w)\, h(t(q,w),x)$.

A gsm is an agsm in which Q'=Q.

Definition 3 :

Given an agsm (resp. a gsm), we define for all $w \in X^*$:

$$g(w) = \begin{cases} h(q_0,w) \text{ if } t(q_0,w) \in Q' \\ \emptyset \text{ otherwise} \end{cases}$$

(resp. $g(w) = h(q_0,w)$), and for all $f \in Y^*$:

$g^{-1}(f) = \{w \in X^* \mid g(w) = f\}$.

We then extend these operations (g and g^{-1}) to operations on languages by :

$\forall L \subset X^* : g(L) = \{g(w) \mid w \in L\}$,

$\forall M \subset Y^* : g^{-1}(M) = \{w \mid g(w) \in M\}$.

In the following we allways denote by g as well the 7-uple (resp. the 6-uple) as the operation on languages for an agsm (resp. a gsm), and g^{-1} will denote the operation of inverse agsm (resp. inverse gsm).

Definition 4 [8] :

An AFDL is a family of languages closed under the set D of operations composed of gsm^{-1}, marked union, marked star, and removal of endmarkers.

Lemma 1 [8] :

An AFDL is closed under intersection with regular sets.

Definition 5 [2] :

A cylinder is a family of languages closed under the set C of operations composed of inverse homomorphism and intersection with regular sets.

Lemma 2 :

The sets of operations $\{agsm^{-1}\}$ and $\{gsm^{-1}$, intersection with regular sets$\}$ have exactly the same power.

Lemma 3 :

The powers of the sets of operations $\{gsm^{-1}\}$ and C are incomparable.

So we have

Proposition 1 :

The diagram in the introduction represents the relative power of the sets of operations involved.

Definition 6 :

Let E be a set of operations on languages, and let $\mathcal{L}$ be a family of languages closed under the operations of E. A subset $\mathcal{R}$ of $\mathcal{L}$ will be called an E-base of $\mathcal{L}$ if every language in $\mathcal{L}$ can be obtained from elements* of $\mathcal{R}$ with the operations in E.

Definition 7 :

In the same conditions, $\mathcal{L}$ will be said E-principal if there exists an E-base of $\mathcal{L}$ reduced to one element only, called an E-generator of $\mathcal{L}$.

2. D-principality and {agsm^{-1}}-principality

In this section we state :

Theorem 1 :

Let $\mathcal{L}$ be an AFDL. $\mathcal{L}$ is D-principal if an only if $\mathcal{L}$ is {agsm^{-1}}-principal.

The sufficient condition is obvious. One has only to prove the necessary condition. A more precise result is :

Proposition 2 :

If $\mathcal{L}$ is a D-principal AFDL with D-generator L, then $\mathcal{L}$ is {agsm^{-1}}-principal with {agsm^{-1}}-generator $(Ld)^*$, d being a marker.

The proof [2] rests on two remarks :

Remark 1 :

If $\mathcal{L}$ is a D-principal AFDL with D-generator L, then $(Ld)^*$ is also a D-generator of $\mathcal{L}$.

Remark 2 :

The least family of languages closed under agsm^{-1} and containing the language $(Ld)^*$ is an AFDL.

3. {agsm^{-1}}-principality and {gsm^{-1}}-principality

In the following, we will allways consider families $\mathcal{L}$ closed under agsm^{-1}.

The next step in this study is to prove :

Theorem 2 :

Let $\mathcal{L}$ be a family of languages closed under agsm^{-1}. $\mathcal{L}$ is {gsm^{-1}}-principal if and only if $\mathcal{L}$ is {gsm^{-1}}-principal.

Together with theorem 1 we get as corollary :

Theorem 3 :

Let $\mathcal{L}$ be an AFDL. $\mathcal{L}$ is D-principal if and only if $\mathcal{L}$ is {gsm^{-1}}-principal.

To establish theorem 2, as the sufficient condition is trivial, one only has to prove :

Proposition 3 :

If $\mathcal{L}$ is a family of languages which is {agsm^{-1}}-principal then $\mathcal{L}$ is {gsm^{-1}}-principal.

Corollary :

If $\mathcal{L}$ is a family of languages closed under agsm^{-1} (resp. is an AFDL) which is C-principal, then $\mathcal{L}$ is {gsm^{-1}}-principal.

The idea of the proof is to construct, from an {agsm^{-1}}-generator of $\mathcal{L}$, a {gsm^{-1}}-generator using a parity argument : from the {agsm^{-1}}-generator we derive a new {agsm^{-1}}-generator in which the length of every word is even, and from any agsm h, we define a gsm h' such that the length of the output word is even if and only if the corresponding state in the agsm h is a final state. This proof, rather technical too, is developed in [3].

4. {gsm^{-1}}-principality and C-principality

The question left is : $\mathcal{L}$ being an {agsm^{-1}}-principal family of languages, is $\mathcal{L}$ C-principal ? The answer, negative, is given by :

Proposition 4 :

The least family of languages closed under agsm^{-1} and containing the language $\{x^n y^n \mid n > 0\}$ is not C-principal.

From a {gsm^{-1}} -generator L of $\mathcal{L}$ we derive a C-base of $\mathcal{L}$ leading to an infinite hierarchy of nested principal cylinders; the problem of proving $\mathcal{L}$ to be C-principal or not becomes proving this hierarchy to be strict or not :

Let $\mathcal{L}$ be a family of languages, closed under agsm^{-1}, {agsm^{-1}}-principal. Let L be then a {gsm^{-1}}- generator of $\mathcal{L}$ over an alphabet X, and let a, c and d be three symbols not in X. Let $Y = X \cup \{a,c,d\}$ and e be the homomorphism from Y^* to X^* that erases these three letters.

We now consider the following regular sets :

$\forall k > 0 \ S_k = \{f \in Y^* \mid c^{k+1} \text{ is not a subword of } f\} = Y^* \setminus Y^* c^{k+1} Y^*$,

$\forall k > 0 \ T_k = Y^* \setminus Y^* d(X^* a)^{k+1} Y^*$,

and $R = (d(aX^*c^+)^+)^*$.

A word f in R can be decomposed in a product $f = g_1 g_2 \dots g_n$ where $g_i \in d(a X^* c^+)^+$. Such a subword g_i of f will be called a bloc (g_i is the bloc of rank i). A bloc can be decomposed itself in a product $d\, h_1 h_2 \dots h_p$ where $h_i \in a X^* c^+$. Such a subword h_i of a bloc will be called an elementary factor (h_i is the ith elementary factor of the bloc).

For a word f in R we will call path in f, and denote $p(f)$, the word obtained as product of one elementary factor in each bloc of f, if there exists one, satisfying the two conditions :

1. In the first bloc the first elementary factor is chosen .
2. If the elementary factor chosen in the kth bloc is $a\,w\,c^{r}$ with $w \in X^*$, the elementary factor chosen in the $1+k^{th}$ bloc is the rth elementary factor of this bloc.

We define the families of languages :

$\forall k > 0 \quad H_k = \{f \in R \cap S_k \cap T_k \mid e(p(f)) \in L\}$,

$\forall k > 0 \quad H'_k = \{f \in R \cap S_k \mid e(p(f)) \in L\}$.

We can state the following proposition :

<u>Proprosition 5</u> :

The family $\{H_k \mid k > 0\}$(resp. $\{H'_k \mid k > 0\}$) is a C-base of .

As $H_k = H'_k \cap T_k$ for all k, one has to prove that for every $k > 0$ H'_k belongs to $\mathcal{L}$ and that for every language M in $\mathcal{L}$, one can find a language H_k such that M can be obtained from H_k by means of operations of C.

Now, if we call $H_\infty = \{f \in R \mid e\ (p(f)) \in L\}$, we have for all k :$H'_k = H_\infty \cap S_k$, so if H_∞ belongs to $\mathcal{L}$, $\mathcal{L}$ is C-principal with H_∞ as C-generator.

Such a language H_∞ can be defined in the same way starting from any language L, so we will call it H_∞ (L) in the following.

We get the sufficient condition :

<u>Theorem 4</u> :

Let $\mathcal{L}$ be a family of languages closed under $agsm^{-1}$ which is $\{agsm^{-1}\}$-principal with $\{agsm^{-1}\}$-generator L. If $H_\infty (L) \in \mathcal{L}$, then $\mathcal{L}$ is C-principal with H_∞ (L) as C-generator.

As a corollary, we have :

<u>Theorem 5</u> :

Let $\mathcal{L}$ be a D-principal AFDL with D-generator L. If $H_\infty\ ((Ld)^*) \in \mathcal{L}$, then is C-principal with $H_\infty\ ((Ld)^*)$ as C-generator.

<u>5. Conclusions and open problem</u>

Let y, $\bar{y}$ and z be three new symbols not in X, and $X' = X \cup \{y,\bar{y},z\}$. Let b be the bijection between X'^* and Y^* defined by :

$\forall x \in X : b(x) = x, \quad b(y) = c, \quad b(\bar{y}) = a, \quad b(z) = d$.

Let $\bar{b} = b^{-1}$. b and $\bar{b}$ are homomorphisms.

We consider the context-free language A generated by the grammar :

$$v_o \to v_1 + \varepsilon$$

$$v_1 \to y \;\; v_2 \; \bar{y} + v_3 \;\; z + z$$

$\forall x \in X \cup \{y\}$ $\qquad v_2 \to v_2 \; x \;\; + v_1$

$\forall x \notin X \cup \{y,\bar{y}\}$ $\qquad v_3 \to v_3 \; x \;\; + \bar{y}$

and the regular set $B = y + (\bar{y}\, X^* \, y^+)^*$.

We define a tranformation which is a kind of substitution :

For any language L over an alphabet X, T(L) is defined by :

$T(L) = \{z \, \bar{y} \, f_1 \; A \;\; f_2 A \, f_3 \, \ldots\ldots \, A f_{n-1} \;\; A f_n \;\; B \mid f_1 \, f_2 \, \ldots\ldots \, f_n \in L\}$.

Clearly $T(L) = b^{-1}(H_\infty(L))$ and $H_\infty(L) = \bar{b}^{-1}(T(L))$.

So we can state theorems 4 and 5 with T(L) instead of H_∞ (L).

The following theorem makes it more convenient.

<u>Theorem 6 a</u> :

Let $\mathcal{L}$ be an $\{agsm^{-1}\}$-principal family of languages with $\{agsm^{-1}\}$-generator L. Then $T(L) \in \mathcal{L}$ if and only if $\mathcal{L}$ is closed under the transformation T.

<u>Theorem 6 b</u> :

Let $\mathcal{L}$ be a D-principal AFDL with D-generator L. Then $T((Ld)^*) \in \mathcal{L}$ if and only if $\mathcal{L}$ is closed under the transformation T.

So the next two theorems are equivalent respectively to theorems 4 and 5.

<u>Theorem 7</u> :

If $\mathcal{L}$ is a family of languages which is $\{agsm^{-1}\}$-principal and is closed under the transformation T, then $\mathcal{L}$ is C-principal.

<u>Theorem 8</u> :

If $\mathcal{L}$ is a D-principal AFDL closed under the transformation T, then $\mathcal{L}$ is C-principal.

Alg, the family of all context-free languages, Nge, the largest rational cone strictly included in Alg, Det, the family of deterministic context free languages, as well as Psi, the family of realtime deterministic context-free languages, are families which are closed under the transformation T.

Lin, the family of linear languages, and Ocl, the family of one-counter languages, are not closed under this transformation.

The former are D-principal (resp. $\{agsm^{-1}\}$-principal, or $\{gsm^{-1}\}$-principal) if and only if they are C-principal. So it is not so surprising if the two proofs of D-principality (for Alg [16] and for Psi [2,4]) are in fact proofs of C-principality of these families.

On the other side, though Lin and Ocl are not C-principal [7,1] they may be $\{agsm^{-1}\}$-principal. We conjecture however, after S.A. Greibach [16], that they are not, and also :

Conjecture 1 :

Let $\mathcal{L}$ be an $\{agsm^{-1}\}$-principal family of languages with $\{agsm^{-1}\}$-generator L. $\mathcal{L}$ is C-principal if and only if $T(\mathcal{L}) \subset \mathcal{L}$.

This conjecture can be stated equivalently as :

Conjecture 2 :

Among the families of languages closed under $agsm^{-1}$, only the ones closed under transformation T may be C-principal.

REFERENCES

[1] J.M. AUTEBERT, Non-principalité du cylindre des langages à compteur, Mathematical Systems Theory, 11 (1977) p. 157-167.

[2] J.M. AUTEBERT, Cylindres de langages algébriques, Thèse de Doctorat d'Etat, publication du LITP n°78-37,Paris, 1978.

[3] J.M. AUTEBERT, Opérations de cylindre et applications séquentielles gauches inverses, to appear in Acta Informatica.

[4] J.M. AUTEBERT, The AFDL of deterministic realtime context-free languages is principal, submitted for publication.

[5] L. BOASSON, Cônes Rationnels et Familles Agréables de Langages. Application aux Langages à Compteur, Thèse de 3ème cycle, Paris VII, 1971.

[6] L. BOASSON and M. NIVAT, Sur diverses familles de langages fermées par transduction rationnelle, Acta Informatica, 2 (1973) p. 180-188.

[7] L. BOASSON and M. NIVAT, Le cylindre des langages linéaires, Mathematical Systems Theory, 11 (1977) p. 147-155.

[8] W.J. CHANDLER, Abstract families of deterministic languages, Proc. 1st ACM Conf. on Theory of Computing, Marina del Rey, Calif. (1969), p. 21-30.

[9] S. EILENBERG, Communication au Congrès International des Mathématiciens, Nice, 1970.

[10] C.C. ELGOT and J.E. MEZEI, On relations defined by generalized finite automata, IBM Journal of Research and Dev., 9 (1962) p. 47-68.

[11] S. GINSBURG, The Mathematical Theory of Context-free Languages, Mc Graw-Hill, New-York, 1966.

[12] S. GINSBURG and J. GOLDSTINE, On the largest full sub-AFL of an AFL, Mathematical Systems Theory, 6 (1972) p. 241-242.

[13] S. GINSBURG and S.A. GREIBACH, Abstract Families of Languages, in:Studies in Abstract Families of Languages, Memoirs of the Amer. Math. Soc., 87 (1969) p. 1-32.

[14] S. GINSBURG and S.A. GREIBACH, Principal AFL, J. Comput. System Sci,, 4 (1970) p. 308-338.

[15] S.A. GREIBACH, Chains of full AFL's, Mathematical Systems Theory, 4 (1970) p. 231-242.

[16] S.A. GREIBACH, The hardest C.F. languages, SIAM Journal on Computing, 2 (1973) p. 304-310.

[17] S.A. GREIBACH, Jump PDA's and hierarchies of deterministic C.F. languages, SIAM Journal on Computing, 3 (1974) p. 111-127.

[18] M. NIVAT, Transductions des langages de Chomsky, Annales de l'Institut Fourier, 18 (1968) p. 339-456.

[19] A. SALOMAA, Formal Languages, Academic Press, New-York, 1973.

COMPUTABLE DATA TYPES

G. COMYN
G. WERNER
UNIVERSITE DE LILLE-I
I.U.T. DEPARTEMENT INFORMATIQUE
59650 - VILLENEUVE D'ASCQ - FRANCE

I. INTRODUCTION

Data types have been introduced in programming as a unifying concept in the definition of machine independent constructs. Traditionally, several particular aspects of such constructs are used for their definition, emphasizing seemingly independent principles of classification :

- data types as sets of values, such as the integers, the reals, binary trees, etc;
- data types determined by an access function or, more generally, some scheme if implementation; for example, Random Access Memory, arrays, lists or files with a particular mechanism for their access and updating, etc... (note that these data types may be quite arbitrary when considered as a value-set);
- data types considered as a set of functions or operations on an appropriate, but variable value-set; e.g. procedures with parameters of variable type.

Recent research in the theory of data types has been highly influenced by the work of D. Scott [1,2] and by the (related) category-theoretic definitions of datatypes (e.g. [3,4]). In this theory, effective computability has been introduced by the a-posteriori restriction to computable subcategories or to finitary categories, see [4]. Another problem - as pointed out by ADJ [5] - is the derivation of appropriate domain equations for the data types we want to define.

The solution which we are proposing here is quite close to Scott's theory since it is essentially based on a model for the (typed) lambda-calculus, due to Ju. L. Ershov [6,7], who develops this model in the framework of the general theory (category) of numbered sets. Computable data types may then be defined as computably numbered families of computable objects, forming a sub-category closed in particular under exponentiation (function space).

We feel that the above mentioned kinds of data types have different roles which should be distinguishable within a general unifying theory.

In the first place, each data type (data set or function) has a variety of different implementations. Addressing schemes and thereby implementations are closely related to the numberings of the elements of data, therefore we believe that our model well suits the definition of data types together (or without) a specific addressing or implementation - at least to some extent.

Secondly, we think that finite data types have a special significance. Their structures, mathematical as well as computational, have often been extensively studied and the more general types should be definable from them. Therefore our definition is based on the numbered sets of objects which have a recursive basis, i.e., which may be computed with any desired degree of approximation from strongly computable families of finite data objects whose inclusion relation is a recursive partial order.

The basic notions of Ershov's theory and the definition of computable domains on a recursive basis are briefly stated in the first section. Next we give formal definitions of two well-known syntactic data types. These definitions are based on closure operations : the third section contains some basic facts and relations concerning retracts, domains over a recursive basis, effective closure and contraction-operators.

Our data types are not closed under union or intersection, but both union and intersection may be embedded in some data type.

<u>Notations</u> : N the integers, $\mathbb{R}$ the class of recursive functions, $\mathbb{RE}$ the class of r.e. sets; D_x is the finite set $\{x_1,\ldots,x_n\}$ with canonical index $x = 2^{x_1}+\ldots+2^{x_n}$; $\mathcal{D}$ is a canonically r.e. class of finite sets iff $\mathcal{D} = \{D_{f(x)} | x\in N\}$ for some $f\in\mathbb{R}$.

2. EFFECTIVELY COMPUTABLE DOMAINS

We consider numbered sets or, more generally, numbered families of objects : (A,α) is a numbered set if $\alpha: N\to A$ is onto (Ershov [6,7]).

Let $\mathbb{F}$ be any class of functions and let ϕ be a mapping from (A,α) to (B,β). Then ϕ is an $\mathbb{F}$-morphism iff there is an $f\in\mathbb{F}$, $f : N\to N$, such that $\alpha\circ\phi = f\circ\beta$. Let $\mathrm{Mor}((A,\alpha),(B,\beta))$ stand for the class of all $\mathbb{F}$-morphisms from (A,α) to (B,β).

The class of all numbered sets together with their morphisms forms a category N; effectively computable data-types are to be defined as a sub-category closed under product and exponentiation (function-space).

Let (A,α) and (A,α') be two different numberings of A. Then $f\in\mathbb{R}$ <u>reduces</u> α to α' $\alpha \leq \alpha'$, iff $\alpha = f\circ\alpha'$.

Let $R = (\mathbb{RE},\pi)$ be the family of r.e. sets numbered by π, the standard numbering of Post. A numbered set (of sets) $S = (\mathbb{S},\nu)$ is an R-*family* (S,μ), if $\mu : S\to R$ is a monomorphism. Such a family corresponds (up to an equivalence) to a subset $\mathbb{S} \subseteq \mathbb{RE}$, numbered by ν such that $\nu\leq\pi$ with μ the identity-mapping from $\mathbb{S}$ to $\mathbb{RE}$.

An R-family S such that for some g , $g\in\mathbb{R}$,

$$g\circ\mu\circ\nu = \pi \qquad \begin{array}{ccc} \mathbb{S} & \xrightarrow{\ \mu\ } & \mathbb{RE} \\ \nu\uparrow & & \uparrow\pi \\ N & \xleftarrow[\ g\ ]{} & N \end{array}$$

is an *effectively principal* R-family iff :

(i) ν is principal i.e., every numbering ν' of $\mathbb{S}$ is such that $[\nu' \leq \pi \Rightarrow \nu' \leq \nu]$,

(ii) a Gödel number for g may be effectively determined from a Gödel number for a numbering of $\mu(\mathbb{S})$.

For any numbered set $A = (A,\alpha)$, the numbering α induces a quasi-ordering :
$\forall a_0, a_1 \in A,\ a_0 \le_\alpha a_1 \Longleftrightarrow \forall A_0,\ A_0 \subseteq A(\alpha^{-1}(A_0)\text{ r.e. } \& \ a_0 \in A_0 \Rightarrow a_1 \in A_0)$

Let $S = (\mathbb{S},\nu)$ be an effectively principal R-family (S,μ). Then the following properties may hold for μ , ν and g :

(a) $\forall x\ x \in \pi^{-1}\cdot\mu(\mathbb{S})\quad \mu\nu g(x) = \pi(x)$,

(b) $\forall x\ x \in N \qquad \mu\nu g(x) \le_\pi \mu(x)$,

(c) $\exists\bar{\mu}\ \bar{\mu} \in Mor(R,S)$ such that $\mu\circ\bar{\mu}$ is the identity morphism on S.

An R-family satisfying (a) and (b) above is a <u>special standard class</u> (S-class, Lachlan) or <u>sn-family</u> (Ershov). An R-family satisfying (a) and (c) is a <u>R-retract</u> (Ershov).

It can be shown (Lachlan,Ershov) that if (S,μ) is an S-class, then there is a canonically r.e, class of finite sets Φ such that :

(i) $\emptyset \in \Phi$,

(ii) $R \in \mathbb{S} \Longleftrightarrow R \in \mathbb{RE}$ & there is a monotone sequence of sets from Φ,
$F_0 \subseteq F_1 \subseteq \dots \subseteq F_i \subseteq \dots$ such that $R = \bigcup_i\{F_i \mid i \in N\}$,

(iii) $\forall R \in \mathbb{S}$, $\forall F \subseteq N\ (F \subseteq R \ \&\ F \text{ finite} \Rightarrow (\exists F_0 \in \Phi)\ (F \subseteq F_0 \subseteq R))$.

(iv) for each computable chain $(R_i \mid i \in N)$ of elements from $\mathbb{S}$ such that
$R_i \subseteq R_{i+1}$ for all $i \in N$, $R = \bigcup_i\{R_i \mid i \in N\}$ belongs to $\mathbb{S}$.

It follows that the S-classes are partially ordered (p.o.) families of r.e. sets and that they are complete for the lub's of computable chains.

Among the effectively principal S-families we want to distinguish those which have a recursive basis : Let $(D,\le)$ be a po-set; $B \subseteq D$ is a basis for $(D,\le)$ if $\forall x$, $x \in D$, $\exists C \subseteq B$, C an ω-chain of B such that $x = \text{lub } C$; $(D,\le)$ is ω-chain-complete if every denumerable chain has its lub in D . Call an ω-chain a c-chain iff it is computable.

Let $S = (\mathbb{S},\nu)$ be an S-class. If Φ is the system of finite sets of S, numbered by ϕ so that $\phi \le \nu_{fin}$, then Φ is a po-set with smallest element $\emptyset$: $(\Phi,\subseteq,\emptyset)$.

Definition 1. Φ is a recursive basis of S iff $(\Phi,\subseteq,\emptyset)$ is a recursive upper semilattice (partial) iff

(i) the boundedness-predicate $U_\Phi(\phi(x),\phi(y)) \Longleftrightarrow \exists z[\phi(x) \subseteq \phi(z)\ \&\ \phi(y) \subseteq \phi(z)]$ is a recursive predicate over $N\times N$,

(ii) $\forall x\ \forall y[U_\Phi(\phi(x),\phi(y))] \Rightarrow \text{lub}(\phi(x),\phi(y))$ is defined and $\{\langle x,y,z\rangle \mid \phi(z)=\text{lub}(\phi(x),\phi(y))\}$ is recursive. Then we have the following :

Definition 2. A po-domain $\mathbb{D}$ is effectively computable if :

(i) $\mathbb{D}$ has a recursive basis Φ ,

(ii) $\mathbb{D}$ is c-chain complete ,

(iii) there is a numbering δ of $\mathbb{D}$ and a morphism μ such that $((\mathbb{D},\delta),\mu)$ is an S-class.

The retracts of computable domains (according to Definition 2) form a sub-category of N closed under product and exponentiation : let S_0 and S_1 be retracts of effectively computable domains with recursive basis (Φ_0,ϕ_0) and (Φ_1,ϕ_1) respectively and let $Mor(S_0,S_1)$ be the set of all $\mathbb{R}$-morphisms from S_0 to S_1. Then Ershov's theorem [7] states that the numbered set $(Mor(S_0,S_1),\nu)$ is a retract of an effective computable domain generated by a recursive basis (a recursive partial upper semi-lattice).

3. DEFINITION AND SPECIFICATION OF COMPUTABLE DATA TYPES

We now turn to the definition of a certain class of computable data types which have been called syntactic by Mayoh [8], since there is a set of formal expressions E (and a grammar for E) such that the data type is generated by E.

We first state our main definition, which is based on the effectively computable domains of the preceding section, then we discuss two examples of syntactic data types and make more precise our definition of computable data types.

Let Φ be some canonically r.e. class of finite sets such that $(\Phi,\subseteq,\emptyset)$ is a recursive upper semi-lattice (partial) and let $\mathbb{D}$ stand for an effectively computable domain on basis Φ.

Definition 3. An effectively computable data type is a retract of some effectively computable domain $\mathbb{D}$.

In particular, any retract of $(\mathbb{RE},\pi)$ having a recursive basis Φ is a computable data type.

The first example are the binary trees. Viewed as a numbered set of objects, binary trees are closed subsets of a po-set; the set A_f of finite trees form their recursive basis. The numberings of the nodes and of the trees have respectively to be compatible with the basic partial order.

Let DY denote the set of dyadic numerals (integers on the alphabet {1,2}); $d : N \to DY$ is the lexicographical numbering of the dyadic numerals (which is one-one and onto, i.e., a bijection).

Define the inclusion ordering in DY by : $x \subseteq y \iff \exists z \;\; xz = y$ for all $x,y \in DY$ and accordingly $x \wedge y = glb(x,y)$.

Then a binary tree (of dyadic numbers) is a set of dyadic numbers closed downward under the inclusion ordering : $d(D_x)$ is a binary tree $\iff \forall u \, \forall v \;\; u \in d(D_x)$ & $v \in d(D_x) \Rightarrow u \wedge v \in d(D_x)$ & $\forall u \, \forall v \, \forall p \;\; u \in d(D_x)$ & $v \in d(D_x)$ & $u \wedge v \subset p \subseteq u$ *or* $u \wedge v \subset p \subseteq v \Rightarrow p \in d(D_x)$.

Therefore, finite trees may be defined by an effective closure operation $Cl(D_x) := \{d^{-1}(u) \mid \forall p \, \forall q, \; p,q \in d(D_x) \; [p \wedge q \subseteq u \subseteq p$ *or* $p \wedge q \subseteq u \subseteq q]\}$.

Define $$D_{f(x)} := \begin{cases} D_x \text{ if } d(D_x) \text{ is a binary tree} \\ Cl(D_x) \text{ otherwise} \end{cases}$$

Then $(\mathbf{A}_f,\alpha)$ is a recursive upper semi-lattice of dyadic trees (as closure over $P_\omega(N)$ it is even a lattice) and $(\{D_{f(x)} \mid x \in N\},\phi)$ is a recursive upper semi-lattice of integer trees, α and ϕ being defined so that the following diagram commutes :

.../...

$$\begin{array}{ccccc} A_f & \xleftarrow{\ d\ } & Cl(P_\omega(N)) & \xleftarrow{\ Cl\ } & P_\omega(N) \\ \alpha\uparrow & & \phi\uparrow & & \uparrow \nu_{fin} \\ N & \xrightarrow{\ id_N\ } & N & \xrightarrow{\ f\ } & N \end{array} \qquad \begin{array}{l} \phi(x) = Cl(\nu_{fin} f(x)) \\ \text{and} \\ \alpha = \phi \circ d \end{array}$$

Finite binary trees of S , where S is any numbered set, may be defined in a similar manner; infinite trees can be obtained by the lub's of c-chains of finite trees.

It should be interesting that a syntactic definition of the finite dyadic trees may be obtained from a recursive form of the closure operator : Let $u = \min_{p,q \in d(D_x)} p \wedge q$ and let $L(D_x)$ ($R(D_x)$) be the set of left (right) sons of u in $d(D_x)$; put $D_{x_1} := d^{-1}(L(D_x))$ and $D_{x_2} := d^{-1}(R(D_x))$ and define :

$$Cl(D_x) := \begin{cases} \emptyset \text{ if } D_x = \emptyset \\ d^{-1}(u) \cup Cl(\textit{if } D_{x_1} = \emptyset \textit{ then } \emptyset \textit{ else } d^{-1}(u1) \cup D_{x_1}) \\ \qquad\qquad \cup Cl(\textit{if } D_{x_2} = \emptyset \textit{ then } \emptyset \textit{ else } d^{-1}(u2) \cup D_{x_2}) \end{cases}$$

where u1 and u2 designate the immediate successors of u in the dyadic tree of root u.

This (bottom-up) definition of a closure can be used to write down a grammar (with parallel rewriting rules !) for the (top-down) generation of the dyadic trees :

A ::= Λ|DY(Λ,Λ)|DY(DY1(A,A),Λ)|DY(Λ,DY2(A,A))

A ::= DY(DY1(A,A),DY2(A,A))

where Λ denotes the empty tree and DY some dyadic numeral.

This grammar produces the usual parenthesized form of binary trees. Note the difference with the grammar : A ::= Λ|DY(A,A) which corresponds to the equation for binary trees of [2] or [3] : $A \cong \emptyset + DY \times A \times A$.

The solution of this equation is the set of all trees with their nodes numbered in all possible ways. This is not our data type "binary tree", since it lacks the underlying partial order.

Our next example is the usual test-case for syntactic data types, namely the type Stack (S). Consider Stack(DY); let $\mathbb{S}_f$ be the set of finite sequences of dyadic numerals which are the values of this type. At each moment, the stack may be described by a stack-configuration which is \$Λ or \$X, X ∈ DY whenever it is defined and is ω else.

Then the type Stack (DY) (or Stack (D) if D = {1,2}) is an algebra of finite functions from DY to DY, generated from the basic functions by composition :

Stack(D) = (DY;New,Push,Pop,Top).

These basic functions are defined as usually; Push and Top are considered as the input - and output functions, respectively.

Let now SF be a set of formal expressions defined by the grammar :

SF ::= Push(SF,D)|Pop(SF)|Top(SF)|New

D ::= 1|2

where SF stands for "stack-function". If we define the functions Push,Pop and New by :

Push(New,d) = \$d ∀d ∈ D

$Pop(Push(x,d)) = x \quad \forall d \in D\ , \forall x \in \mathbf{C}_f$ (the set of configurations)

$Pop(\emptyset) = \omega$

then a relation $<_s$ is induced in the set of stack-configurations by $\forall x \forall y$ $x <_s y \iff y = Push(x,1)$ *or* $y = Push(x,2)$ *or* $y = Pop(x)$. Each formal expression $f \in SF$ now determines a unique stack-sequence $y \in \mathbf{S}_f$ i.e., a finite sequence of stack-configurations $x_0, x_1, \ldots, x_n$ such that $x_0 = New$ and $\forall i,\ 0 \le i < n,\ x_i <_s x_{i+1}$.

The set of configurations $\mathbf{C}_f$ can be taken as the labels of the nodes of a ternary tree (with repetitions) whose branches are the stack-sequences. Let ν be a numbering of this tree.

Each formal expression $f \in SF$ defines an input-sequence x_f in an obvious way.

A finite (stack-computable) function F is simply a finite set of formal expressions and (equivalently) the finite set of corresponding stack-sequences, subject to the following compatibility-condition :

$\forall f \forall g\ f \in F\ \&\ g \in F\ \&\ f \ne g \Longrightarrow x_f \ne x_g$.

Let D_x be a finite set of indexes of nodes in the tree of stack-configurations; let $\subseteq_t$ denote the inclusion-ordering of the ternary numerals. Then the set :

$$Cl(D_x) = \{\nu^{-1}(p) \mid p \subseteq_t q \text{ for some } q \in \nu(D_x)\} = D_{f(x)}$$

is an index-set for a set of stack-sequences.

Then define $D_{gf(x)} = \begin{cases} D_{f(x)} & \text{if the corresponding stack-sequences are compatible} \\ \text{the smallest compatible extension of } D_{f(x)}, & \text{otherwise} \end{cases}$

It can be shown that $(\{D_{gf(x)} \mid x \in N\}, \subseteq, \emptyset)$ forms a recursive partial upper semilattice of (indexes for) finite stack functions - we omit technical details.

4. COMPUTABLE DOMAINS AND RETRACTS

The preceding examples show that effective closure operations are likely to be a very useful tool for the definition of retracts. Each of these example-types was defined by a closure and also had a recursive basis of finite sets. Do these properties hold for all retracts ?

We investigate here relations between computable domains over a recursive basis and retracts.

Let F be an effective closure operator, i.e., F is extensive, monotone and idempotent and there is a recursive $g \in \mathbb{R}$ such that $F(W_x) = W_{g(x)}$ for all x .

Let $(\mathbb{Q},\nu)$ be the set $\{F(W_x) \mid x \in N\}$ numbered by ν, $\nu \le \pi$ by f, let $1_{\mathbb{Q}}$ be the identity morphism on $\mathbb{Q}$.

Put $\quad W_{f(x)} := \begin{cases} W_x & \text{if } W_x \text{ is a closed set} \\ Cl(W_x) = W_{g(x)} & \text{otherwise} \end{cases}$. Then we have :

Proposition 1. a) There is a closure F such that $((\mathbb{Q},\nu),1_{\mathbb{Q}})$ is not an S-class .

b) If $\forall x\ [F(W_x)-W_x]$ is r.e. then $((\mathbb{Q},\nu),1_{\mathbb{Q}})$ is a retract of $(\mathbf{RE},\pi)$.

<u>Corollary</u> : In general, $(\mathbb{Q},\nu)$ does not have a recursive basis (of finite sets) .

Proof :

a) $$\begin{array}{ccc} \mathbb{Q} & \overset{F}{\underset{}{\rightleftarrows}} & \mathbb{RE} \\ \nu \uparrow & \mu = 1_{\mathbb{Q}} & \uparrow \pi \\ N & \overset{g}{\underset{f}{\rightleftarrows}} & N \end{array}$$

$\mu\nu g(x) = \pi(x)$ for $x \in \pi^{-1}(\mu(\mathbb{Q}))$
and $\mu\nu g(x) \subseteq \pi(x)$ otherwise for an S-class;
$\mu\nu g(x) = \pi f g(x)$, so $W_{fg(x)} = W_{g(x)} \supset W_x$ for $x \notin \pi^{-1}(\mu(\mathbb{Q}))$
which contradicts the above condition.

b) To show : f is a recursive function if $F(W_x) - W_x$ is r.e. for all x .
Let $C_x = W_{g(x)} - W_x$ be r.e.,

put $W_{f(x)} = \begin{cases} W_x & \text{if } C_x = \emptyset \\ W_{g(x)} & \text{if } C_x \neq \emptyset \end{cases}$

which reads : enumerate all of W_x in $W_{f(x)}$ as long as $C_x = \emptyset$, then, if some y appears in C_x , enumerate all of $W_{g(x)}$ in $W_{f(x)}$. Hence f is recursive. □

Now let $(\Phi, \subseteq, \emptyset)$ be the recursive basis of some computable domain and let F be a closure operator mapping finite sets to finite sets. Then, by well-known properties of closure operators :

<u>Fact</u> : An effective closure F of a recursive upper semi-lattice is a recursive upper semi-lattice.

If ϕ is a numbering of Φ, then

$$\mathrm{Sup}_\Phi(\phi(x),\phi(y)) = \phi(x) \cup \phi(y) \in \Phi \Rightarrow \mathrm{Sup}_{F(\Phi)}(F(\phi(x)),F(\phi(y))) = F(F(\phi(x)) \cup F(\phi(y))) \in F(\Phi).$$

We are now able to state a necessary and sufficient condition such that the retract of some computable domain $\mathbb{D}$ is also a computable domain :

Proposition 2. A retract $\mathbb{Q}$ of some computable domain $\mathbb{D}$ has a recursive basis (of finite sets) iff there is some idempotent contraction-morphism $\bar{\mu}$ such that $\bar{\mu}(\mathbb{D}) = \mathbb{Q}$.

Proof: Suppose $\mathbb{Q}$ has a recursive basis (Φ,ϕ) ; let $\mathbb{D}$ be $\mathbb{RE}$.

For all x , define a recursive $g \in \mathbb{R}$ by the following effective procedure : Enumerate W_x in steps and thereby enumerate $W_{g(x)}$ in steps; at step n we have $W^n_{g(x)}$ and $W^n_x - W^n_{g(x)}$; at step n+1 enumerate a new $y \in W_x$; if $\exists \phi(i) \in \Phi, \phi(i) \subseteq (W^n_x \cup \{y\})$ & $\phi(i) \nsubseteq W^n_x$ then enumerate $\phi(i) \cap (W^n_x - W^n_{g(x)} \cup \{y\})$ in $W^{n+1}_{g(x)}$ else put y in $W^{n+1}_x - W^{n+1}_{g(x)}$.

This defines $W_{g(x)}$ as

$$W_{g(x)} = \begin{cases} W_x & \text{if } W_x = \bigcup_i \{\phi(i) \mid i \in W_n\} \text{ for some } n \\ \bigcup_i \phi(i) \mid i \in W_n\} & \text{otherwise.} \end{cases}$$

Note that $W_{g(x)} \subseteq W_x$; now let $\bar{\mu}(W_x) = W_{g(x)}$. $\bar{\mu}$ is a contraction and $\bar{\mu}(\bar{\mu}(W_x)) = \bar{\mu}(W_x)$ by definition. Take $\Phi = \{\bar{\mu}(D_x) \mid x \in N\}$.

Conversely, let $\bar{\mu}(W_x) = W_{g(x)}$ be a contraction morphism (so that g represents $\bar{\mu}$) such that $\bar{\mu}\bar{\mu}(W_x) = \bar{\mu}(W_x)$. Then, for all x, $\mu\nu g(x) \subseteq \pi(x)$, since $\pi f g(x) = \pi g(x)$, $W_{g(x)} \subseteq W_x$ and $\mu\nu g(x) = \pi(x)$ for all x such that $W_x = W_{g(x)}$; f being defined as in Proposition 1.

Hence $\mathbb{Q}$ is an S-class. $\mathbb{Q}$ is also a computable domain, since $\bar{\mu}$ maps all finite sets to finite sets and preserves the recursive order relations. □

5. UNION AND INTERSECTION OF COMPUTABLE DATA TYPES

The definitions of polymorphic functions, of parametrized types and procedures require some rule how data types should be compared, how sub-types, union-types and intersection-types are to be defined. We show that computable data types are not closed under union or intersection.

Let $\mathcal{D} = (\mathbb{D},\nu)$ be an effectively computable data type and (Φ,ϕ) its recursive basis. By definition, $\mathcal{D}$ is the closure of Φ under c-chains; $\mathcal{D}$ is c-chain complete and - inductive.

For some $R \in \mathbb{D}$, define Φ^R by $\Phi^R = \{\phi(i) | \phi(i) \subseteq R\}$. Then we have:

Lemma : Φ^R is a directed set and is an order-ideal in Φ.

Consider two computable data types $\mathcal{D}_0 = (\mathbb{D}_0,\nu_0)$ and $(\mathbb{D}_1,\nu_1)$ with recursive basis (Φ_0,ϕ_0) and (Φ_1,ϕ_1) respectively. Suppose that the intersection $\mathbb{D}_0 \cap \mathbb{D}_1$ contains some infinite set R. The sets Φ_0^R and Φ_1^R generated by R in Φ_0 and in Φ_1 respectively are ideals and if

$R = \bigcup_i\{\phi_0(x) | \phi_0(x) \in \Phi_0^R \;\&\; x \in W_{n_0}\}$ and $R = \bigcup_i\{\phi_1(x) | \phi_1(x) \in \Phi_1^R \;\&\; x \in W_{n_1}\}$,

then $\forall\, \phi_0(x),\ x \in W_{n_0},\ \exists \phi_1(u),\ u \in W_{n_1}\ [\phi_0(x) \subseteq \phi_1(u)]$,

and $\forall\, \phi_1(y),\ y \in W_{n_1},\ \exists \phi_0(v),\ v \in W_{n_0}\ [\phi_1(y) \subseteq \phi_0(v)]$.

This relation may be extended to the ideals of Φ_0 and of Φ_1 :

Definition 4. Let $\mathbb{A}$ and $\mathbb{B}$ be order ideals of Φ_0 and Φ_1 respectively, then $\mathbb{A}$ and $\mathbb{B}$ are cofinal iff $\forall\, \phi_0(x)\ \exists \phi_1(u)\ [\phi_0(x) \in \mathbb{A} \;\&\; \phi_1(u) \in \mathbb{B} \;\&\; \phi_0(x) \subseteq \phi_1(u)]$,

$\forall\, \phi_1(x)\ \exists \phi_0(u)\ [\phi_0(u) \in \mathbb{A} \;\&\; \phi_1(x) \in \mathbb{B} \;\&\; \phi_1(x) \subseteq \phi_0(u)]$.

Proposition 3. The computable data types are not closed under union or intersection.

Proof: The intersection $\mathcal{D}_0 \cap \mathcal{D}_1$ does not contain finite sets if $\mathcal{D}_0 \cap \mathcal{D}_1$ has incomparable, but cofinal basis in Φ_0 and Φ_1 .

Clearly, each $R \in \mathcal{D}_0 \cap \mathcal{D}_1$ is generated by some cofinal $\mathbb{A} \subseteq \Phi_0$, $\mathbb{B} \subset \Phi_1$, but the closures with respect to c-chains of $\mathbb{A}$ and $\mathbb{B}$ respectively are not identical in general: if C is a c-chain of $\mathbb{A}$ and lub $C = R_0 \subset R$ then R_0 is the lub of some c-chain of $\mathbb{B}$ iff there are cofinal ideals $\mathbb{A}' \subset \mathbb{A}$ and $\mathbb{B}' \subset \mathbb{B}$ generated by R_0 in Φ_0 and in Φ_1 respectively. Let $\Phi_0' \subseteq \Phi_0$ and $\Phi_1' \subseteq \Phi_1$ be the greatest cofinal subsets of Φ_0 and Φ_1 respectively. The union $\Phi = \Phi_0' \cup \Phi_1'$ is still a poset with a recursive boundedness-predicate if each of Φ_0' and Φ_1' is, but it is not any longer an upper semi-lattice, since

$U_\Phi(\phi_0(x),\phi_1(y)) \Rightarrow \exists \phi_1(v)[\phi_0(x) \subseteq \phi_1(v)] \;\&\; U_{\Phi_1}(\phi_1(v),\phi_1(y))$, and

$U_\Phi(\phi_0(x),\phi_1(y)) \Rightarrow \exists \phi_0(u)[\phi_1(y) \subseteq \phi_0(u)] \;\&\; U_{\Phi_0}(\phi_0(x),\phi_0(u))$,

so that there were two different lub's : $\text{lub}_{\Phi_1}(\phi_1(v),\phi_1(y))$ and $\text{lub}_{\Phi_0}(\phi_0(x),\phi_0(u))$, a contradiction. Hence the non-closure for the union. □

In order to obtain an upper semi-lattice Φ has to be completed by lub's for all $(\phi_0(x),\phi_1(y))$ belonging to cofinal sets. Different completions can be obtained; the problem if there exists a least completion $\bar{\Phi}$ of Φ or, more generally, a least computable data type $\mathcal{D}$ containing the intersection $\mathcal{D}_0 \cap \mathcal{D}_1$ (or the union $\mathcal{D}_0 \cup \mathcal{D}_1$) seems to be an open problem.

REFERENCES

1. SCOTT D., *Outline of a mathematical theory of computation*. Proc. 4th Ann. Princeton Conf. on Information Sciences and Systems, pp. 169-176.

2. SCOTT D., *Data types as lattices*. SIAM J. Computing, Vol. 5 Nr 3, September 1976.

3. LEHMANN D.J., SMYTH M.B., *Data types*. 18th Ann. Symp. Fondations of Computer Science, 1977, pp. 7-12.

4. SMYTH M.B., PLOTKIN G.D., *The Category-Theoretic Solutions of Recursive Domain Equations*. 18th Ann. Symp. Foundations of Computer Science, 1977, pp. 13-17.

5. THATCHER J.W., WAGNER E.G., WRIGHT J.B., *Data Type Specification; Paramaterization and the Power of Specification Techniques*. Proceedings of the Tenth Ann. ACM Symp. on Theory Computing 1978, pp. 119-132.

6. ERSHOV Ju.L., *Theorie der Numerierungen I.*, Zeitschr. f. math. Logik und Grundlagen d. Math. Bd. 19 (1973), pp. 289-388.

7. ERSHOV Ju.L., *Theorie der Numerierungen II*, same Zeitschr. Bd 21 (1975), pp. 473-584.

8. MAYOH B.H., *Data Types as Functions*, Proceedings MFCS'78, Springer Lecture Notes in Computer Science pp. 56-70.

PROGRAM EQUIVALENCE AND PROVABILITY

G. COUSINEAU
L.I.T.P.
Université PARIS VI
75221 - PARIS 5 - FRANCE

P. ENJALBERT
THOMSON-CSF/LCR
Domaine de Corbeville, B.P.10
91401 - ORSAY - FRANCE

SUMMARY

Given a Hoare-like deduction system in which can be proved partial correctness assertions of the form [P] S [Q], where S is a program and P, Q are first-order formulas, we are interested in the following question : "If $\Vdash$[P] S_1 [Q]([P] S_1 [Q] is provable) and S_2 is equivalent to S_1, does it imply that $\Vdash$[P] S_2 [Q] ?" We prove here that the answer is yes for syntactic or Ianov equivalence although it is no for Paterson equivalence or any semantical equivalence. The proofs are based on a syntactic tree formalism for program semantics (COUSINEAU-NIVAT [5], COUSINEAU [6]) and the existence of a complete proof system for these trees.

INTRODUCTION

Most of the papers published in the last three years concerning Hoare Logic have concentrated on the problem of completeness. Following COOK [3], most of the authors have assumed that the expressivity of the assertion language was a reasonable requirement and studied whether or not it implied completeness in different senses for various classes of programs (see for instance CLARKE [2], HAREL, MEYER, PRATT [10], GORELICK [9]). On the other hand, very little attention has been paid to the cases where the assertion language is not necessarily expressive and consequently incompleteness occurs even in the simple case of flowcharts. Yet many interesting computation domains such as trees, lists etc... have no expressive language. It seems to us that lack of interest comes from the fact that people had no tool to compare different kinds of incompleteness. A Hoare like system can be incomplete for some stupid reason (one has forgotten to put a rule for concatenation into it) or for some much deeper one (see : APT, BERGSTRA, MEERTENS [1], WAND [14]). Here we propose the following method : any Hoare like system induces an equivalence between programs in which two programs are equivalent iff one can prove the same thing about them. (Let us call it equiprovability). The classification of these equivalences will give information about the power of the proof system. In this paper, we begin with comparing this new equivalence with more traditional ones. For the purpose, the syntactic tree formalism that we use seems much more appropriate than the language theory formalism used so far by dynamic logic, as it will appear in the following.

I - PROGRAMS AND PROGRAM TREES

Given a first-order language L defined on sets P, F, V respectively of predicate symbols, function symbols and variables, we denote by Form (L) and Term (L) the formulas and terms of L. We call test any atomic formula and assignment any pair $x \leftarrow u$ where $x \in V$ and $u \in Term(L)$

We denote by Te and As the sets of tests and assignments. Given a set $Ex = \{\omega i / i \in \mathbb{N}\}$ of exit instructions, the programs we consider are expressions built according to the following rules :

- An assignment or an exit instruction is a program .
- If S_1, S_2, S are programs and r is a test, then

  $S_1 ; S_2$, <u>if</u> r then S_1 <u>else</u> S_2 <u>fi</u>, and <u>do</u> S <u>od</u> are programs.

The instruction <u>do</u> <u>od</u> is to be interpreted as a doforever and ωi as instructions that cause an exit of i embedded do loops if i>0. $\omega 0$ is introduced for homogeneity and just passes control to the next instruction. The use of this kind of loop causes some modification in the rule of our proof system compared to that of Hoare but is very rewarding in defining semantics of programs.

Giving respectively arity 0, 1 and 2 to symbols of Ex, As and Te, we call tree program (MANNA [13]) any finite or infinite tree built from these alphabets.

The set of tree programs would be denoted $M^{\infty}(Te \cup As \cup Ex)$ in the notations of COURCELLE-NIVAT [4]. To any program S is associated a tree program t(S) called its syntactic tree which is intuitively the tree obtained by splitting all its junctions and unwinding all its loops. Formally, the syntactic tree t(S) associated to a program S is defined by COUSINEAU-NIVAT [5] in the present symposium.

For our purpose it is convenient to define the semantics of programs via their syntactic trees. To do that we use the Doner formalism [7] in which a tree program is a partial mapping $t: \{1,2\}^* \rightarrow Te \cup As \cup Ex$ satisfying the following conditions :

(i) $\forall m, m' \in \{1,2\}^*$ $m \in dom(t)$ and m' is a left factor of m imply $m' \in dom(t)$,

(ii) $\forall m \in dom(t)$ t(m) belongs to Te, As or Ex according to the fact that the number of $i \in \{1,2\}$ such that $mi \in dom(t)$ is 2, 1 or 0.
(mi denotes the concatenation of the word m and the number i considered as a symbol.)

We shall denote by t/m the subtree of root m in t defined by (a/m)(m')=a(m m'). Let I be an interpretation of the language L, D_I the domain of I and $\{f_I/f\in F\}$ and $\{p_I/p\in P\}$ the interpretation of function and predicate symbols as partial functions and predicates over D_I. To every term $u \in$ Term(L) corresponds some partial function u_I and to every formula $w \in$ Form(L) corresponds some partial predicate w_I. If a program or a tree program uses k variables $x_1,\ldots,x_k$, for any u or atomic w appearing in it, we consider u_I and w_I as k-ary. Given a data (an element $\vec{d} \in D_I^k$) a tree-program t performs a computation $\{(\alpha_i,\beta_i)\}$ $\alpha_i \in$ dom(t), $\beta_i \in D_I^k$ defined as follows :

- $\alpha_0=\varepsilon$ $\beta_0=\vec{d}$
- if $t(\alpha_i)=$'$x_j \leftarrow u$' $\in$ As, then if $u_I(\beta_i)$ is undefined, the computation stops and its result is undefined; otherwise $\alpha_{i+1}=\alpha_i 1$ and β_{i+1} is β_i in which the jth element has been replaced by $u_I(\beta_i)$
- if $t(\alpha_i)=w \in$ Te, then if $w_I(\beta_i)$ is undefined, the computation stops and its result is undefined; otherwise $\beta_{i+1}=\beta_i$ and $\alpha_{i+1}=\alpha_i 1$ or $\alpha_i 2$ depending on whether $w_I(\beta_i)$ is true or false.
- if $t(\alpha_i)=\omega_j$, the computation stops and its result is $t_I(\vec{d})=(\beta_i,j)$. If a program S has been correctly written, t(S) contains only $\omega 0$ but the trees associated to its parts can contain ωi for i>0. That is why the result of a computation is both a value and a level of exit.

So to every tree program using k variables and every interpretation I is associated a k-ary partial function t_I. To every program S and interpretation I is associated $S_I=t(S)_I$.

Now, we can define between programs the following equivalences :

(1) Syntactic equivalence : $S_1 \equiv S_2$ iff $t(S_1)=t(S_2)$

(2) Semantic equivalences : $S_1 \equiv_\rho S_2$ iff $S_{1_I}=S_{2_I}$ for all I belonging to some class ρ of interpretations.

Syntactic equivalence is stronger than any semantic equivalence. Moreover, between semantic equivalences we distinguish Paterson equivalence which is stronger than all the others : $S_1 \equiv_p S_2$ iff $S_{1_I}=S_{2_I}$ for all interpretations I.

It is well known (see : LUCKHAM, PARCK, PATERSON [12]) that in this definition "for all interpretation" can be replaced by "for all free interpretation". A free (or Herbrand) interpretation H is an interpretation in which the domain is Term(L), the functions f_H are symbolic and the data is always taken to be $\vec{x}=(x_1,\ldots,x_k)$. To every program tree t we associate functions $Cond_t$ and F_t which, given any point $m \in$ dom(t), give $Cond_t(m)$=a finite conjunction of atomic formulas such that for any interpretation I satisfying it the computation of t for I will reach the point m and $F_t(m)$=the value of the variables at point m in any free interpretation. We shall also write $Cond_t(m,m')$ and $F_t(m,m')$ for $Cond_{t/m}(m')$ and $F_{t/m}(m')$.

Finally, for $F \in Form(L)$, $u \in Term(L)$, x a variable, F^u_x is the result of substituting u for every free occurrence of x in F.

II - DEDUCTIVE SYSTEMS FOR PROGRAMS AND PROGRAM TREES

A general study of such deductive systems can be found in ENJALBERT [8]. Here we introduce a new one, for so-called "Program trees with assertions", and give some results needed to prove the announced theorems.

PROGRAM TREES WITH ASSERTIONS (p.t.w.a.)

Let $L_1 \subseteq L_2$ be two first order languages. A p.t.w.a. on (L_1,L_2) is a couple $T=\langle t,a\rangle$ where :

- t is a program tree on L_1
- a is a mapping : $dom(t) \to Form(L_2) \cup \{U\}$
  (U=Undefined ; that is : a is a partial mapping : $dom(t) \to Form(L_2)$)

Set dom(T)=dom(t). There is an immediate extension of the magma operations for p.t.w.a. : $\langle r,F\rangle(T_1,T_2)$ and $\langle x\leftarrow u,F\rangle(T_1)$, for r a test, $x\leftarrow u$ an assignment, $F \in Form(L_2)\cup\{U\}$, and T_1, T_2 two p.t.w.a. . Same remark for the syntactic order.

THE DEDUCTIVE SYSTEM FOR p.t.w.a.

In a program tree many exit symbols ω_i may occur. To define partial correctness of program trees we shall not consider a single exit condition but associate one condition to every ω_i.

A set of exit assertions in L is a sequence of formulae in Form(L) : $Q=(Q_i)_{i\in\mathbb{N}}$, a finite number of which are different from True.

Let $L_1 \subseteq L_2$ be two languages. We are interested in triples [P] T [Q] where T is a p.t.w.a. on(L_1,L_2), $P \in Form(L_2)$, and Q is a set of exit assertions in L_2.

The next step is to define two functions W_p (Weakest precondition) and Pr (Premisses) in the following recursive way :

For T, Q as above, T finite, $W_p(T,Q) \in Form(L_2)$ and Pr(T,Q) is a finite set of triples [P'] T' [Q] .

1st case : $T=\langle t,a\rangle$ with $a(\varepsilon)\neq U$

- $W_p(T,Q)=a(\varepsilon)$
- $Pr(T,Q)=\{[a(\varepsilon)]\langle t,a'\rangle[Q]\}$, with $a'(\varepsilon)=U$ and $\forall m\neq\varepsilon\ a'(m)=a(m)$

2nd case : $T=\langle t,a\rangle$ with $a(\varepsilon)=U$

If $t(\varepsilon)=\omega_i$
- $W_p(T,Q)=Q_i$
- $Pr(T,Q)=\emptyset$

If $t(\varepsilon)=x\leftarrow u$, then $T=\langle x\leftarrow u,U\rangle(T_1)$

- $W_p(T,Q)=[W_p(T_1,Q)]^u_x$
- $Pr(T,Q)=Pr(T_1,Q)$

If $t(\varepsilon)=r$, then $T=\langle r,U\rangle(T_1,T_2)$

- $W_p(T,Q)=(r\rightarrow W_p(T_1,Q))\wedge(\neg r\rightarrow W_p(T_2,Q))$
- $Pr(T,Q)=Pr(T_1,Q) \cup Pr(T_2,Q)$

<u>The deductive system τ</u> provides proofs for formulae which are either triples [P] T [Q] or formulae in $Form(L_2)$, relatively to any first order theory T written in L_2.

<u>Axioms</u> : $\mathsf{T} \Vdash_\tau F$ if $F \in Form(L_2)$ and $\mathsf{T} \vdash F$

<u>Rule 1</u> : If T is finite :

$$\frac{\mathsf{T} \Vdash_\tau P\rightarrow W_p(T,Q) \quad \mathsf{T} \Vdash_\tau Pr(T,Q)}{\mathsf{T} \Vdash_\tau [P]\ T\ [Q]}$$

<u>Rule 2</u> : If $(T_k)_{k\in K}$ is a directed family of p.t.w.a. :

$$\frac{\forall k\in K\ \mathsf{T} \Vdash_\tau [P]\ T_k\ [Q]}{\mathsf{T} \Vdash_\tau [P]\ \underset{k\in K}{Lub}(T_k)\ [Q]}$$

The meaning of assertions in p.t.w.a. should be clear from the following Path Theorem. In particular, the reader can convince himself that $\mathsf{T} \Vdash_\tau [P]\langle t,a\rangle[Q]$ iff in any model of T, t is partially correct with respect to P, Q <u>and</u> the intermediate assertions in a.

<u>The Path Theorem</u>

$\mathsf{T} \Vdash_\tau [P]\ T\ [Q]$ $(T=\langle t,a\rangle)$

$\iff \forall m\leq m'\leq m'' \in dom(T)$, $a(m)$ and $a(m')\neq U$, $t(m'')=\omega_i$:

(1) $\mathsf{T} \vdash (P\wedge Cond_t(m)) \rightarrow a(m)(F_t(m))$

(2) $\mathsf{T} \vdash (a(m)\wedge Cond_t(m,m')) \rightarrow a(m')(F_t(m,m'))$

(3) $\mathsf{T} \vdash (a(m')\wedge Cond_t(m',m'')) \rightarrow Q_i(F_t(m',m''))$

<u>THE DEDUCTIVE SYSTEM π FOR PROGRAMS</u>

We consider triples [P] S [Q] as before, except that S is now a program. The system π is quite classical but for rule π-3 which is an extension of the usual while-rule.

<u>Axioms</u> : $\mathsf{T} \Vdash_\tau [P]\ \omega_i\ [Q]$ if $\mathsf{T} \vdash P\rightarrow Q_i$.

Rules :

π-1
$$\frac{T \Vdash_{\pi} [P]S[Q]}{T \Vdash_{\pi} [P^u_x]x\leftarrow u;S[Q]}$$

π-2
$$\frac{T \Vdash_{\pi} [P\wedge r]S_1[Q] \quad T \Vdash_{\pi} [P\wedge\neg r]S_2[Q]}{T \Vdash_{\pi} [P] \underline{\text{If}}\ r\ \underline{\text{then}}\ S_1\ \underline{\text{else}}\ S_2\ \underline{\text{fi}}\ [Q]}$$

π-3
$$\frac{T \Vdash_{\pi} [P]S[(P,Q_0,Q_1,\ldots)]}{T \Vdash_{\pi} [P]\ \underline{\text{do}}\ S\ \underline{\text{od}}\ [(Q_0,Q_1,\ldots)]}$$

π-4
$$\frac{T \Vdash_{\pi} [P]S[Q]}{T \Vdash_{\pi} [P']S[Q]} \quad \text{if} \quad T\vdash P'\rightarrow P$$

π-5
$$\frac{T \Vdash_{\pi} [P]S_1[Q] \quad T \Vdash_{\pi} [Q_0]S_2[Q']}{T \Vdash_{\pi} [P]S_1\ ;\ S_2[Q']} \quad \text{if} \quad \forall_i\neq 0\ Q'_i=Q_i$$

π has the usual properties (see ENJALBERT [8]).For instance, validity is a consequence of the lemma 1 in section III and the Path Theorem.

III - PROGRAM EQUIVALENCE AND PROVABILITY

PATERSON EQUIVALENCE

Theorem 1 : There are two programs $S_1\equiv_p S_2$, and (P,Q) such that : $\Vdash_{\pi}[P]S_1[Q]$ but Not $\Vdash_{\pi}[P]S_2[Q]$

For any assignment A, the trivial program S_1=do A ; ω0 od never halts in any interpretation, and : $\Vdash_{\pi}$[True]S_1[False]. But {S/[True]S[False] is valid} is not r.e. (LUCKHAM, PARK, PATERSON [12]). Thus there is a S_2 which never halts, and therefore $S_2\equiv_p S_1$, while Not $\Vdash_{\pi}$[True]S_2[False].

Of course, theorem 1 is valid replacing Paterson Equivalence by any semantic equivalence.

SYNTACTIC EQUIVALENCE

Theorem 2 :
$$\left.\begin{array}{l} S_1\equiv S_2 \\ T \Vdash_{\pi} [P]S_1[Q] \end{array}\right\} \Rightarrow T \Vdash_{\pi} [P]S_2[Q]$$

Sketch of proof : in 3 lemmas.

Definitions :

- T = <t,a> is saturated iff a is total.
- T = <t,a> is unified iff : ∀m,m' ∈ dom(T), t/m=t/m'⟹a(m)=a(m') (two identical subtrees of t have identical top-assertions).
- T = <t,a> has property U iff T is saturated and : ∀m ∈ dom(T), {a(m')/a/m'=a/m} is finite.

<u>Lemma 1</u> : If S is a program and $T \Vdash_{\pi} [P]S[Q]$, there is a p.t.w.a. $T(S)=<t(S),a>$ such that : (1) $T \Vdash_{\tau} [P]T(S)[Q]$, (2) T(S) has property U.

Given a proof Π of [P]S[Q] and the triples [P']S'[Q'] proved at each stage of Π (the S's are subformulae of S), we build an assertions tree a that keeps memory of the P's.

<u>Lemma 2</u> : If T = <t,a> has property U, there is an a^u such that $<t,a^u>$ is unified and $T \Vdash_{\tau} [P]T[Q] \Rightarrow T \Vdash_{\tau} [P]<t,a^u>[Q]$.

By U, $\forall m \in dom(T) \bigvee_{\substack{a/m \\ =a/m'}} a(m)$ is equivalent to a finite formula. Take $a^u(m)$ to be that formula. We can verify that the three conditions of the Path Theorem are satisfied.

<u>Lemma 3</u> : If S is a program and T=<t(S),a> is a saturated, unified p.t.w.a, then $T \Vdash_{\tau} [P]T[Q] \Rightarrow T \Vdash_{\pi} [P]S[Q]$.

Suppose for instance that S=<u>do</u> S' <u>od</u>, t(S)=*(t(S')). Since T is saturated, a(ε) exists (≠U). Now, if $m \in \{1,2\}^*$ is any occurrence of t(S) in t(S), a(m)=a(ε) because T is unified. a(ε) will be a loop invariant we can use to apply rule π-3.□

<u>IANOV EQUIVALENCE</u>

An intermediate level of equivalence can be defined between syntactic and Paterson equivalence by associating to any program S (or tree program t) a one-variable program $\overline{S}$ (or tree program $\overline{t}$) (IANOV [11]).

For example, if S = z←1 ; <u>do</u> <u>if</u> p(x) <u>then</u> z←f(y,z) ; x←g(x) <u>else</u> ω1 <u>fi</u> <u>od</u> , then $\overline{S}$ = X←A(X) ; <u>do</u> <u>if</u> P(X) <u>then</u> X←F(X) ; X←G(X) <u>else</u> ω1 <u>fi</u> <u>od</u> .

S_1 and S_2 are said to be Ianov-equivalent ($S_1 \equiv_I S_2$) iff $\overline{S}_1 \equiv_p \overline{S}_2$.

It is possible to define canonical tree-programs in such a way that two canonical tree-programs are Ianov-equivalent iff they are identical. This enables us to prove the following.

<u>Theorem 3</u> : $T \Vdash_{\pi} [P]S_1[Q]$ and $S_1 \equiv_I S_2$ imply $T \Vdash_{\pi} [P]S_2[Q]$.

<u>WEAK EQUIVALENCE</u>

Finally, we can study the converse problem : given a class C of formulae, suppose one can prove in π the same specifications (P,Q) in C for two programs S_1 and S_2; what can we say about S_1 and S_2 ? Our last theorem is a first answer to that kind of problem. Let us recall that two programs in L S_1 and S_2 are <u>weakly equivalent</u> ($S_1 \equiv_w S_2$) iff for any interpretation of S_1 and S_2 (structure for L + input), if S_1 and S_2 halt, they compute the same output.

Theorem 4 : Let S_1 and S_2 be two programs in L such that, for every (P,Q) in L without quantifiers, $T \Vdash_{\overline{\pi}} [P]S_1[Q] \Rightarrow T \Vdash_{\overline{\pi}} [P]S_2[Q]$. Then $S_1 \equiv_w S_2$.

Sketch of proof : Let I be any interpretation for S_1 and S_2. Suppose S_1 halts in I ; the computation follows a (finite) path in $t(S_1)$, leading to an occurrence m of an ωi, and : $\Vdash_{\overline{\pi}} [Cond_{t(S_1)}(m) \wedge \overline{x}=\overline{y}]\ S_1\ [\overline{x}=F_{t(S_1)}(m)(\overline{y})]$ ($\overline{x}=(x_1,\ldots,x_k)$ are the variables of S_1 and S_2 ; $\overline{y}=(y_1,\ldots y_k)$ are new variables).

Thus if S_2 halts for I and the hypothesis of theorem 4 is satisfied, S_2 certainly computes the same output. That is exactly $S_2 \equiv_w S_1$.□

CONCLUSION.

In fact, different notions of equiprovability between programs can be defined, according to the power of the assertion language and the theories T considered. One can try to compare the usual syntactic or semantic equivalences with these new ones, and eventualy find equiprovability relations equivalent to them. Research in that direction, prolonging the present paper, is being performed.

REFERENCES.

1. Apt, K. Bergstra, J. and Meerteens L., Recursive assertions are not enough-or are they ? Theor. Comput. Sci. 8, 1 (February 1979).

2. Clarke,E., Programming language constructs for which it is impossible to obtain "good" Hoare like axiom systems. Fourth ACM Symposium on Principles of Programming Languages (1977).

3. Cook, S., Soundness and completeness for program verification, SIAM Journal of Computing 7,1 (1978).

4. Courcelle, B. and Nivat, M., The algebraic semantics of recursive program schemes. MFCS 1978, Springer Lecture Notes n°62.

5. Cousineau, G.,The algebraic structure of flowcharts.This symposium.

6. Cousineau, G., An algebraic definition for control structures. To appear in Theor. Comput. Sci.

7. Doner, Tree acceptors and some of their applications. J. of Comput. and System Sci. 4 (1970).

8. Enjalbert, P., Systèmes de déduction pour les arbres et les schémas de programmes. 4ème Colloque de Lille sur les arbres en algèbre et en programmation (1979).

9. Gorelick, G., A complete axiomatic system for proving assertions about recursive and non-recursive programs. Technical Report n° 75. University of Toronto (January 1975).

10. Harel, D., Meyer, A. and Pratt, V., Computability and completeness in logics of programs. 9th ACM symposium on Theory of Computing. Boulder (1977).

11. Ianov, I., The logical schemes of algorithms. In: Problems of cybernetics, Pergamon Press (1960).

12. Lukham, D., Park, D., and Paterson, M., On formalized computer programs. J. of Comput. and System Sci. 4 (1970).

13. Manna, Z., Mathematical theory of computation, Mc Graw-Hill (1974).

14. Wand, M., A new incompleteness result for Hoare's systems. JACM 25, 1 (1978).

INTERACTIVE L SYSTEMS WITH ALMOST INTERACTIONLESS BEHAVIOUR†

K. Culik II
Department of Computer Science
University of Waterloo
Waterloo, Ontario, Canada
N2L 3G1

and

J. Karhumäki
Department of Mathematics
University of Turku
Turku, Finland

ABSTRACT

A restricted version of interactive L systems is introduced. A P2L system is called an essentially growing 2L system (e-G2L system) if every length-preserving production is interactionless (context-free). It is shown that the deterministic e-G2L systems can be simulated by codings of propagating interactionless systems, and that this is not possible for the nondeterministic version. Some interesting properties of e-GD2L systems are established, the main result being the decidability of the sequence equivalence problem for them.

1. Introduction

The area of L systems has had a rapid growth, see [5], however this is mainly due to their mathematical investigation rather than their biological application. For a biologist the deterministic models are for the most relevant as emphasized by A. Lindenmayer [4]. For this reason and also for their mathematical simplicity the DOL and DIL systems, the deterministic versions of the basic interactionless and interactive systems, are of special importance.

---

† This research was supported by the National Sciences and Engineering Council of Canada, Grant No. A 7403.

Since the latter are much more powerful than the former it seems to us important to investigate systems of intermediate capability. One way to get such systems is if we allow interaction only when cells are dividing but not when they are merely changing states. Quite surprisingly in the case of propagating systems (no cells dying) the behaviour of such systems will be shown to be closer to the interactionless systems rather than to the interactive ones.

Our results seem to be well motivated mathematically, too. It is well known, see Baker's Theorem in [3], that certain restrictions on the form of productions of a context-sensitive grammar make the grammar essentially lose its "context-sensitiveness", i.e. to generate a context-free language. We will introduce a different kind of restriction on the form of productions of a deterministic context-sensitive parallel rewriting system (D2L system) which has essentially the same effect, that is, it makes an interactive (context-sensitive) system behave in an almost interactionless (context-free) manner. Similarly, as in the case of sequential context-sensitive grammars, the restriction seems to be a mild one, and therefore the obtained results are rather surprising. We say that a D2L system is essentially growing (an e-GD2L system) if the system is propagating (nonerasing) and each of its productions which is actually context-sensitive is strictly growing. In other words an e-GD2L system is a PD2L system such that each of its nongrowing productions is actually context independent.

As our basic result we show that every e-GD2L system can be simulated by a coding of a propagating DOL system (CPDOL system). Then it is easy to show that both the languages and the sequences generated by e-GD2L systems are properly between those generated by PDOL and CPDOL systems. Hence each e-GD2L language can be generated by a nondeterministic context-free system with nonterminals (EOL system).

In Section 4 we obtain several applications of the basic simulation result and the method of its proof. First we observe that the length sequence equivalence problem for e-GD2L systems is decidable. Then we demonstrate that despite the fact that e-GD2L growth functions are the same as the PDOL growth functions, it is possible to realize some growth functions with a considerably smaller number of symbols by e-GD2L systems than by PDOL systems.

We show that the "cell number minimization problem", see [7, page 116], is decidable for e-GD2L systems.

We conclude Section 4 with the main result of this paper, namely, the decidability of the sequence equivalence problem for e-GD2L systems. Hence e-GD2L systems are the most complicated type of L systems known for which this important problem is decidable. Our result is somewhat surprising in the view that this problem is undecidable for PD1L systems [8].

In the last section we show that Theorem 1 cannot be extended to nondeterministi e-G2L systems. This extension would mean that e-G2L languages were included in CPOL languages, therefore also in EOL languages, see [5]. However, this is impossible since we will show that each ETOL language can be expressed as $h(L) \cap R$ for some homomorphism h, e-G2L language L and regular set R.

In this paper we omit most proofs, only some outlines will be given. For detailed proofs see [2].

2. Preliminaries and Basic Definitions

A 2L system is a triple $G = \langle \Sigma,P,w \rangle$ where Σ is a non-empty alphabet, w is an element of Σ^* and P is a finite relation from
$V = \{\$\} \times \Sigma \times \{\$\} \cup \{\$\} \times \Sigma^2 \cup \Sigma^2 \times \{\$\} \cup \Sigma^3$ into Σ^* satisfying the following completeness condition: For each $u \in V$ there exists at least one v in Σ^* such that $(u,v) \in P$. An element (u,v) of P is called a production and usually written $u \to v$; the letter $\$$ not in V is called the environmental symbol. The relation $\Rightarrow_G$ (or $\Rightarrow$ in short) on Σ^* is defined as follows. For words x and y $x \Rightarrow y$ holds true if and only if one of the following conditions is satisfied: (i) $x \in \Sigma$ and $(\$,x,\$) \to y \in P$, (ii) $x = x_1x_2$, $y = y_1y_2$, with $x_1,x_2 \in \Sigma$, $y_1,y_2 \in \Sigma^*$, and $(\$,x_1,x_2) \to y_1$, $(x_1,x_2,\$) \to y_2 \in P$, (iii) $x = x_1 \ldots x_n$, $y = y_1 \ldots y_n$, with $n \geq 3$, $x_1,\ldots,x_n \in \Sigma$, $y_1,\ldots,y_n \in \Sigma^*$ and $(\$,x_1,x_2) \to y_1$, $(x_1,x_2,x_3) \to y_2, \ldots, (x_{n-2},x_{n-1},x_n) \to y_{n-1}$, $(x_{n-1},x_n,\$) \to y_n \in P$. Let $\Rightarrow^*$ be the transitive and reflexive closure of $\Rightarrow$. The language generated by G is $L(G) = \{x \mid w \Rightarrow^* x\}$.

In this paper propagating systems, i.e. systems where erasing productions are not allowed, are considered. Moreover, in most cases systems are assumed to be deterministic in the following sense. A 2L system $G = \langle \Sigma,P,w \rangle$ is deterministic (abbreviated a D2L system) if the relation P is a function from V into Σ^*. For a D2L system G the conditions $x \Rightarrow_G y$ and $x \Rightarrow_G y'$ imply $y = y'$, and hence G defines the sequence

$$s(G) = w_0,w_1,\ldots$$

where $w_0 = w$ and $w_i \Rightarrow_G w_{i+1}$ for $i = 0,1,\ldots$. Such sequences are called D2L sequences. In the deterministic case we will also write $\delta(a,b,c) = d$ when $(a,b,c) \to d$.

Let $G = \langle\Sigma,P,w\rangle$ be a 2L system. A production $(x,a,y) \to \alpha$, with $x,y \in \Sigma \cup \{\$\}$, is called <u>context-free</u> if $\{(z,a,v) \to \alpha \mid z,v \in \Sigma \cup \{\$\}\} \subseteq P$. So the abbreviation $a \to \alpha$ for the context-free production $(x,a,y) \to \alpha$ can be used. The productions of G, which are not context-free, are called <u>context-sensitive</u>. In the deterministic case we may also talk about context-free and context-sensitive letters.

For $w \in \Sigma^*$, $|w|$ denotes the length of w, for a set S $|S|$ denotes the cardinality of S. Now we introduce the basic notions of this paper.

<u>Definition</u> A 2L system $G = \langle\Sigma,P,w\rangle$ is <u>strictly growing</u> (an s-G2L system) iff $|v| \geq 2$ for each $u \to v$ in P. System G is <u>essentially growing</u> 2L system (e-G2L system) iff it is propagating and $|v| \geq 2$ for each context-sensitive production $u \to v$ in P. (We will be mainly interested in deterministic e-GL2 systems (e-GD2L systems).)

Every production of an s-G2L system must be length-increasing, while an e-G2L system may have length-preserving productions, if they are context-free. Therefore, any propagating context-free productions are allowed in e-G2L systems and thus the e-GD2L systems include all PD0L systems. Also the s-GD2L systems are a special case of the e-GD2L systems and we introduce them mainly to facilitate the explanation of some proof techniques in a simple setting before a general proof. However, all the results concerning deterministic systems will be proved for the more general case of e-GD2L systems.

Throughout this paper we use the basic notions and results of formal language theory and L systems, we refer the reader, e.g., to [6] and [5]. In particular, by a coding we mean a letter-to-letter homomorphism. Moreover, the maximal prefix (resp. suffix) of a string x not longer than k is denoted by $\mathrm{pref}_k(x)$ (resp. $\mathrm{suff}_k(x)$).

3. The Interactionless Simulation of Restricted Interaction

In this section we consider essentially (strictly) growing D2L systems. We first observe that s-GD2L systems can be simulated by CPD0L systems in the sense that any s-GD2L sequence is obtained as a coding of a PD0L sequence. The result is seen as follows. Let $G = <\Sigma,\delta,w>$ be an s-GD2L system and let $(b,c,d) \to \gamma$ be one of its productions. Now, we consider the letter c in the context ...abcde... for some letters a,b,d and e . We know how to rewrite c in that context (for this purpose the context ...bcd... is sufficient) but we also know what are the neighbours of the result (i.e. γ) of the length two. This follows since they are determined by words $\delta(a,b,c)$ and $\delta(c,d,e)$. So the use of quintuples (a,b,c,d,e) makes it possible to simulate the derivations of G by a PD0L system. We omit the details since the result is only a special case of the following stronger theorem.

Theorem 1 For any e-GD2L system $G = <\Sigma,\delta,w>$ there exist a PD0L system G' and a coding c such that $s(G) = c(s(G'))$ and hence also $L(G) = c(L(G'))$.

We point out the basic idea of a proof. For details see [2]. Essential is the fact that not only the intermediate but all the descendants of a word in a given context are determined by a fixed amount of context. More precisely, one can show the following. Given a in Σ and $x,y \in \Sigma^{2|\Sigma|}$. Define recursively the sequence

$$(*) \qquad x_n\alpha_n y_n \quad , \quad n \geq 0 \quad ,$$

by setting

$$x_0 = x \quad , \quad \alpha_0 = a \quad , \quad y_0 = y$$

and

$$x_{i+1} = \text{suff}_{2|\Sigma|}(\delta(?,x_i,\text{pref}_1(\alpha_i))) \quad ,$$
$$\alpha_{i+1} = \delta(\text{suff}_1(x_i),\alpha_i,\text{pref}_1(y_i)) \quad ,$$
$$y_{i+1} = \text{pref}_{2|\Sigma|}(\delta(\text{suff}_1(\alpha_i),y_i,?)) \quad ,$$

where $\delta(?,z,c)$ (resp. $\delta(c,z,?)$) denotes the word which is obtained from z without knowing its left (resp. right) neighbour. Then (*) is infinite, i.e. the rewriting does not terminate because of the lack of information about neighbours.

To clarify the above idea let us consider an e-GD2L system with the axiom 1011110 and the productions

$$1 \to 0 \quad ,$$
$$(0,0,0) \to 10 \quad ,$$
$$(b,0,c) \to 01 \quad , \quad \text{if } b \neq 0 \text{ or } c \neq 0 \; .$$

Now the sequence (*) starting from the third letter of the axiom and using context of length 4 is as follows:

$$10, 1 ,1100
$001, 0 ,000
1010, 01 ,1010
1001, 010 ,0010
1010, 01001 ,1001
1001,01001010,0010
. . . .

where we use commas "to separate words from their neighbours". It is not difficult to see that the sequence does not terminate. The fact that the left neighbour of length 4 is always obtained follows since the rewritings of 01 and 10 are independent of their right neighbours and since only the three rightmost letters are needed to generate words 1010 and 1001. It is a little more complicated to see that also the right neighbours, i.e. y_i's, are defined for all n. This is, however, clear after the observation that the rewriting of $\mathrm{suff}_2(\alpha_i)$ is context-independent.

Constructing the beginnings of all the (necessary) sequences (*) the productions for the simulating system can be defined. For instance in our example

($$10,1,1110) → ($001,0,000),
($001,0,000) → (1010,0,1101)(0100,1,1010) .

Of course, a coding c takes a triple into its middle component. It is important to note that, unlike in the case of s-GD2L systems, the neighbouring words, i.e. x_i's and y_i's, are not all of the same length.

Above we used context of length 4, i.e. of length $2|\Sigma|$. It can be shown that this amount is always enough. On the other hand this amount is also needed, in general. This follows since if we use in our example only context of length 3 then the sequence (*) will terminate:

$10, 1 ,110
010, 0 ,00
001, 01 ,10
010, 010 ,0
001,01001,?

Later on in (Theorem 7) it will be shown that the determinism is an essential assumption for Theorem 1 to hold. Also Theorem 1 cannot be generalized for D2L systems with strictly growing context-sensitive productions and arbitrary (possibly erasing) context-free productions. This is seen as follows. Let G be any PD2L system. We show that it can be simulated, in a sense, by a D2L system with strictly growing context-sensitive productions and arbitrary context-free productions. Such a system G' is defined in the following way. For any length-preserving production $(a,b,c) \to d$ in G G' contains productions

$$(a,b,c) \to \overline{d}\, a_\lambda \;, \quad \overline{d} \to d \quad \text{and} \quad a_\lambda \to \lambda$$

where $\overline{d}$ denotes the "barred copy" of d . Further for any production

$$(a,b,c) \to b_1 \dots b_n \quad \text{in } G \;, \quad \text{with} \quad n \geq 2 \;,$$

G' contains productions

$$(a,b,c) \to \overline{b}_1 \dots \overline{b}_n \quad \text{and} \quad \overline{b}_i \to b_i \quad \text{for } i = 1,\dots,n \;.$$

Then clearly

$$L(G) = L(G') \cap \Sigma^*$$

where Σ denotes the alphabet of G . Since we may choose L(G) not to be an EOL language we may also choose L(G') not to be an EOL language. Hence G' cannot be simulated by any DOL system in the sense of Theorem 1.

Let us denote by $\mathcal{L}_{\text{e-GD2L}}$ (resp. $S_{\text{e-GD2L}}$) the family of languages (resp. sequences) generated by e-GD2L systems. Then using Theorem 1 we get

<u>Theorem 2</u> $\mathcal{L}_{\text{PDOL}} \subsetneq \mathcal{L}_{\text{e-GD2L}} \subsetneq \mathcal{L}_{\text{CPDOL}}$

and $S_{\text{PDOL}} \subsetneq S_{\text{e-GD2L}} \subsetneq S_{\text{CPDOL}}$.

All inclusions above follow from Theorem 1 and the definition of the e-GD2L system. That the first inclusions are proper follows from Example in the next section. The strictness of the second inclusions, in turn, are seen by considering a suitable language over a* .

□

4. <u>Applications of the Simulation Result</u>

After demonstrating the position of our language family $\mathcal{L}_{\text{e-GD2L}}$ within the hierarchy of L families we now derive some interesting properties of our systems.

From Theorem 1 and the fact that the length sequence equivalence problem is decidable for PD0L systems, see [7], it follows:

Theorem 3 The length sequence equivalence problem for e-GD2L systems is decidable.

One of the consequences of Theorem 1 is that any e-GD2L growth function is a PD0L growth function, too. Hoever, as it is seen in the next example, the number of letters needed to realize a given function by an e-GD2L system may be much smaller than that needed to realize the same function by a PD0L system.

Example Let us define an e-GD2L system (or in fact an s-GD2L) G as follows. Its alphabet equals $\{1,\ldots,k\}$ and its axiom is 11. To define the productions let γ be the function which gives the lexicographic order of the set $\{(i,j) \mid i,j = 1,\ldots,k\ ,\ i < j\}$, i.e. $\gamma(1,2) = 1$, $\gamma(1,3) = 2, \ldots,$ $\gamma(k-1,k) = k(k-1)/2$. Now the productions of G are as follows

$$
\begin{aligned}
&(\$,1,1) \to 12\ , \\
&(1,1,\$) \to 13 \ldots 1k23 \ldots 2k \ldots (k-1)k\ , \\
&(-,i,j) \to ij\ , \qquad\qquad \text{for } i < j\ , \\
&(i,j,-) \to (ij)^{\gamma(i,j)}\ , \quad \text{for } i > j\ ,
\end{aligned}
$$

where - denotes that the element there is arbitrary. So the derivation starts as follows

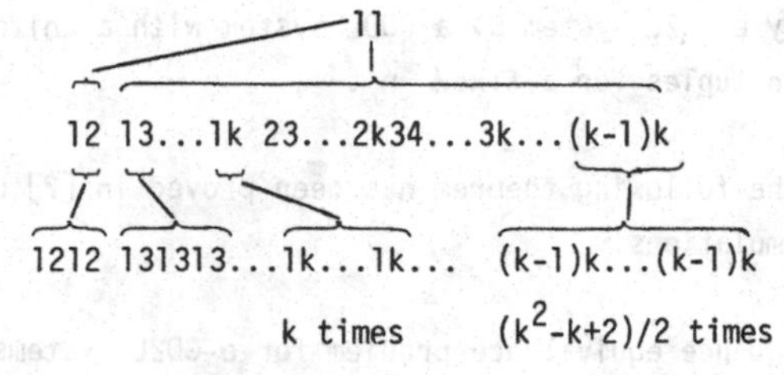

If we denote by f the growth function just defined we get

$$f(n) = 2 \sum_{j=1}^{(k^2-k)/2} (j+1)^n\ , \quad \text{for } n \geq 1\ .$$

This formula implies that any PD0L (or D0L) system generating the growth function of G must contain at least $(k^2-k)/2$ letters. However, f is realized by an e-GD2L system with k letters only.

The above example gives the motivation for the following definition. Let C be a class of deterministic L systems. The cell number minimization problem for C is the following: Given an arbitrary function f realized by a system in C. Is there an algorithm to find a system from C with an alphabet of the minimal cardinality such that its growth function equals f ?

The following theorem has been proved in [2].

Theorem 4 The cell number minimization problem for e-GD2L systems is decidable.

Now, we turn to consider the sequence equivalence problem for e-GD2L systems. In a subcase, i.e. in the case of s-GD2L systems, the decidability of this problem is a consequence of the simulation of these systems by PDOL systems and the decidability of the problem for PDOL systems. Indeed, two s-GD2L sequences $s(G)$ and $s(H)$ are equivalent if and only if the PDOL sequences $s(G')$ and $s(H')$, where G' and H' are the "quintuple PDOL systems" simulating G and H, are equivalent.

For e-GD2L systems the situation is more complicated, mainly due to the fact that the letters of G' in the proof of Theorem 1 are not of uniform length as words of Σ^*. Hence, two e-GD2L systems may be equivalent although the corresponding simulating systems of Theorem 1 are not. Moreover, it seems to be impossible to simulate an arbitrary e-GD2L system by a PDOL system with a uniform "length of letters", i.e. with n-tuples for a fixed n.

Nevertheless, the following theorem has been proved in [2] using a somewhat different type of simulations.

Theorem 5 The sequence equivalence problem for e-GD2L systems is decidable.

The detailed proof (see [2]) is rather long;we describe here only the basic idea behind it. To do so we must introduce some notation. Given an e-GD2L system $G = \langle\Sigma,\delta,w\rangle$, let

$$\Sigma_1 = \{a \in \Sigma \mid a \text{ is context-free}\} ,$$
$$\Sigma_s = \{a \in \Sigma_1 \mid \text{there exists } t > 0 \text{ such that } \delta^t(a) = a\} .$$

Further a subalphabet Δ of G is called unbounded if $\Delta \subseteq \Sigma_s$ and for each natural n and each a in Δ there exists a subword $x \in \Delta^*$ of a word in $L(G)$ such that

$\#_a(x) \geq n$ where $\#_a(x)$ denotes the number of a's in x . The <u>maximal unbounded subalphabet</u> of G is clearly unique and it is denoted by Δ_G .

Now let $G = \langle \Sigma,\delta,w \rangle$ and $H = \langle \Sigma,\nu,w \rangle$ be e-GD2L systems. It can be shown that if $s(G) = s(H)$ then G and H are "structurally similar" in the sense that their maximal unbounded subalphabets coincide, i.e. $\Delta_G = \Delta_H$. Let us refer to this alphabet as the common subalphabet of the pair (G,H) and let us denote it by $\Delta_{G,H}$.

It is not difficult to show that the maximal unbounded subalphabets of G and H can be effectively found. If they are different then the systems are nonequivalent. If they coincide then the pair (G,H) has a common subalphabet and we continue as follows. We define PDOL systems G' and H' (or, in fact, we must decompose G and H into several such systems) simulating G and H in the sense of Theorem 1 (cf. also example there). So each occurrence of a letter in s(G) has as a symbol of G' two-sided context of finite length. Moreover, this context is always either of a fixed length N or of the form ax or xa where $|x| \leq N-1$, $x \in (\Sigma \backslash \Delta_{G,H})^*$ and $a \in \Delta_{G,H}$ (of course, modifications are needed near the edges of the words). So it follows that

$$s(G) = s(H) \quad \text{iff} \quad s(G') = s(H')$$

and hence the result follows from the decidability of the equivalence problem for PDOL systems, see [1].

By Theorems 3, 4 and 5 e-GD2L systems have many favourable properties which general D2L systems do not have. Indeed, the decision problems in Theorems 3 and 5 are undecidable for PD1L systems, see [8]. We also want to point out that e-GD2L systems form the most complicated class of deterministic L systems known to have the decidable equivalence problem. For CPDOL systems the problem is still open.

5. Nondeterministic Case

In this final section we show that Theorem 1 cannot be extended to nondeterministic systems. This is somewhat surprising since one might expect (after Theorem 1) that the strict growth in connection with context-sensitive rules essentially "blocks" the interaction here, too. However, because of the parallelism in the rewriting process, it is possible to use nondeterministic strictly growing context-sensitive rules to control the derivation.

Theorem 6 For any ETOL language L there exist an s-G2L system G , homomorphism h and a regular set R such that

$$L = h(L(G)) \cap R .$$

We show the general construction used in [2] on an example. Consider the TOL system $H = \langle\Sigma, t_1, t_2, a\rangle$ where $\Sigma = \{a,b\}$ and the "tables" t_1 and t_2 are as follows:

$$t_1 : \quad a \to aa \qquad b \to b \qquad\qquad t_2 : \quad a \to ab \mid ba \mid a \qquad b \to b \mid bb$$

It is well known that $L(H) = \{w \in \Sigma^* : \#_a(w) = 2^n ,\ n \geq 0\}$ and that L(H) is not an EOL language. We now give an s-G2L system G , a homomorphism (coding) h and a regular set R so that $L(H) = h(L(G)) \cap R$. The alphabet of G is $\Delta = \{a,b,\bar{a},\#\}$ with $\bar{a}$ being the axiom. The productions are

$$(\$,\bar{a},\$) \to \bar{a}\bar{a} \mid aa \qquad (\bar{a},\bar{a},\bar{a}) \to \bar{a}\bar{a} \mid aa$$
$$(\$,\bar{a},\bar{a}) \to \bar{a}\bar{a} \mid aa \qquad (\$,\bar{a},a) \to \#\#$$
$$(\bar{a},\bar{a},\$) \to \bar{a}\bar{a} \mid aa \qquad (a,\bar{a},\bar{a}) \to \#\#$$
$$(\bar{a},\bar{a},a) \to \#\# \qquad (a,\bar{a},\$) \to \#\#$$
$$(a,\bar{a},a) \to \#\# \qquad a \to ab \mid ba \mid a$$
$$\# \to \#\# \qquad b \to b \mid bb$$

Homomorphism $h : \Delta^* \to \{a,b,\#\}^*$ is defined by $h(\bar{a}) = a$ and $h(\xi) = \xi$ for $\xi \in \Delta - \{\bar{a}\}$, and $R = \{a,b\}^*$. It is easy to verify that $L(H) = h(L(G)) \cap R$.

As a consequence we can demonstrate the impossibility of a simulation similar like in Theorem 1 for the nondeterministic e-G2L systems. Consider any language L in $\mathcal{L}_{ETOL} - \mathcal{L}_{EOL}$. By Theorem 6 we can write $L = h(L_0) \cap R$ where h is a homomorphism, L_0 an e-G2L language and R a regular set. Then, the closure properties of $\mathcal{L}_{EOL}$ imply the following:

Theorem 7 There are s-G2L languages, and therefore also e-G2L languages, which are not in $\mathcal{L}_{EOL}$ (= $\mathcal{L}_{COL}$) .

References

1. Culik II, K. and Fris, I., The decidability of the equivalence problem for DOL-systems, Inf. and Control,35 (1977), 20-39.

2. Culik II, K. and Karhumäki, J., Interactive L systems with almost interactionless behaviour, submitted to Inf. and Control, also Research Report CS-78-47, Department of Computer Science, University of Waterloo, Waterloo, Ontario, Canada. 1978.

3. Harrison, M., Introduction to Formal Language Theory, Addison-Wesley (1978).

4. Lindenmayer, A., Private communication (1977).

5. Rozenberg, G. and Salomaa A., The mathematical theory of L systems, in : J. Tou (ed.), Advances Information Systems Science, 6 (1976), Plenum Press, New York, 161-206.

6. Salomaa, A., Formal Languages, Academic Press (1973).

7. Salomaa, A. and Soittola, M., Automata-Theoretic Aspects of Formal Power Series, Springer-Verlag (1978).

8. Vitanyi, P.M.B., Growth of strings in context dependent Lindenmayer systems, in: L systems, edited by G. Rozenberg and A. Salomaa, Lecture Notes in Computer Science 15 (1974), Springer-Verlag.

ON THE SIMPLIFICATION OF CONSTRUCTIONS IN DEGREES OF UNSOLVABILITY VIA COMPUTATIONAL COMPLEXITY

Robert P. Daley
University of Pittsburgh
Pittsburgh, Pa. 15260, USA

§1. INTRODUCTION:

In this paper we show how some of the infinite injury priority arguments can be simplified by making explicit use of the primitive notions of axiomatic computational complexity theory. An important factor in the simplification is the use of busy beaver sets (see [2]) to provide the basis for the required diagonalizations thereby permitting rather simple and explicit descriptions of the sets constructed. Another is the replacement of the characteristic function χ_A of a set A with the next-element function ν_A.

We devote the remainder of this section to the requisite definitions and notions as well as some preliminary lemmas. A more comprehensive discussion of many of the notions in this section can be found in [2]. Proofs of the lemmas presented here can be found in [3].

For any set X of natural numbers we use $X|s$ to denote the set $\{r|r\in X \text{ and } r \le s\}$. We will assume that $\{\psi_i^X\}$ is a (relativized) acceptable Gödel numbering (see Rogers [5]) and that $\{\Psi_i^X\}$ is a (relativized) computational complexity measure for $\{\psi_i^X\}$ (see Blum [1], Lynch [4], Symes [8]) such that for all X, i, p, and s,

$(\Omega 1)$ $\Psi_i^X(p) \ge \psi_i^X(p)$,

$(\Omega 2)$ $s \ge \Psi_i^X(p) \Rightarrow [\psi_i^{X|s}(p) = \psi_i^X(p)$ and $\Psi_i^{X|s}(p) = \Psi_i^X(p)]$, and

$(\Omega 3)$ $\Psi_i^{X|s} \le s \Rightarrow [\psi_i^{X|s}(p) = \psi_i^X(p)$ and $\Psi_i^{X|s}(p) = \Psi_i^X(p)]$.

---

1) This research was supported by NSF Grant MCS 76-00102

The latter two conditions insure that the computational complexity measure takes into account the members of **X** "used" in a computation, which is an essential notion of relative computations (see Shoenfield [7]).

For any set **X** of natural numbers the next element function of **X**, ν_X is defined by $\nu_X(p) = \min\{q \mid q > p \text{ and } q \in X\}$. We use **deg X** to denote the degree of unsolvability of the set **X**, and clearly

$$\text{(1.1)} \qquad \deg X \le \deg Y \iff (\exists i)[\psi_i^Y = \nu_X].$$

We begin by constructing for each recursively enumerable set W_e a set B_e such that $\deg W_e = \deg B_e$ as follows:

$$b_e(n) = \max\{\Psi_e(p) \mid p \le n \text{ and } \psi_e(p)\downarrow\},$$

$$b_e(n,s) = \begin{cases} \max\{\Psi_e(p) \mid p \le n \text{ and } \Psi_e(p) \le s\}, \\ n, \text{ if no such } p \text{ exists.} \end{cases}$$

$$g_e(p) = \begin{cases} \min\{q \mid q \le p \text{ and } \Psi_e(q) = p\}, \\ p, \text{ if no such } q \text{ exists.} \end{cases}$$

$$B_e = \{b_e(n)\},$$

$$B_e^s = \{b_e(n,s) \mid n \le s\},$$

$$A_e = \overline{B}_e.$$

It should be clear that $b_e(n,s)$ and $g_e(p)$ are total recursive functions and that $b_e(n) \le b_e(n+1)$. The following lemma gives the most important properties of the set B_e.

Lemma 1.1: (a) A_e is recursively enumerable.
(b) B_e is retraceable.
(c) $\deg B_e = \deg A_e = \deg W_e$.
(d) $(\forall n)[B_e \mid b_e(n) = B_e^{b_e(n)}]$.
(e) $(\forall s)[B_e \mid s = B_e^s \iff s \in B_e]$.
(f) $(\forall e)(\forall i)[\psi_i \text{ total and } W_e \text{ non-recursive} \Rightarrow (\exists_n^\infty)[\psi_i(b_e(n)) < b_e(n+1)]]$.

§2. SACKS' DENSITY THEOREM

We illustrate the simplification techniques on the following theorem.

Theorem 2.1: (Sacks) If **C** and **D** are recursively enumerable sets such that **deg D** < **deg C**, then there exists a recursively enumerable set **A** such that **deg D** < **deg A** < **deg C**.

In constructing such a set **A** we will find it more convenient to work with its complement **B**. Let **C** and **D** satisfy the hypotheses of the Theorem 2.1, and let $\mathbf{C} = \mathbf{W}_c$ and $\mathbf{D} = \mathbf{W}_d$. The first step of the simplification is to replace the sets **C** and **D** by their busy beaver equivalents. We define,

$$b_\alpha(n) = b_d(b_c(n)),$$

$$b_\alpha(n,s) = b_d(b_c(n,s),s),$$

$$g_\alpha = g_c(g_d(p)),$$

$$\mathbf{B}_\alpha = \{b_\alpha(n)\},$$

$$\mathbf{B}_\alpha^s = \{b_\alpha(n,s) \mid n \le s\},$$

$$\mathbf{A}_\alpha = \overline{\mathbf{B}}_\alpha.$$

The following is a direct consequence of the above definitions and Lemma 1.2.

Lemma 2.1: (a) $\mathbf{A}_\alpha$ is recursively enumerable.
(b) $\mathbf{deg}\ \mathbf{A}_\alpha = \mathbf{deg}\ \mathbf{B}_\alpha = \mathbf{deg}\ \mathbf{C}$.
(c) $(\forall n)[\mathbf{B}_d | b_\alpha(n) = \mathbf{B}_d^{b_\alpha(n)}]$.
(d) $(\forall i)[\psi_i^{\mathbf{B}_d}$ total $\Rightarrow (\exists_n^\infty)[\psi_i^{\mathbf{B}_d}(b_\alpha(n)) < b_\alpha(n+1)]]$.

From the construction given below for **A** the following will be quite evident.

Lemma 2.2: (a) **A** is recursively enumerable.
(b) $(\forall n)[\mathbf{B} | b_\alpha(n) = \mathbf{B}^{b_\alpha(n)}]$.
(c) $\mathbf{B}_\alpha \subseteq \mathbf{B} \subseteq \mathbf{B}_d$.

Since **A** is recursively enumerable the stage s approximations $\mathbf{A}^s$ and $\mathbf{B}^s$ will be recursive set functions. The conditions of Theorem 2.1 can be split into the following four conditions: (1) $\mathbf{deg}\ \mathbf{B} \le \mathbf{deg}\ \mathbf{B}_\alpha$; (2) $\mathbf{deg}\ \mathbf{B}_d \le \mathbf{deg}\ \mathbf{B}$; (3) $\mathbf{deg}\ \mathbf{B}_\alpha \not\le \mathbf{deg}\ \mathbf{B}$; (4) $\mathbf{deg}\ \mathbf{B} \not\le \mathbf{deg}\ \mathbf{B}_d$. We consider each of these conditions in turn. Condition (1) follows from

Lemma 2.2(b), and condition (2) follows from Lemma 2.2(c) and the retraceability of B_d (Lemma 1.1(b)). In view of (1.1) condition (3) can be replaced by

(R_i^-) $$\psi_i^B \neq \nu_C.$$

A recursive transformation τ for ψ_i^B is defined by

$$\psi_{\tau(i)}^{X}(p) = \min\{s \in X \mid s > p \text{ and } (\forall q \le p)[\Psi_i^{B^s}(q) \le s]\}.$$

If $\psi_{\tau(i)}^{B_d^s}(p) = s$ and $s \in B_d$, then clearly $\psi_{\tau(i)}^{B_d}(p) = s$. If $B|s = B^s$ in addition, then by (Ω1) and (Ω3) we have

$$\psi_i^B(p) < \Psi_i^B(p) = \Psi_i^{B|s}(p) = \Psi_i^{B^s}(p) \le s.$$

Therefore,

(2.1) $$\psi_{\tau(i)}^{B_d^s}(p) = s \text{ and } s \in B_d \text{ and } B|s = B^s \Rightarrow \psi_i^B(p) < \psi_{\tau(i)}^{B_d}(p) = s.$$

During the construction, therefore, stages $s \in B_d$ are sought for which $\psi_{\tau(i)}^{B_d^s}(b_\alpha(n)) = s < b_\alpha(n+1)$, which by Lemma 2.1(d) are known to exist whenever $\psi_i^{B_d}$ is total, and then an attempt is made to arrange that $B|s = B^s$ by restraining these elements from **A**. For this reason (R_i^-) is known as a negative requirement.

Similarly, condition (4) can be replaced by

(R_i^+) $$\psi_i^{B_d} \neq \nu_B,$$

and a recursive transformation σ for $\psi_i^{B_d}$ defined by

$$\psi_{\sigma(i)}^{X}(p) = \min\{s \in X \mid s > p \text{ and } (\forall q \le p)[\Psi_i^{X}(q) \le s]\}.$$

In an analagous manner to (2.1) it can be shown that

(2.2) $$\psi_{\sigma(i)}^{B_d^s}(p) \le s \text{ and } s \in B_d \Rightarrow \psi_i^{B_d}(p) < \psi_{\sigma(i)}^{B_d}(p) \le s.$$

Thus to satisfy (R_i^+) it suffices to find stages $s \in B_d$ such that $\psi_{\sigma(i)}^{B_d^s}(b_\alpha(n)) \le s < b_\alpha(n+1)$, which by Lemma 2.1(d) are known to exist whenever $\psi_i^{B_d}$ is total, and then arrange that $\nu_B(b_\alpha(n)) = b_\alpha(n+1)$ by putting into the set **A** all integers in the interval $(b_\alpha(n), b_\alpha(n+1))$. For this reason (R_i^+) is known as a positive requirement.

However the act of satisfying the positive requirement (R_i^+) may invalidate (injure) some negative requirement (R_j^-) because some integer previously restrained from **A** on behalf of (R_j^-) was put into **A** to satisfy (R_i^+). In finite injury priority constructions such conflicts are resolved by assigning prioritites to the requirements, and then showing that each negative requirement can be injured only finitely many times and so will eventually be satisfied. Unfortunately, here the situation is further complicated when B_d is not recursive and $\psi_{\sigma(i)}^{B_d}$ is not total, because it is possible that $\psi_{\sigma(i)}^{B_d}(p)\uparrow$ and at the same time $\psi_{\sigma(i)}^{B_d^s}(p)\downarrow$ for infinitely many $s\notin B_d$. Moreover, since B_d is not recursive one cannot identify stages $s\in B_d$. Therefore it is possible for a positive requirement to injure a negative requirement of lower priority infinitely many times and possibly prevent the satisfaction of that negative requirement altogether.

Fortunately, such infinite injury can be avoided by searching for stages $s\in B_d$ such that $\psi_{\sigma(i)}^{B_d^s}(r) \le s < b_\alpha(n+1)$, where $r\in B_d^s$ and $r = \psi_{\sigma(i)}^{B_d^s}(b_\alpha(n))$, which again are known to exist if $\psi_i^{B_d}$ is total, and setting $v_B(r) = b_\alpha(n+1)$, where $r = \psi_{\sigma(i)}^{B_d^s}(b_\alpha(n))$, in order to satisfy (R_i^+). If $\psi_{\sigma(i)}^{B_d}$ is not total, then the only way that (R_i^+) can injure a legitimately established negative requirement (R_j^-) (i.e., one where $\psi_{\tau(j)}^{B_d^t}(b_\alpha(n)) = t$, with $t\in B_d$ and $B_t = B|t)$ is for $r < t < s$. Since $t\in B_d$ and $r\in B_d^s$ it follows that $r\in B_d$, and therefore by (2.2) we have that $b_\alpha(n)\in \mathbf{dom}\ \psi_{\sigma(i)}^{B_d}$. But it is clear from the definition of σ that if $\psi_{\sigma(i)}^{B_d}$ is not total, then it has finite domain. Therefore, each positive requirement can cause only finitely many injuries to legitimately established requirements.

We now present an enumeration procedure for the set **A**. The procedure proceeds in stages and uses two types of markers. The markers of the form $\boxed{n}$ are associated with the enumeration procedure for B_e, and the position of marker $\boxed{n}$ at stage s is $b_e(n,s)$, and its final resting place is $b_e(n)$. The markers of the form $\boxed{k|m|u}$ are used to preserve the computations $\psi_{\tau(k)}^{B_d^s}(b_\alpha(m)) = s$, where $s = b_d(u)$. The construction also employs two functions α and β, and $\alpha_s(i)$ $(\beta_s(i))$ specifies the stage s witness hypothesized for (R_i^-) $((R_i^+))$. The functions $\tilde{\alpha}$ and $\tilde{\beta}$ indicate midstage values of α and β.

Stage s:

(1) Place marker $\boxed{s}$ on integer s.

(2) Let $m = g_\alpha(s)$ and $u = g_d(s)$.

For each $i < s$ set

$\tilde{\alpha}_s(i) =$ **if** $\alpha_{s-1}(i) \geq m-1$ **then** 0 **else** $\alpha_{s-1}(i)$,

$\tilde{\beta}_s(i) =$ **if** $\beta_{s-1}(i) \geq m-1$ **then** 0 **else** $\beta_{s-1}(i)$,

and set $\tilde{\alpha}_s(s) = \tilde{\beta}_s(s) = 0$.

(a) Move all markers $\boxed{m}$, $\boxed{m+1}$, ..., $\boxed{s}$ onto integer s.

(b) Remove all markers $\boxed{k|n|v}$ for which $n \geq m$ or $v \geq u$, and set $\tilde{\beta}_s(k) = 0$.

(c) For each $i \leq s$, if i satisfies

(i) $\tilde{\alpha}_s(i) = 0$,

(ii) $\psi_{\sigma(i)}^{B_d^s}(b_\alpha(m-1,s)) < s$,

(iii) $\psi_{\sigma(i)}^{B_d^s}(r) < s$,

(iv) no marker $\boxed{k|n|v}$ for any $k \leq i$ is on any integer in the interval (r,s), where $r = \psi_{\sigma(i)}^{B_d^s}(b_\alpha(m-1,s)$,

then remove all markers in the interval (r,s) and set $\alpha_s(i) = m-1$,

otherwise set $\alpha_s(i) = \tilde{\alpha}_s(i)$.

(d) Define B^s to be the set of all marked integers, and A^s to be the set of all unmarked integers $\leq s$.

(3) (a) For each $i \leq s$, if i satisfies

(i) $\tilde{\beta}_s(i) = 0$,

(ii) $\psi_{\tau(i)}^{B_d^s}(b_\alpha(m-1,s)) < s$,

(iii) $r \in B_d^s$,

(iv) $B^s | r = B_r$,

where $r = \psi_{\tau(i)}^{B_d^s}(b_\alpha(m-1,s))$,

then set $\beta_s(i) = m-1$,

otherwise set $\beta_s(i) = \tilde{\beta}_s(i)$.

(b) For each $i \leq s$ and each $n \leq s$ for which

(i) $\tilde{\beta}_s(i) = 0$,

(ii) $\psi_{\tau(i)}^{B_d^s}(b_\alpha(n,s)) = s$,

place a marker $\boxed{i|n|u}$ on each member of B^s in the interval $(b_\alpha(n,s),s]$.

It should be observed that only during step 2, where markers are removed, are integers enumerated into **A**, and that $\psi_{\tau(i)}^{\mathbf{B}_d^s}(b_e(n,s)) = s$ can be effectively decided from $\mathbf{B}_t$ for $t \le s$, and so involves no circularity. Also, $\mathbf{B}^s$ ($\mathbf{A}^s$) consists of all integers $\le s$ which are marked (unmarked).

The correctness of Lemma 2.2 should now be apparent from the above construction. Tracing the computation of $\alpha_s(i)$, $\tilde{\alpha}_s(i)$, $\beta_s(i)$, and $\tilde{\beta}_s(i)$ we see that

$$\tilde{\alpha}_s(i) \le \alpha_s(i) < g_e(s) \quad \text{and} \quad \tilde{\beta}_s(i) \le \beta_s(i) < g_e(s).$$

The following lemma is an easy consequence of this and Lemma 1.1(d).

Lemma 2.3: (a) For each i, $\alpha(i) = \lim_{s\in \mathbf{B}_\alpha} \alpha_s(i)$ exists, and if $\alpha(i) \ne 0$, then $\alpha(i) = \lim_{s\to\infty} \alpha_s(i)$ and $(\forall s \ge b_\alpha(\alpha(i)+1))[\tilde{\alpha}_s(i) = \alpha_s(i) = \alpha(i)]$.

(b) For each i, $\beta(i) = \lim_{s\in \mathbf{B}_\alpha} \beta_s(i)$ exists, and if $\beta(i) \ne 0$, then $\beta(i) = \lim_{s\to\infty} \beta_s(i)$ and $(\forall s \ge b_\alpha(\beta(i)+1))[\tilde{\beta}_s(i) = \beta_s(i) = \beta(i)]$.

The following lemma is proved by induction using Lemma 2.3 and the ideas given preceding the enumeration procedure.

Lemma 2.4:

(a) $(\forall i)[\alpha(i) \ne 0 \Rightarrow [\psi_{\sigma(i)}^{\mathbf{B}_d}(\psi_{\sigma(i)}^{\mathbf{B}_d}(b_\alpha(\alpha(i)))) < b_\alpha(\alpha(i)+1)$, and $\nu_{\mathbf{B}}(\psi_{\sigma(i)}^{\mathbf{B}_d}(b_\alpha(\alpha(i)))) = b_\alpha(\alpha(i)+1)]]$.

(b) $(\forall i)[\beta(i) \ne 0 \Rightarrow [\psi_{\tau(i)}^{\mathbf{B}_d}(b_\alpha(\beta(i))) < b_\alpha(\beta(i)+1)$, and $\mathbf{B}|r = \mathbf{B}^r$, where $r = \psi_{\tau(i)}^{\mathbf{B}_d}(b_\alpha(\beta(i)))]]$.

(c) $(\forall i)[\alpha(i) = 0 \Rightarrow \psi_{\sigma(i)}^{\mathbf{B}_d}$ is not total],

$(\forall i)[\beta(i) = 0 \Rightarrow \psi_{\tau(i)}^{\mathbf{B}_d}$ is not total].

We can now complete the proof of Theorem 2.1. The proofs of conditions (1) and (2) have already been indicated, so that it remains to verify only conditions (3) and (4). Suppose $\psi_i^{\mathbf{B}_d}$ is total. Then $\psi_{\sigma(i)}^{\mathbf{B}_d}$ is total and by Lemma 2.4 $\alpha(i) \ne 0$. By Lemma 2.4(a), (Ω1), and (2.2),

$$\psi_i^{\mathbf{B}_d}(r) < \Psi_i^{\mathbf{B}_d}(r) \le \psi_{\sigma(i)}^{\mathbf{B}_d}(r) < \nu_{\mathbf{B}}(r),$$

where $r = \psi^{B_d}_{\sigma(i)}(b_\alpha(\alpha(i)))$. Therefore, **deg B $\not\leq$ deg B_d**. Suppose ψ^{B}_{i} is total. Given p, let $s \in B_\alpha$ be such that $s \geq \max\{\Psi^{B}_{i}(q) \mid q \leq p\}$. Then by ($\Omega$2) and Lemma 2.2(b),

$$\Psi^{B^s}_{i}(q) = \Psi^{B|s}_{i}(q) = \Psi^{B}_{i}(q),$$

for all $q \leq p$. Since $B_\alpha \subseteq B_d$, $\psi^{B_d}_{\tau(i)}(p)\downarrow$, and therefore $\psi^{B_d}_{\tau(i)}$ is total. By Lemma 2.4(d) $\beta(i) \neq 0$ and by Lemma 2.4(b) we have $B|r = B^r$ and $\psi^{B_d}_{\tau(i)}(b_\alpha(\beta(i))) < \nu_B(b_\alpha(\beta(i)))$, where $r = \psi^{B_d}_{\tau(i)}(b_\alpha(\beta(i)))$, and then by ($\Omega$2) we have,

$$\Psi^{B|r}_{i}(b_\alpha(\beta(i))) = \Psi^{B^r}_{i}(b_\alpha(\beta(i))) \leq r.$$

Therefore, by (Ω3), and (Ω1) and by combining the above inequalities we have,

$$\psi^{B}_{i}(b_\alpha(\beta(i))) < \Psi^{B}_{i}(b_\alpha(\beta(i))) = \Psi^{B|r}_{i}(b_\alpha(\beta(i))) < \nu_{B_\alpha}(b_\alpha(\beta(i))).$$

Therefore, **deg B_α $\not\leq$ deg B**, and Theorem 2.1 is proved.

REFERENCES

1. Blum, M., "A machine independent theory of the complexity of recursive functions", JACM **14**(1967), 322-336.
2. Daley, R.,"On the simplicity of busy beaver sets", Zeit. fűr Math. Logik und Grund. der Math. **24**(1978), 207-224.
3. Daley, R., "Busy beaver sets and the degrees of unsolvability (revised)", Technical Report 78-2 (1978), Computer Science Department, University of Pittsburgh.
4. Lynch, N., Meyer, A., and Fischer, M., "Relativizations of the theory of computational complexity", Transactions AMS **220**(1976), 243-287.
5. Rogers, H., Theory of Recursive Functions and Effective Computability, McGraw Hill (1967).
6. Sacks, G., "The recursively enumerable sets are dense", Annals of Math. **80**(1964), 300-312.
7. Shoenfield, J., Degrees of Unsolvability, North Holland (1971).
8. Symes, M., "The extension of machine-independent computational complexity theory to oracle machine computation and to the computation of finite functions", Ph. D. Dissertation, Department of Applied Analysis and Computer Science, University of Waterloo (1971).

AN ALGEBRAIC EXTENSION OF THE

CHOMSKY - HIERARCHY

W. Damm

Lehrstuhl für Informatik II, RWTH Aachen

1. Introduction

This paper describes the proof of strictness of the OI-hierarchy of languages, which has been introduced in [2,7,12,17]. One way of defining this hierarchy is to generalize the *fixed point characterizations* of regular, context-free, and macro languages. Taking left concatenation with $a \in V$ and a constant e denoting the empty word as algebraic structure on PV^*, the above characterizations involve taking fixed points over domains of level 0 - PV^* -, level 1 - $PV^* \to PV^*$ - , and level 2 - $(PV^* \to PV^*) \to (PV^* \to PV^*)$, respectively. *Level-n* (OI-) *languages*, then, will be defined by taking *fixed points* on the *n*-th *function space* over PV^*. Strictness of this hierarchy has remained an open problem since conjectured by Wand [17].

The solution of this problem involved two different formalizations of the above intuitive idea. A more algebraic definition is given within a combinatorial framework (see e.g. [3,2]), while the proof of (size-)closure under intersection with regular sets, which is given in this paper, requires a *λ-calculus* oriented model. As the following example illustrates the definition of level-n languages involving the λ-calculus also gives a link between procedures in ALGOL 68 and the OI-hierarchy.

Consider the program

```
begin int input 1, input 2, output;
    proc  EXP = (proc(int)int f, int n, int k) int:
        begin int result;
            if n = o then result := f(k) else result := 2 ↑ EXP(f,n-1,k) fi;
            result
        end of EXP;
    output := EXP(square, input 1, input 2)
end of MAIN.
```

A denotational semantics of this program would involve operations {0,1,2,-,↑, square} dealing with integers, predicates as =, a conditional if, locations {loc, loc 1, loc 2, loc 3} and store-primitives as {content, assign, eval} in a set Ω. Translating to λ-terms gives

assign(loc 3, Y_2 (λ EXP. λ(f,n,k). body-EXP) (square, content(loc1), content(loc
where

body-EXP = eval(if(=(n,o),assign(loc,f(k)), assign(loc,↑(2,EXP(f,-(n,1),k)))),loc) .

The above translation isolates two features of ALGOL 68 procedures:

-non-recursive procedures with *finite modes* induce λ-terms with *typed abstraction* and *typed application* only;

-*recursive* procedures with at most n nested occurences of the proc-declarator in the mode specification part are translated into λ-terms augmented by an atom Y_n for the fixed-point operation on the n-th function space.

To come from ALGOL 68 procedures to language theory, simply view the elements of Ω as *names* for the respective operations, i.e. as *operation symbols*. Then the above λ-term essentially describes a fixed-point on level 2 over the domain of (tree-) languages over the (many-sorted) alphabet Ω, which, by application to arguments of appropriate type, yields a language over Ω. Thus, at least on this abstract level, the hierarchy result indicates that the power of ALGOL 68 procedures increases with their mode depth.

1. Mathematical background

Let I be a set of *base types*. An I-*set* A is a family $(A^i \mid i \in I)$. For I sets A,B we define *union* $A \cup B := (A^i \cup B^i \mid i \in I)$, *inclusion* $A \subseteq B$ iff $\forall i \in I\ A^i \subseteq B^i$, I-*mappings* $f : A \to B$ iff $f = (f^i : A^i \to B^i \mid i \in I)$ pointwise.

We denote by I^* the set of words over I. If $l(w) = r \in \omega$, we write $w = w(1)\cdot \ldots \cdot w(r)$. For $k \in \omega$ we let $[k] := \{1,..,k\}$.

The set $D^*(I)$ of *derived types over* I is defined inductively by $D^0(I) := I$, $D^{n+1}(I) := D^n(I)^* \times D^n(I)$, $D^*(I) := \bigcup_{n\in\omega} D^n(I)$.

We assume that the reader is familiar with the notions *complete partial order* (cpo) - each *directed* subset T has least upper bound $\sqcup T$ - and *continuous* functions - preserves lubs of nonempty directed sets. Any I-cpo A induces canonically the $D^*(I)$-cpo A^* defined by $A^e := \{()\}$, $(\alpha,\nu) \in D^{n+1}(I) \sim A^{\alpha\nu} := A^\alpha \times A^\nu$, $A^{(\alpha,\nu)} := A^\alpha \to A^\nu$, the cpo of all continuous functions from A^α to A^ν.

Let Ω be a D(I)-set of *operations symbols*. A *continuous* Ω-*algebra* $\mathcal{A}$ is a pair $(A,\varphi_A) \in \Delta\text{-}\underline{alg}(\Omega)$ consisting of an I-cpo A and a D(I)-mapping $\varphi_A : \Omega \to (A^{(w,i)} \mid (w,i) \in D(I))$.

An Ω-*homomorphism* $f : \mathcal{A} \to \mathcal{B}$ is a structurepreserving I-mapping $f : A \to B$. It is well known [10] that the Ω-algebra CT_Ω of *infinite trees* over Ω is *initial* in $\Delta\text{-}\underline{alg}(\Omega)$, i.e. for all $\mathcal{A} \in \Delta\text{-}\underline{alg}(\Omega)$ there exists a unique continuous ⊥-preserving Ω-homomorphism $h_{\mathcal{A}} : CT_\Omega \to \mathcal{A}$ $(\perp = \sqcup\ \emptyset)$. We denote by FT_Ω (T_Ω) the restriction of CT_Ω to *finite trees* (without *minimum symbols* $(\perp_i \mid i \in I)$), hence PT_Ω denotes the subset-algebra of *tree languages* over Ω.

With Ω we associate the D(I)-set $\Omega_+ := \Omega \cup (\{+_i\} \mid (ii,i) \in D(I))$. The unique continuous ⊥-preserving Ω_+-homomorphism $CT_{\Omega_+} \to PT_\Omega$ (with "+" denoting union) will be called set.

2. Applicative terms

In this paper we will define level-n languages by a generalization of macro grammars [9]. As rigth hand sides of a production we will allow typed applicative terms over the D(I)-set Ω of *terminal* symbols, and D*(I)-sets of *nonterminals* and *parameters*, denoted X and Y, respectively.

For $\sigma \in (X \cup Y)^{\tau}$ with $\tau \in D^{n}(I)$ we set $\underline{type}(\sigma) := \tau$, $\underline{level}(\sigma) := n$. The D*(I)-set $AT_{\Omega \cup X \cup Y}$ of *applicative terms* over Ω,X, and Y is the smallest D*(I)-set T s.t. $\Omega \cup X \cup Y \subseteq T$ and $t \in T^{(\alpha,\nu)}$, $t_j \in T^{\alpha(j)}, l(\alpha) = r \sim t(t_1,..,t_r) \in T$

Consider a nonterminal x of type $\tau = (\alpha_n,\ldots,(\alpha_o,i)\ldots) \in D^{n+1}(I)$. Then the right hand side of an x-production will only contain parameters "fitting" to τ in a set Y^{τ}. Let, for $\alpha \in D^{n}(I)^{*}$, $Y_{\alpha} := \{ y_{j,\alpha(j)} \mid j \in [l(\alpha)]\}$, thus $Y_e = \emptyset$, and let $y_{\alpha} := (y_{1,\alpha(1)},\ldots,y_{r,\alpha(r)})$. Then $Y^{\tau} := \bigcup_{o \leq j \leq n} Y_{\alpha_j}$.

For $s \in AT_{\Omega \cup X \cup Y^{\tau}}$ and $t_j \in AT^{\alpha_j}_{\Omega \cup X \cup Y}$ we denote by $\underline{sub}_{Y^{\tau}}(s)(t_n)\ldots(t_o)$ the term obtained by simultaneously substituting t_j for y_{α_j} in s.

In case $\tau = (w,i) \in D(I)$ and s contains exactly one occurence of $y_{j,w(j)}$ we abbreviate $\underline{sub}_{Y^{\tau}}(s)(t_o)$ by $s \leftarrow t_o$.

Let $\tau \in D^{n+1}(I)$ be as above. For $t \in AT^{\tau}_{\Omega \cup X \cup Y}$, we denote by $t\downarrow$ the term obtained by giving t all parameters down to level o (for $m \in \{o,\ldots,n+1\}$ $t\downarrow_{n+1} := t$, $t\downarrow_m := t\downarrow_{m+1}(y_{\alpha_m})$, $\downarrow := \downarrow_o$).

We conclude this section by recalling the semantics of applicative terms. Let $A \in \Delta\text{-}\underline{alg}(\Omega)$, and let $\rho : X \to A^{*}$ be a D*(I)-mapping (called *environment*). For $t \in AT^{\nu}_{\Omega \cup X \cup Y^{\tau}}$, the *semantics* of t over A with respect to the environment ρ is the continuous mapping $[\![t,A]\!]\rho : A^{\alpha_n} \to (A^{\alpha_{n-1}} \to \ldots (A^{\alpha_o} \to A^{\nu})\ldots)$ given by

$[\![t,A]\!]\rho(f_n)\ldots(f_o) := t \in \Omega \to \varphi_A(t), t \in X \to \rho(t), t = y_{j,\alpha_r(j)} \to pr_j(f_r)$,

$t = t_o(t_1,\ldots,t_r) \to [\![t_o,A]\!]\rho(f_n)\ldots(f_o)([\![t_1,A]\!]\rho(f_n)\ldots(f_o),\ldots,[\![t_r,A]\!]\rho(f_n)..(f_o))$

3. Level-n schemes

In this section we define syntax and semantics and state two theorems for a class of already "normalized" λ-schemes. By a process similar to the elimination of nested procedure declarations any typed λ-term with fixed point operators can be transformed into a set of equations with typed applicative terms as right hand sides (see [8]). To simplify notation we assume $I = \{i\}$.

Definition 1

Let $X = \{x_o,\ldots,x_N\}$. A *level-n scheme over* Ω is a system of equations $x_j\downarrow = \underline{rhs}(x_j)$ $(j = o,..,N)$ with $\underline{type}(x_o) = i$ s.t. for all j $\underline{rhs}(x_j) \in AT^{i}_{\Omega \cup X \cup Y^{\underline{type}(x_j)}}$ and $\underline{max}\{\underline{level}(x_j) \mid j \in \{o,\ldots,N\}\} = n$. The class of all level-n schemes over Ω will be denoted $n\text{-}\lambda(\Omega)$.

□

In this context, X corresponds to a set of procedure names. The body of x_j is just $\underline{rhs}(x_j)$, and $Y^{\underline{type}(x_j)}$ contains the formal parameters of x_j.

Example 1

The following level-2 scheme corresponds to the sample program of the introduction. Let $i := \underline{\text{int}} \in I$.

$$\text{MAIN} = \text{assign}(\text{loc } 3, \text{EXP}(\text{square})(\text{content}(\text{loc } 1), \text{content}(\text{loc } 2)))$$

$$\text{EXP}(y_{1,(i,i)})(y_{1,i}, y_{2,i}) = \text{eval}(\text{if}(=(y_{1,i}, o), \text{assign}(\text{loc}, y_{1,(i,i)}(y_{2,i})),$$
$$\text{assign}(\text{loc}, \uparrow(2, \text{EXP}(y_{1,(i,i)})(-(y_{1,i}, 1), y_{2,i})))), \text{loc})$$

□

As suggested by the notation, the denotational semantics of S will be obtained by solving the set of mutually recursive procedure declarations, i.e. by taking the least fixed point of a functional canonically induced by S over A.
Note that $[\![\underline{\text{rhs}}(x), A]\!]\rho \in A^{\underline{\text{type}}(x)}$.

Definition 2

Let $A \in \Delta\text{-}\underline{\text{alg}}(\Omega)$, and let S be as above. S induces a functional

$$S_A : A^{\underline{\text{type}}(x_o)} \times \ldots \times A^{\underline{\text{type}}(x_N)} \rightarrow A^{\underline{\text{type}}(x_o)} \times \ldots \times A^{\underline{\text{type}}(x_N)}$$
$$(a_o, \ldots, a_N) \mapsto ([\![\underline{\text{rhs}}(x_o), A]\!](x_j \mapsto a_j), \ldots, [\![\underline{\text{rhs}}(x_N), A]\!](x_j \mapsto a_j))$$

The *semantics of* S *over the interpretation* A is defined by

$$[\![S, A]\!] := pr_1 \; (Y(S_A))$$

□

It should be clear that the semantics of the sample scheme over the particular interpretation described in the introduction coincides with the standard (call by value) denotational semantics of the sample program.

By definition, $0\text{-}\lambda(\Omega)$ equals the class $R(\Omega)$ of *rational schemes* over Ω [10,16], and $1\text{-}\lambda(\Omega)$ equals $RPS(\Omega)$, the class of *recursive program schemes* over Ω [13]. From the equivalence of n-λ-schemes and n-rational schemes $R_n(\Omega)$ shown in [4] we obtain the following Mezei-Wright-like result.

Theorem 1 [2]

Let $T(S) := [\![S, CT_\Omega]\!]$ be the *infinite tree of* S. Then

$$\forall\, A \in \Delta\text{-}\underline{\text{alg}}(\Omega) \quad [\![S, A]\!] = h_A \; (T(S)) \,.$$

□

By the same equivalence we inherit the following normal form theorem from rational schemes [16].

Theorem 2 [4]

For any $S \in n\text{-}\lambda(\Omega)$ we can effectively find an equivalent n-λ-scheme S' over Ω s.t. all right hand sides of S' are of one of the following forms:

(1) $f\!\downarrow$ for $f \in \Omega$, (2) $x\!\downarrow$ for $x \in X$, (3) $y\!\downarrow$ for $y \in Y$,

(4) $y_o(y_1(y_\alpha), \ldots, y_r(y_\alpha))\!\downarrow$ for $y_j, y_{j,\alpha(j)} \in Y$ of appropriate type,

(5) $x_o(x_1(y_\alpha), \ldots, x_r(y_\alpha))\!\downarrow$ for $x_j \in X, y_{j,\alpha(j)} \in Y$ of appropriate type

and $\underline{\text{size}}(S') \leq p(\underline{\text{size}}(S))$ for some polynomial p.

□

We now indicate how to operationally generate approximations of the infinite tree of S, by viewing a level-n scheme as a schematic tree grammar. In this context the $x \in X$ correspond to nonterminals, which are rewritten according to their right hand side using the copy-rule. For our purposes it will be sufficient to consider outermost-innermost derivations.

<u>Definition 3</u>

Let S be as above, and let $t, t' \in AT^{i}_{\Omega \cup X \cup Y}$.

t' *is derivable from* t *in* $S_\perp$ $(t \underset{S_\perp}{\Rightarrow} t')$ iff $\exists\, s \in FT_{\Omega \cup Y_{iw}}$ $\exists\, s_o \in AT^{i}_{\Omega \cup X \cup Y}$ $\exists\, s' \in AT^{w}_{\Omega \cup X \cup Y}$ $s_o = x_j(t_n)\ldots(t_o) \wedge t = s \leftarrow (s_o, s') \wedge (t' = s \leftarrow (\perp_i, s')$ $\vee\ t' = s \leftarrow (\underline{sub}_{Y\underline{type}(x_j)}(\underline{rhs}(x_j))(t_n)\ldots(t_o), s'))$.

The *schematic language generated* by S is defined by

$L(S_\perp) := \{t \in FT_\Omega \mid x_o \overset{*}{\underset{S_\perp}{\Rightarrow}} t\}$. □

<u>Example 2</u>

Define the *integer-types* $\underline{n} \in D^n(I)$ by $\underline{o} := i$, $\underline{n+1} := (\underline{n}, \underline{n})$.
Let $\Omega = \{e, a, b\}$ with types $\underline{o}, \underline{1}, \underline{1}$, and let $X = \{x_o, x_1, x_2, x_3, A, B\}$ with types $\underline{o}, \underline{3}, \underline{2}, \underline{3}, \underline{3}, \underline{2}$, respectively. Let $S_3 \in 3\text{-}\lambda(\Omega_+)$ be given by

$x_o = x_3(x_2)(b)(e)$ $\quad x_1{\downarrow} = y_2(y_2(y_1)){\downarrow}$ $\quad x_2{\downarrow} = y_1(y_1(y_o))$

$x_3{\downarrow} = +(x_3(x_1(y_2))(B(y_1))(A{\downarrow}), A{\downarrow})$ $\quad A{\downarrow} = y_2(a){\downarrow}$ $\quad B{\downarrow} = b(y_1(y_o))$.

Then

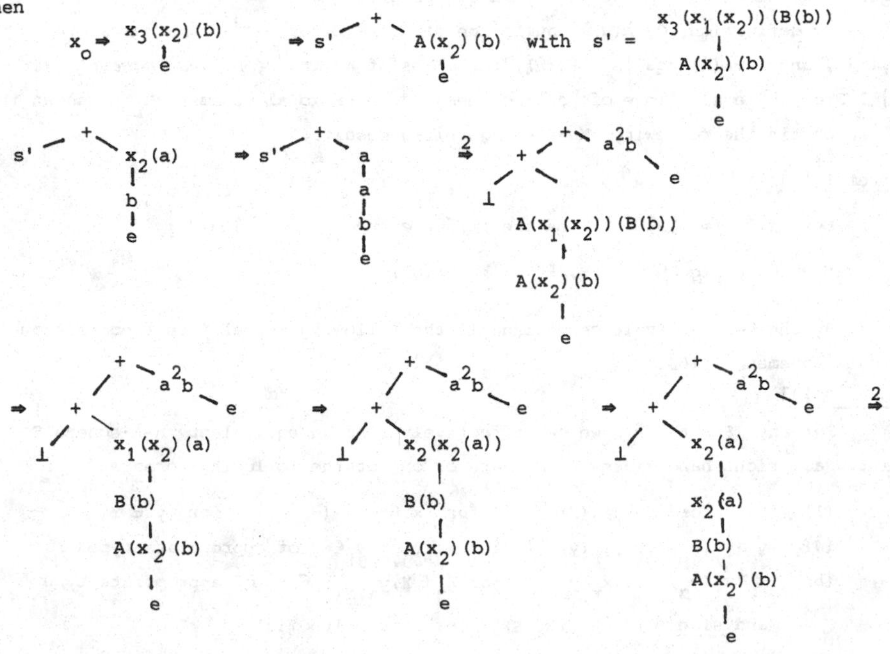

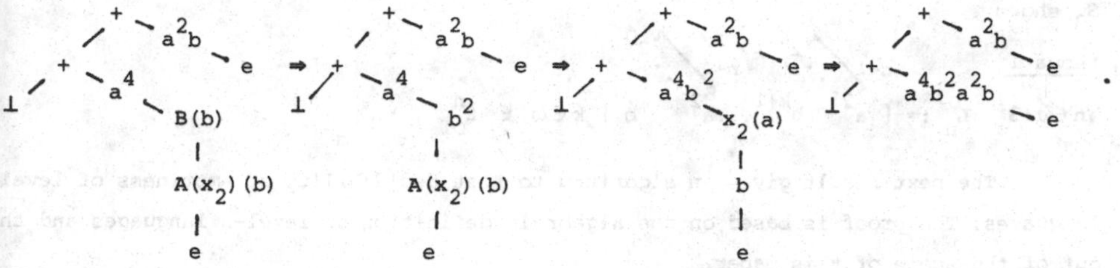

Let $w_o := e$, $w_{k+1} := a^{2^{2^k}} b^{k+1} w_k$. It should be clear that

$$L(S_\perp) = \{+(\overbrace{\ldots,(+}^{k}(\perp, w_k),\ldots,w_1) \mid k \in \omega\} .$$ □

<u>Theorem 3</u> [4]

$L(S_\perp)$ is directed, and $\sqcup L(S_\perp) = T(S)$.

Sketch of proof:

From rational schemes we inherit the property $T(S) = \sqcup K(S)$, where $K(S) : \omega \to FT^i_\Omega$ is the *Kleene-sequence* of S. We then prove $x_o \overset{*}{\Rightarrow} K(S)(n)$ and $x_o \overset{n}{\Rightarrow} t \sim \perp(t) \leqslant K(S)(n)$ by induction on $n \in \omega$ □

In connection with theorem 1 this gives an *operational semantics* for level-n schemes over discrete interpretations.

4. Level-n languages

Consider a string alphabet V. We define Ω_V as the *monadic* alphabet which contains a zero-ary symbol e denoting the empty word, and unary symbols a denoting left concatenation with $a \in V$. In order to generate languages over V we generalize to *level*-n *grammars*, where each nonterminal may be rewritten according to a finite number of definitions. Under the isomorphism between PV^* (with the above algebraic structure) and PT_{Ω_V} concatenation is mapped onto functional composition in $PT_{\Omega_V} \to PT_{\Omega_V}$, hence level-0,1,2 grammars correspond to right-linear, context-free, and macro grammars [9], respectively. Rather than formally introducing level-n grammars we will take the generalization of the fixed point characterizations of the above language classes as the definition of level-n languages.

<u>Definition 4</u>

Let $\Omega = \{e, a_1, \ldots, a_r\}$ be a monadic alphabet, and let $S \in n\text{-}\lambda(\Omega_+)$. The *level*-n *language generated by* S, L(S), is defined as the interpretation of S over PT_Ω. The class of level-n languages over Ω will be denoted $L^n_{OI}(\Omega)$. □

The equivalence to the expected definition by rewriting is an easy consequence of theorems 1 and 3. Since "+" denotes finite union in PT_Ω it simulates nondeterministic choice.

By theorem 1, $L(S) = \underline{set}(T(S))$, hence the scheme S_3 of example 2 generates the language $\{w_k \ldots w_1 \mid k \in \omega\}$. An obvious generalization of the construction of

S_3 shows

<u>Lemma 1</u>

$\forall n \in \omega+3 \quad L_n := \{ a^{\overbrace{2^{\cdot^{\cdot^{2^k}}}}^{n-1}} b^{k+1} \ldots a^{\overbrace{2^{\cdot^{\cdot^{2^0}}}}^{n-1}} b \mid k \in \omega \} \in \mathcal{L}^n_{OI}(\Omega)$. □

The next result gives an algorithm to test decidability of emptiness of level-n languages. The proof is based on the algebraic definition of level-n languages and thus out of the scope of this paper.

<u>Theorem 4 [3]</u>

Let $S \in n\text{-}\lambda(\Omega_+)$. Then

$$L(S) \neq \emptyset \quad \text{iff} \quad \exists w \in L(S) \quad l(w) \leq \overbrace{2^{\cdot^{\cdot^{2^{\underline{size}(S)}}}}}^{n} .$$

□

A crucial result in the proof of the hierarchy is to show size-closure of level-n languages under intersection with regular sets. In a first step, we lift this problem via <u>set</u>.

Let R be any language over Ω. With R we associate the tree language R_+ over Ω_+ by $R_+ := \{t \in FT_{\Omega_+} \mid \underline{set}(t) \subseteq R\}$.

For $L \subseteq FT_{\Omega_+}$, let $R_+ \leq L := \{t \in R_+ \mid \exists\, s \in L \quad t < s\}$, where $t < s$ iff s is obtained from t by replacing some occurences of $\bot$ in t by $e \in \Omega$. It can easily be proved that $R \cap \bigcup \underline{set}(L) = \bigcup \underline{set}(R_+ \leq L)$, hence by theorem 3 we have

<u>Lemma 2</u>

$L(S) \cap R = \bigcup \underline{set}(R_+ \leq L(S_\bot))$.

□

Now assume $R = L(Q)$ for some finite deterministic automaton with states $[k]$. Then it remains to construct some level-n scheme $\overline{S}$ with $L(\overline{S}_\bot) = R_+ \leq L(S_\bot)$. Let S be in normalform. It can easily be seen that $x_o \overset{*}{\underset{S_\bot}{\Rightarrow}} t \rightsquigarrow t \in FT_{\Omega_+ \cup N}$, where $N := AT^i_X$. To motivate the construction of $\overline{S}$, consider the automaton Q working in a top-down fashion on a path of t, and assume Q reaches a leave $l \in \{e, \bot\} \cup N$. Then $\overline{S}$ should

(i) memorize the current state of Q in the top nonterminal of l if $l \in N$;

(ii) simulate $S_\bot$ in case $l = \bot$, since the corresponding path defines $\emptyset \subseteq R$;

(iii) substitute e by $\bot$ in case e is reached in a nonfinal state;

(iv) simulate $S_\bot$ in case e is reached in a final state.

By (i), $\overline{S}$ should have $[k] \times X$ as nonterminals. Since we do not have any information about the states to be attached to $x_1, \ldots x_r$ while simulating a rule $x\!\downarrow \; = \; x_o(x_1(y_\alpha), \ldots, x_r(y_\alpha))\!\downarrow$, we allow all possible states as arguments, and "later on" choose the appropriate state through projections. This implies that the type of nonterminals has to be extended.

<u>Definition 5</u>

(a) Extension of types : for $\tau \in D^n(I)$ we define $\overline{\tau} \in D^n(I)$ by $\overline{i} := i$, $\overline{(\alpha(1)\ \dots\ \alpha(r),\nu)} := (\overline{\alpha(1)}^k \dots \overline{\alpha(r)}^k, \overline{\nu})$. We will identify $[k] \times Y^\tau$ and $Y^{\overline{\tau}}$ with $\tau = (\alpha_{n-1},\dots,(\alpha_o,i)..)$ under the bijection $(q, y_{j,\alpha_r(j)}) \mapsto y_{l+j,\overline{\alpha}_r^k(l+j)}$ where $l := l(\overline{\alpha}_r^{q-1})$.

(b) Extension of terms: for $t \in AT^\tau_{X\cup Y}$ we define $\overline{t} \in AT^{\overline{\tau}^k}_{\overline{X}\cup\overline{Y}}$ by $\overline{x} := (1x,\dots,kx)$, $\overline{y} := (1y,\dots,ky)$, $\overline{t(t_1,\dots,t_r)} := (pr_1(\overline{t})(\overline{t}_1,\dots,\overline{t}_r),\dots,pr_k(\overline{t})(\overline{t}_1,\dots,\overline{t}_r))$.

(c) Let $\overline{N} := \{pr_j(\overline{t}) \mid t \in N \wedge j \in [k]\}$. We define (the inverse of $\overline{\ }$ on trees) $\underline{\ } : FT_{\Omega_+\cup\overline{N}} \to FT_{\Omega_+\cup N}$ to be the unique Ω_+-homomorphism generated by $pr_j(\overline{t}) \mapsto t$.

□

<u>Construction of $\overline{S}$</u>

Let S with nonterminals $X = \{x_o,\dots,x_N\}$ be in normalform, and let $Q = ([k], \delta, 1, F)$ be a Rabin-Scott automaton with transition function δ, starting state 1, and final states F.

$\overline{S}$ has nonterminals $\overline{X} := [k] \times X$ with <u>type</u>(qx) := $\overline{\underline{type}(x)}$, axiom $1x_o$, and equations

(i) if <u>rhs</u>$(x_j) = e$ then $\forall q \in F$ $qx_j = e$,

(ii) if <u>rhs</u>$(x_j) = +\downarrow$ then $\forall q \in [k]$ $qx_j\downarrow = +(qy_{1,i}, qy_{2,i})$,

(iii) if <u>rhs</u>$(x_j) = a_p\downarrow$ then $\forall q \in [k]$ $qx_j\downarrow = a_p(\delta(q,a_p)y_{1,i})$,

(iv) if <u>rhs</u>(x_j) is of the form (2), (3), (4), or (5) then $\overline{x_j\downarrow} = \overline{\underline{rhs}(x_j)}$.

□

<u>Theorem 5</u>

$$L(\overline{S}_\perp) = L(Q)_+ \leqslant L(S_\perp) .$$

Proof:

The proof is based on the following three lemmata which follow by induction on the length of the derivation and case-splitting according to the right-hand-side.

(1) $\overline{S}$ simulates S : $1x_o \overset{*}{\underset{\overline{S}_\perp}{\Rightarrow}} t \curvearrowright x_o \overset{*}{\underset{S_\perp}{\Rightarrow}} \underline{t}$.

(2) $\overline{S}$ simulates Q : Let $1x_o \overset{*}{\underset{\overline{S}}{\Rightarrow}} t \in FT_{\Omega_+\cup\overline{N}}$, and let n be the number of leaves of t, hence $t = s \leftarrow (s_1,\dots,s_n)$ for some $s_j \in \overline{N}\cup\{e,\perp\}$, $s \in T_{\Omega_+\cup Y_i^n}$.

Let <u>path</u>$_j$(s) denote the string in $\{a_1,\dots,a_r\}^*$ leading to s_j. For $t \in \overline{N}$, let <u>state</u>(t) denote the state of the top nonterminal. Then

$$\forall j \in [n] \quad ((s_j = e \curvearrowright \delta^*(\underline{path}_j(s)) \in F) \wedge (s_j \in \overline{N} \curvearrowright \delta^*(\underline{path}_j(s)) = \underline{state}(s_j))) .$$

(3) " $\supset$ " We use the notation of (2). $(x_o \overset{*}{\underset{S_\perp}{\Rightarrow}} s \leftarrow (s_1,\dots,s_n) \wedge s_j \in \{e,\perp\} \cup N$
$\wedge\ q_j = \delta^*(\underline{path}_j(s))) \sim 1x_o \overset{*}{\underset{\overline{S}_\perp}{\Rightarrow}} s \leftarrow (\tilde{s}_1,\dots,\tilde{s}_n) \wedge (s_j = \perp \sim \tilde{s}_j = \perp,$
$s_j \in N \sim \tilde{s}_j = pr_{q_j}(\overline{s}_j),\ s_j = e \sim \tilde{s}_j \in \{\text{if } q_j \in F \text{ then } e \text{ else } \perp, \perp\}$. □

This proves the conjecture in [14] that level-n languages are size-closed under intersection with regular sets. Let $REG_m(\Omega)$ denote the class of languages accepted by m-state Rabin-Scott automata.

<u>Corollary 1</u>

$\forall R \in REG_m(\Omega) \quad \forall S \in n\text{-}\lambda(\Omega_+) \quad \exists \overline{S} \in n\text{-}\lambda(\Omega_+)$:

$$L(\overline{S}) = L(S) \cap R \ \wedge\ \underline{size}(\overline{S}) \leq p(\underline{size}(S),m)$$

for some polynomial $p \in POL$.

□

In [4] we prove that $\mathcal{L}^n_{OI}$ forms a substitution closed AFL.

5. Hierarchies

The results of the previous section are taylored to derive upper bounds for the complexity of level-n languages in terms of the rational index as introduced in [1] .

<u>Definition 6</u>

For $R,L \subseteq T_\Omega$ with $R \cap L \neq \emptyset$, let $d(R,L) := \min\{\,l(w) \mid w \in R \cap L\}$.
The *rational index* of L is the function $g_L : \omega \to \omega$ given by
$g_L(m) := \max\{\,d(R,L) \mid R \in REG_m(\Omega) \wedge R \cap L \neq \emptyset\}$.

Let $F := \{f : \omega \to \mathbb{R}_+\}$. For $f,g \in F$, define $f \leq g$ iff $\exists m_o \in \omega \ \forall m \geq m_o \ f(m) \leq g(m)$. The class of languages over Ω with rational index bounded by some $f \in F$ is denoted $\mathcal{L}(F,\Omega)$. For $F' \subseteq F$,
let $EXP(F') := \{\, f \in F \mid \exists g \in F' \ \ f = m \mapsto 2^{g(m)}\}$.

□

<u>Theorem 6</u>

$$\mathcal{L}^n_{OI}(\Omega) \subseteq \mathcal{L}(EXP^n(POL),\Omega) \ .$$

Proof:

Let $L = L(S) \in \mathcal{L}^n_{OI}(\Omega)$. For any $R \in REG_m(\Omega)$ we have $\underline{size}(\overline{S}_R) \leq p(\underline{size}(S),m)$, hence by theorem 4 and corollary 1

$$d(R,L) \leq \underbrace{2^{\cdot^{\cdot^{\cdot^{2^{p(\underline{size}(S),m)}}}}}}_{n}, \text{ thus } g_L \leq EXP^n(POL).$$

□

It should be obvious that the rational index of L_{n+1} cannot be bounded by a function in $EXP^{n-1}(POL)$. The last exponential increase is due to a trick taken from [15] , by considering intersections with $a^+(b^p)^+a^+(b^{p+1})^+\dots a^+(b^{2p-1})+(a \vee b)^*$.

Lemma 3

$\forall n \in \omega+3 \quad L_n \notin \mathcal{L}(EXP^{n-1}(POL), \{a,b,e\})$.

□

This proves the OI-hierarchy of languages strict.

Since the emptiness-problem for level-n languages is decidable and they are closed under intersection with regular sets, we have strict inclusion in the clan REC of recursive languages. Together with the relation to regular, context-free, and macro languages sketched in section 4 we obtain

Theorem 7

Let Ω be a monadic alphabet containing two unary symbols. Then

$\forall n \in \omega+3 \quad REG(\Omega) \subsetneq CF(\Omega) \subsetneq MAC(\Omega) \subsetneq \mathcal{L}^n_{OI}(\Omega) \subsetneq \mathcal{L}^{n+1}_{OI}(\Omega) \subsetneq REC(\Omega)$. □

Since the above language classes (except for REC) are defined as interpretations of corresponding classes of schemes, the OI-hierarchy induces a hierarchy of program schemes.

Let $\lambda(\Omega)$ denote the class of (untyped) λ-schemes [6].
By results in [5,8], level-n schemes are translatable into λ-schemes, but not vice versa.

Theorem 8

Let Ω contain one constant, one binary and two monadic symbols. Then

$\forall n \in \omega+2 \quad R(\Omega) < RPS(\Omega) < n\text{-}\lambda(\Omega) < n+1 - \lambda(\Omega) < \lambda(\Omega)$.

□

This verifies the conjecture by Indermark [11] that the auxiliary use of higher type procedures augments the power of a programming language.

6. References

[1] BOASSON, B., COURCELLE, B., NIVAT, M., *A new complexity measure for languages*, Proc. conference on Theoretical Computer Science, University of Waterloo, 1977.

[2] DAMM, W., *Languages defined by higher type program schemes*, Proc. 4th international colloquium on Automata, Languages, and Programming, Lecture Notes in Computer Science 52 (1977), pp. 164-179, Springer Verlag.

[3] DAMM, W., *The IO and OI hierarchies*, Schriften zur Informatik und Angewandten Mathematik, 41, RWTH Aachen (1978).

[4] DAMM, W., Dissertation at the RWTH Aachen (1979), to appear.

[5] DAMM, W., FEHR, E., *On the Power of Self-Application and Higher Type Recursion*, Proc. 5th international colloquium on Automata, Languages and Programming, Lecture Notes in Computer Science 62 (1978), Springer Verlag, pp. 177-191.

[6] DAMM, W., FEHR, E., INDERMARK, K., *Higher type recursion and self-application as control structures*, Formal Description of Programming Concepts ed. E. Neuhold, North-Holland Publishing Company (1978), pp. 461-48

[7] ENGELFRIET, J., SCHMIDT E.M., *IO and OI, part I and II*, Journal of Computer an System Sciences Vol. 15, 3 (1977), pp. 328-352, and Vol. 16, 1 (1978), pp. 67-99.

[8] FEHR, E., *On typed and untyped λ-schemes*, Schriften zur Informatik und Angewandten Mathematik 44, RWTH Aachen (1978).

[9] FISCHER, M.J., *Grammars with macro-like productions*, Proc. 9th IEEE conference on Switching and Automata Theory (1968), pp. 131-142.

[10] GOGUEN, J.A., THATCHER, J.W., WAGNER, E.G., WRIGHT, J.B., *Initial Algebra Semantics and Coninuous Algebras*, JACM, Vol. 24, 1 (1977), pp. 68-95.

[11] INDERMARK, K., *Schemes with recursion on higher types*, Proc. 5th conference on Mathematical Foundations of Computer Science, Lecture Notes in Computer Science 45 (1976), Springer Verlag, pp. 352-358.

[12] MASLOV, A.N., *The Hierarchy of Indexed Languages of an Arbitrary Level*, Soviet. Math. Dokl., Vol. 15, 14 (1974), pp. 117o-1174.

[13] NIVAT, M., *On the interpretation of recursive program schemes*, Atti del convegr di informatica teorica, Symposia Mathematica, Vol. 15 (1975), Rome, pp. 256-281.

[14] SCHMIDT, E.M., *Succintness of Description of Context-free, Regular, and Finite Languages*, Datalogisk Afdelning Report, DAIMI PB-84, Aarhus University (1978).

[15] STEYAERT, J.M., *Evaluation des Index Rationnels de quelques Familles de Langages*, IRIA-report No. 261 (1977).

[16] WAND, M., *A Concrete Approach to Abstract Recursive Definitions*, Automata, Languages, and Programming, ed. M. Nivat, North-Holland Publishing Company (1973), pp. 331-341.

[17] WAND, M., *An Algebraic Formulation of the Chomsky-Hierarchy*, Category Theory Applied to Computation and Control, Lecture Notes in Computer Science 25 (1975), pp. 209-213.

BOUNDS ON COMPUTATIONAL COMPLEXITY AND APPROXIMABILITY OF INITIAL SEGMENTS OF RECURSIVE SETS

M.I. Dekhtjar

Institute of Mathematics, Novosibirsk, 630090, USSR

1. Introduction

In this paper we study the relationship between bounds of the computational complexity of recursive sets and the minimal length of programs which compute initial segments of these sets (complexity in the sense of Kolmogorov-Markov).

If some recursive set (r.s.) A has a great lower bound g of the computational complexity, then there seems to be natural an attempt to find such programs that recognize initial segments of A within complexity $f < g$. The growth of the length of these programs (as function of the initial segments length) characterizes the difficulty of approximation of set A by sets whose computational complexity doesn't exceed f.

In Section 3 we establish the lower bound on this function which depends on the "distance" between the upper and the lower bounds of computational complexity of A (Theorems 1,2). This bound "exponentially" improves the bound obtained by the author in [1]. It follows from the speed-up Lemma which shows that any computation of a set which is easy to approximate can be sped up on infinitely many arguments. In Section 4 the examples of the sets which are easy to approximate are presented (Theorem 3). They show that lower bounds obtained in Theorems 1 and 2 can't be essentially improved.

2. Notations and definitions

We denote by the same letter an infinite binary sequence $A = A(0)A(1)\ldots A(x)\ldots$ ($A(x) \in \{0,1\}$) and a set $A=\{ x \mid A(x) = 1\}$. A^x denotes the initial segment of sequence (set) A : $A^x = A(0)A(1)\ldots A(x)$.

We fix a standard Gödel numbering of the one-tape one-head Turing machines (TM) $\{\mathfrak{M}_i\}$ (e.g. as in [2]) and the corresponding numbering of the partial recursive functions $\{\varphi_i\}$. As a basic measure of computational complexity we consider the space complexity $\{s_i\}$ (as usual $s_i(x)$ is the number of tape squares that are required for $\mathfrak{M}_i$ to compute $\varphi_i(x)$). A total recursive function f is said to be tape-constructible if for some i $f = s_i$.

$|x|$ is used to denote the length of the binary representation of number x. We suppose that there exists a universal TM $\mathfrak{M}$ such that for any pair (i,x) $\mathfrak{M}$ computes $\varphi_{\mathfrak{M}}(i,x)=\varphi_i(x)$ and $s_{\mathfrak{M}}(i,x)\leqslant |i|s_i(x)$.

For any set A, number x and total recursive function (t.r.f.) f, $f(y)\geqslant |y|$, we define the complexity of f-approximation of the initial segment A^x as the number

$$M^f(A^x) = \min\{|i| \mid \forall y \leqslant x\,[\,\varphi_i(y)=A(y)\ \&\ s_i(y) \leqslant f(y)\,]\} .$$

This definition can be naturally extended on any other complexity measure. The basic properties of the functions $\lambda x.M^f(A^x)$ and related results may be found in [3].

We shall also use the next notations and abbreviations :

$U = \{f \mid f$ is unbounded non-decreasing t.r.f.$\}$;

$f \preccurlyeq g \iff \varliminf\limits_{x\to\infty} g(x)/f(x) > 0$;

lgx stands for $\log_2 x$;

$\lfloor a \rfloor$ - the whole part of real a ;

$\text{Comp.}A \leqslant h \iff \exists i\,[\,(\varphi_i = A)\ \&\ \overset{\infty}{\forall} x\,(s_i(x) \leqslant h(x))\,]$;

$\text{Comp.}A > g \iff \forall i\,[\,(\varphi_i = A) \Rightarrow \overset{\infty}{\forall} x\,(s_i(x) > g(x))\,]$.

For a pair of t.r.f. h and g such that $h(x) > g(x)$ for almost all x we define an auxiliary function

$$u_{h,g}(x) = \max\{y \mid (y=0) \vee \forall z \leqslant y\,(h(z) \leqslant g(x))\}$$

and let $u^{(0)}_{h,g}(x)=x,\ \dots,\ u^{(n+1)}_{h,g}(x)=u_{h,g}(u^{(n)}_{h,g}(x))$. "Distance" between the functions h and g will be characterized by a function

$$r_{h,g}(x) = \max\{n \mid (n=0) \vee (u^{(n)}_{h,g}(x) > 0)\} .$$

It follows from the definition that $r_{h,g}(x) \leqslant x$ for all x and if $g(x)\geqslant |x|$, then $r_{h,g}$ is unbounded t.r.f.. If $h\in U$ and $g\in U$, then $r_{h,g}\in U$. In this case $r_{h,g}(x)$ is equal to the number of steps of the staircase placed between the graphs of h and g with the top at the point $(x,g(x))$.

3. Lower bounds of approximability

Let t.r.f. g be almost everywhere a lower bound of the computational complexity of A, i.e. $\text{Comp.}A > g$. It follows from the results of [1,4] that there doesn't exist an increasing lower bound of f-approximability of A which only depends on g : for every $e \in U$ there exists such r.s. A that $\mathrm{Comp.}A > g$ and $\overset{\infty}{\forall} x [M^f(A^x) \leq e(x)]$ for every $f(x) \geq |x|$. In case when some upper bound of the computational complexity of A is also known, i.e. $g < \mathrm{Comp.}A \leq h$, it is established in [1] that $\overset{\infty}{\exists} x [M^f(A^x) \geq \lfloor \lg\lg r_{h,g}(x) \rfloor]$ (for $f(x) \leq g(x)/|x|$). This bound can be increased exponentially.

THEOREM 1. Let h and g be tape-constructible functions, A be a r.s. and $g < \mathrm{Comp.}A \leq h$. Then for every tape-constructible $f \in U$ such that $|z| f(z) < g(z)$ for all z, and for infinitely many x, $M^f(A^x) \geq \lfloor \lg r_{h,g}(x) \rfloor$.

The proof of this theorem follows immediately from the next Lemma on speed up computations of sets that are easy to approximate.

LEMMA. Let for tape-constructible h,g and $f \in U$ and for r.s. A the next conditions hold :

i) $\overset{\infty}{\forall} x [h(x) > g(x) > |x| f(x)]$;

ii) $\mathrm{Comp.}A \leq h$;

iii) $\overset{\infty}{\forall} x [M^f(A^x) < \lfloor \lg r_{h,g}(x) \rfloor]$.

Then there exists a program i such that $\varphi_i = A$ and $\overset{\infty}{\exists} x [s_i(x) \leq g(x)]$ (i.e. computation A(x) can be sped up in infinitely many arguments x from h(x) to g(x)).

Sketch of the proof. We fix at first programs i_1 and i_2 and argument $\tilde{x}$ such that $\forall x [(\varphi_{i_1}(x) = A(x) \ \& \ s_{i_1}(x) \leq h(x)) \& (\varphi_{i_2}(x) = r_{h,g}(x) \ \& \ s_{i_2}(x) \leq g(x)) \ \& \ (x \geq \tilde{x} \Rightarrow M^f(A^x) \leq \lfloor \lg r_{h,g}(x) \rfloor - 1)]$. Program i contains as subprograms i_1 and i_2 and on input x computes as follows :

a) if $x < \tilde{x}$ then write output A(x) using the inner memory;

b) if $x \geq \tilde{x}$ then compute the number $k(x) = \lfloor r_{h,g}(x)/2 \rfloor$;

c) find such minimal $\ell \leq k(x)$ that $\forall z \leq x [s_\ell(z) \leq f(z)]$ and for every $j \leq k(x)$ if $\tilde{z} = \min\{ z \mid s_j(z) \leq f(z) \ \& \ \varphi_j(z) \neq \varphi_\ell(z) \}$, then $\varphi_\ell(\tilde{z}) = \varphi_{i_1}(\tilde{z}) = A(\tilde{z})$; compute $\varphi_\ell(x)$ and write it as output.

It follows immediately from iii) and the choice of k(x) in b) that $\varphi_i(x) = A(x)$ for all x. Introduce an auxiliary notation :

$v_y(x) = \max\{\tilde{z} \mid (\tilde{z} \leq x) \& \exists \ell, j [(\ell \leq k(y)) \& (j \leq k(y)) \& \ \tilde{z} = \min\{z \mid (s_\ell(z) \leq f(z)) \& (s_j(z) \leq f(z)) \ \& \ (\varphi_\ell(z) \neq \varphi_j(z))] \}$.

Then by the definitions and the construction of i the next claims hold.

Claim 1. For every y $|\{v_y(x) \mid x \leq y\}| \leq k(y) \leq r_{h,g}(y)/2$.

Claim 2. There exist infinitely many x such that $v_x(x) < u_{h,g}(x)$

Claim 3. For every x $s_i(x) \leq \max\{g(x), |k(x)| f(x), \max\{h(z) \mid z \leq v_x(x)\}\}$.

Now the definition of $u_{h,g}(x)$, condition i) and claims 2 and 3 imply that $\exists^{\infty} x\, [s_i(x) \leq g(x)]$.

COROLLARY 1.1 Let h be an optimal (up to a constant factor) space complexity of A (i.e. Comp.A $\leq$ h and for any g if CompA $\leq$ g then g $\succcurlyeq$ h) and let there exist c such that $\lim_{x\to\infty} h(x)/h(x+c) = 0$. Then, for $f(x) = h(x)/|x|$, $\exists d\, \exists^{\infty} x\, [M^f(A^x) \geq \lg x - d]$.

COROLLARY 1.2 For any $f \in U$, $H \in U$ there exists $g \in U$ such that for any A if $g <$ Comp.A $\leq H \circ g$, then $\exists c\, \exists^{\infty} x\, [M^f(A^x) \geq \lg x - c]$.

Below we present some more examples of lower bounds of f-approximability obtained by Theorem 1.

$f(x) \leq$	$g(x)$	$h(x)$	$\lg r_{h,g}(x) \succcurlyeq$
$x/\|x\|$	x	$\|x\| x$	$\lg\|x\|$
$x/\|x\|$	x	x^2	$\lg\lg\|x\|$
$\|x\|^k$	$\|x\|^{k+1}$	$\|x\|^{k+1+p}, p>0$	$\lg\lg\lg\|x\|$
$2^x/\|x\|$	2^x	2^{2x}	$\lg\|x\|$
$2^{2^x}/\|x\|$	2^{2^x}	$2^{2^{x+1}}$	$\|x\|$

REMARK 1. Theorem 1 can't be improved by changing quantification $\exists^{\infty} x$ into $\forall^{\infty} x$. This is due to the results of [5] where "sparse" sets with optimal computations have been constructed. The results imply that for $s(x) = |x|$, every tape-constructible h and every (arbitrarily slowly increasing) $e \in U$ there exists a r.s. A such that

i) h is an optimal space complexity of A ;

ii) $\exists^{\infty} x\, [M^s(A^x) \leq e(x)]$.

REMARK 2. In conditions of Theorem 1 the "honesty" requirements on complexity bounds h and g can be omitted. Then the assertion of Theorem 1 will hold after the change of $r_{h,g}(x)$ into any t.r.f. r(x) such that r(x) is computable within space g(x) and $r(x) \leq r_{h,g}(x)$ for almost all x.

REMARK 3. Theorem 1 with slight changes remains valid for the tin and other "natural" complexity measures. For instance, it will hold for the time if we change $u_{h,g}$ in the definition of $r_{h,g}$ into function $\tilde{u}_{h,g}(x) = \max\{y \mid (y = 0) \vee \forall z \leq y\ (\ x \cdot h(z) < g(x)\ )\}$ and choose f such that $g(x) > x^2 f(x) \lg f(x)$ for all x.

We supposed above that the complexity of any computation of **A** is greater than g(x) for almost all x. There are a lot of natural problems A that have not great enough lower bounds of complexity on almost all arguments but, nevertheless, for any TM $\mathfrak{M}_i$ computing A $\max\{s_i(x) \mid |x| = n\}$ increases quickly as function of n. Theorem 2 below generalizes Theorem 1 and allows us to obtain lower bounds of f--approximability for such kind of problems.

<u>THEOREM 2.</u> Let b(n) be any tape-constructible function such that b(0) = 0 and b(n) < b(n+1) for all n. Let for tape-constructible functions h, g, f ∈ U and for r.s. A the following conditions hold:

(i) $\overset{\infty}{\forall} n\ [h(n) > g(n) > |b(n+1) - 1|\, f(b(n+1) - 1)$;

(ii) $\exists i[(\varphi_i = A)\ \&\ \overset{\infty}{\forall} n(\ \max\{s_i(x) \mid b(n) \leq x < b(n+1)\} \leq h(n)\ )]$;

(iii) $\forall j[(\varphi_j = A) \Rightarrow \overset{\infty}{\forall} n(\ \max\{s_j(x) \mid b(n) \leq x < b(n+1)\} > g(n)\ )]$.

Then

$$\overset{\infty}{\exists} n \exists x\ [\ (\ b(n) \leq x < b(n+1)\ )\ \&\ (\ M^f(A^x) \geq \lfloor \lg r_{h,g}(n) \rfloor\ )].$$

The proof of this theorem is similar to the proof of Theorem 1 and therefore is omitted.

Note that Theorem 1 follows from Theorem 2 under b(n) = n. Under $b(n) = 2^{n-1}$ the next analogy of Corollary 1.1 may be obtained.

<u>COROLLARY 2.1</u> Let h be a tape-constructible function such that for some c $\lim_{n \to \infty} h(n)/h(n+c) = 0$ and A be a r.s. such that

(i) $\exists i[(\varphi_i = A)\ \&\ \overset{\infty}{\forall} n\ (\ \max\{s_i(x) \mid |x| = n\} \leq h(n)\ )]$;

(ii) $\forall j[(\ \varphi_j = A\ ) \Rightarrow (\ \max\{s_j(x) \mid |x| = n\} \succcurlyeq h(n)\ )]$.

Then for $f(x) = h(|x|)/|x|$ there exists such constant d that

$$\overset{\infty}{\exists} x(\ M^f(A^x) \geq \lg\lg x - d\ ).$$

<u>REMARK 4.</u> Theorem 2 may be applied, for example, to complete in **EXPSPACE** problems to obtain lower bound O(lglgn) on their polynomial approximability. But for such kind of problems exponential lower bounds on their polynomial approximability were established in [6]. On the other hand, Theorems 1 and 2 are applicable to any set within given bounds of computational complexity, not only to a complete one.

4. Sets that are easy to approximate

In this section we establish the upper bounds of f-approximability for sets with fixed bounds of the computational complexity. They show that the lower bounds of f-approximability obtained by Theorem 1 and 2 are exact enough and can't be essentially increased. The next theorem provides examples of easy to approximate sets within any quickly enough increasing bounds of computational complexity.

THEOREM 3. Let h, g and f be tape-constructible functions from U such that $g(x) = o(g(x+1))$, $h(x) = o(h(x+1))$, $h(x) > g(x+1)$ for all x and $r_{h,g}(x)$ is computable within space f(x). Then for any (arbitrarily slowly increasing) $e \in U$ there exists a r.s. A such that

(i) $g <$ Comp.A $\leqslant h$;

(ii) $\overset{\infty}{\forall} x\,[M^f(A^x) \leqslant e(x)\ \lg r_{h,g}(x)]$.

Sketch of the proof. At first we define a sequence $\{y_k\}$: $y_0 = 0$, $y_{2n} = \min\{z \mid r_{h,g}(z) = n\}$, $y_{2n+1} = \lfloor (y_{2n} + y_{2n+2})/2 \rfloor$, and an auxiliary function $\tilde{g}$: $\tilde{g}(x) = h(y_n)$ for $x \in [y_n, y_{n+1})$. Let $b \in U$ be such computable in linear space function that $b(x) \leqslant \sqrt{e(x)}$ for all x. A set A will be defined by $\tilde{g}$, b and $\{y_n\}$ with the help of a standard diagonalization. After stage n of construction A(x) will be defined for all $x \in [0, y_{n+1})$. In addition some finite set L_n of programs "cancelled" after stage n will be defined.

Stage 0. Put $A(x) = 0$ for $x \in [0, y_1)$, $L_0 = \emptyset$.

Stage n+1. Let $i = \min\{j \mid (j = \infty) \vee [\,(j \notin L_n) \,\&\, (j \leqslant b(y_n))\ \&$ $\&\ \exists z\,[\,(z \in [y_{n+1}, y_{n+2}))\,\&\,(|j| \cdot s_j(z) \leqslant \tilde{g}(z))]]\}$.

If $i = \infty$ then put $A(x) = 0$ for $x \in [y_{n+1}, y_{n+2})$, and $L_{n+1} = L_n$.

If $i < \infty$ then find $\tilde{z} = \min\{z \mid z \in [y_{n+1}, y_{n+2})\ \&\ (|i|\, s_i(z) \leqslant \tilde{g}(z)\,)\}$ and put $A(x) = 1 \dot{-} \varphi_i(z)$ for $x \in [y_{n+1}, y_{n+2})$, and $L_{n+1} = L_n \cup \{i\}$.

It is not difficult to verify that A(x) can be computed within space g(x). Therefore Comp.A $\leqslant \tilde{g} \leqslant h$. By the standard arguments it may also be established that for every i if $\varphi_i = A$, then $\overset{\infty}{\forall} x\,(s_i(x) > \tilde{g}(x)/|i|\,)$. The conditions on the growth of h and g and the definition of $\tilde{g}$ imply that $g(x) = o(\tilde{g}(x))$. Hence, Comp. A $> g$ and assertion (i) holds.

To prove (ii) we note that a sequence A^x, $x \in [y_n, y_{n+1})$, can be restored by a set $I_n = \{k \mid k \leqslant n\ \&\ A(y_k) = 1\}$. It follows from the construction of A that $|n| \leqslant \lg r_{h,g}(x) + 1$ and $|I_n| \leqslant b(y_n) \leqslant b(x)$. Therefore, the length of the program recognizing A^x within space f doesn't exceed, for some constant c , $c\, b(x) \cdot \lg r_{h,g}(x) \leqslant e(x)\ \lg r_{h,g}($

for almost all x.

COROLLARY 3.1. Let $r \in U$ range over all numbers. Then there exist $f \in U$ and such arbitrarily great t.r.f. h and g that $r_{h,g} = r$ and for every $e \in U$ there exists r.s. A such that

(i) $g < \text{Comp. } A \leq h$;

(ii) $\overset{\infty}{\forall} x \; [\, M^f(A^x) \leq e(x) \lg r(x) \,]$.

It is unknown whether unbounded function e(x) may be avoided in statement (ii) of Theorem 3 (and Corollary 3.1) or at least be decreased to a constant.

REFERENCES

1. Dekhtjar, M.I., On the complexity of approximation of recursive sets, EIK, 12 (1976), 3, 115-122.
2. Trakhtenbrot, B.A., Complexity of algorithms and computations, Novosibirsk State University, Novosibirsk, 1967.
3. Agafonov, V.N., Complexity of algorithms and computations, II, Novosibirsk State University, Novosibirsk, 1975.
4. Freidzon, R.I., Regular approximation of recursive predicates, Zap. nauchn. sem. Leningradskogo otd. Mat. Inst. Steklova, 20 (1971), 220-223.
5. Trakhtenbrot, B.A., Optimal computations and frequency phenomenon of Jablonsky, Algebra and logic, 4 (1965), 5, 79-93.
6. Dekhtjar, M.I., Complexity spectra of recursive sets and approximability of initial segments of complete problems, EIK, 15 (1979), 1/2, 11-32.

ON THE WEIGHTED PATH LENGTH OF BINARY SEARCH TREES FOR UNKNOWN ACCESS PROBABILITIES

Thomas Fischer
Technical University of Dresden
Department of Mathematics
DDR - 8027 Dresden
German Democratic Republic

INTRODUCTION

Given a set of m records with an ordering on their keys, and giv[en] a fixed distribution p of their access probabilities, it is well know[n] that a binary search tree can then be constructed such that its weigh[t]ed path length is less than H(p) + 2, where H(p) is the Shannon entropy of the distribution p. Algorithms for constructing search trees according to given underlying access probabilities have been proposed in a large number of papers, see e.g. Gilbert and Moore [1], Knuth [2, 3], Fredman [4], and Mehlhorn [5]. One of the major shortcomings of the theory, however, is that in practice access probabilities are often not known precisely. Moreover, it may be possible that the true access probabilities are not only unknown in advance but also varying over time. The construction of an optimal search tree, however, and the calculation of its weighted path length require precise knowledge of the actual distribution p. Optimal or nearly optimal binary search trees designed for a given known probability distribution are therefore not applicable to such cases. Three questions thus naturally arise:

(i) If a binary search tree is designed for an access probability distribution q, but the actual distribution is p, how much performance might be lost due to this mismatch?

(ii) What is the minimum attainable weighted path length if we must design a tree for an unknown member of a given class P of distribu-

tions?
(iii) Is there an algorithm for constructing a binary search tree that is in some sense well suited for any member of P?

These three questions are answered in Theorem 1 and 2 and in the Corollary below.

A similar problem has previously been attacked in papers by Allan and Munro [6] and Mehlhorn [7] by investigating techniques for adapting the search tree to the unknown or time varying access probabilities (self-organizing and dynamic binary search trees, respectively). The idea of the present paper is to design a static tree that performs well for any member of a whole class of probability distributions. It is shown that under certain conditions there is always a single tree such that, for each particular probability distribution $p \in P$, its weighted path length does not exceed the bound $\sup_{p \in P} H(p) + 2$. Binary search trees of such a kind are called universal with respect to P. An universal binary search tree could thus be used without specific knowledge of which distribution was given.

WEIGHTED PATH LENGTH AND INACCURACY

Suppose we are given m records represented by their names (keys) $N_1, N_2, \ldots, N_m$ in lexicographical order and a probability vector $p = (p_1, \ldots, p_m, p_0', p_1', \ldots, p_m')$, i. e. $p_i \geq 0$, $i = 1, \ldots, m$, $p_j' \geq 0$, $j = 0, \ldots, m$, and $\sum_{i=1}^{m} p_i + \sum_{j=0}^{m} p_j' = 1$. Let $p_1, \ldots, p_m$ be the probabilities of a successful search for the records $N_1, \ldots, N_m$, while p_j' denotes the probability of requesting for a name that belongs to the interval (N_j, N_{j+1}), $j = 1, \ldots, m-1$. Consequently, the probabilities of searching for names less than N_1 and greater than N_m are p_0' and p_m', respectively. An important method for retrieving information by its name is to store the names in a binary tree. Let l_i denote the path length from the root of the tree to the internal node representing the name N_i, l_j' denotes the corresponding path length of the external node labelled with (N_j, N_{j+1}). Then the weighted path length of such a binary search tree T is defined by

$$L_p(T) = \sum_{i=1}^{m} p_i (l_i + 1) + \sum_{j=0}^{m} p_j' l_j',$$

cf. with Knuth [3].

Now suppose we are given another probability distribution $q = (q_1, \dots, q_m, q_0', q_1', \dots, q_m')$, where $q_i > 0$, $i = 1, \dots, m$, $q_j' > 0$, $j = 0, \dots, m$, which may be interpreted as a hypothesis on the true but unknown distribution p. It is well known that it is possible to construct a binary search tree such that

$$l_i < -\mathrm{ld}\, q_i\,, \quad i = 1, \dots, m,$$

and

$$l_j' < -\mathrm{ld}\, q_j' + 2\,, \quad j = 0, \dots, m,$$

cf. with Fredman [4], Mehlhorn [5], or for a special case see Knuth [3], pp. 445 - 446. Let us denote such a tree by T(q) and consider its weighted path length with regard to the actual distribution p. Thus we have established the following result:

Theorem 1.

$$L_p(T(q)) < H(p,q) + \sum_{i=1}^{m} p_i + 2\sum_{j=0}^{m} p_j'\,,$$

where

$$H(p,q) = -\sum_{i=1}^{m} p_i\, \mathrm{ld}\, q_i - \sum_{j=0}^{m} p_j'\, \mathrm{ld}\, q_j'.$$

The latter function, which is usually called inaccuracy, is well established in information theoretic literature, see Kerridge [8]. It is easy to see that this function is a simple generalization of the usual Shannon entropy H(p), which can be obtained from inaccuracy by setting q = p. Some important properties of inaccuracy are listed in the lemma below. Here and in the following the convention $0 \cdot \mathrm{ld}\, 0 = 0$ is used.

Lemma.

For any pair of probability vectors $p = (p_1, \dots, p_n)$ and $q = (q_1, \dots, q_n)$ the function

$$H(p,q) = -\sum_{i=1}^{n} p_i\, \mathrm{ld}\, q_i$$

is nonnegative, greater than or equal to Shannon's entropy H(p), and unbounded to above. Furthermore, H(p,q) is continuous and linear in p and convex in q. If $q_i > 0$, or $q_i = 0$ implies $p_i = 0$, $i = 1, \dots, n$, it is also continuous in q and always finite.

BINARY SEARCH TREES BASED ON FREQUENCY DISTRIBUTIONS AND UNIVERSAL BINARY SEARCH TREES

At first consider the case in which the actual distribution of access probabilities, p, is unknown in advance but not varying over time. Then it might be possible to observe the frequency of requests for the particular records before designing the final search tree. Suppose p_n is such a frequency distribution based on a sample of size n. Then the weighted path length of a tree T_n designed according to p_n is upperbounded by $H(p,p_n) + 2$, cf. with Theorem 1. Since p_n is a consistent estimate of p, for sufficiently large samples, $H(p,p_n)$ converges to the optimum value, H(p), with probability 1.

Next consider the case that the actual access probabilities are not observable, perhaps because they are varying over time. If thus the true distribution p is completely unknown, the only we can do is to use the uniform probability distribution for constructing the tree. If q is the uniform distribution, i. e. $q_i = q_j' = 1/(2m+1)$, i = 1, ... , m, j = 0, ... , m, we get $H(p,q) = \text{ld}\,(2m+1)$. Hence, when using the uniform probability distribution we get a binary search tree having a weighted path length less than ld (2m+1) + 2. If, in the opposite case, the actual distribution p is completely known and the tree constructing algorithm is applied to p, we have the well known upper bound H(p) + 2.

In the following the situation between these two extreme cases is investigated. Suppose it is known that the true access probability distribution belongs to a certain class P of probability distributions corresponding to a fixed set of names, $N = \{N_1, \dots, N_m\}$. Let T_N denote the set of all possible binary search trees for N and consider the minimum attainable weighted path length of trees $T \in T_N$ for the worst distribution $p \in P$,

$$L_N(P) = \min_{T \in T_N} \sup_{p \in P} L_p(T).$$

It is well known that for any particular distribution p the weighted path length of an optimal binary search tree T_o satisfies

$$L_p(T_o) = \min_{T \in T_N} L_p(T) \leq H(p) + 2,$$

cf. with Theorem 1 for $q = p$. Now we are interested in a corresponding upper bound for $L_N(P)$.

Theorem 2.

If P is convex and compact, then

$$L_N(P) < \sup_{p \in P} H(p) + 2.$$

Proof:

Let P be compact and convex and consider the minimum weighted path length of a tree $T = T(q)$ related to the worst distribution $p \in P$,

$$L'_N(P) = \inf_q \sup_{p \in P} L_p(T(q)).$$

From Theorem 1 it immediately follows that

$$L'_N(P) < \inf_q \sup_{p \in P} H(p,q) + 2,$$

and since $H(p,q)$ is linear in p and convex in q, we get

$$\inf_q \sup_{p \in P} H(p,q) = \sup_{p \in P} \inf_q H(p,q)$$

$$= \sup_{p \in P} H(p),$$

see Golstein [9], Theorem 1. Therefore

$$L'_N(P) < \sup_{p \in P} H(p) + 2,$$

and since $L_N(P) \leq L'_N(P)$, the theorem is proved.

It should be pointed out that the bound of Theorem 2 could be sharpened by using the value $\sup H(p) + \sup \sum_{j=0}^{m} p'_j + 1$, cf. with the bound of Theorem 1. However, we prefer the weaker bound used above since it is easier to work with it. Thus the result of Theorem 2 leads to the following definition:

A binary search tree T is called universal for P, if for each particular probability distribution $p \in P$ its weighted path length satisfies

$$L_p(T) < \sup_{p \in P} H(p) + 2.$$

In the proof of Theorem 2 it is also shown that for convex and compact classes P a universal binary search tree can be found by applying the usual tree constructing algorithm to a suitable distribution q. Hence, for any such convex and compact class P we have to find this best distribution. A simple solution for this problem can be derived from the following lemma proved in a previous paper [10]:

Lemma.

Let P be a convex and compact class of probability distributions and let p_o be such that $H(p_o) = \sup_{p\in P} H(p)$. Then, for each $p \in P$,

$$H(p,p_o) \leq H(p_o).$$

Now let $T(p_o)$ be a binary search tree constructed in accordance to the entropy maximizing distribution p_o. Then, from Theorem 1, it follows that

$$L_p(T(p_o)) < H(p,p_o) + 2,$$

for each $p \in P$, and the Lemma above yields

$$L_p(T(p_o)) < \sup_{p\in P} H(p) + 2.$$

Thus we have derived the following result:

Corollary.

The tree $T(p_o)$ is universal for P.

This result is illustrated in the following example, in which the special case $p_1 = \dots = p_m = 0$ is considered. The corresponding binary search trees can therefore be obtained by the use of the Gilbert - Moore algorithm, cf. with Knuth [3], p. 445. Furthermore, for any real numbers a and b the class $P_{[a;b]}$ is defined as the set of all probability vectors $(p_o', \dots, p_m')$ with $p_o' = \varrho$ and $p_1' = \dots = p_m' = (1-\varrho)/m$ where ϱ is a real number such that $a \leq \varrho \leq b$. The entropy of such probability vectors, $H(p_o',\dots,p_m')$, can then be written as a function of the real parameter ϱ only,

$$\begin{aligned} H(p_o',\dots,p_m') &= H(\varrho) \\ &= \mathrm{ld}\, m - \varrho\, \mathrm{ld}\, \varrho - (1-\varrho)\, \mathrm{ld}\,(1-\varrho). \end{aligned}$$

The graphical representation of $H(\varrho)$ is given below.

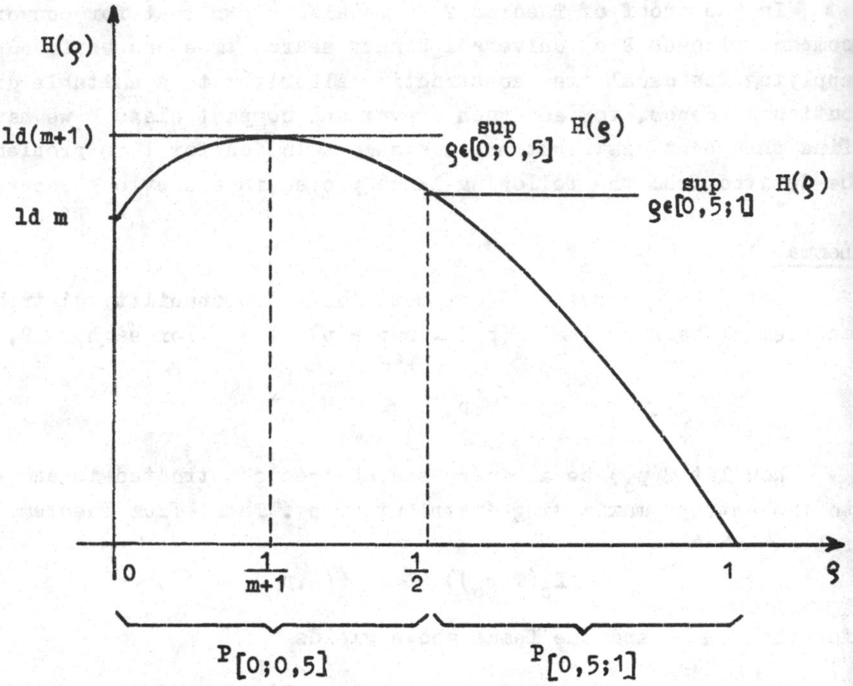

From the shape of the $H(\varrho)$ curve it can be easily seen that for the class $P_{[0;0,5]}$ as well as for $P_{[0;1]}$ the entropy maximizing distribution is the uniform one, which leads to a tree having a weighted path length between ld(m+1) and ld(m+1) + 2, whereas for the class $P_{[0,5;1]}$ a considerable reduction of weighted path length would be possible by using the distribution obtained by setting $\varrho = 0,5$.

CONCLUDING REMARKS

The algorithm for the construction of universal binary search trees can therefore be described as follows:

(i) Given a convex and compact class P of access probability distributions choose a distribution p_o that maximizes the Shannon entropy in P.

(ii) Construct a binary search tree by applying the usual algorithms mentioned above to p_o. Then the tree obtained is universal for P in the sense of the definition above.

Thus the problem of constructing a universal binary search tree has been reduced to the problem of finding a probability distribution that maximizes the Shannon entropy within a convex and compact set of distributions. Note that, since the Shannon entropy H(p) is a concave function of the probability vector p, a solution may be found by using the methods of convex optimization theory.

Clearly, if P is the class of all probability distributions on some ordered set of names, the entropy maximizing distribution is always the uniform one. Moreover, the uniform probability distribution is the maximizing one whenever it belongs to P. This is the worst case of our considerations. Thus the efficiency of the method presented here essentially depends on the nature of the given class P.

REFERENCES

1. Gilbert, E.N. and Moore, E.F., Variable-length binary encodings. Bell System Tech. J. , 38 (1959), 933 - 968.
2. Knuth, D.E., Optimum binary search trees. Acta Informatica, 1 (1971), 14 - 25.
3. ————, The Art of Computer Programming, Vol. 3: Sorting and Searching. Addison - Wesley Publ. Co., Reading, Mass., 1973.
4. Fredman, M.L., Two applications of a probabilistic search technique: Sorting X+Y and building balanced search trees. Proc. 7th Annual ACM Symposium on Theory of Computing, 1975.
5. Mehlhorn, K., A best possible bound for the weighted path length of binary search trees. SIAM J. Comput. 6 (1977), 235 - 239.
6. Allan, B. and Munro, I., Self-organizing binary search. Proc. 17th Symp. on Foundations of Computer Science, 1976.
7. Mehlhorn, K., Dynamic binary search. Lecture Notes Computer Science, Vol. 52. Springer - Verlag, Berlin etc., 1977, 323 - 336.
8. Kerridge, D.F., Inaccuracy and inference. J. Royal Stat. Soc., B 23 (1961), 184 - 194.
9. Golstein, E.G., Konvexe Optimierung. Akademie - Verlag, Berlin, 1973.
0. Fischer, Th., Über universelle Codierungen und Suchprozesse. Schriftenreihe Weiterbildungszentrum MKR, Heft 30 (1978), Technical University of Dresden.

COMPUTATIONAL COMPLEXITY OF APPROXIMATION ALGORITHMS FOR COMBINATORIAL PROBLEMS

G.V. Gens
E.V. Levner
Central Economic and Mathematical Institute
USSR Academy of Sciences, Moscow 117333

1. INTRODUCTION

Combinatorial optimization problems such as allocation, traveling salesman and knapsack have obvious applications in computer systems, for example in efficient organization of data processing, assignment of computer information to physical blocks of storage, resource allocation in multiprocessor systems. The apparent computational intractability of these problems has stimulated research into the possibilities of designing efficient approximation algorithms which, while not always finding optimal solutions, do guarantee solutions that are close to the optimal.

In this paper for the allocation and traveling salesman problems it is shown that to find approximate solutions with guaranteed accuracy is, in a sense, as difficult as to find optimal ones. Various forms of the knapsack problem are considered and fast approximation algorithms with the guaranteed accuracy are obtained for them.

The knapsack problems considered in the paper are formulated as follows.

Problem P_1 (min-knapsack)

$$\text{Minimize} \sum_{i=1}^{n} c_i x_i ,$$

$$\text{subject to} \sum_{i=1}^{n} a_i x_i \geq b,\ x_i \in \{0, 1\}, c_i \geq 0,\ a_i \geq 0 (i=1,\ldots,n).$$

Problem P_2 (sum of subset)

$$\text{Maximize} \sum_{i=1}^{n} a_i x_i ,$$

$$\text{subject to} \sum_{i=1}^{n} a_i x_i \leq b,\ x_i \in \{0, 1\},\ a_i \geq 0 (i=1,\ldots,n) .$$

Problem P_3 (min-triangular-knapsack, or min-job-sequencing-with-due-dates)

$$\text{Minimize } \sum_{i=1}^{n} c_i x_i ,$$

$$\text{subject to } \sum_{i=1}^{j} a_i x_i \geqslant b_j \ (j=1,\dots,n),$$
$$x_i \in \{0,1\},\ c_i \geqslant 0,\ a_i \geqslant 0 (i=1,\dots,n) .$$

Problem P_4 (max-multiple-choice-knapsack)

$$\text{Maximize } \sum_{i=1}^{m} \sum_{j=1}^{k_i} c_{ij} x_{ij} ,$$

$$\text{subject to } \sum_{i=1}^{m} \sum_{j=1}^{k_i} a_{ij} x_{ij} \leqslant b, \sum_{j=1}^{k_i} x_{ij} \leqslant 1,$$
$$x_{ij} \in \{0,1\},\ c_{ij} \geqslant 0,\ a_{ij} \geqslant 0 (j=1,\dots,k_i;\ i=1,\dots,m).$$

Problem P_4' (min-multiple-choice-knapsack)

$$\text{Minimize } \sum_{i=1}^{m} \sum_{j=1}^{k_i} c_{ij} x_{ij} ,$$

$$\text{subject to } \sum_{i=1}^{m} \sum_{j=1}^{k_i} a_{ij} x_{ij} \geqslant b, \sum_{j=1}^{k_i} x_{ij} \leqslant 1,$$
$$x_{ij} \in \{0,1\},\ c_{ij} \geqslant 0,\ a_{ij} \geqslant 0 \ (j=1,\dots,k_i,\ i=1,\dots,m) .$$

Problem P_5 (min-obligatory-multiple-choice-knapsack)

$$\text{Minimize (Maximize)} \sum_{i=1}^{m} \sum_{j=1}^{k_i} c_{ij} x_{ij} ,$$

$$\text{subject to } \sum_{i=1}^{m} \sum_{j=1}^{k_i} a_{ij} x_{ij} \underset{(\leqslant)}{\geqslant} b, \ \sum_{j=1}^{k_i} x_{ij} = 1,$$

$$x_{ij} \in \{0,\ 1\},\ c_{ij} \geqslant 0,\ a_{ij} \geqslant 0 \ (j=1,\dots,k_i,\ i=1,\dots,m).$$

Problem P_6 (continuous fixed-charge-knapsack)

$$\text{Minimize (Maximize)} \sum_{i=1}^{n} (c_i x_i + d_i \operatorname{sgn} x_i) ,$$

$$\text{subject to } \sum_{i=1}^{n} (a_i x_i + b_i \operatorname{sgn} x_i) \geqslant b$$
$$\left(\sum_{i=1}^{n} (a_i x_i + b_i \operatorname{sgn} x_i) \leqslant b \right)$$

$$0 \leqslant x_i \leqslant 1,\ a_i,\ d_i > 0,\ c_i,\ b_i \geqslant 0,\ i=1,\dots,n .$$

Table 1 summarizes the $\mathcal{E}$-approximation algorithms which have recently been devised for the problems P_1 - P_6, N standing for $\sum_{i=1}^{m} k_i$, logarithm considered base 2.

Table 1. Complexity of Approximation Algorithms

Problem	Time Complexity	Space Complexity	Reference
P_1	$O(n^4/\varepsilon)$	$O(n^4/\varepsilon)$	Babat [1]
	$O(n^3/\varepsilon)$	$O(n^3/\varepsilon)$	Gens and Levner [8]
	$O(n^2/\varepsilon)$	$O(n^2/\varepsilon)$	h.l.
P_2	$O(n/\varepsilon^2)$	$O(n+{}^1/\varepsilon^3)$	Ibarra and Kim [5]
	$O(n+{}^1/\varepsilon^3)$	$O(n+{}^1/\varepsilon^2)$	Lawler [7]
	$O(n/\varepsilon)$ or $O(n+1/\varepsilon^2)$	$O(n/\varepsilon)$ or $O(n+1/\varepsilon^2)$	Gens and Levner [4, 8]
	$O(n/\varepsilon+1/\varepsilon^3)$	$O(n+1/\varepsilon)$	h.l. h.l.
P_3	$O(n^3/\varepsilon)$	$O(n^3/\varepsilon)$	Gens and Levner [4, 8]
	$O(n^2\log n+{}^{n^2}/\varepsilon)$	$O(n^2/\varepsilon)$	h.l.
P_4	$O(Nm^2/\varepsilon)$	$O(Nm^2/\varepsilon)$	Lawler [7]
	$O(Nm/\varepsilon)$	$O(Nm/\varepsilon)$	h.l.
P'_4	$O(N\log m+Nm\log m+{}^{Nm}/\varepsilon)$	$O(N+{}^{m^2}/\varepsilon)$	h.l.
$P_5(\max)$	$O(Nm/\varepsilon)$	$O(N+m^2/\varepsilon)$	h.l.
$P_5(\min)$	$O(N\log m+Nm\log m+{}^{Nm}/\varepsilon)$	$O(N+m^2/\varepsilon)$	h.l.
$P_6(\min)$	$O(n^4/\varepsilon+n/\varepsilon^4(\sum_{i=1}^{n}\log(1+{}^{c_i}/_{d_i}))^3)$	$O(n^3/\varepsilon)$	Babat [2]
$P_6(\max)$	$O(n^3/\varepsilon)$	$O(n^2/\varepsilon)$	h.l.
	$O(n^2/\varepsilon)$	$O(n^2/\varepsilon)$	h.l.

2. ANALYSIS OF APPROXIMATION ALGORITHMS

Formally we define approximation algorithms with guaranteed accuracy as follows.

Definition 1. An algorithm will be said to be an ε- approximation one for a problem P if for every instance I_p of P and for a given $\varepsilon > 0$ we have $|f(I_p, \bar{x}) - f^*(I_p)| \leq \varepsilon \cdot f^*(I_p)$, where $f^*(I_p)$ is the optimum value of the object function $f(I_p,x)$ in the problem I_p, $f(I_p,\bar{x})$ is the approximate solution value obtained. The solution $\bar{x}$ will be said to be ε- approximate.

Definition 2. An algorithm will be said to be (ν) - approximation algorithm for a problem P, if for every instance I_P of P and for a given $\nu>0$ $|f(I_P,\bar{x})-f(I_P)|\leq\nu\cdot(\max_{x\in X}f(I_P,x)-\min_{x\in X}f(I_P,x))$, X being the set of feasible solutions of I_P. The solution $\bar{x}$ will be said to be (ν) - approximate.

In the definition 1 the guaranteed upper bound on possible error is measured in terms of the desired optimum, while in the definition 2 this bound is measured in terms of the maximum possible error.

Another approach to measuring the accuracy in discrete optimization may be as follows. Let P be a combinatorial problem, say, a maximization problem with $f(x)$ being an object function. Let all the feasible solutions of the problem P be arranged according to nonincreasing value of $f(x)$:

$$f(x_1)\geq f(x_2)\geq f(x_3)\geq\ldots \qquad (1)$$

All the x's which have the same value of $f(x)$ are assumed to be of the same group. Thus, the sequence (1) gives us a sequence of groups $B_1,B_2,B_3,\ldots$, such that

$$f(x\,|x\in B_1)>f(x\,|x\in B_2)>f(x\,|x\in B_3)>\ldots \qquad (2)$$

Definition 3. The k th solution in the sequence (1) is called the k th best solution. An approximate solution is called a k - solution if it belongs to any of the first k groups in the sequence (2).

It is clear that the k th best solution is, also, a k - solution; in this sense, to find a k - solution is easier than to find the k th best solution.

Definition 4. An ε - approximation (respectively, (ν) - approximation) algorithm will be said to be fast if it has time and space bounded by a polynomial in both problem size and $1/\varepsilon$ (or $1/\nu$). For example, if an algorithm operates in time bounded by $O(n^3+n^2/\varepsilon)$ then it is fast; if the bound is $O(n^{1/\varepsilon})$, it is not fast.

Theorem 1. For the knapsack problem to find a k -solution is a NP-hard problem for any k being bounded by a polynomial function of n.

To prove this fact we consider the following problem G:

$$\text{Maximize } g(x)=2k\cdot\sum_{i=1}^{n}p_ix_i+x_{n+1}+\ldots+x_{n+k-1},$$

subject to $g(x)\leq k.\sum_i p_i+k-1,\quad x_i\in\{0,1\}.$

Let A be a polynomial in time algorithm which gives a k - solution of the knapsack problem, and, hence, of the problem G. Then we see that either A gives a solution x such that $g(x) \geqslant k\sum_i p_i$ and then the sum of subset problem F, stated in terms of recognition, has a solution, or A gives a solution x such that $g(x) < k\sum_i p_i$ and then the problem F has no solution. Since F is the well-known NP - complete problem, it follows that to find A is a NP- hard problem in the sense of Karp [6].

A similar fact is valid for many other combinatorial problems such as traveling salesman, clique cover, set cover, graph coloring, etc.

Now consider the allocation problem A, in the formulation given by Cornuejols, Fisher and Nemhauser [3], and the sum of subset problem S, formulated in terms of the absolute value of a difference between two disjoint sets:

<u>Problem A</u>. Maximize $\sum_{i=1}^{m}\sum_{i=1}^{n} c_{ij}\, x_{ij} - \sum_{j=1}^{n} d_j\, y_j$

subject to $\sum_{j=1}^{n} x_{ij}=1,\ i=1,\ldots,m,\quad 1 \leqslant \sum_j y_j \leqslant K,\quad 0 \leqslant x_{ij} \leqslant y_j \leqslant$

$x_{ij},\ y_j \in \{0,\ 1\}$.

<u>Problem S</u>. Minimize $\left|\sum_{i=1}^{n} p_i x_i - \sum_{i=1}^{n} p_i(1-x_i)\right|,\ x_i \in \{0,\ 1\}$

<u>Theorem 2</u>. To find ε - approximate solutions of the problems A and S is NP - hard.

This fact is proved by reducing the well-known NP - complete problems such as allocation and sum of subset in terms of recognition, to finding ε - approximate solutions of the problems A and S. Thus, Theorem 2 extends the list of such problems originally presented by Sahni and Gonzales [10] .

Note that if a problem P has the optimum value f_P equal to 0, it implies that to find its ε - approximate solution is equivalent to finding the optimal one; however, it does not imply yet the NP - hardness of finding ε - approximate solutions since the zero f_P value in the problem P might be found in polynomial time.

<u>Theorem 3</u>. For the traveling salesman problem to find the fast (ν) - approximation or ε- approximation algorithms is NP - hard, even if the distances obey the "triangle inequality".

In order to prove Theorem 3 it is sufficient to show that the existence of the fast (ν) - approximation or ε - approximation algorithm implies the existence of a polynomial in time algorithm for the NP - complete Hamiltonian cycle problem.

It is easy to show that a similar fact is valid for many other combinatorial problems, in particular, for graph problems with the object function being bounded by a polynomial in the problem size.

3. FAST ε - APPROXIMATION ALGORITHMS

We suggest a fast algorithm $A(P_2)$ for the sum of subset problem P_2 which is a modified version of our earlier algorithm $E(P_2)$, described in [4,8]. The main idea here is to improve the space bounds using the trade-off between the running time and the space.

In the preliminary step of $A(P_2)$ we use the algorithm $E(P_2)$ which in $O(n+1/\varepsilon^2)$ time computes an ε-approximate solution value $f(\bar{x})$. However in this step we do not try to obtain the solution $\bar{x}=(\bar{x}_1,\ldots,\bar{x}_n)$ itself, so the space needed is $O(n+1/\varepsilon)$. The components of solution $\bar{x}$ are determined successively, in s steps, where s is a number of variables not belonging to the list Small in the algorithm $E(P_2)$.

In the first step of $A(P_2)$ we set some $x_1=1$ ($x_1 \bar{\in}$ Small) and run the algorithm $E(P_2)$ once more with the x_1 value fixed. If it yields a solution value not worse than the $f(\bar{x})$ obtained we can fix $\bar{x}_1=1$ in the solution $\bar{x}$ we are seeking for. Otherwise, we can fix $\bar{x}_1=0$. Then we set in turn $x_2=1$, $x_3=1,\ldots,x_s=1$, run $E(P_2)$ each time and fix the values $\bar{x}_2, \bar{x}_3,\ldots,\bar{x}_s$ successively. Next we determine the rest $\bar{x}_{s+1},\ldots,\bar{x}_n$ values in just the same way as in the algorithm $E(P_2)$. In the latter $s \leq 2/\varepsilon$, thus we obtain the bounds stated in the Table 1.

This technique easily generalizes to other knapsack type problems. However, we shall not consider it here.

The techniques we suggest to obtain fast algorithms for the problems P_1, P_3-P_6 consist of using the digit truncation, similar to that described by Sahni [9], and constructing bounds $\hat{f}$, satisfying $\hat{f} \leq f^* \leq c\hat{f}$, where f^* is the optimal solution and c is a constant, usually, $2 \leqslant c \leqslant 8$.

We present two methods for obtaining the bounds $\hat{f}$: (i) sophisti ted "greedy" procedures, and (ii) an iterative dichotomous search.

The first method may be illustrated on the problem P_1. An algorithm yielding in $O(n \log_2 n)$ time the $\hat{f}$ value such that $1/2\ \hat{f} \leqslant f^*(P_1) \leqslant \hat{f}$ takes the following form:

Step 1. Sort $S=\{1,2,\dots,n\}$ according to nondecresing outlay density, c_i/a_i. Set $\hat{f}=\sum_{i\in S} a_i$.

Step 2. Fill the knapsack by the elements of S in the order obtained untill $\sum_i a_i x_i \geqslant b$. (We assume that $\sum_{i=1}^{n} a_i x_i \geqslant b$, otherwise the problem has no solutions). Find $k=\min(j \mid \sum_{i\leqslant j} a_i \geqslant b)$. Set $L=\sum_{i\leqslant k} c_i$; $\hat{f} \leftarrow \min(\hat{f},L)$.

Step 3. Set $S \leftarrow S \setminus k$. If $\sum_{i\in S} a_i > b$, then go to step 2 else stop.

We illustrate the second method of finding $\hat{f}$ on the problem P_3.

Begin by constructing an algorithm $V(f^{\circ})$ which finds a value f° such that $f^{\circ}/n \leqslant f^*(P_3) \leqslant f^{\circ}$:

Step 1. Sort $S=\{1,2,\dots,n\}$ according to nondecreasing outlay, c_i. Denote the elements x_i entering the j th constraint in P_3, i.e. $a_j(x) \geqslant b_j$, by J_j.

Step 2. Select k smallest elements such that $\sum_{i\in J_j}^{k}{}_{i=1} a_i \geqslant b_j$ for any $j=1,\dots,n$ and $\sum_{i\in J_j}^{k-1}{}_{i=1} a_i < b_j$ for some j.

It is clear that the value $k\cdot a_{i_k}$ is the desired f°.

Let p be a positive number. We next devise a fast algorithm A(p) which for every p and $f^*(P_3)$ finds that either $f^*(P_3) \leqslant p$ or $f^*(P_3) \geqslant 3/4\cdot p$. The algorithm A(p) consists of n steps, its construction being similar to that described by the authors in [8], its running time being $O(n^2)$.

So the algorithm for finding the $\hat{f}$ value for the problem P_3 such that $1/2 \cdot \hat{f} \leq f^*(P_3) \leq \hat{f}$ takes the following form:

<u>Step 1</u>. Using the algorithm $V(f^o)$, find the f^o value, such that $f^o/n \leq f^*(P_3) \leq f^o$. Set $f = a_{i_k}$.

<u>Step 2</u>. Set $p=2f$ and run the above algorithm $A(p)$. If $f^*(P_3) < p$, set $\hat{f} \leftarrow p$; the $\hat{f}$ desired is found. If $f^*(P_3) \geq 3/4 \cdot p$, set $3/4 \cdot p = 3/2 \cdot f \rightarrow$ $\rightarrow f$ and go to Step 2.

Since $f^o/n \leq f^*(P_3) \leq f^o$, Step 2 is to be executed no more then $\log_{3/2} n$ times. Thus, the running time to obtain the $\hat{f}$ value is $O(n^2 \log_{3/2} n) = O(n^2 \log n)$.

The technique for obtaining ε-approximate solution of the problem P_3, the $\hat{f}$ value being found, is similar to that of Sahni [9] and demands $O(n^2/\varepsilon)$ time and space.

Note that if we have already devised a fast ε-approximation algorithm, it easily yields the bound $\hat{f}$. For example, consider $O(n^3/\varepsilon)$ algorithm developed by the authors for the min - job - sequencing problem P_3 (without using a bound $\hat{f}$) [4,8]. If we put $\varepsilon = \varepsilon_0 = 1/2$, then in $O(n^3/\varepsilon_0) = O(n^3)$ time we obtain ($\varepsilon = 1/2$) - approximate solution value, $f(\bar{x})$, which, evidently, satisfies $f(\bar{x}) \leq f \leq 2f(\bar{x})$. Thus, using the $\hat{f} = f(\bar{x})$ value and $O(n^2/\varepsilon)$ digit truncation technique, we obtain an $O(n^3 + n^2/\varepsilon)$ algorithm for the said problem. (Above we suggest a better algorithm for this problem).

Now consider the second method solving the problem P_4. It is evident that the optimal value, $f^*(P_4)$, in the problem P_4 satisfies:

$$\max_{ij} c_{ij} \leq f^*(P_4) \leq m \max_{ij} c_{ij} \qquad (3)$$

Let p be a positive number. We firstly present an algorithm $B(p)$ which for every p and f^* finds that either $f^*(P_4) < p$ or $f^*(P_4) > p/4$. In each of the groups $1, 2, \ldots, m$ we choose the x_{ij_i} with maximal c_{ij} value among those elements of a group which have $f_{ij} = c_{ij}/a_{ij} \geq p/2b$. In those groups where $\max_j f_{ij} < p/2b$ we do not choose the x_{ij}.

It can be shown that either $\sum_i c_{ij_i} > p/2$ and then $f(P_4) < p/4$, or $\sum_i c_{ij_i} \leq p/2$ and then $f^*(P_4) > p$.

Algorithm for finding the $\hat{f}$ value in the problem P_4, such that $1/8\ \hat{f} \leq f^*(P_4) \leq \hat{f}$, takes the following form:

Step 1. Find $r = \max_{ij} c_{ij}$ and $R = mr$. Set $\hat{f} \leftarrow R$.

Step 2. Set $p \leftarrow \hat{f}/2$. Using the algorithm B(p), we find that eithe $f^*(P_4) > \hat{f}/8$, or $f^*(P_4) < \hat{f}/2$. In the former case we have $\hat{f}/8 < f^*(P_4) <$ (the desired $\hat{f}$ value is found). In the latter case go to Step 3.

Step 3. Set $\hat{f} \leftarrow p$ and go to Step 2.

Using the techniques described above we can obtain ε - approximation algorithms for the problems P_4'-P_6, their time and space being as shown in the Table 1.

REFERENCES

1. Babat, L.G., Linear function on the N-dimensional unit cube. Dokl. Akad. Nauk SSSR, 222 (1975), 761-762.
2. Babat, L.G., A fixed-charge problem. Izv. Akad. Nauk SSSR, Engeneering Cybernetics, 3, (1978), 25-31.
3. Cornuejols G., Fisher M.L., Nemhauser G.L., Location of bank accounts to optimize floats. Manag.Sci., 23 (1977), 789-810.
4. Gens G.V., Levner E.V. Approximate algorithms for NP-hard scheduling problems. Izv. Akad. Nauk SSSR,Engineering Cybernetics 6, (1978), 38-43.
5. Ibarra O.H., Kim C.E., Fast approximation algorithms for the knapsack and sum of subset problems. J.ACM, 22 (1975), 463-468.
6. Karp R.M., The probabilistic analysis of some combinatorial search algorithms. In: Algorithms and Complexity (ed. Traub J.F.), (1976) 1-19.
7. Lawler E.L., Fast approximation algorithms for knapsack problems. In: Interfaces between Computer Science and Operations Research, Amsterdam, Mathematical Centre Tracts, 99, (1978).
8. Levner E.V., Gens G.V. Discrete Optimization Problems and Efficien Approximation Algorithms. Moscow, Central Economic and Mathematica Institute, USSR Academy of Sciences, (1978) (in Russian).
9. Sahni S. Algorithms for scheduling independent tasks. J.ACM, 23 (1976), 114-127.
10. Sahni S., Gonzales T. P-complete approximation problems. J.ACM, 23 (1976), 555-565.

A REDUCT-AND-CLOSURE ALGORITHM FOR GRAPHS

Alla Goralčíková, Václav Koubek
Computational Centre, Charles University
Malostranské nám. 25, 118 00 Praha 1, Czechoslovakia

By a graph we understand a pair (V,E), where V is a finite set of vertices and $E \subset V \times V$ is a set of edges. A directed path $p:x \rightarrow y$ of length $\|p\| = k$ in (V,E) is a sequence of vertices $x=v_0, v_1, \ldots, v_k = y \in V$ such that $(v_{i-1}, v_i) \in E$ for $i = 1,\ldots,k$.

The transitive closure of (V,E) is the graph $\mathrm{Clos}(V,E) = (V,C)$ where $(x,y) \in C$ iff there is a path $p:x \rightarrow y$ in (V,E) with $\|p\| > 0$.

The transitive reduct $\mathrm{Red}(V,E)$ of an acyclic graph (V,E) is the least graph (V,R) with $\mathrm{Clos}(V,R) = \mathrm{Clos}(V,E)$.

The number of elements of a finite set X will be denoted throughout by $|X|$.

If Alg is an algorithm processing graphs, $\mathrm{Time}(\mathrm{Alg}(V,E))$ denotes the number of steps taken by Alg to process (V,E). We write $\mathrm{Time}(\mathrm{Alg}(V,E)) \leq O(f(V,E))$, for a function $f:K \rightarrow N$ from some class of graphs to the non-negative integers N, if there exists $c \in N$ such that the inequality $\mathrm{Time}(\mathrm{Alg}(V,E)) \leq c.f(V,E)$ holds for all $(V,E) \in K$. We then also say that Alg needs $O(f(V,E))$ time to process a member (V,E) of K, or that Alg has the time complexity $O(f)$ on K.

The main objective of the present note is to describe an algorithm Recl computing simultaneously both $\mathrm{Red}(V,E)$ and $\mathrm{Clos}(V,E)$, for (V,E) acyclic, with the time complexity $\mathrm{Time}(\mathrm{Recl}(V,E)) \leq O(|V|.|R| + |E|)$ comparing favorably with the algorithms described in [1], [2], [5], [6]. We also describe some useful modifications of this basic algorithm.

An essential preliminary part of the algorithms we are going to present here is the computation of a suitable height function on an acyclic graph (a notion adopted from [3]).

A homomorphism h of (V,E) onto $([s], \leq)$, the set of integers $[s] = \{0,1,\ldots,s\}$ naturally ordered by $\leq$, is a height function if $h(x) \neq h(y)$ for every $(x,y) \in E$ with $x \neq y$, and is a Jordan-Dedekind height function, shortly JD-function, if it is also a homomorphism of Red(V,E) onto Red([s], $\leq$) .

For a given height function $h: (V,E) \rightarrow ([s], \leq)$ the integer $h(x)$, $x \in V$, is called the <u>height</u> of x relative to h, the set $L(i) = \{x \in V \; ; \; h(x) = i\}$, $i \in [s]$, is called the <u>i-th h-level</u> of (V,E) .

We first give a simple algorithm Lev constructing recursively a particular height function h for an acyclic graph (V,E) . The ide is that $L(0) = \mathrm{Min}(V,E) = \{x \in V \; ; \; (y,x) \in E \rightarrow y = x\}$ while $L(i) = \mathrm{Min}(V(i),E(i))$, where $V(i) = V - \{x \in V \; ; \; h(x) < i\}$, $E(i) = E \cap (V($ $\times V(i))$. The graph (V,E) will always be assumed, for the algorithmic purposes, to be given in the form of a list of two families of se $\{xE \; ; \; x \in V\}$ and $\{Ey \; ; \; y \in V\}$, where $xE = \{y \; ; \; (x,y) \in E\}$ and $Ey = \{x \; ; \; (x,y) \in E\}$. The algorithm Lev will be simultaneously listing both h-levels and heights of vertices of (V,E).

```
    Initialize  L(i) = ∅  for all  i .
    Procedure  Lev(V,E)
L1: for each  x∈V  do
        if  Ex⊂{x}  then
        begin  L(0):= L(0) ∪ {x} , h(x):= 0  end
    end
    i:= 0
L2: while  L(i) ≠ ∅  do
        for each  x∈L(i) do
            for each  y∈xE  do
                Ey:= Ey - {x}
                if  Ey⊂{y}  then
                begin  L(i+1):= L(i+1) ∪ {x} , h(x):= i+1  end
            end
        end
        i:= i+1
    end
    s:= i-1
    return {L(i) ; i=0,1,...,s} , {h(x) ; x∈V}
    end  Lev
```

It is easy to see that Lev constructs a height function as indicated. As for the time complexity, step L1 is repeated $|V|$ times. Since each step in the innermost cycle of L2 corresponds to a singl edge $(x,y) \in E$ and uses at most six operations, we conclude that $\mathrm{Time}(\mathrm{Lev}(V,E)) = O(|V| + |E|)$.

In the sequel we shall always assume that the representation of a graph (V,E) in the form of the families $\{xE \,;\, x \in V\}$ and $\{Ey \,;\, y \in V\}$ has been completed by list of h-levels and heights of elements of (V,E) (for example furnished by Lev).

Our algorithm Recl will be based on the following easy

<u>Lemma 1</u> . Let (V,E) be an acyclic graph and let $h:(V,E) \to ([s], \leq)$ be the height function computed by Lev . Then for all $x \in V$,

(a) $L(h(x)+1) \cap xR = L(h(x)+1) \cap xC = L(h(x)+1) \cap xE$,

and for all $j > h(x)+1$,

(b) $L(j) \cap xR = L(j) \cap xE - A(x,j)$,

(c) $L(j) \cap xC = (L(j) \cap xR) \cup A(x,j)$,

where $A(x,j) = \bigcup\{L(j) \cap yC \,;\, y \in L(k) \cap xR \,,\, h(x) < k < j\}$.

Here is a description of Recl:

```
Procedure Recl(V,E)
i:= 1
while i ≤ s do
   for each j ≤ s-i do
      for each x ∈ L(j) do
         A:= ∪ {L(j+i) ∩ yC ; y ∈ L(k) ∩ xR , j < k < j+i}
         L(j+i) ∩ xR = L(j+i) ∩ xE - A
         L(j+i) ∩ xC = (L(j+i) ∩ xR) ∪ A
      end
   end
   i:= i+1
end
return {L(j) ∩ xR ; h(x) < j < s , x ∈ V}
       {L(j) ∩ xC ; h(x) < j < s , x ∈ V}
end Recl
```

If the sets $L(j) \cap xE$ and $L(j) \cap xR$ are given by the singly linked lists, while the sets $L(j) \cap xC$ and the auxiliary sets A are determined by their characteristic functions on L(j) , then the innermost cycle of Recl requires $O(|L(j+i) \cap xE|)$ time for $L(j+i) \cap xR$, $O(|L(j+i) \cap xR|)$ time for $L(j+i) \cap xC$, and $O(|L(i+j)| \cdot \sum\{|L(k) \cap xR| \,;\, j < k < j+i\})$ time for A . Since for a given vertex $x \in V$ these computations repeat for each k , $h(x) < k \leq s$, the constructions of the families $\{L(k) \cap xR \,;\, h(x) < k \leq s\}$ and $\{L(k) \cap xC \,;\, h(x) < k \leq s\}$ require $O(\sum\{|L(k) \cap xE| \,;\, h(x) < k \leq s\}) = O(|xE|)$ time, while the constructions of the sets A require $O(\sum\{|L(h(x)+j)|.\sum\{|L(k)\cap xR| ; h(x)< k<h(x)+j\} ; 0<j\leq s-h(x)\})\leq O(\sum\{|L(h(x)+j)|.\sum\{|L(k)\cap xR| ; h(x)< k\leq s\} ; 0<j\leq s-h(x)\}) = O(\sum\{|L(h(x)+j)|.|xR| ; 0<j\leq s-h(x)\})\leq O(|xR|.\sum\{|L(k)| ; 0\leq k\leq s\}) = O(|xR|.|V|)$ time . Thus the whole algorithm requires at most $O(\sum\{|xE|+|xR|.|V| ; x\in V\}) = O(|R|.|V|+|E|)$ time. We have proved

<u>Theorem 1</u> . For the class of acyclic graphs,

$$\mathrm{Time}(\mathrm{Recl}(V,E)) \leqslant O(|R|.|V| + |E|) .$$

<u>Remark.</u> The most time consuming in Recl are the constructions of the sets A . The other operations require $O(|E|+|V|)$ time and it is clear that this estimate cannot be improved. Unfortunately, A is generally not a disjoint union, so we cannot use here any of the fast set-union algorithms (see[2]). For this reason it seems not to be an easy task to lower the estimate of the time complexity attained by Recl .

The transitive closure Clos(V,E) has been defined for an arbitrary (not only acyclic) graph (V,E) . We claim that it is possible to construct Clos(V,E) with aid of Recl , however, Recl cannot be applied directly to (V,E) but to the factorgraph (W,F) of (V,E) by the strong connectedness of (V,E) . (A pair $(x,y)\in V\times V$ is strongly connected iff there exist directed paths both from x to y and from y to x . Strong connectedness is an equivalence on V partitioning V into the strong components of (V,E) and yielding the biggest acyclic factorgraph (W,F) of (V,E) .) If $f:(V,E)\rightarrow(W,F)$ is the canonical projection, then clearly $(x,y)\in \mathrm{Clos}(V,E)$ iff $(f(x),f(y))\in \mathrm{Clos}(W,F)$. We can use the algorithm described in [7] to find the strong components of (V,E) in $O(|V|+|E|)$ time. It is then easy to pass to (W,F) also in $O(|V|+|E|)$ time. We continue by Recl and get Clos(W,F) in $O(|W|.|S|+|F|)$ time, where S are the edges in (W,S) = Red(W,F) .

Putting all the pieces together we get

<u>Theorem 2</u> . There exists an algorithm, on the class of all graphs (V,E) , computing the transitive closure Clos(V,E) of (V,E) with the time complexity $O(|V|.|S|+|E|)$, where S is the set of edges in the transitive reduct of the biggest acyclic factorgraph of (V,E)

The usefulness of a height function $h:(V,E)\rightarrow([s],\leq)$ for algorithms computing the transitive reduct Red(V,E) = (V,R) of an acyclic graph (V,E) is rooted in the simple observation that if h(y) = = h(x) + 1 for $(x,y) \in E$ then $(x,y) \in R$. In this regard the best possible height function is the Jordan-Dedekind one, since then $(x,y) \in R$ iff $(x,y) \in E$ and h(y) = h(x)+1 .

If a graph (V,E) has a JD-function h , then it satisfies the Jordan-Dedekind chain condition (also a notion adopted for graphs from [3]): any two paths $p,q: x \to y$ in Red(V,E) with the same endpoints have the same length $\|p\| = \|q\|$. (Note that this condition is generally not sufficient for the existence of a JD-function.)

Let JDH denote the class of acyclic graphs (V,E) possessing a Jordan-Dedekind height function $h:(V,E) \to ([s], \leq)$.

Lemma 2 . If $(V,E) \in JDH$ then for any four paths in Red(V,E) = = (V,R) of the form $p: z \to x$, $q: z \to y$, $r: v \to x$, $s: v \to y$ it holds

$$\|p\| - \|r\| = \|q\| - \|s\|$$

Proof. $\|p\| - \|r\| = h(x) - h(z) - (h(y) - h(z)) = h(x) - h(y)$,
$\|q\| - \|s\| = h(x) - h(v) - (h(y) - h(v)) = h(x) - h(y)$.

This lemma provides a justification for the following algorithm Jodeh operating on JDH and computing a JD-function for $(V,E) \in JDH$.

```
     Procedure  Jodeh(V,E)
J1:  for each  x∈V  do
       if  Ex⊂{x}  then
       begin  M(0):= M(0)∪{x} , H:= H∪{x}, k(x) = (x,0)  end
     end
     i:= 0
J2:  while M(i) ≠ ∅  do
       for each  x∈M(i)  do
          for each  y∈xE  do
            Ey:= Ey - {x}
            if  k(x) = (t,n)  then
               if (t,m)∈K(y)  for some  m  then
                  replace (t,m)  by  (t,max{m,n+1})  in K(y)
                  else  add  (t,n+1)  to  K(y)
            if  Ey⊂{y}  then
            begin  M(i+1):= M(i+1)∪{y} , choose any  (t,n)∈K(y)
                   such that t∈H , k(y):= (t,n) ,
                   for each  (u,m)∈K(y) , u∈H , u ≠ t  do
                      insert  (t,n) = (u,m)  into the stack  S
                      H:= H - {u}
```

```
                    end
                end
            end
        end
        i:= i+1
    end
    s:= 0 , v:= 0
J3: for each x∈H do
        h(x):= 0
    end
    while S ≠ ∅ do
        if (t,n) = (u,m) ∈ S then
        begin delete (t,n) = (u,m) from S , H:= H ∪{u},
              h(u):= h(t)+n-m , v:= min{v,h(u)} , s:= max{s,h(u)}
        end
    end
J4: for each x∈V-H do
        if k(x) = (t,n) then
        begin h(x):= h(t)+n , s:= max{s,h(x)} end
    end
    return {L(i) ; v≤i≤s , L(i) ={x ; h(x) = i}}
    end Jodeh
```

Clearly Jodeh stops. If $(x,y)\in E$ and $k(x) = (t,n)$ then $(t,p)\in K(y)$ for some $p\geqslant n+1$ such that $(u,m) = (t,p)$ is in S for $k(y) = (u,m)$, hence $h(t)+p = h(u)+m = h(y)\geqslant h(t)+n+1 > h(t)+n =$ $= h(x)$. Since $(u,m) = (t,p)$ had been in S , there are paths $r:t\to$ $\to x$ and $s:t\to y$ in $Red(V,E) = (V,R)$ with $\|r\| = n$, $\|s\| = p$. I $(x,y)\in R$ we get a path $q:t\to y$ in $Red(V,E)$ with $\|q\| = n+1$, he $p = n+1$. Thus $h(y) = h(u)+m = h(t)+n+1 = h(x)+1$, which proves h to be a JD-function.

As for the time complexity, J1 requires $O(|V|)$ time, by a si milar argument as for L1 in Lev . If we use a structure on $K(x)$ enabling us to find an element (t,n) of $K(x)$ by its first component, then $K(x)$ is actualized in a constant time, thus J2 require $O(|V|+|E|)$ time, by the same reason as L2 in Lev . Since there is a one-to-one correspondence between the data in S and the elements of $M(0)$, established by the assignment $(t,n) = (u,m)\mapsto u$, J3 requires $O(|V|)$ time. Since J4 requires $O(|V|)$ time, too, we get

Theorem 3 . There exists an algorithm on the class JDH of grap computing Red(V,E) with the time complexity $O(|V|+|E|)$.

Remark . If we write to the stack all the data from K(y) then we can easily modify Jodeh so as to decide whether a graph (V,E) fulfilling the Jordan-Dedekind chain condition belongs to JDH .

The height function h computed earlier by Lev proves helpful also in the task of finding infima $x \wedge y$ for couples of points of a finite poset $(P, \leq)$. An algorithm Inf computing infima is based on the easy observation that $h(x \wedge y) \geqslant h(z)$ for any $z \in P$ with $z \leqslant x$, $z \leqslant y$. Since the computation of a single infimum can be done in $O(|P|)$ time, we can easily obtain an algorithm deciding whether a given poset $(P, \leq)$ is a lattice in $O(|P|^3)$ time .

The algorithms described in this paper have found their applications in [4] .

References

1. A. V. Aho, M. R. Garey, J. D. Ullman, The transitive reduction of a directed graph, SIAM J. Comput., 1:2(1972), 131-137.
2. A. V. Aho, J. E. Hopcroft, J. D. Ullman, The design and analysis of computer algorithms, Addison-Wesley, 1974.
3. G. Birkhoff, Lattice theory, Amer. Math. Soc., vol. XXV (1967).
4. M. Demlová, J. Demel, V. Koubek, Several algorithms for finite algebras, to appear in FCT 1979.
5. M. J. Fischer, A. R. Meyer, Boolean matrix multiplication and transitive closure, IEEE 12th Annual Symposium on Switching and Automata Theory (1971), 129-131.
6. H. B. Hunt III, T. G. Szymanski, J. D. Ullman, Operations on sparse relations, Comm. ACM 20:3(1977), 172-176.
7. R. E. Tarjan, Depth first search and linear graph algorithms, SIAM J. Comput.,1:2(1972), 146-160.

SMALL UNIVERSAL MINSKY MACHINES

Ľudmila Gregušová

Department of Cybernetics, Faculty of Electrical Engineering
of Slovak Tech. Univ., Vazovova 5, 880 19 Bratislava

Ivan Korec

Department of Algebra, Faculty of Sciences of Comenius Univ.
Mlynská Dolina, 816 31 Bratislava

Some authors looked for universal Turing machines with the smallest possible state-symbol product, i. e. consisting of the smallest possible number of instructions. By Minsky [6], for example, Ikeno exhibited a six-symbol, ten-state universal Turing machine and Watanabe a five-symbol, eight-state machine. The best result, proved by Minsky, was a four-symbol, seven-state universal Turing machine. On the other hand, it was proved that no two-symbol, two-state Turing machine can be universal.

We solve analogical problem for the Minsky machines. However, the activity of a Minsky machine in one step is simpler than that of a Turing machine. Therefore greater numbers of instructions would be expected. For example, no Minsky machine consisting of at most eight instructions is universal (see Korec [4]). M. Gregušová constructed universal Minsky machines consisting of 85 and 76 instructions (see [1], [2]). Here we present the universal Minsky machine U consisting of 37 instructions; U uses no artificial input and output coding. If we code x by 2^x then we can obtain a universal Minsky machine consisting of 32 instructions.

N denotes the set of nonnegative integers. Under a /partial/ function we always mean a /partial/ function on the set N. We recall some notions and notations of Minsky machines.

<u>Definition 1.</u> a) Ordered quadruples of the forms

(1) $(q_iS_jPq_k)$, $(q_iS_jMq_k)$, $(q_iS_jq_mq_n)$

where i, j, k, m, n $\in$ N, will be called Minsky instructions (briefly: M-instructions).

b) A Minsky machine (briefly: M-machine) is a finite set of M-instructions which does not contain two different M-instructions with the same first elements.

The symbols S_0, S_1, S_2, ... denote counters. Each of them can contain an arbitrarily large nonnegative integer. The content of S_j is denoted $\langle S_j \rangle$. The symbols q_0, q_1, q_2, ... are states, q_1 is the initial state and q_0 the final state. The symbol P, M in (1) means +1 and $\dot{-}1$ (with respect to $\langle S_j \rangle$). The last M-instruction in (1) means jump to q_m if $\langle S_j \rangle \neq 0$ and jump to q_n if $\langle S_j \rangle = 0$.
The n-ary partial function computed by an M-machine Z is denoted ϕ_Z^n. If Z computes $\phi_Z^n(x_1,\ldots,x_n)$ then Z starts with the numbers $x_1,\ldots, x_n$ stored in S_1, ..., S_n; other counters contain zero. The value $y = \phi_Z^n(x_1, \ldots, x_n)$ is defined if and only if Z halts in the final state q_0; then $y = \langle S_0 \rangle$.

Let S be a set of unary partial functions and F be a binary partial function. F is said to be a universal partial function for the set S if

a) for every a $\in$ N the unary partial function g, g(x) = F(a,x) belongs to S, and

b) for every partial function g $\in$ S there is a $\in$ N such that g(x) = F(a,x) for all x $\in$ N.

<u>Definition 2.</u> An M-machine Z will be called universal if the partial function ϕ_Z^2 is universal for the set of all unary partial recursive functions.

The universal M-machine U, which is constructed in this paper, simulates the so-called modified Minsky machines. They are defined as follows.

<u>Definition 3.</u> a) Ordered triples of the forms

(2) (q_iPq_j), $(q_iS_0q_j)$, $(q_iS_2q_j)$, $(q_iS_3q_j)$

where i, j $\in$ N, will be called modified Minsky instructions (briefly: M3-instructions).

b) A modified Minsky machine (briefly: M3-machine) is a finite set of M3-instructions which does not contain two different M3-instructions with the same first element.

The denotation M3 indicates that only three counters are used : S_0 S_2 and S_3. The M3-instruction (q_iPq_j) means addition of 1 to $\langle S_0 \rangle$, $\langle S_2 \rangle$ and $\langle S_3 \rangle$. The M3-instruction $(q_iS_nq_j)$ means subtraction of 1 from $\langle S_n \rangle$ and jump to q_j if $\langle S_n \rangle \neq 0$; otherwise it means jump to q_{j+1}. The counter S_1 is left for the first argument of the universal partial function Φ^2_U, i. e. for the number of a simulated machine. Ordered quadruples $(q_i;a_0,a_2,a_3)$ where i, a_0, a_2, $a_3 \in$ N are called M3-configurations; q_i is a state and a_0, a_2, a_3 are interpreted as contents of registers S_0, S_2, S_3, respectively.

Definition 4. Let V be an M3-machine. We shall write

$$(q_i;a_0,a_2,a_3) \rightarrow (q_j;b_0,b_2,b_3) \quad (V)$$

if there is an m $\in \{0,2,3\}$ such that one of the following conditions hold

(i) $(q_iS_mq_j) \in V$, $b_m = a_m \dot{-} 1$ and $b_n = a_n$ for $n \in \{0,2,3\} - m$;

(ii) $(q_iS_mq_{j-1}) \in V$, $b_m = a_m = 0$ and $b_n = a_n$ for $n \in \{0,2,3\} - m$;

(iii) $(q_iPq_j) \in V$ and $b_n = a_n + 1$ for $n \in \{0,2,3\}$.

If V is an M3-machine and X, Y are M3-configurations then we write $X \Longmapsto Y$ (V) if and only if there are M3-configurations $X = X_0$, X_1, ..., $X_n = Y$ such that $X_{i-1} \rightarrow X_i$ (V) for all i = 1,...,n , and there is no M3-configuration X_{n+1} such that $X_n \rightarrow X_{n+1}$ (V). The smallest non-zero index of the first elements of the M3-instructions of V will be denoted init(V). The state $q_{init(V)}$ will be the initial state of V.

Definition 5. Let V be an M3-machine and let f be a unary partial function. We shall say that V computes f, and write $f = \phi^1_V$, if for every x, y $\in$ N $f(x) = y$ if and only if there are b, c $\in$ N such that

$$(q_{init(V)};0,x,0) \Rightarrow (q_0;y,b,c) \quad (V).$$

It is known (see e.g. Mal'cev [5]) that every unary partial recursive function is computable by an M-machine which uses only three counters. This result easily implies

<u>Theorem.</u> For every unary partial function f there is an M3-machine V such that $f = \phi^1_V$.

We may assume that the state $q_{init(V)}$ does not occur at the third place of any M3-instruction of V and that V never halts when $\phi^1_V(x)$ is not defined.

Now we show how to associate a number y to any M3-machine V in such a way that $\phi^1_V(x) = \phi^2_U(y,x)$ for all x $\in$ N. Let v be the maximal index of the states of V. Denote by p_r the r-th prime, i. e. $p_0 = 2$, $p_1 = 3$ etc. By Sierpiński [7] there is a positive integer r such that

$$p_r > 4v + 3 \quad \text{and} \quad p_{r+v} < 2p_r.$$

Let us choose such an r, e.g. the least one. To each M3-instruction $(q_i X q_j)$ of V we associate the congruence

(3) $$y \equiv 4j + |X| \pmod{p_{r+w}} ,$$

where $|X| = 0,1,2,3$ if $X = S_0, S_2, S_3, P$, respectively, and w = 1 if i = init(V), w = i otherwise. To every prime p_{r+u}, $1 < u \leq v$, which is not used as a modul in (3) we associate the congruence

(4) $$y \equiv 4u \pmod{p_{r+u}}.$$

Finally, to every prime p_k, $k \leq r$, we associate the congruence

(5) $$y \equiv 0 \pmod{p_k^n} ,$$

where $p_k^n < p_{r+v} < p_k^{n+1}$. The Chinese remainder theorem implies that the system of congruences (3), (4), (5) is solvable. Every its positive solu tion y (e. g. the least one) can be used as a number of the M3-machine V Notice that y is divisible by all composed numbers less than p_{r+v}.

Let y be a number of V and X be an M3-instruction of V with the fir element q_i. If we want to know the second and the third element of X, we divide the number y by the numbers 1, 2, 3 etc., and we find the i-th (i $i = \mathrm{init}(V)$ then the first) non-zero remainder in these divisions. This remainder can be uniquely represented in the form $4j + k$, $0 \leq k \leq 3$, $j, k \in N$. Then the third element of X is q_j and the second one is S_0, S_2 S_3, P if $k = 0,1,2,3$, respectively. The same principle is used in the universal M-machine U.

The M-machine U uses the counters $S_0, S_1, \ldots, S_7$. Their role is shown in Figure 1.

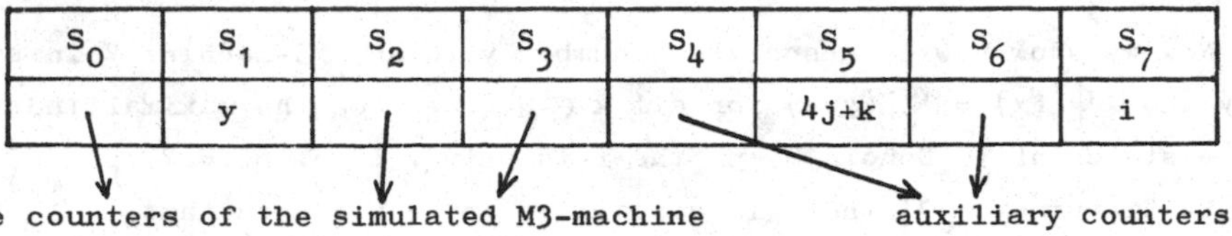

Figure 1.

The auxiliary counter S_4 helps to store y; y will be preserved as the sum $\langle S_1 \rangle + \langle S_4 \rangle$. Analogously, the latest divisor will be preserved as $\langle S_5 \rangle + \langle S_6 \rangle$.

The M-machine U will consist of a decoder D, a working block W and three further instructions:

$$U = D \cup W \cup \{(q_{20}S_7q_{21}q_0),(q_{21}S_6q_{22}q_{23}),(q_{22}S_6Mq_{21})\}.$$

It is illustrated in Figure 2. The decoder

$$D = \{(q_{23}S_1q_{24}q_{26}),(q_{24}S_1Mq_{25}),(q_{25}S_4Pq_{23}),(q_{26}S_5Pq_{27}),(q_{27}S_6q_{28}q_{29}),$$
$$(q_{28}S_6Mq_{26}),(q_{29}S_6Pq_{30}),(q_{30}S_5Mq_{31}),(q_{31}S_4Mq_{32}),(q_{32}S_1Pq_{33}),$$
$$(q_{33}S_4q_{34}q_{35}),(q_{34}S_5q_{29}q_{27}),(q_{35}S_5q_{36}q_{23}),(q_{36}S_7Mq_{37}),(q_{37}S_7q_{23}q_1)\}$$

finds the number $4j + k$ from the number i by a sequence of succesive

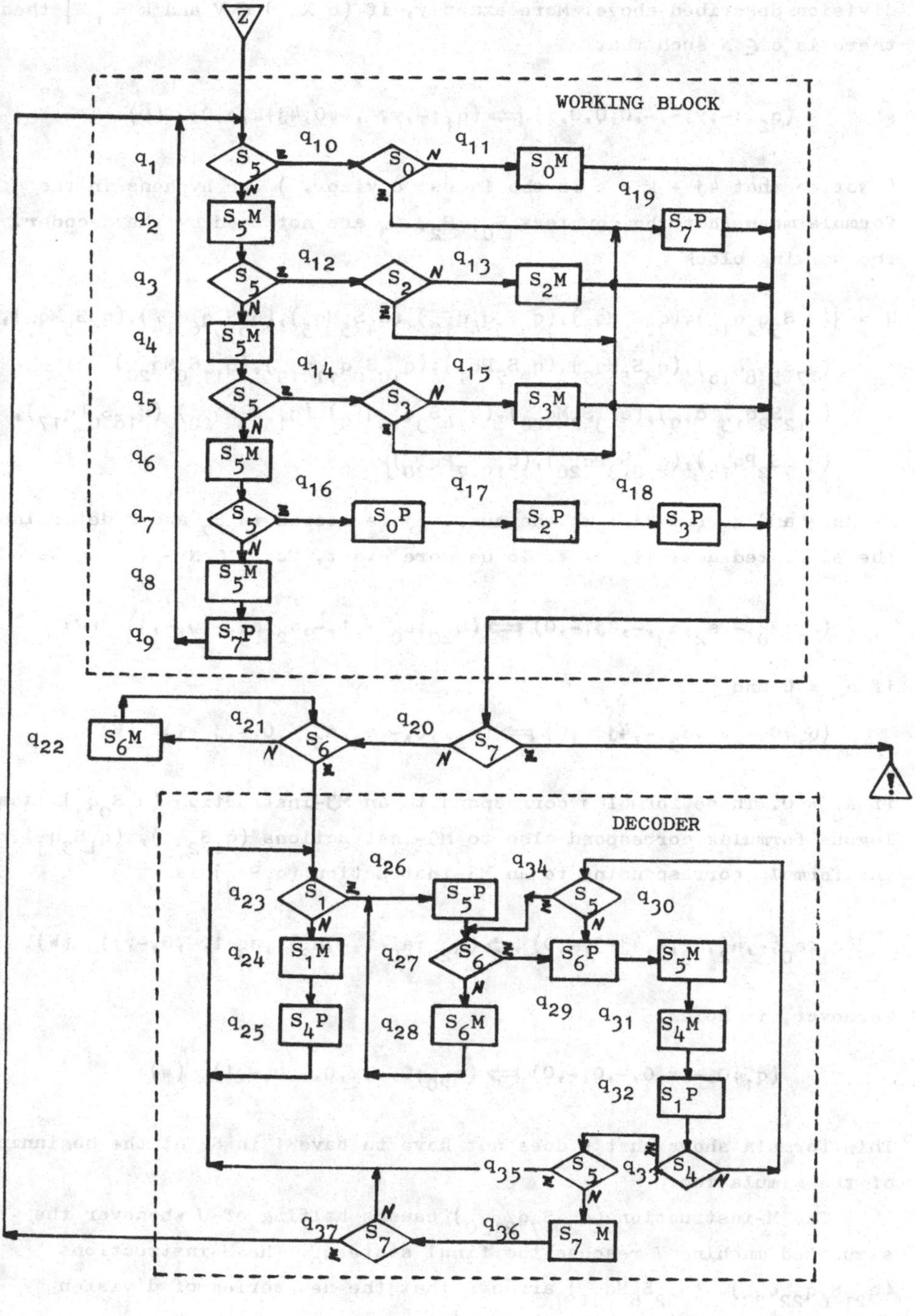
Z
WORKING BLOCK
q1
q2
q3
q4
q5
q6
q7
q8
q9
q10
q11
q12
q13
q14
q15
q16
q17
q18
q19
S5
S5M
S0
S0M
S7P
S2
S2M
S3
S3M
S0P
S2P
S3P
q20
q21
q22
S6
S6M
S7
!
DECODER
q23
q24
q25
q26
q27
q28
q29
q30
q31
q32
q33
q34
q35
q36
q37
S1
S1M
S4P
S5P
S6
S6P
S6M
S5M
S4M
S1P
S4
S5
S7
S7 M
N
Z

Figure 2.

divisions described above. More exactly, if $(q_iXq_j) \in V$ and $k = |X|$ then there is $c \in N$ such that

$$(q_{23};-,y,-,-,0,0,0,i) \models\!\Rightarrow (q_1;-,y,-,-,0,4j+k,c,0) \quad (D).$$

(Notice that $4j + k + c$ is the latest divisor.) The hyphens in the formula mean that the counters S_0, S_2, S_3 are not used by the decoder. The working block

$$W = \{(q_1S_5q_2q_{10}),(q_2S_5Mq_3),(q_3S_5q_4q_{12}),(q_4S_5Mq_5),(q_5S_5q_6q_{14}),(q_6S_5Mq_7),$$
$$(q_7S_5q_8q_{16}),(q_8S_5Mq_9),(q_9S_7Pq_1),(q_{10}S_0q_{11}q_{19}),(q_{11}S_0Mq_{20}),$$
$$(q_{12}S_2q_{13}\ q_{19}),(q_{13}S_2Mq_{20}),(q_{14}S_3q_{15}q_{19}),(q_{15}S_3Mq_{20}),(q_{16}S_0Pq_{17}),$$
$$(q_{17}S_2Pq_{18}),(q_{18}S_3Pq_{20}),(q_{19}S_7Pq_{20})\}$$

finds j and k from $4j + k$; the number j is stored in S_7 and k determines the simulated activity of V. To be more exact, for $j \in N - \{0\}$ we have

$$(q_1;a_0,-,a_2,a_3,-,4j,-,0) \models\!\Rightarrow (q_{20};a_0 - 1,-,a_2,a_3,-,0,-,j) \quad (W)$$

if $a_0 \neq 0$ and

$$(q_1;0,-,a_2,a_3,-,4j,-,0) \models\!\Rightarrow (q_{20};0,-,a_2,a_3,-,0,-,j+1) \quad (W)$$

if $a_0 = 0$. These formulas correspond to an M3-instruction $(q_iS_0q_j)$. Analogous formulas correspond also to M3-instructions $(q_iS_2q_j)$, $(q_iS_3q_j)$. The formula corresponding to an M3-instruction (q_iPq_j) is

$$(q_1;a_0,-,a_2,a_3,-,4j+3,-,0) \models\!\Rightarrow (q_{20};a_0+1,-,a_2+1,a_3+1,-,0,-,j) \quad (W).$$

Moreover, it holds

$$(q_1;0,-,x,0,-,0,-,0) \models\!\Rightarrow (q_{20};0,-,x,0,-,0,-,1) \quad (W).$$

This formula shows that U does not have to have 1 in S_7 at the beginning of the simulation.

The M-instruction $(q_{20}S_7q_{21}q_0)$ causes halting of U whenever the simulated machine V reaches the final state q_0. The M-instructions $(q_{21}S_6q_{22}q_{33})$, $(q_{22}S_6Mq_{21})$ arrange that the new series of division starts again with the divisor 1.

So the M-machine U simulates the computation of V. More exactly, if y is a number of an M3-machine V then for every $x \in N$ there are m, n, c $\in$ N such that

$$(q_1;0,y,x,0,0,0,0,0) \Rightarrow (q_0;\phi^1_V(x),y,m,n,0,0,c,0) \quad (U)$$

if $\phi^1_V(x)$ is defined. If $\phi^1_V(x)$ is not defined then the simulation is infinite. Hence for every unary partial recursive function f there is $y \in N$ such that $f(x) = \phi^2_U(y,x)$ for all $x \in N$. All the partial functions $g(x) = \phi^2_U(a,x)$, $a \in N$, are obviously partial recursive, hence U is a universal Minsky machine. It consist of 37 instructions.

Until now we have not used any special input and output coding. However, such coding is usually permitted in the definition of universal Turing machines (see Maľcev [5]). If we would similarly change the definition of universal M-machines we could reduce the attained number of 37 instructions by:

1) coding the input x by 2^x and the output y by 2^y;
2) using only S_0, S_2 instead of S_0, S_2, S_3;

   instead of M3-machines we consider M2-machines with the instructions of the forms

$$(q_iS_0q_j),(q_iS_2q_j),(q_iPq_j); \tag{6}$$

3) saving 5 instructions in the working block; to each M2-instruction of the form (6) we attach the remainder $3j + k$, $k = 0, 1, 2$.

So we would get a Minsky machine, universal in the sense of Maľcev [5] and consisting of 32 instructions.

References

1. Gregušová, Ľ., The Construction of "Small" Universal Minsky Machine. Journal of Electrotechnical Faculty of STU, ALFA Bratislava, 1979, (in print, in Slovak).

2. Gregušová, Ľ., Doctoral Thesis, Faculty of Sciences of Comenius University, Bratislava, 1978, (in Slovak).

3. Korec, I., A Complexity Valuation of the Partial Recursive Functions

Following the Expectation of the Time... Acta F. R. N. Univ. Comen. - Mathematica XXIII, 1969, 53-112.

4. Korec, I., Decidability (Undecidability) of Equivalence of Minsky Machines with Components Consisting of at most seven (eight) Instructions. Proceedings of MFCS 1977, Lecture Notes in Computer Science 53 324-332.

5. Mal'cev, A. I., Algorithms and Recursive Functions. Izd. NAUKA, Moscow 1965 (in Russian).

6. Minsky, M. L., Computation, Finite and Infinite Machines, Prentice Hall, New Jersey, 1967.

7. Sierpiński, W., Theory of Numbers. Warszawa - Wroclaw, 1950, (in Poli

PARALLEL AND TWO-WAY RECOGNIZERS OF DIRECTED ACYCLIC GRAPHS

(extended abstract)

Tsutomu Kamimura
Giora Slutzki
Department of Computer & Information Sciences
University of Delaware
Newark, Delaware 19711

1. Introduction.

Finite state automata operating on trees in parallel fashion have been studied extensively both as recognition devices and as transformation devices [1,3,4,5,11,12,15]. On the other hand, 2-way tree walking automata operating sequentially were treated in [1,5] only as transducers. The connections between these two types of automata with respect to their transformation power were first studied in [1] and then very systematically in [5].

In this paper we extend both kinds of automata to operate as recognizers on more general graphs, called dags. Dags are special directed ordered acyclic graphs which model derivations of phrase-structure grammars analogously to the way that trees model derivations of context-free grammars. A rather complete characterization of the relative power of the following features is obtained (when relevant): parallel versus sequential, deterministic versus nondeterministic and finite state (i.e. bounded memory) versus a (restricted type of) push-down store, see Fig. 6. Interesting and new results concerning trees follow as special cases.

Some results about dags and (only parallel) dag automata (in slightly less general form) were reported in [2,6]; however these papers focused on derivation structures of phrase-structure grammars (which are dags) and their recognition.

The paper is organized as follows. Section 2 gives the formal definition of dags. Section 3 introduces the top-down and the bottom-up varieties of parallel automata and their deterministic and nondeterministic versions. Although these definitions were already given in [10], they are repeated for completeness. The results are summarized in Fig. 4. It is rather interesting to note that nondeterminism buys us no additional power in the bottom-up case. In Section 4 we define two-way dag walking automata: with and without the push-down facility. Here determinism and nondeterminism turn out to be equivalent in the presence of the push-down, but nonequivalent without it. The results are shown in Fig. 5. The last section compares the parallel automata with the 2-way machines, see Fig. 6, and briefly summarizes some properties of the classes of dag languages definable by our automata.

2. Dags.

For basic graph terminology the reader is referred to [16].

Dags are graphs having the following properties: they are labeled, directed, acyclic, ordered, planar and connected. The labels are symbols out of a doubly ranke alphabet which is a set $\Sigma = \bigcup_{i,j} \Sigma_{ij}$ where each Σ_{ij} is a finite set and only for a finit number of i and j $\Sigma_{ij} \neq \emptyset$. An element $\sigma \in \Sigma_{ij}$ has head-rank i and tail-rank j. Also, we define $\Sigma_{*j} = \bigcup_i \Sigma_{ij}$ and $\Sigma_{i*} = \bigcup_j \Sigma_{ij}$. Then (rooted) dags are defined inductively (as in the tree case) along with the concept of "leaves".

2.1. Definition. Let Σ be a doubly ranked alphabet. The set of partial dags over Σ, denoted by P_Σ, is defined as follows:

(i) If $a \in \Sigma_{0*}$, then $a \in P_\Sigma$; leaves(a)=a (when convenient we will identify the node with its label).

(ii) Let $d \in P_\Sigma$ with leaves(d)=$a_1 \ldots a_n$ and $a_i \in \Sigma_{*m}$; let $b_1, \ldots, b_m \in \Sigma_{1*}$. Then d' of Fig. 1(a) is in P_Σ and leaves(d')=$a_1 \ldots a_{i-1} b_1 \ldots b_m a_{i+1} \ldots a_n$.

(iii) Let $d \in P_\Sigma$ with leaves(d)=$a_1 \ldots a_n$ and $a_i, a_{i+1}, \ldots, a_j \in \Sigma_{*1}$ for some $1 \leq i \leq j \leq n$. Let $b \in \Sigma_{(j-i+1)*}$; then d' of Fig. 1(b) is in P_Σ with leaves(d')=$a_1 \ldots a_{i-1} b a_{j+1} \ldots a_n$. ⊠

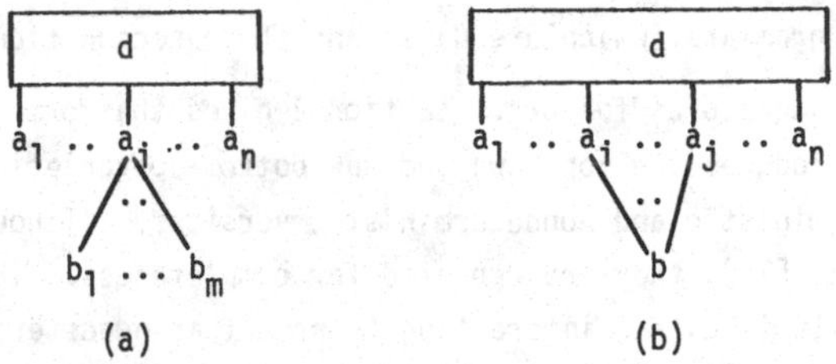

Fig. 1.

The set of dags over Σ is then $D_\Sigma = \{d \in P_\Sigma \mid \text{leaves}(d) \in \Sigma_{*0}^*\}$.

3. Parallel Dag Automata.

For a doubly ranked alphabet Σ we define a companion alphabet $\Sigma'=\{\sigma' \mid \sigma\in\Sigma\}$ such that σ and σ' have precisely the same head and tail ranks.

3.1. Definition. A finite dag automaton is a construct $A=(Q,\Sigma,R)$ where Q is a finite set of states, Σ is a doubly ranked alphabet and R is a finite set of rules of the form $r: \alpha\to\beta$. α and β are respectively the left-hand side and the right-hand side of r. A is deterministic if two different rules have different left-hand sides; otherwise A is nondeterministic. A being top-down or bottom-up depends on the form of α and β above as follows.

(a) A is top-down if the rules in R are of the form

$$[p_1\ldots p_n]\sigma\to\sigma'(q_1\ldots q_m)$$

(b) A is bottom-up if the rules in R are of the form

$$\sigma(q_1\ldots q_m)\to[p_1\ldots p_n]\sigma'$$

for some $\sigma\in\Sigma_{nm}$ and $p_1,\ldots,p_n,q_1,\ldots,q_m\in Q$. ⊠

The reason for introducing the primes in the right-hand sides of rules is to signify that σ has been processed and to prevent repeated reprocessing of the same dag.

A configuration of the finite dag automaton $A=(Q,\Sigma,R)$ is a dag over the doubly ranked alphabet $\Delta=\Sigma\cup\Sigma'\cup Q$ with $Q\subseteq\Delta_{11}$. Let d_1 and d_2 be two configurations of A. Then the (direct computation) relation $\vdash_A$ is defined as follows:

(i) If A is top-down, then $d_1 \vdash_A d_2$ if d_1 contains a subdag of Fig. 2(a), R contains the rule $[p_1\ldots p_n]\sigma\to\sigma'(q_1\ldots q_m)$ and d_2 is obtained from d_1 by replacing the subdag of Fig. 2(a) by the subdag of Fig. 2(b).

(ii) If A is bottom-up, then $d_1 \vdash_A d_2$ if d_1 contains a subdag of Fig. 2(c), R has a rule $\sigma(q_1\ldots q_m)\to[p_1\ldots p_n]\sigma'$ and d_2 is obtained from d_1 by replacing the subdag of Fig. 2(c) by the subdag of Fig. 2(d).

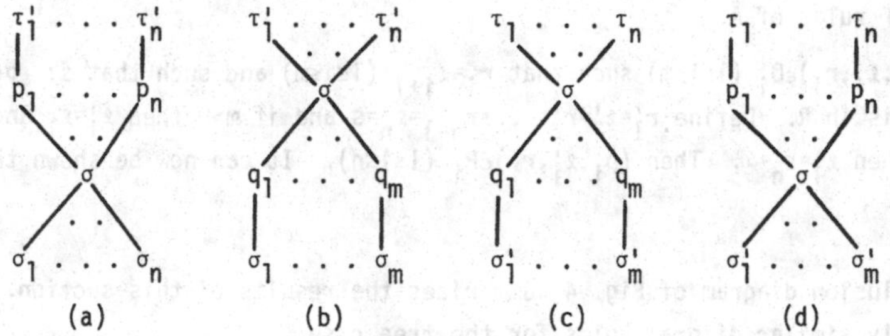

Fig. 2.

Given $\vdash_A$, $\overset{*}{\vdash}_A$ is the reflexive-transitive closure of $\vdash_A$. The dag language recognized by A is $L(A)=\{d\in D_\Sigma \mid d \overset{*}{\vdash}_A d'\}$ where $d'\in D_{\Sigma'}$ is the dag resulting from $d\in D_\Sigma$ by priming all the labels of d.

NT and DT denote respectively the set of all nondeterministic and deterministic top-down dag automata and similarly NB and DB in the bottom-up case. For a class K of dag automata $\mathcal{L}(K)=\{L(A)\mid A\in K\}$ is the class of dag languages defined by automata in K. Languages in $\mathcal{L}(NB)$ are said to be recognizable, and $\mathcal{L}(NB)$ will also be denoted by $RECOG_D$. If a dag automaton instead of priming the labels of the processed dag consistently relabels them by symbols of another doubly ranked alphabet then we obtain a device called finite state relabeling, cf. [3,4]. The formal definition is left to the reader. A relabeling is just a (total) single state relabeling. Let T be a finite state relabeling from Σ to Δ and let $L\in D_\Sigma$; then $T(L)=\{g\in D_\Delta \mid d \overset{*}{\vdash}_T g$ for some $d\in$

We are now ready to compare the machines defined above. The first of the next two results is easy while the second follows from a similar result for trees, see e.g. [4].

3.2. Theorem. $\mathcal{L}(NT)=\mathcal{L}(NB)$. ⊠

3.3. Theorem. $\mathcal{L}(DT)\subsetneq\mathcal{L}(NT)$. ⊠

In view of Theorem 3.3 the next result is rather surprising (even though for trees it is easy).

3.4. Theorem. $\mathcal{L}(DB)=\mathcal{L}(NB)$.

Proof. We have to show $\mathcal{L}(NB)\subseteq\mathcal{L}(DB)$. Unlike the tree case, a simpleminded subset construction does not work here because of the "non-tree-like" rules $\sigma(q_1\ldots q_m)\to[p_1\ldots p_n]\sigma'$ $(n>1)$ of A in NB. Let $\hat{A}\in DB$ be the automaton to be constructed.

The states of $\hat{A}$ are $\hat{Q}=\mathcal{P}(\{(q,\ell,r)\mid q\in Q,\ell,r\in R\cup\{\phi\}\})$. Let $\sigma\in\Sigma_{nm}$ and $Q_1,\ldots,Q_m\in\hat{Q}$. We explain the construction of $P_1,\ldots,P_n\in\hat{Q}$ such that $\sigma(Q_1\ldots Q_m)\to[P_1\ldots P_n]\sigma'$ will be in $\hat{R}$, the set of rules of $\hat{A}$.

Let $(q_i,\ell_i,r_i)\in Q_i$ $(1\le i\le m)$ such that $r_i=\ell_{i+1}$ $(1\le i\le m)$ and such that s: $\sigma(q_1\ldots q_m)\to[p_1\ldots p_n]\sigma'$ is in R. Define $r_1'=\ell_2'=r_2'=\ldots=r_{n-1}'=\ell_n'=s$ and if $m>0$ then $\ell_1'=\ell_1$ and $r_n'=r_m$ but if $m=0$ then $\ell_1'=r_n'=\phi$. Then $(p_i,\ell_i',r_i')\in P_i$ $(1\le i\le n)$. It can now be shown that $L(A)=L(\hat{A})$. ⊠

The inclusion diagram of Fig. 4 summarizes the results of this section. Note that precisely similar diagram holds for the tree case.

$$\mathcal{L}(NT)=\mathcal{L}(DB)=\mathcal{L}(NB)=RECOG_D$$

$$|$$

$$\mathcal{L}(DT)$$

Fig. 4.

4. Two-way Dag Walking Automata.

Our sequential two-way dag automata have a single pointer into the input dag. They are allowed to walk freely on the dag except that when moving up they must go through the edge by which they came down. This requirement is especially meaningful when the automaton is allowed to temporarily write some information on the dag (compare to remarks at the end of [1] and also to [5]). Thus, the automata which are allowed to use the input as temporary storage are restricted to have a synchronized push-down store between the root of the dag and the current location of the pointer; see [5] for the tree transducer case.

4.1. Definition. A two-way push-down dag automaton is a construct $A=(Q,\Sigma,\Gamma,\delta,q_0,F)$ where Q is a finite set of states, Σ is a doubly ranked alphabet, Γ is the push-down alphabet, $q_0\in Q$ is the initial state of A and $F\subseteq Q$ is the subset of final states. δ is a mapping from $Q\times\Sigma\times\Gamma$ to finite subsets of $Q\times D$ where $D=\{-i\,|\,i\geq 1\}\cup\{(i,\gamma)\,|\,\gamma\in\Gamma,\ i\geq 1\}$. ☒

A configuration of a two-way push-down dag automaton $A=(Q,\Sigma,\Gamma,\delta,q_0,F)$ on a dag $d\in D_\Sigma$ is a quadruple $<q,d,\alpha,\beta>$ where $q\in Q$, α is a path in d, $\alpha=(n_1,n_2,\ldots,n_k)$ with n_1 the root of d and n_k the node currently scanned, and $\beta=\gamma_1\ldots\gamma_k\in\Gamma^*$ is the contents of the push-down store. Let $C_1=<q,d,(n_1,\ldots,n_k),\beta_1\gamma>$ and $C_2=<p,d,(n_1,\ldots,n_\ell),\beta_2>$ be two configurations with n_k labeled by σ and $\gamma\in\Gamma$. Then $C_1 \vdash_{\overline{A}} C_2$ if either of the following holds.

1) $(p,(j,\gamma'))\in\delta(q,\sigma,\gamma)$, $\beta_2=\beta_1\gamma\gamma'$, $\ell=k+1$ and n_ℓ is the jth son of n_k.

2) $(p,-j)\in\delta(q,\sigma,\gamma)$, $\beta_2=\beta_1$, $\ell=k-1$ and n_ℓ is the jth father of n_k. For $k=1$, $<q,d,(n_1),\gamma> \vdash_{\overline{A}} <p,d,(\ ),\lambda>$.

A is deterministic if the following conditions are satisfied: (i) $(p,(j,\gamma))\in\delta(q,\sigma,\gamma')$ implies $\delta(q,\sigma,\gamma')=\{(p,(j,\gamma))\}$ (ii) $(p_1,-j_1),(p_2,-j_2)\in\delta(q,\sigma,\gamma)$ implies $j_1\neq j_2$. Note that in general a deterministic δ is not a partial function from $Q\times\Sigma\times\Gamma$ to $Q\times D$ as in the tree case [5]. This is because such a restriction would force the automaton to move up in the same direction and the same state each time it comes to some node in the same state (with the same push-down symbol) and is about to move up. This would make our automata unduly awkward. On the other hand note that the above restrictions force a deterministic behavior for any input dag. The relation $\vdash^{*}_{\overline{A}}$ is the reflexive-transitive closure of $\vdash_{\overline{A}}$ and the dag language recognized by A is $L(A)=\{d\in D_\Sigma\,|\,(q_0,d,(n_1),\gamma) \vdash^{*}_{\overline{A}} (p,d,(\ ),\lambda)$ with some $p\in F$ and $\gamma\in\Gamma\}$; n_1 is the root of d.

A two-way push-down dag automaton is <u>finite state dag walking</u> automaton if its push-down alphabet contains a single symbol. In this case the push-down is redundant and the configurations will be $\langle q,d,\alpha\rangle$ with $q \in Q$, $d \in D_{\Sigma}$ and α a path from the root of d to some node.

2N-PD and 2D-PD denote respectively the set of all nondeterministic and deterministic two-way push-down dag automata; similarly, 2N and 2D are the versions without the push-down facility.

The inclusion relationships between the two-way dag walking automata are summarized next.

<u>4.2. Theorem.</u> The following inclusion diagram (Fig. 5) holds.

$$\mathcal{L}(2N\text{-}PD)=\mathcal{L}(2D\text{-}PD)$$
$$|$$
$$\mathcal{L}(2N)$$
$$|$$
$$\mathcal{L}(2D)$$

<u>Fig. 5.</u> ⊠

<u>4.3. Remarks.</u> The following remarks could help to understand Theorem 4.2.

(i) The equality $\mathcal{L}(2N\text{-}PD)=\mathcal{L}(2D\text{-}PD)$ is proved using the method of transition tables [7,13] extended to dags.

(ii) The proper inclusion $\mathcal{L}(2N)\subsetneq\mathcal{L}(2N\text{-}PD)$ is proved by showing that the set of derivation dags (trees) D_G of the grammar $G=(\{S,A\},\{a,b\},\{S\to b|A, A\to aa|aA|Aa|AA\},S)$ cannot be recognized by any finite state dag walking automaton while it can be easily recognized using the push-down facility.

(iii) The complement of D_G of (ii) with respect to its doubly ranked alphabet $\Sigma=\{S,A,a,b\}$ with $\Sigma_{01}=\{S\}$, $\Sigma_{12}=\{A\}$, $\Sigma_{10}=\{a,b\}$ can be recognized by a nondeterministic automaton in 2N but by no automaton in 2D. ⊠

Let us note in passing that the diagram of Fig. 5 holds unmodified for the tree case (in fact our counterexamples were tree languages).

<u>5. Comparison, Properties and Problems.</u>

The facilities of (one-way) parallelism and two-way movement capability are compared. It turns out that the synchronized push-down store is precisely what is needed to handle the parallelism present in NB. On the other hand, the parallelism in DT cannot be handled by 2N and the two-way motion capability cannot be handled by DT.

Thus the inclusion diagram of Fig. 6 holds.

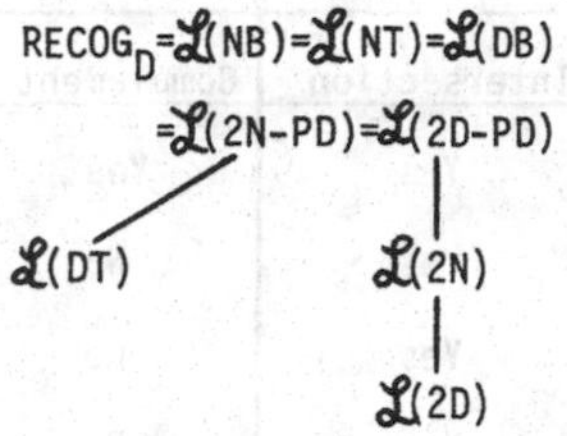

Fig. 6.

5.1. Remarks.

(i) The proof of RECOG$_D \subseteq \mathcal{L}$(2N-PD) is (more or less) straightforward.

(ii) Given any A$\in$2N-PD, the equivalent $\hat{A}\in$2D-PD constructed in the proof of Remark 4.3(i) enables an easy construction of B$\in$NB, equivalent to $\hat{A}$. This shows that $\mathcal{L}$(2N-PD)$\subseteq$RECOG$_D$.

(iii) The language { A(a, b) , A(b, a) } is not in $\mathcal{L}$(DT) but can be easily accepted by an automaton in 2D.

(iv) The language D_G of Remark 4.3(ii) is not in $\mathcal{L}$(2N) but can be recognized by a deterministic top-down automaton. ⊠

The next theorem characterizes the difference between RECOG$_D$ on the one hand and DT), $\mathcal{L}$(2N) and $\mathcal{L}$(2D) on the other.

5.2. Theorem. Let K be any of DT, 2N or 2D. For every L$\in$RECOG$_D$ there exists a nguage L'$\in\mathcal{L}$(K) and a relabeling h such that L=h(L'). Moreover, there is a finite ate relabeling T such that L'=T(L). ⊠

Theorem 5.2, the fact that derivation structures of phrase-structure grammars e (dags) in RECOG$_D$ [2,6,9,10] and the undecidability of the emptiness problem for rase-structure grammars yield the next result.

5.3. Theorem. The emptiness problem is unsolvable for any of the classes of g languages in Fig. 6. ⊠

This situation should be compared with the tree case where the emptiness problem solvable [4,15].

Finally we summarize the closure properties of our classes of languages in Table 1. o questions are left open and we conjecture that the answer to both is negative.

	Union	Intersection	Complement	Finite State Relabeling
$RECOG_D$	Yes	Yes	Yes	Yes
$\mathcal{L}(DT)$	No	Yes	No	No
$\mathcal{L}(2N)$	Yes	Yes	No	No
$\mathcal{L}(2D)$	?	Yes	?	No

Table 1

References.

1. Aho, A.V. and Ullman, J.D., Translations on a Context Free Grammar, Info. and Control 19(1971), 439-475.

2. Buttlemann, H.W., On the Syntactic Structure of Unrestricted Grammars, I and II, Info. and Control 29(1975), 29-101.

3. Engelfriet, J., Bottom-up and Top-down Tree Transformations - a comparison, Math. Syst. Th. 9(1975), 193-231.

4. Engelfriet, J., Tree Automata and Tree Grammars, Lecture Notes DAIMI, FN-10, University of Aarhus, Denmark, 1975.

5. Engelfriet, J., Rozenberg, G. and Slutzki, G., Tree Transducers, L-systems and Two-way Machines, Proc. 10-th Annual ACM STOC, San Diego, CA, 1978, 66-74.

6. Hart, J.M., Acceptors for the Derivation Languages of Phrase-Structure Grammars, Info. and Control 25(1974), 75-92.

7. Hopcroft, J.E. and Ullman, J.D., Nonerasing Stack Automata, JCSS 1(1967), 166-186

8. Hopcroft, J.E. and Ullman, J.D., Formal Languages and their Relation to Automata, Addison-Wesley, Reading, Mass., 1969.

9. Kamimura, T., Automata on Directed Acyclic Graphs, Ph.D. dissertation, in preparation.

10. Kamimura, T. and Slutzki, G., DAGs and Chomsky Hierarchy, to appear in Proc. of 6th Coll. of EATCS on Automata, Languages and Programming, July 1979, Graz, Austria.

11. Levy, L.S. and Joshi, A.K., Some Results in Tree Automata, Math. Syst. Th. 6(1972 336-342.

12. Ogden, W.F. and Rounds, W.C., Composition of n Tree Transducers, Proc. 4th Annual ACM STOC, Denver, CO, 1972, 198-206.

13. Shepherdson, J.C., The Reduction of Two-way Automata to One-way Automata, IBM J. Res. Devel. 3(1959), 198-200.

14. Thatcher, J.W., Characterizing Derivation Trees of Context-Free Grammars through a Generalization of Finite Automata Theory, JCSS 1(1967), 317-322.

15. Thatcher, J.W. and Wright, J.B., Generalized Finite Automata Theory with an Application to a Decision-Problem of Second Order Logic, Math. Syst. Th. 2(1968), 57-81.

16. Wilson, R., Introduction to Graph Theory, Oliver and Boyd, Edinburgh, 1972.

FULLY EFFECTIVE SOLUTIONS OF RECURSIVE DOMAIN EQUATIONS

Akira Kanda
Department of Computer Studies
University of Leeds
Leeds, LS2 9JT
U.K.

ABSTRACT This paper studies how we can exclude noncomputable elements of effectively given domains to obtain effective domains, and how we can obtain an effective domain which is an effectively initial solution to a recursive domain equation.

1. Introduction

Recursive domain equations play essential roles in denotational semantics of programming languages (See Scott-Strachey[4] and Tennent [8]). Also recently Lehmann-Smyth [3] showed the importance of them in abstract data type specification. Systematic ways of obtaining noneffective initial solutions of them have been known (See Scott [5,6], Smyth-Plotkin [7]). Kanda-Park [2] showed a method of obtaining initial "semi-effective" solutions which are effectively given domains.

This paper concerns fully effective initial solutions, which are effective domains. A preliminary report on this problem could be found in Kanda [1].

Throughout this paper, we assume a fixed acceptable indexing $<\phi_i>$ and $<W_i>$ of partial recursive functions and r.e. sets s.t. W_i = range (ϕ_i).

2. Effectively Given Domains

The following terminology seems to be widely accepted. A poset is directed (bounded) complete iff every directed (bounded) subset has a least upper bound (lub). A cpo is a directed complete poset with a least element (called bottom). An element x of a poset D is compact iff for every directed subset $S \subseteq D$ with a lub, $x \sqsubseteq \sqcup S$ implies $x \sqsubseteq s$ for some $s \in S$. A directed complete poset D is countably algebraic iff the set E_D of all compact elements of D is countable and for every $x \in D$, $J_x = \{e \in E_D | e \sqsubseteq x\}$ is directed and $x = \sqcup J_x$. E_D will be called the basis of D. If D is a countably alegbraic cpo then for every cpo Q, every monotone $m: E_D \to Q$ has a unique continuous extension $\bar{m}: D \to Q$ given by $\bar{m}(x) = \sqcup m(J_x)$. A poset has bounded joins iff every bounded finite subset has a lub. It can readily be seen that a countably algebraic cpo has bounded joins iff its basis has bounded joins iff it is bounded complete.

Definition 2.1 (1) Let D be a countably algebraic cpo with bounded joins (countably algebraic domain). A total indexing $\varepsilon:N\to E_D$ is effective iff there is a pair of recursive predicates (b,ℓ) called the characteristic pair of ε, s.t.:

$b(x)$ iff $\varepsilon(f_s(x))$ is bounded in E_D, and

$\ell(k,x)$ iff $\varepsilon(k) = \sqcup\varepsilon(f_s(x))$,

where f_s is the standard enumeration of finite subsets of N.

(2) An indexed domain is an ordered pair (D,ε) where D is a countably algebraic domain and $\varepsilon:N\to E_D$ is a total indexing. An indexed domain is an effectively given domain iff ε is effective. We will write D^ε for (D,ε). In case D^ε is effectively given, the characteristic pair of D^ε is the characteristic pair of ε.

(3) Given an effectively given domain D^ε, $x\in D$ is computable w.r.t. ε (or in D^ε) iff for some r.e. set W, $\varepsilon(W)$ is directed and $x=\sqcup\varepsilon(W)$. We say that an r.e. set W is ε-directed if $\varepsilon(W)$ is directed. The set of all computable elements of D^ε is denoted by $Comp(D^\varepsilon)$.

(4) Given effectively given domains D^ε, and $D'^{\varepsilon'}$, a function $f:D\to D'$ is computable w.r.t. $(\varepsilon,\varepsilon')$ iff the graph of f which is $\{ \langle n,m\rangle \mid \varepsilon'(m)\sqsubseteq f.\varepsilon(n)\}$ is r.e.

In case D^ε and $D'^{\varepsilon'}$ have the same characteristic pair, D^ε is merely a renaming of $D'^{\varepsilon'}$. More formally, there is an order isomorphism $f:D\to D'$ s.t. $f\cdot\varepsilon=\varepsilon'$. We denote this relation by $D^\varepsilon \overset{r}{\equiv} D'^{\varepsilon'}$. To within $\overset{r}{\equiv}$ we can introduce the following partial indexing $\overline{\xi}$ called the acceptable indexing of the class of effectively given domains s.t. $\overline{\xi}(\langle i,j\rangle)$ is the effectively given domain whose characteristic pair is (ϕ_i,ϕ_j). Notice that if τ is a partial function then we write $\tau(x)$ iff it is defined.

Given an effectively given domain D^ε, there is a recursive function d_ε (called the directing function) s.t. for every $j\in N$, $W_{d_\varepsilon(j)}$ is ε-directed and if W_j is already ε-directed then $\sqcup\varepsilon(W_j)=\sqcup\varepsilon(W_{d_\varepsilon(j)})$. Therefore we have a total indexing δ_ε (called the directed indexing) of $Comp(D^\varepsilon)$ s.t. $\delta_\varepsilon(i)=(\sqcup\varepsilon W_{d_\varepsilon(i)})$.

Definition 2.2 Let D^ε and $D'^{\varepsilon'}$ be indexed domains.

(1) $D^\varepsilon\times D'^{\varepsilon'} \overset{df}{\equiv} (D\times D', \varepsilon\times\varepsilon')$, (2) $D^\varepsilon+D'^{\varepsilon'} \overset{df}{\equiv} (D+D',\varepsilon+\varepsilon')$,

(3) $[D^\varepsilon\to D'^{\varepsilon'}] \overset{df}{\equiv} ([D\to D'],[\varepsilon\to\varepsilon'])$.

where $\varepsilon\times\varepsilon'$ and $\varepsilon+\varepsilon'$ are evidently induced indexings of $E_D\times E_{D'}$ and $E_D+E_{D'}$, respectively. $[\varepsilon\to\varepsilon']$ is defined by $[\varepsilon\to\varepsilon'](n)=$ if $\sigma(n)$ has a lub then $\sqcup\sigma(n)$ else $\perp$, and $\sigma(n)=\{[\varepsilon(i),\varepsilon(j)] \mid \langle i,j\rangle\in P(n)\}$ where P is the standard enumeration of finite subsets of $N\times N$ and $[e,e']$ is defined by $[e,e'](x)=$ if $x\sqsupseteq e$ then e' else $\perp$.

It is well-known that "effectively giveness" is invariant under x,+, and $\rightarrow$. Furthermore these domain constructors are effective operators w.r.t. the acceptable indexing $\overline{\xi}$.

A function $f:D\rightarrow D'$ is computable w.r.t. $(\varepsilon,\varepsilon')$ iff $f\in Comp([D^{\varepsilon}\rightarrow D'^{\varepsilon'}])$. This equivalence in fact is "effective" uniformly in D^{ε} and $D'^{\varepsilon'}$.

Fact 2.3 (1) A function from an effectively given domain to another is computable w.r.t. their bases iff it maps computable elements to computable elements recursively in directed indices.

(2) The composition of computable functions is recursive in directed indices uniformly in the ranges and domains of functions to be composed.

Kanda-Park [2] showed that order isomorphic relation is not sufficient to identify two effectively given domains. They proposed a notion of "effective isomorphism" for this purpose.

Definition 2.4 (1) Given indexed domains D^{ε} and $D'^{\varepsilon'}$, a function $f:E_D\rightarrow E_{D'}$ is an effective imbedding from ε to ε' (in symbols $f:\varepsilon\rightarrow\varepsilon'$) iff

1. f is injective,
2. $f\cdot\varepsilon=\varepsilon'\cdot r_f$ for some recursive function $r_f:N\rightarrow N$,
3. $\{\varepsilon(i_1),\ldots,\varepsilon(i_n)\}$ is bounded iff $\{f\cdot\varepsilon(i_1),\ldots,f\cdot\varepsilon(i_n)\}$ is bounded,
4. $f(\sqcup\{\varepsilon(i_1),\ldots,\varepsilon(i_n)\}) = \sqcup\{f\cdot\varepsilon(i_1),\ldots,f\cdot\varepsilon(i_n)\}$

(2) We say ε and ε' are effectively isomorphic (in symbols $\varepsilon \overset{e}{\cong} \varepsilon'$ or $D^{\varepsilon} \overset{e}{\cong} D'^{\varepsilon'}$) iff there exists an effective embedding $f:\varepsilon\rightarrow\varepsilon'$ s.t. $f^{-1}:\varepsilon'\rightarrow\varepsilon$.

If $f:\varepsilon\rightarrow\varepsilon'$ then $\overline{f}:D\rightarrow D'$ is an embedding with the adjoint $g:D'\rightarrow D$ given by $g(y)=\sqcup\{e\in E_D \mid f(e)\sqsubseteq y\}$. We will call $\overline{f}$ an effective embedding. In case $r_f=\phi_j$ we say that f (or $\overline{f}$) has a recursive index j.

A pair-wise computable (p-computable) embedding is an embedding which is computable as well as its adjoint. The next fact establishes the "effective" equivalence of effective embeddings of effectively given domains and p-computable embeddings.

Fact 2.5 (1) We can effectively go from a recursive index of an effective embedding of effectively given domains to directed indices of itself and its adjoint.

(2) Given directed indices of a p-computable embedding and its adjoint, we can effectively obtain a recursive index of the p-computable embedding.

If $D^{\varepsilon} \overset{e}{\cong} D'^{\varepsilon'}$ and either of them is effectively given then both of them are effectively given and $Comp(D^{\varepsilon}) \cong Comp(D'^{\varepsilon'})$.

An indexed poset is a pair (E,ε) where E is a countable poset with a bottom and bounded joins and ε is a total indexing of E. If ε is effective, i.e. it has a characteristic pair, E^{ε} is called an effective poset. The (algebraic) completion of an indexed poset (E,ε) is an indexed domain $(\overline{E},\overline{\varepsilon})$ where $\overline{E}$ is the algebraic completion of E and $\overline{\varepsilon} : N\rightarrow\tau(E)$ is given by $\overline{\varepsilon}(n)=\tau.\varepsilon(n)$ where $\tau:E\rightarrow\overline{E}$ is given by $\tau(x) = \{e \in E \mid e \sqsubseteq x\}$.

Fact 2.6 (1) The completion of an effective poset is an effectively given domain.

(2) Given an effectively given domain D^{ε}, E_D^{ε} is an effective poset and $(\bar{E}_D, \bar{\varepsilon})$ is equal to (D, ε) (to within $\overset{r}{\cong}$).

(3) An indexed domain is an effectively given domain iff it is (effectively isomorphic to) the completion of an effective poset.

An ω-sequence $\langle D_m^{\varepsilon}m, f_m\rangle$ of effective embeddings of effectively given domains is effective iff there exists a recursive function $q{:}N \to N$ s.t. $\pi_1 \cdot q(m)$ is a recursive index of f_m and $\pi_2 \cdot q(m)$ is an acceptable index of $D_m^{\varepsilon}m$ iff there is a recursive function q s.t. $\pi_1 \cdot \pi_1 \cdot q(m)$ is a directed index of f_m, $\pi_2 \cdot \pi_1 \cdot q(m)$ is a directed index of the adjoint g_m of f_m, and $\pi_2 \cdot q(m)$ is an acceptable index of $D_m^{\varepsilon}m$. The inverse limit of an ω-sequence $\langle D_m^{\varepsilon}m, f_m\rangle$ (in symbols $\lim\langle D_m^{\varepsilon}m, f_m\rangle$) is an indexed domain $(D_\infty, \varepsilon_\infty)$, where D_∞ is the inverse limit of $\langle D_m, f_m\rangle$ and ε_∞ is defined by :

$$\varepsilon_\infty(0) = f_{0\infty} \cdot \varepsilon_0(0) \qquad \varepsilon_\infty(1) = f_{0\infty} \cdot \varepsilon_0(1)$$
$$\varepsilon_\infty(2) = f_{1\infty}, \varepsilon_1(0) \qquad \varepsilon_\infty(3) = f_{0\infty} \cdot \varepsilon_0(2)$$

....

where $\langle f_{m\infty}\rangle$ is the universal cone of $\langle D_m, f_m\rangle$.

Fact 2.7 Let $\langle D_m^{\varepsilon}m, f_m\rangle$ be an effective sequence of effective embeddings of effectively given domains. Then $(D_\infty, \varepsilon_\infty)$ is an effectively given domain. Also $f_{m\infty}: D_m \to D_\infty$ is an effective embedding from ε_m to ε_∞. Furthermore there are recursive functions λ_d, ρ_d s.t. $\lambda_d(m)$ and $\rho_d(m)$ are directed indices of $f_{m\infty}$ and $g_{\infty m}$ respectively where $g_{\infty m}$ is the adjoint of $f_{m\infty}$.

3. Effective Domains

For a theory of computation, it is at least desirable to handle only computable objects. In this section, we will see that $(\mathrm{Comp}(D^{\varepsilon}), \varepsilon)$ functions perfectly well as a domain (called effective domain).

An effectively algebraic (ef-algebraic) domain is a pair (X, ε) where X is a poset and ε is a total indexing of E_X s.t.:

(1) E_X has bounded joins.

(2) If $\varepsilon(W_j)$ is directed then $\bigsqcup \varepsilon(W_j) \in X$. We call such r.e. set W_j ε-directed.

(3) For every $x \in X$, there is a ε-directed r.e. set W s.t. $x = \bigsqcup \varepsilon(W)$.

We call ε (or E_X^{ε}) the basis of X.

Definition 3.1 An ef-algebraic domain (X, ε) is an effective domain iff there exists a pair (called the characteristic pair of ε) of recursive predicates (b, ℓ) s.t.: $b(x)$ iff $\varepsilon(f_s(x))$ is bounded by E_X, and $\ell(k,x)$ iff $\varepsilon(k) = \bigsqcup \varepsilon(f_s(x))$, where f_s is the standard enumeration of the finite subsets of N. In this case we call ε an effective indexing of E_X (or an effective basis of X).

Notice that to within the renaming relation $\overset{r}{\cong}$, a characteristic pair uniquely determines an effective domain. In case (ϕ_i,ϕ_j) is a characteristic pair, then we write $\tilde{\zeta}$ (<i,j>) to denote the effective domain determined by (ϕ_i,ϕ_j). We say that <i,j> is an acceptable index of $\tilde{\zeta}$ (<i,j>).

The "directing function" does exist for every effective domain. Thus we can introduce the "directed indexing" χ_ε to every effective domain (X,ε) as we did for the set of all computable elements of each effectively given domains. By an effectively directed (ef-directed) subset of an effective domain X^ε we mean a directed subset $Z \subseteq X$ s.t. $Z = \chi_\varepsilon(W)$ for some r.e. set W. We say that this W is χ_ε-directed.

Theorem 3.2 An effective domain is ef-directed complete, i.e. every ef-directed subset has a lub.

Proof Let $Z \subseteq X$ be ef-directed s.t. $Z = \chi_\varepsilon(W)$ for some χ_ε-directed W. For every x in Z we have an ε-directed r.e. set W^x s.t. $x = \bigsqcup \varepsilon(W^x)$. Let $Y = \bigcup_x \varepsilon(W^x)$. It can readily be seen that $Y = \varepsilon(W')$ for some ε-directed r.e. set W'. Furthermore $\bigsqcup Z = \bigsqcup Y$.

Definition 3.3 Let X^ε and X'^{ε} be effective domains. A function $f : X \to X'$ is fully computable (f-computable) w.r.t. $(\varepsilon,\varepsilon')$ iff

(1) the graph of f is r.e.,

(2) f is effectively continuous w.r.t. $(\varepsilon,\varepsilon')$, i.e. for every χ_ε-directed r.e. set W, $f(\chi_\varepsilon(W))$ is effectively directed and $f(\bigsqcup\chi_\varepsilon(W)) = \bigsqcup f(\chi_\varepsilon(W))$.

4. Effective Isomorphism

By the same reason as for the effectively given domains, we need the notion of effective isomorphism as a criterion for identifying effective domains.

We define effective embeddings and effective isomorphisms among the bases of ef-algebraic domains exactly as we did in 2.4. Given ef-algebraic domains X^ε and $X'^{\varepsilon'}$ and an effective embedding $f : \varepsilon \to \varepsilon'$, let $\tilde{f} : X \to X'$ be the following extension : $\tilde{f}(\bigsqcup\varepsilon(W)) = \bigsqcup f(\varepsilon(W))$ for every ε-directed W. Notice $\tilde{f}$ is well-defined since f is an effective embedding. We call f an effective embedding from X^ε to $X'^{\varepsilon'}$ (or from ε to ε').

For an effective embedding f of ef-algebraic domains, we can not expect more than monotonicity. In fact it could not be even an embedding even though it is called an "effective embedding". But effective embeddings of effective domains enjoy much more interesting properties.

Definition 4.1 Let X^{ε} and $X'^{\varepsilon'}$ be effective domains. We say that a function $f:X\to X'$ is pairwise fully computable (pf-computable) embedding from ε to ε' iff f is f-computable w.r.t. $(\varepsilon,\varepsilon')$ and there exists a unique f-computable w.r.t. $(\varepsilon',\varepsilon)$ map $g:X'\to X$ s.t. $f\cdot g\sqsubseteq id_{X'}$, and $g\cdot f= id_X$.

Theorem 4.2 Let X^{ε} and $X'^{\varepsilon'}$ be effective domains. A map $f:X\to X'$ is an effective embedding from ε to ε' iff it is a pf-computable embedding from ε to ε'.

Despite discouragingly poor character of effective embeddings of ef-algebraic domains, effective isomorphisms of them are quite interesting.

Lemma 4.3 Let X^{ε} and $X'^{\varepsilon'}$ be ef-algebraic domains s.t. $f:\varepsilon\to\varepsilon'$ is an effective isomorphism. Then $E_X \cong E_{X'}$ via f and $X \cong X'$ via f.

Proof By the injectiveness of f and f^{-1}.

In case either X^{ε} or $X'^{\varepsilon'}$ is an effective domain and $\varepsilon \overset{e}{\cong} \varepsilon'$ via $f:\varepsilon\to\varepsilon'$, then both of them are effective domains and $X \cong X'$ via a pf-computable isomorphism f. We will denote this fact by $X^{\varepsilon} \overset{e}{\cong} X'^{e'}$. Note that a pf-computable isomorphism is an isomorphism which is f-computable as well as its adjoint.

5. Effective Completion

By an effective completion of an effective poset (B,ε), we mean a poset $(\tilde{B}^{\varepsilon},\sqsubseteq)$ together with a total indexing $\bar{\varepsilon}:N\to\tau(B)$ s.t. $\bar{\varepsilon}$ is as in section 2 (between 2.5 and 2.6) and $\tilde{B}^{\varepsilon} = \{\varepsilon(W) \mid W:\varepsilon\text{-directed}\}$. We will write $\tilde{B}$ for $\tilde{B}^{\varepsilon}$ if ε is evident from the context.

Remember that for each effective poset there is a unique characteristic pair. If two effective posets (B,ε) and (B',ε') have the same characteristic pair then (B,ε) is a renaming of (B',ε'), i.e. there is an isomorphism $f:B\to B'$ s.t. $f\cdot\varepsilon= \varepsilon'$. Thus they are the same. Notice that the characteristic pairs of effective(ly given) domains are special cases of this argument since every effective basis is an effective poset. To within $\overset{r}{\cong}$ we can introduce a partial indexing (called an acceptable indexing) ξ of effective posets s.t. $\xi(\langle i,j\rangle)$ denotes the effective poset determined by the characteristic pair (ϕ_i,ϕ_j).

Theorem 5.1 Given an effective poset (B,ε) we have:

(1) $\tau(B) = E_{\tilde{B}}$ and $(\tau(B),\bar{\varepsilon}) = (B,\varepsilon)$,

(2) for every $x\in\tilde{B}$, there is a $\bar{\varepsilon}$-directed r.e. set W^x s.t. $x=\bigsqcup\bar{\varepsilon}(W^x)=\bigsqcup\tau\cdot\varepsilon(W^x)$,

(3) $(\tilde{B},\bar{\varepsilon}) = (Comp(\bar{B},\bar{\varepsilon}),\bar{\varepsilon})$,

(4) $(\tilde{B},\bar{\varepsilon})$ is an effective domain.

Theorem 5.2 Let (X,ε) be an effective domain. Then (E_X,ε) is an effective poset and $(\tilde{E}_X,\bar{\varepsilon}) = (X,\varepsilon)$.

Proof Define $\phi:X\to\tilde{E}_X$ by $\phi(x)= \{e\in E_X \mid e\sqsubseteq x\}$. Evidently ϕ is an isomorphism and $\phi\cdot\varepsilon = \bar{\varepsilon}$. Therefore $\bar{\varepsilon} \overset{r}{\cong} \varepsilon$.

Corollary 5.3 An ef-algebraic domain is an effective domain iff it is (effectively isomorphic to) the effective completion of an effective poset iff it is (effectively isomorphic to) $(\mathrm{Comp}(D^{\varepsilon}),\varepsilon)$ for some effectively given domain D^{ε}.

Notice that we have observed that every effective basis of an effective (ly given) domain is an effective poset and every effective poset is an effective basis of some effective(ly given) domain. In fact we have :

<u>Theorem 5.4</u> (<u>The Acceptable Indexing Theorem</u>)

(1) $\tilde{\zeta}(i)=\widetilde{\widetilde{\zeta}(i)}=\mathrm{Comp}(\bar{\zeta}(i))$,

(2) $\overline{\zeta}(i)=\overline{\zeta(i)}$.

6. Domain Constructors

Let X^{ε} and $X'^{\varepsilon'}$ be effective domains. We define $X^{\varepsilon}\times X'^{\varepsilon'}$ and $X+X'^{\varepsilon'}$ by : $X^{\varepsilon}\times X'^{\varepsilon'} \overset{\mathrm{df}}{=} (X\times X')^{\varepsilon\times\varepsilon'}$ and $X^{\varepsilon}+X'^{\varepsilon'} \overset{\mathrm{df}}{=} (X\times X')^{\varepsilon+\varepsilon'}$ where $\varepsilon\times\varepsilon'$ and $\varepsilon+\varepsilon'$ are the same as in 2.2. By 5.4, $X^{\varepsilon} = \mathrm{Comp}(\overline{E}_X^{\bar{\varepsilon}})$. But evidently $\mathrm{Comp}(\overline{E}_X^{\bar{\varepsilon}})\times(\overline{E}_{X'}^{\bar{\varepsilon}'}) = \mathrm{Comp}(\overline{E}_X^{\bar{\varepsilon}'}\times\overline{E}_{X'}^{\bar{\varepsilon}'})$. Therefore $X^{\varepsilon}\times X'^{\varepsilon'}$ is an effective domain. Similarly $X^{\varepsilon}+X'^{\varepsilon'}$ is an effective domain.

The problem of function space is not so straightforward because effective domains are not necessarily cpo's.

<u>Definition 6.1</u> Let X^{ε} and $X'^{\varepsilon'}$ be effective domains. Define the function space $(X^{\varepsilon}\to X'^{\varepsilon'})$ to be $(X\to X')^{(\varepsilon\to\varepsilon')}$ where $(X\to X')$ is the set of all f-computable (w.r.t. $(\varepsilon,\varepsilon')$) functions with the pointwise ordering, and $(\varepsilon\to\varepsilon')$ is the following total indexing of $E_{(X\to X')}$: $(\varepsilon\to\varepsilon')(n)$ = <u>if</u> $\sigma(n)$ has a lub <u>then</u> $\sqcup\sigma(n)$ <u>else</u> $\bot$, where $\sigma(n) = \{(\varepsilon(i),\varepsilon(j))\mid\langle i,j\rangle\in P(n)\}$ and $(e,e'):X\to X'$ is a step function s.t. $(e,e')(x)$ = <u>if</u> $x\sqsupseteq e$ <u>then</u> e' <u>else</u> $\bot$. Evidently $(e,e') = [e,e']\upharpoonright X$ where $[e,e']$ is a step function $\overline{E}_X\to\overline{E}_{X'}$.

<u>Lemma 6.2</u> Let X^{ε} and $X'^{\varepsilon'}$ be effective. $h:X\to X'$ is f-computable w.r.t. $(\varepsilon,\varepsilon')$ iff it is the restriction to $X=\mathrm{Comp}(\overline{E}_X^{\bar{\varepsilon}})$ of a function $E_X\to E_{X'}$ which is computable w.r.t. $(\bar{\varepsilon},\bar{\varepsilon}')$.

<u>Proof</u> Necessity is trivial. We prove sufficiency. Assume $h:\mathrm{Comp}(\overline{E}_X^{\bar{\varepsilon}})\to\mathrm{Comp}(E_{X'}^{\varepsilon'})$ is f-computable w.r.t. $(\bar{\varepsilon},\bar{\varepsilon}')$. Evidently $h\upharpoonright\tau(E_X):\tau(E_X)\to\overline{E}_{X'}$ is monotone. Thus $\Phi_h:\overline{E}_X\to\overline{E}_{X'}$ s.t. $\Phi_h\sqcup\bar{\varepsilon}(W)=\sqcup h.\bar{\varepsilon}(W)$ for all ε-directed W, is the unique continous extension of $h\upharpoonright\tau(E_X)$. Since h is f-computable, Φ_h is computable w.r.t. $(\bar{\varepsilon},\bar{\varepsilon}')$. Also $\Phi_h(\sqcup\bar{\varepsilon}(W)) = \sqcup h.\bar{\varepsilon}(W)= h(\sqcup\bar{\varepsilon}(W))$ for every ε-directed W. Notice that the second equality is due to the fact that we can effectively go from effective indexing $\bar{\varepsilon}$ to the directed indexing $\chi_{\bar{\varepsilon}}$. Thus Φ_h is the unique computable extension of h.

<u>Theorem 6.3</u> Let X^{ε} and $X'^{\varepsilon'}$ be effective domains. We have $(X^{\varepsilon}\to X'^{\varepsilon'}) = (\mathrm{Comp}([\overline{E}_X^{\bar{\varepsilon}}\to\overline{E}_{X'}^{\bar{\varepsilon}'}]),[\bar{\varepsilon}\to\bar{\varepsilon}'])$. Therefore $(X^{\varepsilon}\to X'^{\varepsilon'})$ is an effective domain.

<u>Proof</u> Define $\alpha:\mathrm{Comp}([\overline{E}_X^{\bar{\varepsilon}}\to\overline{E}_{X'}^{\bar{\varepsilon}'}])\to(\mathrm{Comp}(\overline{E}_X^{\bar{\varepsilon}})\to\mathrm{Comp}(\overline{E}_{X'}^{\bar{\varepsilon}'}))$ by $\alpha(h)=h\upharpoonright\mathrm{Comp}(\overline{E}_X^{\bar{\varepsilon}})$. Then α is an isomorphism with the adjoint β s.t. $\beta(h)=\Phi_h$. Evidently $\alpha[\bar{\varepsilon}\to\bar{\varepsilon}'] = (\bar{\varepsilon}\to\bar{\varepsilon}')$. Therefore $(X^{\varepsilon}\to X'^{\varepsilon'}) \overset{r}{\cong} (\mathrm{Comp}([\overline{E}_X^{\bar{\varepsilon}}\to\overline{E}_{X'}^{\bar{\varepsilon}'}]),[\bar{\varepsilon}\to\bar{\varepsilon}'])$.

Notice that x,+, and → are effective operators w.r.t. acceptable indices. Furthermore the directed indexing $\delta_{[\bar{\varepsilon}\to\bar{\varepsilon}']}$ of $(\mathrm{comp}([\bar{E}^{\varepsilon}_{X}\to\bar{E}^{\varepsilon'}_{X'}]), [\bar{\varepsilon}\to\bar{\varepsilon}'])$ is equivalent to the directed indexing $\chi_{(\varepsilon\to\varepsilon')}$ of $(X^{\varepsilon}\to X'^{\varepsilon'})$ in such a sense as $\alpha(\delta(i)) = \chi(i)$. Therefore from 2.3 we immediately have :

<u>Lemma 6.4</u> (1) Every f-computable function maps recursively in directed indices. (2) The composition of f-computable functions is recursive in directed indices uniformly in the ranges and domains of the functions to be composed.

Also we can introduce the recursive indices of effective embeddings as we did for effectively given domains, and we can observe the same result as 2.5 for effective domains.

7. Effective Inverse Limit

We define the effectiveness of an ω-sequence of effective embeddings of effective domains in the same manner as we did for effectively given domains. Even-though an ω-sequence of embeddings is effective, the usual inverse limit construction gives us a poset which is uncountable. Therefore we need a notion of "effective inverse limit" which will cut down the cardinal of limits to ≤ω.

<u>Definition 7.1</u> The <u>effective inverse limit</u> of an effective sequence $\langle X^{\varepsilon}_{m}m, j_m\rangle$ of effective embeddings of effective domains is a pair $(X_\infty, \varepsilon_\infty)$ where X_∞ is a poset $\{\langle x_m\rangle \mid x_m = g_m(x_{m+1})$, there is a recursive function s:N→N s.t. s(m) is a direct index of $x_m\}$ with a coordinatewise ordering, and ε_∞ is defined as in section 2. We write $\text{ef-}\varprojlim\langle X^{\varepsilon}_{m}m, f_m\rangle$ for $(X_\infty, \varepsilon_\infty)$. Evidently $\text{ef-}\varprojlim\langle X^{\varepsilon}_{m}m, f_m\rangle$ is ef-algebraic.

<u>Theorem 7.2</u> Let $\langle X^{\varepsilon}_{m}m, f_m\rangle$ be an effective sequence, then $\text{ef-}\varprojlim\langle X^{\varepsilon}_{m}m, f_m\rangle$ is an effective domain.

<u>Proof</u> (outline) We can show $\text{ef-}\varprojlim\langle X^{\varepsilon}_{m}m, f_m\rangle = \text{el-}\varprojlim\langle \mathrm{Comp}(\bar{E}^{\varepsilon}_{X_m}m)^{\bar{\varepsilon}}m, f_m\rangle = \mathrm{Comp}(\varprojlim\langle \bar{E}^{\varepsilon}_{X_m}m, \Phi_{f_m}\rangle)^{\bar{\varepsilon}_\infty}$ where Φ_{f_m} is the unique computable extension of f_m.

8. Effective Categories and Effective Functors

In this section we will study a general framework for providing initial solutions to recursive domain equations.

<u>Definition 8.1</u> An <u>E-category</u> is a category <u>K</u> together with an object indexing κ and a morphism indexing ∂(K,K') : N→Hom(K,K') for each pair (K,K') of objects s.t. there is a recursive function ∂-compose satisfying:

$$\partial(\kappa(i),\kappa(k))(\partial\text{-compose }(i,j,k,l,m)) = \partial(\kappa(j),\kappa(k))(m).\partial(\kappa(i),\kappa(j))(l)$$

<u>Definition 8.2</u> (1) <u>ω</u> is the category 0≤1≤2≤... .

(2) An <u>effective codiagram</u> (with a codiagram index j) in an E-category (<u>K</u>,κ,∂) is a functor G : <u>ω</u>→<u>K</u> s.t. ϕ_j is recursive and

$G(n) = \kappa(\pi_1.\phi_j(n))$,

$G(n\le n+1) = \partial(G(n), G(n+1))(\pi_2.\phi_j(n))$.

(3) An <u>effective cocone</u> (with a cocone index <i,j>) of an effective diagram G is

a cocone $<\lambda_n : G(n) \to \kappa(i)>$ s.t. $\lambda_n = \partial(G(n),\kappa(i))\ \phi_j(n))$ where ϕ_j is recursive.
(4) An <u>effective colimiting cocone</u> of an effective codiagram $G:\underline{\omega}\to\underline{K}$ is an effective cocone $<\partial_n : G(n) \to K>$ of G s.t. there exists a recursive function C-Med s.t. for every effective cocone$<\lambda_n : G(n) \to K'>$ with a cocone index $<i,j>$, $\alpha=\partial(K,K')$ (C-Med(i,j)) is the unique morphism which mediates from $<\partial_n>$ to $<\lambda_n>$. We will denote such K by ef-colim G.

<u>Definition 8.3</u> Given E-categories $(\underline{K},\kappa,\partial)$ and $(\underline{K}',\kappa',\partial')$, a functor $F:\underline{K}\to\underline{K}'$ is <u>semi effective</u> (w.r.t. $<(\kappa,\partial),(\kappa',\partial')>$) iff there are recursive functions f_{ob} and f_{mr} s.t. $F(\kappa(n)) = K'(f_{ob}(n))$ and $F(\partial(\kappa(i),\kappa(j))(n)) = \partial'\kappa'(\pi_1.\pi_1 f_{mr}(n)),\ \kappa'(\pi_2.\pi_1.f_{mr}(n))(\pi_2.f_{mr}(n))$. Such F is <u>effective</u> (w.r.t. $<(\kappa,\partial',(\kappa',\partial',>$) iff it preserves all existing effective colimiting cocones.

<u>Definition 8.4</u> Given an E-category $(\underline{K},\kappa,\partial)$ and an endo functor $F:\underline{K}\to\underline{K}$, an <u>F-algebra</u> is a pair $<A,\alpha>$ where $\alpha:FA\to A$. An <u>F-homomorphism</u> from an F-algebra $<A,\alpha>$ to another $<b,\beta>$ is a <u>K</u>-morphism $f:A\to B$ s.t.: $f.\alpha=\beta.Ff$.
If $A=\kappa(a)$ and $\alpha=\partial(FA,A)(i)$, we say that $<A,\alpha>$ has an <u>algebra index</u> $<a,i>$. $\underline{A}^F$ will denote the category of F-algebras and F-homomorphisms.

<u>Definition 8.5</u> The <u>effective initial F-algebra</u> is an F-algebra $<I,\iota>$ s.t. there exists a recursive function Int : $N\times N\to N$ s.t. for every F-algebra $<A,\alpha>$ with an algebr index $<a,i>$, $\partial(I,A)(Int(a,i))$ is the unique F-homomorphism from $<I,\iota>$ to $<A,\alpha>$.

<u>Lemma 8.6</u> Effective initial F-algebra, if exists is an isomorphism.

<u>Theorem 8.7</u> <u>(The Fundamental Theorem)</u>

Let $(\underline{K},\kappa,\partial)$ be an E-category with an initial object $\perp$. Let $F : \underline{K}\to\underline{K}$ be semi effective. Evidently $\Delta : \perp \overset{\perp_{F\perp}}{\to} F\perp \overset{F(\perp_{F\perp})}{\to} F^2\perp \to \ldots$ is an effective codiagram in $(\underline{K},\kappa,\partial)$, where $\perp_{F\perp}$ is the unique morphism from $\perp$ to $F\perp$. Assume $\mu = <\mu_n : F^n\perp \to A>$ is an effective colimiting cocone of Δ. Also assume that $F.\Delta$ has an effective colimiting cocone $F\mu = <F\mu_n : F^{n+1}\perp \to FA>$. Then the effective initial F-algebra exists.

<u>Corollary 8.8</u> Let $(\underline{K},\kappa,\partial)$ be an E-category with an initial object $\perp$ and $F:\underline{K}\to\underline{K}$ be an effective functor s.t.

$$\Delta : \perp \overset{\perp_{F\perp}}{\to} F\perp \overset{F(\perp_{F\perp})}{\to} F^2\perp \to \ldots.$$

has an effective colimit, then effective initial F-algebra exists.

<u>Definition 8.9</u> An <u>effective category</u> is an E-category with an initial object s.t. every effective codiagram has an effective colimiting cocone.

<u>Corollary 8.10</u> Let $(\underline{K},\kappa,\partial)$ be an effective category and $F : \underline{K}\to\underline{K}$ be an effective functor, then the effective initial F-algebra exists.

It should be noted that the notion of effective universality, like effective colimit and effective initiality, implies effective existence of the unique universal morphisms. In this sense our theory here is more appropriate than the previous developments of Kanda [1] and Kanda & Park [2].

Given two effective categories (K,κ,∂) and $(\underline{K}',\kappa',\partial')$ the product category $\underline{K}\times\underline{K}'$ together with the evidently induced object indexing $\kappa\times\kappa'$ and morphism indexing $\partial\times\partial'$ is an effective category which will be denoted by $(\underline{K},\kappa,\partial)\times(\underline{K}',\kappa',\partial')$. Evidently $F : (\underline{K},\kappa,\partial)\times(\underline{K}',\kappa',\partial')\to(\underline{K}'',\kappa'',\partial'')$ iff F is effective in both $\underline{K}$ and $\underline{K}'$. Also notice that the composition of two effective functors is again effective.

Theorem 8.11 The category of effective(ly) domains and effective embeddings together with $\tilde{\zeta}(\bar{\zeta})$ as an object indexing and the recursive indexing as a morphism indexing is an effective category. The effective diagrams are effective sequences and effective colimits are effective inverse (inverse) limits of effective sequences. We will denote this effective category $E^E(ED^E)$ without being explicit about indexing.

Given an effective embedding f of effective(ly given) domains with the adjoint g, we call (f,g) an effective embedding pair. If f and g have directed indices i and j respectively, we say that the effective embedding pair has a directed index <i,j>.

Theorem 8.12 The category of effective(ly given) domains and effective embedding pairs together with $\tilde{\zeta}(\bar{\zeta})$ as an object indexing and direct indexing as a morphism indexing is an effective category. The effective diagrams are effective sequences and the effective colimits are effective inverse (inverse) limits. We denote this category by $E^{EP}(ED^{EP})$ without explicitly mentioning indexings.

It can readily be seen that $\times,+,$ and $\to$ induce bi-functors which are effective of these effective categories. Thence for every recursive domain equation which involves these domain constructors we can obtain an effectively initial solution which is an effective(ly given) domain. In fact this solution is to within effective isomorphism.

The following theorem says that ED^E and ED^{EP} are redundant version of E^E and E^{EP}.

Theorem 8.13 $ED^E \simeq E^E \simeq ED^{EP} \simeq E^{EP}$ within the category of effective categories and effective functors.

Theorem 8.14 An effective domain together with the directed indexing as an object indexing and the evident morphism indexing is an effective category and all f-computable functions are effective functors.

ACKNOWLEDGEMENT I am very grateful to M. Smyth and D. Park for very helpful discussions and criticisms, I also am benefitted from discussions with M. Paterson, M. Beynon and W. Wadge. This research was supported in part by the SRC grant GR/A66772 under the direction of M. Smyth. I thank J. Ferguson for her patience to type this difficult to type manuscript.

REFERENCES

[1] Kanda, A, Data types as effective object, Theory of Computation Report No. 22, Warwick University, (1977).

[2] Kanda-Park, When two effectively given domains are identical, Proc. of the 4th GI Theoretical Computer Science Conference, Lecture Notes in Computer Science No. 67, (1979).

[3] Lehmann-Smyth, Data Types, Proc. of the 18th IEEE FOCS Conference, (1977).

[4] Scott-Strachey, Towards a mathematical semantics of computer languages, Proc. of Symposium on Computer and Automata, Polytechnical Inst. of Brooklyn (1971).

[5] Scott D., Data types as lattices, Unpublished lecture notes, Amsterdam, (1972).

[6] Scott D., Data types as lattices, SIAM J. on Computing, Vol. 5, (1976).

[7] Smyth-Plotkin, The categorical solution of recursive domain equations, Proc. of the 18th IEEE FOCS Conference, (1977).

[8] Tennent R., The denotational semantics of programming languages, CACM, Vol. 19, No. 18, (1977).

[9] Egli-Constable, Computability concepts for programming language semantics, Theoretical Computer Science, Vol. 2, (1976).

[lo] Rosen-Markowsky, Bases for chain complete posets, IBM J. of R&D, Vol. 5, No. 3, (1976).

[11] Smyth M., Effectively given domains, Theoretical Computer Science, Vol. 5, (1977).

A NOTE ON COMPUTATIONAL COMPLEXITY OF A STATISTICAL DEDUCIBILITY TESTING PROCEDURE

Ivan Kramosil

Institute of Information Theory and Automation, Czechoslovak Academy of Sciences, Prague, Czechoslovakia

1. INTRODUCTION

Automated theorem proving has been taken for a common part of artificial intelligence and computer science since the birth of these two branches of applied science. Usually, theorem proving is investigated as an autonomous domain in the sense that a decision taken about a tested formula is not followed by some further decision making or action. Cf., e. g., Chang and Lee [1] as a surveyal monography on theorem proving. However, some recent results dealing with applications of mathematical logic and proof theory in artificial intelligence proved an interesting possibility of theorem proving when used as a tool for, say, robot plan formation, cf. Štěpánková and Havel [7].

The application-oriented approach to theorem proving forces also an appropriate change of criteria with respect to which a theorem-proving procedure can be classified. In the classical, pure mathematics and also in the case of an autonomous theorem-proving procedure we have always the possibility to give up our effort to prove or disprove a formula before reaching an answer, supposing the length or complexity of the desired decision is beyond our powers. Such a resignation is not taken as an error and is always strictly preferred to a wrong decision. However, in the case of, say, robot plan formation we must decide somehow, hence, using the proof theory apparatus, we must take a decisive answer whether the corresponding formalized proof does or does not exist. In such a situation an approximative, perhaps statistical, deducibility testing program may be of a great significance.

Probably the first statistically based deducibility testing procedure was that proposed by A. Špaček [6]. Here a rather modified version of this method will be briefly presented, a more detailed explanation can be found in Kramosil [2] or in some special papers. Our aim in this paper will be to study this statistical deducibility testing method from the viewpoint of its computational complexity.

Consider a formalized theory $<L,T>$; L is a formalized language, $T \subset L$ is the set of theorems. Usually the property of theoremhood is not decidable, i. e., T is not a recursive subset of L.

Let $T_0 \subset T$ be a recursive (and often finite) set of theorems such that for every sentence $x \in L$ we are able to decide effectively and within some a priori given time and space limitations whether x belongs to T_0 or not. Considering a theorem proving program, say, a resolution -based one, together with necessary time and storage limitations, then T_0 is the set of theorems provable by this program within these limitations. We shall suppose, in the rest of this paper, that:

(a) There is given a complexity measure s defined over the set of all formulas of the formalized theory $<L,T>$ and taking its values in the set $N = \{0,1,2,...\}$. This mapping will be called, in what follows, the <u>size</u> of a formula or sentence. Suppose that for each $n \in N = \{0,1,2,...\}$, $n>0$, there is at least one sentence $x \in L$ such that $s(x) = n$.

For example, the length of formula or its syntactical depth can serve as such size .

(b) For each pair of sentences $x,y \in L$ the implication $x \rightarrow y$ belongs to T_0 only if $|s(x) - s(y)| \leq K = K(T_0)$, where $K(T_0)$ is an a priori given positive integer (but not all implications with this property belong, in general, to T_0).

If we take the syntactical depth to play the role of size we can see that the demand $|s(x) - s(y)| \leq K$ is satisfied if $x \rightarrow y$ is provable and possesses a proof consisting of K or less formulas. The same will hold if $s(x) = min\{\ell(x),n\}$ for $x \in T$, $s(x) = n$ for $x \in L - T$, where $\ell(x)$ is the length of a shortest proof of x and $n \in N$ is an a priori given threshold value.

2. A STATISTICAL DEDUCIBILITY TESTING PROCEDURE

Consider the following decision procedure. Let $n(0)$, $m(0) \in N$, $1 \leq m(0) \leq n(0)$, let $x \in L$ be a sentence which is to be tested for theoremhood. Sample at random, statistically independently and from the same probability distribution, $n(0)$ sentences $a_1, a_2, ..., a_{n(0)}$ from L and test, for $i = 1,2,...,n(0)$, whether $\ulcorner a_i \rightarrow x \urcorner \in T_0$. If this is the case for at least $m(0)$ values of i, proclaim x to be a theorem, in the other case proclaim x to be a non-theorem. In order to find under which conditions such a decision procedure has a sense we have to formalize it as follows.

Let $<\Omega,S,P>$ be a probability space, hence, Ω is a nonempty set, S is a σ-field of subsets of Ω, and P is a normalized set measure on S.

et $x, a_1, a_2, \ldots, a_{n(0)}$ be random variables (i. e., measurable mappings) efined on $\langle\Omega, S, P\rangle$ and taking their values in the set L of sentences; uppose that $a_1, a_2, \ldots, a_{n(0)}$ are mutually independent and equally istributed. Set $I(x) = 1$ for $x \in T_0$, $I(x) = 0$ for $x \in L - T_0$. The decision rule above can be expressed as follows

$$(x) = d(x(\omega), m(0), \{a_i(\omega)\}_{i=1}^{n(0)}) = d_1 \text{ iff } \sum_{i=1}^{n(0)} I(a_i(\omega) \to x(\omega)) \geq m(0),$$
$$(x) = d_2 \text{ otherwise,} \tag{1}$$

here d_1, d_2 are possible decisions and the random event $\{d(x(\omega))=d_1\}$ s taken as proclaiming $x(\omega)$ to be a theorem, $\{d(x(\omega))=d_2\}$ as proclaimng $x(\omega)$ to be a non-theorem. The assumption that the tested formula s sampled at random is necessary in order to be able to classify the tatistical properties of the testing procedure globally, i. e., without respect to a particular tested formula. When denoting random events elements of S) we write simply $\{A(\omega)\}$ instead of $\{\omega : \omega \in \Omega, A(\omega)\}$.

Denote by p, q the two following conditional probabilities

$$p = P(\{\ulcorner a_1(\omega) \to x(\omega)\urcorner \in T_0\} \mid \{x(\omega) \in T\}),$$
$$q = P(\{\ulcorner a_1(\omega) \to x(\omega)\urcorner \in T_0\} \mid \{x(\omega) \in L-T\}). \tag{2}$$

Theorem 1. If

$P(\{x(\omega) \in T_0\} \mid \{x(\omega) \in T\}) > P(\{a_1(\omega) \in L-T\})$, then $p > q$.

Proof. A simple consequence of Theorem 5.3, Kramosil [2]. This theorem offers a weaker condition for the inequality $p>q$ to hold.

Calling an $a \in L$ "successful" (w.r. to $x \in L$ and T_0) if $\ulcorner a \to x \urcorner \in T_0$, the original decision problem (whether $x(\omega) \in T$ or not) can be transformed into this one: whether the probability of sampling a successful $a_i(\omega)$ equals p or q. This is nothing else than the usual case of parametric hypothesis testing theory - simple hypothesis $H : P_0=p$ against a simple alternative $A : P_0=q$. If $p>q$, then the decision function (1) is adequate and $m(0)$ can be chosen in such a way that (1) is optimum among all rules resulting from (1) by $m(0)$ ranging from 1 to $n(0)$. Here optimality is taken either in the sense of minimal sum P_I+P_{II} or as minimal P_I (P_{II}, resp.) under the condition that P_{II} (P_I, resp.) is kept below a given threshold value. Let us recall the two corresponding probabilities of errors:

$$P_I = P(\{d(x(\omega)) = d_1\} \mid \{x(\omega) \in L-T\}),$$
$$P_{II} = P(\{d(x(\omega)) = d_2\} \mid \{x(\omega) \in T\}). \tag{3}$$

A possible asymptotic solution for $m(0)$ can be found in Kramosil [2] or Kramosil and Šindelář [3], here we shall not investigate the statistical properties of the proposed decision function. The value $n(0)$ can be either given a priori as a free parameter of the decision function d or chosen appropriately in order to keep the probabilities of error under some limit. A way of setting $n(0)$ must be taken as a part of the decision procedure described above, below we shall make this step precise.

3. A SIZE-RELATIVIZED DEDUCIBILITY TEST

When choosing the parameters of the testing procedure only on the basis of the values p, q and prescribed upper bounds for P_I or P_{II} we do not take into consideration the size $s(x(\omega))$ of the tested formula $x(\omega)$. It is why the criteria of the test have just a global and average character and the quality of this test may significantly differ when relativized to, say, the class of sentences of the same given size. Hence, we would like to modify the testing procedure in such a way that, first, $x(\omega)$ is sampled, then $s(x(\omega))$ is computed or estimated, and finally, the parameters $n(0) = n(0)(s(x(\omega)))$, $m(0) = m(0)(s(x(\omega)))$ are chosen in such a way that the relativized probabilities of error satisfied given limitations. Denote, for each $i \in N$,

$$p_i = P(\{\ulcorner a_1(\omega) \rightarrow x(\omega)\urcorner \in T_0\} \mid \{x(\omega) \in T,\ s(x(\omega)) = i\}),$$
$$q_i = P(\{\ulcorner a_1(\omega) \rightarrow x(\omega)\urcorner \in T_0\} \mid \{x(\omega) \in L-T,\ s(x(\omega)) = i\}). \tag{4}$$

Let us suppose, for the sake of simplicity, that all these conditional probabilities exist. Denote, moreover,

$t_i = P(\{s(a(\omega)) = i\}),\ i \in N$

and remember that

$$p = \sum_{i \in N} p_i \cdot P(\{s(x(\omega)) = i\} \mid \{x(\omega) \in T\}),$$
$$q = \sum_{i \in N} p_i \cdot P(\{s(x(\omega)) = i\} \mid \{x(\omega) \in L-T\}).$$

Theorem 2.

(1) $p_i \leq \sum_{j=i-K}^{i+K} t_j q_i \leq \sum_{j=i-K}^{i+K} t_j$, $i \in N$,

(2) $\lim p_i = 0$, $\lim q_i = 0$, $i \rightarrow \infty$.

Proof. If $s(x(\omega)) = i$ then a necessary condition for $\ulcorner a_1(\omega) \rightarrow x(\omega)\urcorner \in T_0$ is that $|s(a_1(\omega)) - s(x(\omega))| \leq K$, i.e.,

$$p_i = P(\{\ulcorner a_1(\omega) \rightarrow x(\omega)\urcorner \in T_0\} \mid \{x(\omega) \in T,\ s(x(\omega)) = i\}) \leq$$
$$\leq P(\{s(a_1(\omega)) \in \langle i-K, i+K\rangle\}) = \sum_{j=i-K}^{i+K} P(\{s(a_1(\omega)) = j\}) = \sum_{j=i-K}^{i+K} t_j\ ,$$

and similarly for q_j. $\sum_{j=1}^{\infty} t_j = 1$, hence, $t_j \rightarrow 0$, so $\sum_{j=i-K}^{i+K} tj \leq$ $\leq (2K+1)\ max\ \{tj.: i-K \leq j \leq i+K\} \rightarrow 0$. Q.E.D.

Theorem 2 implies that the most simple case of p_i, q_i, i.e., $p_i = p$, $q_i = q$, $i \in N$ does not meet the other conditions. In order to simplify the situation enough to be able to arrive at some explicit results concerning the computational complexity let us adopt the two following simplifying assumptions concerning the values p_i, q_i.

(C1) $p_i = (2K+1)p.t_i$, $i \in N$,

(C2) $p_i/q_i = p/q$ hence, $q_i = (2K+1)q.t_i$, $i \in N$.

To explain intuivitely (C1) let us remark that in the case of monotonous random variable a_1 when $t_i \geq t_{i+1}$, $i \in N$, we can write $(2K+1)t_{i-K} \geq \sum_{j=i-K}^{i+K} t_j \geq (2K+1)t_{i+K}$, and, roughly, $\sum_{j=i-K}^{i+K} t_j \approx (2K+1)t_i$. Now, (C1) asserts that, under the condition $x(\omega) \in T$, the ratio of successful $a_1(\omega)$'s satisfying the demand $|s(a_1(\omega)) - s(x(\omega))| \leq K$ to all $a_1(\omega)$'s coping with this demand is approximately (in the sense of $\approx$) the same as the ratio of all successful $a_1(\omega)$'s and this latter ratio is nothing else than p. So $p_i | (2K+1)t_i = p/1$, which gives (C1). (C2) asserts that the ratio p_i/q_i equals p/q, i. e., the relativization of our statistical deducibility problem to formulas of complexity i does not change the ratio of the two corresponding probabilities. We would like to emphasize the fact that we realize the unprecise and approximative character of these assumptions and we accept them just with the aim to obtain some explicit results within the limited extend of this paper.

Let us modify the statistical deducibility testing procedure above in such a way that, having p, q and $x(\omega)$, we compute $i = s(x(\omega))$, and try to find an optimum (in the given sense) test of the hypothesis $P_0 = p_i$ against the alternative $P_0 = q_i$. Suppose that we want to have $P_I + P_{II} \leq \varepsilon > 0$ given a priori and that we look for $n(0), m(0)$ which meet this demand using the Tchebyshev inequality. Namely, if $s(x(\omega)) = i$ then $\{\ulcorner a_j(\omega) \rightarrow x(\omega) \urcorner \in T_0\}$ is a random event occuring for each $j \in N$ independently and with the same probability, say P_0. Tchebyshev inequality then reads (cf. Rényi [5], e. g.)

$$P(\{| I_n^*(\omega) - P_0| \geq \varepsilon\}) < \frac{D(I)}{n\varepsilon^2}, \ n \in N,$$

where $I_n^*(\omega) = (\sum_{j=1}^{n} I(a_j(\omega) \rightarrow x(\omega)))n^{-1}$, $D(I)$ is the dispersion of the random variable I. We make precise the decision function d by setting: $d(x(\omega)) = d_1$, if

$$|I_n^*(\omega) - p_i| \leq |I_n^*(\omega) - q_i|, \tag{5}$$

$d(x(\omega)) = d_2$ otherwise.

Random variable I takes the value 1 with probability $P_0 (= p_i$ or $q_i)$ and 0 with probability $1 - P_0$, so $D(I) = P_0(1 - P_0)$. (C2) and $p \geq q$ implies $p_i \geq q_i$, so under the condition $p_i \leq 1/2$, $D(I) \leq p_i(1 - p_i) = K_1 t_i p(1 - K_1 t_i p)$, setting $K_1 = 2K+1$.

The decision procedure can lead to an error only under the condition that the relative frequency $I_n^*(\omega)$ differs from P_0 by more than $1/2|p_i - q_i|$. Hence,

$$P_I + P_{II} \leq P(\{|I_n(\omega) - P_0| \geq 1/2|p_i - q_i|\}) \leq \frac{K_1 t_i p(1 - K_1 t_i p)}{n(1/2|p_i - q_i|)^2} = \frac{4p(1 - K_1 p t_i)}{nK_1 t_i (p-q)^2} < \varepsilon, \tag{6}$$

using (C1). With the aim to meet the condition $P_I + P_{II} \leq \varepsilon$ we set $n(0)$ as the minimal n satisfying the right side of (6) and compute $m(0)$ using this $n(0)$ and (5). Namely, $m(0) = Int((1/2)n(0)(p_i - q_i)) + 1$. Till the end of this paper we shall consider this way of choosing $m(0)$ and $n(0)$ as fixed and integral part of the investigated statistical deducibility testing procedure to which the following computational complexity estimates are related. We leave outside the optimality problem of this decision rule as well as the possibility of its replacement by another way of choosing $m(0)$ and $n(0)$.

4. RESULTS CONCERNING TIME AND SPACE COMPLEXITIES

Consider a time-complexity measure TC and a space-complexity measure SC of the characteristic function I of the set $T_0 \in L$. Formally, TC: $P(L) \times L \times L \to N$; TC$(T_0, x, y)$ expresses the time or the number of some uni operations necessary for deciding whether $\ulcorner x \to y \urcorner \in T_0$ or not. Analogously SC: $P(L) \times L \times L \to N$ measures the storage demands of the same procedure. Let these two complexity measures be also able to quantify the time and sp ce pretentions connected with the process of sampling at random sentences $x(\omega)$, $a_1(\omega)$, $a_2(\omega)$,... (random sampling can be formalized, fro the computational point of view, as a Turing Machine equipped with oracle). We shall suppose, for the sake of simplicity, that the time (space, resp.) complexity connected with sampling at random the tested formula $x(\omega)$ equals a constant tc_x (sc_x, resp.); the same is supposed to be valid for the random sample of formulas $a_j(\omega)$, $j \leq n(0)$, with tc_a (sc_a, resp.). Denote by $b_t(x)$ the time complexity (measured by TC) of the oper tion consisting in (1) enlarging by one the value of an auxiliary vari le v enregistering the number of sampled $a_j(\omega)$'s, (2) enlarging by one the value of an auxiliary variable w enregistering the number of successful $a_j(\omega)$'s, if the last sampled $a_j(\omega)$ was such that $\ulcorner a_j(\omega) \to x(\omega) \urcorner \in T_0$, and finally, (3) deciding whether $w \geq m(0)$ or $v \geq n(0)$, in order to stop the testing procedure. Denote by $b_t(x)$ the time complexity (measured by TC) connected with the computation of $s(x)$ and suppose that $b_t(x) = b_t(y)$ if $s(x) = s(y)$, so we shall use the notation $b_t(i)$, $i \in N$, instead of $b_t(x)$. Symbols b_{os}, $b_s(x)$ and $b_s(i)$ play analogous roles with respect to SC. Moreover, we may suppose that there exist two constants $B_t(T_0)$ and $B_s(T_0)$ such that

$$sup\{TC(T_0, x, y) : x, y \in L\} \leq B_t(T_0),$$
$$sup\{SC(T_0, x, y) : x, y \in L\} \leq B_s(T_0).$$

This assumption is not too restrictive, as the set T_0 is just chosen or defined in such a way that the problem whether an implication $x \to y$ belongs to T_0 or not was decidable "easily and effectively", i.e., within some time and storage limitations. Finally, denote by $TC(x)$ and $SC(x)$ the time and space complexities of the statistical deducibility testing procedure defined above, set

$$TC(i) = sup\{TC(x): x \in L,\ s(x)=i\},$$

and similarly for $SC(i)$. Our aim in the rest of this paper will be to express or estimate the values $TC(i)$ and $SC(i)$ as functions of i, tc_x, tc_a, ... and the other input parameters. The specific character of the complexities TC or SC and the appropriateness of their choice will not be considered in this paper.

Theorem 3. $$TC(i) \leq K_1 \cdot \frac{1}{t_i} + K_2 + b_t(i), \tag{7}$$

where $K_1 = Int(\frac{4p}{(2K+1)(p-q)^2\varepsilon})(B_t(T_0) + tc_a + b_{ot})$

and $K_2 = B_t(T_0) + tc_a + tc_x + b_{ot}$

are constants, i. e., do not depend on x.

Proof. In order to test $x(\omega)$ we need (1) to sample it and to compute $s(x(\omega)) = i$, i.e., $tc_x + b_t(i)$ time units, then (2) to repeat at most $n(0)$-times the process of sampling $a_j(\omega)$, verifying $\ulcorner a_j(\omega) \to x(\omega)\urcorner \in T_0$ and enregistering the result; these three operations request $B_t(T_0)+tc_a+b_{ot}$ time units, so

$$TC(i) \leq n(0)(B_t(T_0) + tc_a + b_{ot}) + tc_x + b_t(i),$$

where, by (6)

$$n(0) = Int(\frac{4p}{(2K+1)t_i(p-q)^2\varepsilon}) + 1. \tag{8}$$

An easy calculation gives the result. Q.E.D.

Theorem 4.

$$SC(i) \leq sc_x + sc_a + B_s(T_0) + b_{os} + b_s(i) + 4c_1 log_2(Int\frac{4p}{(2K+1)t_i(p-q)^2\varepsilon}) + 1), \tag{9}$$

where c_1 is the space complexity (measured by SC) of one-bit unit of storage (hence, if SC counts just the bits, then $c_1=1$). The right side of (9) can be asymptotically expressed as

$$K_1^* + K_2^* log_2 (1/t_i) + b_s(i), \tag{10}$$

where $K_2^* = 4c_1$ and

$$K_1^* = sc_x + sc_a + B_s(T_0) + b_{os} + 4c_1\ log_2(\frac{4p}{(2K+1)(p-q)^2\varepsilon}).$$

Proof. The value sc_x (sc_a, resp.) expresses the storage necessary for sampling and inscribing $x(\omega)$ ($a_j(\omega)$, resp.), $B_s(T_0)$ is the space necessary for deciding whether $\ulcorner a_j(\omega) \to x(\omega)\urcorner \in T_0$, and b_{os} is the space occupied by the procedure modifying the values v, w and comparing them with $m(0)$, $n(0)$. $b_s(i)$ space units are necessary for computing $s(x(\omega)) = i$. Finally, we need some space to enregister the four integers v, w, $m(0)$,

$n(0)$. As none of them exceeds $n(0), 4.\log_2(n(0))$ bits will do. Using (8) we obtain (9). (10) follows immediately. Q.E.D.

Theorem 5. Let there exist two functions f, g, such that $t_i^{-1} \leq f(i) = \sum_{K=0}^{m} d_K i^K$, $b_t(i) \leq g(i) = \sum_{K=0}^{m} \beta_K i^K$, $m, n, \alpha_0, \ldots, \alpha_m$, $\beta_0, \ldots, \beta_m$, $i \in N$.
Then there exists a function h polynomial in i and such that $TC(i) \leq h(i)$, $i \in N$.

Proof. Set $h(i) = (Int\ K_1+1)f(i) + (Int\ K_2+1)g(i)$ Q.E.D.

Roughly said, Theorem 5 shows that under some conditions the time complexity of the investigated statistical deducibility testing procedure is of polynomial type. This result can serve as an interesting counterpart of the well-known fact that the computational complexity of deterministic theorem-proving procedures is at least of exponential type. This complexity reduction has been achieved, of course, by admitting a positive probability with which the decision taken about the tested formula can be wrong. Such a fact can be seen as an illustrative example verifying the well-known M. O. Rabin's thesis [4] that the admission of the possibility of an error can reduce substantially (in the sense of type reduction the computional complexity of problem-solving algorithms. Let us note that a probability distribution $\{t_i\}$ coping with the demand of Theorem 5 can be easily obtained; set e.g., $t_0=0$, $t_i=c/i^2$, $c=\sum_{i=1}^{\infty}(1/i)^2=\pi^2/6$, $i=1,2,\ldots$. On the other hand, e.g., geometric distribution with $t_i=(1-\lambda)\lambda^i$, $0<\lambda<1$, $i \in N$ does not satisfy the condition.

Theorem 6.

(a) Let f, g be as in Theorem 5, let $t_i^{-1} \leq f(i)$, $b_t(i) \leq \log_2 g(i)$, $i \in N$. Then there exist $c_2, c_3 \in N$ such that $SC(i) \leq c_2 \log_2 i + c_3$.

(b) Let $t_i^{-1} \leq \gamma_1 exp(\gamma_2 i)$, $b_s(i) \leq \delta_1 exp(\delta_2 i)$, $\delta_j, \gamma_j \in N$, $j=1,2$, $i \in N$. Then there exist $c_4, c_5 \in N$ such that $SC(i) \leq c_4 i + c_5$, $i \in N$.

Proof. Simple substitutions into Theorem 4. Q.E.D.

Hence, the space complexity reduction is still more remarkable then the time reduction. This result seems to be interesting namely in the light of the fact that just the space complexity problems (population explosion) have challenged the question of practical applicability of deterministic theorem-proving algorithms.

References:

1. Ch. L. Chang, R. C. T. Lee: Symbolic Logic and Mechanical Theorem Proving. Academic Press, New York, London, 1973.

2. I. Kramosil: Statistical Approach to Proof Theory. Appendix to Kybernetika (Prague), vol. 15, 1979.

3. I. Kramosil, J. Šindelář: Statistical Deducibility Testing with Stochastic Parameters. Kybernetika 14 (1978), 6, 385-396.

4. M. O. Rabin: Theoretical Impediments to Artificial Intelligence. In: Information Processing 1974, North Holland Publ. Comp., Amsterdam, 1974, pp. 615-619.

5. A. Rényi: Probability Theory. Akadémiai Kiadó, Budapest, 197o.

6. A. Špaček: Statistical Estimation of Provability in Boolean Logics. In: Transactions of the Second Prague Conference on Information Theory,..., Prague, 1959, NČSAV (Publishing House of the Czech. Acad. of Sci.), 196O, pp. 6O9-626.

7. O. Štěpánková, I. M. Havel: A Logical Theory of Robot Problem Solving. Artificial Intelligence 7 (1976), pp. 129-161.

CONTEXT FREE NORMAL SYSTEMS

Manfred Kudlek
Fachbereich Informatik
Universität Hamburg
Schlüterstraße 70
D-2000 Hamburg 13

1. Introduction

Post Normal Systems (or Normal Systems, abbreviated by NS) are rewriting systems with rules of the form $aw \rightarrow wb$ where $a \in V^+$, $b \in V^*$. They have been introduced by E. Post (1) who also proved that such systems generate the same class of languages as Semi-Thue Systems (STS) do which have rules of the form $w_1aw_2 \rightarrow w_1bw_2$ namely the class of recursively enumerable languages (RE).
With a similar proof it may be shown that Normal Systems with $|a| \leq |b|$ (called Context Sensitive Normal Systems, CSNS) generate the class of Context Sensitive languages (CS). Context Free Normal Systems (CFNS) however, generate a larger class than Context Free languages (CF). CFNS's have very simple rewriting rules since derivations are performed at the first symbol of a word. They also have some close relations to Lindenmayer Systems. The generated class is lying between EOL and ETOL languages.

2. Definitions and Basic Properties

It is assumed that the notations of L systems are known (2,3).

D1 Normal System

A Normal System (NS) is a formal system $G=(V,A,R)$ with a finite alphabet V, an axiom $A=\{\omega\} \in V^*$, and a finite set R of rules of the form $r : aw \rightarrow wb$ (written $a \rightarrow b$) with $a \in V^+$, $b \in V^*$ and w arbitrary.

A NS is called Context Sensitive (CSNS) if $|a| \leq |b|$ or $S \rightarrow \varepsilon$ where S is the axiom and doesn't occur on any right side of a rule.

A NS is called Context Free (CFNS) if $|a| = 1$. Note that ε-rules are allowed in CFNS's, and that $x \rightarrow x$ is not a trivial rule.

A derivation is defined as usually by

$w \Rightarrow w'$ iff $w=av$, $w'=vb$, $a \to b \in R$

$w \overset{n}{\Rightarrow} w'$ iff $\exists\, w=w_o,w_1,\dots,w_n=w' \in V^* : w_{i-1} \Rightarrow w_i$, $0 \leq i < n$

$w \overset{*}{\Rightarrow} w'$ iff $w=w'$ or $\exists\, n \in \mathbb{N} : w \overset{n}{\Rightarrow} w'$.

As for L systems the letters E, P, D, F will be used to denote extended, propagating, deterministic, and finite axiom set systems respectively :

E : $V = V_N \cup V_T$, $V_N \cap V_T = \emptyset$ (notation $G = (V_N,V_T,A,R)$)

P : $|a| \leq |b|$

F : $|A| \geq 1$

D : if $x \to b_1$, $x \to b_2 \in R$ then $b_1 = b_2$ (for CFNS's only)

The language generated by such systems is defined by

$L(G) = \{w \in V^* \mid \exists w_o \in A : w_o \overset{*}{\Rightarrow} w\}$ for systems without E

$L(G) = \{w \in V^* \mid \exists w_o \in A : w_o \overset{*}{\Rightarrow} w\} \cap V_T^*$ for systems with E

The classes of languages generated by such systems will be denoted by underlining.

<u>E1</u> $G = (\{a,b\},\{a\},\{a\},\{a \to bb, b \to a\}) \in$ EPDCFNS

with $L(G) = \{a^{2^n} \mid n \geq 0\} \in$ <u>EPDCFNS</u> .

<u>D2</u> Cyclic permutation

If $L \subset V^*$, then $Cyc(L) = \{w \in V^* \mid \exists u,v \in V^* : w = uv \text{ , } vu \in L\}$

is the set of all cyclic permutations of words from L .

For completeness it will be shown in the following theorems that <u>ENS</u> = <u>ESTS</u> = <u>RE</u> and <u>ECSNS</u> = <u>ECSSTS</u> = <u>CS</u> . All rules are written in the short form $a \to b$ for NS's as well as for STS's.

<u>T1</u> <u>ENS</u> $\subset$ <u>ESTS</u> .

Proof : Let $G = (V_N,V_T,A,R) \in$ ENS with $A = \{S\}$.

Define $G' = (V_N',V_T,A',R') \in$ ESTS by $A' = \{S'\}$,

$V_N' = V_N \cup \{S',A,B,C,D,E,F,G,H\} \cup \{T_i \mid 1 \leq i \leq |R|\}$, and $R' = R_1' \cup R_2'$ with

$R_1' = \{S' \to ESF,\ Ea_i \to AT_i\ (\ a_i \to b_i \in R \text{ , } 1 \leq i \leq |R|\ ),\ T_i x \to xT_i\ (\ 1 \leq i \leq |R| \text{ , } x \in V\ ),$
$T_iF \to Bb_iF\ (\ 1 \leq i \leq |R|\ ),\ xB \to Bx\ (\ x \in V\ ),\ AB \to E\}$

$R_2' = \{E \to CD,\ Dx \to xD\ (\ x \in V_T\ ),\ Dx \to Gx\ (\ x \in V_N\ ),\ DF \to H,\ xH \to Hx\ (\ x \in V_T),$
$DH \to \varepsilon,\ xG \to Gx\ (\ x \in V\ ),\ DG \to E\}$

Then R_1' generates all and only such $w \in L_{SF}(G) = \{w \in V^* \mid S \overset{*}{\Rightarrow} w\}$ in the form EwF . R_2' produces from these all and only such $w \in L(G)$. Therefore $L(G') = L(G)$, hence <u>ENS</u> $\subset$ <u>ESTS</u> .

T2 ESTS ⊂ ENS .

Proof : Let $G = (V_N, V_T, A, R) \in ESTS$ with $A = \{S\}$.

Define $G' = (V_N', V_T, A', R') \in ENS$ by $A' = \{S'\}$,

$V_N' = \bar{V} \cup \{S', A, B, C, K, L, M, N, P\}$ and $R' = R_1' \cup R_2' \cup R_3'$ with

$R_1' = \{\bar{x} \to \bar{x}\ (x \in V),\ B \to B\}$,

$R_2' = \{S' \to E\bar{S},\ E \to BA,\ A\bar{x} \to \bar{x}A\ (x \in V),\ A\bar{a} \to \bar{b}C\ (a \to b \in R),\ AB \to E,$
$C\bar{x} \to \bar{x}C\ (x \in V),\ CB \to E\}$,

$R_3' = \{E \to PK,\ P \to L,\ K\bar{x} \to xMK\ (x \in V_T),\ LxM \to LxM\ (x \in V_T),\ xM \to xM\ (x \in V_T),$
$KLx \to Nx\ (x \in V_T),\ Mx \to x\ (x \in V_T),\ MN \to \varepsilon\}$.

Then R_1', R_2' generate all and only such $w \in L_{SF}(G)$ in the form $E\bar{w}$, and R_3' transforms all and only such $w \in L(G)$ from $E\bar{w}$ into w. Therefore $L(G') = L(G)$, hence ESTS ⊂ ENS .

T3 ECSNS ⊂ ECSSTS .

Proof : Let $G = (V_N, V_T, A, R) \in ECSNS$ with $A = \{S\}$.

Define $G' = (V_N', V_T, A', R') \in ECSSTS$ by $A' = \{S_{EF}\}$,

$V_N' = V_N \cup V_{EF} \cup V_E \cup V_F \cup V_A \cup V_B \cup \bar{V} \cup \{C_r \mid r \in R\}$, and $R' = R_1' \cup R_2'$ with

$R_1' = \{x_{EF} \to y_{EF}\ (x \to y \in R),\ x_{EF} \to y_E a z_F\ (x \to yaz \in R),$
$x_E a y_F \to u_E b z_F\ (xay \to ubz \in R),\ x_E a y_F \to y_E b z_F\ (xa \to bz \in R),$
$x_E a y \to y_A C_r^{n(r)}\ (r : xa \to bz \in R,\ |bz| = n(r)),$
$C_r^{n(r)} x \to x C_r^{n(r)}\ (r \in R,\ x \in V),\ C_r^{n(r)} x_F \to x_B b z_F\ (r \in R,\ x \in V),$
$x y_B \to x_B y\ (x, y \in V),\ x_A y_B \to x_E y\ (x, y \in V)\}$,

$R_2' = \{x_{EF} \to x\ (x \in V_T),\ x_E y_F \to xy\ (x, y \in V_T),\ x_E y \to x\bar{y}\ (x, y \in V_T),$
$\bar{x} y \to x\bar{y}\ (x, y \in V_T),\ \bar{x} y_F \to xy\ (x, y \in V_T)\}$.

Then R_1' generates all and only such $w \in L_{SF}(G)$ in the forms x_{EF} or $x_E v y_F$. R_2' produces from these all and only such $w \in L(G)$. Therefore $L(G') = L(G)$, hence ECSNS ⊂ ECSSTS .

T4 ECSSTS ⊂ ECSNS .

Proof : Let $G = (V_N, V_T, A, R) \in ECSSTS$ with $A = \{S\}$.

Define $G' = (V_N', V_T, A', R') \in ECSNS$ by $A' = \{S_E\}$,

$V_N' = V_N \cup \bar{V} \cup V_A \cup V_B \cup V_C \cup V_E$, and $R' = R_1' \cup R_2' \cup R_3'$ with

$R_1' = \{\bar{x} \to \bar{x}\ (x \in V),\ x_A \to x_A\ (x \in V)\}$,

$R_2' = \{x_E \bar{a} \to y_E \bar{b}\ (xa \to yb \in R),\ x_E \bar{y} \to x_A y_B\ (x, y \in V),\ x_B \bar{y} \to \bar{x} y_B\ (x, y \in V),$
$x_B y_A \to \bar{x} y_E\ (x, y \in V),\ x_B \bar{a} \to y_C \bar{b}\ (xa \to yb \in R),\ x_C \bar{y} \to \bar{x} y_C\ (x, y \in V),$
$x_C y_A \to \bar{x} y_E\ (x, y \in V)\}$,

$R_3' = \{x_E \to x\ (x \in V_T),\ \bar{x} \to x\ (x \in V_T)\}$.

Then R_1', R_2' generate all and only such $w \in L_{SF}(G)$ in the form $x_E \bar{v}$, and R_3' transforms all and only such $w \in L(G)$ from those into xv .

Therefore $L(G') = L(G)$ and hence ECSSTS ⊂ ECSNS .

<u>T5</u> <u>ENS</u> = <u>ESTS</u> = <u>RE</u> and <u>ECSNS</u> = <u>ECSSTS</u> = <u>CS</u> .

Proof : By <u>T1</u> to <u>T4</u>

<u>T6</u> <u>EFXNS</u> = <u>EXNS</u> ($X \in \{\epsilon, CS, CF\}$).

Proof : If $G = (V_N, V_T, A, R)$ and $|A| \geq 1$ define $G' = (V_N \cup \{S\}, V_T, \{S\}, R')$ with $R' = R \cup \{S \rightarrow a \mid a \in A\}$. Then $G' \in$ EXNS and $L(G') = L(G)$.

From now on Context Free Normal Systems will be considered only.
The next theorem shows relations between CFNS derivations and OL derivations.

<u>T7</u> To each CFNS derivation $w_o \overset{*}{\Rightarrow} w$ (CFNS) by $G \in$ XCFNS (X arbitrary) there exists a uniquely determined integer k and another derivation using for the first k steps the rules R of G as XOL rules ($w_o \overset{*}{\Rightarrow} w_k$ (XOL), $w_k = ab$) and after that using them as XCFNS rules for $m = |a| < |ab|$ steps ($ab \overset{m}{\Rightarrow} bc$ (XCFNS), $w = bc$).

Proof : If XCFNS rules are applied $|u|$ times on a word $u \in V^*$, then this may also be achieved by one XOL step.

Now, if $w_o \overset{*}{\Rightarrow} w$ (XCFNS) then there exist uniquely determined $w_o, w_1, \ldots, w_k \in V^*$ with $w_{i-1} \Rightarrow w_i$ (XOL) , $w_{i-1} \overset{m(i)}{\Rightarrow} w_i$ (XCFNS) ,

$m(i) = |w_{i-1}|$ ($1 \leq i \leq k$) , $w_k \overset{m(k)}{\Rightarrow} w$ (XCFNS) ,

$m(k) = |a| < |ab|$, $w_k = ab$, $w = bc$.

This fact is shown by the following figure :

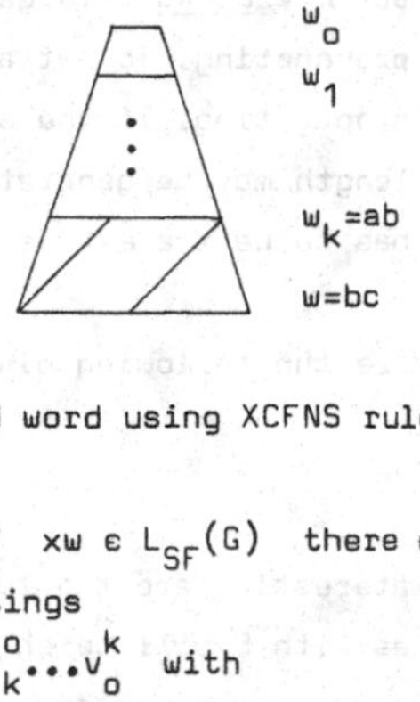

The next theorem traces back the first symbol of a derived word using XCFNS rules.

<u>T8</u> Let $G \in$ XCFNS with arbitrary X. To each derivation of $xw \in L_{SF}(G)$ there exist $w_o \in A$, an integer k, and a unique sequence of splittings

$w_o = u_o^o x_o v_o^o$, $w_1 = u_o^1 u_1^o x_1 v_1^o v_o^1$, ... , $w_k = u_o^k \ldots u_k^o x_k v_k^o \ldots v_o^k$ with

$x_i \in V$, $u_i^j \Rightarrow u_i^{j+1}$ (XOL) , $v_i^j \Rightarrow v_i^{j+1}$ (XOL) , $x_i \Rightarrow u_{i+1}^o x_{i+1} v_{i+1}^o$ (XOL) and

$u_o^k \ldots u_k^o x_k v_k^o \ldots v_o^k \overset{m}{\Rightarrow} x_k v_k^o \ldots v_o^k u_o^{k+1} \ldots u_k^1$ (XCFNS) , $m = |u_o^k \ldots u_k^o|$,

$xw = x_k v_k^o \ldots v_o^k u_o^{k+1} \ldots u_k^1$.

Proof : This is an immediate consequence from T7.

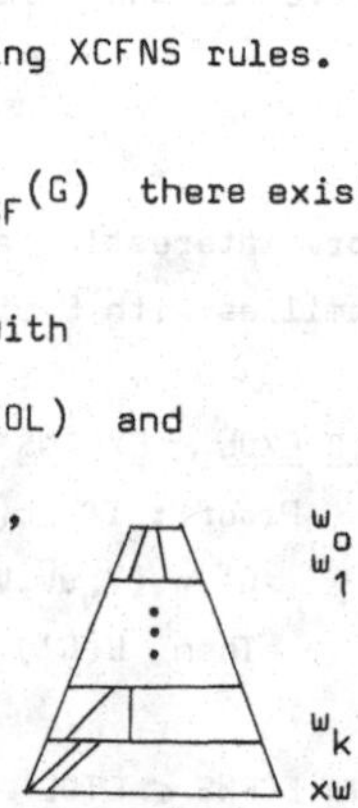

<u>T9</u> If $\epsilon \in L_{SF}(G)$ then to each derivation of ϵ there exist $w_o \in A$, an integer k, and a unique sequence of splittings $w_o = u_o^o x_o v_o^o$, ... ,

$w_{k-1} = u_{k-1}^o x_{k-1} v_{k-1}^o \ldots v_o^{k-1}$, $w_k = x_k v_k^o \ldots v_o^k$ with

$u_i^o \Rightarrow \epsilon$, $x_i \Rightarrow u_{i+1}^o x_{i+1} v_{i+1}^o$, $v_i^j \Rightarrow v_i^{j+1}$, $w_k \Rightarrow \epsilon$ (XOL) .

Proof : As in <u>T8</u>.

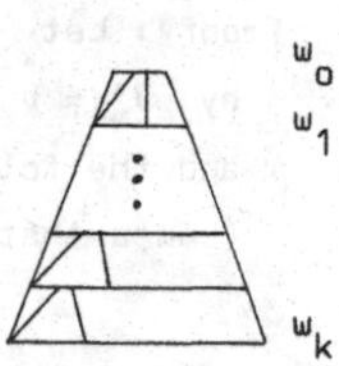

3. Hierarchies and Closure Properties

T10 For $G \in XCFNS$ with arbitrary X it is decidable whether $\varepsilon \in L(G)$.

Proof : Define $M_0 = \{x \in V \mid x \to \varepsilon \in R\}$ and

$M_{i+1} = M_i \cup \{y \in V \mid \exists u \in M_i^* : y \to u \in R\}$ inductively.

Then $M = M_k = M_{k+1}$ for all $l \in \mathbb{N}$ with $k = |V|$ and $x \in M$ iff $x \overset{*}{\Rightarrow} \varepsilon$.

Now, if $a = x_1 \ldots x_m \in A$ then $a \overset{*}{\Rightarrow} \varepsilon$ iff $x_i \overset{*}{\Rightarrow} \varepsilon$ ($1 \le i \le m$) .

Therefore it is decidable whether $\varepsilon \in L(G)$.

Only a few results concerning systems without E will be shown.

T11 DCFNS $\not\subset$ PCFNS .

Proof : $L = \{\varepsilon, a\} \in$ DCFNS by $G = (\{a\},\{a\},\{a \to \varepsilon\}) \in$ DCFNS .

But $L \notin$ PCFNS .

T12 PCFNS $\not\subset$ DCFNS .

Proof : $L = \{a,b\}^+ \in$ PCFNS by $G = (\{a,b\},\{a\},\{a \to a, a \to b, a \to aa, b \to a, b \to b\})$

But $L \notin$ DCFNS . To get words of arbitrary length one rule has to be strictly propagating. To get all words of one length not both rules can be strictly propagating. If the second rule is length preserving at most $|w|$ words of that length may be generated. But $2^{|w|}$ such words exist. Therefore the second rule has to be the ε-rule generating ε in contradiction to L.

Therefore the following diagram holds :

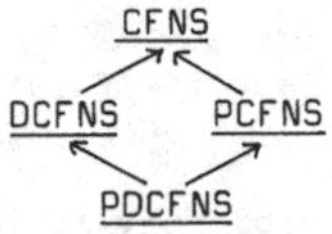

More interesting are the language families with E. Their relations to OL families with E will be shown.

T13 EXOL $\subset$ EXCFNS for $X \in \{\varepsilon, P, D, F, PD, PF, DF, PDF\}$.

Proof : If $L(G) \in$ EXOL with $G = (V_N, V_T, A, R)$ define $G' \in$ EXCFNS by

$G' = (V_N \cup \bar{V}, V_T, A, \bar{R} \cup R')$ with $\bar{R} = \{x \to \bar{a} \mid x \to a \in R\}$, $R' = \{\bar{x} \to x \mid x \in V\}$.

Then $L(G') = L(G)$.

T14 ECFNS $\subset$ ETOL .

Proof : Let $G = (V_N, V_T, \{S\}, R) \in$ ECFNS . Define $G' = (V_N', V_T, \{S'\}, \mathcal{T}') \in$ ETOL

by $V_N' = \hat{V} \cup \bar{V} \cup \bar{\bar{V}} \cup V_N \cup \{S', A\}$

and the following tables where for shortness all trivial rules are omitted :

(note that $\varepsilon \in L(G)$ is decidable by T10)

$\tau' = \{ \boxed{S' \to \hat{S}A} \}$

$\cup \{ \boxed{\hat{x} \to \hat{y}\bar{b} \; , \; A \to \bar{\bar{a}}A \; , \; \bar{z} \to \bar{c} \; (z \to c \in R) \; , \; \bar{\bar{z}} \to \bar{\bar{c}} \; (z \to c \in R)} \; |$

$\quad x \to ayb \in R \; , \; y \in V$ (each such splitting) $\}$

$\cup \{ \boxed{\hat{x} \to x \; (x \in V) \; , \; \bar{x} \to x \; (x \in V) \; , \; \bar{\bar{y}} \to a \; (y \to a \in R) \; , \; A \to \varepsilon} \}$

$\cup \{ \boxed{\hat{x} \to \varepsilon \; (x \in V) \; , \; \bar{x} \to \varepsilon \; (x \in V) \; , \; \bar{\bar{x}} \to \varepsilon \; (x \in V) \; , \; A \to \varepsilon} \}$ if $\varepsilon \in L(G)$.

The first table is the starting table, and the last one (or two if $\varepsilon \in L(G)$) are the terminal tables.

By T8 to each derivation of xw there exists a sequence $S = w_0$, $w_1 = u_1^0 x_1 v_1^0$, ... , $w_k = u_1^{k-1} \ldots u_k^0 x_k v_k^0 \ldots v_1^{k-1}$, $xw = x_k v_k^0 \ldots v_1^{k-1} u_1^k \ldots u_k^1$.

Using the tables the following sequence may be constructed :

S' , $\hat{S}A$, $\hat{x}_1 \bar{v}_1^0 \bar{\bar{u}}_1^0 A$, ... , $\hat{x}_k \bar{v}_k^0 \ldots \bar{v}_1^{k-1} \bar{\bar{u}}_1^{k-1} \ldots \bar{\bar{u}}_k^0 A$, $x_k v_k^0 \ldots v_1^{k-1} u_1^k \ldots u_k^1$

On the other hand, to each derivation of w using tables there exists a corresponding derivation in the way of T8. Therefore $L(G') = L(G)$.

If $\varepsilon \in L(G)$ the same holds using one of the two terminal tables and T9.

Hence ECFNS $\subset$ ETOL . The derivations are illustrated by the following two diagrams :

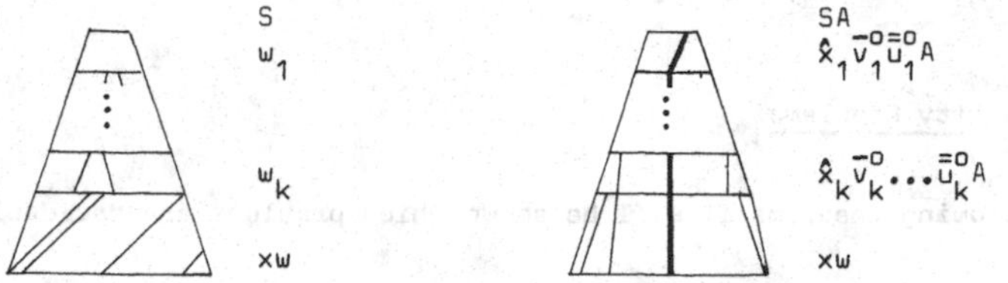

15 EPCFNS $\subset$ EPTOL .

Proof : Let $G = (V_N, V_T, \{S\}, R) \in$ EPCFNS . Define $G' = (V_N', V_T, \{S'\}, \tau') \in$ EPTOL by $V_N' = \tilde{V} \cup \hat{V} \cup \bar{V} \cup \bar{\bar{V}} \cup \check{V} \cup \mathring{V} \cup V_N \cup \{S'\}$

and the following tables, where the trivial rules are omitted again :

$\tau' = \{ \boxed{S' \to \tilde{S}} \}$

$\cup \{ \boxed{\tilde{x} \to \tilde{y}} \; | \; x \to y \in R \}$

$\cup \{ \boxed{\tilde{x} \to \hat{y}\bar{b}\bar{\bar{c}}\check{z} \; (\text{if } a = cz) \; , \; \tilde{x} \to \hat{y}\bar{d}\mathring{z} \; (\text{if } a = \varepsilon \; , \; b = dz)} \; |$

$\quad x \to ayb \in R \; , \; y \in V$ (each such splitting) $\}$

$\cup \{ \boxed{\begin{array}{l} \hat{x} \to \hat{y}\bar{b} \; , \; \check{z} \to \check{z} \; (\text{if } a = \varepsilon) \; , \; \check{z} \to \bar{\bar{z}}\bar{\bar{c}}\check{u} \; (\text{if } a = cu) \; , \; \mathring{z} \to \mathring{z} \; (\text{if } a = \varepsilon) \; , \\ \mathring{z} \to \bar{\bar{z}}\bar{\bar{c}}\check{u} \; (\text{if } a = cu) \; , \; \bar{z} \to \bar{a} \; (z \to a \in R) \; , \; \bar{\bar{z}} \to \bar{\bar{a}} \; (z \to a \in R) \end{array}} \; |$

$\quad x \to ayb \in R \; , \; y \in V$ (each such splitting) $\}$

$\cup \{ \boxed{\hat{x} \to x \; (x \in V) \; , \; \bar{x} \to x \; (x \in V) \; , \; \bar{\bar{z}} \to a \; (z \to a \in R) \; , \; \check{z} \to a \; (z \to a \in R)} \}$

Again, the first table is the starting table and the last one the terminating one. As in T14 it follows that $L(G') = L(G)$, hence EPCFNS $\subset$ EPTOL .

T16 EDCFNS $\subset$ EDTOL and EPDCFNS $\subset$ EPDTOL .

Proof : This is shown exactly as in T14 and T15 .

The next theorems show closure properties of classes with E. The classes without E will not be considered here. Most of those classes are not closed under AFL operatior

T17 EPCFNS is closed under ε-free homomorphism $h_\varepsilon : V_1^* \to V_2^*$.

Proof : Let $G = (V_N,V_T,A,R) \in$ EXCFNS . Then define $G' = (\bar{V}_1,V_2,\bar{A},R')$ by $V_1 = V_N \cup V_T$, and $R' = \{\bar{x}\to\bar{a} \mid x\to a \in R\} \cup \{\bar{x}\to h_\varepsilon(x) \mid x \in V_T\}$

Then $L(G') = L(G)$.

T18 EXCFNS is closed under union $\cup$ for $X \in \{\varepsilon,P\}$.

Proof : If $G_i \in$ EXCFNS with $G_i = (V_{Ni},V_{Ti},A_i,R_i)$ define $G \in$ EXCFNS by $G = (V_N,V_T,A,R)$ with $V_N = \bar{V}_1 \cup \bar{\bar{V}}_2 \cup \{S\}$, $V_T = V_{T1} \cup V_{T2}$, $A = \{S\}$,
$R = \{S\to\bar{a} \mid a \in A_1\} \cup \{S\to\bar{\bar{a}} \mid a \in A_2\} \cup \{\bar{x}\to\bar{a} \mid x\to a \in R_1\} \cup \{\bar{\bar{x}}\to\bar{\bar{a}} \mid x\to a \in R_2\} \cup$
$\cup \{\bar{x}\to x \mid x \in V_{T1}\} \cup \{\bar{\bar{x}}\to x \mid x \in V_{T2}\}$.

Then $L(G) = L(G_1) \cup L(G_2)$.

4. Decidability Problems

In the following theorems it will be shown which problems are decidable and which not

T19 The membership problem is decidable for ECFNS ($w \in L(G)$) .

Proof : This is a consequence from T14 and ETOL c CS .

T20 The emptiness problem is decidable for ECFNS ($L(G) = \emptyset$) .

Proof : This is also a consequence from T14 and ETOL c IND (3).

T21 The finiteness problem is decidable for ECFNS ($L(G)$ finite) .

Proof : This too is a consequence from T14 and ETOL c IND (3).

T22 The language equivalence problem is not decidable for CFNS ($L(G_1) = L(G_2)$) .

Proof : This may be shown by a proof similar to that for OL in (2).

Let $(a_1,\dots,a_n)$, $(b_1,\dots,b_n)$ with $a_i,b_i \in V^*$ an instance of a Post Correspondence Problem (PCP). Let $\overleftarrow{w}$ denote the mirror word to w .

Then define $G_1 = (V_1,\{S\},R_1)$ by $V_1 = V \cup \{A,B,C,D,E,F,H,S,T\}$,

$R_1 = \{S\to TH,\ T\to a_iE\overleftarrow{b}_i\ (1\leq i\leq n),\ T\to xAx\ (x \in V),\ T\to xBy\ (x,y \in V,\ x\neq y),\ A\to A,$
$A\to xAx\ (x \in V),\ A\to xBy\ (x,y \in V,\ x\neq y),\ A\to xC\ (x \in V),\ A\to Fx\ (x \in V),$
$B\to xB\ (x \in V),\ B\to Bx\ (x \in V),\ B\to B,\ C\to xC\ (x \in V),\ B\to D,\ C\to C,\ C\to D,$
$D\to D,\ E\to E,\ E\to a_iE\overleftarrow{b}_i\ (1\leq i\leq n),\ F\to F,\ F\to Fx\ (x \in V),\ F\to D,\ H\to H,\ x\to x\}$

and $G_2 = (V_2, \{S\}, R_2)$ by $V_2 = V_1$, $R_2 = R_1 \cup \{E \to D\}$. Then

$$L(G_1) = \{S, TH\} \cup$$

$$\cup\ Cyc(\ \{ A\overleftarrow{w}Hw \mid w \in V^+\} \cup \{D\overleftarrow{u}Hw \mid w,u \in V^+,\ w \neq u\}$$

$$\cup \{ Bub\overleftarrow{w}Hwaz \mid w,u,z \in V^*,\ a,b \in V,\ a \neq b\}$$

$$\cup \{ C\overleftarrow{w}Hwz \mid w,z \in V^+\} \cup \{Fz\overleftarrow{w}Hw \mid w,z \in V^+\}$$

$$\cup \{ E\overleftarrow{u}Hw \mid w=a_{i(1)} \cdots a_{i(t)}\ ,\ u=b_{i(1)} \cdots b_{i(t)}\ ,\ t \geq 1\ ,\ 1 \leq i(j) \leq n \}\ ),$$

$$L(G_2) = L(G_1) \cup Cyc(\ \{ D\overleftarrow{u}Hw \mid w=a_{i(1)} \cdots a_{i(t)}\ ,\ u=b_{i(1)} \cdots b_{i(n)}\ ,\ t \geq 1,\ 1 \leq i(j) \leq n\}\ )$$

and $L(G_1) \neq L(G_2)$ iff PCP has a solution.

Therefore the language equivalence problem is undecidable for CFNS.

23 If $L_1, L_2 \in$ CFNS it is undecidable whether $L_1 \cap L_2 = \emptyset$.

Proof : Let $(a_1, \ldots, a_n)$, $(b_1, \ldots, b_n)$ be an instance of a PCP.

Then define $G_1 = (V_1, \{S\}, R_1)$ by $V_1 = V \cup \{1, \ldots, n\} \cup \{S, A, C\}$

$R_1 = \{S \to A,\ A \to iAa_i,\ A \to iCa_i,\ i \to i\ (\ 1 \leq i \leq n\ ),\ x \to x\ (x \in V),\ A \to A,\ C \to C\}$

and $G_2 = (V_2, \{T\}, R_2)$ by $V_2 = V \cup \{1, \ldots, n\} \cup \{T, B, C\}$

$R_2 = \{T \to B,\ B \to iBb_i,\ B \to iCb_i,\ i \to i\ (\ 1 \leq i \leq n\ ),\ x \to x\ (x \in V),\ B \to B,\ C \to C\}$

Then $L(G_1) = \{S, A\} \cup Cyc(\ \{ uAv \mid u=i(t) \ldots i(1),\ v=a_{i(1)} \cdots a_{i(t)},\ t \geq 1\}$
$\cup \{ uCv \mid u=i(t) \ldots i(1),\ v=a_{i(1)} \cdots a_{i(t)},\ t \geq 1\}\ )$

and $L(G_2) = \{T, B\} \cup Cyc(\ \{ uBv \mid u=i(t) \ldots i(1),\ v=b_{i(1)} \cdots b_{i(t)},\ t \geq 1\}$
$\cup \{ uCv \mid u=i(t) \ldots i(1),\ v=b_{i(1)} \cdots b_{i(t)},\ t \geq 1\}\ )$

Therefore $L(G_1) \cap L(G_2) \neq \emptyset$ iff PCP has a solution.

Hence $L_1 \cap L_2 = \emptyset$ is undecidable for CFNS.

5. Open problems

Several problems concerning ECFNS are left open. Among them are closure properties under catenation, catenation closure, inverse homomorphism, and intersection with regular sets. It is also unknown if there exists a Normal Form theorem as the Chomsky Normal Form for CF. Further open problems are the corresponding automata.

6. References

(1) E. L. Post, Formal reduction of the general combinatorial decision problem, AJM 65 (1943), pp. 197-215 .

(2) G. T. Herman, G. Rozenberg, Developmental Systems and Languages, North Holland, Amsterdam , 1975 .

(3) L Systems, LNCS 15, Springer, Berlin , 1974 .

NEW PROOFS FOR JUMP DPDA´S

Matti Linna

Martti Penttonen

Department of Mathematics, University of Turku

20500 Turku 50, Finland

INTRODUCTION

The deterministic context-free languages (see e.g. Harrison [5]) constitute an important subfamily of the context-free languages. The need of good parsing algorithms and the intriguing equivalence problem have lead to the investigation of various subfamilies and various characterizations of this family. The deterministic pushdown automata with jumps, jump dpda´s in short, establish a useful device of characterizing the deterministic languages.

A jump dpda is a dpda with the additional facility of erasing the pushdown memory until a symbol from a set specified by the transition is found, all this at one step. Thus, in erasing situations, these jump transitions can speed up the computation. It is easy to see that, if no restrictions on the computation time are settled, the jump transitions do not increase the generating capacity. But it has been shown in [3] and will be shown here, that all deterministic languages can be accepted by jump dpda´s in realtime. The usefulness of jump dpda´s has been demonstrated e.g. by the simplified regularity algorithm of deterministic languages [2].

In spite of the importance of jump dpda´s, the proofs of the basic results cannot be found in the litterature. Some results can be found in [4] but without proofs, and [3] is unpublished. In [1] it is proved that all deterministic languages can be accepted in realtime by somewhat different jump dpda´s. Various results can be found in [2] including Theorem 2 of this paper. Here we give a short, self-contained and, we hope, transparent proof of that theorem.

We shall show that all prefix-free deterministic context-free languages can be accepted in realtime by one-state jump dpda´s whose jump sets contain only a single symbol. This is what our Theorem 2 says. It was first proved in [2]. The proof of

Theorem 2 is strongly dependent on the constructions of Theorem 1. Theorem 1 would follow from Proposition 1 (1), proved in [3], and Proposition 2 of [2], but unfortunately the proof of Proposition 2 is not correct. The erroneous construction does not simulate a jump dpda, where the erasing is extended until and including a symbol of the set given in the jump transition, but rather a modification of the jump dpda, where the symbol of the jump transition is saved. We feel that there is no easy way to correct the proof. Essentially, one has to prove our Theorem 1, which starts from the "conventional" dpda directly. However, the proof of the main result, our Theorem 2, is independent of Proposition 2 and is correct in [2].

1. PRELIMINARIES

In this section we fix our terminology and prove two normal form lemmas about deterministic pushdown automata.

The empty word over an arbitrary alphabet is denoted by e. The cardinality of a set S is denoted by $|S|$. If f is a partial function, $f(a)\downarrow$ and $f(a)\uparrow$ mean that f is, respectively is not, defined on the argument a.

Definition. A dpda is a 7-tuple $M = (Q,\Sigma,\Gamma,\delta,q_0,Z_0,F)$, where Q, Σ and Γ are the finite sets of the states, inputs and stack symbols, $q_0 \in Q$, $Z_0 \in \Gamma$, $F \subseteq Q$, and $\delta: Q \times (\Sigma \cup \{e\}) \times \Gamma \to Q \times \Gamma^*$ is a partial function satisfying the determinism condition: if $\delta(q,e,Z)\downarrow$, then for all $a \in \Sigma$, $\delta(q,a,Z)\uparrow$. The relation $\vdash_M$ on $Q \times \Sigma^* \times \Gamma^*$ is defined as follows.

$$(q,ax,Z\gamma) \vdash_M (p,x,\gamma'\gamma) \quad \text{iff} \quad \delta(q,a,Z) = (p,\gamma').$$

Moreover, $\vdash_M^*$ is the reflexive, transitive closure of $\vdash_M$. The language L(M) accepted by M by empty stack and final state is the set

$$L(M) = \{w \mid (q_0,w,Z_0) \vdash_M^* (q_f,e,e),\ q_f \in F\}.$$

For a more detailed introduction to the terminology, consult [5].

Lemma 1. For every dpda $M = (Q,\Sigma,\Gamma,\delta,q_0,Z_0,F)$ there exists a dpda $M' = (Q,\Sigma,\Gamma',\delta',q_0,[Z_0],F)$ such that

(i) $L(M) = L(M')$,

(ii) if $\delta'(q,a,Z) = (p,\gamma)$ then $|\gamma| \leq 2$ ($|\gamma|$ is the length of γ),

(iii) if $\delta'(q,a,Z) = (p,e)$ then $a = e$,

(iv) if $\delta'(q,e,Z) = (p,\gamma)$ then $\gamma = e$.

Proof. Assume that M is loop-free (see [5]). There exists a constant k such that for all $q,p \in Q$, $\eta \in \Gamma^+$ and $a \in \Sigma$, if $(q,a,\eta) \vdash_M^* (p,e,\gamma)$ then $|\gamma| \leq |\eta| + k$. It can also be assumed that M pops only on e, because otherwise one can replace $(q,a,Z) \vdash_M (p,e,e)$ by $(q,a,Z) \vdash_M (p_E,e,E) \vdash_M (p,e,e)$, where E is a new stack letter and p_E is a new state. The stack alphabet of M' is $\Gamma' = \{[q] \mid q \in Q\} \cup \{[\gamma] \mid \gamma \in \Gamma^+, 1 \leq |\gamma| \leq k\}$ and the transition function δ' is defined for all $p,q \in Q$, $a \in \Sigma$, and $\eta \in \Gamma^*$ such that $1 \leq |\eta| \leq k$, as follows.

1. If $(q,a,\eta) \vdash_M^+ (p,e,\gamma)$ is the maximal (unique) computation then

$$\delta'(q,a,[\eta]) = \begin{cases} (p,[\gamma_1][\gamma_2]) \text{ if } |\gamma| > k,\ \gamma = \gamma_1\gamma_2,\ |\gamma_2| = k, \\ \qquad\qquad (\text{Note that } |\gamma| \leq 2k) \\ (p,[\gamma]) \text{ if } 1 \leq |\gamma| \leq k, \\ (p,[p]) \text{ if } \gamma = e, \end{cases}$$

and

$$\delta'(q,e,[q]) = (q,e) \text{ for all } q \in Q.$$

2. If $(q,e,\eta) \vdash_M^+ (p,e,e)$ then

$$\delta'(q,e,[\eta]) = (p,e).$$

Otherwise δ' is undefined.

It is easy to verify that L(M) = L(M').

We shall now give another normal form that simplifies constructions. It affirms that the stack is changed only by pushing new letters into it or popping existing ones, but never changing the topmost letter to another. This property is preserved in the constructions of Theorem 1 and Theorem 2.

<u>Lemma 2</u>. For every dpda $M = (Q,\Sigma,\Gamma,\delta,q_0,Z_0,F)$ there exists a dpda $M' = (Q',\Sigma,\Gamma',\delta',[q_0,Z_0],Z_0',F')$ such that

(i) L(M) = L(M'), and

(ii) each transition of M' is of one of the following three types:

$$\delta'(q,e,Z) = (p,e),$$
$$\delta'(q,a,Z) = (p,Z),$$
$$\delta'(q,a,Z) = (p,YZ).$$

<u>Proof</u>. Let M be in the normal form of Lemma 1, $Z_0' \notin \Gamma$, $\Gamma' = \Gamma \cup \{Z_0'\}$, $Q' = Q \times \Gamma'$ and $F' = F \times \Gamma'$. The transition function δ' is defined for all $q,p \in Q$, $X,Y,Z \in \Gamma$, $U \in \Gamma'$ and $a \in \Sigma$ as follows.

$$\delta'([q,Z],e,U) = ([p,U],e) \text{ if } \delta(q,e,Z) = (p,e),$$
$$\delta'([q,Z],a,U) = ([p,Y],U) \text{ if } \delta(q,a,Z) = (p,Y),$$
$$\delta'([q,Z],a,U) = ([p,X],YU) \text{ if } \delta(q,a,Z) = (p,XY).$$

Note that M' differs from M only having the topmost symbol in the finite state memory. Thus an input w empties the stack of M iff it empties the stack of M'. The correspondence between F and F' implies that L(M) = L(M').

Remark 1. Without loss of generality we may assume that the first transition in every computation, unless it is a popping transition, changes only the state, i.e. is of the form $\delta(q_0,a,Z_0) = (q,Z_0)$. If necessary, the k in the proof of Lemma 1 is increased.

2. JUMP DPDA

Definition. A jump dpda is a 7-tuple $M = (Q,\Sigma,\Gamma,\delta,q_0,Z_0,F)$, where all other components are as in the definition of the dpda, but δ is now a partial function $\delta: Q \times (\Sigma \cup \{e\}) \times \Gamma \to (Q \times \Gamma^*) \cup (\{J\} \times Q \times 2^\Gamma)$ satisfying the determinism condition. The relation $\vdash_M$ is defined as follows.

$$(q,ax,Z\gamma) \vdash_M (p,x,\gamma')$$

iff either $\delta(q,a,Z) = (p,\gamma'')$ and $\gamma' = \gamma''\gamma$, or $\delta(q,a,Z) = (J,p,S)$ and $Z\gamma = \gamma''Y\gamma'$ for some $Y \in S$ and $\gamma'' \in (\Gamma-\{S\})^*$. M is a single jump dpda if in all transitions $\delta(q,a,Z) = (J,q,S)$, the set S consists of a single element only.

The specialty of the jump dpda is the ability to pop arbitrarily many letters from the stack at one step. By Lemma 1, we may assume that all e-transitions of a dpda are popping transitions. In the next theorem we shall show that all the e-transitions can be combined to a single jump transition.

To illustrate the correspondence between the computations of a dpda M and the simulating jump dpda M', we define the relation $\models_M$ as follows:

(1) $$(q,a,Z\gamma) \models_M (p,e,\gamma')$$

iff $\delta(q,e,Z)\uparrow$, $(q,a,Z\gamma) \vdash_M^* (p,e,\gamma')$, and either $\gamma' = e$ or $\gamma' = Z'\gamma''$ and $\delta(p,e,Z')\uparrow$. As usual, $\models_M^*$ means the reflexive, transitive closure of $\models_M$.

In the next theorem, a jump dpda M' is constructed such that the computation (1) is simulated by a single step of M'.

Theorem 1. For every dpda $M = (Q,\Sigma,\Gamma,\delta,q_0,Z_0,F)$ there exists a realtime single jump dpda $M' = (Q',\Sigma,\Gamma',\delta',q_0',0,F)$ such that

(i) L(M) = L(M'),

(ii) the transitions of M' are of the form

$\delta'(q,a,Z) = (J,p,Y)$,

$\delta'(q,a,Z) = (p,Z)$,

$\delta'(q,a,Z) = (p,XYZ)$.

Proof. Let M be of the form of Lemma 2, $Q = \{q_0,\ldots,q_n\}$, $Q' = Q \cup \{q_0'\}$, and $\Gamma' = \{B\} \cup \Gamma_{adr} \cup \Gamma_{push}$, where

$\Gamma_{adr} = \{0,1,\ldots,n+2\}$,

$\Gamma_{push} = \{[Z,\lambda] \mid Z \in \Gamma$ and $\lambda: Q \to Q \times \Gamma_{adr}$ is a partial function$\}$

and B is a special symbol needed only in a transition to satisfy the length condition (ii). The initial symbol 0 appears only as the bottommost symbol.

The jump dpda M' will be defined in such a way that a computation $(q_i,a,Z_mZ_{m-1}\ldots Z_0) \models_M (q_j,e,Z_m',\ldots Z_0')$ is simulated by a single transition of M'. The configuration $(q_i,Z_mZ_{m-1}\ldots Z_0)$ of M is simulated by the configuration $(q_i,[Z_m,\lambda_m]l_{m-1}[Z_{m-1},\lambda_{m-1}]\ldots l_0[Z_0,\lambda_0]B0)$ of M'. The integers $l_{m-1},\ldots,l_0,0 \in \Gamma_{adr}$, called jump addresses, and the partial functions $\lambda_m,\ldots,\lambda_0$ contain the popping information as follows: If M begins popping in the state q_i on Z_m, (to be exact, now we are considering the second configuration of a macro-step $\models_M$) and ends popping in the state q_j with Z_k on the top, then $\lambda_m(q_i) = (q_j,l)$, where $l = l_k$ and $l \neq l_r$ for $r = m,m-1,\ldots,k+1$. Note that, because the popping transitions of M are e-transitions, the point where popping ends depends only on the state and the contents of the stack.

Assume now that M pushes by the transition $\delta(q_i,a,Z) = (q_j,YZ)$, denoting $Z = Z_m$ and $\lambda = \lambda_m$. Then the next address is calculated by the function adr: $\Gamma_{push} \to \Gamma_{adr}$, and the new popping information by the function push: $Q \times \Sigma \times \Gamma_{push} \to \Gamma_{push}$ as follows:

$$\mathrm{adr}([Z,\lambda]) = \min\{t \mid t > 0 \text{ and for no } j,k,\ \lambda(q_j) = (q_k,t)\}, \text{ and}$$

$$\mathrm{push}(q_i,a,[Z,\lambda]) = [Y,\lambda'],$$

where $\delta(q_i,a,Z) = (q_j,YZ)$ and for $k = 0,\ldots,n$,

$$\lambda'(q_k) = \begin{cases} \text{undefined if } \delta(q_k,e,Y)\uparrow, \\ (q_r,\mathrm{adr}([Z,\lambda])) \text{ if } \delta(q_k,e,Y) = (q_r,e) \text{ and } \lambda(q_r)\uparrow, \\ \lambda(q_r) \text{ if } \delta(q_k,e,Y) = (q_r,e) \text{ and } \lambda(q_r)\downarrow. \end{cases}$$

Note that for all Z and λ, $\mathrm{adr}([Z,\lambda]) \leq n+2$. The initial popping information is given by

$$\lambda_0(q_k) = \begin{cases} \text{undefined if } \delta(q_k,e,Z_0)\uparrow, \\ (q_r,0) \text{ if } \delta(q_k,e,Z_0) = (q_r,e). \end{cases} \qquad (k = 0,\ldots,n)$$

The transition function δ' is defined for all q_i, q_j in Q, Z, Y in Γ and a in Σ as follows:

1. If $\delta(q_0,a,Z_0) = (q_j,Z_0)$ then

$$\delta'(q_0',a,0) = \begin{cases}(J,q_k,0) \text{ if } \delta(q_j,e,Z_0) = (q_k,e), & \text{(a)}\\ (q_j,[Z_0,\lambda_0]B0) \text{ if } \delta(q_j,e,Z_0)\uparrow. & \text{(b)}\end{cases}$$

2. If $\delta(q_i,a,Z) = (q_j,Z)$ then

$$\delta'(q_i,a,[Z,\lambda]) = \begin{cases}(q_j,[Z,\lambda]) \text{ if } \lambda(q_j)\uparrow, & \text{(a)}\\ (J,q_r,l) \text{ if } \lambda(q_j) = (q_r,l). & \text{(b)}\end{cases}$$

3. If $\delta(q_i,a,Z) = (q_j,YZ)$ then

$$\delta'(q_i,a,[Z,\lambda]) = \begin{cases}(q_j,\text{push}(q_i,a,[Z,\lambda])\text{adr}([Z,\lambda])[Z,\lambda]) \text{ if } \delta(q_j,e,Y)\uparrow, & \text{(a)}\\ (q_k,[Z,\lambda]) \text{ if } \delta(q_j,e,Y) = (q_k,e) \text{ and } \lambda(q_k)\uparrow, & \text{(b)}\\ (J,q_r,l) \text{ if } \delta(q_j,e,Y) = (q_k,e) \text{ and } \lambda(q_k) = (q_r,l). & \text{(c)}\end{cases}$$

Otherwise δ' is undefined.

If a word of length one empties the stack of M, then it also empties the stack of M' by the transition 1(a). Otherwise the computation is initiated by 1(b), by which the stack, particularly λ_0, is properly determined. The most important part in the proof is to assure that in the transition 3(a) the popping information λ' in $[Y,\lambda'] = \text{push}(q_i,a,[Z,\lambda])$ and the address adr$([Z,\lambda])$ are correctly formed. This follows immediately by an inductive argument from the definitions of the functions push and adr.

The correspondence between the computations of M and M' is very easily illustrated. The transitions 2(a) and 2(b) correspond, denoting $Z_m = Z$, to the moves

$$(q_i,a,Z_mZ_{m-1}\ldots Z_0) \models_M (q_j,e,Z_mZ_{m-1}\ldots Z_0) \text{ and}$$
$$(q_i,a,Z_mZ_{m-1}\ldots Z_0) \models_M (q_r,e,Z_sZ_{s-1}\ldots Z_0),$$

where $l = l_s$ but $l \neq l_t$ for $t = m,m-1,\ldots,s+1$. The transitions 3(a), 3(b) and 3(c) correspond to the moves

$$(q_i,a,Z_mZ_{m-1}\ldots Z_0) \models_M (q_j,e,YZ_mZ_{m-1}\ldots Z_0),$$
$$(q_i,a,Z_mZ_{m-1}\ldots Z_0) \models_M (q_k,e,Z_mZ_{m-1}\ldots Z_0), \text{ and}$$
$$(q_i,a,Z_mZ_{m-1}\ldots Z_0) \models_M (q_r,e,Z_sZ_{s-1}\ldots Z_0),$$

where $l = l_s$ but $l \neq l_t$ for $t = m,m-1,\ldots,s+1$.

Remark 2. For easy reference we summarize here some properties of the jump dpda $M' = (Q',\Sigma,\Gamma',\delta',q_0',0,F)$:

(i) $\Gamma' = \Gamma_{push} \cup \Gamma_{adr} \cup \{B\}$ and adr is a function $\Gamma_{push} \to \Gamma_{adr}$.

(ii) The initial symbol $0 \in \Gamma_{adr}$ appears as the bottommost symbol of the stack and after the first step it is reached only at the last step with a jump.

(iii) The transitions are of the form

1. $\delta'(q_0',a,0) = (J,p,0)$,
2. $\delta'(q_0',a,0) = (p,ZB0)$, where $Z \in \Gamma_{push}$,
3. $\delta'(q,a,Z) = (p,XYZ)$, where $X,Z \in \Gamma_{push}$ and $Y = adr(Z)$,
4. $\delta'(q,a,Z) = (p,Z)$, where $Z \in \Gamma_{push}$,
5. $\delta'(q,a,Z) = (J,p,Y)$, where $Z \in \Gamma_{push}$ and $Y \in \Gamma_{adr}$.

(iv) In two successive jumps the jump addresses are different.

Theorem 2. For every dpda M there exists a realtime one state single jump dpda M" such that

(i) $L(M) = L(M")$,

(ii) the transitions of M" are of the form

$\delta"(a,Z) = (J,Y)$,

$\delta"(a,Z) = Y_1 \ldots Y_n Z$, where $n \geq 0$.

Proof. Let $M' = (Q',\Sigma,\Gamma',\delta',q_0,0,F)$ be the realtime single jump dpda of Theorem 1 such that $L(M) = L(M')$. For convenience we have changed the notation for the initial state, and denote $Q' = \{q_0,\ldots,q_n\}$. Let

$$\Gamma" = \{[q_i,Z] \mid i = 0,\ldots,n,\ Z \in \Gamma',\ Z \neq 0,B\} \cup \{0,B\}.$$

By the property (iii) of Remark 2, the configurations of M' are of the form $(q_i,X_mY_{m-1}X_{m-1}\ldots Y_1X_1B0)$, where each $X_i \in \Gamma_{push}$ and $Y_i = adr(X_i)$. We shall construct a realtime one state single jump dpda $M" = (\Sigma,\Gamma",\delta",0)$ such that this configuration is simulated by the stack

$$[q_i,X_m][q_{i-1},Y_m][q_{i-1},X_m]\ldots[q_0,Y_m][q_0,X_m][q_n,Y_{m-1}][q_n,X_{m-1}]\ldots$$
$$[q_0,Y_{m-1}][q_0,X_{m-1}]\ldots[q_n,Y_1][q_n,X_1]\ldots[q_0,Y_1][q_0,X_1]B0.$$

Thus the current state is stored in the topmost letter of the stack of M".

Formally, $\delta"$ is defined for all q_i,q_j in Q', X,Y,Z in Γ' and a in Σ as follows. (Compare with Remark 2)

1. If $\delta'(q_0,a,0) = (J,q_j,0)$ and $q_j \in F$ then
   $\delta"(a,0) = (J,0)$. (In this case a is accepted)

2. If $\delta'(q_0,a,0) = (q_j,ZB0)$ then
   $\delta"(a,0) = [q_j,Z][q_{j-1},Y][q_{j-1},Z]\ldots[q_0,Y][q_0,Z]B0$, where $Y = adr(Z)$.

3. If $\delta'(q_i,a,Z) = (q_j,XYZ)$ then

$$\delta''(a,[q_i,Z]) = [q_j,X][q_{j-1},U][q_{j-1},X]\ldots[q_0,U][q_0,X][q_n,Y][q_n,Z]\ldots$$
$$[q_i,Y][q_i,Z], \text{ where } U = adr(X).$$

4. If $\delta'(q_i,a,Z) = (q_j,Z)$ then

$$\delta''(a,[q_i,Z]) = \begin{cases} [q_j,Z][q_{j-1},Y][q_{j-1},Z]\ldots[q_i,Y][q_i,Z] \text{ if } j \geqq i \ (Y = adr(Z)) & \text{(a)} \\ (J,[q_j,Y]) \text{ if } j < i \ (Y = adr(Z)) & \text{(b)} \end{cases}$$

5. If $\delta'(q_i,a,Z) = (J,q_j,Y)$ then

$$\delta''(a,[q_i,Z]) = \begin{cases} (J,[q_j,Y]) & \text{if } Y \neq 0 & \text{(a)} \\ (J,0) & \text{if } Y = 0 \text{ and } q_j \in F. & \text{(b)} \end{cases}$$

Otherwise δ'' is undefined.

In the transitions 3 and 4 it is important that the address of the topmost letter is uniquely determined by the function adr. Therefore the chains will never be mixed. In the transition 5(a), note that the property (iv) of Remark 2 assures that the jump goes deep enough. By induction on the length of w it is easy to verify that for all w in Σ^*,

$$(q_0,w,0) \vdash^*_{M'} (q_i,e,X_mY_{m-1}X_{m-1}\ldots Y_1X_1B0)$$

iff

$$(w,0) \vdash^*_{M''} (e,[q_i,X_m][q_{i-1},Y_m][q_{i-1},X_m]\ldots[q_0,Y_m][q_0,X_m][q_n,Y_{m-1}][q_n,X_{m-1}]\ldots$$
$$[q_0,Y_{m-1}][q_0,X_{m-1}]\ldots[q_n,Y_1][q_n,X_1]\ldots[q_0,Y_1][q_0,X_1]B0),$$

where $Y_m = adr(X_m)$.

Moreover, the property (ii) of Remark 2 and the jump transitions 1 and 5(b) assure that M" empties the stack iff M' empties the stack in a final state. Hence L(M") = L(M') = L(M).

ACKNOWLEDGEMENT

The authors are very grateful to Bruno Courcelle and Tero Harju for many useful comments and discussions and for careful reading of the manuscript.

REFERENCES

1. Cole, S.N., Deterministic pushdown store machines and real-time computation. Journal of the Association for Computing Machinery, 14(1971), 453-460.

2. Courcelle, B., On jump-deterministic pushdown automata. Mathematical Systems Theory, 11(1977), 87-109.

3. Greibach, S., Jump-pda´s, deterministic context-free languages, principal AFDL´s and polynomial time recognition. Unpublished report.

4. Greibach, S., Jump-dpda´s and hierarchies of deterministic context-free languages. SIAM Journal of Computing, 3(1974), 111-127.

5. Harrison, M.A., Introduction to Formal Language Theory. Addison-Wesley, 1978.

SYNCHRONIZATION AND MAXIMALITY FOR VERY PURE SUBSEMIGROUPS OF A FREE SEMIGROUP

Aldo de Luca

and

Antonio Restivo

Laboratorio di Cibernetica del C.N.R.,80072,Arco Felice,Napoli.Italy

0.INTRODUCTION

A *very pure* subsemigroup P of a free semigroup A^+ is a subsemigroup of A^+ satisfying the condition : for all $u, v \in A^+$ $uv, vu \in P \Rightarrow u, v \in P$.

The notion of very pure subsemigroup of a free semigroup plaies an important role in some problems of algebra, information theory and language theory. In particular the bases of very pure subsemigroups, called *very pure codes*, have been considered by M.P. Schützenberger in the factorizations of free monoids and in the construction of the bases of free Lie algebras [17-18] . Further very pure subsemigroups and codes have remarkable synchronizing properties which are of relevant interest in the theory of information transmission [2,10,12,13] . Recently J.Pin [11] has shown that a characterization of the variety of *locally testable languages* [6] can be obtained in terms of the notion of very pure subsemigroups.

The aim of this paper is to present some new results which are mainly concerned with the properties of synchronization and maximality of very pure subsemigroups and codes. In section 2 a brief account of the synchronizing properties of very pure subsemigroups, generalizing some previously published results, is given. In section 3 two different notions of maximality for a very pure code are introduced: one with respect to code condition and the other with respect to the property of being "very pure". The main result of this section states that the two notions are indeed equivalent under the hypothesis that the code is *nondense* . Several corollaries are derived. One in particular, shows that a maximal very pure code has to be infinite. Finally in section 4 some results concerning the smallest very pure subsemigroup containing a given set $X \subseteq A^+$ are proven.

1.PRELIMINARIES

Let A^+ be the free semigroup generated by a finite *alphabet* A. In the following

$A^* = A^+ \cup \{1\}$ will denote the *free monoid* generated by A ,where 1 is the identity element of A^* or *empty word* .For any $f \in A^*$, $|f|$ denotes the *length* of f where conventionally, $|1| = 0$.

A classical result of Schützenberger [15] states that a subsemigroup P of A^+ is free if and only if for all $s \in A^+$, $p, q \in P$ $sp, qs \in P \Rightarrow s \in P$. From the definition one has that a very pure subsemigroup P of A^+ is free. In fact if $sp, qs \in P$ then $spq, pqs \in P$ so that since P is very pure $s \in P$.

The base $X = P \setminus P^2$ of a free subsemigroup P of A^+ is usually called *code* since any word of P can be *uniquely* factorized in terms of the elements of X (*code words*). When P is very pure X is called *very pure code*.

Let P be a subsemigroup of A^+ and $X = P \setminus P^2$ the (unique) minimal set of generators of P .One can then introduce in P the following relation (X-conjugation) defined for all $f, g \in P$ as f X-conj g if there exist $u, v \in X^*$ such that $f = uv$, $g = vu$. When u,v are different from the empty word f and g are said *strictly* X-conjugate. It is clear that X-conjugation implies A-conjugation whereas the converse is not generally true.

Proposition 1.1. Let P be a subsemigroup of A^+, $X = P \setminus P^2$ and $P^1 = P \cup \{1\}$. The following conditions are equivalent:

i. P is very pure.

ii. for all $h \in A^+$ $hP \cap Ph \neq \emptyset \Rightarrow h \in P$.

iii. for all $u, v \in A^+$ $uv \in X \Rightarrow vP^1u \cap P = \emptyset$.

iv. for all $u \in A^*$, $v \in A^+$ $uv \in X$, $x \in X$ and $|v| < |x| \Rightarrow vP^1u \cap xP^1 = \emptyset$.

v. P is a free subsemigroup of A^+ and for all $u, v \in P$ u and v are strictly X-conjugate if and only if they are strictly A-conjugate.

The previous proposition whose proof can be found in [3] and [5] ,gives then different characterizations of very pure subsemigroups and co des .We remark that condition iii. and iv. express that any word of P can be *circularly* factorized in a unique way in code words. Condition v. is the one used by Schützenberger in the factorization of free monoids [17]. Condition ii.,at last, has been considered by Levenstein in order to characterize the family of finite codes having a bounded synchronization delay [8].

In the following we shall denote by *VP* (*resp. VC*) the families of very pure subsemigroups (*resp.* very pure codes) of A^+. Let us prove the following closure properties of *VP* and *VC* that we shall use in the next:

Proposition 1.2. *VP* is closed under intersection.

Proof: Let $\{ P_\gamma \mid \gamma \in \Gamma \}$ be any subfamily of *VP* and set $P = \bigcap_{\gamma \in \Gamma} P_\gamma$ .For any $h \in A^+$ one has that $hP \cap Ph \subseteq hP_\gamma \cap P_\gamma h$ for all $\gamma \in \Gamma$. If $hP \cap Ph \neq \emptyset$ then for all $\gamma \in \Gamma$ $hP_\gamma \cap P_\gamma h \neq \emptyset$ so that from proposition 1.1 it follows $h \in P_\gamma$ for all $\gamma \in \Gamma$ and then $h \in \bigcap_{\gamma \in \Gamma} P_\gamma = P$.

Proposition 1.3. *VP* and *VC* have the Zorn property.

Proof : Let γ be a chain of $P_\alpha \in VP$. One has that $P = \bigcup_\gamma P_\alpha \in VP$. In fact let $u, v \in A^+$ be such that uv, $vu \in P$. There exist then two elements of the chain P_{α_1} and P_{α_2} for which $uv \in P_{\alpha_1}$ and $vu \in P_{\alpha_2}$. Let us suppose $P_{\alpha_1} \subseteq P_{\alpha_2}$. Since P_{α_2} is very pure it follows that $u, v \in P_{\alpha_2} \subseteq P$. Let now γ be a chain of elements of *VC* and be $X = \bigcup_\gamma X_\alpha$. Since, for any $X_\alpha \in VC$, X_α^+ is very pure, one has a chain of very pure subsemigroups so that from the previous result $X^+ = (\bigcup_\gamma X_\alpha)^+ = \bigcup_\gamma X_\alpha^+ \in VP$. At last it is easy verify that X is just the base of X^+ so that $X \in VC$.

Let us remark that propositions 1.2 and 1.3 hold true also for the families of free subsemigroups and codes of A^+. Further properties of very pure subsemigroups and codes are the following:

1. The property *VP* is preserved under the formation of inverse homomorphic images.

2. If $X \in VC$ then any subset $X' \subseteq X$ belongs to *VC*.

Let X and Y be two codes over the alphabet A . If $X \subseteq Y^+$ then the image of X in Y^+ is still a code (over the alphabet Y), say Z . X is said to be obtained by *composition* of Z and Y and denoted by $X = Z \otimes Y$. X is called *indecomposable* if it admits trivial decompositions only.

3. The composition of two very pure codes is a very pure code.

A subset M of A^+ is *recognizable* (by a finite automaton) if its *syntactic semigroup* S(M) is finite [6] . In the following we shall denote by σ the canonical epimorphism $\sigma\colon A^+ \to S(M)$.

If T is a subset of A^+, a word $u \in A^+$ is said to be *completable* in T if u is factor of some word of T , i.e. if there exist $v, w \in A^*$ such that $vuw \in T$. Otherwise u is said to be *incompletable* in T . If T and S are subsets of A^+, we say that T is *dense with respect to* S if any word of S is completable in T . If $S = A^+$, we simply say that T is *dense*.

2. SYNCHRONIZING PROPERTIES OF VERY PURE CODES

Let P be a subsemigroup of A^+. We recall the following definition. A pair $(u,v) \in P \times P$ is a *synchronizing pair* for P if for all $s, t \in A^*$ $suvt \in P$ implies $su, vt \in P$. The subsemigroup P is called *synchronizing* if it has at least a synchronizing pair.

If P is a synchronizing free subsemigroup of A^+ its base $X = P \setminus P^2$ is said to be a *synchronizing code* . It is clear from the definition that *if X is a synchronizing code then X is nondense with respect to* X^+. In the case of very pure subsemigroups of A^+ it holds the following [5] :

Proposition 2.1. If P is a very pure subsemigroup of A^+ such that its base is nondense with respect to P then P is synchronizing.

The condition that $X = P \setminus P^2$, where P is any free subsemigroup of A^+, is non dense with respect to P is verified in the particularly interesting case of recognizable codes [2] so that from proposition 2.1 one derives:

Proposition 2.2. A recognizable very pure subsemigroup of A^+ is synchronizing.

In the case of recognizable codes, the notion of very pure is however closely related to a stronger synchronization property. Let us give the following definition: A subsemigroup P of A^+ has a *bounded synchronization delay* if a positive integer k exists such that all the pairs of $P^k \times P^k$ are synchronizing. The least integer for which this condition is verified is called the *synchronization delay* of P.

If P is a free subsemigroup of A^+ having a bounded synchronization delay, then a positive integer p has to exist such that the following condition, called F(p), holds: $A^* X^p A^* \cap X = \emptyset$, with $X = P \setminus P^2$. A theorem of Restivo [13], proved by means of combinatorial techniques, relates the concepts of very pure subsemigroup and free subsemigroup having a bounded synchronization delay. Moreover an algebraic characterization of the syntactic semigroup S(P) of a very pure subsemigroup P of A^+ has been given by the authors Perrin and Termini [4] in the case in which P is finitely generated and extended by de Luca [2] to more general cases. For a *recognizable* subsemigroup P of A^+ the following proposition holds:

Proposition 2.3. Let P be a recognizable free subsemigroup of A^+. The following conditions are equivalent:

i. P has a bounded synchronization delay.

ii. P is very pure and satisfies the condition F(p) for some p.

iii. The 0-minimal ideal J of the syntactic semigroup S(P) has trivial *H*-classes and all the idempotents of $P\sigma$ belong to J.

iv. For all the idempotents e of $P\sigma$, $eS(P)e \subseteq \{e, 0\}$.

If P is finitely generated then the condition F(p) is obviously verified and *the class of free subsemigroups having a bounded synchronization delay coincides with the class of very pure subsemigroups*. Moreover from a result of Restivo [12] it follows that these two classes coincide with the class of *locally testable free subsemigroups* of A^+ (see also [7]).

The results reported in this section show the close relation existing between the notion of "very pure" and synchronizing properties. However, we stress that there are very pure subsemigroups which *do not admit synchronizing pairs*. From proposition 2.1 one has that a very pure subsemigroup P of A^+ is not synchronizing if and only if its base is dense with respect to P. In [5] we have shown that this is the case for the *restricted Dyck's language* D_n over an alphabet A of 2n letters, for all $n \geq 1$. In fact, by making use of condition iv. of proposition 1.1, one can prove that D_n is a very pure subsemigroup of A^+. The base Δ_n of D_n is the set of *prime Dyck's words* (a Dyck's word is prime if it cannot be written as the product of two or more nonempty Dyck's words). One can show that Δ_n is dense with respect to D_n. Moreover Δ_1 is dense.

3. MAXIMAL VERY PURE CODES AND SUBSEMIGROUPS

Let $X \subseteq A^+$ be a code. X is called *complete* , or maximal (as a code), if it does not exist another code X' such that $X \subset X'$. A *completion* $\bar{X}$ of X is any complete code $\bar{X}$ such that $\bar{X} \supseteq X$. Since the family of codes of A^+ has the Zorn property then from the *Hausdorff maximal principle* it follows that for any code X there exists at least a completion $\bar{X}$ of X . Restivo [14] has recently proven the existence of *finite* codes which *do not admit a finite completion*. The following basic characterization of complete codes has been given by Schützenberger [16] (see also [6], [9]):

Theorem 3.1. Let X be a code which is not dense. X is complete if and only if X^+ is dense.

Let us now consider the family *VC* of very pure codes. We say that $X \in VC$ is *complete as very pure* if it does not exist $X' \in VC$ such that $X' \supset X$. A very pure completion $\bar{\bar{X}}$ of $X \in VC$ is any very pure code which is complete as very pure and such that $\bar{\bar{X}} \supseteq X$. Since from proposition 1.3 the family *VC* has the Zorn property it follows that any $X \in VC$ has a very pure completion. It is clear that if X is complete as code then X is complete as very pure. Our main result is the following proposition [5] :

Proposition 3.1. Let $X \in VC$. If X is complete as very pure then X^+ is dense.

(Outline of the proof). In the case Card A = 1 the proof is trivial. Let us then suppose Card A > 1. If $P = X^+$ is not dense then there would exist a word $f \in A^+$ such that $A^* f A^* \cap P = \emptyset$. Let us then consider the word $f' = f^2 b^{|f|}$, where b is a letter different from the first letter of f . Since f' is not a sesquipower $X' = X \cup \{f'\}$ is a code. Moreover we can prove that X' is very pure by showing (cf. proposition 1.1) that for all $u, v \in A^+$ if $uv \in X'$ then $v (X')^* u \cap (X')^+ = \emptyset$. The proof is obtained by contradiction considering two cases. In the first case $uv \in X$ and in the second $uv = f'$. If $uv \in X$ and $vgu = g'$ with $g \in (X')^*$, $g' \in (X')^+$, one has that the only possibility is that $g, g' \in (X')^* \setminus (X)^*$ so that the previous equation can be written as:

$$vh f^2 b^{|f|} wu = h' f^2 b^{|f|} w',$$

with $h, h' \in X^*$, $w, w' \in (X')^*$. One can then prove that $vh = h'$ and $wu = w'$. This implies, since $uv \in X$, that $u, v \in (X')^+$ and $X' \cap [(X')^+]^2 \neq \emptyset$, which is absurd since X' is a code. If $uv = f^2 b^{|f|}$ one has to consider the following three subcases: 1. $|u|, |v| > |f|$. 2. $|v| > |f|$, $|u| \leq |f|$. 3. $|v| \leq |f|$. In all these cases one contradicts either that f begins with a letter $\neq b$, or that X' is a code, or, at last, that $A^* f A^* \cap X^+ = \emptyset$.

Let us now examine some corollaries of the previous proposition the proof of which we omit for the sake of brevity.

Corollary 3.1. Let X be a nondense very pure code. If X is complete as very pure then X is complete as code.

We explicit ly observe that the hypothesis that X is nondense is certainly verified when X is recognizable.

Corollary 3.2. If X is a *finite* very pure code, different from the alphabet A, then X is not complete as very pure.

Corollary 3.3. Let X be a recognizable code having a bounded synchronization delay. If X is maximal as regard to this property, then X is complete as code.

Let P be a free subsemigroup of A^+. P is said *maximal as free subsemigroup* if it does not exist a free subsemigroup Q such that $P \subset Q \subset A^+$, i.e. P is indecomposable. It is clear that a maximal free subsemigroup of A^+ is generated by a complete code. The converse, however, is not generally true.

Similarly we say that $P \in VP$ is *maximal as very pure subsemigroup* if it does not exist $Q \in VP$ such that $P \subset Q \subset A^+$. If P is maximal as very pure then its base has to be a complete very pure code so that from proposition 3.1, P has to be dense. However differently from what occurs in the case of very pure codes (cf. corollary 3.1) one has that *if* P *is a maximal very pure subsemigroup of* A^+ *, such that its base is not dense, then* P *is not, in general, maximal as free subsemigroup of* A^+. This fact is shown by the following example. Consider the alphabet $A = \{ a,b \}$ and the subsemigroup $P = \{ a^* b \}^+$. P is very pure since its base has a bounded synchronization delay (cf. proposition 2.3) and maximal since how one can easily prove by lemma 4.1 it cannot exist a very pure subsemigroup Q such that $P \subset Q \subset A^+$. However P is not maximal as free subsemigroup, since $P \subset \{ a^2, ab, b \}^+ \subset A^+$.

A consequence of corollary 3.2 is that the family of finitely generated very pure subsemigroups of A^+, ordered by inclusion, does not contain maximal elements.

4. AUXILIARY RESULTS

In the section 1 we have seen that the family of very pure subsemigroups of A^+ is closed under intersection (cf. proposition 1.2). Thus for any $X \subseteq A^+$ one can define *the smallest very pure subsemigroup* P *containing* X, i.e. P is the meet of all very pure subsemigroups of A^+ containing X. Extending a similar result shown in [1] in the case of free subsemigroups of A^+, we can prove the following:

Lemma 4.1. Any element of the base Y of P is a left factor (right factor) of a word of X at least.

Proof : The proof is by contradiction. Suppose that a $y \in Y$ exists such that $yY^* \cap X = \emptyset$. We can then consider the set $Z = (Y \setminus \{y\})\, y^*$. Z looked as code over the alphabet Y is very pure. Since Y is also a very pure code over the alphabet A , from the closure property of very pure codes by composition (cf. section 1) it follows that Z is a very pure code over the alphabet A . Further since obviously $X \subseteq Z^+$, one concludes that $X \subseteq Z^+ \subset P$, which contradicts the fact that P is the smallest very pure subsemigroup containing X .

As a consequence of the previous lemma the following holds :

Proposition 4.1. If X is finite then Card Y $\leq$ Card X .

Proof: Since $X \subseteq Y^+$ and Y freely generates Y^+, we can introduce the map $\alpha : X \to Y$ defined for all $x \in X$ as $\alpha(x) = y$ if and only if $x \in yY^*$. From lemma 4.1 it follows that α is surjective so that in the finite case Card Y $\leq$ Card X .

For any subset X of A^+ let us define $\mathscr{P}(X) = \{ h \in A^+ \mid hX \cap Xh \neq \emptyset \}$. If X is a subsemigroup P of A^+ then $\mathscr{P}(P) \supseteq P$. By using condition ii. of proposition 1.1 one has that a subsemigroup P of A^+ is very pure if and only if $\mathscr{P}(P) = P$.

Let us now introduce for any $X \subseteq A^+$ the following sequence P_n of subsemigroups of A^+:

$$P_o = X^+ \quad , \quad P_{n+1} = [\mathscr{P}(P_n)]^+ \quad , n \geq 0 .$$

It is clear that $P_n \subseteq P_{n+1}$,$n \geq 0$. Defining $P(X) = \bigcup_{n \geq o} P_n$ one easily derives the following proposition (see, Spehner [19]) the proof of which we omit:

Proposition 4.2 . P(X) is the smallest very pure subsemigroup of A^+ containing X .

Let us remark that the interest of the previous proposition lies in the fact that one can obtain from it, at least in some particular cases, algorithms in order to construct, starting from a given set X , the base Y of the smallest very pure subsemigroup containing X.

In conclusion we observe that the results obtained in this section for very pure subsemigroups of A^+ cannot, in general, be extended to the family of free subsemigroups having a *bounded synchronization delay* and to that of *locally testable* free subsemigroups of A^+, which reduce themselves to the family *VP* in the finitely generated case. As shown by the examples reported below both these families are not closed under intersection so that is not, in general, possible define the smallest free subsemigroup having a bounded synchronization delay and the smallest locally testable free subsemigroup of A^+ containing a given set.

Examples . Let us consider, for any positive integer n , the free subsemigroup of A^+, $A = \{a,b\}$, generated by the set $X_n = \{ a^{2k} b \mid k \leq n \} \cup a^{2n} a^+ b$. The subsemigroup X_n^+ is strictly locally testable and moreover $X_{n+1}^+ \subset X_n^+$. The meet of all

X_n^+ is the free subsemigroup generated by the set $X = (a^2)^* b$ and X^+ obviously is not locally testable.

For free subsemigroups having a bounded synchronization delay let us give the following example. Consider, for any positive integer n, the free subsemigroup generated by the set $X_n = \{ a^k b a^{k+1} b \mid k \leq n \} \cup a^{n+1} a^* b$. X_n^+ has a bounded synchronization delay and moreover $X_{n+1}^+ \subset X_n^+$. The meet of all X_n^+ is generated by the set $X = \{ a^k b a^{k+1} b \mid k \geq 0 \}$. The subsemigroup X^+ has an infinite delay of synchronization (cf. [13]) .

REFERENCES

1. Berstel, J., Perrin, D. , Perrot, J.F. and A. Restivo, Sur le théorème du défaut, J. of Algebra, (1979) , *in press*.
2. De Luca, A., On some properties of the syntactic semigroup of a very pure subsemigroup, RAIRO, I.T. (1979), *in press*.
3. De Luca, A., A note on "very pure" subsemigroups of a free semigroup, *in* "Non commutative structures in Algebra and Geometric Combinatorics". Proceedings of the Conference held at Laboratorio di Cibernetica del CNR, Arco Felice, Napoli, July 1978 (to appear).
4. De Luca, A., Perrin, D., Restivo, A. and S. Termini, Synchronization and simplification, Discrete Mathematics (1979), *in press*.
5. De Luca, A. and A. Restivo, On some properties of very pure codes, Theoretical Computer Science, (1979) , *in press*.
6. Eilenberg, S., Automata, Languages and Machines, Academic Press, New York, vol. A (1974), vol. B (1976).
7. Hashiguchi, K. and N. Honda, Properties of code events and homomorphisms over regular events, J. Comput. System Sci. 12 (1976) , 352-367.
8. Levenstein, V. I., On some properties of encoding and on self-tuning automatons for the decoding of information, Problemy Kibernetiki 11 (1964), 63-121.
9. Perrin, D., Codes et combinatoire du monoide libre, Cours de DEA 1976-77, Université de Paris VII (1978).
10. Perrin, D. and M.P. Schützenberger, Un problème élémentaire de la théorie de l'Information. Colloques Internationaux du C.N.R.S. n. 276-Théorie de l'Information. Cachan, 4-8 juillet 1977, pp. 249-260.
11. Pin J. , Une caractérisation de trois variétés de langages bien connues, 4-th G.I. Conference on Theoretical Computer Science, 26-28 Mars 1979, Aachen, Lecture Notes in Computer Science, Springer -Verlag.
12. Restivo, A. , On a question of McNaughton and Papert, Information and Control 25 (1974), 93-101.
13. Restivo, A. A combinatorial property of codes having finite synchronization delay, Theoretical Computer Science 1 (1975), 95- 101.
14. Restivo, A., On codes having no finite completions, Discrete Mathematics, 17 (1977), 309-316.
15. Schützenberger, M.P., On an application of semigroup methods to some problems in coding, IRE Trans. Information Theory, 2 (1956), 47-60.
16. Schützenberger, M.P., Codes à longueur variables , Cours à l'Ecole d'Eté de l'OTAN sur les méthodes combinatoires en théorie du codage, Royan, France (1965) *unpublished*

7 . Schützenberger,M.P. ,On a factorization of free monoids,Proc.of AMS , 16 (1965) , 21-24.

8 . Schützenberger ,M.P.,Sur une propriété combinatoire des algèbres de Lie libres pouvant etre utilisée dans un problème de mathématique appliquée,Séminaire Dubreil-Pisot (Algèbre et théorie des nombres),exposé n.1, 1958/59,Paris.

9. Spehner,J.C., Sur les codes synchronisants, C.R. Acad.Sc. Paris,t.281 (1975), Série A, 779-782.

ON THE SETS OF MINIMAL INDICES
OF PARTIAL RECURSIVE FUNCTIONS

G. B. Marandžjan

Computing Center,

Academy of Sciences of the Armenian SSR,

P. Sevak st. 1, Erevan 375200, USSR

Let $\{\varphi_i\}$ be a uniform effective numbering of the family of partial recursive functions of one variable. Denote by M_φ the set

$$\{x \mid (Ay)(\varphi_x = \varphi_y \Rightarrow x \leqslant y)$$

of minimal indices. A. Meyer [1] showed that M_φ is of Turing degree $0''$ and asked whether $M_\varphi \equiv_m M_{\varphi'}$ for all uniform effective numberin φ and φ' or whether $M_\varphi \equiv_{tt} M_{\varphi'}$. P. Young [1] showed that M_φ and $M_{\varphi'}$ are not 1 - equivalent in the general case. The first question of A. Meyer was negatively answered by J. Kinber [2] and the present author [3] independently and in different ways. Namely, J. Kinber proved the existence of such a pair of numberings φ and φ' that M_φ and $M_{\varphi'}$ are incomparable with respect to bounded trut table reducibility. The present author proved the existence of such a pair of numberings φ and φ' that M_φ and $M_{\varphi'}$ are incomparable with respect to c - reducibility (i. e. conjunctive reducibility). Further these results were extended to other reducibilities by the same authors. However, the second question remains open.

Denote by W_i the domain of a partial recursive function φ_i.

Denote by M_w the set $\{x \mid (Ay)(W_x = W_y \Rightarrow x \leq y)\}$.

Proposition 1. M_w is of Turing degree $0''$ and belongs to $\Sigma_2 - \Pi_2$.

Proof is a slight modification of A. Meyer's proof [1] with the following main differences:

(1) Replace the set Z in Lemma 6 by $Z = \{i \mid W_i = N\}$;

(2) Define f in the Theorem 7 as follows

$f(n) = \max\{t \mid (Ek)(k \leq n \;\&\; t \in W_k \;\&\; i = i_1 \;\&\; i \in M_w \;\&\; t$ is the first element in a standard enumeration of $W_k)\}$.

Proposition 2. There exist φ and φ' such that $M_w = M_{w'}$ but M_φ and $M_{\varphi'}$ are of incomparable c - degrees.

The proof is too long to be presented here, and so we confine ourselves to a short sketch of the main idea of the proof.

In [3] two acceptable numberings Ξ and Ψ are constructed by concatenating finite segments defined in such a way that the impossibility of c - reduction of Ξ to Ψ via φ_n was ensured by the (2n + 2)-th segments of these numberings. Let fix an arbitrary acceptable numbering $\{\psi_i\}$ called the initial one. Now we construct the needed numberings alternating segments just like above mentioned with segments obtained from the corresponding segments of the initial numbering by replacing all the values of functions (in their domain) by zeroes. These segments of the initial numbering are choosen in such a way that any of them contains, at least, one index j satisfying the following property:

$(Ai)(i < j \Rightarrow W_i = W_j \;\&\; (Ey)(\psi_j(y)$ is defined and $\psi_j(y) \neq 0))$.

The segments of the first kind provide $M_\varphi \mid_c M_{\varphi'}$ whereas the second ones provide $M_w = M_{w'}$. The remaining needed modifications are of secondary character.

REFERENCES

1. Meyer, A., Program size in restricted programming languages. Information and Control, 21 (1972), 382 - 394.
2. Kinber, J., On btt - degrees of sets of minimal numbers in Gödel numberings. Zeitschrift für mathematische Logik und Grundlagen der Mathematik, 23 (1977), 201 - 212.
3. Маранджян,Г.Б., С-степени множеств минимальных индексов алгоритмов. Известия АН Армянской ССР, серия "Математика", XII,(1977), 130-137.

SOME REMARKS ON BOOLEAN SUMS

Kurt Mehlhorn
Fachbereich 10 - Angewandte
Mathematik und Informatik
Universität des Saarlandes
6600 Saarbrücken, BRD

Abstract

Neciporuk, Lamagna/Savage and Tarjan determined the monotone network complexity of a set of Boolean sums if any two sums have at most one variable in common. Wegener then solved the case that any two sums have at most k variables in common. We extend his methods and results and consider the case that any set of h+1 distinct sums have at most k variables in common. We use our general results to explicitly construct a set of n Boolean sums over n variables whose monotone complexity is of order $n^{5/3}$. The best previously known bound was of order $n^{3/2}$. Related results were obtained independently by Pippenger.

I. Introduction, Notations and Results

We consider the monotone network complexity of sets of Boolean sums $f = (f_1,\ldots,f_m) : \{0,1\}^n \to \{0,1\}^m$ with

$$f_i = \bigvee_{j \in F_i} x_j \quad \text{and} \quad F_i \subseteq \{1,\ldots,n\}.$$

Sets of Boolean sums were also considered by Neciporuk, Lamagna/Savage, Tarjan, Wegener and Pippenger.

$C_B(f)$ denotes the network complexity of f over the basis B; we will consider $B = \{\vee\}$ and $B = \{\vee,\wedge\}$. A set of Boolean sums is called <u>(h,k)-disjoint</u> if for all pairwise distinct $i_0,i_1,i_2,\ldots,i_h$: $|F_{i_0} \cap F_{i_1} \cap \ldots \cap F_{i_h}| \le k$. It is possible to represent a set of Boolean sums $f : \{0,1\}^n \to \{0,1\}^m$ by a bipartite graph with inputs $\{x_1,\ldots,x_n\}$

and outputs $\{f_1,\dots,f_m\}$. The edge (x_j,f_i) is present if and only if $j \in F_i$. Then (h,k)-disjointness is equivalent to saying that the associated bipartite graph does not contain $K_{k+1,h+1}$ (= complete bipartite graph with k+1 inputs and h+1 outputs).

<u>Theorem 1:</u> Let $f : \{0,1\}^n \to \{0,1\}^m$ be a (h,k)-disjoint set of Boolean sums. Then

$$C_{\wedge,\vee}(f) \geq \sum_{i=1}^{m} (|F_i|/k - 1)/h \cdot \max(1,h-1)$$

Neciporuk, Lamagna/Savage, Tarjan proved the theorem in the case $h = 1 = k$. Wegener extended their results to the case $h = 1$ and arbitrary k. The first three authors used their result to explicitly construct sets of n Boolean sums over n variables whose monotone network complexity is $\Omega(n^{3/2})$.
We explicitly construct sets of Boolean sums

$$f : \{0,1\}^n \to \{0,1\}^m \quad \text{such that}$$

$C_{\wedge,\vee}(f) = \Omega(n^{5/3})$. This result was independently obtained by Pippenger.

II. Proofs

Our proof of theorem 1 is based on two lemmas. In these lemmas we will make use of complexity measure C_B^*. $C_B^*(f)$ is the network complexity of f over the basis B under the assumption that all sums $\bigvee_{j\in F} x_j$ with $|F| \leq k$ are given for free, i.e. the sums $\bigvee_{j\in F} x_j$ can be used as additional inputs.
Measure C_B^* was introduced by Wegener.

<u>Lemma 1:</u> Let $f : \{0,1\}^n \to \{0,1\}^m$ be a (h,k)-disjoint set of Boolean sums.
Then
a) $C_\vee^*(f) \leq \max\{1,h-1\}\ C_{\wedge,\vee}^*(f)$
b) $C_\vee(f) \leq \max\{1,h-1,k-1\}\ C_{\wedge,\vee}(f)$

<u>Proof:</u> a) Let N be an optimal $*$-network for f over the basis $\{\vee,\wedge\}$. Then N contains s $\vee$-gates and t $\wedge$-gates, $s+t = C_{\vee,\wedge}^*(f)$.

For $i = 0,1,\dots,t$ we show the existence of a $*$-network N_i for f with $\le t-i$ $\wedge$-gates and $\le s+(h-1)\cdot i$ $\vee$-gates.

We have $N_0 = N$. Suppose now N_i exists. If N_i does not contain an $\wedge$-gate then we are done. Otherwise let G be a last $\wedge$-gate in topological order, i.e. between G and the outputs there are no other $\wedge$-gates. Let g be the function computed by G, g_1 and g_2 the functions at the input lines of G. Then

$$g = s_1 \vee \dots \vee s_p \vee t_1 \vee \dots \vee t_q ,$$

where s_i is a variable and t_j is of length at least 2, is the monotone disjunctive normal form of g.

Case 1: $p \le k$. The sum $s_1 \vee \dots \vee s_p$ comes for free. By theorem I of Mehlhorn/Galil g may be replaced by $s_1 \vee \dots \vee s_p$ and an equivalent circuit is obtained. This shows the existence of network N_{i+1} with $\le t-i-1$ $\wedge$-gates and $\le s+(h-1)(i+1)$ $\vee$-gates.

Case 2: $p > k$: There are some outputs, say $f_1, f_2, \dots, f_\ell$, depending on G. Between G and the output f_j there are only $\vee$-gates and hence $f_j = g \vee u_j$. Since f_j is a boolean sum, u_j is not the constant 1. Hence $\{s_1,\dots,s_p\} \subseteq F_j$ for $j = 1,\dots,\ell$. Since f is (h,k)-disjoint we conclude $\ell \le h$.

Claim: For every j, $1 \le j \le \ell$: either $f_j = g_1 \vee u_j$ or $f_j = g_2 \vee u_j$.

Proof: Since $g = g_1 \wedge g_2$ and $f_j = g \vee u_j$ we certainly have $f_j \le g_1 \vee u_j$ and $f_j \le g_2 \vee u_j$. Suppose both inequalities are proper. Then there are assignments $\alpha_1, \alpha_2 \in \{0,1\}^n$ with $f_j(\alpha_1) = 0 < 1 = (g_1 \vee u_j)(\alpha_1)$ and $f_j(\alpha_2) = 0 < 1 = (g_2 \vee u_j)(\alpha_2)$.
Let $\alpha = \max(\alpha_1,\alpha_2)$. Since f_j is a boolean sum $f_j(\alpha) = 0$ and since $g_1 \vee u_j$ and $g_2 \vee u_j$ are monotone $(g_1 \vee u_j)(\alpha) = (g_2 \vee u_j)(\alpha) = 1$. Hence either $u_j(\alpha) = 1$ or $g_1(\alpha) = g_2(\alpha) = 1$ and hence $g(\alpha) = 1$. In either case we conclude $f_j(\alpha) = (g \vee u_j)(\alpha_j) = 1$. Contradiction. □

We obtain circuit N_{i+1} equivalent to N_i as follows.

1) Replace g by the constant 0. This eliminates $\wedge$-gate G and at least one $\vee$-gate. After this replacement the output line corresponding to

f_j, $1 \le j \le \ell$, realizes function u_j.

2) For every output f_j, $1 \le j \le \ell$, we use one $\vee$-gate to sum u_j and g_1 (resp. g_2). This adds $\ell \le h$ $\vee$-gates.

Circuit N_{i+1} has $\le s+(h-1)(i+1)$ $\vee$-gates and $\le t-i-1$ $\wedge$-gates.

In either case we showed the existence of $*$-network N_{i+1}.
Hence there exists a $*$-network realizing f and containing at most $s+(h-1)\cdot t \le \max\{1,h-1\}(s+t) = \max\{1,h-1\}\cdot C_{\wedge,\vee}(f)$ $\vee$-gates and no $\wedge$-gates
This ends the proof of part a.

b) In order to prove b) we only have to observe that in case 1) above (i.e. $p \le k$) we can explicitely compute $s_1 \vee \ldots \vee s_p$ using at most k-1 $\vee$-gates. Hence N_{i+1} contains at most (k-1) additional $\vee$-gates. □

Lemma 1 has several interesting consequences. Firstly it shows that $\wedge$-gates can reduce the monotone network complexity of sets of (h,k)-disjoint Boolean sums by at most a constant factor. Secondly, the proof of lemma 1 shows that optimal circuits for (1,1)-disjoint sums use no $\wedge$-gates and that there is always an optimal monotone circuit for (2,2)-disjoint sums without any $\wedge$-gates.

<u>Lemma 2:</u> Let $f : \{0,1\}^n \to \{0,1\}^m$ be a (h,k)-disjoint set of Boolean sums. Then

$$C_\vee(f) \ge C_\vee^*(f) \ge \sum_{i=1}^{m} (\lceil |F_i|/k \rceil - 1)/h$$

<u>Proof:</u> Let S be an optimal $*$-network over the basis $B = \{\vee\}$.
Since $f_i = \bigvee_{j \in F_i} x_j$ and input lines represent sums of at most k variable output f_i is connected to at least $\lceil |F_i|/k \rceil$ inputs.
Let G be any gate in S. Since S is optimal G realizes a sum of > k variables and hence at most h outputs f_i depend on G (cf. the discussion of case 2 in the proof of Lemma 1).
For every gate G let n(G) be the number of outputs f_i depending on G. Then $n(G) \le h$ and hence

$$\sum_{G \in S} n(G) \le h \cdot C_\vee^*(f)$$

Next consider the set of all gates H connected to output f_i, $1 \le i \le m$. This subcircuit must contain a binary tree with $\lceil |F_i|/k \rceil$ leaves, (corresponding to the input lines connected to f_i) and hence contains at least $\lceil |F_i|/k \rceil -1$ gates. This shows

$$\sum_{G \in S} n(G) = \sum_{i=1}^{m} \text{ number of gates connected to output } f_i$$

$$\geq \sum_{i=1}^{m} (\lceil |F_i| / k \rceil -1). \qquad \square$$

Wegener proved Lemmas 1 and 2 for the case h = 1. This special case is considerably simpler to prove. Pippenger proved Lemma 2 by a more complicated graph-theoretic approach.

Theorem 1 is now an immediate consequence of Lemmas 1 and 2. Namely,

$$C_{\vee,\wedge}(f) \geq C^*_{\vee,\wedge}(f) \qquad \text{by definition of } C^*_{\vee,\wedge}$$

$$\geq C^*_{\vee}(f)/\max(1,h-1) \qquad \text{by Lemma 1a}$$

$$\geq \sum_{i=1}^{m} (|F_i|/k-1)/h\cdot\max(1,h-1) \quad \text{by Lemma 2}$$

III. Explicit construction of a "hard" set of Boolean sums

Brown exhibited bipartite graphs with n inputs and outputs, $\Omega(n^{5/3})$ edges, and containing no $K_{3,3}$.

His construction is as follows. Let p be an odd prime and let d be a non-zero element of GF(p) (the field of integers modulo p), such that d is a quadratic non-residue modulo p if $p \equiv 1$ modulo 4, and a quadratic residue modulo p if $p \equiv 3$ modulo 4. Let H be a bipartite graph with $n = p^3$ inputs and outputs. The inputs (and outputs) are the triples (a_1,a_2,a_3) with $a_1,a_2,a_3 \in GF(p)$. Input (a_1,a_2,a_3) is connected to output (b_1,b_2,b_3) if

$$(a_1-b_1)^2+(a_2-b_2)^2+(a_3-b_3)^3 = d \text{ modulo } p$$

Brown has shown that bipartite graph H has $p^4(p-1)$ edges and that it contains no copy of $K_{3,3}$.

By the remark in the introduction a bipartite graph corresponds in a natural way to a set of boolean sums. Here we obtain a set of boolean sums over $\{x_1,\ldots,x_n\}$ with $\sum_{i=1}^{n} |F_i| = \Omega(n^{5/3})$.

Furthermore, this set of boolean sums is (2,2)-disjoint.
Theorem 1 implies that the monotone complexity of this set of boolean sums is $\Omega(n^{5/3})$.

Bibliography

Brown, W.G.: On graphs that do not contain a Thompson graph. Can. Math. Bull, 9 (1966), pp. 281-285.

Lamagna, E.A. / Savage, J.E.: Combinatorial complexity of some monotone functions, 15th SWAT Conference, New Orleans, 14o-144, 1974.

Mehlhorn, K. / Galil, Z.: Monotone switching circuits and Boolean matrix product, Computing 16, 99-111, 1976.

Neciporuk, E.I.: On a Boolean matrix, Systems Research Theory, 21, 236-239,1971.

Pippenger, N.: On another boolean matrix, IBM Research Report 69/4, Dec. 1977.

Wegener, I.: A new lower bound on the monotone network complexity of boolean sums, Preprint, Dept. of Mathematics, University of Bielefeld, 1978.

ON THE PROPOSITIONAL ALGORITHMIC LOGIC

Grażyna Mirkowska
Institute of Mathematics, University of Warsaw
00-901 Warsaw, POLAND

ABSTRACT

The aim of propositional algorithmic logic PAL is to investigate properties of program connectives and to develop tools useful in the practice of proving properties of program schemes. The tautologies of PAL become tautologies of algorithmic logic after replacing program variables by programs and propositional variables by formulas.

INTRODUCTION

We shall investigate here properties of program connectives: begin ... end, if ... then ... else ..., while ... do ..., either ... or ... (the connective of nondeterministic choice). We are also interested in tautologies, i.e. expressions which are true by virtue of their syntactical composition, independently of various interpretations which can be associated with signs occurring in them.

The first result in propositional algorithmic logic belongs to Yanov[9]. In 1972 Grabowski[2] proved that zero-order algorithmic logic with propositional variables only is decidable. In 1977 Fisher and Ladner [1] constructed Propositional Dynamic Logic which contains program variables apart of propositional variables. The set of well-formed expressions splits into two sets : of schemes of programs and of formulas. This turned out to be a proper continuation of schematology as originated by Yanov. Formulas of PDL are schemes of statements about termination, correctness, equivalence and other properties of programs. In [1] Fisher and Ladner proved that PDL is decidable and estimated the complexity of the algorithm which recognizes the set of tautologies of PDL. Next Segerberg [8] , Parikh [5] and Pratt [6] gave three different proofs of completeness of axiomatic systems for

We intend here to study PAL which differs from PDL in syntax and semantics. Syntax - schemes of programs in PAL are constructed from propositional formulas and program variables by means of program connectives. This is in contrast with PDL where program expressions are regular expressions over tests, propositional formulas and program variables. The minor difference is that we consider two kinds of formulas $\Delta K\alpha$ and $\nabla K\alpha$. The formula $\nabla K\alpha$ corresponds to PDL's $\langle K\rangle\alpha$ where $\Delta K\alpha$ corresponds to $[K^{+}]\alpha$ of DL^{+}. The semantics introduced in PAL assumes a richer structure, namely, we introduce in PAL valuations of propositional variables explicitly. Interpretation of a program variable will be then a relation between valuations.

The goals of PAL go beyond those of PDL, we are able to characterize the set of tautologies, we can prove its decidability, moreover we see further questions which should be answered.

a/ There are possible several variants of the semantics. The meaning of program variables atomic programs may be restricted to the class of : total functions from the set W of valuations into itself, partial functions from W into W, total relations, relations in W. Each of these semantics gives another set of tautologies and may be considered separately.

<u>REMARK</u> There are two sources of nondeterminism in a language of programs. First, we can admit the program connective either ... or .. of nondeterministic choice. Second, we can admit nondeterministic atomic programs . Obviously they can be combined simultaneously as in DL and PDL or they can be considered separately.

b/ We wish to study not only deuctive system but also propositional theories with specific axioms in order to obtain the equality:

The set of theorems of a consistent theory is equal to the set of all formulas valid in every model of the set of axioms.

In view of noncompactness of propositional logics of programs it is important to be able to recognize those inference rules of PAL which can be used in the practice of proving properties of programs.

c/ We would like to indicate that the semantics proposed here allows us to consider not only program variables but also propositional assignments of the form

$\langle$propositional variable$\rangle$:= $\langle$propositional formula$\rangle$

With this in mind we are able to assert at least three useful facts:

1. Every program scheme can be normalized.
It means that for every program scheme M there exists an equivalent one of the form

begin M1 ; while α do M2 end

where the schemes M1 and M2 do not contain the while connective.
2. The language of extended PAL, i.e. with propositional assignments, is at least as expressive as the language of PDL; this follows from

$K^* \sim$ begin[p: = 0 or p:= 1]; while p do [K or p:=0] end

3. All schemes of axioms of PAL are also schemes of axioms of AL. In other words the axiomatization of AL arises from the axiomatization of PAL with propositional assignments simply by adding rules for introduction and elimination of quantifiers.
d/ The tautologies of PAL are schemes of predicate tautologies of algorithmic logic.

§1. SYNTAX AND SEMANTICS

We consider here the formalized language - an extension of the language of classical propositional calculus in which there are two kinds of variables (propositional and program variables) and two kinds of connectives (usual propositional connectives and program connectives).

Let V_o denote the set of all propositional variables and V_p the set of all program variables. Let F_o be the set of all propositional classical formulas composed in the usual way by means of propositional connectives: disjunction $\cup$, conjunction $\cap$, negation $\neg$ and implication $\Rightarrow$ and two logical constants 1 and 0 from propositional variables.

By a scheme of program we understand any element of the set Π of expressions which is the least set containig V_p and is closed under following rules:

- if M, N are schemes of programs, then [begin M ;N end] and [either M or N] are schemes of programs
- if γ is a classical formula, i.e. $\gamma \in F_o$, and $M, N \in \Pi$, then [while γ do M], [if γ then M else N] are schemes of programs.

Now, we define the set of all formulas as the least set containing F_o and such that

- if $\alpha \in F$ and $M \in \Pi$, then $\Delta M\alpha \in F$ and $\nabla M\alpha \in F$
- if α, β are formulas, then $\neg\alpha, (\alpha\cup\beta), (\alpha\cap\beta), (\alpha\Rightarrow\beta)$ are in the set F.

To define the semantics of the language L_0 we shall consider th two-element Boolean algebra $B_0 = \langle \{0, 1\}, \vee, \wedge, -, \rightarrow \rangle$ and valuations of propositional variables in B_0, i.e. functions $v: V_0 \rightarrow B_0$. Let W denote the set of all valuations of propositional variables.

By an interpretation of the program variables we shall mean any function I which to every program variable assigns a binary relation in W, $I : V_p \rightarrow 2^{W \times W}$. For every program variable K, K_I denotes a relation which is assigned to K by the interpretation I.

Values of classical propositional formulas are defined depending on valuations of propositional variables [cf. Rasiowa-Sikorski 7].

The notion of interpretation may be extended to the set of all programs. Let us introduce first some definitions.

By a configuration we shall understand an ordered pair $\langle v, M_1; ..; M_n \rangle$ where v is a valuation of propositional variables an $M_1; ..; M_n$ is a list of programs (may be empty).

For a given interpretation I of the program variables let $\mapsto$ denote the binary relation in the set of all configurations such that

1. $\langle v, M_1; ..; M_n \rangle \mapsto \langle v', M_2; ..; M_n \rangle$ where M_1 is an atomic program $M_1 \in V_p$ and $\langle v, v' \rangle \in M_{1I}$

2. $\langle v, \text{either } M_1 \text{ or } M_2; M_3; ..M_n \rangle \mapsto \langle v, M_1; M_3; ..M_n \rangle$
   $\langle v, \text{either } M_1 \text{ or } M_2; M_3; ..M_n \rangle \mapsto \langle v, M_2; M_3; ..M_n \rangle$

3. $\langle v, \text{begin } M_1; M_2 \text{ end } ; M_3, ..M_n \rangle \mapsto \langle v, M_1; M_2; M_3; ..M_n \rangle$

4. $\langle v, \text{if } \gamma \text{ then } M_1 \text{ else } M_2; M_3; ..M_n \rangle \mapsto \begin{cases} \langle v, M_1; M_3; ..M_n \rangle \text{ iff } \gamma(v) = 1 \\ \langle v, M_2; M_3; ..M_n \rangle \text{ iff } \gamma(v) = 0 \end{cases}$

5. $\langle v, \text{while } \gamma \text{ do } M_1 ; M_2; ..M_n \rangle \mapsto \begin{cases} \langle v, M_1; \text{while } \gamma \text{ do } M_1; M_2; .. \rangle \text{ iff } \gamma(v) = 1 \\ \langle v, M_2; M_3; .. M_n \rangle \text{ in the opposite case} \end{cases}$

The sequence of configurations will be called computation of the program scheme M in the interpretation I at the initial valuation v, iff $c_1 = \langle v, M \rangle$ and for all i, $c_i \mapsto c_{i+1}$.
If the computation is a finite sequence $c_1, \ldots, c_n$ and the last configuration c_n is of the form $\langle v', \rangle$ (i.e. the second part of the configuration is the empty sequence), then the computation will be called successful. The valuation v' in the successful computation will be called result of the computation of the program M.
The set of all results of the program M in the interpretation I at the initial valuation v will be denoted by $M_I(v)$.

Hence, for a given interpretation I, to every program M we can assign a binary relation M_I such that $v \, M_I v'$ iff $v' \in M_I(v)$.

Now we are ready to define the value of any formula for a given interpretation of the program variables and a given valuation of the propositional variables.

$(\nabla M\alpha)_I(v) = \mathbb{1}$ iff there exists asuccessful computation of the program M at the valuation v in the interpretation I such that its value satisfies the formula α.

$(\Delta M\alpha)_I(v) = \mathbb{1}$ iff all computations of the program M at the valuation v in the interpretation I are successful and all the results satisfy the formula α.

By a tautology we shall mean a formula α such that for every interpretation I and every valuation v, $\alpha_I(v) = \mathbb{1}$.

EXAMPLE

For every program scheme and every formula α the following schemes are tautologies of propositional algorithmic logic

$\Delta M\alpha \Rightarrow \nabla M\alpha)$, $(\Delta M \neg\alpha \Rightarrow \neg\nabla M\alpha)$, $(\neg\Delta M\alpha \cup \neg\Delta M\neg\alpha)$ □

§2. SEMANTICAL CONSEQUENCE OPERATION

By a model of the set of formulas Z we shall mean a pair $\langle W_0, I\rangle$ where I is an interpretation and W_0 is a subset of the set of all valuations W such that for every program variable K and for any valuation $v \in W_0$, $K_I(v) \in W_0$ and for all formulas α from the set Z, $\alpha_I(v) = \mathbb{1}$, i.e. α is valid in $\langle W_0, I\rangle$.

We shall say that α is a semantic consequence of the set of formulas Z, $Z \models \alpha$, iff for every pair $\langle W_0, I\rangle$, if $\langle W_0, I\rangle$ is a model for Z, then $\langle W_0, I\rangle$ is a model for α.

The semantical consequence operation $\models$ has the same properties as the classical one excluding the compactness property.

Lemma There exists a set of formulas Z and a formula α such that $Z \models \alpha$ and for every finite subset Z_0 of Z there exists a model for Z_0 which is not a model for α.

Proof.

Assume that $V_0 = \{a_0, a_1, \dots\}$. Let $Z = \{\Delta[\text{begin } K_1; K_2^{\,i} \text{ end}]\, a_0\}_{i\in\omega}$ and $\alpha = \neg(\nabla[\text{begin } K_1; \text{ while } a_0 \text{ do } K_2 \text{ end}]\,\mathbb{1})$, where K_1, K_2 are program variables and a_0 is a propositional variable.

It is easy to show that $Z \models \alpha$.

For every subset X of ω, let us construct an interpretation I in the following way : for every valuations v, v', v'' of propositional variables

$vK_{1I}\ v'$ iff $v'(a_i) = \mathbb{1}$ for $i \in X$, and $v'(a_i) = \mathbb{0}$ for $i \notin X$

$v'K_{2I}\ v''$ iff $v''(a_i) = v'(a_{i+1})$ for $i = 0,1,\ldots$

The pair $\langle V_0, I \rangle$ is a model for the set $Z_0 = \{\Delta[\text{begin } K_1; K_2^i \text{ end}]\, a_0\}_{i\in}$ but it is not a model for α. $\square$

§3. SYNTACTICAL CONSEQUENCE OPERATION

We shall give a syntactical characterization of the semantics consequence operation described above. Assume the following schemes as axioms :

All axioms of classical propositional calculus and

$$\nabla K\, \mathbb{1} \Rightarrow \Delta K \mathbb{1} \qquad \Delta M\alpha \Rightarrow \nabla M\alpha$$

$$✡[\text{begin } M_1 ; M_2 \text{ end}]\,\alpha \equiv (✡M_1(✡M_2\alpha))$$

$$✡[\text{if } \gamma \text{ then } M_1 \text{ else } M_2]\alpha \equiv (\gamma \cap ✡M_1\alpha) \cup (\neg\gamma \cap ✡M_2\alpha)$$

$$✡[\text{while } \gamma \text{ do } M_1]\,\alpha \equiv (\neg\gamma \cap \alpha) \cup (\gamma \cap ✡M_1(✡[\text{while } \gamma \text{ do } M_1]\,\alpha))$$

$$\nabla[\text{either } M_1 \text{ or } M_2]\alpha \equiv (\nabla M_1\alpha \cup \nabla M_2\alpha)$$

$$\Delta[\text{either } M_1 \text{ or } M_2]\alpha \equiv (\Delta M_1\alpha \cap \Delta M_2\alpha)$$

$$\Delta M(\alpha \cap \beta) \equiv (\Delta M\alpha \cap \Delta M\beta)$$

$$\nabla M(\alpha \cup \beta) \equiv (\nabla M\alpha \cup \nabla M\beta)$$

$$\Delta M \neg\alpha \Rightarrow \neg\nabla M\alpha \qquad \Delta M\mathbb{1} \Rightarrow (\neg\nabla M\alpha \Rightarrow \Delta M\neg\alpha)$$

$$\neg ✡M\, 0$$

We assume the following rules of inference :

$$\frac{\alpha, (\alpha \Rightarrow \beta)}{\beta} \qquad \frac{(\alpha \Rightarrow \beta)}{✡M\alpha \Rightarrow ✡M\beta}$$

$$\frac{\{(✡[\text{if } \gamma \text{ then } M]^i(\alpha \cap \neg\gamma) \Rightarrow \beta)\}_{i\in\omega}}{(✡[\text{while } \gamma \text{ do } M]\,\alpha \Rightarrow \beta)}$$

In all the above schemes K denotes a program variable, M, M_1, M_2 denote schemes of programs, γ a classical formula and α, β arbitrary formulas from F. The formula $\alpha \equiv \beta$ is used as an abbreviation for two formulas $(\alpha \Rightarrow \beta)$ and $(\beta \Rightarrow \alpha)$. Every formula containing the sign ✡

should be read twice: as a formula with Δ in all places where ✡ occurs and then as a formula with ∇ in all places where ✡ occurs.

The set of all axioms and inference rules defines in a usual way the syntactical consequence operation C. The system $\langle L_0 , C \rangle$ consisting of the language L of propositional algorithmic logic and of syntactic consequence operation C will be called propositional algorithmic logic PAL .

By an easy verification we have the following lemmas .

<u>Lemma</u> The propositional algorithmic logic is consistent . ▯

<u>Lemma</u> For every formula α , if α is a theorem of PAL then α is a tautology. ▯

§4. DETERMINISTIC, TOTAL INTERPRETATION OF ATOMIC PROGRAMS

In this paragraph we shall consider the class of interpretations of L_0 which to every program variable K assigns a total function in the set of all valuations W. We shall call them functional .

Let us extend the set of axioms defined in §3 by the axioms of the following two schemes :

$\Delta K \mathbb{1}$ for $K \in V_p$

$\nabla K \alpha \Rightarrow \Delta K \alpha$ for $K \in V_p$ and $\alpha \in F$.

Denote the new consequence operation by C_f and the corresponding propositional calculus by PAL_f .

We shall say that α is valid in a functional interpretation I_f if for every valuation v, $\alpha_{I_f}(v) = \mathbb{1}$.

By an easy verification we have the following adequacy theorem

<u>Theorem</u> For every formula α, if α is a theorem of PAL_f then α is valid in every functional interpretation . ▯

The two further facts are the main results of this section.

<u>Theorem</u> A theory $T = \langle L_0, C_f, A \rangle$ based on PAL_f is consistent iff T has a model . ▯

<u>Theorem</u> For any consistent theory T based on PAL_f the following conditions are equivalent

(i) α is a theorem of T,

(ii) α is valid in every model for T .

Proof.

(ii)→(i) by the adequacy theorem.

To prove that (i) implies (ii) assume that α is not a theore of T. Consider the Lindenbaum algebra of the theory and denote its el ments by $\|\alpha\|$ for $\alpha \in F$. As a consequence of the assumption, $\|\neg\alpha\| \neq 0$ Hence, there exists Q-filter $\mathfrak{F}$ in the Lindenbaum algebra such that $\|\neg\alpha\| \in \mathfrak{F}$. We shall construct a model for T such that α is not vali

Let ζ denote the set of all finite sequences of program vari bles with the signs Δ or ∇ , e.g. $\Delta K_1 \nabla K_2 \nabla K_3$. By v_c for $c \in \zeta$ we denote a valuation such that

$$v_c(a) = \begin{cases} 1 & \text{iff } \|ca\| \in \mathfrak{F} \\ 0 & \text{if } \|ca\| \notin \mathfrak{F} \end{cases} \qquad v_\emptyset(a) = \begin{cases} 1 & \text{if } \|a\| \in \mathfrak{F} \\ 0 & \text{if } \|a\| \notin \mathfrak{F} \end{cases} \quad \text{for all } a \in V$$

Let I be an interpretation of the program variables which to e ry program variable K assigns function K_I such that $K_I(v_c) = v_{c\Delta K}$

By induction on the complexity of the formula we can prove that for every c, $\|c\alpha\| \in \mathfrak{F}$ if and only if $\alpha_I(v_c) = 1$.

Since $\|\neg\alpha\| \in \mathfrak{F}$, we have $\alpha_I(v) = 0$. ▯

§5. SUBSTITUTION THEOREM

In this section we aim to indicate that the tautologies of prop sitional algorithmic logic are schemes of tautologies of nondeterm nistic algorithmic logic , Mirkowska [3] .

Let α be a formula of PAL and let s be a substitution of the

$$[a_1/\alpha_1, \ldots, a_n/\alpha_n, K_1/M_1, \ldots, K_m/M_m],$$

where $a_i \in V_0$ for $i=1,..n$, $K_j \in V_p$ for $j=1,..m$, M_j is a program of no deterministic algorithmic logic NAL and α_i is a formula of NAL .

By $\overline{s\alpha}$ we shall mean an expression that arrises from the fo mula α by the simultanous replacement of any variable a_i by the fo mula α_i respectively, and any program variable K_j by program M_j .

Theorem For every formula α of PAL and for every substitution s su that $\overline{s\alpha}$ is a well-formed formula of NAL, if α is a tautology of PAL then $\overline{s\alpha}$ is a tautology of NA

§6. FINAL REMARKS

In the subsequent papers we shall study

- relational interpretations of the language of PAL and the connections between modal logics and PAL
- propositional algorithmic logic of schemes of concurrent programs, in order to find properties of the program connective of parallelism cobegin ... coend .

REFERENCES

1. Fisher,M.,Ladner,R., Propositional dynamic logic of regular programs, Proc. 9th Ann. ACM Symp.,Boulder,Colorado,1977.
2. Grabowski,M., The set of tautologies of zero-order algorithmic logic is decidable, Bull.Acad.Pol.Sci. Ser.Math. 20 (1972), 575-582.
3. Mirkowska,G., On formalized systems of algorithmic logic,ibid. 19, (1971),421-428.
4. Mirkowska,G., Algorithmic logic with nondeterministic programs, to appear in Fundamenta Informaticae.
5. Parikh,R., A completeness result for PDL, MFCS,september 1978.
6. Pratt,V.R., A practical decision method for propositional dynamic logic, Tenth ACM symposium on Theory of Computing ,1978.
7. Rasiowa,H.,Sikorski,R., The mathematics of metamathematics, PWN, Warsaw 1963.
8. Segerberg,K., A completeness theorem in the modal logic of programs, Notices of the ACM, 24.6. A-552,1977.
9. Yanov,J., On equivalence of operator schemes, Problems of Cybernetics 1,1-100,1959 .

Ch(k) GRAMMARS: A CHARACTERIZATION OF LL(k) LANGUAGES

Anton Nijholt
Department of Mathematics
Vrije Universiteit, Amsterdam
The Netherlands.

Eljas Soisalon-Soininen
Department of Computer Science
University of Helsinki
Finland.

1. INTRODUCTION

From the point of view of parsing the LL(k) grammars constitute a very attractive class of context-free grammars. For each LL(k) grammar a top-down parsing algorithm can be devised which is essentially a one-state deterministic push-down automaton. From a more theoretic point of view LL(k) grammars are attractive as well. It is well-known, for example, that it is decidable whether two LL(k) grammars are equivalent. Also the hierarchy of LL(k) languages with regard to the length k of the look-ahead is a characteristic property.

The class of LL(k) grammars is properly contained in the class of LR(k) grammars, and even the family of LL(k) languages is properly contained in the family of LR(k) languages. If we focus on the "gap" between LL(k) and LR(k) grammars the following points are of interest.

(i) There is the obvious difference in grammar definition.

(ii) The generating capacities are different.

(iii) Apart from the difference between LR(0) and LR(1) languages the length k of the look-ahead does not play a role for LR(k) languages.

(iv) Every LL(k) grammar is both left parsable and right parsable but there are LR(k) grammars which are not left parsable [1].

We consider the present paper as a contribution to the research which tries to clarify the differences between LL(k) and LR(k) grammars. Research in this area has been reported e.g. in Rosenkrantz and Lewis [7], Brosgol [2], Hammer [4], Soisalon-Soininen and Ukkonen [9], Demers [3] and Soisalon-Soininen [8]. In this paper we introduce the class of so called Ch(k) grammars (pronounced "chain k grammars"). This class of grammars is properly contained in the class of LR(k) grammars and it properly contains the LL(k) grammars. However, the family of Ch(k) languages coincides with the family of LL(k) languages. Nevertheless, the parsing properties of Ch(k) grammars are quite different from the parsing properties of LL(k) grammars. The class of Ch(k) grammars can be considered as a generalization of the class of simple chain grammars [6] in the same sense as the class of LL(k) grammars is a generalization of the class of simple LL(1) grammars.

The present paper is organized as follows. In Section 2 we define the necessary background concerning context-free grammars and parsing. The Ch(k) grammars are defined in Section 3 where also some basic properties of Ch(k) grammars are proved. In Section 4 we demonstrate that the well-known transformation process of left factoring the given grammar will always produce an LL(k) grammar from a Ch(k) grammar and that, in fact, this process cannot produce an LL(k) grammar from a non-Ch(k) grammar. This result implies the equality of the classes of LL(k) and Ch(k) languages, and it also clarifies the relationship of Ch(k) grammars with some other classes of grammars.

2. BACKGROUND

In this section we review various commonly known definitions (cf. [1]) and give some notations. A quadruple $G = (N,\Sigma,P,S)$ is a *context-free grammar* (*grammar* for short) if N and Σ are finite disjoint sets, P is a finite subset of the product $N \times (N \cup \Sigma)^*$ and S is an element of N. Elements of the set N are called *nonterminals* and denoted by capital Latin letters from the beginning of the alphabet A,B,C,...,S. Elements of the set Σ are called *terminals* and denoted by small Latin letters from the beginning of the alphabet a,b,c,...,s. By X, Y and Z we denote elements which are either in N or in Σ. The elements (A,ω) of P are called *productions* and denoted by $A \to \omega$. The symbol S is called the *start symbol* of the grammar.
Terminal strings, i.e. strings in Σ^*, are denoted by small Latin letters from the end of the alphabet t,u,v,...,z, whereas small Greek letters $\alpha,\beta,\gamma,\dots,\omega$ denote strings in $(N \cup \Sigma)^*$. The *empty string* is denoted by ε. The *derives* relation $\Rightarrow$ of G on the set $(N \cup \Sigma)^*$ is defined by the condition $\alpha A\beta \Rightarrow \alpha\omega\beta$ if α and β are strings in $(N \cup \Sigma)^*$ and $A \to \omega$ is a production in P. If here α is required to be a terminal string, then we get the definition of the *leftmost* derives relation of G, denoted by $\underset{L}{\Rightarrow}$, and if β is required to be a terminal string, then we get the definition of the *rightmost* derives relation of G, denoted by $\underset{R}{\Rightarrow}$.
A sequence $Q_1,Q_2,\dots,Q_n$ of strings Q_i in $(N \cup \Sigma)^*$ is called a *leftmost derivation* (respectively *rightmost derivation*) of Q_n from Q_1 in the grammar G if $Q_i \underset{L}{\Rightarrow} Q_{i+1}$ (respectively $Q_i \underset{R}{\Rightarrow} Q_{i+1}$) holds in G for each $i = 1,\dots,n-1$ whenever $n > 1$. The sequence of productions used in a leftmost derivation of a string Q from the start symbol S is a *left parse* of Q in G, and the reverse of the sequence of productions in a rightmost derivation of Q from S is a *right parse* of Q in G.
Let $G = (N,\Sigma,P,S)$ and $G' = (N',\Sigma,P',S')$ be grammars and let $h: P'^* \to P^*$ be a homomorphism. We say that G' *left-to-right covers* G with respect to the homomorphism h, if the following two conditions hold:

(i) if π' is a left parse of a terminal string w in the grammar G', then $h(\pi')$ is a right parse of w in the grammar G;

(ii) if π is a right parse of a terminal string w in the grammar G, then there is a left parse π' of w in G' such that $h(\pi') = \pi$.

If, in these conditions, "left" (resp. "right") is replaced by "right" (resp. "left"), then, under the conditions, G' *right-to-right covers* (resp. *left-to-left covers*) the grammar G with respect to the homomorphism h. The shorthand "right cover" is often used for "right-to-right cover", as well as the shorthand "left cover" is used for "left-to-left cover". If the grammar G' left-to-right covers, right covers or left covers the grammar G, then we say that G' *covers* G. Observe that the *language* $L(G') = \{x \in \Sigma^* \mid S' \Rightarrow^* x\}$ generated by the grammar G' equals the language L(G) if G' covers G.

Let $G = (N,\Sigma,P,S)$ be a grammar and let k be a nonnegative integer. If α is string in $(N \cup \Sigma)^*$ then we denote by $k{:}\alpha$ the first k symbols of α whenever the length $|\alpha|$ of α is greater than or equal to k, and α otherwise. By $FIRST_k(\alpha)$ we denote the set of all strings $k{:}u$ such that u is a terminal string and α derives u. The grammar G is said to be LL(k) if, for a terminal string w, a nonterminal A and strings γ, δ_1 and δ_2 in $(N \cup \Sigma)^*$ such that $A \to \delta_1$ and $A \to \delta_2$ are distinct productions of G, the condition

$$S \underset{L}{\Rightarrow^*} wA\gamma$$

implies that

$$FIRST_k(\delta_1\gamma) \cap FIRST_k(\delta_2\gamma) = \emptyset.$$

The definition of an LL(k) grammar immediately implies some properties of LL(k) grammars. For example, each LL(k) grammar is *unambiguous*, i.e. each terminal string in the language has exactly one left parse. In addition, if an LL(k) grammar is *reduced*, i.e. every production is used in a left (or right) parse of some terminal string, then it is not *left-recursive*, i.e. it has no nonterminal A such that $A \Rightarrow^+ A\alpha$ for a general string α.

3. DEFINITION OF Ch(k) GRAMMARS

In order to intuitively characterize the class of grammars to be defined and its relationship with other classes of grammars, we first illustrate the deterministic top-down, bottom-up and left-corner parsing algorithms; i.e. the parsing algorithms that apply to LL(k), LR(k) and LC(k) [7,8] grammars, respectively. Consider the derivation tree shown in Figure 1.

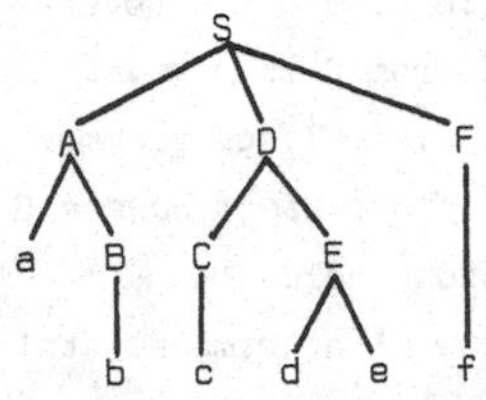

Figure 1. Derivation tree

n the top-down parsing algorithm for LL(k) grammars the productions for the non-erminals in the tree are recognized in the order S,A,B,D,C,E,F. Each production in he tree is recognized before its descendants and its right siblings and their escendants. In the bottom-up parsing algorithm for LR(k) grammars the productions or the nonterminals in the tree are recognized in the order B,A,C,E,D,F,S. ach production in the tree is recognized after its descendants but before its ncestors and its right siblings and their descendants. In the left-corner parsing lgorithm for LC(k) grammars the productions are recognized in the order A,B,S,C,D, ,F. Each production is recognized after its left corner but before any of the iblings of the left corner (or their descendants).

he Ch(k) grammars can now be characterized as LR(k) grammars for which the left-and sides of the productions can be recognized in the same order as the whole roductions in top-down parsing, but the right-hand sides are recognized in the rder of the bottom-up parse. This method for constructing the derivation tree "node by node" corresponds to the way in which the well-known recursive descent-arser constructs the tree. For example, when the top-down parsing algorithm has ecognized the productions for S,A,B and D in the derivation tree of Figure 1, then n the case of Ch(k) grammars the left-hand sides S,A,B and D are determined but he whole productions only for B and A.

n a sense, the Ch(k) grammars constitute a dual of the PLR(k) grammars [8,9] in a imilar way as the LL(k) grammars constitute a dual of LR(k) grammars as regards he construction of the derivation tree. The PLR(k) grammars are LR(k) grammars for hich the left-hand sides of the productions can be recognized in the same order as he whole productions in left-corner parsing, but the right-hand sides are ecognized in the order of the bottom-up parse. Thus the PLR(k) grammars are those or which deterministic "node by node" parsing bottom-up is possible, whereas the h(k) grammars are those for which deterministic "node by node" parsing top-down is ossible.

EFINITION 3.1. Let k be a non-negative integer. A grammar $G = (N,\Sigma,P,S)$ is said o be a Ch(k) *grammar* if, for a terminal string w, a nonterminal A and strings γ, , δ_1 and δ_2 in $(N \cup \Sigma)^*$ such that $A \to \alpha\delta_1$ and $A \to \alpha\delta_2$ are distinct productions of and α is the longest common prefix of $\alpha\delta_1$ and $\alpha\delta_2$, the condition

$$S \underset{L}{\overset{*}{\Rightarrow}} wA\gamma$$

mplies that

$$\mathrm{FIRST}_k(\delta_1\gamma) \cap \mathrm{FIRST}_k(\delta_2\gamma) = \emptyset. \qquad \square$$

bserve the obvious difference with the definition of LL(k) grammars. In that case he implication $\mathrm{FIRST}_k(\alpha\delta_1\gamma) \cap \mathrm{FIRST}_k(\alpha\delta_2\gamma) = \emptyset$ is used. Thus in the case of Ch(k) rammars it is not necessary to consider the terminal strings which can be derived rom the longest common prefix α of the right-hand sides of two distinct productions $A \to \alpha\delta_1$ and $A \to \alpha\delta_2$. These observations imply immediately

THEOREM 3.2. Every LL(k) grammar is a Ch(k) grammar.

Proof. Assume that a grammar $G = (N,\Sigma,P,S)$ is not a Ch(k) grammar. Then there exist a terminal string w, a nonterminal A and strings γ, α, δ_1 and δ_2 in $(N \cup \Sigma)^*$ such that $A \to \alpha\delta_1$ and $A \to \alpha\delta_2$, where α is the longest common prefix of $\alpha\delta_1$ and $\alpha\delta_2$, are two distinct productions of G, $S \overset{*}{\underset{L}{\Rightarrow}} wA\gamma$ and $FIRST_k(\delta_1\gamma) \cap FIRST_k(\delta_2\gamma) \neq \emptyset$. But then also $FIRST_k(\alpha\delta_1\gamma) \cap FIRST_k(\alpha\delta_2\gamma) \neq \emptyset$, which means that the grammar G is not LL(k), as desired. □

As an informal description of the definition of Ch(k) grammars and of its relationship with the definitions of other classes of grammars consider the following situation. There exist terminal strings w, x, y and z, a nonterminal A, a symbol X which is a nonterminal or a terminal, and a general string α such that $A \to X\alpha$ is a production and

$$S \overset{*}{\Rightarrow} wAz, \quad X \overset{*}{\Rightarrow} x, \quad \alpha \overset{*}{\Rightarrow} y.$$

Consider then the terminal string wxyz. The production $A \to X\alpha$ in question in the derivation tree of wxyz can be recognized with certainty after scanning

(i) w and k:xyz if the grammar is LL(k),

(ii) wx and k:yz if the grammar is LC(k),

(iii) wxy and k:z if the grammar is LR(k), PLR(k) or Ch(k).

However, if the grammar is PLR(k) then the left-hand side A of the production $A \to X\alpha$ is recognized after scanning wx and k:yz, and if the grammar is Ch(k) then the left-hand side A is recognized after scanning w and k:xyz.

As we remarked in the introduction the Ch(k) grammars can be considered as a generalization of simple chain grammars [6]. A grammar $G = (N,\Sigma,P,S)$ is said to be a simple chain grammar if G is ε-free (i.e. P contains no production of the form $A \to \varepsilon$), P is prefix-free (i.e. there are no two productions $A \to \alpha\beta$ and $A \to \alpha$ in P) and for any pair of productions $A \to \alpha X\beta$ and $A \to \alpha Y\gamma$, X and Y are in $N \cup \Sigma$, such that $X \neq Y$, we have $FIRST_1(X) \cap FIRST_1(Y) = \emptyset$. For left part grammars [5] the requirement that P is prefix-free is dropped. The following theorem is an immediate consequence of the above discussion.

THEOREM 3.3. A grammar $G = (N,\Sigma,P,S)$ is a simple chain grammar if and only if P is prefix-free and G is an ε-free Ch(1) grammar.

For example, the grammar with productions

$$S \to a, \quad S \to ab$$

is not a simple chain grammar, because these two productions do not constitute a prefix-free set of productions. However, this grammar is Ch(1). Thus we conclude by Theorem 3.3 that the class of simple chain grammars is properly contained in the class of Ch(k) grammars whenever $k > 0$. Further, since there exist simple chain grammars which are not LL(k) for any k [6], we conclude by Theorems 3.2 and 3.3 that the class of LL(k) grammars is properly contained in the class of Ch(k) grammars whenever $k > 0$. In the case $k = 0$ both classes contain only the grammars that derive at most one terminal string.

The following two theorems are also immediate consequences of the Ch(k) definition. The proofs are analogous to corresponding ones in the LL(k) case and therefore omitted.

THEOREM 3.4. Each Ch(k) grammar is unambiguous.

THEOREM 3.5. A reduced Ch(k) grammar is not left-recursive.

4. PROPERTIES OF Ch(k) GRAMMARS

In this section our primary interest is to relate the Ch(k) grammars and LL(k) grammars by a grammatical transformation. In fact, we shall show that the Ch(k) grammars are exactly those which can be transformed into LL(k) grammars by *left-factoring* the grammar until it has no two productions of the form $A \rightarrow \alpha\beta$ and $A \rightarrow \alpha\gamma$ where $\alpha \neq \varepsilon$. This implies, in particular, that the language generated by a Ch(k) grammar is always an LL(k) language, and thus, by Theorem 3.2, the family of Ch(k) languages equals the family of LL(k) languages. Furthermore, this result implies the interesting property that Ch(k) grammars are PLR(k) grammars [8,9]. Since the left factoring process yields an LL(k) grammar if and only if the given grammar is Ch(k), we can perform the test whether a grammar is Ch(k) by left-factoring the grammar and then testing the LL(k)-property. However, the LL(k) parser of the resulting grammar cannot be used to produce left or right parses in the original grammar. That is, left-factoring can distort the structure of the grammar such that no left-to-left nor left-to-right cover is obtained [6]. Nevertheless, we can give a simple modification of the left-factoring process such that the above mentioned properties are preserved except that, as an additional bonus, the transformed grammar left-to-right covers the original grammar.

We begin by defining that a grammar is in the *left-factored* form, if it has no two productions $A \rightarrow \alpha\beta$ and $A \rightarrow \alpha\gamma$ such that α is not the empty string. The definitions of LL(k) and Ch(k) grammars imply immediately

THEOREM 4.1. A grammar in the left-factored form is LL(k) if and only if it is Ch(k).

The process of left-factoring can be regarded as a transformation which is composed by consecutive steps of "factoring" two distinct productions $A \to \alpha\beta$ and $A \to \alpha\gamma$, $\alpha \neq \varepsilon$, into productions $A \to \alpha A'$, $A' \to \beta$ and $A' \to \gamma$, where A' is a new nonterminal. These steps are performed, in an arbitrary way, until the grammar is in the left-factored form. It should be noticed that the above specification of the process does not define the "left-factored" grammar uniquely. However, our results are independent of the particular way in which the individual steps and their order in the left-factoring process are chosen.

THEOREM 4.2. The grammar obtained by the left-factoring process is LL(k) if and only if the original grammar is Ch(k).

Sketch of proof. By Theorem 4.1 it is enough to show that the process of left-factoring does not affect the Ch(k)-ness of a grammar and that the process of left-factoring cannot produce a Ch(k) grammar from a non-Ch(k) grammar. It is clear by the definition that this is true as regards one individual step in the left-factoring process. Since the whole process is just a consecutive sequence of these individual steps, we thus conclude the theorem. □

COROLLARY 4.3. The families of Ch(k) and LL(k) languages are identical.

The PLR(k) grammars [8,9] are exactly those grammars which can be transformed into LC(k) grammars by the left-factoring process [8]. Thus, since the class of LL(k) grammars is properly contained in the class of LC(k) grammars [8], we conclude by Theorem 4.2 that the class of Ch(k) grammars is properly contained in the class of PLR(k) grammars. This implies further that Ch(k) grammars are LR(k) grammars, since PLR(k) grammars are LR(k) grammars [8].
The inclusion of the Ch(k) grammars in the class of PLR(k) grammars is an interesting property because PLR(k) grammars can be transformed into LL(k) grammars such that the transformed grammar left-to-right covers the original grammar [8]. This is thus true also for Ch(k) grammars. However, the transformation involved is rather complicated, and it is thus desirable to find out easier possibilities.
Let therefore $G_1 = (N_1,\Sigma,P_1,S)$ be a grammar and let $G_1' = (N_1',\Sigma,P_1',S)$ where $N_1' = N_1 \cup \{[A\alpha] \mid A \to \alpha \text{ is in } P\}$ and $P_1' = \{A \to \alpha[A\alpha] \mid A \to \alpha \text{ is in } P_1\} \cup \{[A\alpha] \to \varepsilon \mid A \to \alpha \text{ is in } P_1\}$. Further let $G_2 = (N_2,\Sigma,P_2,S)$ be a grammar obtained by the left-factoring process from G_1', and let $h\colon P_2^* \to P_1^*$ be a homomorphism defined by the conditions

$$h([A\alpha] \to \varepsilon) = A \to \alpha, \text{ and}$$
$$h(A \to \alpha) = \varepsilon.$$

LEMMA 4.4. The grammar G_2 left-to-right covers the grammar G_1 with respect to the homomorphism h.

<u>Sketch of proof</u>. The appearance of a production of the form $[A\alpha] \to \varepsilon$ in a left parse of a terminal string x in G_1' means that a substring y of x for which $A \Rightarrow \alpha \Rightarrow^* y$ in G_1 has been analyzed. Thus it can be shown by any easy induction that the sequence of these productions in a left parse of x in G_1' defines a right parse of x in G_1. Further, this is also true when G_1' is replaced by the grammar G_2, since the process of left-factoring does not affect the order in which the productions of the form $[A\alpha] \to \varepsilon$ are recognized. By formalizing the above discussion we can conclude that the grammar G_2 left-to-right covers the grammar G_1 with respect to the homomorphism h. □

Since the grammar G_2 obtained by the left-factoring process from G_1' is LL(k) if and only if any grammar obtained by the left-factoring process from G_1 is LL(k), we conclude by Theorem 4.2 and Lemma 4.4 that the following theorem holds.

<u>THEOREM 4.5.</u> Each Ch(k) grammar of size[*)] n can be left-to-right covered by an LL(k) grammar of size O(n).

REFERENCES

1. A.V. Aho and J.D. Ullman, The Theory of Parsing, Translation and Compiling, Vols 1 and 2, Prentice Hall, Englewood Cliffs, N.J., 1972 and 1973.
2. B.M. Brosgol, Deterministic translation grammars, Proc. 8th Princeton Conf. on Information Sciences and Systems 1974, pp. 300-306.
3. A.J. Demers, Generalized left corner parsing, Conf. Record of the 4th ACM Sympos. on Principles of Programming Languages 1977, pp. 170-182.
4. M. Hammer, A new grammatical transformation into LL(k) form, Conf. Record of the 6th Annual ACM Sympos. on Theory of Computing 1974, pp. 266-275.
5. A. Nijholt, A left part theorem for grammatical trees, Discrete Mathematics 25 (1979), pp. 51-63.
6. A. Nijholt, Simple chain grammars and languages, to appear in Theoretical Computer Science.
7. D.J. Rosenkrantz and P.M. Lewis, Deterministic left-corner parsing, IEEE Conf. Record of the 11th Annual Sympos. on Switching and Automata Theory 1970, pp. 139-152.
8. E. Soisalon-Soininen, Characterization of LL(k) languages by restricted LR(k) grammars, Ph.D. Thesis, Report A-1977-3, Department of Computer Science, University of Helsinki.
9. E. Soisalon-Soininen and E. Ukkonen, A characterization of LL(k) languages, in: Automata, Languages and Programming (eds. S. Michaelson and R. Milner), Third Colloquium, Edinburgh University Press, Edinburgh, 1976, pp. 20-30.

[*)] The size $|G|$ of a grammar G is defined by $|G| = \sum_{A\to\alpha\in P} |A\alpha|$.

A UNIFORM APPROACH TO BALANCED BINARY AND MULTIWAY TREES

Th. Ottmann
Institut für Angewandte Informatik und Formale Beschreibungsverfahren, Universität Karlsruhe, W-Germany

D. Wood
Dept. of Applied Mathematics, McMaster University, Hamilton, Ontario L8S 4K1, Canada

1. Introduction

Balanced search trees such as AVL-trees, 2-3 trees and B-trees are among the most popular data structures used to implement dictionaries. These structures guarantee O(log N) performance for the three operations searching, insertion and deletion. While AVL-trees and 2-3 trees are used to manipulate sets of keys in main memory, B-trees have been, until recently, the only choice for implementing dictionaries which use backup store. Kwong and Wood [KW] introduced a relaxed variant of B-trees, f(m)-trees, which seem to be an attractive alternative to B-trees. In both cases (B-trees and f(m)-trees) one node corresponds to a page of fixed size which can hold up to m-1 keys and m pointers to sons. The insertion of a new key ensues by the recursive strategy of splitting an "over-full" node having m keys into two (both with $\lceil m/2 \rceil -1$ keys in the case of B-trees and at least $f(m)-1 \geq 1$ keys in the case of f(m)-trees) and then moving one of its keys upwards. Though this strategy is applicable to ternary trees, cf. the insertion procedure for 2-3 trees of Hopcroft [AHU], its extension to the binary case is not obvious. In fact the standard AVL insertion procedure (for binary height-balanced trees) comprises a quite different strategy. Further, if the keys are inserted in ascending order into an initially empty B-trees, the pages of the resulting tree are only half filled. An attempt to solve this "sparsity problem" for B-trees is the main idea of "overflow" introduced by Bayer and McCreight [BM] which leads to the notion of B*trees, see Knuth [K], p. 477 ff. Instead of splitting an over-full node look first at its left (or right) brother. Say the immediate right brother has only j keys and j+1 sons where $j < m-1$. Then the over-full node "flows over" into its brother node. If the brother node is already full, then split both nodes and create three new nodes, each about two-thirds full. The result of this modification is an increase in storage utilization. At least two-thirds of the available space is utilized.

We might even look at both immediate left and right brothers before splitting nodes. However, this also does not yield trees whose nodes are maximally filled when keys are inserted in ascending order.

We show how trees can be made as dense as possible by looking at all brothers of a node before splitting it. We introduce classes of "dense" m-ary trees for arbitrary $m \geq 2$ and design insertion procedures for these classes in a general framework. Though we look at all brothers before splitting nodes, surprisingly it turns out that in the "weakly dense" case the amount of work to be done is proportional to log N where N is the number of stored keys. Further iterative insertion of keys in ascending order produces trees whose nodes have as many sons as possible.

It would be desirable also to estimate the average storage utilization after a sequence of random insertions for dense m-ary trees. But it turns out that Yao's analysis [Y] for B-trees cannot be applied except for the special case $m = 2$ (cf. [OW]). The reason for this fact is that (except for $m = 2$) looking at brothers before splitting nodes makes it impossible to predict the effect of a (random) insertion for subtrees of a given small height.

When we allow multiple users to search and update a tree in parallel looking at brothers before splitting nodes turns out to be very messy. To look only upwards, as in the B-tree and f(m)-tree case, seems to be the better philosophy. That, nonetheless, node-splitting does not yield degenerate trees is assured in [BM], respectively [KW] by the requirement that each node must have at least $\lceil m/2 \rceil$, respectively $f(m) \geq 2$ sons. In a final section we discuss the question of whether this requirement can be weakened in order to include also the case of binary trees into the general framework.

2. Dense m-ary trees

We define classes of m-ary trees where $m \geq 2$ is a fixed natural number. Each internal node in the tree may have j sons where $1 \leq j \leq m$. A node which has the maximal number m of sons or which is a leaf is said to be saturated, otherwise it is unsaturated.

An m-ary tree T is said to be r-dense, where r is a natural number with $1 \leq r \leq m-1$ iff (1)...(5) hold.

(1) The root of T is at least binary.

(2) Each unsaturated node different from the root has either only saturated brothers and at least one such brother or at least r saturated brothers.

(3) All leaves have the same depth.

(4) Each node with i+1 sons has i keys ($0 \leq i < m$).

(5) The i keys $k_1, \ldots, k_i$ of a node p with i+1 sons $\sigma_1 p, \ldots, \sigma_{i+1} p$ are ordered such that for all j, $1 \leq j \leq i$, the following holds: The keys in the subtree with root $\sigma_j p$ are less than k_j, which in turn is less than the keys in the subtree with root $\sigma_{j+1} p$.

A class of m-ary trees is called dense if it is a class of r-dense m-ary trees for some r. In particular, we speak of weakly dense m-ary trees and strongly dense m-ary trees, respectively, if we have in mind the classes of 1-dense and (m-1)-dense m-ary trees, respectively. Observe that there is only one class of dense binary trees. This class coincides with the class of 1-2 brother trees [OW].

There are two classes of dense ternary trees. Figure 1 shows an example of a strongly dense ternary tree and an example of a weakly dense ternary tree each with 11 leaves and 10 keys.

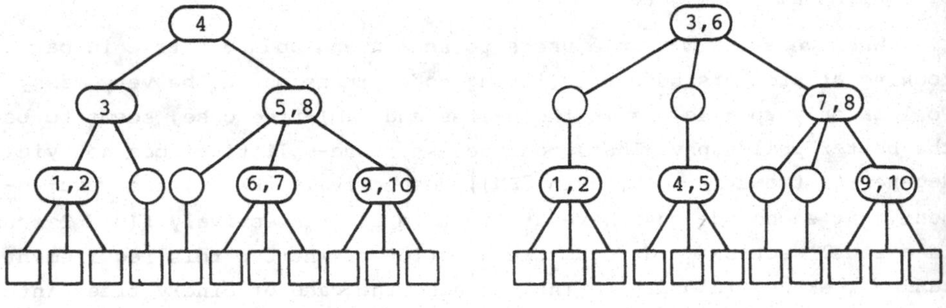

Figure 1

It is clear that for the given height h the complete m-ary tree (where each internal node has m sons) has the maximal number of leaves of all r-dense m-ary trees of height h.

We can easily derive a recurrence formula for the minimal number of leaves of an r-dense m-ary tree of height h, $L_{min}(r,m,h)$; this recurrence formula can be solved using standard methods (cf.[COW] for the details) and gives

$$L_{min}(r,m,h) = \begin{cases} 2; & \text{if } h = 1 \\ A_1 \cdot \alpha_1^{h+1} + A_2 \cdot \alpha_2^{h+1}; & \text{if } h \geq 2 \end{cases}$$

where

$$\alpha_1 = \frac{1}{2}\,(r + \sqrt{4(m-r) + r^2}\,)$$

$$\alpha_2 = \frac{1}{2}\,(r - \sqrt{4(m-r) + r^2}\,)$$

$$A_1 = \frac{1-\alpha_2}{\alpha_1 - \alpha_2}, \quad A_2 = \frac{\alpha_1 - 1}{\alpha_1 - \alpha_2}$$

This gives in the special case of binary trees ($m = 2$, $r = 1$) the result that the minimal number of leaves of dense binary trees, (i.e. of brother trees) of height h is about $1.171(1.618)^h$. Further it is easy to show by induction on h that for every r-dense m-ary tree of height h $L_{min}(r,m,h) \geq m^{\lfloor h/2 \rfloor}$.

Hence, every r-dense m-ary tree with N keys and N+1 leaves is of height $h \leq 2 \cdot \log_m(N+1)+1$. This shows that r-dense m-ary trees cannot degenerate into linear lists.

3. Brother-tree splitting for dense m-ary trees

It is clear how to search in a dense m-ary tree. In order to insert a new key into a dense m-ary tree we firstly search for the key in the tree. Since the key is not found we insert it in a node p at the father of leaves level and we create also a new leaf q (representing the new key-interval in between the sons of p). If the node p now contains fewer than m keys the insertion is completed. Otherwise (i.e. if p contains m keys) we call _up_ (p,q).

The upwards restructuring procedure _up_ will be designed in a uniform way for all r-dense m-ary trees, where r is arbitrary in the allowed interval. _Up_ may eventually call the subroutines _right-shift_ or _left-shift_ which we explain first:

right-shift (p,q);
On entry q is a right brother of p or $p = q$; p is at least binary.

Case 1 [$p \neq q$]

Let p be the i-th son of its father φp and p' the (i+1)st son of φp.

Let r be the rightmost son of p. Make r the leftmost son of p' and call right-shift (p',q)

Case 2 [p = q]

FINISH.

<u>left-shift</u> is defined similarly.

In the above and all below designed procedures we pass over the question of how to locally transfer keys from one node to the other.

However, we mention the crucial point in the insertion-procedure for dense m-ary trees: The shifting of nodes may destroy the property of being an r-dense tree. Consider, for example, the following strongly dense ternary tree:

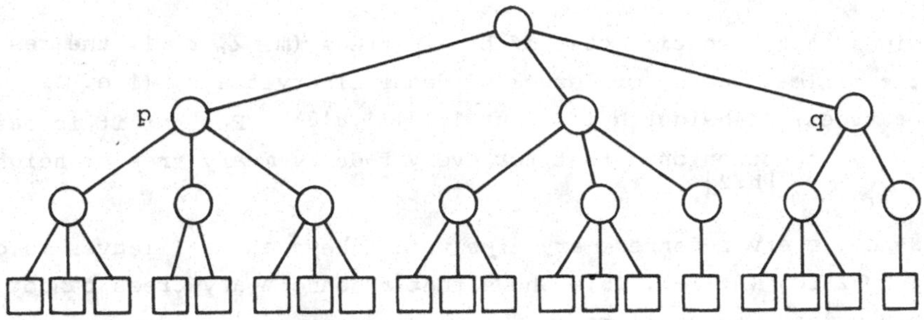

The result of performing <u>right-shift</u> (p,q) is no longer a strongly dense ternary tree, since q obtains two unary sons. Therefore additiona work has to be done in order to keep the tree in strongly dense form. W will return to this problem later and assume for the moment that <u>right-shift</u> and <u>left-shift</u> always yield r-dense trees when applied to these trees.

Let us call a node a <u>⊗-node</u> if it is either the saturated root of an r-dense m-ary tree or a leaf.

<u>up</u> is a recursive procedure with the following invariant condition Whenever <u>up</u>(p,q) is called then

- p and all of p's sons are ⊗-nodes,
- q is either a leaf or has a single son σq which is a ⊗-node,
- q lies to the right of the leftmost son of p and to the left of the rightmost son of p.

<u>up</u>(p,q);

Case 1 [p has an unsaturated brother]

Let βp be an unsaturated brother of p which is nearest to p. (That means, all brothers occurring in between p and βp are saturated). Make q an additional son of p and call <u>right-shift</u> (p,βp), if p is a right brother of p, and call <u>left-shift</u> otherwise; FINISH.

Case 2 [p has only saturated brothers]

Case 2.1 [p has an unsaturated father]

Make q an additional son of p; remove the leftmost son ℓ of p and make it the only son of a newly created node u; make u an additional son of the father φp of p and FINISH.

Case 2.2 [p has no father, i.e. p is the root]

Then create a new root φp with p as its only son and proceed as in Case 2.1.

Case 2.3 [The father φp of p is saturated]

Assume first that p is not the rightmost son of its father φp. Create a new immediate right brother q' of p which obtains as its only son the rightmost son of p; make q an additional son of p; call <u>up</u>(φp,q'). If p is the rightmost son of φp proceed analogously by creating a new immediate left brother of p.

(Observe that the invariant condition is maintained.)

<u>end of up</u>

It is clear that in the worst case h (recursive) calls of <u>up</u> can occur where h is the height of the tree before insertion. Thus, the performance time of <u>up</u> is dominated by h and the time to carry out <u>right-shift</u> (resp. <u>left-shift</u>). If the shift procedures are of the above specified form where no further restructuring is required then their performance time is constant. Simple shifting without further restructuring is sufficient for the case of binary and weakly dense ternary trees. For in these cases any call of <u>left-shift</u> or <u>right-shift</u> again yields a dense binary and a weakly dense ternary tree, respectively. Thus, for these two classes total performance time of <u>up</u> is O(h)=O(log N), where N is the number of leaves (or stored keys). It can be shown by quite involved combinatorial arguments that the same holds for any class of weakly dense m-ary trees: (Cf. [COW] for the proof.)

<u>Theorem 1</u>:

It is possible to insert a new key into an N-key weakly dense m-ary tree in time at most $O(m^3 \log N)$. Moreover, the insertion procedure maintains the order of keys and retains the weakly dense tree structure.

We do not know exactly for which other classes of dense m-ary tree insertion can be carried out in time O(log N) for fixed $m \geq 3$. It seems that in general almost the whole tree has to be restructured in order to maintain the r-dense tree structure after insertion.

Very little is known about deletion of keys in dense m-ary trees. However, in the binary case deletion can be carried out in time O(log N see [OW], and in the case of strongly dense ternary trees an $O(\log^2 N)$ deletion procedure is available. But the general case remains open.

4. Storage utilization under brother tree splitting

As already mentioned in Section 1 iterative insertion of keys in ascending order into the initially empty tree using the splitting-procedure for B-trees or f(m)-trees yields trees which are very "sparse"; The storage utilization, i.e. the ratio of the number of storage cells (which is (m-1) times the number of internal nodes of the resulting tree) divided by the number of stored keys is about 50 % (or even less in the case of f(m)-trees). In contrast to this the above specified insertion procedure for r-dense m-ary trees yields trees with the maximal possible storage utilization:

Theorem 2:

Iterative insertion of (m^h-1) keys in strictly ascending order beginning with the one-leaf tree of height 1 and with no key yields the complete m-ary tree of height h.

The proof is by induction on h, cf. [COW], and holds uniformly for every class of dense m-ary trees. Furthermore, the "critical" shifting of nodes to unsaturated brothers passing several saturated nodes never applies when inserting keys in ascending order.
We conjecture that Theorem 2 can be strengthened as follows: The tree which results by iteratively inserting a given sequence of keys approximates the complete m-ary tree as better as the given sequence is more presorted.

In order to estimate the average storage utilization after a sequence of random insertions Yao [Y] has developed a method for 2-3 and B-trees. The very heart of his method is to compute precisely the number of all subtrees of a given small height (say 1,2, or 3) in the resulting tree after N random insertions and to estimate the nodes and keys on heigher levels. This approach requires that the effect of an insertion into a given small subtree be predicted precisely. It is a fortuitious circumstance that Yao's method is still applicable to dense binary trees, see [OW], where an average storage utilization which is better than for 2-3 trees is obtained. Unfortunately Yao's approach already fails for dense ternary trees. Consider, for example, the saturated root of a height 1 subtree in a dense ternary tree:

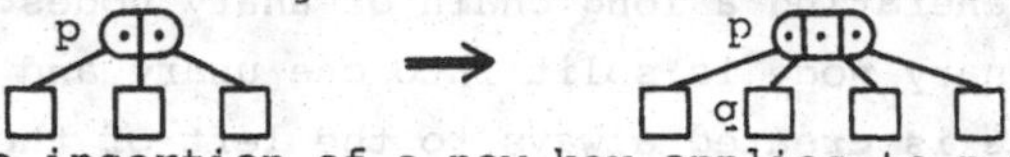

and up(p,q)

If the insertion of a new key applies to p we insert the new key into p, create a new leaf q as an additional son of p and call up(p,q). The effect of up now depends on whether or not the brothers of p are saturated or not. If all brothers of p are saturated p will be split and a new node on level 1 will be created. But no new node will be created, if p has an unsaturated brother, because one of p's sons is shifted to this unsaturated brother. Hence, we are unable to say what are the resulting subtrees of height 1 after an insertion into the subtree with root p when considering only this subtree. A possible way out of these difficulties seems to be the following one: We define a modified algorithm by performing brother-tree splitting as described in Section 3 up to a certain level ℓ above the leaves. At level ℓ we do a B-tree splitting (cf. Section 5) i.e. we do not look at brothers at all before splitting nodes. Independent of what kind of splitting we choose at the levels higher (nearer to the root) than ℓ it seems to be plausible that the thus modified algorithm keeps at least as many keys at low levels as the original one on the average.

Now we can compute the average number of keys at levels $\leq \ell$ for the modified algorithm by Yao's method which gives us a lower bound for the original algorithm.

5. B-tree splitting for binary trees

In [KW] the B-tree splitting procedure has been modified by allowing an arbitrary split of an over-full node (with m keys and m+1 sons) rather than a split into two equal parts as in the B-tree case. The resulting trees do not degenerate, because in [KW] the two new nodes must have at least one key and two (pointers to) sons. Clearly, this relaxed variant of B-trees still does not cover the case of binary trees. In order to obtain a splitting procedure which is also applicable in the binary case and which follows the strategy of only looking upwards (to the father of the actual node) and not to brothernodes we allow an even more liberal splitting mechanism.

If a node of an m-ary tree has m keys and m+1 (pointers to) sons it can be split into two nodes with s-1 and m-s keys respectively, where $1 \leq s \leq \lceil m/2 \rceil$.

As it stands this splitting procedure does not prevent degenerate trees.

For, obviously, the insertion of one new key may lead to a sequence of node-splittings generating a long chain of unary nodes. If, for example, an overfull binary node is split into one unary and one binary such that the unary node is created always to the left of the binary node, iterative insertion of N keys in ascending order yields binary trees of height N.

But degenerate trees cannot occur if the following condition holds

(c) There is a fixed natural number $k \geq o$ such that for both new nodes p there is a $k' \leq k$ such that $\sigma^{k'}p$ is at least binary.

Here $\sigma^{k'}p$ is defined recursively by:

$\sigma^{o}p = p$

$\sigma^{n+1}p =$ the only son of $\sigma^{n}p$, if $\sigma^{n}p$ is unary.

It should be clear that condition (c) is fulfilled for the case of m-ary trees where $m \geq 3$, since the splitting procedure of [KW] trivially fulfills it. The only problem is to find a strategy, which ensures condition (c) also holds in the binary case. The intended strategy, of course, should not include looking at brothers! An obvious strategy of this kind is to split an overfull binary, i.e. a ternary node into one unary and one binary node such that the unary node has a binary descendant after the minimal possible number of generations.

Let us call a node a $\otimes$-node if it is either a leaf or binary. Then it is easy to see that iterative insertion may lead to a ternary node p (which has to be split) only if at least one of its three sons is a $\otimes$-node. The above strategy now recommends a split of p such that the unary node obtains a $\otimes$-node as its only son whenever possible. If unfortunately, both the rightmost and leftmost sons of p are unary make that son of p the only son of the unary fragment node which has a $\otimes$-no as descendant in shorter distance.

It is now easy to see that iterative insertion of keys in ascendin order no longer yields degenerate trees. We conjecture that the resulting number of internal nodes and keys at the lowest h levels after a sequence of N random insertions does not differ very much from the corresponding number in brother trees (cf. Section 4 and [OW]).

Unfortunately however, we do not know how this dynamically defined class of binary trees (obtainable by iterative insertions according to the above splitting strategy) behaves. In particular we do not know whether the class contains degenerate trees.

eferences

AHU] Aho,A.V., Hopcroft,J.E., and Ullman,J.D.: The Design and Analysis of Computer Algorithms, Addison-Wesley, Reading, Mass., (1974).

BM] Bayer,R. and McCreight,E.: Organization and Maintenance of Large Ordered Indexes, Acta Informatica 1, 1972, 173-189.

COW] Culik II,K., Ottmann,Th., and Wood,D.: Dense multiway trees, Report 77, Institut für Angewandte Informatik und Formale Beschreibungsverfahren, Universität Karlsruhe, 1978.

K] Knuth,D.: The Art of Computer Programming, Vol. 3: Sorting and Searching, Addison - Wesley, Reading, Mass., (1973).

KW] Kwong,Y.S. and Wood,D.: T-Trees: A variant of B-trees, Computer Science Technical Report 78-CS-18, McMaster University, Hamilton, Ontario, 1978.

OW] Ottmann,Th. and Wood,D.: 1-2 brother trees or AVL trees revisited, to appear in Computer Journal (1979).

Y] Yao,A.: On random 2,3 trees, Acta Informatica 9, 1978, 159-170.

ON THE GENERATIVE CAPACITY OF SOME CLASSES OF GRAMMARS WITH REGULATED REWRITING

Gheorghe Păun

University of Bucharest, Division of Systems Studies
Str. Academiei 14, Bucharest, R-7olo9 Romania

INTRODUCTION

Since the computational power of context-free grammars is too weak and the context-sensitive ones "shed little light on the problem of attaching meanings to sentences" [3], context-free grammars with restrictions in derivation have been introduced. Generally, such devices generate intermediate families of languages between the context-free and the context-sensitive language families ([9]). So far, about twenty types of such grammars are known.

A major problem in this area is that of the generative capacity of these grammars, compared to that of the unrestricted Chomsky grammars and to the computational power of other similar devices. The purpose of the present paper is exactly to compare the generative capacity of some well-known variants of matrix grammars: simple matrix grammars and right linear simple matrix grammars [4], matrix grammars of finite index [2],[6], scattered context grammars [3]. The index of matrix and of simple matrix grammars, the degree of simple matrix and of right linear simple matrix grammars, combined with the use of λ-rules, induce eight hierarchies of matrix languages. All these hierarchies are infinite (the result is known for some of them). We prove that the λ-rules can be eliminated from right linear simple matrix grammars and from finite index matrix grammars, but cannot be removed from simple matrix grammars (of degree n or of index n). Consequently, there are only six distinct hierarchies. The relations between the families on the same level of these hierarchies are investigated too.

Another main result of the paper is the proof of the strict inclusion of the family of simple matrix languages (generated with or without using λ-rules) in the family of λ-free generated scattered context languages. (Although in [4] it is claimed - without proof - that any simple matrix language is a matrix language, we feel that this is not true; see arguments in [6].)

The results of the paper are summarized in the next diagram where the directed arrows denote strict inclusions (and $n \geq 4$). The families in this diagram are: $\mathcal{SM}$ = simple matrix languages, $\mathcal{SM}(n)$ = simple matrix languages of degree no greater than n, $\mathcal{SM}(\mathrm{ind}\ n)$ = simple matrix languages of index not greater than n, $\mathcal{SMR}(n)$ = right linear simple matrix languages of degree not greater than n, $\mathcal{SMR}$ = right linear simple matrix languages, $\mathcal{M}(\mathrm{ind}\ n)$ = matrix languages of index not greater than n, $\mathcal{SM}_f$ = simple matrix languages of finite index, $\mathcal{M}_f$ = matrix languages of finite index, $\mathcal{S}$ = scattered context languages. The superscript λ indicates that λ-rules are allowed. The definitions of simple matrix grammars can be found in [4], that of matrix grammars in [9], the index of grammars and languages is defined in [2] and the scattered context grammars in [3]. The monograph [7] contains almost all the known results in the area of matrix and matrix-like restrictions in derivation.

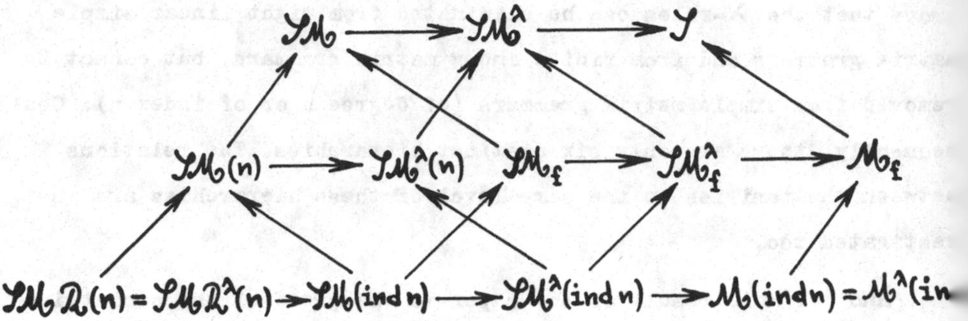

RESULTS

We assume the reader to be familiar with the basic results in regulated rewriting, particularly with the papers quoted above. Thus we omit all the definitions but that of the index.

Let $G = (V_N, V_T, S, M)$ be a matrix grammar. We denote by $N(x)$ the length of the string obtained by erasing all terminal symbols in x, $x \in (V_N \cup V_T)^*$. Let D be a derivation according to G,

$$D : S = w_1 \Rightarrow w_2 \Rightarrow \ldots \Rightarrow w_n.$$

We define

$$\mathrm{ind}(D,G) = \max \{ N(w_i) \mid i = 1,2,\ldots,n \}.$$

For $x \in L(G)$ we put

$$\mathrm{ind}(x,G) = \min \{ \mathrm{ind}(D,G) \mid D \text{ is a derivation of } x \text{ according to the grammar } G \}.$$

Then, the index of G is

$$\mathrm{ind}(G) = \sup \{ \mathrm{ind}(x,G) \mid x \in L(G) \}$$

and, for a language $L \in \mathcal{M}^\lambda$, we define

$$\mathrm{ind}(L) = \inf \{ \mathrm{ind}(G) \mid L = L(G) \}.$$

In a similar way we define the index for simple matrix languag of any type.

In what follows we present the lemmas and theorems without proofs (the proofs - some of them are very tedious - will be published elsewhere).

The inclusion $\mathcal{L} \subseteq \mathcal{L}^\lambda$ is trivial for any family $\mathcal{L}$ in the above list. In [4] it is proved that $\mathcal{L}^\lambda(n) \subseteq \mathcal{L}^\lambda(n+1)$, for any $\mathcal{L} \in \{\mathcal{SM}, \mathcal{SMR}\}$. Similarly, $\mathcal{L}(n) \dot{\subset} \mathcal{L}(n+1)$, for these families. ($\subset$ denotes a proper inclusion.) The inclusions $\mathcal{SM}(n) \subset \mathcal{SM}$, $\mathcal{SM}(\text{ind } n) \subseteq \mathcal{SM}_f$, and $\mathcal{M}(\text{ind } n) \subseteq \mathcal{M}_f$ are obvious. From the proofs in [5] it follows that $\mathcal{SM}(\text{ind } n) \subset \mathcal{M}(\text{ind } n)$, $\mathcal{SM}^\lambda(\text{ind } n) \subset \mathcal{M}^\lambda(\text{ind } n)$. These inclusions are proper since the language $\{a^n b^n c^n \mid n \geq 1\}^*$ belongs to $\mathcal{M}(\text{ind } 2)$ but not to $\mathcal{SM}^\lambda$ [4].

An algorithmic procedure can be constructed which eliminates the λ-rules from right linear simple matrix grammars, that is, we have

<u>Theorem 1</u>. $\mathcal{SMR}(n) = \mathcal{SMR}^\lambda(n)$, $n \geq 1$.

Such a result is not true for context-free simple matrix grammars. (This problem is not investigated in [4].) Indeed, the language

$$L = \{a^n b^n a^{k_1} b^{k_1} a^{k_2} b^{k_2} \ldots a^{k_m} b^{k_m} a^n b^n \mid n \geq 1,\ m \geq 1,\ k_i \geq 1 \text{ for all } i\}$$

belongs to $\mathcal{SM}^\lambda(2)$ but not to $\mathcal{SM}$. Consequently, we have

<u>Theorem 2</u>. $\mathcal{SM}(n) \subset \mathcal{SM}^\lambda(n)$, $n \geq 1$, and $\mathcal{SM} \subset \mathcal{SM}^\lambda$.

As $L \in \mathcal{SM}^\lambda(\text{ind } 4)$, we have $\mathcal{SM}(\text{ind } n) \subset \mathcal{SM}^\lambda(\text{ind } n)$ too for $n \geq 4$.

Clearly, $\mathcal{SM}(\text{ind } n)$ and $\mathcal{SM}^\lambda(\text{ind } n)$ define infinite hierarchies and $\mathcal{SM}(\text{ind } n) \subseteq \mathcal{SM}(n)$, $\mathcal{SM}^\lambda(\text{ind } n) \subseteq \mathcal{SM}^\lambda(n)$. The last inclusions are proper since we have

<u>Theorem 3</u>. The family $\mathcal{SM}(1)$ contains infinite index languages (according to simple matrix grammars of any degree).

Such a language is

$$L' = \{c\}(D\{c\})^+$$

where D is the Dyck language over $\{a,b\}$ (D has an infinite index with respect to context-free grammars [8]).

Therefore, $\mathcal{SM}_f \subset \mathcal{SM}$ and $\mathcal{SM}_f^\lambda \subset \mathcal{SM}^\lambda$ are proper inclusions.

Although the λ-rules cannot be removed from context-free simple matrix grammars, the λ-matrices (the matrices containing only λ-rules) and the vocabularies of nonterminals generating only one (null or non-null) string can be algorithmically eliminated. In this aim we use the "pure form" theorem from [5]. This theorem is also used in proving the inclusion $\mathcal{SM}^\lambda \subset \mathcal{S}$. (Let $G = (V_1,\ldots,V_n,V_T,S,M)$ be a simple matrix grammar and let $x = x_1x_2\ldots x_n$, $x_i \in (V_i \cup V_T)^*$ be a string which can be derivated from S in the grammar G. Let A_i be the leftmost nonterminal in each string x_i, $i = 1,2,\ldots,n$. The grammar G is said to be pure if for any matrix $(B_1 \longrightarrow y_1,\ldots,B_n \longrightarrow y_n)$ in M, either $(A_1,\ldots,A_n) = (B_1,\ldots,B_n)$ or $A_i \neq B_i$ for every $i = 1,2,\ldots,n$. As it was proved in [5], for any simple matrix grammar an equivalent pure grammar can be algorithmically constructed.)

Lemma 1. Any language in $\mathcal{SM}^\lambda(n)$ is the homomorphic image of the intersection of n languages in $\mathcal{SM}^\lambda(2)$.

The homomorphism used is a linearly erasing one.

Lemma 2. $\mathcal{SM}^\lambda(2) \subset \mathcal{S}$.

Combining with the results in [3] concerning the closure properties of the family $\mathcal{S}$, we obtain

Theorem 4. $\mathcal{SM}^\lambda \subset \mathcal{S}$.

The inclusion is proper since, contrarily to $\mathcal{SM}^\lambda$, the family $\mathcal{S}$ contains languages with non-semi-linear Parikh image.

In [6] it was proved that $\mathcal{M}_f = \mathcal{M}_f^\lambda$. By a suitable reformulation of the proof of Theorem 4 in [6], we obtain

Theorem 5. $\mathcal{M}(\text{ind } n) = \mathcal{M}^\lambda(\text{ind } n)$, $n \geq 1$.

The main step of the proof is

Lemma 3. Let G be a matrix grammar of index n. There is an equivalent matrix grammar G' of index not greater than n+1, $G' = (V_N, V_T, S,$ with $V_N = V_N^1 \cup V_N^2 \cup \{S\}$, $V_N^1 \cap V_N^2 = \emptyset$ and the matrices in M are of the following forms

a) $(S \longrightarrow AB)$, $A \in V_N^1$, $B \in V_N^2$,

b) $(A_1 \longrightarrow x_1, \ldots, A_u \longrightarrow x_u, B \longrightarrow C)$, $A_i \in V_N^1$, $x_i \in$ $\in (V_N^1 \cup V_T)^+$, $B, C \in V_N^2$,

c) $(B \longrightarrow \lambda)$, $B \in V_N^2$,

d) $(S \longrightarrow x)$, $x \in V_T^*$.

(Then, attacking the symbols in V_N^2 to nonterminals in V_N^1, we obtain a λ-free matrix grammar of the same index as G or smaller.)

Finally, in order to prove that $\mathcal{M}$(ind n), $n \geq 1$, is an infinite hierarchy, the following pumping lemma is used.

Lemma 4. For any language $L \in \mathcal{M}$(ind n), $L \subseteq V^*$, there exists an integer p such that for any $z \in L$, $|z| > p$, there is a string $z' \in L$ such that

a) $z' = y_1 z_1 y_2 z_2 \ldots y_t z_t y_{t+1}$, $z = z_1 z_2 \ldots z_t$, $y_i, z_i \in V^*$ for all i,

b) $z' = u_1 v_1 w_1 x_1 u_2 v_2 w_2 x_2 u_3 \ldots u_k v_k w_k x_k u_{k+1}$, $k \leq n$, $u_i, v_i, w_i, x_i \in V^*$ for all i and

i) $|v_1 x_1 \ldots v_k x_k| > 0$,

ii) $u_1 v_1^s w_1 x_1^s u_2 \ldots u_k v_k^s w_k x_k^s u_{k+1} \in L$ for any $s \geq 1$.

($|x|$ denotes the length of the string x.)

Theorem 6. $\mathcal{M}$(ind n) $\subset \mathcal{M}$(ind (n+1)) is a strict inclusion for any $n \geq 1$.

(The language $L_n = \{ b(a^i b)^{2n} \mid i \geq 1 \}$ is in $\mathcal{M}$(ind n) $- \mathcal{M}$(ind (n-1)).)

One can prove that every family $\mathcal{M}$(ind n), $n \geq 1$, is a full semi-AFL not closed under concatenation, iteration +, substitution, intersection and complementation. Moreover, $\mathcal{M}$(ind n), $n \geq 1$, have the "prefix implies regular" property defined in [1], as well as the families $\mathcal{SM}$(n), $n \geq 1$.

OPEN PROBLEMS

In the field of regulated rewriting there are a lot of significan unsolved problems. Many such problems are related to the above discussed families of languages. We formulate only some of them:

P1) Which is the relation between $\mathcal{M}_f$ and $\mathcal{SM}^{\lambda}_f$? (More specifically, are there languages in $\mathcal{SM}^{\lambda}_f$ which are not in $\mathcal{M}_f$?) (We believ that $D \notin \mathcal{M}_f$.)

P2) Is the family $\mathcal{SM}^{\lambda}$ included in the family of matrix languages (We don't think so since $\mathcal{SM}^{\lambda}$ is closed under doubling, but we feel that there are context-free languages L such that $\{xx \mid x \in L\}$ is not a matrix language.)

P3) Closure properties of $\mathcal{SM}$(ind n), $\mathcal{SM}^{\lambda}$(ind n), $\mathcal{SM}_f$, $\mathcal{SM}^{\lambda}_f$.

REFERENCES

1. BOOK, R., On languages with a certain prefix property, Math. Syst. Theory, lo, 3(1976/77), 229-237.
2. BRAINERD, B., An analog of a theorem about context-free languages, Inform. Control, 11 (1968), 561-568.
3. GREIBACH, S., HOPCROFT, J., Scattered context grammars, J. of Computer and System Sci., 3 (1969), 233-247.
4. IBARRA, O., Simple matrix grammars, Inform. Control, 17 (197o), 359-394.
5. PĂUN, GH., On the generative capacity of simple matrix grammars of finite index, Inform. Proc. Letters, 7, 2(1978), loo-lo2.
6. PĂUN, GH., On the family of finite index matrix languages, J. of Computer and System Sci., (In press).
7. PĂUN, GH., Matrix grammars. Applications, Ed. Ştiinţifică şi Enciclopedică, Bucureşti, 1979 (In Roum.).
8. SALOMAA, A., On the index of context-free languages, Inform. Control, 14 (1969), 474-478.
9. SALOMAA, A., Formal languages, Academic Press, New York and London 1973.

VALIDITY TEST FOR FLOYD'S OPERATOR-PRECEDENCE PARSING ALGORITHMS

Peter Ružička

Computing Research Center
Dúbravská 3, 885 31 Bratislava
Czechoslovakia

ABSTRACT

The classes of languages definable by operator-precedence grammars and by Floyd's operator-precedence algorithms are studied. Operator-precedence languages are shown to be a proper superclass of languages accepted by Floyd's operator-precedence parsing algorithms. An algorithm is developed to decide equivalence of an operator-precedence grammar and the underlying Floyd's operator-precedence parsing algorithm, a result of possible practical significance. As a consequence a necessary and sufficient condition for an operator-precedence grammar to be valid grammatical characterization for the underlying Floyd's operator-precedence parsing algorithm is obtained.

1. INTRODUCTION

Recent work on the improvement in both the running time and the size of bottom-up parsing algorithms has been oriented in several directions. Much effort has been devoted to develop optimizing transformations to reduce the size and/or the time of parsing algorithms and in addition modified techniques have been proposed to produce parsing algorithms of practical size. The research on the validity of shift-reduce parsing algorithms also belongs to this area.

The shift-reduce operator-precedence parsing algorithm is conceptually very simple and hereby a very effective technique for syntactical analysis. It is sufficiently general to handle the parsing of a variety of the programming-language constructs. However, the main impediment for using this technique is that Floyd's operator-precedence parsing algorithm might accept strings not in the language of the underlying operator-precedence grammar. This is a consequence of using different means in the definition of both concepts. While operator-precedence grammars use nonterminal symbols as well as unique operator-precedence relations to derive valid input strings, Floyd's parsing algorithms use only operator-precedence relations to recognize valid input strings. On the other hand, the great advantage of this method is the size efficiency because a copy of the grammar need not be kept to know which reduction to make by Floyd's operator-precedence parsing algorithms.

Other parsing techniques are known in which it is necessary to overcome the same problem of invalid strings being recognized. Well known are canonical precedence parsing algorithms of Gray [5] , or skeletal LR parsing algorithms of El Djabri [3] and Demers [2] , or Knuth's [7] top-down parsing algorithms with partial back-up used in Mc Clure's translator writing system TMG . Knuth has proved that in the case of top-down parsing algorithms with partial back-up it is not algorithmically decidable whether a grammar generates the same language as it is accepted by the intended parsing algorithm. Demers presented an algorithm to decide whether a skeletal LR parsing algorithm accepts exactly the language of the underlying grammar. Demers further claimed that his validity test applies to skeletal canonical precedence parsing algorithms as well. We consider a similar problem for operator-precedence parsing algorithms and we solve it by presenting a decision test for the equivalence of an operator-precedence grammar and the underlying Floyd's operator-precedence parsing algorithm.

In this paper we first show that languages accepted by Floyd's operator-precedence parsing algorithms form a proper subclass of operator-precedence languages. Then we present a validity test for Floyd's operator-precedence parsing algorithms. Finally we discuss the consequence of this result for the problem of Levy [8] . The problem of Levy is to state conditions imposed on the precedence grammar in order to eliminate the problem "of losing one's place in a parse". Our validity test states necessary and sufficient condition to ensure that an operator-precedence grammar which satisfies this condition can be parsed in a complete way by an operator-precedence parsing algorithm.

2. BACKGROUND

In this section we present some basic definitions and terminology in the area of context-free languages and parsing theory.

A context-free grammar is a quadruple $G = \langle N,T,P,S \rangle$, where N is the finite set of nonterminal symbols, T is the finite set of terminal symbols, S is the start symbol from N , and $P \subseteq N \times V^*$ is a set of productions. Instead of (A,α) we write $A \longrightarrow \alpha$. A single production is one of the form $A \longrightarrow B$, where A and B are nonterminal symbols. A grammar is backwards-deterministic if no two productions have the same right-hand side.

Unless specified we use Roman capitals $(A,B,C,\dots,X,Y,Z)$ to denote nonterminal symbols, lower-case Roman letters at the beginning of the alphabet $(a,b,c,\dots)$ to denote terminal symbols, lower-case Roman letters near the end of the alphabet $(x,y,z,w,\dots)$ to denote terminal strings, lower-case Greek letters $(\alpha,\beta,\dots)$ to denote strings of terminal and nonterminal symbols. We use e for the empty string.

We write $\alpha \Longrightarrow \beta$ in G if $\alpha = \alpha_1 A \alpha_3$, $\beta = \alpha_1 \alpha_2 \alpha_3$, and $A \longrightarrow \alpha_2$ is a production in P . The relation $\overset{*}{\Longrightarrow}$ is the reflexive and transitive closure of $\Longrightarrow$, and $\overset{+}{\Longrightarrow}$ is the transitive closure. The context-free language for the grammar G is $L(G) = \{ w \in T^* \mid S \overset{*}{\Longrightarrow} w \}$. A derivation $\alpha_1 \Longrightarrow \alpha_2 \Longrightarrow \dots \Longrightarrow \alpha_k$ in G is called leftmost (denoted as $\underset{L}{\Longrightarrow}$) if at the i-th step the leftmost nonterminal symbol of α_i is replaced according to some production of G to yield α_{i+1} .

A grammar G is ambiguous if there is some string in L(G) for which there are two distinct leftmost derivations. G is otherwise unambiguous. Two leftmost derivations are similar if the corresponding derivation trees are identical up to relabeling of interior nodes. We say that G_1 is structurally equivalent to G_2 if every derivation in G_1 is similar to some derivation in G_2 and vice versa. We say that G_1 is equivalent to G_2 if $L(G_1) = L(G_2)$.

An e-free grammar is a context-free grammar in which no right-hand side of production is e . An operator grammar is a context-free grammar in which no production has a pair of adjacent nonterminal symbols on its right-hand side. Operator-precedence relations $\lessdot$, $\doteq$, $\gtrdot$ are defined in the following manner:

For $a,b \in T$

i. $a \doteq b$ if there is $A \longrightarrow \alpha a B b \beta$ in P for B = e or $B \in N$

ii. $a \lessdot b$ if $A \longrightarrow \alpha a B \beta$ in P and $B \overset{+}{\Longrightarrow} C b \gamma$ for $C = e$ or $C \in N$

iii. $a \gtrdot b$ if $A \longrightarrow \alpha B b \beta$ in P and $B \overset{+}{\Longrightarrow} \gamma a C$ for $C = e$ or $C \in N$

An operator-precedence grammar is an e-free operator grammar in which operator-precedence relations $\lessdot, \doteq, \gtrdot$ are disjoint.

An operator-precedence parsing algorithm works using an input string of terminal symbols, a parsing stack and the precedence relations from some operator-precedence grammar in the following way:
Initially the input contains $a_1 a_2 \ldots a_n \dashv$ and the parsing stack contains $\vdash$.

```
begin
  repeat forever
    if topstack = ⊢ and current-input-symbol = ⊣
    then ACCEPT
    else
      if topstack = a and current-input-symbol = b
      then
        select
          a <· b or a ≐ b : shift b from the input onto the stack;
          a ·> b : repeat pop the stack and put this topstack symbol
                          to X
                   until the topstack <· X ;
          otherwise : ERROR
        end ⟨select⟩
      fi
    fi
  end ⟨repeat⟩
end
```

Floyd's operator-precedence parsing algorithm is $\langle \mathcal{A}, M \rangle$, where $\mathcal{A}$ is an operator-precedence parsing algorithm driven by precedence relations M from some operator-precedence grammar.

We presume here that the reader is familiar with the LR(k) parsing as described e.g. in Aho and Ullman [1] . We use notations and definitions given in this reference concerning LR(k) parsing only in proofs of Lemma 2 and Lemma 5.

3. RELATION BETWEEN LANGUAGE CLASSES

In this section we investigate the relationship between operator-precedence languages and languages accepted by Floyd´s operator-precedence parsing algorithms. The main result is that operator-precedence grammars have more generative power than is the acceptance power of Floyd´s parsing algorithms.

It has been already remarked by Fischer [4] that operator-precedence grammars suffer one serious drawback, namely that though Floyd´s shift-reduce operator-precedence parsing algorithm accepts all input strings of an operator-precedence grammar, there is no guarantee that such a parsing algorithm will not also accept invalid input strings of that grammar. To illustrate this disadvantage a simple grammar G with the following set of productions is chosen: $S_0 \longrightarrow \vdash S \dashv$, $S \longrightarrow B\ a\ B$, $B \longrightarrow b$. We see that $L(G) = \{ \vdash b\ a\ b \dashv \}$. However, if one parses the string $\vdash b \dashv$, one finds that $\vdash \lessdot b$ and $b \gtrdot \dashv$ and so the input string $\vdash b \dashv$ is reduced to $\vdash B \dashv$. This sentential form now has $\vdash \doteq \dashv$ and if no check is performed to ensure that both terminal and nonterminal symbols match the right-hand side of some production, $\vdash b \dashv$ will be accepted as a valid string of the language. Thus, we summarize our first observation about grammars.

Fact 1

There are Floyd´s operator-precedence parsing algorithms which can also accept input strings not valid for the underlying operator-precedence grammar.

Now, consider an arbitrary finite language L over one-letter alphabet $\{ a \}$ containing a string of the length longer than one. It is certainly an operator-precedence language. Each Floyd´s operator-precedence parsing algorithm accepting L has to be ruled by operator-precedence relations containing one of $a \lessdot a$, $a \doteq a$, or $a \gtrdot a$, and thus it accepts an infinite language $\{ a^n \mid n \geqslant 0 \}$ which properly contains L . Hence, it holds

Lemma 1

There are operator-precedence languages which cannot be accepted by any Floyd´s operator-precedence parsing algorithm.

It is not so obvious that language accepted by a Floyd´s operator-precedence parsing algorithm is an operator-precedence language. We can see that it holds

Lemma 2
Given any Floyd's operator-precedence parsing algorithm $\mathcal{A}_M$ driven by operator-precedence relations M , we can find an equivalent backwards-deterministic operator-precedence grammar G . Moreover, the start symbol S of G does not appear on the right-hand side of any production of G and if $A \longrightarrow B$ is a single production of G then $A = S$.

We present here only a rough sketch of the proof. Let us give some helpful notations first. Consider T to be a set of terminal symbols (operators) , M to be operator-precedence relations over T , and $R = \{ (a,b) \mid M(a,b) = \doteq \} \subseteq T \times T$. (a,b) from R can be graphically represented as an oriented edge from a node a to a node b . Thus, R can be viewed as a digraph G_R with possibly many starting and ending nodes. A path in G_R is a sequence of consecutive nodes in G_R . A cycle C is the path of the length greater than one in which from any node in C one can reach an arbitrary node in C . Each edge entering a node in C which does not belong to C points to the beginning of the cycle C . End node of the cycle C is a node from which starts an edge entering a node not in C . We say that M contains a $\doteq$ - cycle if there is a cycle in G_R .

The idea of the proof is firstly to modify operator-precedence relations M to M_1 and then to give an appropriate grammatical characterization G_{M_1} of the operator-precedence parsing algorithm $\mathcal{A}_M$.

If b is the beginning of a cycle and a is the end of the same cycle, then replace $a \doteq b$ in M by the relation $a \lessdot b$. We call the resulting set M_1 . Then M_1 is the set of unique operator-precedence relations. We now let $G' = \langle \{S'\}, T, P', S' \rangle$ where S' is the start symbol and

$P' = \{ S' \longrightarrow S_1 a_1 S_1 a_2 S_1 \dots S_1 a_k S_1 \mid S_1 \in \{S', e\}$ and either there are $b,c \in T$ such that $M(b,a_1) = \lessdot$, $M(a_k,c) = \gtrdot$, $M(a_i,a_{i+1}) = \doteq$ for $1 \leqslant i < k$ and $a_1,\dots,a_k$ does not contain a cycle,or $a_1,\dots,a_k$ is a path between two consecutive ends of cycle (or between consecutive start and end/end of cycle or between consecutive end of cycle and end) $\}$.

In general, this grammar generates a superclass of $L(\mathcal{A}_M)$ and it is often ambiguous, but it is still possible to generate a valid set of LR(1) tables for this grammar. We can begin by constructing the canonical collection of sets of LR(0) items for G'. The second step is to construct a set of SLR(1) tables from these sets of items. Since the grammar is often ambiguous, the parsing action is not always uniquely defined for all lookaheads and for all tables. We use the operator-precedence relations M_1 to resolve these conflicts. In other words, the action can always be uniquely defined by consulting operator-precedence relations M_1. Now, we can use machine description grammar [2] G_1 to obtain an operator-precedence grammar which is equivalent to $\mathcal{A}_M$. Finally, following Fischer [4] for any operator-precedence grammar G_1 one can find an equivalent backwards-deterministic operator-precedence grammar G_{M_1} in which the start symbol S does not appear on the right-hand side of any production in G_{M_1} and if $A \longrightarrow B$ is a single production, then $S = A$. This gives the proof of Lemma 2.

Now, considering Lemma 1 and Lemma 2 we have the second observation about languages.

<u>Theorem 1</u>
Languages accepted by Floyd's operator-precedence parsing algorithms form a proper subclass of operator-precedence languages.

4. VALIDITY TEST

It was already mentioned in the previous section that Floyd's operator-precedence parsing algorithm may accept a proper superset of the intended language. In this section we present an algorithm to decide whether an operator-precedence parsing algorithm accepts exactly the language generated by the underlying grammar. The principal idea behind this decision algorithm is the grammatical characterization of the operator-precedence parsing algorithm and the reduction of our test to the equivalence problem for two backwards-deterministic operator-precedence grammars with the same operator-precedence relations. The latter problem is shown to be algorithmically decidable.

We begin with a technical assertion that a single production removal does not affect the operator-precedence property.

<u>Lemma 3</u>
Given an arbitrary operator-precedence grammar, one can find a structurally equivalent operator-precedence grammar without single productions.

The key idea of our test algorithm is obtained in the following assertion.

Lemma 4

Given two backwards-deterministic operator-precedence grammars G_1 and G_2 with equal operator-precedence relations, there is an algorithm to decide whether it holds $L(G_1) = L(G_2)$.

Proof:

Let G_1 and G_2 be two backwards-deterministic operator-precedence grammars with equal operator-precedence relations. We first show that $L(G_1) = L(G_2)$ if and only if G_1 is structurally equivalent to G_2.

1. Let $L(G_1) = L(G_2) = L$. By Lemma 3 there are structurally equivalent operator-precedence grammars G_1' and G_2' without single productions. We show that an arbitrary input string in L has the same derivation structure. Consider an arbitrary string $w \in L$. Each string $w = a_1 a_2 \ldots a_n$ can be derived in G_1' and G_2' in a unique way. We can distinguish two cases:

i. Derivation in G_1':

$$S_1' \overset{*}{\underset{L}{\Longrightarrow}} a_1 \ldots a_i A_1 \beta_1 \underset{L}{\Longrightarrow} a_1 \ldots a_k B_1 \gamma_1 \overset{*}{\underset{L}{\Longrightarrow}} a_1 \ldots a_n$$

Derivation in G_2':

$$S_2' \overset{*}{\underset{L}{\Longrightarrow}} a_1 \ldots a_i A_2 \beta_2 \underset{L}{\Longrightarrow} a_1 \ldots a_s B_2 \gamma_2 \overset{*}{\underset{L}{\Longrightarrow}} a_1 \ldots a_n ,$$

where $n \geqslant s > k \geqslant i \geqslant 1$. From the derivation of w in G_1' it holds $a_k \lessdot a_{k+1}$, but from the derivation of w in G_2' it must be $a_k \doteq a_{k+1}$ a contradiction with the assumption that G_1' and G_2' have equal operator-precedence relations.

ii. Derivation in G_1':

$$S_1' \overset{*}{\underset{L}{\Longrightarrow}} uA\beta_1 \underset{L}{\Longrightarrow} uv\beta_1 \overset{*}{\underset{L}{\Longrightarrow}} a_1 \ldots a_n$$

Derivation in G_2':

$$S_2' \overset{*}{\underset{L}{\Longrightarrow}} uB\beta_2 \overset{*}{\underset{L}{\Longrightarrow}} a_1 \ldots a_n ,$$

where β_1 equals to β_2 up to renaming of nonterminal symbols. However, G_1' and G_2' are backwards-deterministic grammars and have equal operator-precedence relations and thus G_1' and G_2' are structurally equivalent.

2. If G_1 is structurally equivalent to G_2, then $L(G_1) = L(G_2)$.

In order to characterize Floyd's operator-precedence parsing algorithm grammatically it is not possible to apply the construction from Lemma 2 because this construction does not preserve operator-precedence relations. Therefore, another construction is given. More precisely we show how for a Floyd's parsing algorithm a grammar can be constructed which will generate the same language and in what cases it is backwards-deterministic operator-precedence grammar. We make the following claim

Lemma 5

Given any Floyd's operator-precedence parsing algorithm $\mathcal{A}_M$ ruled by operator-precedence relations M without $\doteq$ - cycle, we can find an equivalent backwards-deterministic operator-precedence grammar G with the same operator-precedence relations M .

Again we give only an idea of the proof. We construct a grammar $\bar{G} = \langle \{S\}, T, P, S \rangle$, where S is the start symbol and

$$P = \{ S \longrightarrow S_1 a_1 S_1 \ldots S_1 a_k S_1 \mid S_1 \in \{S, e\} \text{ and there are } b,c \in T \text{ such that } M(b,a_1) = \lessdot , M(a_k, c) = \gtrdot , M(a_i, a_{i+1}) = \doteq \text{ for } 1 \leqslant i < k \} .$$

It holds $L(\bar{G}) \supseteq L(\mathcal{A}_M)$. Following the proof of Lemma 2 we can obtain a backwards-deterministic operator-precedence grammar G with operator-precedence relations M generating $L(\mathcal{A}_M)$.

For any Floyd's operator-precedence parsing algorithm ruled by operator-precedence relations M containing $\doteq$ - cycle there does not exist an equivalent operator-precedence grammar with equal operator-precedence relations.

Now, we are prepared to formulate the main theorem of this section.

Theorem 2

For any backwards-deterministic operator-precedence grammar G with operator-precedence relations M and for the Floyd's operator-precedence parsing algorithm $\mathcal{A}_M$ ruled by operator-precedence relations M there is an algorithm to decide whether it holds $L(G) = L(\mathcal{A}_M)$.

Proof:

If M contains $\doteq$ - cycle, then $L(G) \neq L(\mathcal{A}_M)$. If M does not contain $\doteq$ - cycle, then following Lemma 5 we can find an equivalent backwards-deterministic operator-precedence grammar G_1 . Using Lemma 4 given two backwards-deterministic operator-precedence grammars G and G_1 with the same operator-precedence relations, there is an algorithm to decide whether it holds $L(G) = L(G_1)$.

5. CONCLUSION

Operator-precedence grammars are suitable especially for specifying a variety of programming language constructs using an information about the precedence and associativity of operators. However, operator-precedence parsing algorithms possess the curious property that one can accept inputs that are not in the language of the underlying grammars. We have answered natural questions concerning two classes of languages which are definable using the operator-precedence grammars and the Floyd´s operator-precedence parsing algorithms and concerning the decidability of the equivalence problem for these two models. The latter result can be reformulated as the necessary and sufficient condition for an operator-precedence grammar to be valid grammatical characterization for Floyd´s operator-precedence parsing algorithms. This result solves the problem stated by Levy [8] . We have not yet looked at a modification of operator-precedence relations in order to obtain valid characterization of Floyd´s parsing algorithms. A partial solution of this problem has been got by Henderson and Levy [6] by defining extended operator-precedence relations.

6. REFERENCES

1. Aho, A.V., and J.D. Ullman, The Theory of Parsing, Translation and Compiling. Vol. I : Parsing. Englewood Cliffs, N.J. , Prentice-Hall, 1972.
2. Demers, A.J., Skeletal LR Parsing. IEEE Conference Record of 15th Annual Symposium on Switching and Automata Theory, 185 - 198, 1974.
3. El Djabri, N., Extending the LR Parsing Techniques to Some Non-LR Grammars. TR - 121 , Department of AACS, Princeton University, Princeton, New Jersey, 1973.
4. Fischer, M.J., Some Properties of Precedence Languages. Proc. ACM Symposium on Theory of Computing, 181 - 190, 1969.
5. Gray, J.N., Precedence Parsers for Programming Languages. Ph. D. Thesis, Department of Computer Science, University of California, Berkeley, 1969.
6. Henderson, D.S., and M.R. Levy, An Extended Operator-Precedence Parsing Algorithm. Computing Journal 19 : 3, 229 - 233 , 1976.
7. Knuth, D., Top Down Syntax Analysis. Acta Informatica 1 : 2, 79 - 110, 1971.
8. Levy, M.R., Complete Operator Precedence. IPL 4 : 2, 38 - 40, 1975.

ON THE LANGUAGES OF BOUNDED PETRI NETS

Peter H. Starke
Sektion Mathematik
der Humboldt-Universität zu Berlin
DDR-1086 Berlin, PSF 1297.

Introduction

In this paper we investigate Petri nets under a prescribed token capacity of their places working with a transition rule which ensures that the token capacity of the places is not exceeded. We introduce the families of languages representable as the free terminal languages (of firing sequences) of such nets having the token capacity k in each place. By implementing the bounded transition rule to ordinary Petri nets it is shown that these families form a proper hierarchy. We characterize them algebraically by their closure properties. Finally, we give two remarks on the regularity of Petri net languages.

1. Definitions

In this section we define the basic notions to be dealt with in the sequel and give some examples and remarks. Following HACK [1] we call $N = (P,T,F,m_o)$ a <u>Petri net</u> iff P and T are finite disjoint sets (of <u>places</u> and <u>transitions</u> resp.) with $P \cup T \neq \emptyset$, F is a mapping from the set $(P \times T) \cup (T \times P)$ into the set $\mathbb{N}$ of all natural numbers (including 0) and m_o is the <u>initial marking</u>, i.e. a mapping from P into $\mathbb{N}$.

For markings $m,m' \in \mathbb{N}^P$ we define the markings $m+m'$, $m-m'$ and the relation $m \leq m'$ componentwise (pointwise). To each transition $t \in T$ we adjoin two markings t^- and t^+ as follows:

$t^-(p) = F(p,t)$ (the number of tokens, t takes from p)

$t^+(p) = F(t,p)$ (the number of tokens, t sends to p).

The so-called <u>transition rule</u> describes the change of the markings of a net under the firing of transitions, and, it can be given by a partial function of $\mathbb{N}^P \times T$ into $\mathbb{N}^P$. We here are going to deal with two kinds of transition rules, namely the usual one given by

$$\delta_N(m,t) = \begin{cases} m - t^- + t^+, \text{ if } m \geq t^-, \\ \text{not defined, else;} \end{cases} \qquad (m \in \mathbb{N}^P,\ t \in T)$$

where a transition t has concession under the marking m iff $m \geq t^-$, and the c-<u>bounded</u> <u>transition</u> <u>rule</u> (where $c \in \mathbb{N}^P$ is fixed) given by

$$\delta_N^c(m,t) = \begin{cases} m - t^- + t^+, \text{ if } t^- \leq m \leq c \text{ and } m - t^- + t^+ \leq c, \\ \text{not defined, else.} \end{cases}$$

The number c(p) is understood as the token capacity of the place p which should not be exceeded during the working of the net.

Let us remark that in our net in Fig. 1 the transition t has concession iff $c(p) \geq 2$.

Fig. 1.

Both functions, δ_N and δ_N^c, can be extended in the natural way to partial functions from the set $\mathbb{N}^P \times W(T)$ into $\mathbb{N}^P$ where W(T) denotes the free semigroup with identity e generated by T. The domain (of definition) of the extended functions is denoted by D_N, D_N^c resp. Then the sets of markings reachable in N and the languages represented by N are given as follows:

$$R_N(m) = \{\delta_N(m,q) : (m,q) \in D_N\}; \qquad R_N^c(m) = \{\delta_N^c(m,q) : (m,q) \in D_N^c\},$$

$$L_N(m) = \{q : (m,q) \in D_N\}; \qquad L_N^c(m) = \{q : (m,q) \in D_N^c\},$$

$$L_N(m,M) = \{q : (m,q) \in D_N \ \&\ \delta_N(m,q) \in M\};$$

$$L_N^c(m,M) = \{q : (m,q) \in D_N^c \ \&\ \delta_N^c(m,q) \in M\},$$

where M is a (in general finite) set of markings.

<u>Corollary</u> 1

1. δ_N^c is a restriction of δ_N.
2. $m \in R_N^c(m_o) \longrightarrow m \in R_N(m_o) \ \&\ m \leq c$.
3. $L_N^c(m) \subseteq L_N(m) \ \&\ L_N^c(m,M) \subseteq L_N(m,M)$.

The converse of 1.2 does not hold as one can see easily from the net in Fig. 2. The marking m = (0,1) (i.e. $m(p_1)=0$, $m(p_2)=1$) is reachable from $m_o = (1,0)$ by the firing sequence abbb but for c = (1,1) m is not c-reachable.

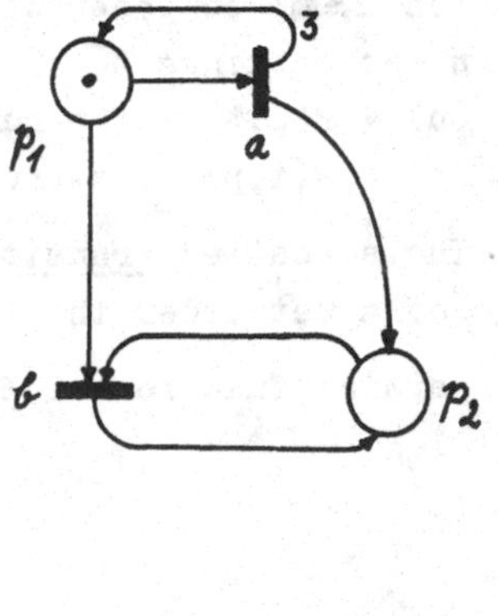

Fig. 2

We call the Petri net $N = (P,T,F,m_o)$ to be c-bounded ($c \in \mathbb{N}^P$) iff for all (reachable) markings $m \in R_N(m_o)$ we have $m \leq c$. Now we can show:

Theorem 2

The following conditions are equivalent:

(a) $L_N^c(m_o) = L_N(m_o)$ (b) $R_N^c(m_o) = R_N(m_o)$ (c) N is c-bounded.

Proof. The implications (a) → (b) and (b) → (c) directly follow from Corollary 1. To show (c) → (a) we assume that there exists a word $q \in L_N(m_o) \setminus L_N^c(m_o)$ which is minimal with respect to the initial segment relation $\sqsubseteq$. By $e \in L_N^c(m_o)$ we obtain $e \neq q = rt$ with $t \in T$, $r \in L_N(m_o) \cap L_N^c(m_o)$. Let $m = \delta_N(m_o,r) = \delta_N^c(m_o,r)$. From $m \in R_N(m_o)$ and condition (c) we obtain $m \leq c$ and by $(m,t) \in D_N$ we have $t^- \leq m \leq c$. Moreover, $\delta_N(m,t) = (m-t^-+t^+) \in R_N(m_o)$, thus, $m-t^-+t^+ \leq c$, and therefore, $(m,t) \in D_N^c$, which is in contradiction with $q = rt \notin L_N^c(m_o)$.

Remark. The so-called safe transition rule gives concession to a transition t in an ordinary Petri net (i.e. $F:(P \times T) \cup (T \times P) \rightarrow \{0,1\}$) iff all the preconditions p of t (i.e. $F(p,t)=1$) are fullfilled (marked) and no postcondition p' ($F(t,p')=1$) of t is fullfilled. Consider the net in Fig. 3. Obviously t has concession under our 1-bounded transition rule δ_N^1 but not under the safe transition rule δ_N^s. Hence we have

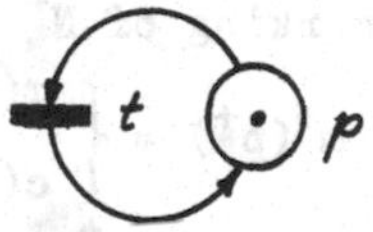

Fig. 3

$$L_N(m_o) = \{t^i : i \in \mathbb{N}\} \neq L_N^s(m_o) = \{e\},$$

while $R_N(m_o) = \{m_o\} = R_N^s(m_o)$. Thus Theorem 2 does not hold true under the safe transition rule.

2. Implementation of the bounded transition rule

By adding to the given net N, for each place p, a so-called co-place $\bar{p}$ which controls the number of tokens in p, we can construct a net N^+ which is bounded and which under a certain initial marking behaves under δ_{N^+} in the same way as N under δ_N^c for a given capacity c. In this way the bounded transition rule can be implemented to an Petri net working under the usual transition rule.

Let be $N = (P,T,F,m_o)$ and $N' = (P',T',F',m_o')$ Petri nets and

$c \in \mathbb{N}^P$, $c' \in \mathbb{N}^{P'}$. We call N and N' to be <u>equivalent with respect to</u> (c,c') iff $\{L_N^c(m_o,M) : M \subseteq \mathbb{N}^P\} = \{L_{N'}^{c'}(m_o',M') : M' \subseteq \mathbb{N}^{P'}\}$.

<u>Theorem 3</u>

For every Petri net $N = (P,T,F,m_o)$ and every capacity marking $c \in \mathbb{N}^P$ with $m_o \leq c$ there exist a Petri net $N^+ = (P^+,T,F^+,m_o^+)$ and a marking $c^{++} \in \mathbb{N}^{P^+}$ such that

(i) N and N^+ are equivalent with respect to (c,c^{++}),

(ii) N^+ is c^{++}-bounded,

(iii) the number of tokens in N^+ is constant.

The construction of N^+ and c^{++} is done as follows. Let be $\bar{P} = \{\bar{p} : p \in P\}$ (with $\bar{P} \cap P = \emptyset$). We call $\bar{p}$ the co-place of p. For $\bar{p} \in \bar{P}$, $t \in T$ we put

$$\bar{F}(\bar{p},t) = \begin{cases} F(t,p) - F(p,t), & \text{if } F(t,p)-F(p,t) > 0, \\ 0, & \text{else}, \end{cases}$$

$$\bar{F}(t,\bar{p}) = \begin{cases} F(p,t) - F(t,p), & \text{if } F(p,t)-F(t,p) > 0, \\ 0, & \text{else}. \end{cases}$$

Moreover, we set $P^+ = P \cup \bar{P}$, $F^+ = F \cup \bar{F}$, and, for $m \in \mathbb{N}^P$, let be m^+ the marking of N^+ with

$$m^+(p^+) = \begin{cases} m(p), & \text{if } p^+ = p \in P, \\ c(p)-m(p), & \text{if } p^+ = \bar{p} \in \bar{P}. \end{cases}$$

In this way, m_o^+ is defined. For any set $M \subseteq \mathbb{N}^P$ we put $M^+ = \{m^+ : m \in M\}$. Finally, we set

$$c^{++}(p^+) = c(p) \quad \text{for} \quad p^+ \in \{p,\bar{p}\}.$$

Now one can show the following assertions:

(3.1) $m \in \mathbb{N}^P$ & $(m,t) \in D_N^c \longrightarrow (m^+,t) \in D_{N^+}$ & $(\delta_N^c(m,t))^+ = \delta_{N^+}(m^+,t)$,

(3.2) $q \in W(T)$ & $(m_o,q) \in D_N^c \longrightarrow (m_o^+,q) \in D_{N^+}$ & $(\delta_N^c(m_o,q))^+ = \delta_{N^+}(m_o^+,q)$,

(3.3) $q \in W(T)$ & $(m_o^+,q) \in D_{N^+} \longrightarrow (m_o,q) \in D_N^c$ & $(\delta_N^c(m_o,q))^+ = \delta_{N^+}(m_o^+,q)$,

(3.4) $L_{N^+}(m_o^+) = L_N^c(m_o)$ & $R_{N^+}(m_o^+) = (R_N^c(m_o))^+$,

(3.5) $m' \in R_{N^+}(m_o^+)$ & $p \in P \longrightarrow m'(p) + m'(\bar{p}) = c(p)$.

Obviously, (3.5) implies (ii) and (iii), and, from (3.2), (3.3) and (3.4) we obtain (i) and

$$L_N^c(m_o,M) = L_{N^+}(m_o^+,M^+) \qquad (M \subseteq \mathbb{N}^P).$$

3. A hierarchy of FL's

In this section we introduce, for every $k \in \mathbb{N}$, the family of all (regular) languages which are representable by a Petri net the places of which all have the token capacity k. We shall show that these FL's form a proper hierarchy.

Let be $k \in \mathbb{N}$. Then $\bar{k}$ denotes (for any P) the marking which assigns k tokens to every $p \in P$. By $\mathcal{T}_{PN}^{(k)}$ we denote the family of all languages L such that there exist a Petri net $N = (P,T,F,m_o)$ and a finite set $M \subseteq \mathbb{N}^P$ of markings with $m_o \leq \bar{k}$ and $L = L_N^{\bar{k}}(m_o,M)$. Then

$$\mathcal{T}_{PN}^{*} = \bigcup_{k \in \mathbb{N}} \mathcal{T}_{PN}^{(k)}$$

is the family of all languages which are representable by bounded Petri nets. $\mathcal{T}_{PN}^{*}$ is properly contained in $\mathcal{T}_{PN}$, the family of all (free terminal) Petri net languages. Obviously, we have

$$\mathcal{T}_{PN}^{(0)} = \{\emptyset, \{e\}, \langle X \rangle : \mathrm{card}(X) \in \mathbb{N}\},$$

where $\langle\rangle$ denotes the forming of the catenation closure. (In a net, representing $\langle X \rangle$ all the places are isolated.)

Theorem 4

$$\mathcal{T}_{PN}^{(0)} \subset \mathcal{T}_{PN}^{(1)} \subset \ldots \subset \mathcal{T}_{PN}^{(k)} \subset \mathcal{T}_{PN}^{(k+1)} \subset \ldots \subset \mathcal{T}_{PN}^{*} \subset \mathcal{T}_{PN}.$$

Proof. Let be $L = L_N^{\bar{k}}(m_o,M) \in \mathcal{T}_{PN}^{(k)}$. By Theorem 3 we can construct the net N^+ such that

$$L = L_N^{\bar{k}}(m_o,M) = L_{N^+}(m_o,M^+) = L_{N^+}^{\overline{(k+1)}}(m_o,M^+) \in \mathcal{T}_{PN}^{(k+1)},$$

hence $\mathcal{T}_{PN}^{(k)} \subseteq \mathcal{T}_{PN}^{(k+1)}$ and $\mathcal{T}_{PN}^{*} \subseteq \mathcal{T}_{PN}$. To show that the inclusions are proper ones we consider the one-element languages $\{a^i\}$ ($i \in \mathbb{N}$). Obviously, $\{a^i\} \in \mathcal{T}_{PN}^{(i)}$. Assume that $\{a^{i+1}\} \in \mathcal{T}_{PN}^{(i)}$, $\{a^{i+1}\} = L_N^{\bar{i}}(m_o,M)$ for certain N,m_o,M. Then $(m_o,a) \in D_N^{\bar{i}}$ and $m_o \neq \delta_N^{\bar{i}}(m_o,a)$ since we have $\delta_N^{\bar{i}}(m_o a^i) \notin M$ but $\delta_N^{\bar{i}}(m_o,a^{i+1}) \in M$. This implies that $\Delta a = -a^- + a^+ \neq 0$. Now, for $p \in P$ with $\Delta a(p) \geq 1$ we have

$$\delta_N^{\bar{i}}(m_o,a^{i+1})(p) = m_o(p) + (i+1)\Delta a(p) \geq i+1$$

contradicting $\delta_N^{\bar{i}}(m_o,a^{i+1})(p) \leq i$, and, for $p \in P$ with $\Delta a(p) \leq -1$ we obtain by $m_o(p) \leq i$ the contradiction $\delta_N^{\bar{i}}(m_o,a^{i+1})(p) < 0$.

It is verified easily that there exists no one-letter language

$L \in \mathcal{T}_{PN} \setminus \mathcal{T}^{*}_{PN}$, since a free terminal Petri net language with one letter only is either finite or the full semigroup generated by that letter.

By $D_{a,b}$ we denote the DYCK-language D_1 over $\{a,b\}$ and by $D^{(i)}_{a,b}$ the sublanguage of $D_{a,b}$ containing all the formulae of depth less or equal i. Hence

$$D^{(0)}_{a,b} = \{e\}, \quad D^{(i+1)}_{a,b} = \langle\{a\}\cdot D^{(i)}_{a,b}\cdot\{b\}\rangle, \quad D_{a,b} = \bigcup_{i\in\mathbb{N}} D^{(i)}_{a,b} \, .$$

One can see without difficulties that $D^{(i+1)}_{a,b} \in \mathcal{T}^{(i+1)}_{PN} \setminus \mathcal{T}^{(i)}_{PN}$ and $D_{a,b} \in \mathcal{T}_{PN} \setminus \mathcal{T}^{*}_{PN}$.

4. Closure properties

In this section we give without proof some results on the closure properties of the FL's defined above.

Theorem 5

All the families $\mathcal{T}^{(i)}_{PN}$ and $\mathcal{T}^{*}_{PN}$ are closed under intersection but are not closed under union and complementation.

Theorem 6

All the families $\mathcal{T}^{(i)}_{PN}$ $(i \geq 1)$ and $\mathcal{T}^{*}_{PN}$ are not closed under any of the following operations: catenation, catenation closure, homomorphisms.

Theorem 7

All the families $\mathcal{T}^{(i)}_{PN}$ and $\mathcal{T}^{*}_{PN}$ are closed under inverse homomorphisms, restriction, left derivatives and right quotient by finite languages.

Remark. For a family of languages which is closed under inverse homomorphisms closure under restriction and closure under intersection are equivalent.

5. Characterization

From our previous results it follows that $\mathcal{T}^{(i)}_{PN}$ contains the language $D^{(i)}_{a,b}$ and is closed under intersection, restriction, inverse homomorphisms, left derivatives and right quotients by finite languages. We shall prove:

Theorem 8

$\mathcal{T}^{(i)}_{PN}$ is the least family of languages containing $D^{(i)}_{a,b}$ and which is closed under intersection, inverse homomorphisms, left derivatives and right quotients by finite languages.

Proof. If L is a language let be X(L) the least alphabet such that $L \subseteq W(X(L))$. First we recall a usefull lemma from [2]:

Lemma 8.1: Let for j=1,2 $N_j = (P_j, T_j, F_j, m_o^j)$ be two place-disjoint Petri nets without isolated transitions and let M_1, M_2 be finite subsets of $\mathbb{N}^{P_1}$, $\mathbb{N}^{P_2}$ such that $X(L_{N_j}(m_o^j, M_j)) = T_j$. Finally let be $N = N_1 \cup N_2$ the Petri net $(P_1 \cup P_2, T_1 \cup T_2, F_1 \cup F_2, m_o^1 \cup m_o^2)$. Then

$$L_{N_1}(m_o^1, M_1) \circledR L_{N_2}(m_o^2, M_2) = L_N(m_o^1 \cup m_o^2, \{m_1 \cup m_2 : m_1 \in M_1 \ \&\ m_2 \in M_2\}).$$

Now let us consider an arbitrary Petri net $N = (P, T, F, m_o)$, a finite set $M \subseteq \mathbb{N}^P$ and a natural number i. We are going to show that $L^I_N(m_o, M)$ is an element of every family $\mathcal{L}^{(i)}$ which contains $D^{(i)}_{a,b}$ and which is closed under restriction, inverse homomorphisms, left derivatives and right quotients by finite languages.

In the first step we erase from N all the transitions $t \in T$ such that $t \notin X(L^I_N(m_o, M))$. This is possible in an constructive way since $L^I_N(m_o, M)$ is regular. In the resulting net $N' = (P, T', F', m_o)$ we have

$$L^I_{N'}(m_o, M) = L^I_N(m_o, M) \quad \& \quad T' = X(L_{N'}(m_o, M)).$$

If $T' = \emptyset$ then $L^I_{N'}(m_o, M) \in \{\emptyset, \{e\}\} \subseteq \mathcal{L}^{(i)}$ since $\emptyset = \partial_b D^{(i)}_{a,b}$ and $\{e\} = h^{-1}(D^{(i)}_{a,b})$ where h is the empty homomorphism $h: \emptyset \to W(\{a,b\})$.

Consider the case $T' \neq \emptyset$ and denote by T_o the set of all isolated transitions of N'. If $T_o = \emptyset$ we denote N' by N" else we denote by N" the net obtained from N' by erasing all the isolated transitions. Clearly,

$$L^I_{N'}(m_o, M) = L^I_{N''}(m_o, M) \circledR W(T_o).$$

Since $W(T_o) = h^{-1}(D^{(i)}_{a,b})$ where $h(t) = e$ for all $t \in T_o$, it is sufficient to prove that the languages $L^I_{N''}(m_o, M)$ is an element of $\mathcal{L}^{(i)}$ where $N'' = (P, T'', F'', m_o)$ contains no isolated transitions and $T'' = X(L^I_{N''}(m_o, M))$ is fullfilled.

Now for $p \in P$ consider the subnet $N_p = (\{p\}, T_p, F_p, m_o(p))$ of N" consisting of the place p and all the transitions connected to p:

$T_p = \{t : t \in T'' \;\&\; F''(p,t)+F''(t,p) > 0\}$.

One shows without difficulties that

$T_p = X(L^{\bar{I}}_{N_p}(m_o(p),\{m(p) : m \in M\}))$.

Now, applying the construction of Theorem 3 to N" and all the N_p's we obtain by $N'' = \bigcup_{p \in P} N_p$ and Lemma 8.1 $(N'')^+ = \bigcup_{p\in P} N_p^+$ and

$$L^{\bar{I}}_{N''}(m_o,M) = L_{(N'')^+}(m_o,M^+) = \textcircled{R}_{p\in P} L_{N_p^+}((m_o(p),i-m_o(p)),\{(m(p),i-m(p) : m \in M\})$$

$$= \textcircled{R}_{p\in P} L^{\bar{I}}_{N_p}(m_o(p),\{m(p) : m \in M\}).$$

Therefore it is enough to prove that all the "one-place" languages are elements of $\mathcal{L}^{(i)}$.

Let be $N = (\{p\},T,F,k)$ a one-place Petri net without isolated transitions, $M = \{j_1,\ldots,j_l\} \subseteq \mathbb{N}$, $k \leq i$. Then one can see by straightforward calculations that

$$L^{\bar{I}}_N(k,M) = h^{-1}(\partial_{a^k}(D^{(i)}_{a,b}/\{b^{j_1},\ldots,b^{j_k}\})) \in \mathcal{L}^{(i)}$$

where $h: T \longrightarrow W(\{a,b\})$ is the homomorphism with $h(t) = b^{t^-(p)}a^{t^+(p)}$.

6. Final remark

Obviously, if a language L is representable by a bounded Petri net, i. e. $L \in \mathcal{T}^*_{PN}$, then L is regular and $L \in \mathcal{T}_{PN}$. But the converse is true too.

<u>Theorem 9</u>: $\mathcal{T}_{PN} \cap \mathrm{REG} = \mathcal{T}^*_{PN}$

Proof. Let be $L = L_N(m_o,M) \in \mathcal{T}_{PN}$ and $\mathrm{card}(\{\partial_q L : q \in W(T)\}) = k \in \mathbb{N}$. By M' we denote the set of all markings of N which are reachable from m_o and from which a marking from M is reachable:

$M' = \{m' : m' \in R_N(m_o) \;\&\; R_N(m') \cap M \neq \emptyset\}$.

It is sufficient to show that M' is finite since in this case there exists a number $i \in \mathbb{N}$ such that $m' \leq \bar{i}$ for all $m' \in M'$ which implies that $L_N(m_o,M) = L^{\bar{I}}_N(m_o,M) \in \mathcal{T}^*_{PN}$.

It is easy to verify that the following assertions hold:

(i) $M' = \{\delta_N(m_o,q) : \exists r \; qr \in L\}$,

(ii) $(m_o,q) \in D_N \longrightarrow L_N(\delta_N(m_o,q),M) = \partial_q L_N(m_o,M)$

(iii) $\{L_N(m',M) : m' \in M'\} = \{\partial_q L : q \in W(T)\} \setminus \{\emptyset\}$.

Markings m',m" from M' are called to be equivalent ($m' \sim m''$) iff $L_N(m',M) = L_N(m'',M)$. $M'/\sim$ denotes the partition of M' induced by the equivalence relation $\sim$, $[m]$ is the equivalence class containing m.

Now, from (iii), we obtain $card(M'/\sim) \leq k$. To complete the proof we show that, for $m \in M'$, $card([m]) \leq card(M)$ from which it follows that $card(M') \leq k \cdot card(M) \in \mathbb{N}$.

Let $m_1, \ldots, m_l$ be l pairwise different markings from $[m]$ and $r \in L_N(m,M)$. Then, for $j = 1, \ldots, l$, $m'_j = \delta_N(m_j, r) = m_j + \Delta r \in M$. Since $m'_i \neq m'_j$ iff $m_i \neq m_j$ iff $i \neq j$ we obtain $l \leq card(M)$.

Corollary 10

$L_N(m_o, M) \in REG$ iff for every $m \in M$, $L_N(m_o, \{m\}) \in REG$.

If $L_N(m_o, M)$ is regular, then the set M' defined above is finite, therefore, $M'(m) = \{m' : m' \in R_N(m_o) \ \& \ m \in R_N(m')\} \subseteq M'$ is finite from which the regularity of $L_N(m_o, \{m\})$ follows.

Theorem 11

$L_N(m_o) \in REG \longrightarrow \forall m (m \in \mathbb{N}^P \longrightarrow L_N(m_o, \{m\}) \in REG)$

Proof. Valk [3] has shown that, if $L_N(m_o)$ is regular, then there exists a marking c such that

$\forall m \forall m' (m \in R_N(m_o) \ \& \ m' \in R_N(m) \longrightarrow m' \geq m - c)$.

This implies for $m' \in M'(m)$ that $m \geq m' - c$, i.e. $m' \leq m + c$. Hence $M'(m)$ is finite and, therefore, $L_N(m_o, \{m\})$ is regular.

The converse of Theorem 11 is not true.

7. References

[1] Hack, M.T.: Petri Net Languages. MAC-TM 59, MIT 1975.

[2] Starke, P.H.: Free Petri Net Languages. Seminarberichte der Sektion Mathematik der HU Berlin, No. 7, 1978.

[3] Valk, R.: Petri Nets and Regular Languages. Preprint 1978.

DYCK LANGUAGE D_2 IS NOT ABSOLUTELY PARALLEL.

Miron Tegze
Faculty of Mathematics and Physics,
Charles University, Prague, Czechoslovakia.

INTRODUCTION.

It is known that the family of absolutely parallel languages (APL) is not contained in the family of context-free languages (CFL), since the language $\{a^i b^i c^i \; ; \; i \geqslant 0\}$, which is an APL, is not a CFL. In the present paper we show that the language D_2 (which is a CFL) is not an APL, hence CFL and APL are incomparable by inclusion.

Moreover, it is known that the family of CFL equals the family of languages generated by stack transducers. Rajlich [1] has shown that the family of APL coincides with the family of languages generated by two-wa finite-state transducers. Thus the result of the paper concerns the generative power of these machines, too.

This paper is divided into three sections. In the first one we present some definitions and preliminary results. The second section involves our main result. The proof of Lemma 3, which is omitted in the first section, is placed in the third one. The reader, who is not interested in technical details, can omit this proof.

1. PRELIMINARY RESULTS.

DEFINITION of CFG and CFL. A context-free grammar (CFG) is any 4-tuple $H = (N,T,S,P)$, where N and T are disjoint finite sets of symbols, $S \in N$, and P is a finite set of productions of the form $p = A \rightarrow y$ with $A \in N$, $y \in (N \cup T)^*$. $w \xRightarrow[H]{p} w'$ iff $w = uAv$ and $w' = uyv$, where $uv \in (N \cup T)^*$. $w \Rightarrow_H w'$ iff there exists $p \in P$ such that $w \xRightarrow[H]{p} w'$. Then $\xRightarrow[H]{*}$ is the reflexive and transitive closure of $\Rightarrow_H$. The language generated by H is $L(H) = \{ w \in T^* \; ; \; S \xRightarrow[H]{*} w \}$.

The family of context-free languages is $CFL = \{L(H); \text{ H is a CFG}\}$.

<u>DEFINITION of D_2.</u> The Dyck language D_2 is the language generated by a CFG $E = (\{S\}, \{],[,),(\}, S, \{S \to \Lambda, S \to (S), S \to [S], S \to SS\})$.

Thus D_2 is the language of "well-founded parenthetical expressions" constructed from two types of parenthesis.

<u>DEFINITION.</u> Let $w \in T^*$, $T = \{],[,),(\}$. We say "in w all [,]-brackets are complete" iff :

1. (the number of] in w) = (the number of [in w).
2. Let w = uv. Then (the number of [in u) $\geqslant$ (the number of] in u).

Let $w = u[_1 v]_2 x$. We say "in $[_1 \ldots]_2$-brackets of w all [,]-brackets are complete" iff in v all [,]-brackets are complete. We say "the brackets $[_1 \ldots]_2$ are spoilt" iff in $[_1 \ldots]_2$ of w all [,]-brackets are complete and the numbers of) and of (in $[_1 \ldots]_2$ are different.

The following lemma is easily verified :

<u>LEMMA 1.</u> Let $w \in T^*$, where $T = \{],[,),(\}$. Suppose w contains some spoilt brackets $[_1 \ldots]_2$: $w = u[_1 v]_2 x$. Then $w \notin D_2$.

<u>DEFINITION of APG and APL.</u> An absolutely parallel grammar (APG) is any 4-tuple $G = (N,T,S,P)$, where N and T are disjoint finite sets of symbols, called non-terminal and terminal alphabet, respectively. Denote $V = N \cup T$. $S \in N$ is the start-symbol and P is a finite set of productions of the form $q = (A_1, A_2, \ldots, A_m) \to (y_1, \ldots, y_m)$, where $A_i \in N$, $y_i \in V^*$. Then $w \overset{q}{\Longrightarrow}_G w'$ iff $w = u_1 A_1 u_2 \ldots A_m u_{m+1}$, $w' = u_1 y_1 u_2 \ldots y_m u_{m+1}$ with $u_i \in T^*$. $w \Longrightarrow_G w'$ iff there exists $q \in P$ such that $w \overset{q}{\Longrightarrow}_G w'$. Then $\overset{*}{\Longrightarrow}_G$ is the reflexive and transitive closure of $\Longrightarrow_G$. The language generated by G is $L(G) = \{w \in T^*;\ S \overset{*}{\Longrightarrow}_G w\}$. The family of absolutely parallel languages is $APL = \{L(G);\ G \text{ is an APG}\}$.

<u>ILLUSTRATION.</u> We can imagine the "work" of G as follows ($u_i \in T^*$) :

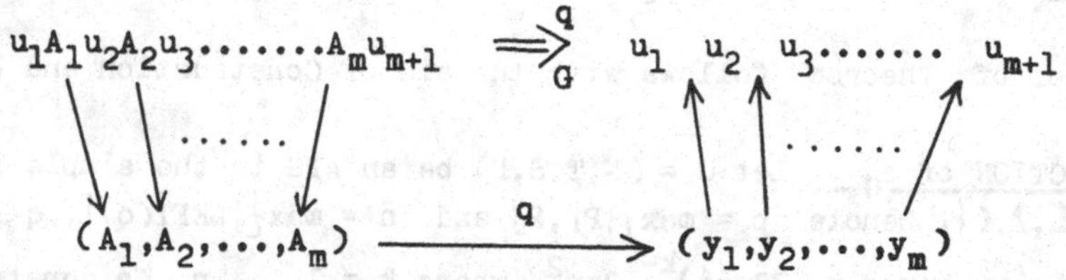

<u>NOTE.</u> We usually omit the symbol G in $\Longrightarrow_G$. Instead of $w \overset{q}{\Longrightarrow} w'$ we also write $w' = q(w)$, similarly instead of $w_1 \overset{q_1}{\Longrightarrow} w_2 \overset{q_2}{\Longrightarrow} w_3 \overset{q_3}{\Longrightarrow} \ldots\ldots \overset{q_{r-1}}{\Longrightarrow} w_r$ we also write $w_r = \{q_1, \ldots, q_{r-1}\}(w_1)$. The left side of the production we call LEFT : $LEFT(q) = (A_1, \ldots, A_m)$. The ordered set of non-terminals of a word we call NOT : $NOT(w) = (A_1, \ldots, A_m)$.

DEFINITION. Let $G=(N,T,S,P)$ be an APG. We say "G is in the simple form" iff for any $q \in P$ the number of terminals on the right side of q is at most one. (This means that any $q \in P$ writes at most one terminal.)

LEMMA 2. There is an effective procedure to find, for any APG H, an APG G in the simple form such that $L(H) = L(G)$.

Proof (outline, for details see [2]). Each "wrong" production replace by a finite set of new productions as follows : Let for example

$$q = (A_1, A_2, A_3) \longrightarrow (d_1 d_2 B_1 d_3 B_2, d_4 d_5, B_3 B_4 d_6),$$

where $A_i, B_i \in N$ and $d_i \in T$. Replace this production by 6 new productions $q_1, \dots, q_6$ with help of 5 new non-terminals $Y_1, \dots, Y_5$:

$$\begin{aligned}
q_1 &= (A_1, A_2, A_3) \longrightarrow (Y_1, d_1 A_1, A_2, A_3) \\
q_2 &= (Y_1, A_1, A_2, A_3) \longrightarrow (Y_2, d_2 B_1 A_1, A_2, A_3) \\
q_3 &= (Y_2, B_1, A_1, A_2, A_3) \longrightarrow (Y_3, B_1, d_3 B_2, A_2, A_3) \\
q_4 &= (Y_3, B_1, B_2, A_2, A_3) \longrightarrow (Y_4, B_1, B_2, d_4 A_2, A_3) \\
q_5 &= (Y_4, B_1, B_2, A_2, A_3) \longrightarrow (Y_5, B_1, B_2, d_5, B_3 B_4 A_3) \\
q_6 &= (Y_5, B_1, B_2, B_3, B_4, A_3) \longrightarrow (\Lambda, B_1, B_2, B_3, B_4, d_6)
\end{aligned}$$

Then obviously (for any $w, w' \in V^*$) $w \overset{q}{\Longrightarrow} w'$ iff $w \overset{q_1}{\Longrightarrow} \dots\dots \overset{q_6}{\Longrightarrow} w'$. The new non-terminals guard that $L(G) = L(H)$.

2. THE MAIN RESULT.

THEOREM. The Dyck language D_2 is not absolutely parallel.

The proof of Theorem follows with the aid of Construction and Lemma 3.

CONSTRUCTION of c_G. Let $G = (N,T,S,P)$ be an APG in the simple form and $T \supseteq \{],[,),(\}$. Denote $p = \max\{|P|, 2\}$ and $n = \max\{|LEFT(q)|;\ q \in P\}$. Let s_k be the number $s_k = (33pn^4)^{k-1} 2pn^2$, where $k = 1, \dots, n$. We construct the words $c_1, \dots, c_k, \dots, c_n = c_G$ by induction on k :

<u>$k = 1$.</u> Let $a = e = [((\dots((,\ d = h =))\dots))]$, where $|a| = |d| = |e| = |h| = s_1 + 1$. Then define $c_1 = adeh$. Obviously $c_1 \in D_2$.

<u>$k > 1$.</u> Let c_{k-1} be already defined. Let $a = e = [((\dots(([,\ d = h =]))\dots))]$, where $|a| = |d| = |e| = |h| = s_k + 2$. Let $b = c = f = g = c_{k-1}$. Then define $c_k = abcdefgh$:

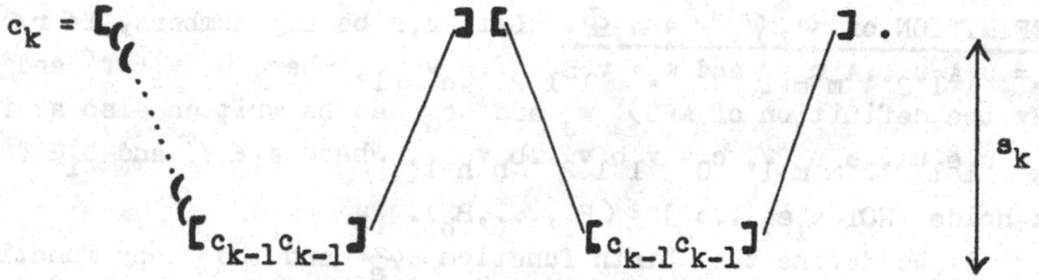

Obviously it holds $c_k \in D_2$.

<u>Finally</u> denote $c_G = c_n$. Clearly it holds $c_G \in D_2$.

<u>LEMMA 3.</u> Let $G=(N,T,S,P)$ be an APG in the simple form. If $c_G \in L(G)$, then there exists $w \in L(G)$ such that w contains some spoilt [...]-brackets.

We postpone the proof of Lemma 3 to the third section of the paper.

<u>Proof of the theorem.</u> Suppose an APG G in the simple form can derive all the words of the language D_2 : $L(G) \supseteq D_2$. Construct the word c_G. It holds $c_G \in D_2$, hence c_G can be derived by G. By Lemma 3 there exists some $w \in L(G)$ such that w contains some spoilt [...]-brackets. By Lemma 1 it holds $w \notin D_2$, hence $L(G) \supsetneqq D_2$. Hence none APG in the simple form can generate exactly the language D_2.

By Lemma 2 each APG is equivalent to some APG in the simple form, hence none APG can generate exactly the language D_2.

It means that D_2 is not absolutely parallel : $D_2 \notin APL$.

3. PROOF OF LEMMA 3.

Let $c_G \in L(G)$. Let δ be a derivation of the word c_G in the APG G :

$$\delta : S \overset{q_0}{\Longrightarrow} w_1 \overset{q_1}{\Longrightarrow} w_2 \overset{q_2}{\Longrightarrow} \dots\dots \overset{q_t}{\Longrightarrow} c_G \in T^* .$$

(In this section δ is always this particular derivation). We shall make precise concepts "this symbol of c_G arises from this non-terminal of w_s" or "this symbol of c_G is in w_s already written and it is written on this place" or "this non-terminal of w_s arises from this non-terminal of w_r (where $r < s$)" or "the production q_s writes this symbol of c_G". Therefore we introduce "birth function", "copy function" and "change function".

<u>Important note</u> : Whenever we say "the symbol (subword) of the word c_G", we mean an occurrence of the symbol (subword) implicitly.

The definition of new concepts follows :

DEFINITION of φ, ρ, ψ and Φ. Let r,s be any numbers, $1 \le r \le s \le t$. Let $w_r = u_1A_1u_2..A_mu_{m+1}$ and $w_s = v_1B_1v_2..B_hv_{h+1}$, where $u_i, v_i \in T^*$ and $A_i, B_i \in N$. (By the definition of APG) w_s and c_G can be written also as follows : $w_s = u_1a_1u_2..a_mu_{m+1}$, $c_G = v_1b_1v_2..b_hv_{h+1}$, where $a_i \in V^*$ and $b_i \in T^*$. Clearly it holds $NOT(a_1a_2...a_m) = (B_1,...,B_h)$.

1. We define the birth function φ_s and the copy function ρ_s as follows : Let z be any symbol of the word c_G. Then

If z is contained in some v_i, then $\varphi_s(z)=\Omega$ and $\rho_s(z)$=(the index of this v_i). We say "z is in w_s (already) written" .

If z is contained in some b_i, then $\rho_s(z)=\Omega$ and $\varphi_s(z)$=(the index of this b_i). We say "z is in w_s not yet written" .

We say "the production q_s writes z" iff $\varphi_s(z)\neq\Omega$ and $\varphi_{s+1}(z)=\Omega$.

2. We define the change function $\psi_s^r : \langle 1,...,m\rangle \rightarrow \langle 1,...,h\rangle$ as follows : $\psi_s^r(i)= j$ iff the i-th non-terminal of w_s is contained in the subword a_j of w_s. It means that B_i arises from A_j.

3. We define the birth function Φ_s for subwords : Let $x=x_1...x_k$ be a subword of c_G, $x_i \in T$. Then $\Phi_s(x)= \{ \varphi_s(x_i) ; \varphi_s(x_i)\neq\Omega$ and $1 \le i \le k \}$.

See also the example described in the following figure :

FIGURE.

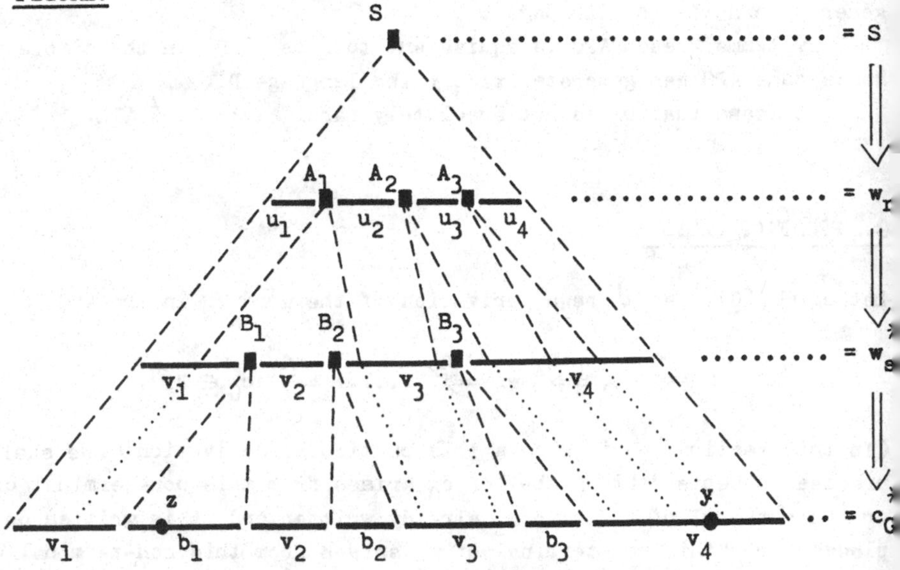

In this example it holds $\varphi_s(z)=1$, $\rho_s(z)=\Omega$, $\varphi_s(y)=\Omega$, $\rho_s(y)=4$, $\varphi_r(z)=1$, $\varphi_r(y)=3$, $\psi_s^r(1)=1$, $\psi_s^r(3)=2$, $\Phi_r(v_3)=\{1,2\}$, $\Phi_r(v_4)=\{2,3\}$, $\Phi_s(y)=\emptyset$, etc.

In the following Fact we state (without proof - for proof see [2]) some consequences of the above definition.

FACT. Consider the situation described in Definition of φ, ρ, ψ and Φ. Let z,y be symbols of c_G and x a subword of c_G. Then it holds :

1. If $\varphi_s(y) \neq \Omega \neq \varphi_s(z)$ and (y is in c_G written on the left of z), then $\varphi_s(y) \leq \varphi_s(z)$.
2. If $\varphi_s(y) = \Omega = \varphi_s(z)$ then (y is in c_G on the left of z) iff (y is in w_s on the left of z).
3. If $\varphi_s(y) = \Omega \neq \varphi_s(z)$ then (y is in c_G on the left of z) iff $\rho_s(y) \leq \varphi_s(z)$ iff (y is in w_s on the left of the $\varphi_s(z)$-th non-terminal).
4. If $\varphi_s(z) = \Omega$ then also $\varphi_r(z) = \Omega$.
5. If $x = uv$ then $\Phi_s(x) = \Phi_s(u) \cup \Phi_s(v)$.
6. If $x = uzv$ and $\varphi_s(z) = \Omega$, then $\Phi_s(u) \cap \Phi_s(v) = \emptyset$.
7. If $\varphi_s(z) \neq \Omega$, then $\varphi_r(z) = \psi_s^r(\varphi_s(z))$.
8. If $\varphi_s(y) \neq \Omega$ for any symbol y of x, then $|\Phi_r(x)| \leq |\Phi_s(x)|$.
9. If $\varphi_s(z) = \varphi_r(z) = j \neq \Omega$, then $\psi_s^r(j) = j$.
10. If $\psi_s^r(j) = j$, then $a_j = a_{j1} B_j a_{j2}$, where $a_{j1}, a_{j2} \in V^*$.
11. If $NOT(w_s) = (A_1, \ldots, A_m) = NOT(w_r)$ and $\rho_r(z) = \rho_s(z) = j \leq n$, then in both the words w_r and w_s, A_j is the first non-terminal, which is written on the right of the symbol z.

The own proof of Lemma 3 is divided into three sections : In the first two sections Induction assertion V_k is proved for any k, $1 \leq k \leq n$. In the third one we show that V_n implies Lemma 3.

Induction assertion V_k. Denote c_k' one fixed occurrence of c_k in c_G. If $|\Phi_i(c_k')| \leq k$ for any i , $1 \leq i \leq t$, then there exists a derivation δ' : $S \Rightarrow^* w \in T^*$ such that w contains some spoilt [...]-brackets.

I. Basis k = 1. We omit the proof of Basis, since it is the particular case of the proof of Induction step. (See also notes 1,2,3) .

II. Induction step $1 < k \leq n$. Assume that V_{k-1} is true. Let c_k' = abcdefgh (cf. Construction). We distinguish two cases dependent on what happens during deriving the word c_G by the derivation δ :

A. At least one symbol of de arises earlier than writing of one of the parts a, h is finished.

B. One of the parts a, h is finished before any symbol of de arises.

Roughly speaking, in the case A the new symbol separates some two subwords c_{k-1} involved in bc and fg. Then V_{k-1} can be used, since it is possible to prove $|\Phi_i(c'_{k-1})| \leq k-1$ for some c_{k-1} of c_G and for any i, $1 \leq i \leq t$. In the case B the APG G is not able to count all the symbols ❫ resp. ❪ involved in h resp. a .

Note 1 : If k = 1 , then the case A can be eliminated.

CASE A.

Distinguish two possibilities : The first symbol arisen in bcdefg is:

(A-i)) in cdef .

(A-ii)) in b or in g .

(A-i) : Denote this arisen symbol z. Denote c'_k = uzv. Obviously b is a subword of u and g is a subword of v. We can assume that the part b arises earlier than g .

Roughly speaking, before z arises, nothing in b is written. After z arises, at least one non-terminal is separated in the part g and before it can make itself free, b is finished. Since for the subword c'_k there are at disposal always at most k non-terminals, for the subword b there are at disposal always at most k-1 non-terminals, which is the sense of the assumption $|\Phi_i(b)| \leq k-1$. And hence V_{k-1} can be used on b .

More formally , let w_r be the first word of δ in which z is written : $r = \min\{i \,;\, \varphi_i(z) = \Omega\}$. Let $s = \min\{i \,;\, \Phi_i(b) = \emptyset\}$. Then :

If $s \leq i \leq t$ then $\Phi_i(b) = \emptyset$.

If $r \leq i \leq s$ then $\Phi_i(b) \cap \Phi_i(g) = \emptyset$, $\Phi_i(b) \cup \Phi_i(g) = \Phi_i(c'_k)$ and $\Phi_i(g) \neq \emptyset$. Hence $|\Phi_i(b)| \leq k-1$.

If $1 \leq i \leq r$ then also $|\Phi_i(b)| \leq k-1$, since $|\Phi_r(b)| \leq k-1$ and since it holds $\varphi_r(y) \neq \Omega$ for any symbol y of b .

Hence $|\Phi_i(b)| \leq k-1$ for any i, $1 \leq i \leq t$, and V_{k-1} can be used on b = c'_{k-1}

We omit the proof of (A-ii), since it is similar.

CASE B.

Suppose that a is written earlier than h. Let $\alpha = \min\{i \,;\, \Phi_i(a) = \emptyset\}$. Denote $[_1$ the left [of a , $[_2$ the right [of a , $]_3$ the left] of d and $]_4$ the right] of d. It holds $\varphi_i(y) \neq \Omega$ for any symbol y of d and for any i, $1 \leq i \leq \alpha$. We distinguish 4 possibilities :

(i) At least $\frac{s_k}{3}$ symbols ❪ of a have arisen before $[_1$ or $[_2$ arises.

(ii) $[_2$ had arisen before $\frac{s_k}{3}$ symbols ❪ of a arose; then at least $\frac{s_k}{3}$ symbols ❪ of a have arisen before $[_1$ arises.

(iii) The same as (ii) with changed $[_1$ and $[_2$.

(iv) Both $[_1$ and $[_2$ have arisen before $\frac{2s_k}{3}$ symbols ❪ of a arise.

Note 2: In the Basis there is only one possibility derived from (iii).

The proof of all the cases is similar. We present only the outline of the proof of (ii).(By (ii)) in the part a, G writes the symbol $[_2$, then it writes symbols ("long enough" and then it writes $[_1$. From B we know that during this part of δ no symbol of de arises. This enables us to choose the numbers r,s , $1 \leq r \leq s \leq \alpha$, so that :

$$\delta : S \overset{Q}{\Longrightarrow}^* w_r \overset{R}{\Longrightarrow}^* w_s \overset{K}{\Longrightarrow}^* c_G ,$$

where $Q,R,K \in P^*$, and so that it holds :

α) the production q_r writes some (into the part a .

β) $NOT(w_r) = NOT(w_s)$, $1 \leq \varphi_r([_1) = \varphi_s([_1) \leq \wp_r([_2) = \wp_s([_2) \leq \varphi_r(]_3) = \varphi_s(]_3) \leq \leq \varphi_r(]_4) = \varphi_s(]_4) \leq n$.

γ) during the sequence of productions R no symbol of bc arises.

Now we repeat the sequence R in the derivation δ and define :

$$\delta' : S \overset{Q}{\Longrightarrow}^* w_r \overset{R}{\Longrightarrow}^* w_s \overset{R}{\Longrightarrow}^* w_s' \overset{K}{\Longrightarrow}^* w \in T^* .$$

From the definition of APG it follows that c_G and w_r differ similarly as w and w_s. The "stabilization" of φ and $\wp$ warrants that the terminals of w arise on the "wanted" places. Hence if we denote $[_1, [_2,]_3$ and $]_4$ the symbols of w which are written in w on the places corresponding to the places of $[_1, [_2,]_3$ and $]_4$ in c_G, then it holds :

(a) in the part $[_1 \ldots [_2$ of w there is at least one (in addition to the number of (in the part $[_1 \ldots [_2$ of the word c_G.

(b) in the part $[_1 \ldots [_2$ of w occurs none of the symbols) ,] or [.

(c) the number of) resp. (resp.] resp. [is the same in both the parts $[_2 \ldots]_3$ of the word w, and $[_2 \ldots]_3$ of the word c_G.

(d) The part $]_3 \ldots]_4$ is the same in both the words w and c_G .

The (a) follows from α, β , the (b) follows from β, the (c) follows from β, γ and the (d) follows from β with the aid of Fact 1,3,9,10,11 .

Finally there are two possibilities :

1. If in $[_1 \ldots]_4$ of w all [,]-brackets are complete, then the brackets $[_1 \ldots]_4$ are spoilt, since the numbers of) and of (in $[_1 \ldots]_4$ differ.
2. Otherwise there exists $]_5$ in $[_1 \ldots]_4$ such that in $[_1 \ldots]_5$ of w all [,]-brackets are complete. Since in $[_1 \ldots [_2$ of w and in $]_3 \ldots]_4$ of w there is no], this $]_5$ is contained in $[_2 \ldots]_3$. Hence $[_1 \ldots]_5$ contains at most $2|c_{k-1}| = |bc|$ symbols) and at least $s_k = |a|$ symbols (. Since $s_k > 2|c_{k-1}|$, in $[_1 \ldots]_5$ of w the numbers of) and of (differ. Hence $[_1 \ldots]_5$ is spoilt.

Note 3. In the proof of Basis this possibility 2 can be eliminated.

Hence w involves some spoilt [,]-brackets in both the cases 1 and 2.

This completes the proof of Case B and of Induction step, too.

It follows from Basis and Induction step that V_n holds.

III. V_n implies Lemma 3. Since $|\mathrm{LEFT}(q)| \leq n$ for any $q \in P$, it holds $|\Phi_i(c_G)| \leq n$ for any i, $1 \leq i \leq t$. Since $c_n = c_G$, the assumptions of V_n are fulfilled. Applying V_n we obtain a derivation δ' of a word w such that w contains some spoilt [...]-brackets.

This completes the proof of Lemma 3.

ACKNOWLEDGEMENT.

I wish to thank my advisor Professor Michal Chytil for help and encouragement.

REFERENCES.

1. Rajlich, V., Absolutely parallel grammars and two-way finite-state transducers. J. Comput. Syst. Sci. 6, (1972), 324-342.
2. Tegze, M., Some properties of functions, realized by different types of automata (in Czech). Dissertation, Charles University, (1978).
3. Chytil, M.P. and Jákl, V., Serial composition of two-way finite-state transducers and simple programs on strings. Proceedings of ICALP, Turku, Lecture Notes in Comput.Sci.52, Springer-Verlag, Berlin, (1977)

FIXED POINTS IN THE POWER-SET ALGEBRA OF INFINITE TREES

/ abstract /

Jerzy Tiuryn

RWTH Aachen, Lehrstuhl für Informatik II

and

Warsaw University, Institute of Mathematics

Introduction

In this paper we are studying two notions of continuity, both weaker than the well known and broadly applied notion of *ω-continuity*. Examples grouped in this paper show that there are natural situations when one is dealing with monotone maps, defined on "sufficiently" complete carriers which are not ω-continuous but which still have a very pleasant property allowing to solve fixed point systems of equations within *ω steps of iterations*.

The first above mentioned generalization of ω-continuity is obtained simply by requiring solvability of fixed point systems of equations with *arbitrary parameters* from a carrier, within ω steps of iterations (i.e. by taking the least upper bound of ω iterations, which is assumed to exist). This is the notion of a *regular algebra* introduced by the author in Tiuryn [1977] . However the only examples of regular algebras which are not ω-continuous, known to the author till 1978, were based on the fact that a carrier is not ω-complete but still all operations are ω-continuous (e.g. the algebra of regular trees). This paper provides an example of a regular algebra which is not ω-continuous due to essentially different reasons - the carrier is a complete lattice but operations are not ω-continuous.

But then a new phenomenon arises. It turns out that in some cases even the property of being regular is too strong, and the reason lies in "too arbitrary" parameters. This gives rise to a weaker notion than that of regularity. So the second generalization of ω-continuity is obtained by restricting solvability within ω steps of iterations, only to fixed point systems *without parameters*. This is the idea laying behind the notion of a *semiregular algebra*, originally introduced in this paper. It can be easily seen (the proof from Tiuryn [1977] applies to this case as well) that the algebra of regular trees is still initial in the category of semiregular algebras. However, we conjecture that there does *not* exist a free semiregular algebra generated by nonempty set of free generators. On the other hand there are confusingly many natural examples (as we hope to show in this paper) of semiregular algebras which are not regular, proving this notion to be useful.

The paper provides a useful reduction of questions about regularity (or semiregularity) of a given algebra to questions about a certain kind of continuity of basic operations which seems to link much better algebraic and order aspects of structures. This is done in the first part of the paper. The second part uses this results to investigate the properties of power-set algebras of finite and/or infinite trees, from the continuity point of view.

There is another aspect of interest in infinite trees. This is connected with semantics of nondeterministic recursive program schemes (or equivalently with infinitistic behaviours of context-free tree grammars). In fact the author took the inspiration for writing this paper from the paper Arnold, Nivat [1977] . Theorem 4.5 fill a gap in the above mentioned paper in removing the assumption of being a Greibach scheme, which was used to prove that the greatest fixed point of an operator associat with a given grammar is obtainable as an intersections of ω iterations (when startin with the top element). This justifies the author's interest in the present paper to the order dual to inclusion. Theorem 4.5 has been obtained independently in Arnold, Nivat [1978] for the case of OI substitution.

Results obtained in this paper are grouped in a table at the end.

1. Regular and semiregular algebras

As the power set algebras of finite and/or infinite trees form many-sorted structures - we formulate our definitions in this section for the heterogenuous case. The reader may compare this with homogenuous notion of a regular algebra given in Tiuryn [1977] or in Tiuryn [1978].

1.1 Let S be a nonempty set, called the set of sorts. The set S will be fixed throughout this section. Denote by S^* the set of all finite words over S, the empt word will be denoted by λ. Concatenation of two words $w,v \in S^*$ is denoted by wv.

An S-sorted signature Σ is an indexed family $<\Sigma_{w,s}>_{(w,s)} \in S^* \times S$ of disjoint sets. Elements of a $\Sigma_{w,s}$ are called function symbols of arity w, and type

An S-sorted set A is an indexed family $<A_s>_{s \in S}$. If $w = s_1 \ldots s_n \in S^*$, then by A^w we denote the set $A_{s_1} \times \ldots \times A_{s_n}$.
In particular A^s for $s \in S$ denotes A_s, and A^λ is a singleton.

1.2 Let Σ be an S-sorted signature, and $w = s_1 \ldots s_n \in S^*$. Denote by $T_\Sigma(w)$ the S-sorted set of all finite terms over Σ with variables among $\{\phi_1^{s_1},\ldots,\phi_n^{s_n}\}$. The variable $\phi_i^{s_i}$, $o < i \leq n$, is supposed to range over the elements of sort s_i.

1.3 Let Σ be an S-sorted signature. A Σ-algebra $\mathcal{A}$ is an S-sorted set A (the carrier of $\mathcal{A}$) with interpretation of every function symbol $\sigma \in \Sigma_{w,s}$ as a function $\sigma_{\mathcal{A}} : A^w \to A_s$. If $w,v \in S^*$, $t \in T_\Sigma(w)^v$ (i.e. t is a v-vector of w-ary terms) and $\mathcal{A}$ is a Σ-algebra, then $t_{\mathcal{A}} : A^w \to A^v$ is the meaning of t in $\mathcal{A}$. A map $p : A^w \to A^v$ is called a polynomial in $\mathcal{A}$ if there is $t \in T_\Sigma(w)^v$ such that $p = t_{\mathcal{A}}$. A map $f : A^w \to A^v$ is called an algebraic map in $\mathcal{A}$ if there exist $u \in S^*$, a polynomial $p : A^{wu} \to A^v$, and $a \in A^u$, such that for all $x \in A^w$, $f(x) = p(x,a)$.

1.4 We will deal in this paper exclusively with ordered algebras, i.e. the carrier of sort s is equipped with a partial order $\leq_s$ and a bottom element $\perp_s$, and all operations are monotone. We frequently omit the subscript s when it does not lead to a confusion. The same concerns products: A^w $(w = s_1 \dots s_n \in S^*)$ – the order relation in A^w (defined component wise) will be denoted by $\leq$, and the bottom element in A^w will be denoted by $\perp$. The least upper bound of a set X, if it exists, will be denoted by $\sup X$.

1.5 Let $\mathcal{A}$ be an ordered Σ-algebra, $w \in S^*$, and let $f\colon A^w \to A^w$ be an algebraic map in $\mathcal{A}$. Denote by L_f the set $\{f^i(\perp) : i \in \omega\}$ where $f^0 = id_{A^w}$, and $f^{i+1} = f \circ f^i$ for $i \in \omega$. Let $s \in S$. A subset $E \subseteq A_s$ is called an iteration if there is $w \in S^*$, and an algebraic map $f\colon A^{sw} \to A^{sw}$ such that $E = e_1^{sw}(L_f)$. Here $e_1^{sw} : A^{sw} \to A_s$ is the projection on the first component. A subset $E \subseteq A_s$ is called a pure iteration if $E = e_1^{sw}(L_p)$ for some polynomial $p\colon A^{sw} \to A^{sw}$. Therefore pure iterations are iterations without parameters from the carrier A.

1.6 An ordered Σ-algebra $\mathcal{A}$ is said to be a regular algebra iff for every $w \in S^*$, and for every *algebraic map* in $\mathcal{A}$, $f\colon A^w \to A^w$, all the following hold:

(1.6.1) $\sup L_f$ exists in A^w.

(1.6.2) $f(\sup L_f) = \sup L_f$.

From the above definition it follows that $\sup L_f$ is the least fixed point of f. It means that regular algebras are precisely those algebras where one can solve every finite system of fixed point equations with parameters, getting the least solution as the least upper bound of *ω-iterations*. Every ω-continuous algebra (i.e. with ω-complete carrier and ω-continuous operations) is a regular one, but the converse in general is not true. For example, the algebra of regular trees (cf. Tiuryn [1977]) is a regular algebra which is not ω-continuous due to the fact that the carrier is not ω-complete (the ω-operations are still ω-continuous). In this paper we shall see another reason for not being ω-continuous - there exist natural examples of algebras with complete carriers and monotone operations which are not ω-continuous. It will turn out that when considering equations with arbitrary parameters (as it has been done in definition 1.6), there are natural examples when in general ω steps of iterations are not enough to get solutions, while without parameters ω steps are always enough. This leads to the following generalization of the concept of regular algebra.

1.7 An ordered Σ-algebra $\mathcal{A}$ is said to be a semiregular algebra iff for every $w \in S^*$, and for every *polynomial* in $\mathcal{A}$, $p\colon A^w \to A^w$, all the following hold:

(1.7.1) $\sup L_p$ exists in A^w.

(1.7.2) $p(\sup L_p) = \sup L_p$.

1.8 Observe that if the carrier of $\mathcal{A}$ is ω-complete then $\mathcal{A}$ is a semiregular algebra iff (1.8.1) for every polynomial $p\colon A^w \to A^w$, $\sup\{p^n(\perp) : n \in \omega\}$ is a fixed point of p.

There are some situations (cf. for example Arnold, Nivat [1977]) where one is forced to prove (1.8.1) while polynomials are not ω-continuous and the induction on the length of terms cannot be applied due to great complexity. For this reason we introduce a weaker notion of continuity, showing that the above mentioned problem can b reduced to a usually simpler problem concerning only basic operations.

To introduce not too many definitions we consider in this paper only the case o algebras with a Δ-complete carrier (in these algebras every iteration has a least upper bound).

1.9 Let A be an ordered Σ-algebra with a Δ-complete carrier, let $w = s_1 .. s_n \in S^*$, and $s \in S$. A map $f: A^w \to A_s$ is said to be algebraically continuc iff for arbitrary iterations $E_1, \ldots, E_n$ with $E_i = \{ a_k^i : k \in \omega \} \subseteq A_{s_i}$ for $i = 1, \ldots, n$ there exists an iteration $E \subseteq A_s$ such that

(1.9.1) $\{ f(a_k^1, \ldots, a_k^n) : k \in \omega \} \subseteq E.$

(1.9.2) $f(\sup E_1, \ldots, \sup E_n) = \sup \{ f(a_k^1, \ldots, a_k^n) : k \in \omega \}.$

A map f is said to be algebraically semicontinuous if it has the above prope ty when restricting only to pure iterations. The importance of the above introduced notions lies in the following results.

1.10 Theorem

Let A be an ordered Σ-algebra with a Δ-complete carrier. The following conditions are equivalent:

(1.10.1) A is a *regular* algebra .

(1.10.2) All Σ-operations in A are *algebraically continuous*.

1.11 Theorem

Under the same assumptions about A as in 1.10., the following conditions are equivalent:

(1.11.1) A is a *semiregular* algebra .

(1.11.2) All Σ-operations in A are *algebraically semicontinuous*.

1.12 As we will deal in the sequel also with properties of greatest fixed points, it will be convenient, for the sake of uniformity of presentation, to treat them at least fixed points in dual order.

Therefore, if α is a property stating something about a poset $\underline{P} = (P, \leqslant)$, then we shall say that $\underline{P}$ has property co-α iff α is true in the dual poset $\underline{P}^* = (p, \leqslant^*)$ where $p \leqslant^* q$ iff $q \leqslant p$. For example, $\underline{P}$ is co-ω-complete iff every down-directed subset of P has greatest lower bound. Thus one has definitions of a co-regular alg bra, of an algebraic co-continuity, of a co-iteration , etc.

2. The power-set algebras: PT_Σ and PT_Σ^∞

2.1 Let Σ be a *one sorted* signature, and let $n \in \omega$. Denote by $T_\Sigma(n)$ (resp. by $T_\Sigma^\infty(n)$) the set of all *finite* (resp. of all *finite* or *infinite*) *trees* with variables among $\{ x_1, \ldots, x_n \}$.

2.2 Denote by $D(\Sigma)$ an ω-sorted signature (derived alphabet of Σ, cf. also Engelfr Schmidt [1977]) defined as follows. For each $n \geqslant 1$ and for each i, $1 \leqslant i \leqslant n$, let Π_i^n be a new symbol (the i-th projection symbol of sort n); for each $n \in \omega$, let U_n be a new symbol (the <u>join symbol</u> of sort n); and let for each $n,k \in \omega$, $c_{n,k}$ be a new symbol (the (n,k)-th <u>composition symbol</u>). We assume that all the symbols introduced above do not belong to Σ. Then $D(\Sigma)$ is determined by the following conditions

(2.2.1) $D(\Sigma)_{\lambda,o} = \Sigma_o$,

(2.2.2) for $n \geq 1$, $D(\Sigma)_{\lambda,n} = \Sigma_n \cup \{\Pi_i^n : 1 \leq i \leq n\}$,

(2.2.3) for $n \in \omega$, $n \neq 1$, $D(\Sigma)_{nn,n} = \{U_n\}$,

(2.2.4) for $n,k \in \omega$, $n \neq 1$ or $k \neq 1$, $D_{n\underbrace{k..k}_{n\text{-times}}k} = \{c_{n,k}\}$,

(2.2.5) $D(\Sigma)_{11,1} = \{U_1, c_{1,1}\}$
(here 11 is a word over ω of length 2),

(2.2.6) $D(\Sigma)_{w,n} = \emptyset$ otherwise .

The signature introduced above should be called perhaps a *nondeterministic derived alphabet* of Σ in contrast to a *deterministic* one, introduced in Engelfried and Schmidt [1977] .

<u>2.3 PT_Σ</u>–

PT_Σ is going to be a $D(\Sigma)$-algebra. Let X be a set, by $P(X)$ we denote the set of all subsets of X. The carrier of sort n, $PT_{\Sigma,n}$ is $P(T_\Sigma(n))$. This is naturally ordered by inclusion relation. The meaning of $\sigma \in \Sigma_n$, for $n \geq 0$, is the one-element set $\{\sigma(x_1 \ldots x_n)\}$ (for $n = 0$ this is simply the set $\{\sigma\} \in P(T_\Sigma(0))$). Π_i^n is interpreted as $\{x_i\}$. U_n is interpreted as binary set union (one for each sort). $c_{n,k}$ is interpreted as a substitution of tree languages. In this paper we are only interested in IO and OI substitutions (cf. Engelfriet and Schmidt [1977] for definitions). Therefore, accordingly which substitution has been chosen, one gets $D(\Sigma)$-algebras PT_Σ^{IO} and PT_Σ^{OI} .

<u>2.4 PT_Σ^∞</u>–

The $D(\Sigma)$-algebra PT_Σ^∞ is defined similarly to 2.3. The carrier of sort n, $PT_{\Sigma,n}^\infty$, is $P(T_\Sigma^\infty(n))$, the order relation is inclusion. $D(\Sigma)$-constant symbols and U_n's are interpreted the same as in 2.3. Extension of OI substitution to infinite trees the reader may find in Arnold, Nivat [1977], an extension of IO substitution can be defined similarly. The $D(\Sigma)$-algebra, obtained in this way, we denote by $PT_\Sigma^{\infty OI}$ and $PT_\Sigma^{\infty IO}$.

<u>2.5</u> It can be easily checked that all the $D(\Sigma)$-operations in algebras PT_Σ and PT_Σ^∞ (for OI and IO) are monotone, and U_n's for $n \in \omega$, are Δ-continuous and co-ω-continuous. Therefore by Theorem 1.10 and 1.11 it follows that the global properties of the above defined power-set algebras depend on properties of substitution.
For $z \in \{OI, IO\}$ by $\underset{z}{\leftarrow}$ we denote the z-substitution.

<u>2.6 Theorem</u>

(2.6.1) For every signature Σ the $D(\Sigma)$-algebra PT_Σ^{IO} is ω-continuous and co-ω-continuous.

(2.6.2) The same holds for PT_Σ^{OI}.

3. Examples

3.1 Example (*$PT_\Sigma^{\infty OI}$ is not regular for a polyadic signature Σ*)

Let $\Sigma_o = \{a\}$, $\Sigma_1 = \{s\}$, $\Sigma_2 = \{\alpha\}$, $\Sigma_n = \emptyset$ for $n > 2$.

Consider the following tree $t \in T_\Sigma^\infty(1)$:

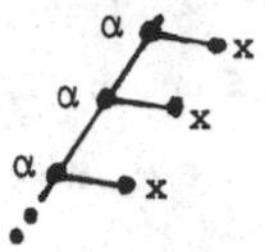

It is easy to check that the sequence $\{A_n, n \in \omega\}$, where $A_n = \{a, s(a), \ldots, s^n(a)\}$, is an iteration. Then the tree t' :

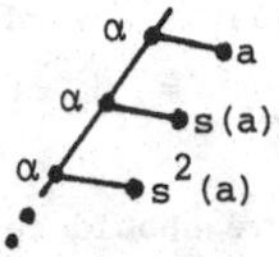

belongs to $\{t\} \underset{OI}{\leftarrow} \bigcup_{n\in\omega} A_n$, but it does not belong to $\bigcup_{n\in\omega} (\{t\} \underset{OI}{\leftarrow} A_n)$.

This example shows that the algebraic mapping $f : P(T_\Sigma^\infty(o)) \to P(T_\Sigma^\infty(o))$ defined by $f(L) = \{t\} \underset{OI}{\leftarrow} L$, is *not* algebraically continuous. Therefore, by Theorem 1.10 $PT_\Sigma^{\infty OI}$ is not regular. It can be easily checked that the system

$$\phi_1^{(o)} = \{ s(x) \} \underset{OI}{\leftarrow} (\phi_1^{(o)} \cup \{a\}) ,$$

$$\phi_2^{(o)} = \{t\} \underset{OI}{\leftarrow} \phi_1^{(o)} ,$$

needs exactly $\omega+1$ steps of iterations to "achieve" the solution. One can easily prove the following

3.2 Proposition

(3.2.1) $PT_\Sigma^{\infty OI}$ is always semiregular.

(3.2.2) $PT_\Sigma^{\infty IO}$ is always ω-continuous.

(3.2.3) For a monadic signature Σ, $PT_\Sigma^{\infty OI}$ is ω-continuous.

3.3 Example (*$PT_\Sigma^{\infty OI}$ and $PT_\Sigma^{\infty IO}$ are not co-regular for a polyadic signature Σ*)

Let Σ be a signature like in Example 3.1, for every $n\in\omega$ let $t_n \in T_\Sigma^{\infty(1)}$ denote the following tree:

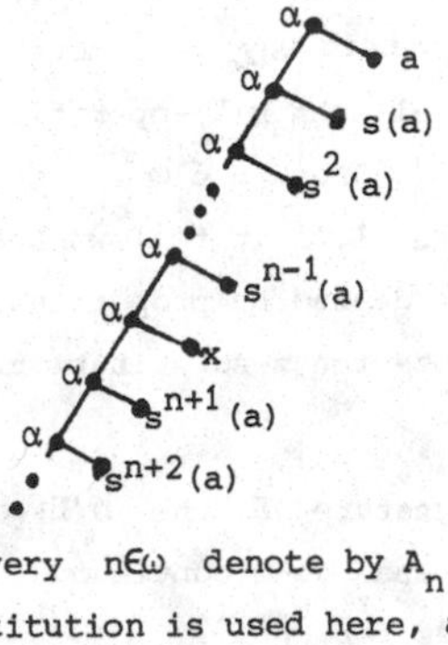

Let $L = \{t_n : n\in\omega\}$. For every $n\in\omega$ denote by A_n the set $\{s^n(x)\} \leftarrow T_\Sigma^\infty(o)$ (we do not specify which kind of substitution is used here, as A_n is the same for OI as well

as for IO substitution). $\{A_n : n\in\omega\}$ is obviously a co-iteration. Observe that the tree t' from Example 3.1 belongs to $\bigcap_{n\in\omega}(L\leftarrow A_n)$, but it does not belong to $L \leftarrow \bigcap_{n\in\omega} A_n = L \leftarrow \{s^{\omega}\}$ (since in each t_n the variable x occurs exactly once, substitutions OI and IO in the above formulas coincide). The above example also shows that the system

$$\phi_1^{(o)} = \{s(x)\} \leftarrow \phi_1^{(o)},$$
$$\phi_2^{(o)} = L \leftarrow \phi_1^{(o)},$$

provides exactly $\omega+1$ steps of co-iterations.

3.4 Example (*$PT_{\Sigma}^{\infty OI}$ and $PT_{\Sigma}^{\infty IO}$ are not co-regular for a monadic signature Σ containing at least two unary operation symbols*)

Let $\Sigma_1 = \{s,g\}$, $\Sigma_n = \phi$ for $n \neq 1$. Let $t_n = sgs^2gs^3g\ldots s^ng(x)$, for $n\in\omega$. Define $A_n = \{s^n(x)\} \leftarrow T_{\Sigma}^{\infty}(o)$ for $n\in\omega$. Let $L = \{t_n : n\in\omega\}$. Then the infinite word $sgs^2gs^3g\ldots$ belongs to $\bigcap_{n\in\omega}(L\leftarrow A_n)$ but does not belong to $(L \leftarrow \bigcap A_n) = L \leftarrow \{s^{\omega}\}$.
Again the system

$$\phi_1^{(o)} = \{s(x)\} \leftarrow \phi_1^{(o)},$$
$$\phi_2^{(o)} = L \leftarrow \phi_1^{(o)},$$

produces $\omega+1$ steps of co-iterations.

However in the case of one element monadic signature the situation looks different.

3.5 Theorem

If $\Sigma_1 = \{s\}$, $\Sigma_n = \phi$ for $n > 1$, then PT_{Σ}^{∞} is co-regular *not* co-ω-continuous algebra.

As the reader may guess, the essential reason for the phenomena produced by examples 3.3 and 3.4 lies in "too arbitrary parameters". The rest of the paper is devoted to investigate for what parameters co-iterations "stabilize" in ω steps. In particular we show that PT_{Σ}^{∞} is a co-semiregular algebra for OI and IO substitutions, and for arbitrary signature Σ. This we will do using topological methods.

4. Fixed point systems of equations with closed parameters

For simplicity of exposition we assume here that we are dealing with a finite signature Σ. However, all results carry over for arbitrary signatures when adopting *Mycielsky-Taylor topology* on T_{Σ}^{∞} which is always compact (cf. Mycielski, Taylor [1976]).

For a finite signature Σ we use a distance function d in the set T_{Σ}^{∞} defined by $d(t,t') = 0$, if $t = t'$; else $2^{-|w|}$, where $w\in\omega^*$ is the shortest path such that $t(w) \neq t'(w)$. It is well known that this function makes T_{Σ}^{∞} into a complete metric space, which is compact if and only if the signature Σ is finite.

Results which we state here are valid for IO and OI substitutions, thus we will omit subscripts and denote substitution by $\leftarrow$.

4.1 Theorem (the main result)

Let $n,k\in\omega$.

(4.1.1) If $L \subseteq T_{\Sigma}^{\infty}(n)$ is a closed subset and for every $i = 1,..,n$, $\{A_m^{(i)} : m\in\omega\}$ is a decreasing sequence of closed subsets in $T_{\Sigma}^{\infty}(k)$, then

$$L \leftarrow (\bigcap_m A_m^{(1)},\ldots,\bigcap_m A_m^{(n)}) = \bigcap_m (L \leftarrow (A_m^{(1)},\ldots,A_m^{(n)})).$$

(4.1.2) If $\{L_m : m\in\omega\}$ is decreasing sequence of closed subsets in $T_\Sigma^\infty(n)$, and $A_1,\dots,A_n \subseteq T_\Sigma^\infty(k)$ are closed subsets, then

$$(\bigcap_m L_m) \leftarrow (A_1,\dots,A_n) = \bigcap_m (L_m \leftarrow (A_1,\dots,A_n)).$$

Proof In both cases it is enough to prove $\supseteq$-inclusion. We give a proof of both cases at once as it is based on the same argument. Let ξ belong to the set standing on the right hand side of equality ((4.1.1) or (4.1.2)). Define

$F_m = \{t \in L : \xi \in t \leftarrow (A_m^{(1)},\dots,A_m^{(n)})\}$ in the first case, and

$F_m = \{t \in L_m : \xi \in t \leftarrow (A_1,\dots,A_n)\}$ in the second case. Then we need the following technical lemmas.

4.2 Lemma

Let $n,k\in\omega$. If $L \subseteq T_\Sigma^\infty(n)$, $A_1,\dots,A_n \subseteq T_\Sigma^\infty(k)$ are closed subsets, then $L \leftarrow (A_1,\dots,A_n)$ is a closed subset of $T_\Sigma^\infty(k)$.

4.3 Lemma

Let $n,k\in\omega$. If $\xi \in T_\Sigma^\infty(k)$, $A_1,\dots,A_n \subseteq T_\Sigma^\infty(k)$ are closed subsets, then the set $\{t \in T_\Sigma^\infty(n) : \xi \in t \leftarrow (A_1,\dots,A_n)\}$ is closed in $T_\Sigma^\infty(n)$.

4.4 Lemma

Let $n,k\in\omega$, $t \in T_\Sigma^\infty(n)$, and let $\{A_m^{(i)} : m\in\omega\}$ be a decreasing chain of subsets of $T_\Sigma^\infty(k)$, for $i = 1,\dots,n$, such that $\bigcap_m A_m^{(i)} \neq \phi$ for $i = 1,\dots,n$.

Then $t \leftarrow (\bigcap_m A_m^{(1)},\dots,\bigcap_m A_m^{(n)}) = \bigcap_m (t \leftarrow (A_m^{(1)},\dots,A_m^{(n)}))$.

Continuation of 4.1

Obviously each F_m is nonempty and $F_m \supseteq F_{m+1}$ for $m\in\omega$. By 4.2 and 4.3 each F_m is closed. Hence by compactness there exists $t \in \bigcap_m F_m$.

In the first case (4.1.1) it means that $\xi \in \bigcap_m (t \leftarrow (A_m^{(1)},\dots,A_m^{(n)}))$.

Again by compactness, $\bigcap_m A_m^{(i)} \neq \phi$ for $i = 1,\dots,n$. By 4.4 $\xi \in t \leftarrow (\bigcap_m A_m^{(1)},\dots,\bigcap_m A_m^{(n)})$, what completes the proof of 4.1.1 .

In the second case it means that $t \in \bigcap_m L_m$ and thus $\xi \in (\bigcap_m L_m) \leftarrow (A_1,\dots,A_n)$, completing 4.1.2. From 1.11 and 4.1 one immediately derives the following result (for the OI substitution it has been obtained by Arnold and Nivat [1978]).

4.5 Corollary

For every signature Σ, the algebras $PT_\Sigma^{\infty OI}$ and $PT_\Sigma^{\infty IO}$ are co-semiregular.

Denote by CT_Σ^{OI} (resp. by CT_Σ^{IO}) a $D(\Sigma)$-algebra of all closed subsets, with OI-substitution (resp. with IO-substitution). By 4.2 we see that CT_Σ^{OI} is a $D(\Sigma)$-subalgebra of PT_Σ^{OI} (CT_Σ^{IO} is a $D(\Sigma)$-subalgebra of $PT_\Sigma^{\infty IO}$). By 4.1 one gets immediately the following corollary.

4.6 Corollary

For every signature Σ the algebras CT_Σ^{OI} and CT_Σ^{IO} are co-ω-continuous.

To make the situation described in this paper more visible the results are grouped together and presented in the table below.

<table>
<tr><th></th><th>PT_{Σ}^{OI}</th><th>PT_{Σ}^{IO}</th><th>$PT_{\Sigma}^{\infty OI}$</th><th>$PT_{\Sigma}^{\infty IO}$</th><th>CT_{Σ}^{OI}</th><th>CT_{Σ}^{IO}</th></tr>
<tr><td rowspan="2">Monadic, one element signature</td><td colspan="2">ω-continuous</td><td colspan="2">ω-continuous</td><td colspan="2">ω-continuous</td></tr>
<tr><td colspan="2">co-ω-continuous</td><td colspan="2">co-regular, and not co-ω-continuous</td><td colspan="2">co-ω-continuous</td></tr>
<tr><td rowspan="2">Monadic ≤ 2-element signature</td><td colspan="2">ω-continuous</td><td colspan="2">ω-continuous</td><td colspan="2">ω-continuous</td></tr>
<tr><td colspan="2">co-ω-continuous</td><td colspan="2">co-semiregular, and not co-regular</td><td colspan="2">co-ω-continuous</td></tr>
<tr><td rowspan="2">Polyadic signature</td><td colspan="2">ω-continuous</td><td>semiregular not regular,</td><td>ω-continuous</td><td>semiregular not regular</td><td>ω-continuous</td></tr>
<tr><td colspan="2">co-ω-continuous</td><td>co-semiregular, not co-regular</td><td>co-semiregular not co-regular</td><td>co-ω-continuous</td><td>co-ω-continuous</td></tr>
</table>

References

Arnold, A., Nivat, M. [1977] Non deterministic recursive program schemes. Proceedings FCT'77 , Lecture Notes in Computer Science No.56, Karpinski, M. (ed.), Springer Verlag.

Arnold, A., Nivat, M. [1978] The metric space of infinite trees. Algebraic and topological properties. Rapport de Recherche No. 323, IRIA.

Engelfriet, J., Schmidt E.M. [1977] IO and OI. I. J. Comput. System Sci., 15.

Mycielski, J., Taylor, W. [1976] A compactification of the algebra of terms. Algebra Universalis, 6.

Tiuryn, J. [1977] Fixed points and algebras with infinitely long expressions. Part I. Regular algebras. Proceedings MFCS'77, Lecture Notes in Comp. Sci. No. 53,Gruska, J. (ed.), Springer Verlag. The full version will appear in Fundamenta Informaticae.

Tiuryn, J. [1978] On a connection between regular algebras and rational algebraic theories. 2 nd Workshop Meeting on: Categorical and Algebraic Methods in Computer Science and System Theory, Dortmund.

ON RELAXATION RULES IN ALGORITHMIC LOGIC

B.A.Trakhtenbrot

Institute of Mathematics, Siberian Branch,
the USSR Academy of Sciences, Novosibirsk 630090, USSR

§ 1. INTRODUCTION

Besides subtle analysis of program constructs many facilities that do not concern penetration in program-structure are widely used in the logic of algorithms. To such facilities that are rather of general set-theoretical and logical nature belong as well the rules we call here relaxation rules, e.g.

$$\frac{P\{G\}Q,\ P'\supset P,\ Q\supset Q'}{P'\{G\}Q'} \qquad (1)$$

$$\frac{P\{G\}Q,\ G'\subseteq G}{P\{G'\}Q} \qquad (2)$$

$$\frac{P\{G\}Q,\ \ x\notin \text{glob } G}{\exists x\, P\{G\}\ \exists x Q} \qquad (3)$$

Rules 1-2 are sound on the propositional level. Namely, one may interpret the metaprogram variables G, G' as state transformers (binary relations) on an arbitrary set Σ, and pre-, post-conditions P, Q, P', Q' as arbitrary one-place predicates on Σ. Unlike 1 and 2, the soundness of rule 3, and even its very formulation are confined to the functional level, i.e. structural states s, t, ... are only considered, which map the set V of program identifiers into the semantical object-domain D. At the same time, to each condition P (to each transformer G) a finite subset glob $P \subseteq V$ (glob $G \subseteq V$) is related, such that the behaviour of P (of G) is completely determined by the values of the states on glob P (glob G).

The status of the relaxation rules manifests itself by means of some specific strong statements. E.g. on propositional level the partial correctness statements $P\{G\}Q$ belong to them, where P is the weakest precondition of Q, or Q is the strongest postcondition of P. To some extent on the functional level a similar role is played by the statements we call accurate (§ 4). As a matter of fact, the technique of accurate statements is a usual tool in axiomatizing algorithmic logic (see [2], [3], [5], [6]); nevertheless, its explicit characterization is seemingly lacking. Maybe that circumstance explains some lack of coordination and also some inaccuracies in different versions of algorithmic logic. This paper is intended as a contribution to the explicit presentation of the accurate statements technique. In § 2 notations and terminology are explained. The main facts and comments to them are in §§ 3,4.

§ 2. TERMINOLOGY, NOTATIONS

Let $z=z_1,\ldots,z_k$ be an identifier vector, P - a condition, G - a transformer. Then, glob $P \subseteq z$, glob $G \subseteq z$ (verbally - "z is a support for P","z is a support for G") are defined as below:

$$\text{glob } P \subseteq z =_{def} \text{ there exists } p: D^k \to (\text{true},\text{false}) \text{ s.t.}$$

$$\forall s \in \Sigma \qquad (P(s) \leftrightarrow p(z(s))),$$

$$\text{glob } G \subseteq z =_{def} \text{ there exists } g: D^{2k} \to (\text{true},\text{false}) \text{ s.t.}$$

$$\forall st \in \Sigma \qquad (G(s,t) \leftrightarrow g(z(s), z(t))).$$

The predicates $p =_{def} \lambda z^s p(z^s)$, $g =_{def} \lambda z^s z^t, g(z^s, z^t)$ from the definitions above, are called the slice of P (respectively of G) over the support z. To avoid cumbersome notations we shall take some liberty in them hoping that will not lead to misunderstanding. The following examples are explanatory. Suppose, for instance, that glob $P \subseteq zw$ (where z and w abbreviate identifier vectors $z_1 \ldots z_k$ and $w_1 \ldots w_k$) and that the corresponding slice is $\lambda z^s w^s p(z^s, w^s)$; then $p(z,w)$ will be the notation for P. Particularly, such notations are implied below in a) - b) where constructs are described that may be characterized respectively as information transfer from conditions to transformer and vice versa.

Besides partial correctness statements (shortly - H-statements), $P\{G\}Q$ correct accomplishment statements (shortly - M-statements) $P\langle G\rangle Q$ are considered as well, with the following semantics:

$$P\langle G\rangle Q =_{def} \forall s(P(s) \supset \exists t\ (G(s,t) \wedge Q(t))) \qquad (4)$$

That is why we use everywhere below H and M-mnemonics to emphasize the situations when — though sometimes implicitely — H-statements, or M-statements are concerned.

a)

Given $P =_{def} p(z,w)$ and $Q =_{def} q(z,w)$ their H-transfer over the support z consists in constructing the transformer G with support z, such that

$$g(z^s,z^t) =_{def} \forall w^s(p(z^s,w^s) \supset q(z^t,w^s));$$

PQZ is the notation for that transformer. Similarly, the M-transfer over z yields the transformer denoted $\overline{PQz}$ with slice g defined thus:

$$g(z^s,z^t) =_{def} \exists w^s(p(z^s,w^s) \wedge q(z^t,w^s))$$

b)

Given transformer G with support $z = z_1 \ldots z_k$ and with slice $\lambda z^s z^t g(z^s,z^t)$, its H-transfer conditions are P_H, Q_H with support zw, where $w = w_1 \ldots w_k$ and, besides, $w \cap z = \emptyset$. The slices p_H and q_H are defined by

$$p_H(z^s,w^s) =_{def} z^s=w^s, \quad q_H(z^s,w^s) =_{def} g(w^s,z^s)$$

Similarly, the pair of conditions $P_M Q_M$ is declared as the result of M-transfer, where

$$p_M(z^s,w^s) =_{def} g(z^s,w^s), \quad q_M(z^s,w^s) =_{def} z^s=w^s$$

§ 3. RELAXATION RULES

Given two pairs of conditions PQ and P'Q' with support z it may occur that

$$\forall G(\text{glob } G \subseteq z \wedge P\{G\}Q \rightarrow P'\{G\}Q') \tag{5}$$

If so, the pair P'Q' is said to be a H-relaxation of the pair PQ over the support z. Similarly, M-relaxations of the pair PQ over the support z are considered as well, and respectively notations H(PQ, P'Q', z) and M(PQ, P'Q', z) are used. In these terms what the soundness of the rule (1) (and its analogy for M-statements) really means is just the following: for arbitrary P, Q, P', Q', z

$$P' \supset P \wedge Q \supset Q' \rightarrow H(PQ,P'Q',z) \tag{6-H}$$

$$P' \supset P \wedge Q \supset Q' \rightarrow M(PQ,P'Q',z) \tag{6-M}$$

What other relaxation rules are possible? The answer essentially depends on the additional requirements one may claim concerning the support of the conditions.

THEOREM 1. If conditions P, Q, P', Q', are only considered with support z , then

$$H(PQ,P'Q',z) \leftrightarrow (P' \supset P \wedge Q \supset Q') \vee (P' = \text{false}) \vee (Q' = \text{true}) \tag{7-H}$$

$$M(PQ,P'Q',z) \leftrightarrow (P' \supset P \wedge Q \supset Q') \vee (P' = \text{false}) \vee (P = \text{true} \wedge Q = \text{false}) \tag{7-M}$$

Hence in the special situation above implications (6) are almost reversible. However, generally speaking, different relaxation rules exist that are precisely characterized in Theorem 2 below. Preliminarily, let us note that (unlike partial correctness) in the case of correct accomplishment for arbitrary PQ the existence of G with support z such that $P\langle G\rangle Q$ is not obligatory. Hence, the following M-consistency criterion must be taken into consideration.

$$M(PQ,z) =_{def} \forall z^s w^s (p(z^s,w^s) \supset \exists z^t q(z^t,w^s)) \quad (8)$$

Below, for definiteness, the pairs under comparison are assumed to be:

$$P =_{def} p(z,w),\ Q =_{def} q(z,w),\ P' =_{def} p'(z,v),\ Q' =_{def} q'(z,v)$$

THEOREM 2.

$$H(PQ,P'Q',z) \leftrightarrow \forall z^s v^s (p'(z^s,v^s) \supset$$
$$\supset \underline{\forall z^t \exists w^s ((p(z^s,w^s) \supset q(z^t,w^s)) \supset q'(z^t,v^s)))} \quad (9\text{-H})$$

$$M(PQ,P'Q',z) \leftrightarrow \forall z^s v^s (p'(z^s,v^s) \supset$$
$$\supset \left(M(PQ,z) \supset \underline{\exists w^s \forall z^t (p(z^s,w^s) \wedge (q(z^t,w^s) \supset q'(z^t,v^s))))}\right) \quad (9\text{-M})$$

Let us denote H(PQ,Q',z) and M(PQ,Q',z) the subformulas underlined in (9-H) and (9-M). It is easy to realize that the predicate $\lambda z^s v^s H(PQ,Q',z)$ is the slice of the weakest precondition R such that R {G} Q holds for all G, where glob G ⊆ z and P{G}Q. The meaning of M(PQ,Q',z) is similar if the additional requirement (8) is fulfilled for the pair PQ.

Preserving the same notations H(PQ,Q',z) and M(PQ,Q',z) for the weakest preconditions defined above and taking into account that if the premises P ⟨G⟩ Q and glob G ⊆ z are true, so is M(PQ,z), we can formulate the general relaxation rules as follows:

$$\frac{\text{glob } G \subseteq z,\ P\{G\}Q,\ P' \supset H(PQ,Q',z)}{P'\{G\}Q'} \quad (10\text{-H})$$

$$\frac{\text{glob } G \subseteq z,\ P\langle G\rangle Q,\ P' \supset M(PQ,Q',z)}{P'\langle G\rangle Q'} \quad (10\text{-M})$$

REMARKS

1°. The proof of Theorem 1 shows in fact that on the propositional level the general H-relaxation rule becomes

$$\frac{P\{G\}Q,\ (P'\supset P\wedge Q\supset Q')\vee(P'=\text{false})\vee(Q'=\text{true})}{P'\ \{G\}\ Q'},$$

i.e. only slightly differs from the consequence rule (1). The same remark holds for M-relaxation. However, on the functional level from (9-10) many special rules may be derived that are common in algorithmic logic, for instance, rule (3). The following rule

$$\frac{\text{glob } G\subseteq z,\ p(z,w)\{G\}q(z,w)}{z{=}u\ \{G\}p(u,w)\supset q(z,w)} \qquad (11)$$

and its reverse are also worth noting. Clearly, they suggest why initialization of global variables is frequently used as precondition in partial correctness statements.

2°. For the first time general H-relaxation occurs actually (though in some other terms and notations) in Hoare's adaptation rule [1] Unfortunately, the explicit expression for what we denote H(PQ,Q',z) is incorrect in [1]. (See [6]; we only learned later from [3] that this incorrectness was pointed out long ago by Morris in unpublished notes). Incidentally, Hoare's expression coincides with that we referred for M(PQ,Q',z); however, we have not so far come across any considerations of M-relaxation.

§ 4. ACCURATE STATEMENTS ABOUT PARTIAL CORRECTNESS AND CORRECT ACCOMPLISHMENT

On the propositional level the H-statement P{G}Q, where P is the weakest precondition of Q, is the strongest possible for given G and Q. On the functional level a similar rule is played by the following statements:

$$H_1(P,G,Q,z) =_{def} \text{glob } G\subseteq z \ \wedge\ \forall P'Q'(P'\{G\}Q' \leftrightarrow \\ \leftrightarrow H(PQ,P'Q',z)) \qquad (12\text{-}H)$$

$$M_1(P,G,Q,z) =_{def} \text{glob } G \subseteq z \ \wedge \ \forall P'Q'(P' \langle G \rangle Q' \leftrightarrow M(PQ,P'Q',z)) \tag{12-M}$$

An alternative approach in pointing out "strong" statements is based on the monotonicity of H and M-statements, which on the propositional level is reflected in rule (2) and its M-analogy.

$$\frac{P \ \langle G \rangle \ Q, \ G' \supseteq G}{P \ \langle G' \rangle \ Q} \tag{2-M}$$

These are the precise definitions:

$$H_2(P,G,Q,z) =_{def} \text{glob } G \subseteq z \wedge \ \forall G'(\text{glob } G' \subseteq z \ \rightarrow (P\{G'\} Q \leftrightarrow G' \subseteq G)) \tag{13-H}$$

$$M_2(P,G,Q,z) =_{def} \text{glob } G \subseteq z \ \wedge \ \forall G'(\text{glob } G' \subseteq z \ \rightarrow (P \langle G' \rangle Q \leftrightarrow G' \supseteq G)) \tag{13-M}$$

Note that in Theorems 3-H and 3-M below the constructs $\underline{PQz}$ and $\overline{PQz}$ from § 2 are used.

<u>THEOREM 3-H</u>. For each support z the following three statements are equivalent:

$H_1(P,G,Q,z)$, $H_2(P,G,Q,z)$, and $(G = \underline{PQz})$

In what concerns M-statements, there are some additional peculiarities. While in Theorem 2 the M-consistency requirement (8) was taken into account, now a stronger one

$$M!(PQ,z) =_{def} \ \forall z^s w^s (p(z^s,w^s) \supset \ \exists! \ z^t q(z^t,w^s)) \tag{8'}$$

will work. Given a pair PQ, condition P_Q is defined:

$$P_Q(z^s,w^s) =_{def} p(z^s,w^s) \wedge \exists!\ z^t q(z^t,w^s)$$

Obviously:

$$P_Q \subseteq P \wedge M!(P_Q Q,z),\ M!(PQ,z) \leftrightarrow (P_Q = P)$$

THEOREM 3-M. For each support z the following three statements are equivalent:

$$M_1(P,G,Q,z),\ M_2(P,G,Q,z),(G = \overline{P_Q Qz}) \wedge P \langle P_Q Qz \rangle\ Q$$

COROLLARY. If $M!(PQz)$ holds, then the equivalence of

$$M_1(P,G,Q,z),\ M_2(P,G,Q,z),\ (G = \overline{PQz})$$

is implied.

Suppose $H_1(P,G,Q,z)$ holds and, hence, $H_2(P,G,Q,z)$ and $G = \underline{PQz}$ hold as well; then the pair PQ is said to be a z-accurate H-pair for G, and $P\{G\}Q$ is said to be a z-accurate H-statement. The similar terminology is used in what concerns accurate M-pairs and accurate M-statements - Theorems 3-H, 3-M, and accurate pairs may be useful in axiomatizing algorithmic logic because they allow us to confine to the inference of only accurate H and M-statements; indeed, it is the relaxation rules (10-H, 10-M) that provide the inference of all other true H-statements and M-statements. To justify this approach the following two questions are to be answered preliminarily:

(1) For what transformers G with support z z-accurate H-pairs (M-pairs) PQ do exist?

(2) What is the estimate for the supports of these conditions P,Q one may guarantee?

THEOREM 4. For each transformer G with glob $G \subseteq z$ there exist z-accurate H-pairs and z-accurate M-pairs with support zw ($z=z_1 \ldots z_k$, $w=w_1 \ldots w_k$).

In particular, the pairs $P_H Q_H$ and $P_M Q_M$ in which the information of G is transferred (see § 2) do the job.

Is it possible to improve the estimate for the support of the pair PQ? In some cases - yes, but in general obstacles arise that become clear by considering the following question:

(3) Let glob G $\subseteq$ z; in what cases $\overset{\text{do}}{\text{z}}$-accurate pairs exist with support z? (unlike pairs $P_H Q_H$ and $P_M Q_M$ that require extra variables w).

To confine to H-pairs, we remark that since G = <u>PQz</u> (see § 2)

$$\forall z^s z^t (g(z^s, z^t) \leftrightarrow (p(z^s) \supset q(z^t))) \tag{14}$$

holds, that is the same as

$$\forall st(G(s,t) \leftrightarrow ((P(s) \supset Q(t)) \wedge \ s = t \text{ outside of } z\,)) \tag{14'}$$

But the analysis of (14') obviously shows that this is possible only in very special (and trivial) situations. For instance, if $\neg P(s_o)$ and $Q(t_o)$ hold, then

$$\forall t(t = s_o \text{ outside of } z \rightarrow G(s_o, t));$$

$$\forall s(s = t_o \text{ outside of } z \rightarrow s(s - t_o \ \text{ outside of } \ z \quad G(s, t_o))$$

In particular, no single valued transformer (well, but just such transformers are described by deterministic programs) allows z-accurate pairs without extra variables. This is the reason why the analysis of partial correctness on the propositional level did not result in the notion of accurate H-statements though just the same analysis discovered"strong" statements $P\{G\}Q$, with the weakest precondition P. On the other hand, the functional level allows the use of extra (relative to glob G) variables and by that the existence of accurate statements. Theorems 3,4 give evidence that in such statements the point is in transferring information there and back from transformer to conditions and from conditions to transformer.

It is worth mentioning that the role played by extra variables is actually - though in some other terms - clarified in [4], and that in axiomatizing procedures all three characteristics of accurate statements are used ([2], [3], [6]) .

REFERENCES

1. C.A.R.Hoare, Procedures and parameters: An axiomatic approach, Lecture Notes in Mathematics, v. 188, Symposium on Semantics of Algorithmic Languages, 1971, 102-116.
2. D.Harel, A.Pnueli, J.Stavi, A Complete Axiomatic System for Proving Deductions about Recursive Programs. Proceedings 9-th ACM Symp. Theory of Computing, 249-260, 1977.
3. J.Guttag, J.Horning, R.London, A Proof Rule for Euclid Procedures. Proceedings of the IFIP-Conference on Formal Description of Programming Concepts, North-Holland 1978, 211-220.
4. K.R.Apt, L.G.L.T.Meertens, Completeness with finite systems of intermediate assertions for recursive schemes. Preprint IW 84/77. Stichting Mathematisch Centrum, Amsterdam, 1977.
5. R.Cartwright, D.Oppen, Unrestricted Procedure Calls in Hoare's Logics, Conference Record of the Fifth Annual ACM Symposium on Principles of Programming Languages, 1978, 131-140.
6. Б.А.Трахтенброт, "О полноте алгоритмической логики", Кибернетика, № 2, 1979.

L-FUZZY FUNCTORIAL AUTOMATA

Věra Trnková

Math. Institute of Charles University

Praha 8., Sokolovská 83, 18600, Czechoslovakia

I. Introduction.

Finite sequential non-deterministic automata accept the same languages as the deterministic ones. This well-known fact was generalized by Santos in [1]. He proved that finite stationary maximin sequential automata accept the same languages again. Another kind of generalization is given by Thatcher and Wright in [2], where finite non-deterministic tree automata are shown to accept the same languages as the deterministic ones. More generally, Eilenberg and Wright [3] investigate automata in varieties, but they prove the coincidence of the classes of languages only in varieties without equations, i.e. for the tree automata again. Maximin tree automata, more generally, maximin automata in varieties, form a straightforward generalization including all the above cases.

In the present paper we investigate L-fuzzy automata in the varieties where the defining equations are in basic operations. Our notion of fuzziness is that of Zadeh [4,5] and Goguen [6,7] and includes the maximin and minimax stationary automata. We characterize all those varieties in which finite maximin automata accept the same languages as the deterministic ones. We prove that this is true only in rather special varieties: if redundant operations are omitted, then only a type of commutativity is permitted.

In the paper, categorical language is used. Varieties with equations in basic operations are described as categories of F-algebras with F:Set ⟶ Set being a superfinitary functor (see III.c)). This categorical description makes the positive results extremely simple and lucid (see VII.). Nevertheless, the proof of the negative result (only tree-group fuzzy automata accept the same languages as the deterministic ones) requires a little routine in manipulating set functors.

II. Arbib-Manes machines.

A unified approach to many kinds of automata has been introduced by Arbib and Manes [8]. Let us recall their basic notions. If $\mathcal{K}$ is a category and $F:\mathcal{K}\longrightarrow\mathcal{K}$ a functor, an F-algebra (in $\mathcal{K}$) is any pair (Q,δ) where $\delta:FQ\longrightarrow Q$ is a $\mathcal{K}$-morphism. A homomorphism $h:(Q,\delta)\longrightarrow(Q',\delta')$ is a $\mathcal{K}$-morphism such that $\delta\cdot h=F(h)\cdot\delta'$. A free F-algebra over an object I consists of an F-algebra $(I^{\#},\varphi)$ and a $\mathcal{K}$-morphism $\eta:I\longrightarrow I^{\#}$ such that for every F-algebra (Q,δ) and every $\mathcal{K}$-morphism $i:I\longrightarrow Q$ there exists a unique homomorphism $i^{\#}\ (I^{\#},\varphi)\longrightarrow(Q,\delta)$ with $\eta\cdot i^{\#}=i$ ($i^{\#}$ is called the free extension of i). If free F-algebras exist over any object I, F is called a varietor (an input process in the terminology of [8]). An Arbib-Manes machine M consists of an F-algebra (Q,δ) (Q is the state object of M, δ its transition), an initial morphism $i:I\longrightarrow Q$ and an output morphism $y:Q\longrightarrow Y$. We indicate $M=(I,i,Q,\delta,Y,y)$. If F is a varietor, then the behaviour beh M is defined as $i^{\#}\cdot y:I^{\#}\longrightarrow Y$, where $i^{\#}$ is the free extension of the initial morphism i. As a minor modification, let us define an F-acceptor (see [9]) as $\mathcal{A}=(I,i,Q,\delta,T,t)$, where I,i,Q,δ are as above and $t:T\longrightarrow Q$ is a subobject of Q. The language $\mathcal{L}(\mathcal{A})$ of $\mathcal{A}$ is $(i^{\#})^{-1}(T)$ (where the preimage is categorically modelled by a pullback).

III. Examples.

a) Let us denote by Set the category of all sets and mappings. If Σ is a set, the functor $F_{\Sigma}:\mathrm{Set}\longrightarrow\mathrm{Set}$ is defined by

$$F_{\Sigma}X = X\times\Sigma\ ,\quad F_{\Sigma}f = f\times 1_{\Sigma}$$

(where $1_{\Sigma}:\Sigma\longrightarrow\Sigma$ denotes the identical mapping). F_{Σ}-acceptors are precisely (complete deterministic) sequential automata with Σ being the set of inputs. For, Σ^* with the concatenation $\varphi:\Sigma^*\times\Sigma\longrightarrow\Sigma^*$ is a free F_{Σ}-algebra over a one point $\{\Lambda\}$ (see [8]), $i:\{\Lambda\}\longrightarrow Q$ sending the empty string Λ to an initial state q_0 extends to $i^{\#}:\Sigma^*\longrightarrow Q$ by the well-known formula $i^{\#}(s\sigma)=\delta(i^{\#}(s),\sigma)$, $T\subset Q$ is the set of terminal states ($t:T\longrightarrow Q$ is the inclusion), hence $(i^{\#})^{-1}(T)$ is the set of all accepted strings.

b) Let Ω be a ranked alphabet (= type or signature), i.e. Ω consists of a set Σ and an arity function $ar:\Sigma\longrightarrow\{\text{cardinals}\}$. The functor F_Ω :Set $\longrightarrow$ Set is defined by

$$F_\Omega X = \coprod_{\sigma \in \Sigma} X^{ar(\sigma)}, \quad F_\Omega f = \coprod_{\sigma \in \Sigma} f^{ar(\sigma)},$$

where $\coprod$ denotes a disjoint union of sets (or mappings, resp.). Then F_Ω -acceptors are precisely (complete deterministic) Ω -tree automata in the sense of Thatcher and Wright [2]. Hence, let us call the functor F_Ω Ω -tree functor or simply tree functor. In [2], Σ and every $ar(\sigma)$ are always finite. $\Omega = \{\Sigma, ar\}$ with this property is said to be superfinitary. If there is no danger of confusion, we write $F_{\{n\}}$ instead of F_Ω whenever $\Omega = \{\Sigma, ar\}$, Σ consists of a single letter σ and $ar(\sigma)=n$.

c) A functor F:Set $\longrightarrow$ Set is called superfinitary if there is an epitransformation $\nu : F_\Omega \longrightarrow F$ with Ω superfinitary. Thus, F-algebras are precisely universal algebras of the type Ω which satisfy equations determining the epitransformation. E.g. commutative groupoids are F-algebras with $\nu : F_{\{2\}} \longrightarrow F$ prescribed by $(x,y)=(y,x)$. In this case, languages of F-acceptors, being subsets of free F-algebras, are sets of equivalence classes of binary trees, where the equivalence is generated by the exchanging of branches.

IV. L-fuzzy relations.

Let us recall (see Goguen [6,7]) that L is said to be a closg (= complete lattice ordered semigroup) if it is a complete lattice ($\bigvee$ denotes supremum and $\bigwedge$ infimum in L) and a monoid (with respect to an operation $\cdot$) which is sup-distributive, i.e.

$$(\bigvee_i a_i) \cdot (\bigvee_j b_j) = \bigvee_{i,j} (a_i \cdot b_j).$$

Examples of closg's: a) If L is a completely distributive complete lattice, then it is a closg with $a \cdot b$ being the infimum $a \wedge b$. Throughout the paper, L_o denotes the distributive lattice $\{0,1\}$ with $\cdot$ being the infimum.

b) The unit interval $\langle 0,1 \rangle$ with its usual order forms a complete lattice. It can be turned in a closg in at least three natural ways, namely $\overset{1}{\cdot}$ being the infimum, $\overset{2}{\cdot}$ being the supremum and $\overset{3}{\cdot}$ being the multiplication.

If L is a closg, then L-fuzzy relation from a set A into a set B is a mapping $m:A \times B \longrightarrow L$. The value $m(a,b)$ is interpreted as the grade that $a \in A$ is related to B. Let us indicate $m:A \twoheadrightarrow B$ an L-fuzzy relation from A into B. If $m_1:A \twoheadrightarrow B$ and $m_2:B \twoheadrightarrow C$ are L-fuzzy relations, their composition $m_1 \cdot m_2 : A \twoheadrightarrow C$ is defined by

$$(m_1 \cdot m_2)(a,c) = \bigvee_{b \in B} m_1(a,b) \cdot m_2(b,c).$$

All sets and all their L-fuzzy relations form a category (see Goguen [8]). We denote it by L Rel. Clearly, for every sets A,B, the set of all L-fuzzy relations from A to B is partially ordered by

$$m_1 \leq m_2 \text{ iff } m_1(a,b) \leq m_2(a,b) \text{ for all } (a,b) \in A \times B.$$

Let us notice that L_0 Rel is the category of sets and their binary relations as morphisms. The category Set of all sets and mappings is naturally embedded in L Rel such that any mapping $f:A \to B$ is turned in an L-fuzzy relation $m_f:A \twoheadrightarrow B$ by the rule

$$m_f(a,f(a)) = 1, \quad m_f(a,b) = 0 \text{ otherwise.}$$

This embedding preserves the composition of mappings (i.e. $m_{f\cdot g} = m_f \cdot m_g$) whenever $L_0 \subset L$. It is fulfilled in all the above examples and, in what follows, we always suppose it.

V. Fuzzy automata.

Let $F:\mathrm{Set} \to \mathrm{Set}$ be a varietor and L a closg. An L-fuzzy F-automaton is $\mathcal{A} = (I,i,Q,\sigma,Y,y)$, where $i:I \twoheadrightarrow Q$, $\sigma:FQ \twoheadrightarrow Q$, $y:Q \twoheadrightarrow Y$ are L-fuzzy relations. It can be considered as an acceptor as well. The inverse L-fuzzy relation $y^{-1}:Y \twoheadrightarrow Q$, given by the rule $y(a,b)=y^{-1}(b,a)$, determines it.

Examples: a) L_0-fuzzy F_Ω -automata are non-deterministic Ω-tree automata in the sense of Thatcher and Wright [2]. More generally, L_0-fuzzy F-automata are relational automata investigated by Trnková [10, 11].

b) If $L = \langle 0,1 \rangle$ with $a \cdot b$ being the minimum of a and b, then L-fuzzy F_Σ -automata are stationary maximin automata in the sense of Santos [1] or fuzzy automata in the sense of Mizumoto, Toyoda and Tanaka [12].

In analogy with the deterministic case and in accordance with th above examples, we expect the behaviour of an L-fuzzy automaton $\mathcal{A} = (I,i,Q,\sigma,Y,y)$ being an L-fuzzy relation $i^\# \cdot y:I^\# \to Y$, where $(I^\#,\varphi)$ and $\eta:I \to I^\#$ form a free F-algebra over I and $i^\#:(I^\#,\varphi) \twoheadrightarrow (Q,\sigma)$ is a free extension of $i:I \twoheadrightarrow Q$. We recall that the free extension $i^\#$ is defined by the equalities

$$\varphi \cdot i^\# = F(i^\#) \cdot \sigma \quad , \quad \eta \cdot i^\# = i.$$

The second equation causes no difficulty but, in the first one, there is the symbol $F(i^\#)$ which has not yet a meaning. It is necessary to extend $F:\mathrm{Set} \to \mathrm{Set}$ to $\overline{F}:\mathrm{L\ Rel} \to \mathrm{L\ Rel}$, in a standard way, if possible In [11], any $F:\mathrm{Set} \to \mathrm{Set}$ is extended to $\overline{F}:L_0\ \mathrm{Rel} \to L_0\ \mathrm{Rel}$. In the nex part, we show a standard construction how to extend any superfinitary F:Set$\to$Set to $\overline{F}$:L Rel$\to$L Rel for an arbitrary commutative closg L. But in both the cases, the extension $\overline{F}$ does not satisfy all the requirements for being a functor: instead of the equation $\overline{F}(\alpha\cdot\beta) = \overline{F}(\alpha)\cdot\overline{F}(\beta)$, only the weaker condition $\overline{F}(\alpha\cdot\beta)\leq\overline{F}(\alpha)\cdot\overline{F}(\beta)$. On the other hand, if $\alpha\leq\gamma$ then always $\overline{F}(\alpha)\leq\overline{F}(\gamma)$. Let us call an extension with these properties a semifunctor. Semifunctors are quite sufficient for the purpose to introduce the behaviour of L-fuzzy automata. This follows from the theorem below.

Theorem. Let F:Set$\to$Set be a varietor, let $\overline{F}$:L Rel$\to$L Rel be a semifunctor which extends F. Then for any pair of L-fuzzy relations i: :I$\twoheadrightarrow$Q, σ:FQ$\twoheadrightarrow$Q there exists a unique L-fuzzy relation $i^{\#}:I^{\#}\twoheadrightarrow Q$ such that $\varphi\cdot i^{\#}=\overline{F}(i^{\#})\cdot\sigma$ and $\eta\cdot i^{\#}=i$ (where $(I^{\#},\varphi)$ and $\eta:I\to I^{\#}$ is a free F-algebra over I in Set).

Proof. Set is coreflective in L Rel, so the embedding Set$\to$L Rel preserves colimits. Then use VI.1. in [14].

VI. Behaviour of fuzzy automata.

Let L be a commutative closg, F:Set$\to$Set be a superfinitary functor. We show the standard way how to extend F to $\overline{F}$:L Rel$\to$L Rel.

a) For Ω with precisely one letter σ and ar(σ)=n, let us write $F_{\{n\}}$ instead of F_Ω. First, we extend $G=F_{\{n\}}$ as follows. If m:A$\twoheadrightarrow$B, we define G(m):A$\times\ldots\times$A$\twoheadrightarrow$B$\times\ldots\times$B by

$$[G(m)]((a_o,\ldots,a_{n-1}),(b_o,\ldots,b_{n-1}))=m(a_o,b_o)\cdot\ldots\cdot m(a_{n-1},b_{n-1}).$$

b) For arbitrary $\Omega=\{\Sigma,ar\}$, we put $\overline{F}_\Omega=\coprod_{\sigma\in\Sigma}\overline{F^{ar(\sigma)}}$.

c) If F is a superfinitary functor, express it as $\nu:F_\Omega\to F$ with a minimal possible Ω (i.e. if for some Ω' there is also an epitransformation $\nu':F_{\Omega'}\to F$, then $\nu'=\lambda\cdot\nu$ for an epitransformation λ: :$F_{\Omega'}\to F_\Omega$). If m:A$\twoheadrightarrow$B, define $\overline{F}$(m):FA$\twoheadrightarrow$FB such that for every a$\in$FA, b$\in$FB,

$$[\overline{F}(m)](a,b)=\bigvee[(\overline{F}_\Omega(m)](x,y),$$

where the supremum is taken over all the pairs (x,y) with $\nu(x)$=a, $\nu(y)$=b.

In all these cases, $\overline{F}$ can be easily shown to be a semifunctor(in the cases a) and b) even a functor). Thus, the above Theorem can be applied, and we define the behaviour b of an L-fuzzy F-automaton $\mathcal{A}$ = =(I,i,Q,σ,Y,y) as

$$b = i^{\#} \cdot y : I^{\#} \twoheadrightarrow Y.$$

Let us notice that in the special cases mentioned in V. (i.e. investigated in [1,2,9,10,11,12] the new definition of the behaviour coincides with the old ones.

VII. Languages of fuzzy automata.

Let L be a commutative closg, F a superfinitary set functor, $\mathcal{A} = (I,i,Q,\delta,Y,y)$ an L-fuzzy automaton with Q finite and $b=i^{\#} \cdot y:I^{\#} \twoheadrightarrow Y$ its behaviour. As usual, for any threshold $\ell \in L$, put

$$\mathcal{L}(\mathcal{A},\ell) = \{x \in I^{\#} \mid b(x,z) > \ell \text{ for some } z \in Y\}.$$

For L-fuzzy F_{Σ}-automata with $L=\langle 0,1\rangle$ and $a \cdot b$ being the minimum, every $\mathcal{L}(\mathcal{A},\ell)$ is proved to be regular, in Santos [1]. In [2], the analogous result is proved for L_o-fuzzy F_{Ω}-automata (here, there are only the possibilities $\mathcal{L}(\mathcal{A},1)=\emptyset$ and $\mathcal{L}(\mathcal{A},0)$, the latter being the language of the non-deterministic Ω-tree automaton $\mathcal{A}$). We show that both these situations are special cases of the following simple idea. For every set X, denote by $e_X : L^X \twoheadrightarrow X$ the evaluation, i.e. $e_X(f,x) = f(x)$. For any L-fuzzy relation $m:A \twoheadrightarrow B$ denote by $\overline{m}:A \to L^B$ the mapping $\overline{m}(x)=m(x,-)$. If $F:\mathrm{Set} \to \mathrm{Set}$ is extended to $\overline{F}:L\ \mathrm{Rel} \to L\ \mathrm{Rel}$ as a functor, then $\mathcal{A} =(I,i,Q,\delta,Y,y)$ can be replaced by the F-automaton $\mathcal{A}_1=(I,i_1,L^Q,\delta_1,Y,y_1)$ with $i_1=\overline{i}$, $\delta_1=\overline{(F(e_Q)\cdot\delta)}$, $y_1=\overline{e_Q \cdot y} \cdot e_Y$. A simple running in the picture below shows that beh $\mathcal{A}$ = beh $\mathcal{A}_1$.

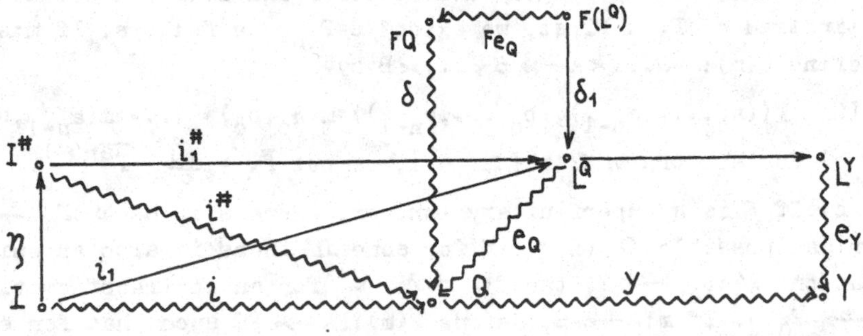

(The above situation, generalizing the classical power automata construction, can be applied on every embedding of Set as a coreflective subcategory.) The automaton $\mathcal{A}_1$ is very near to being a deterministic F-acceptor: its initial morphism and its transition are already mappings; hence to obtain a deterministic F-acceptor with the language $\mathcal{L}(\mathcal{A},\ell)$, it is sufficient to replace the output of $\mathcal{A}_1$ by the set T of terminal states with $T=\{t \in L^Q \mid y_1(t,z) > \ell$ for some $z \in Y\}$.

Unfortunately, $\mathcal{A}_1$ is far from being finite unless L is a finite closg. But if a threshold $\ell \in L$ is given and if there is a homomorphism $h:L \to L'$ with respect to $\cdot$ and $\vee$ such that L' is finite and $h(\ell') > h(\ell)$ iff $\ell' > \ell$, we can replace the L-fuzzy relations by L'-fuzzy relations (by the rule sending any $m:A \twoheadrightarrow B$ to $m':A \twoheadrightarrow B$ with $m'(a,b)=h(m(a,b))$) and $\mathcal{L}(\mathcal{A},\ell)$ is proved to be accepted by a finite deterministic F-acceptor.

Let us call a closg L <u>finitely reducible</u> if, for every $\ell \in L$, there exists a homomorphism with the above properties. This is rather restrictive condition. Nevertheless, every linearly ordered complete lattice with $\cdot$ being the minimum is finitely reducible (hence, the result of Santos [1] for maximin automata is included; minimax automata are obtained by the investigation of $h: \langle 0,1 \rangle \to \langle 0,1 \rangle$, given by $h(x)=1-x$).

VIII. Tree-group automata.

Let $F:\mathrm{Set} \to \mathrm{Set}$ be a superfinitary functor. We say that L-fuzzy F-automata have the same capability as the deterministic F-acceptors iff, for every finite L-fuzzy F-automaton $\mathcal{A}$ and every threshold $\ell \in L$, the language $\mathcal{L}(\mathcal{A},\ell)$ is accepted by a finite deterministic F-acceptor. In this part we describe all the superfinitary functors F for which the capability of the L-fuzzy F-automata (with L being finitely reducible) is the same as the capability of the deterministic F-acceptors. There are precisely tree group functors described below.

Let $n=\{0,\dots,n-1\}$ be a natural number and G a group of permutations of n. For any set X denote by $X^{(n,G)}$ the set $X^n/\sim$, where the equivalence is defined by

$$\alpha \sim \beta \quad \text{iff } \alpha = g \cdot \beta \quad \text{for some } g \in G.$$

Since G is a group, $\sim$ is really an equivalence. Clearly, for every mapping $f:X \to Y$, the mapping $f^n:X^n \to Y^n$ can be factorized on $f^{(n,G)}: X^{(n,G)} \to Y^{(n,G)}$.

Now, let $\Omega = \{\Sigma, ar\}$ be a <u>group-ranked alphabet</u>, i.e. Σ is a set and the arity function ar assigns to every $\sigma \in \Sigma$ a pair $ar(\sigma) = (n_\sigma, G_\sigma)$, where n_σ is a natural number and G_σ is a group of permutations of n_σ. (If every G_σ is trivial, Ω is, up to a formal difference, a ranked alphabet in the usual sense.) We define again

$$F_\Omega X = \coprod_{\sigma \in \Sigma} X^{ar(\sigma)}, \quad F_\Omega f = \coprod_{\sigma \in \Sigma} f^{ar(\sigma)}.$$

The F_Ω is called a <u>tree-group functor</u> and the F_Ω-automaton (deterministic, L-fuzzy) are called <u>tree group automata</u>.

Theorem.

Let $F:\mathrm{Set}\longrightarrow\mathrm{Set}$ be a superfinitary functor. The following assertions are equivalent.

(a) For every finitely reducible commutative closg L, L-fuzzy F-automata have the same capability as the deterministic F-acceptors.

(b) L_o-fuzzy F-automata have the same capability as the deterministic F-acceptors.

(c) F is a tree-group functor.

Proof. (c) $\Longrightarrow$ (a). The extension $\overline{F}:\text{L-Rel}\longrightarrow\text{L-Rel}$ of any tree-group functor F (see VI a),b)) is a functor again, so (a) follows by VII.

(a) $\Longrightarrow$ (b) is trivial.

(b) $\Longrightarrow$ (c): L_o-fuzzy relations are precisely usual relations(let us indicate them like multivalued mappings) and L_o-fuzzy automata are relational automata. Since F is supposed to be superfinitary, there exists an epitransformation $\nu:\coprod_{i=1}^{N} F_{\{n_i\}}\longrightarrow F$ with N and all the n_i's minimal (see VII.c)), i.e., for every $i=1,\dots,N$, $d_i=\nu(1_{n_i})$ fulfils

$$(1)\qquad d_i\in Fn_i\setminus\bigcup_{\substack{j=1\\ j\neq 1}}^{N}\nu(F_{\{n_j\}}n_i).$$

α) Let us suppose that ν glues together a part of $F_{\{n_1\}}$ and of $F_{\{n_2\}}$, i.e. there exists a set X and $f\in F_{\{n_1\}}X=X^{n_1}$, $g\in F_{\{n_2\}}X=X^{n_2}$, such that $\nu_X(f)=\nu_X(g)$. We may suppose that $n_1\geq n_2$ and $f(n_1)=g(n_2)$ has precisely one element. If X is a set, let us denote by $(x_o,\dots,x_{n_1-1})^1$ (or $(x_o,\dots,x_{n_2-1})^2$) the $f\in F_{\{n_1\}}X$ (or the $g\in F_{\{n_2\}}X$) sending i to x_i. Hence, we have

(2) for every $x\in X$, we have $\nu(x,\dots,x)^1=\nu(x,\dots,x)^2$.

We construct a relational F-automaton $\mathcal{A}=(I,i,Q,\delta,Y,y)$ as follows $I=Y=Q=n_1$, i sends every element of I to the whole Q, $y=1_Q$, $\delta:FQ\twoheadrightarrow Q$ sends d_1 (see (1)) to the whole Q, $\delta(x)=\emptyset$ otherwise. Let $i^{\#}:(I^{\#},\varphi)\twoheadrightarrow(Q,\delta)$ be the free extension of i. Then, for every $x\in I^{\#}$, either $i^{\#}(x)$ is the whole Q (and then $x\in\mathcal{L}(\mathcal{A})=\mathcal{L}(\mathcal{A},0)$) or $i^{\#}(x)=\emptyset$. Moreover,

(3) if $x_o,\dots,x_{n_1-1}$ are distinct elements of $L(\mathcal{A})$, then $\varphi(\nu(x_o,\dots,x_{n_1-1})^1))\in L(\mathcal{A})$ while $\varphi(\nu(x_o,\dots,x_{n_2-1})^2))\notin L(\mathcal{A})$

For, if we denote $c_1=\nu(x_o,\dots,x_{n_1-1})^1$, $c_2=\nu(x_o\dots,x_{n_2-1})^2$, we have

$(F(i^{\#}))(c_1) \ni d_1$ while $(F(i^{\#}))(c_2) \subset \nu(F_{\{n_2\}}Q)$ so that $\sigma((F(i^{\#}))(c_1)=$ $=Q$ while $\sigma((F(i^{\#}))(c_2))=\emptyset$. (3) also implies that $\mathcal{L}(\mathcal{A})$ is infinite. Now, let $\bar{\mathcal{A}}=(I,j,\bar{Q},\bar{\sigma},T,t)$ be a finite deterministic F-acceptor such that $\mathcal{L}(\bar{\mathcal{A}})=\mathcal{L}(\mathcal{A})$. Since $\bar{Q}$ is finite, there exist infinite $Z \subset \mathcal{L}(\mathcal{A})$ such that $j^{\#}$ maps Z on the same state $q_o \in \bar{Q}$. Choose the $x_o,\dots,x_{n_1-1}$ in Z. Then $\varphi(c_1) \in \mathcal{L}(\bar{\mathcal{A}})$ iff $\varphi(c_2) \in \mathcal{L}(\bar{\mathcal{A}})$ because (2) implies $(F(j^{\#}))(c_1)=(F(j^{\#}))(c_2)$. This contradicts (3).

β) By α), we may suppose that $F=\coprod_{i=1}^{N} \nu(F_{\{n_i\}})$. Denote by G_1 the set of all $g \in F_{\{n_1\}}n_1$ such that $\nu(g)=\nu(1_{n_1})$. The minimality of n_1 implies that every $g \in G_1$ is a permutation of n_1 (otherwise, $\nu(F_{\{n_1\}})$ could be expressed as some $\nu'(F_{\{m\}})$ with $m<n_1$). Since ν is a transformation, G_1 is a group. We are going to prove that $\nu(F_{\{n_1\}})=F_{\{(n_1,G_1)\}}$. Let us suppose the contrary, i.e.

(4) there exist $f,h:n_1 \to n_1$ such that $\nu(f)=\nu(h)$ but $f \neq g\cdot h$ for all $g \in G_1$.

$\beta1$) Let us suppose that $f(n_1) \setminus h(n_1) \neq \emptyset$. Hence, we may suppose $h(n_1)=\{0\}$, $f(n_1)=\{0,1\}$. More precisely, $n_1=A\cup B$, $A=\{0,\dots,k-1\}$, $B=\{k,\dots,n_1-1\}$, $k\geq 1$ and f sends A to 1 and B to 0. We construct a relational F-automaton $\mathcal{A}=(I,i,Q,\sigma,Y,y)$ as follows: $I=n_1\times n_1$, $Q=n_1$, $Y=$ $=A$; i sends every $a\in n_1\times A$ to A and every $b\in n_1\times B$ to B; $\sigma:FQ \twoheadrightarrow Q$ sends d_1 (see (1)) to Q and every $x \in FQ\setminus\{d_1\}$ to B; y sends every $a\in A$ to itself and every $b \in B$ to $\emptyset$. We show that $\mathcal{L}(\mathcal{A})$ is not accepted by any finite deterministic F-acceptor. Let $i^{\#}:(I^{\#},\varphi)\to(Q,\sigma)$ be the free extension of i. Then for every $x\in I^{\#}$, either $i^{\#}(x)$ is Q (and then $x \in$ $\in \mathcal{L}(\mathcal{A})$ or $i^{\#}(x)=B$ (and then $x \notin \mathcal{L}(\mathcal{A})$). Moreover

(5) if $x_o,\dots,x_{n_1-1}$ are distinct elements of $I^{\#}\setminus\mathcal{L}(\mathcal{A})$ and $z_o,\dots,z_{k-1}$ are distinct elements of $\mathcal{L}(\mathcal{A})$ then $\varphi(\nu(z_o,\dots,z_{k-1},x_k,\dots,x_{n_1-1}))\in\mathcal{L}(\mathcal{A})$ but $\varphi(\nu(x_o,\dots,x_{n_1-1}))\in$ $\in \mathcal{L}(\mathcal{A})$.

For, if $c_1=\nu(z_o,\dots,z_{k-1},x_k,\dots,x_{n_1-1})$, $c_2=\nu(x_o,\dots,x_{n_1-1})$, then $d_1\in(F(i^{\#}))(c_1)$ while $d_1\notin(F(i^{\#}))(c_2)$ because $i^{\#}(x_j)=B$ for all $j=0,\dots$ $\dots,n_1-1$ and card $B<n_1$. Clearly, both $\mathcal{L}(\mathcal{A})$ and $I^{\#}\setminus\mathcal{L}(\mathcal{A})$ are infinite. Let us suppose that there is a finite deterministic F-acceptor $\bar{\mathcal{A}}=(I,j,\bar{Q},\bar{\sigma},T,t)$ such that $\mathcal{L}(\bar{\mathcal{A}})=\mathcal{L}(\mathcal{A})$. Since $\bar{Q}$ is finite, there exist infinite $Z_1\subset\mathcal{L}(\mathcal{A})$ and $Z_o\subset I^{\#}\setminus\mathcal{L}(\mathcal{A})$ such that $j^{\#}$ maps Z_1 on a state $q_1\in\bar{Q}$ and Z_o on a state $q_o\in\bar{Q}$. Choose the distinct $z_o,\dots,z_{k-1}$ in Z_1 and $x_o,\dots,x_{n_1-1}$ in Z_o. Then $\varphi(c_1)\in\mathcal{L}(\mathcal{A})$ iff $\varphi(c_2)\in\mathcal{L}(\bar{\mathcal{A}})$ because $(F(j^{\#}))(c_1)=(F(j^{\#}))(c_2)$. This contradicts (5).

$\beta 2$) If $h(n_1)\setminus f(n_1)\neq\emptyset$, interchange h and f in $\beta 1$).

$\beta 3$) Finally, let us suppose $h(n_1)=f(n_1)$. Denote $B=f(n_1)$. Clearly, card $B\geq 2$ (otherwise f=h). Since $h(n_1)$ is also B, there exists $\ell: n_1\to n_1$ such that $\ell\cdot f=h$. We construct a relational F-automaton $\mathcal{A}=(I,i,Q,\delta,Y,y)$ as follows: $I=n_1\times n_1$, $Q=(n_1\times n_1)\cup\{\xi\}$, $Y=\{\xi\}$ (where ξ is supposed not in $n_1\times n_1$); $i:I\twoheadrightarrow Q$ sends every (k,j) to $n_1\times f^{-1}(f(j))$; $y:Q\twoheadrightarrow Y$ sends ξ to itself, $y(q)=\emptyset$ otherwise; now, we define $\delta: FQ\twoheadrightarrow Q$ as follows: for every $z\in FQ$ choose $x\in\coprod_{i=1}^{N} F_{\{n_1\}}Q$ with $\nu(x)=z$; if $x=(a_0,\dots,a_{n_1-1})\in F_{\{n_1\}}Q$ with $a_0,\dots,a_{n_1-1}$ distinct, then put

$$\delta(z) = n_1\times f^{-1}(b) \text{ whenever } a_j\in n_1\times f^{-1}(b) \text{ for some } b\in B \text{ and all } j=0,\dots,n_1-1;$$

$$\delta(z) = \xi \text{ whenever } a_j\in n_1\times f^{-1}(f(g(j))) \text{ for some } g\in G_1 \text{ and all } j=0,\dots,n_1-1;$$

$$\delta(z) = \emptyset \quad \text{in all other cases;}$$

(hence, in fact, δ is independent on the choice of $x\in\nu^{-1}(z)$). We show that $\mathcal{L}(\mathcal{A})$ is not accepted by any finite deterministic F-acceptor. Let $i^{\#}:(I^{\#},\varphi)\twoheadrightarrow(Q,\delta)$ be the free extension of i. One can prove that $I^{\#}$ is decomposed into the following disjoint sets:

$$\mathcal{L}(\mathcal{A}) = \{z\in I^{\#}\mid i^{\#}(z)=\xi\}$$
$$Z_b = \{z\in I^{\#}\mid i^{\#}(z)=n_1\times f^{-1}(b)\},\ b\in B,$$
$$Z_\emptyset = \{z\in I^{\#}\mid i^{\#}(z)=\emptyset\}.$$

All the sets Z_b are infinite. (For, the sets $n_1\times f^{-1}(b)$ are large enough - this is the rôle of the multiplying by the first factor n_1 - thus if $z_0,\dots,z_{n_1-1}\in Z_b$, then $\varphi(\nu(z_0,\dots,z_{n_1-1}))\in Z_b$.) Hence $\mathcal{L}(\mathcal{A})$ is also infinite because the following assertion is fulfilled.

(6) If $z_0,\dots,z_{n_1-1}$ are distinct elements of $I^{\#}$ such that $z_j\in Z_{f(j)}$ for all $j=0,\dots,n_1-1$, then $\varphi(\nu(z_0,\dots,z_{n_1-1}))\in\mathcal{L}(\mathcal{A})$ but $\varphi(\nu(z_{\ell(0)},\dots,z_{\ell(n_1-1)}))\notin\mathcal{L}(\mathcal{A})$

(The last fact follows from the definition of ℓ and δ. For, if $a_0,\dots,a_{n_1-1}$ are distinct elements of $F_{\{n_1\}}Q$ such that $\nu(a_0,\dots,a_{n_1-1})\in (F(i^{\#}))(\nu(z_{\ell(0)},\dots,z_{\ell(n_1-1)}))$, then necessarily $a_j\in n_1\times f^{-1}(f(\ell(g(j$ for some $g\in G_1$ and all $j=0,\dots,n_1-1$. But $\ell\cdot f=h$ and $g\cdot h\neq g'\cdot f$ for each $g'\in G_1$, by (4). Hence, for each $g'\in G_1$, there exists j such that $f^{-1}(f(g'(j)))\cap f^{-1}(h(g(j)))=\emptyset$. Consequently, never $\delta(\nu(a_0,\dots,a_{n_1-1}))=\xi$.) Now, let $\bar{\mathcal{A}}=(I,j,\bar{Q},\bar{\delta},T,t)$ be a finite deterministic

F-acceptor with $\mathcal{L}(\bar{\mathcal{A}})=\mathcal{L}(\mathcal{A})$. Since $\bar{Q}$ is finite, there exist infinite $\bar{Z}_b \subset Z_b$, $b \in B$, such that $j^{\#}$ maps each $\bar{Z}_b$ on a unique state $q_b \in \bar{Q}$. Choose $z_i \in \bar{Z}_{f(i)}$. Then, for $c=\nu(z_0,\dots,z_{n_1-1})$, $\bar{c}=\nu(z_{\ell(0)},\dots,z_{\ell(n_1-1)})$, we have $\varphi(c) \in \mathcal{L}(\mathcal{A})$ iff $\varphi(\bar{c}) \in \mathcal{L}(\mathcal{A})$ because $(F(j^{\#}))(c)=(F(j^{\#}))(\bar{c})$. This contradicts (6).

References

1. Santos, E.S., Maximin automata, Information and Control, 13(1968), 363-377.
2. Thatcher, J.W. and Wright, J.B., Generalized finite automata theory with an application to a decision problem of second-order logic, Math. Syst. Theory 2(1968), 57-81.
3. Eilenberg, S. and Wright, J.B., Automata in general algebras, Information and Control 11(1967), 52-70.
4. Zadeh, L.A., Fuzzy sets, Information and Control 8(1965), 338-353.
5. Zadeh, L.A., Fuzzy sets and systems, Proceedings of the Symposium on System Theory, Polytechnic Press of the Polytechnic Institute of Brooklyn, New York 1965.
6. Goguen, J.A., L-fuzzy sets, J. of Math. Anal. and Appl. 18(1967), 145-174.
7. Goguen, J.A., Concept Representation in Natural and Artificial Languages: Axioms, Extensions and Applications for Fuzzy Sets, Internat. J. Man-Machine Studies 6(1974), 513-561.
8. Arbib, M.A. and Manes, E.G., Machines in a category, an expository introduction, SIAM Review 16(1974), 163-192.
9. Adámek, J. and Trnková, V., Recognizable and regular languages in a category, Lect. Notes in Comp. Sci. 56(FCT '77 Proceedings), Springer-Verlag 1977, 199-205.
10. Trnková, V., Relational automata in a Category and their Languages, Lect. Notes in Comp. Sci. 56(FCT '77 Proceedings), Springer-Verlag 1977, 340-355.
11. Trnková, V., General theory of relational automata, Fund. Informaticae, to appear.
12. Mizumoto, M., Toyoda, J. and Tanaka, K., Some consideration on fuzzy automata, J. of Computer and System Sciences 3(1969), 409-422.

SCHEMATICS OF STRUCTURAL PARALLEL PROGRAMMING AND ITS APPLICATIONS

G.E. Tseytlin
Institute of Cybernetics
Ukrainian Academy of Sciences
252207 Kiev 207 USSR

INTRODUCTION

The advent of multiprocessor computing systems and elaboration of methods of parallel computations stimulated research on structural parallel programming in connection with the development of large para lel programs, in particular, systems of multiprocessor software. Sinc 1965 the Institute of Cybernetics of the Ukrainian Academy of Science has been engaged in developing a formal apparatus suggested by V.M.Gl shkov [1] to solve a number of important problems of automation of the logical computer structures design and programming. The basis for the given apparatus is the notion of a system of algorithmic algebras (SAA) which is in a complete agreement with the concept of structural programming.

This communication is devoted to the theory of multiprocessing-oriented SAA which may be the basis for schematology of the structura parallel programming [1-2]. Principal results are solutions of problems of axiomatization and equivalence of regular schemes in $\mathcal{S}$-algebras oriented towards formalization of asynchronous parallel computations. Theorems from 2 to 5 formulated below are a natural generalization of results obtained in [2]. The considered schematology of th structural parallel programming found practical application in the solution of problems of artificial intelligence, of parallel translation, of the development of a set of tools oriented towards the desig of parallel programs [1].

SYSTEMS OF ALGORITHMIC ALGEBRAS ORIENTED TO MULTIPROCESSING

The SAA $<\mathfrak{U},\mathfrak{B}>$ is a many-sorted algebraic system [1] and consis of two main sets: a set of operators $\mathfrak{U}$ and a set of three-valued logical conditions $\mathfrak{B}$. Over the sets $\mathfrak{U}$ and $\mathfrak{B}$ is defined a signature of operations $\Omega = \Omega_1 \cup \Omega_2$ which consists of a system Ω_1 of logical operations, assuming values in $\mathfrak{B}$, and a system Ω_2 with operations assuming values in $\mathfrak{U}$.

Among logical operations of the system Ω are the generalized Boolean operations [1] : disjunction ($\vee$), conjunction ($\wedge$), negation ($^{-}$). The system Ω_1 has also an operation of left multiplication of a condition α by an operator A which generates a new condition $\beta = A\alpha$ which coincides with the value of the condition α after application of the operator A.

The set $\tilde{\mathfrak{B}}$ of logical functions, being superpositions of the operations entering into Ω_1, forms an algorithmic logic associated with SAA $\langle \mathfrak{U}, \mathfrak{B} \rangle$.

The system Ω_2 contains the following operations: composition $A \times B$ consisting in subsequent application of operators $A, B \in \mathfrak{U}$; α--disjunction $[\alpha](A \vee B)$ which realizes the conditional transfer with respect to $\alpha \in \mathfrak{B}$ to one of the operators $A, B \in \mathfrak{U}$; α-iteration $\{A\}_\alpha$ consisting in subsequent application of the operator A for $\alpha = 0$ until the condition α is true.

Representations in SAA $\langle \mathfrak{U}, \mathfrak{B} \rangle$ of operators from $\mathfrak{U}$ by superpositions of the enumerated operations are called the regular schemes.

<u>Theorem 1</u> [1] . An arbitrary algorithm or a program are representable in SAA $\langle \mathfrak{U}, \mathfrak{B} \rangle$ by an equivalent regular scheme. A constructive procedure of regularization (of reducing to the regular scheme) is developed for arbitrary algorithms and programs.

It should be emphasized that the SAA apparatus is in a complete accordance with the concept of structural programming and can be the basis of the theory of schemes of structured algorithms and programs. To orient SAA towards formalization of parallel computations and in connection with the development of an apparatus of identical transformations the following means of parallelism [1-2] were introduced into the signature of SAA:

<u>filtration</u> which generates filtres-operators

$$\underline{\alpha} = \begin{cases} E, & \text{if } \alpha = 1 \\ N, & \text{otherwise,} \end{cases}$$

where E is an identical operator and N is an indefinite operator;

<u>synchronous disjunction</u> $A \vee B$ which is a partly defined operation consisting in the simultaneous application of the operators A and B to a state $\tilde{m} \in \mathfrak{M}$; as this takes place $A(\tilde{m}) = B(\tilde{m})$ or one of the operators in the state $\tilde{m}$ interrupts the computation and then the

transformation in $\widetilde{\mathfrak{M}}$ is executed by another operator;

asynchronous disjunction $A \dot{\vee} B$ consisting in an independent parallel execution of operators A and B on various substructures of an operating structure P ;

superpositions of operations, belonging to Ω , which generate operators from $\mathfrak{U}$ and are called the parallel regular schemes (PRS).

Relations which form the basis of identical transformations of PRS are characterized in terms of the introduced constructions [1] .

Let us fix in PRS $F(\widetilde{A};\widetilde{\mathfrak{X}}) \in \mathfrak{U}$, with

$$\widetilde{A} = \{A_1, A_2, \ldots, A_r\}, \quad \widetilde{\mathfrak{X}} = \{x_1, x_2, \ldots, x_s\}$$

sets of elementary operators and conditions, respectively, a certain collection of check points (spots at joints where operators occur in the scheme). Each check point T is associated with the condition α which is false until the process of computations reaches the point T true from the instant at which the given point T is reached, and uncertain in the presence of emergency halts in the way leading to the point T of the given scheme. This condition α will be called the condition of synchronization associated with the point T which hereafter will be denoted by $T(\alpha)$. The condition of synchronization is used in PRS as filtering and iterative conditions. Let E^q be a local identical operator functioning over a substructure P_q such that $E^q(\widetilde{m}_q) = \widetilde{m}_q$ for any condition $\widetilde{m}_q \in \mathfrak{M}$ of the substructure P_q . Then, by definition, assume that the operator E^q operates in one time quantum $t_{E^q} = 1$ while the check of conditions of synchronization, the passage through filtres and check points, do not require time expenditures (they are performed in time $t = 0$).

By synchronizer is meant an α-iteration $S^q(\alpha) = \{E^q\}_{\alpha}$, where α is a logical function depending on conditions of synchronization associated with some check points of PRS. The synchronizer installed in some place of PRS delays computations in the given place of the scheme F up to the instant when its condition of synchronization (associated with the passage through the corresponding check points) becomes true.

Theorem 2. In SAA with an extended signature of operations the synchronous- and asynchronous-type parallel processes and their various combinations are representable in the form of PRS.

Thus, the apparatus of SAA oriented towards formalization of parallel computations may serve as the basis for schematology of the structural parallel programming.

CANONICAL FORMS IN ALGORITHMIC LOGIC

We now turn to the discussion of the algorithmic algebra of logic $\widetilde{\mathfrak{B}}$ being a component of SAA $\langle \mathfrak{U}, \mathfrak{B} \rangle$. Let $\varphi(\widetilde{\mathfrak{X}};\widetilde{A})$ be a logical condition represented by the superposition of operations of the algebra $\widetilde{\mathfrak{B}}$, where $\widetilde{\mathfrak{X}}=(x_1,\dots,x_m)$, $\widetilde{A}=(A_1,\dots,A_n)$ are sets of elementary conditions and variable operators, respectively. By ε-conjunction we shall mean the conjunction of the form

$$\mathcal{U}_i = k x_{i_1}^{\psi_{i_1}} x_{i_2}^{\psi_{i_2}} \dots x_{i_k}^{\psi_{i_k}} \quad (1 \le k \le n),$$

where k may be equal to μ or 1, $\psi_j \in \{-1,0,1\}$, and

$$x_{i_j}^{\psi_{i_j}} = \begin{cases} x_{i_j}, & \text{if } \psi_{i_j}=1 \\ \bar{x}_{i_j}, & \text{if } \psi_{i_j}=0 \\ x_{i_j}\cdot\bar{x}_{i_j}, & \text{if } \psi_{i_j}=-1 \quad j=1,2,\dots,k. \end{cases}$$

The conjunction $\mathcal{U}_i$ is elementary if $k=1$ and $\psi_{i_j} \in E_2=\{0,1\}$ for all $j=1,2,\dots,K$. A conjunction of the form

$$\widetilde{\mathcal{U}} = \mathcal{U}_0 \wedge B_1\mathcal{U}_1 \wedge B_2\mathcal{U}_2 \wedge \dots \wedge B_\ell \mathcal{U}_\ell$$

will be called the Δ-conjunction if $\mathcal{U}_0$ is the ε-conjunction; $\mathcal{U}_i=0$ or $\mathcal{U}_i$ is the ε-conjunction; B_i is a composition of operators from $\widetilde{A}$. The Δ-conjunction $\widetilde{\mathcal{U}}$ is elementary if all its ε-conjunctions are elementary. Non-elementary Δ-conjunctions have the coefficient $k=\mu$ as a conjunctive co-factor. If for the Δ-conjunctions $\widetilde{\mathcal{U}}$ and $\widetilde{\mathcal{U}}'$ the equality $\widetilde{\mathcal{U}} \wedge \widetilde{\mathcal{U}}' = \widetilde{\mathcal{U}}'$ holds, the Δ-conjunction $\widetilde{\mathcal{U}}$ absorbs the Δ-conjunction $\widetilde{\mathcal{U}}'$.

By q-polynominal G is meant a disjunction of the finite number of various and mutually non-absorbing Δ-conjunctions

$$G = G^1 \vee G^\mu,$$

where $G^1 = \bigvee_{i=1}^{r} \widetilde{\mathcal{U}}'$ is a disjunction of elementary Δ-conjunctions occurring in the q-polynomial G, $G^\mu = \bigvee_{j=1}^{r'} \widetilde{\mathcal{U}}_j^\mu$ is a disjunction of the rest of Δ-conjunctions of the given q-polynomial.

Lemma 1. Any logical condition $\varphi(\widetilde{\mathfrak{X}};\widetilde{A})$ representable in the algebra $\widetilde{\mathfrak{B}}$ is reducible to the q-polynomial G; $\varphi(\widetilde{\mathfrak{X}};\widetilde{A})=G$. By ε-constituent $\mathcal{K}^0(\mathcal{K}^{-1})$ is meant the ε-conjunction $\mathcal{U}^0(\mathcal{U}^{-1})$ containing all variables from the set $\widetilde{\mathfrak{X}}$.

By a generalized constituent is meant a Δ-conjunction $\widetilde{\mathcal{K}}^\mu = \mathcal{U}_0 \wedge B_1\mathcal{U}_1 \wedge \dots \wedge B_t\mathcal{U}_t$, where $\mathcal{U}_0 = \mathcal{K}_0^{\psi_0}$, $\mathcal{U}_i \in \{0, \mathcal{K}_i^{\psi_i}\}$ $(\psi_0, \psi_i \in \{0,-1\}, i=1,\dots,t)$, $\mathcal{K}_0^{\psi_0}, \mathcal{K}_0^{\psi_i}$ are ε-constituents, $\widetilde{B}=\{B_i \mid i=1,2,\dots,t\}$ is an ordered collection of all operator compositions occurring in the q-polynomial G. The q-polynomial $G_c = G_1 \vee G^\mu$ such that $\widetilde{G}^\mu = \bigvee_{j=1}^{r'} \widetilde{\mathcal{K}}_j^\mu$, where $\widetilde{\mathcal{K}}_j^\mu$ is a generalized constituent of the q-polynomial G for any $j=1,2,\dots,r'$, will be called the perfect polynomial.

Lemma 2. Any q-polynomial G can be transformed to the equivalent perfect q-polynomial G_c, $G=G_c$. Valid is the following assertion.

Theorem 3. Any logical condition $\varphi(\tilde{\mathfrak{X}};\tilde{A})$ of the algebra $\tilde{\mathfrak{B}}$ is naturally representable in the form of the perfect q-polynomial $G_c(\varphi)=G^1 \vee \tilde{G}^{\mu}$. The q-polynomial $G_c(\varphi)=G_1$ representing the disjunctio of elementary Δ-conjunctions will be called the simple polynomial.

Corollary 1. If the condition φ in the algebra $\tilde{\mathfrak{B}}$ is represen table by the simple q-polynomial $G_c(\varphi)=G_c^1$, such representation is unique.

A special case of the simple q-polynomials is disjunctive norma form (dnf) $G_c^1=\bigvee_{i=1}^{K}\mathcal{U}_i^1$, where $\mathcal{U}_i^1$ are elementary conjunctions.

Corollary 2. If the condition φ in the algebra $\tilde{\mathfrak{B}}$ is representable in dnf, such representation is unique.

This fact discriminates, in principle, between the representability in dnf of conditions from the class $\tilde{\mathfrak{B}}(E_3)\subset\mathfrak{B}$ of three-valued functions and the similar results of algebra of logic. In particular, for the class $\tilde{\mathfrak{B}}(E_3)$ the problem of dnf minimization is unfeasible in view of the single-valued representability in dnf of functions of the given class. For the algorithmic logic $\tilde{\mathfrak{B}}$ the finite complete axiomatics with the unique rule of inference - traditional substitution - is constructed and the problem of identities is solved.

Let $\varphi(\tilde{\mathfrak{X}};\tilde{A})$ be a function in the algorithmic logic $\tilde{\mathfrak{B}}$. The function $\varphi^*(\tilde{\mathfrak{X}};\tilde{A})$ is dual to the function $\varphi(\tilde{\mathfrak{X}};\tilde{A})$ if it can be obtained from the function φ as a result of substitution of:

1. Each occurrence of the operation $\vee$ in the function φ by the operation $\wedge$, and vice versa;
2. Each occurrence of the symbol 1 in the function φ by the symbol 0, and vice versa.

Theorem 4. (Duality principle). Let $\varphi=\varphi(\tilde{\mathfrak{X}},\tilde{A})$ be an arbitrary identity in the algorithmic logic $\tilde{\mathfrak{B}}$ composed of conditions and operators occurring in collections $\tilde{\mathfrak{X}}$ and $\tilde{B}$, respectively. Then the expression $\varphi^*=\varphi^*(\tilde{\mathfrak{X}};\tilde{A})$ is also the identity in $\tilde{\mathfrak{B}}$.

On the basis of this principle methods of representation of logical conditions by conjunctive forms dual to q-polynomials can be developed for the algorithmic logic $\tilde{\mathfrak{B}}$. To provide for the functional completeness of logical means of SAA in the three-valued logic and in connection with the necessity of a more flexible control of computing processes the unary operations $\alpha^{a(c)}$ such that

$$q^a=\begin{cases} C, & \text{if } q=\alpha \\ \bar{C} & \text{otherwise,}\end{cases}$$

$$(q,\alpha=0,\mu,1;\ C=0,1)$$

were introduced [1] .

IDENTICAL TRANSFORMATIONS IN S-ALGEBRAS

In connection with the formalization of problems of system multiprocessing the study of S-algebras, i.e. the modified SAA with closed logical conditions [2] , is of interest. Let $\mathfrak{M}=\{\tilde{m}\}$ be an information set on which the operators and conditions of SAA $\langle\mathfrak{A},\mathfrak{B}\rangle$ are defined. Each condition $\alpha\in\mathfrak{B}$ we shall put into correspondence with subsets $M^0(\alpha), M^1(\alpha)\subseteq\mathfrak{M}$ such that $M^0(\alpha)=\{\tilde{m}\,|\,\alpha(\tilde{m})\neq\mu\}$, $M^1(\alpha)=\{\tilde{m}\,|\,\alpha(\tilde{m})=1\}\subseteq M^0(\alpha)$. The set $M^0(\alpha)$ we shall call the domain of definiteness, and $M^1(\alpha)$ the truth domain of the condition α .

Let us introduce into the set $\mathfrak{M}$ a specific "devil" state $w\in\mathfrak{M}$ such that $A(\tilde{m})=w$ if the operator A is not defined in the state $\tilde{m}\in\mathfrak{M}$; here $A(w)=w$ for any $A\in\mathfrak{A}$. Assume that $w\bar{\in}\bigcup_{\alpha\in\mathfrak{B}}M^0(\alpha)$, i.e. $\alpha(w)=\mu$ for any $\alpha\in\mathfrak{B}$. The subset $M^1\in\mathfrak{M}$ is closed if for any $A\in\mathfrak{A}$ $A(\tilde{m})\in M^1$ (for any $\tilde{m}\in M^1$ such that $A(\tilde{m})\neq w$) holds. The set M^1 is isolated if its complement $\neg M^1=\mathfrak{M}\setminus M^1$ is closed.

The logical condition $\alpha\in\mathfrak{B}$ will be called closed if the set $M^0(\alpha)$ is isolated and $M^1(\alpha)$ is closed. In other words, the closed conditions characterize situations whose existence does not depend on the further development of the process. This manifests itself in the fact that the closed conditions, being true at some instant, continue to be true later on. The notion of the closed logical condition is very close to the property of monotonicity of operators.

The apparatus of closed conditions may be used when organizing the problem interaction in operational systems [1] .

Thus the condition

$$\alpha_i=\begin{cases}1, & \text{if the } i \text{ th problem is solved;}\\ 0, & \text{if the } i \text{ th problem is in the process of solution;}\\ \mu, & \text{when in the process of solution of the } i \text{ th problem the emergency halts arise}\end{cases}$$

is associated with the i th problem.

The condition α_i is closed since its truth at the instant t of the i th problem does not depend on the state of a computing system at succeeding instants $t'>t$. Closed are also the conditions of synchronization, considered earlier, associated with check points in the asynchronous PRS. Thus, the closed conditions may be used as landmarks when organizing control schemes in large programs oriented towards multiprocessing.

Let $\langle \mathfrak{U}, \mathfrak{B} \rangle$ be SAA with a system of generators $\Sigma = \Sigma_1 \cup \Sigma_2$, where $\Sigma_1 \in \mathfrak{U}$ are elementary operators, $\Sigma_2 \in \mathfrak{B}$ are elementary logical conditions. By S-algebra is meant SAA $\langle \mathfrak{U}, \mathfrak{B} \rangle$ in which the subsystem $\Sigma_2 \subset \Sigma$ consists of closed logical conditions. For S-algebras the problem of axiomatization and the problem of equivalence are solved.

The method of reducing an arbitrary PRS to the standard parallel polynomial $D_c = \bigcup_{i=1}^{s} \widetilde{w}_i$, where $\widetilde{w}_i = \bigvee_{j=1}^{n} w$ is a Π-branch consisting of asynchronously interacting compositions $w_j = C_{j_1} C_{j_2} \ldots C_{j k_j}$, $C_{j_{\varkappa}}$ being an elementary operator or the standard α-iteration $(\varkappa = 1, 2, \ldots, k_j)$, underlies the obtained results.

Among intermediate results which themselves are of interest is the development of an apparatus of canonical representations of PRS, a criterion of clinch, of asynchronous branches in PRS, a criterion of fictitiousness of iterative structures, etc.

The axiomatic system constructed for S-algebras contains, along with the traditional substitution, the rule of inference connected with the solution of equations by analogy with axiomatics for the algebra of regular events [3] . The distinguishing feature of the solution of the problem of axiomatization of S-algebras is the use of the method of localization of the given rule which is used only when establishing auxiliary identical transformations and which does not influence the proof of the main result about reduction of arbitrary PRS to their canonical representation. Thus, the validity of the following important assertion is stated.

<u>Theorem 5.</u> For S-algebras a complete axiomatics is built which characterizes the set of all true identities in the given algebras and contains only the substitution as a rule of inference. Here the given axiomatics contains schemes of axioms which parametrically depend on the natural numbers $n = 1, 2, \ldots$.

It should be emphasized that the problem of axiomatization of the algebra of regular events with the use of substitution as the only rule of inference is one of the most complex and interesting problems of automata theory.

The SAA apparatus was in use when solving the following important problems of theory and practice of computing science [1] .

<u>Formalization of Semantics of Programming Languages.</u> Semantic penetration of the programming language into algorithmic algebra is connected with the search for a basis and a system of generators oriented towards representation of its programs. Base operators are interpreted as machine facilities in terms of which the principal language constructions entering into the system of generators of algorithmic

algebras are represented in the regular form.

Semantics of the language penetrated into algorithmic algebras is defined as closure of the corresponding system of generators by superposition of operator functional and logical structures entering into it. The given approach has been used when constructing the semantic penetration for address language.

Artificial Intelligence. On the basis of the developed apparatus of identical transformations and axiomatization of SAA oriented towards formalization of parallel computations some experiments were performed on automation of proof of theorems (identities) in appropriate automated systems. In particular, for a MIR-2 computer a program package ANALIST is developed which performs automatic proof in S-algebras by reducing the regular schemes composing the left and right parts of the identity to a polynomial form. The package consists of a control program, a system of subprograms which perform canonization of left and right parts of the theorem being proved,and an identity archives used when proving the theorems in S-algebras. The modularity of structure and the possibility of parametrization of individual components of the package permit its further extension and reorientation towards various classes of SAA.

System Parallel Programming. Clear definition of mechanisms of checking the control conditions and execution of operators in PRS make it possible to accomplish level-by-level transfer to families of programs defined by the given schemes. This level-by-level transfer may be also formalized by varying the system of generators of SAA in the direction of elaboration (enforcement) of elementary operators and conditions according to the concept of up-to-down (down-to-up) programming.

The SAA apparatus forms the basis for a set of tools of system parallel programming which serves for the development of systems of parallel translation of distribution of free resources of multiprocessors, control of flows of problems being solved, etc.

REFERENCES

1. Glushkov V.M., Tseytlin G.E., Yushchenko E.L., Theory of language processors and parallel computations. Kibernetika, 1(1979), 1-19.
2. Tseytlin G.E., Problem of identical transformations of structured program schemata with closed logical conditions. Parts I and II, Kibernetika, 3(1978), 50-57, 4(1978), 10-18.
3. Salomaa A., Two complete axiom systems for the algebras of regular events. J. Assoc. Comput. Mach., 13, 1, 1966, 158-169.

ON AXIOMATIZATION OF DETERMINISTIC PROPOSITIONAL DYNAMIC LOGIC

M.K.Valiev
Institute of Mathematics, Novosibirsk-90, USSR

1. The language of propositional dynamic logic (PDL) with Kripke type semantics for it was introduced by Fischer and Ladner [1], following Pratt [2]. Many assertions about nondeterministic single variable programs (equivalence, correctness, etc.) may be expressed in PDL. In [1] the decidability of satisfiability problem for PDL is proved and the problem of "good" axiomatization of PDL is posed. Segerberg [3] proposed a Hilbert type axiom system for PDL. The completeness of this system was proved by Parikh [4], Gabbay [5] and Segerberg [6]. A Gentzen type axiomatization of PDL was proposed by Pratt [7], however, this system and the proof of its completeness contain some flows and inaccuracies. The object of this note is to give the corrected version of Pratt's system for PDL and to construct on its base a complete Gentzen type axiom system AxD for deterministic variant of PDL (DPDL). The language and semantics for DPDL coincide with the ones for PDL except for semantics of atomic programs which are interpreted as functional relations (partial functions from states to states) rather than as arbitrary binary relations as in PDL A Hilbert type axiom system for DPDL may be easily deduced from AxD, as indicated by Pratt [7] for PDL. It should be noted that the proof of the completeness of PDL given in [4], [6] (we do not know the contents of [5]) substantially use nondeterminateness of atomic programs. In the concluding part of the paper we briefly consider an extension of PDL by some asynchronous programming constructions and show that this extension cannot have a complete axiomatization.

The usual programming constructs " if p then a else b ", "while p do a " may be readily expressed in the language of PDL (DPDL). Therefore, the single variable fragment AL_1 (without the construction $\bigcap K\alpha$) of the algorithmic logic (AL) of Salwicki [8] may be imbedded in DPDL. A finitary axiom system for AL_1 (rather than infinitary system given by Mirkowska [9] for the full AL) may be easily derived from AxD .

A language G which resembles much Al_1 has been earlier introduced by Glushkov [10] (cf. also [11]). However, propositional variables in G are three-valued (may have undefined value), therefore the axiom system for DPDL does not imply any natural axiom system for G. The decidability of G was proved by Semjonov [12], and also follows from the decidability of DPDL (it was pointed out by B.A.Trachtenbrot in his lectures on the algorithmic logic) which may be easily derived from [7]. My interest to DPDL is due to the listening of the above-mentioned course of Professor Trachtenbrot, whom I would like to express my gratitude. I also thank M.I.Dekhtjar for useful discussions.

2. Let us define the language of PDL (DPDL). It contains two sorts of variables: propositional variables P,Q,... and program variables (which we call atomic programs as well) A,B,.... The notions of formula and program are defined by simultaneous recursion.

Definition 1. Every atomic program is a program, every propositional variable is a formula.

2. If a and b are programs, p is a formula, then a; b, $a \cup b$, a*, p? are programs, $\neg$p, [a] p are formulas.

We next use letters a, b,... to denote programs, and letters p, q ... to denote formulas.

Here we give the informal semantics of programs and formulas only. An interpretation of PDL(DPDL) is defined by using a set M (of states). Atomic programs are interpreted for PDL as binary relations over M and as partial functions over M for DPDL. a;b denotes the composition of programs a and b, $a \cup b$ denotes nondeterministic choice of a or b for executing, $a^* = Id \cup a \cup a^2 \cup \dots$, where Id denotes the identity program. p? is the program which defines the identity transformation of a state s, if p is true in s, and undefined, otherwise. The formula [a]p is true in a state s, iff p is true in all the states which may be reached from s by the program a.

It follows immediately from the definitions that [p?] q is equivalent to $p \supset q$, and, consequently, disjunction and conjunction may be also expressed in PDL. We also have, "if p then a else b" and "while p do a" are equivalent to p?; $a \cup \neg$p?;b and (p?;a)*;$\neg$p?, respectively.

3. The complete Gentzen type axiom system Ax for PDL has the following rules (it differs from the system given in [7] in two small details).

$$\frac{}{p \to p}$$

$$\frac{\to p}{\neg p \to} \qquad\qquad \frac{p \to}{\to \neg p}$$

$$\frac{\to p \quad q \to}{[p?]q \to} \qquad\qquad \frac{p \to q}{\to [p?]q}$$

$$\frac{[a]p, [b]p \to}{[a \cup b]p \to} \qquad\qquad \frac{\to [a]p \quad \to [b]p}{\to [a \cup b]p}$$

$$\frac{[a][b]p \to}{[a;b]p \to} \qquad\qquad \frac{\to [a][b]p}{\to [a;b]p}$$

$$\frac{p, [a][a^*]p \to}{[a^*]p \to} \qquad\qquad \frac{U \to p, V \quad p \to [a]p \quad p \to q}{U \to [a^*]q, V} \ \text{(Ind)}$$

$$\frac{p_1, \dots, p_k \to q}{U, [A]p_1, \dots, [A]p_k \to [A]q, V} \ \text{(M)}$$

Let us explain the notation. The rules are written in the form "from zero or more premise sequents infer a conclusion sequent". A sequent $U \to V$, where U and V are sets of formulas $p_1, \dots, p_k$ and $q_1, \dots, q_l$, respectively, is equivalent to the formula $p_1 \& p_2 \& \dots \& p_k \supset$ $\supset q_1 \vee q_2 \vee \dots \vee q_l$. All the rules except for (M) and (Ind) must be thought as abbreviations for schemes of rules which may be obtained by adding sets U (and V, respectively) to the antecedents (and, consequents, respectively) of the conclusion and all the premises of the rule considered.

The original Pratt's system contains the rule $\frac{}{P \to P}$ instead of $\frac{}{p \to p}$ (it may be easily shown that in this case e.g. the sequent $[a^*]p \to [a^*]p$ cannot be deduced). The first premise of the Pratt's rule (Ind) does not contain V (in this case, e.g., $\to [(\neg p?)^*]\neg p, p$ fails to be deduced). The necessity of the first correction of Ax was also observed by Pratt himself (private communication, cf. also [13]).

4. The rules of the axiom system AxD for DPDL coincide with the rules of Ax except for the rule (M) which must be replaced by the rule

(MD)
$$\frac{p_1,\ldots,p_k \to q_1,\ldots,q_l}{U,\ [A]p_1,\ldots,\ [A]p_k \to [A]\ q_1\ ,\ldots,\ [A]\ q_l,\ V},$$

where $k \geqslant 0$, $l \geqslant 1$.

It may be easily deduced from the completeness of AXD that the axiom system for PDL from [3] supplemented by the scheme of axioms

$$[A]\,(p \vee q) \supset [A]\ p \vee [A]\ q$$

is complete for DPDL. We announced this axiom system for DPDL in [14].

5. The proofs of the completeness of Ax and AxD are similar, and we sketch here the proof for AxD only. The first part of the proof deals with the concept of tableau and coincides in essence with the analogical part of the proof in [7].

Definition [7]. A structure is a triple $\langle M, \pi, \tau \rangle$, where M is a set (of states), π is a function which associates to any s in M a set of formulas which are true in s (we write $s \models p$, if p is true in s), τ is a function which associates to any atomic program A a binary relation over M. A structure is deterministic, if for any A $\tau(A)$ is a (partial) function.

The concepts of model and satisfiability of formulas are defined as usual.

Definition [7]. A structure is a tableau, if the following properties hold:

(1) $s \models \neg p \Rightarrow$ non $s \models p$ (2) $s \models \neg\neg p \Rightarrow s \models p$
(3) $s \models [a \cup b]p \Rightarrow s \models [a]p$ and $s \models [b]p$ (4) $s \models \neg[a \cup b]p \Rightarrow s \models \neg[a]p$ or $s \models \neg[b]p$
(5) $s \models [a;b]p \Rightarrow s \models [a][b]p$ (6) $s \models \neg[a;b]p \Rightarrow s \models \neg[a][b]\ p$
(7) $s \models [p?]q \Rightarrow s \models \neg p$ or $s \models q$ (8) $s \models \neg[p?]q \Rightarrow s \models p$ and $s \models \neg q$
(9) $s \models [a^*]p \Rightarrow s \models p$ and $s \models [a][a^*]p$ (10) $s \models \neg[a^*]p \Rightarrow s \models \neg p$ or $s \models \neg[a][a^*]p$
(11) $s \models [A]p \Rightarrow \forall s'\ ((s,s') \in A \supset s' \models p)$
(12) $s \models \neg[A]p \Rightarrow \exists s'\ ((s,s') \in A\ \&\ s' \models \neg p)$
(13) $s \models \neg[a^*]p \Rightarrow \exists s'\ ((s,s') \in a^*\ \&\ s' \models \neg p)$.

The following proposition is essentially contained in [7].

Lemma 1. A formula p of PDL (DPDL) is satisfiable iff there exists a tableau (a deterministic one) which contains p.

6. Definition [1]. A set U of formulas is closed if the following properties hold:

(1) if $\neg p \in U$, then $p \in U$,

(2) if $[a \cup b]p \in U$, then $[a]p \in U$, $[b]p \in U$,
(3) if $[a; b]p \in U$, then $[a][b]p \in U, [b] p \in U$,
(4) if $[a^*] p \in U$, then $p \in U, [a][a^*] p \in U$,
(5) if $[p?]q \in U$, then $p \in U$, $q \in U$,
(6) if $[A]p \in U$, then $p \in U$.

<u>Lemma 2</u> [1]. For any set U there exists a finite closed set $U' \supseteq U$

<u>7</u>. In the main part of the proof we restrict ourselves to the case when formulas have the following property: for any expression of form a^* the regular set which corresponds to expression a does not contain the empty word and words which consist of tests solely. For such a formula we say it does not have the empty word property (e.w.p.) This restriction is adopted because 1) it gives an essential simplification of the proof, 2) any formula is equivalent to a formula without e.w.p., 3) constructions "if" and "while" are naturally expressed by such expressions. In the concluding part of the proof we briefly discuss alterations and complementations which are needed for the general case.

<u>8</u>. For any set U of formulas we denote by U_0 the set of program-free formulas of U, by U'_A - the subset of all the formulas of the form $[A]p, \neg[A]p$ in U, and the subset $\{p_1,\dots,p_k, \neg q_1,\dots,\neg q_l : [A]p_i \in U, \neg[A]q_j \in U\}$ by U_A. U'_A is essential, if it contains at least one formula of the form $\neg[A]q$. A set U is clean if $U = U_0 \cup \bigcup_A U'_A$. We shall often identify a set $\{p_1,\dots p_k\}$ with conjunction $p_1 \& p_2 \& \dots$

<u>Lemma 3</u>. Any formula p without e.w.p. is equivalent to a disjunction of clean conjunctions p_i. Moreover, any sequent of form $U, p \rightarrow V$ may be deduced in AxD from the sequents $U, p_i \rightarrow V$.

Proof (sketch). Consequently applying, as far as possible, equivalences which correspond to properties (2)-(10) in the tableau definition (such that $[a \cup b]p \equiv [a]p \& [b]p, \neg[a^*]q \equiv \neg q \vee \neg[a][a^*]q$) we construct a tree whose nodes are marked by sets (conjunctions) of formulas. From the absence of e.w.p. for p it follows that this tree is finite and its leaves are marked by clean conjunctions p_i. From this tree we easily obtain the needed deduction of $U, p \rightarrow V$.

<u>Lemma 4</u>. For any U the following properties hold:

1) any essential set U'_A is satisfiable iff U_A is satisfiable,

2) if U is clean, then U is satisfiable iff U_0 and all essential U'_A are satisfiable;

3) if U is clean, then there exists a proof of sequent $U \to$ from $U_0 \to$ or from $U_A \to$ for a program A for which U'_A is essential.

The proof easily follows from the tableau definition and (MD).

The following proposition is also useful.

Lemma 5. The following rules are derivable in PDL (DPDL):

$$(1)\ \frac{p \to [a]q}{p \to [a](q \vee r)} \qquad (2)\ \frac{p \to q, r \quad p \to [a][a^*]q, r}{p \to [a^*]q, r}$$

Rule (1) may be proved by induction on complexity of a, rule (2) follows from (1) and (Ind) by choosing of $(p \& \neg r) \vee [a^*]q$ as an invariant for a. Notice that rule (2) is used in the proof of lemma 3.

9. For any set U of formulas we define a tree T(U) whose nodes are labelled by sets (conjunctions) of formulas by the following inductive process. At stage 0 we label the root of T(U) by U. At stage 2i+1, $i \geq 0$, we apply the following procedure to all the leaves of the tree constructed. Let s be a leaf, its label L(s) be $p = \bigvee p_j$, where p_j are clean conjunctions (cf. lemma 3), α_j be the path which leads to p_j in the tree of lemma 3, and q_j be set of all formulas which appear in α_j. Then, for any j such that q_j does not contain formulas of forms $r, \neg r$ simultaneously, we add a node s_j with label q_j to the tree and connect it to s by an edge. At stage 2i, $i > 0$, for any leaf s of the tree constructed with L(s)=q and any A such that q'_A is essential we add a node s' with label q_A to the tree and connect s' to s by an edge with label A.

Introduce the following definitions, having in mind similar definitions of [13]. A node of T(U) is full if it appears at a stage 2i+1, $i \geq 0$. All other nodes are partial (notice that if a partial node s is a leaf of T(U), then L(s) ≡ false). A fat path is a subtree Z of T(U) with following properties: 1) if a full node appears in Z, then all its (immediate) successors appear in Z, 2) if an internal partial node appears in Z, then exactly one its successor appears in Z.

Considering the set of all full nodes of Z as a set of states, we see that Z defines a structure which satisfies conditions (1)-(12) of tableau definition. Then Lemma 1 may be reformulated as

Lemma 6. U is satisfiable iff there exists a fat path in T(U) which defines a structure for which condition (13) holds.

10. Concept of reachability of a node in T(U) from another one by a word in the alphabet of atomic programs and tests is defined naturally (in particular, s is reachable from s by p? iff $p \in L(s)$).

Let U={W, ¬[a]q} and U′ = {W, ¬[a]R}, where R is a new propositiona letter. We denote by M(W,a) the set of all the full nodes s in T(U′) with following properties: 1) L(s) contains ¬R, 2) any predecessor of s does not contain ¬R. It may be shown easily that if s is in M(W, then s is reachable from the root by a word w in a.

For any s in T(U′) we denote by W(a,s) conjunction of all the formulas in L(s) except for ones which contain R, and by W(a) disjunc tion of all the W(a,s), s in M(W,a).

Lemma 7. For any program a and formula W sequent W → [a]W(a) is provable in AxD.

Proof. Proceeds by induction on complexity of a.

a=A. Let $W = \vee W_i$, where W_i are clean. Then, by lemma 3 W→[A]W(A) is reduced to W_i → [A]W(A), and by (MD), to W_{iA} → W(A) which is provabl

a=p?. We omit a simple proof of W → [p?]W(p?).

a = b ∪ c. By induction hypothesis, sequents W → [b]W(b), W→[c]W(c are provable. Then, since W(b ∪ c) contains W(b) and W(c), we obtain provability of W → [b ∪ c]W(b ∪ c) using rule (1) of lemma 5 and ∪-rule.

a=b;c. In this case we proceed by induction on complexity of b with assumption that b is not of the form d;e. Consider the case a=b*; c only (other cases are simple). Apply to W → [b*][c]W(b*;c) rul (Ind) by choosing W(b*) as an invariant for b.

The first premise of rule is provable easily. The second premise may be reduced to sequents W(b*,s) → [b]W(b*), s in M(W,b*). Denote W(b*,s) by V. By induction hypothesis we have provability of V→[b]V(b Then, the second premise would be proved if we prove that for any node x in M(V,b) there exists such a node x′ in M(W,b*) that V(b,x) coincides with W(b*,x′).

Let y be the immediate predecessor of x in T(V, ¬[b]R), and z be the successor of the root connected to y by a path α. Let z′ be a nod in T(W, ¬[b*]R) such that 1) the clean part of L(z′) may be obtained from the clean part of L(z) by replacement of occurrences of R by [b*] 2) s and z′ have a common immediate predecessor (such a node exists). Then, going along path α from z′ we access such a node y′ that label of y′ contains ¬[b*] R, and V(b,y) = W(b*,y′) (notice that since b does not contain empty word, formula ¬R appears in y). Then, a successor of y′ satisfies conditions needed.

The third premise is trivial if c is void. This gives induction step for a=b*. In the general case by induction hypothesis we have provability of V → [c]V(c), where V is W(b*,s), s in M(W,b). Then, since V(c) is contained in W(b*,c), we obtain provability of V → [c]W(b*;c) which shows provability of the third premise of (Ind).

Remark. Idea of lemma 7 is due to [7], however, its realization in [7] contains a number of flows and incorrectnesses.

11. Lemma 8. If sequent $W \to [a]q$ is not provable, then a full node s exists in $T(W, \neg[a]q)$ with following properties: 1)s is reachable from the root by a word w in a, 2 L(s) has the form $\{V, \neg q\}$, 3) $V \to q$ is not provable, 4) for any s' on the path leading from the root to s $\neg L(s')$ is not provable.

Proof. First we show by induction on complexity of a the existence of s with properties 1)-3). Cases a= A, p?, $b \cup c$, b;c are simple. Let $a=b^*$. Apply rule (Ind) to $W \to [b^*]q$ with the invariant $W(b^*)$. The first and second premises are provable (see proof of lemma 7). Since $W \to [b^*]q$ is not provable, then a sequent $W(b^*, s) \to q$, s in $M(W, b^*)$, is not provable. Then, a node s' in $T(W, \neg[b^*]q)$ corresponds to the node s in $T(W, \neg[b^*]R)$ which satisfies properties 1)-3).

Let T' be a (finite) subtree of $T(W, \neg[a]q)$ whose leaves correspond to nodes of $T(W, \neg[a]R)$ from $M(W,a)$. Then existence of a node s with properties 1)-4) may be proved by induction on the height of T'. Namely, let $W \& \neg[a]q \equiv \vee p_j$, where $p_j = \{W_j, \neg[c]q\}$ are clean, $c \leq a$, and p_i is not provable. Then we may choose as s the node in T' with properties 2)-4) and reachable from the node with label p_i by a word in c (such a word exists by induction hypothesis).

12. Now we get to the proof of completeness of AxD for formulas without e.w.p. Further we assume for simplicity of exposition that AxD contains an additional rule $\neg p, q \to r \vdash q \to p, r$. Hence, we may restrict ourselves to proving formulas of form $\neg p$. Applications of this rule may be eliminated from the proof (the strenghtened form of the rule (Ind) is needed for this elimination, namely).

Let p be a formula such that $\neg p$ is not provable. Then we show that p is satisfiable. Thus, our proof of completeness is not constructive in the sense that it gives no proof for a given valid formula. We think that it is essential, in contrary to the failing attempt of [7] to give a constructive proof.

In order to show that p is satisfiable we define the following inductive process of constructing a fat path D in T(p) which satisfies condition (13) of the tableau definition. At first stage we include in D an edge leading from the root to a node s such that $\neg L(s)$ is not provable (see lemma 3) and all the edges which lead from s.

Let a (finite) subtree D_1 of T(p) be constructed up to a moment. Then, for any statement $s \models \neg[a^*]q$, s in D_1, we add to D_1 a path $\alpha(s)$

from s to a node s' with properties 1)-4) of lemma 8 (maybe $\alpha(s)$ li yet in D_1) with all the (immediate) successors of full nodes in $\alpha(s)$, and after this apply the procedure of the first stage to all the leav of the subtree constructed.

It is clear from the description that the fat path D obtained as the limit of this construction satisfies conditions (1)-(13) of the tableau definition, i.e. p is satisfiable.

13. Now we discuss briefly provability in AxD of formulas with empty word property. This may be shown by induction on k(p), where k(p) denotes cardinality of positive occurrences of * in a formula p such that there exist occurrences of * with e.w.p. which are subordinate to the first occurrences. First, the proof given above may be generalized to formulas with k(p)=0. The general case (k(p) > 0) may be reduced to provability of sequents of form $U \rightarrow [a^*]q, V$, where $[a^*]q$ has e.w.p. $[a^*]q$ is equivalent to a formula r without e.w.p. Then, induction step is established by applying rule (Ind) to $U \rightarrow [a^*]q$, V with the invariant r.

14. Let $\mathfrak{A}$ be an extension of language of regular expressions by adding operations +, # where $a+b=\{v_1w_1v_2w_2\ldots v_kw_k : v_1v_2\ldots v_k \in a, w_1w_2\ldots w_k \in b\}$, and $a^{\#}= Id \cup a \cup (a+a) \cup (a+a+a) \cup \ldots$ These operations were used by Kimura in [15] for describing asynchronous programs. The following proposition may be proved.

Proposition. The set of all the expressions in $\mathfrak{A}$ which are equivalent to $(A \cup B)^*$ is not recursively enumerable.

This shows that an extension of PDL (DPDL) by adding of +, # is not axiomatizable. Moreover, the finitary system of equivalent transformations of expressions in $\mathfrak{A}$ proposed in [15] (as any finitary system, certainly) cannot be complete.

A more weak result, namely, that PDL(DPDL) supplemented by +, # has not finite model property, follows from the fact that formula

$$[(AB)^{\#}]P \,\&\, [(A\cup B)^*-(AB)^{\#}]\neg P \,\&\, [(A\cup B)^*]\neg([A]\text{false} \vee [B]\text{false})$$

is satisfiable, however, has no finite model (notice that $(A\cup B)^*-(AB$ may be expressed in $\mathfrak{A}$). A similar formula for context-free PDL was proposed by R.Ladner.

REFERENCES

1. Fischer M.J., Ladner R.E., Propositional modal logic of programs. Proc. 9th ACM Symp. on Theory of Computing, 1977, 286-294.

2. Pratt V.R., Semantical considerations on Floyd-Hoare logic. Proc. 17th IEEE Symp. on Found. of Comp. Sci., 1976,109-121.

3. Segerberg K.,A completeness theorem in the modal logic of programs. Notices of the AMS, 24 (1977), 6, A-552.

4. Parikh R., The completeness of propositional dynamiic logic. Lecture Notes in Computer Science, 64(1978), 403-413.

5. Gabbay D., Axiomatizations of logics of programs. Manuscript, under cover dated Nov. 1977 (the reference taken from [7]).

6. Segerberg K., A completeness theorem in the modal logic of programs. Manuscript, 1978.

7. Pratt V.R., A practical decision method for propositional dynamic logic. Proc. 10th ACM Symp. on Theory of Computing, 1978, 326-337.

8. Salwicki A., Formalized algorithmic languages. Bull. Acad. Pol. Sci., Ser. Sci. Math. Astr. Phys. 18(1970),No.5.

9. Mirkowska G., On formalized systems of algorithmic logic. Bull. Acad. Pol. Sci. Math. Astr. Phys. 19(1971), No.6.

10. Glushkov V.M., Automata theory and formal transformations of microprograms. Kybernetika, 1965, 5 (in Russian).

11. Glushkov V.M., Zeitlin G.E., Yushchenko E.L., Algebra, languages, programming. Kiev, 1974 (in Russian).

12. Semjonov A.L., Some algorithmic problems for systems of algorithmic algebras. Doklady AN SSSR, 239(1978), 5, 1063-1066 (in Russian).

13. Pratt V.R., On near-optimal method for reasoning about actions. MIT/LCS/TM-113, 1978.

14. Valiev M.K., Complete axiomatization of deterministic propositional dynamic logic. All-Union Symp. on Artificial Intelligence and Automatization of Math. Research. Kiev, 1978, 3-4 (in Russian).

15. Kimura T., An algebraic system for process structuring and interprocess communication. Proc. 8th ACM Symp. on Theory of Computing, 1976, 92-100.

BOUNDED RECURSION AND COMPLEXITY CLASSES

Klaus Wagner

Sektion Mathematik der Friedrich-Schiller-Universität
DDR - 69 Jena, Universitätshochhaus

1. INTRODUCTION

Already in the sixties the interest in machine-independent characterizations of important complexity classes by means of recursion theory arose (see Ritchie [1] and Cobham [2]). Up to now there is a series of such results (see further Thompson [3] and Monien [4]) but these results concern only single complexity classes There are no results on general relations between complexity classes and classes of functions generated by several types of bounded recursion. In this paper we present such general relations for bounded syntactic recursion, bounded weak recursion, bounded primitive recursion and bounded recursion by two values (Werteverlaufsrekursion). For the proof of these results we generalize some methods used in the literature for the corresponding single results. The most important step is to find a very simple arithmetization by means of the present type of bounded recursion. As special cases of these results we get the known single results as well as a series of new characterizations of time and tape complexity classes. Because of the limitation on the number of pages all proofs are omitted.

2. Definitions

Let $\mathbb{N} = \{0,1,2,\dots\}$ be the set of natural numbers. We use in what follows the dyadic presentation of natural numbers, i.e. the presentation given by: empty word $\rightarrow$ o, x1 $\longrightarrow$ 2x+1 and x2 $\longrightarrow$ 2x+2. Thus we have a one-one mapping from the set $\{1,2\}^*$ of all words over the alphabet $\{1,2\}$ to $\mathbb{N}$. For $x \in \mathbb{N}$ let $|x|$ be the length of the dyadic presentation of x. We generalize: $|x|_o =_{df} x$, $|x|_{k+1} =_{df} ||x|_k|$. If f is a function from A to B we write $f: A \longmapsto B$. We define the functions O, I^n_m, S, $[\frac{x}{2}]$, $\dot{-}$ and M by $O =_{df} o$, $I^n_m(x_1,\dots,x_n) =_{df} x_m$ for $1 \leq m \leq n$ (by I we denote the set of all theses I^n_m), $S(x) =_{df} x+1$, $[\frac{x}{2}]$

$=_{df}$ the largest natural number not greater than $\frac{x}{2}$, $x \dot{-} y =_{df} \begin{cases} x-y, & \text{if } x \geq y \\ o, & \text{otherwise} \end{cases}$, and $M(x,y) =_{df} \max(x,y)$. We renounce Church's λ-notation of functions. Let Lin be the set of all unary linear functions, and let Pol be the set of all unary polynomials. For $t,t':\mathbb{N}\mapsto\mathbb{N}$ we define $t \leq_{ae} t' \Leftrightarrow_{df} \bigvee_x \bigwedge_{y \geq x} t(y) \leq t'(y)$. For $t:\mathbb{N}\mapsto\mathbb{N}$ let $t^o(x) =_{df} x$, $t^{m+1}(x) =_{df} t(t^m(x))$. The algebraic closure operator, given by the partial operations $O_1,\ldots,O_r$ on $\{f; \bigvee_n f:\mathbb{N}^n \mapsto \mathbb{N}\}$, is denoted by $\Gamma_{O_1,\ldots,O_r}$; for short: $\Gamma_{O_1,\ldots,O_r}(f_1,\ldots,f_s) =_{df} \Gamma_{O_1,\ldots,O_r}(\{f_1,\ldots,f_s\})$. The complexity class SPACE(s) (SPACE-TIME(s,t), TIME(t,s)) consists of all functions computable by a multitape Turing machine which works within space s(n) (space s(n) and time t(n), time t(n) and length of result not greater than s(n)), where n is the length of the input. For $A,B \subseteq \mathbb{N}^{\mathbb{N}}$ let $\mathrm{SPACE}(A) =_{df} \bigcup_{s \in A} \mathrm{SPACE}(s)$, $\mathrm{SPACE\text{-}TIME}(A,B) =_{df} \bigcup_{\substack{s \in A \\ t \in B}} \mathrm{SPACE\text{-}TIME}(s,t)$, and $\mathrm{TIME}(A,B) =_{df} \bigcup_{\substack{t \in A \\ s \in B}} \mathrm{TIME}(t,s)$. Further, let $P =_{df} \mathrm{TIME}(\mathrm{Pol},\mathrm{Pol})$, LINSPACE $=_{df}$ SPACE(Lin), and POLSPACE =SPACE(Pol). The n-th Grzegorczyk class is denoted by E^n and we choose unary functions f_n in such a way that $E^n = \Gamma_{SUB,BPR}(O,I,S,M,f_n)$ and $f_1(x) = 2x$, $f_2(x) = x^2$ and $f_3(x) = 2^x$.

3. Several types of bounded recursion

First we define the relevant operations.

The substitution (SUB):

$h=SUB(f,g_1,\ldots,g_k) \Leftrightarrow_{df}$ there is an $n \in \mathbb{N}$ such that

a) $f:\mathbb{N}^k \mapsto \mathbb{N}$,

b) $g_1,\ldots,g_k: \mathbb{N}^n \mapsto \mathbb{N}$,

c) $h:\mathbb{N}^n \mapsto \mathbb{N}$ is defined by

$$h(x_1,\ldots,x_n)=f(g_1(x_1,\ldots,x_n),\ldots,g_k(x_1,\ldots,x_n)).$$

The usual type of bounded primitive recursion (BPR):

$h=BPR(f,g,k) \Leftrightarrow_{df}$ there is an $n \in \mathbb{N}$ such that

a) $f:\mathbb{N}^n \mapsto \mathbb{N}$,

b) $g:\mathbb{N}^{n+2} \mapsto \mathbb{N}$,

c) $h:\mathbb{N}^{n+1} \mapsto \mathbb{N}$ is defined by

$$h(x_1,\ldots,x_n,o)=f(x_1,\ldots,x_n),$$
$$h(x_1,\ldots,x_n,y+1)=g(x_1,\ldots,x_n,y,h(x_1,\ldots,x_n,y)),$$

d) $k: \mathbb{N}^{n+1} \mapsto \mathbb{N}$ with

$h(x_1,\ldots,x_n,y) \leq k(x_1,\ldots,x_n,y)$.

The bounded weak recursion (BWR). Here the length of the argument, not its value, determines the number of steps of the recursion:

$h = BWR(f,g,k) \Leftrightarrow_{df}$ there is an $n \in \mathbb{N}$ and a function $h': \mathbb{N}^{n+1} \mapsto \mathbb{N}$ such that

a) $h' = BPR(f,g,k)$,

b) $h: \mathbb{N}^{n+1} \mapsto \mathbb{N}$ is defined by

$h(x_1,\ldots,x_n,y) = h'(x_1,\ldots,x_n,|y|)$.

The bounded syntactic recursion (BSR) uses the dyadic presentation of the numbers. Thus also in this case the length of the argument determines the number of steps of the recursion:

$h = BSR(f,g_1,g_2,k) \Leftrightarrow_{df}$ there is an $n \in \mathbb{N}$ such that

a) $f: \mathbb{N}^n \mapsto \mathbb{N}$,

b) $g_1, g_2: \mathbb{N}^{n+2} \mapsto \mathbb{N}$,

c) $h: \mathbb{N}^{n+1} \mapsto \mathbb{N}$ is defined by

$h(x_1,\ldots,x_n,o) = f(x_1,\ldots,x_n)$,

$h(x_1,\ldots,x_n,y+1) = g_1(x_1,\ldots,x_n,y,h(x_1,\ldots,x_n,y))$,

$h(x_1,\ldots,x_n,y+1) = g_2(x_1,\ldots,x_n,y,h(x_1,\ldots,x_n,y))$,

d) $k: \mathbb{N}^{n+1} \mapsto \mathbb{N}$ with

$h(x_1,\ldots,x_n,y) \leq k(x_1,\ldots,x_n,y)$

The bounded recursion by two values (BVR, Werteverlaufsrekursion) takes into consideration not only the last value but also an earlier value of the computed function:

$h = BVR(f,g,s,k) \Leftrightarrow_{df}$ there is an $n \in \mathbb{N}$ such that

a) $f: \mathbb{N}^n \mapsto \mathbb{N}$,

b) $g: \mathbb{N}^{n+3} \mapsto \mathbb{N}$,

c) $s: \mathbb{N}^{n+1} \mapsto \mathbb{N}$ with

$s(x_1,\ldots,x_n,y) \leq y$,

d) $h: \mathbb{N}^{n+1} \mapsto \mathbb{N}$ is defined by

$h(x_1,\ldots,x_n,o) = f(x_1,\ldots,x_n)$,

$h(x_1,\ldots,x_n,y+1) = g(x_1,\ldots,x_n,y,h(x_1,\ldots,x_n,y),$
$h(x_1,\ldots,x_n,s(x_1,\ldots,x_n,y)))$,

e) $k: \mathbb{N}^{n+1} \mapsto \mathbb{N}$ with

$h(x_1,\ldots,x_n,y) \leq k(x_1,\ldots,x_n,y)$.

Between the several types of bounded recursion we have the following relations.

<u>Lemma 1</u> If $0,I,S \in \Gamma_{SUB,BSR}(A)$, then

$$\Gamma_{SUB,BWR}(A) \subseteq \Gamma_{SUB,BSR}(A) .$$

<u>Lemma 2</u> If $0,I,S,M,2x,[\frac{x}{2}], \dot{-} \in \Gamma_{SUB,BWR}(A)$, then

$$\Gamma_{SUB,BSR}(A) \subseteq \Gamma_{SUB,BWR}(A) .$$

Consequently, the BSR and the BWR can do the same, if only some simple functions can be generated. However, the BPR seems to be more powerful, because we have

<u>Lemma 3</u> If $0,I,S,M,2x \in \Gamma_{SUB,BPR}(A)$, then

$$\Gamma_{SUB,BWR}(A) \subseteq \Gamma_{SUB,BPR}(A) ,$$

but for the converse inclusion we need the fast increasing function 2^x:

<u>Lemma 4</u> If $2^x \in \Gamma_{SUB,BWR}(A)$, then

$$\Gamma_{SUB,BPR}(A) \subseteq \Gamma_{SUB,BWR}(A).$$

Similarly, the BVR seems to be more powerful than the BPR:

<u>Lemma 5</u> If $I \in \Gamma_{SUB,BVR}(A)$, then

$$\Gamma_{SUB,BPR}(A) \subseteq \Gamma_{SUB,BVR}(A).$$

<u>Lemma 6</u> If $0,I,M,S,2^x \in \Gamma_{SUB,BPR}(A)$ and every function of A can be bounded by a monotonic function from $\Gamma_{SUB,BPR}(A)$, then

$$\Gamma_{SUB,BVR}(A) \subseteq \Gamma_{SUB,BPR}(A).$$

<u>Corollary 1</u> If $0,I,M,S,2x, [\frac{x}{2}], \dot{-} , 2^x \in A$, then

$$\Gamma_{SUB,BSR}(A) = \Gamma_{SUB,BWR}(A) = \Gamma_{SUB,BPR}(A) = \ _{SUB,BVR}(A) .$$

<u>4. Bounded syntactic recursion (BSR) and bounded weak recursion(BWR)</u>

Generalizing methods used by Cobham [2] and Thompson [3] we get

<u>Theorem 1</u> If a) $t: \mathbb{N} \mapsto \mathbb{N}$ monotonic,

b) there is a $c > 1$ such that $t(n) \geq c \cdot n$,

c) there is a $m \in \mathbb{N}$ such that

$$t \in \text{SPACE-TIME}(t^m(2^n), \text{Pol}(t^m(2^n))),$$

then

$$\text{SPACE-TIME}(\Gamma_{SUB}(t), \text{Pol}(\Gamma_{SUB}(t))) = \Gamma_{SUB,BSR}(0,I,S,+,2^{t(|x|)}) =$$

$$= \Gamma_{SUB,BWR}(0,I,S,M,2x,[\tfrac{x}{2}],\dot{-},2^{t(|x|)}) .$$

Theorem 2 If a) $t: \mathbb{N} \mapsto \mathbb{N}$ monotonic,

b) there is an $r > 1$ such that $t(x) \geq_{ae} x^r$,

c) $t \in \text{SPACE-TIME}(\Gamma_{SUB}(|t(2^n)|), \text{Pol}(\Gamma_{SUB}(|t(2^n)|)))$,

then

$$\text{SPACE-TIME}(\Gamma_{SUB}(|t(2^n)|), \text{Pol}(\Gamma_{SUB}(|t(2^n)|))) = \Gamma_{SUB,BSR}(0,I,S,+,t) = \Gamma_{SUB,BWR}(0,I,S,M,2x,[\tfrac{x}{2}], \dot{-}, t).$$

Theorem 2 implies directly

Corollary 2 1. (THOMPSON [3])

$$\text{SPACE-TIME}(\text{Lin},\text{Pol}) = \Gamma_{SUB,BSR}(0,I,S,+,x^2) = \Gamma_{SUB,BWR}(0,I,S,M,2x,[\tfrac{x}{2}], \dot{-}, x^2).$$

2. (COBHAM [2])

$$P = \text{SPACE-TIME}(\text{Pol},\text{Pol}) = \Gamma_{SUB,BSR}(0,I,S,+,x^{|x|}) = \Gamma_{SUB,BWR}(0,I,S,M,2x,[\tfrac{x}{2}], \dot{-}, x^{|x|}).$$

3. $\text{SPACE-TIME}(n \cdot \text{Pol}(\log^k n), \text{Pol}) = \Gamma_{SUB,BSR}(0,I,S,+, x^{|x|_{k+1}})$

$\Gamma_{SUB,BWR}(0,I,S,M,2x,[\tfrac{x}{2}], \dot{-}, x^{|x|_{k+1}})$.

5. Bounded primitive recursion (BPR)

Generalizing methods used by Ritchie [1] and Thompson [3] we get

Theorem 3 If a) $t: \mathbb{N} \mapsto \mathbb{N}$ monotonic,

b) there is a $c > 1$ such that $t(n) \geq c \cdot n$,

c) there is an $m \in \mathbb{N}$ such that $t \in \text{SPACE}(t^m(2^n))$,

then

$$\text{SPACE}(\Gamma_{SUB}(t)) = \Gamma_{SUB,BPR}(0,I,S,M,2^{t(|x|)}).$$

Theorem 4 If a) $t: \mathbb{N} \mapsto \mathbb{N}$ monotonic,

b) there is an $r > 1$ such that $t(x) \geq_{ae} x^r$,

c) $t \in \text{SPACE}(\Gamma_{SUB}(|t(2^n)|))$,

then

$$\text{SPACE}(\Gamma_{SUB}(|t(2^n)|)) = \Gamma_{SUB,BPR}(0,I,S,M,t).$$

Theorem 4 implies directly

Corollary 3 1. (RITCHIE [1]) $\text{LINSPACE} = \Gamma_{SUB,BPR}(0,I,S,M,x^2)$.

2. (THOMPSON [3]) $\text{POLSPACE} = \Gamma_{SUB,BPR}(0,I,S,M,x^{|x|})$.

3. $\text{SPACE}(n \cdot \text{Pol}(\log^k n)) = \Gamma_{SUB,BPR}(0,I,M,x^{|x|_{k+1}})$.

4. (RITCHIE [1] for $n = 3$) $SPACE(E^n) = SPACE(\Gamma_{SUB}(f_n)) = \Gamma_{SUB,BPR}(0,I,S,M,f_n) = E^n$ for $n \geq 3$.

6. Bounded recursion by two values (BVR)

Generalizing methods of Monien [4] (note that there are considered characteristic functions only) we get:

Theorem 5 If a) $t: \mathbb{N} \mapsto \mathbb{N}$ monotonic,

b) there is a $c > 1$ such that $t(n) \geq c \cdot n$,

c) there is an $m \in \mathbb{N}$ such that $t \in TIME(2^{t^m(2^n)}, t^m(2^n))$,

then

$$TIME(2^{\Gamma_{SUB}(t)}, \Gamma_{SUB}(t)) = \Gamma_{SUB,BVR}(0,I,S,M,2^{t(|x|)}).$$

Theorem 6 If a) $t: \mathbb{N} \mapsto \mathbb{N}$ monotonic,

b) there is an $r > 1$ such that $t(x) \geq_{ae} x^r$,

c) $t \in SPACE(\Gamma_{SUB}(|t(2^n)|))$,

then

$$TIME(2^{\Gamma_{SUB}(|t(2^n)|)}, \Gamma_{SUB}(|t(2^n)|)) = \Gamma_{SUB,BVR}(0,I,S,M,t).$$

Theorem 6 implies directly

Corollary 4 1. (MONIEN [4]) $TIME(2^{Lin}, Lin) = \Gamma_{SUB,BVR}(0,I,S,M,x^2)$.

2. $TIME(2^{Pol}, Pol) = {}_{SUB,BVR}(0,I,S,M,x^{|x|})$.

3. $TIME(2^{n \cdot Pol(\log^k n)}, n \cdot Pol(\log^k n)) = \Gamma_{SUB,BVR}(0,I,S,M,x^{|x|}{}_{k+1})$.

4. $TIME(E^n) = TIME(E^n, E^n) = {}_{SUB,BVR}(0,I,S,M,f_n) = E^n$ for $n \geq 3$.

Taking into consideration the known relations between time and space for Turing machine computations, the theorems 2, 4 and 6 imply

Corollary 5 If a) $t: \mathbb{N} \mapsto \mathbb{N}$ monotonic,

b) $t(x) \geq 2^x$,

c) $t \in SPACE\text{-}TIME(\Gamma_{SUB}(|t(2^n)|), Pol(\Gamma_{SUB}(|t(2^n)|)))$,

then

$$\begin{aligned} \Gamma_{SUB,BWR}(0,I,S,M,2x,[\tfrac{x}{2}], \dot{-}, t) &= \Gamma_{SUB,BSR}(0,I,S,+,t) \\ &= \Gamma_{SUB,BPR}(0,I,S,M,t) \\ &= \Gamma_{SUB,BVR}(0,I,S,M,t) \end{aligned}$$

7. Conclusions

The types of bounded recursion considered in this paper are not sensitive enough to characterize time complexity classes for which the set of names is not closed under polynomials, for example such classes like TIME(Lin) and TIME(n^2). For this one has to use more restricted types of recursion and substitution. Investigations in this spirit are done by Monien [5]. Furthermore, it would be interesting to study the power of such types of bounded recursion like bounded summation and bounded product with respect to complexity classes. For a single result in this direction see Constable [6].

8. Acknowledgement

I am grateful to Dr. Bernhard Goetze, Jena, for some interesting discussions.

9. References

1. Ritchie, R.W., Classes of predictably computable functions, Trans. Amer. Math. Soc. 106 (1963), 137-173.
2. Cobham, A., The intrinsic computational complexity of functions, Proc. 1964 Intern. Congr. on Logic, Math. and Philos. of Science, 24-30.
3. Thompson, D.B., Subrecursiveness: Machine-Independent Notions of Computability in Restricted Time and Storage, MST 6 (1972),3-15.
4. Monien, B., A recursive and a grammatical characterization of the exponentional-time languages, TCS 3 (1977), 61-74.
5. Monien, B., Komplexitätsklassen von Automatenmodellen und beschränkte Rekursion, Bericht Nr. 8, Institut für Informatik, Universität Hamburg, 1974.
6. Constable, Robert.L., Type two computational complexity, Proc. of 5th Ann. ACM Symp. on Theory of Comp., 1973, 108-121.

CHARACTERIZATION OF RATIONAL AND ALGEBRAIC POWER SERIES

Wolfgang Wechler
Sektion Mathematik
Technische Universität Dresden
DDR-8027 Dresden, German Democratic Republic

1. Introduction and basic notions

Let R be a semiring and let M be a monoid. A mapping s from M into R is called a (formal) power series and s itself is written as formal sum

$$s = \sum_{m \in M} (s,m)m ,$$

where (s,m) is the image of $m \in M$ under the mapping s. The values (s,m) are referred to as coefficients of s. R<<M>> denotes the set of all such mappings. The support supp(s) of a power series s is the set $\mathrm{supp}(s) = \{m \in M \mid (s,m) \neq 0\}$. Any power series with a finite support is called a polynomial. The set of all polynomials is denoted by R<M>.

In this paper a new characterization of algebraic power series will be presented. The known characterization of rational power series shall also be established in our framework in order to emphasize the analogy of both kinds of characterization. To prepare such a characterization some necessary concepts from the representation theory of monoids and from an appropriate generalization of module theory must be introduced.

A commutative monoid A with the operation + is called an R-semimodule if, for each r of a given semiring R and each a of A, a scalar product ra is defined in A such that the usual axioms are satisfied: $r(a + a') = ra + ra'$, $(r + r')a = ra + r'a$, $(r \cdot r')a = r(r'a)$, $1a = a$, $0a = 0$ and $r0 = 0$ for $r,r' \in R$, $a,a' \in A$. Obviously, R<<M>> forms an R-semimodule with respect to the usual operations.

An R-semimodule is an algebraic structure in the sense of universal algebra. Therefore, the notions of an R-subsemimodule, of a generated R-subsemimodule and of an R-homomorphism are fixed in that sense and we can skip their definitions here. In order to define one of our main concepts we have to consider the regular representation of a monoid M. Let $Hom_R(R\langle\langle M\rangle\rangle, R\langle\langle M\rangle\rangle)$ denote the set of all R-homomorphisms from $R\langle\langle M\rangle\rangle$ into itself. The regular representation of M is a monoid homomorphism from M into $Hom_R(R\langle\langle M\rangle\rangle, R\langle\langle M\rangle\rangle)$ assigning a mapping ϱ_m to each m of M, where ϱ_m is defined as follows

$$\varrho_m s = \sum_{n \in M} (s, m \cdot n) n$$

for all s of $R\langle\langle M\rangle\rangle$.

We say that an R-subsemimodule A of $R\langle\langle M\rangle\rangle$ is invariant if, for each m of M, $s \in A$ implies $\varrho_m s \in A$. By means of invariant R-subsemimodules of $R\langle\langle M\rangle\rangle$ rational power series can be characterized (cf. [2]. To do the same for algebraic power series invariant R-subsemialgebras are needed. An R-semimodule A is said to be an R-semialgebra if A is additionally a semiring. $R\langle\langle M\rangle\rangle$ forms an R-semialgebra provided a (Cauchy) product of power series can be defined. For that reason it has to be assumed that each m of M possesses only finitely many factorizations $m = m_1 \cdot m_2$. This condition is satisfied for a monoid with length-function, which is a mapping λ from M into the natural numbers $\mathbb{N}$ such that, for each $n \in \mathbb{N}$, $\lambda^{-1}(n)$ is a finite set, $\lambda^{-1}(0) = \{1\}$ (1 denotes also the unit element of M), and $\lambda(m \cdot m') \geqq \lambda(m) + 1$ as wel as $\lambda(m' \cdot m) \geqq \lambda(m) + 1$ for all $m \in M$ and $m' \in M - \{1\}$ [1]. Clearly, the free monoid generated by a finite set is a monoid with length-function. If M is a monoid with length-function, then $R\langle\langle M\rangle\rangle$ is an R-semialgebra with respect to usual operations. An R-subsemialgebra of $R\langle\langle M\rangle\rangle$ is called invariant if it is invariant as R-subsemimodule.

2. Matrix representations of a monoid

In this section we intend to introduce representations of a monoid by matrices over a semiring. Let n be a natural number. The set of all n × n matrices $A = (a_{ij})$ with $a_{ij} \in R$, $i,j = 1,\dots,n$, is denoted by $(R)_n$. Obviously, $(R)_n$ forms a monoid with respect to matrix multiplication.

Definition. Let M be a monoid and let n be a natural number. A homomorphism δ from M into $(R)_n$ is called a finite matrix representation of M.

We are now going to consider infinite matrices. Let N be a set. $(R)_N$ denotes the set of all mappings from N × N into R, which will be written as (possibly infinite) matrices $A = (a_{ij})$ with $a_{ij} = A(i,j)$

for $i,j \in N$. In order to generalize the matrix multiplication in that case we state the following requirement: in each row of A there are all but finitely many coefficients equal to 0. Under this assumption $(R)_N$ forms a monoid with respect to (generalized) matrix multiplication.

Definition. Let M be a monoid with length-function and let N be a finitely generated monoid. A homomorphism δ from M into $(R)_N$ is called a locally finite matrix representation of M if the associated mapping [ ]: $N \to R\langle\langle M\rangle\rangle$ defined by

$$[n] = \sum_{m \in M} \delta(m)_{n,e} m \qquad \text{(e is the unit element of N)}$$

is a homomorphism from N into the multiplicative monoid of $R\langle\langle M\rangle\rangle$.

Since matrix representations shall be used as acceptors the question arises: Under which conditions is a locally finite matrix representation determined by finitely many datas. Assume that M and N are free monoids generated by finite sets X and P, resp. For each pair (x,p) of $X \times P$ let a finite set of elements $d(x,p,q)$ of R be chosen, where $q \in P^*$. Define a mapping $\delta: X \to (R)_{P^*}$ by

$$\delta(x)_{q_1,q_2} = \begin{cases} d(x,p,q) & \text{if there are } p \in P \text{ and } q,q' \in P^* \\ & \text{such that } q_1 = pq' \text{ and } q_2 = qq' \\ 0 & \text{otherwise.} \end{cases}$$

The unique extension $\delta^*: X^* \to (R)_{P^*}$ of δ is a locally finite matrix representation. It is easily seen that the associated mapping [ ]: $P^* \to R\langle\langle X^*\rangle\rangle$ is a homomorphism. On this basis generalized acceptors for algebraic power series are introduced in [3].

3. Recognizable power series

Let M be a monoid and let R be a semiring. Our first aim is the definition of two kinds of recognizable power series using finite resp. locally finite matrix representations. It will be shown that recognizable power series can be characterized by means of invariant R-subsemimodules resp. R-subsemialgebras of $R\langle\langle M\rangle\rangle$.

Definition. A power series s of $R\langle\langle M\rangle\rangle$ is said to be f-recognizable (or shortly recognizable) if there exist a finite matrix representation $\delta: M \to (R)_n$, $n \geq 1$, a row vector $\alpha = (\alpha_1,\ldots,\alpha_n)$ and a column vector $\beta = (\beta_1,\ldots,\beta_n)^T$ such that

$$(s,m) = \alpha \cdot \delta(m) \cdot \beta \qquad \text{for all } m \in M.$$

If $s \in R\langle\langle M\rangle\rangle$ is recognizable, then the n power series s_i defined by $(s_i,m) = (\delta(m)\cdot\beta)_i$ for $i = 1,\dots,n$ generate an R-subsemimodule

$$A = \sum_{i=1}^{n} Rs_i$$

which contains s because of $s = \alpha_1 s_1 + \dots + \alpha_n s_n$. An easy calculation shows that A is invariant. Therefore, we get

Theorem 1. Each recognizable power series s of $R\langle\langle M\rangle\rangle$ belongs to an invariant finitely generated R-subsemimodule of $R\langle\langle M\rangle\rangle$. □

In the case of free monoids the derived condition is necessary and sufficient.

Theorem 2 [2]. Let M be a free monoid generated by a finite set. A power series s of $R\langle\langle M\rangle\rangle$ is recognizable if and only if s belongs to an invariant finitely generated R-subsemimodule of $R\langle\langle M\rangle\rangle$. □

Next, we are going to introduce lf-recognizable power series, which will be characterized in a similar way.

Definition. Let M be a monoid with length-function. A power series s of $R\langle\langle M\rangle\rangle$ is said to be lf-recognizable if there exist a locally finite matrix representation $\delta: M \to (R)_N$, a row vector $\alpha = (\alpha_n)_{n\in N}$ with all but finitely many coefficients equal to 0, and a unit column vector $\beta = (\beta_n)^T_{n\in N}$ with $\beta_e = 1$ and $\beta_n = 0$ for $n \neq e$ such that

$$(s,m) = \alpha\cdot\delta(m)\cdot\beta \qquad \text{for all } m \in M.$$

Theorem 3. Let M be a monoid with length-function. Then each lf-recognizable power series s of $R\langle\langle M\rangle\rangle$ belongs to an invariant finitely generated R-subsemialgebra of $R\langle\langle M\rangle\rangle$.

Proof. Assume that s is an lf-recognizable power series of $R\langle\langle M\rangle\rangle$. Then there exist a locally finite matrix representation $\delta: M \to (R)_N$ and finitely many $\alpha_n \in R$ such that

$$(s,m) = \sum_{n \in N} \alpha_n\cdot\delta(m)_{n,e}.$$

By definition, N is finitely generated. Thus, the R-subsemialgebra A of $R\langle\langle M\rangle\rangle$ consisting of all finite sums

$$s = \sum_{n \in N} r_n[n] \qquad \text{where } r_n \in R$$

is finitely generated too. Take into consideration that [ ] is a homomorphism. Evidently, $s = \sum \alpha_n[n]$ is contained in A.

It remains to show that A is invariant. Let a be an element of A. For an arbitrary element m of M we derive

$$(\varrho_m a, m') = \sum_{n \in N} r_n(\varrho_m[n], m') = \sum_{n \in N} r_n([n], m\cdot m')$$
$$= \sum_{n' \in N} \Big(\sum_{n \in N} r_n \delta(m)_{n,n'} \Big)([n'], m'),$$

where all sums are finite. Hence

$$\varrho_m a = \sum_{n \in N} r'_n[n] \qquad \text{with} \qquad r'_n = \sum_{n' \in N} r_{n'}\delta(m)_{n',n}$$

and, consequently, $\varrho_m a \in A$. □

<u>Theorem 4</u>. Let M be a free monoid generated by a finite set. A power series s of R<<M>> is lf-recognizable if and only if s belongs to an invariant finitely generated R-subsemialgebra of R<<M>>.

Proof. By Theorem 3 it suffices to show that each element of an invariant finitely generated R-subsemialgebra A of R<<M>> is an lf-recognizable power series provided M is the free monoid X* generated by a finite set X. Assume that A is generated by $\{s_p | p \in P\}$, where P is a finite set. Since A is invariant we get

$$\varrho_x s_p = \sum_{q \in P^*} r_{pq}(x) q[p/s_p] \qquad \text{with } r_{pq}(x) \in R\ ,$$

whereby the sum is finite. $q[p/s_p]$ denotes the substitution of each p of P by the corresponding s_p in $q \in P^*$. Now we define a mapping $\delta: X \to (R)_{P^*}$ by the rule

$$\delta(x)_{q_1,q_2} = \begin{cases} r_{pq}(x) & \text{if there are } p \in P \text{ and } q,q' \in P^* \\ & \text{such that } q_1 = pq' \text{ and } q_2 = qq' \\ 0 & \text{otherwise.} \end{cases}$$

We assert that the unique extension $\delta^*: X^* \to (R)_{P^*}$ of δ is a locally finite matrix representation of X*. For that reason it must be shown that the associated mapping [ ] from P* into the multiplicative monoid of R<<X*>> defined by

$$[q] = \sum_{w \in X^*} \delta^*(w)_{q,e} w \qquad \text{for all } q \in P^*$$

is a homomorphism. Evidently,

$$[e] = \sum_{w \in X^*} \delta^*(w)_{e,e} w = 1$$

by definition. In order to show $[qq'] = [q]\cdot[q']$ for $q,q' \in P^*$ we use the statement

$$\delta^*(w)_{qq',q''} = \sum_{\substack{uv=w \\ u \neq 1}} \sum_{q_1 q_2 = q''} \delta^*(u)_{q,q_1} \cdot \delta^*(v)_{q',q_2}$$

for all non-empty words w over X and all $q, q', q'' \in P^*$, which may be proved by induction over the length of w. Let q and q' be elements of P^*. Now we conclude

$$[qq'] = \sum_{w \in X^*} \delta^*(w)_{qq',e} w = \sum_{w \in X^*} \Big(\sum_{uv=w} \delta^*(u)_{q,e} \cdot \delta^*(v)_{q',e} \Big) w$$

$$= [q] \cdot [q'].$$

Without loss of generality we suppose that $(s_p, 1) = 0$ for all p of P. We are now going to prove that each element a of A is lf-recognizable by that δ^*. First we state

$$s_p = [p] \qquad \text{for } p \in P.$$

Obviously, it holds

$$(s_p, x) = (\varrho_x s_p, 1) = \sum_{q \in P^*} r_{pq}(x)(q[p/s_p], 1) = r_{pe}(x)$$

$$= ([p], x)$$

for all $x \in X$. Let $w \in X^*$. Then

$$(s_p, xw) = (\varrho_x s_p, w) = \sum_{q \in P^*} r_{pq}(x)(q[p/s_p], w).$$

Together with

$$(q[p/s_p], w) = \delta^*(w)_{q,e} \tag{$*$}$$

we get

$$(s_p, xw) = \sum_{q \in P^*} \delta^*(x)_{p,q} \cdot \delta^*(w)_{q,e} = \delta^*(xw)_{p,e}$$

$$= ([p], xw),$$

that means $s_p = [p]$. Each a of A can be represented as follows

$$a = \sum_{q \in P^*} r_q q[p/s_p] \ ,$$

whereby the sum is finite. Because of ($*$) we obtain

$$(s, w) = \sum_{q \in P^*} r_q (q[p/s_p], w) = \sum_{q \in P^*} r_q \delta^*(w)_{q,e}$$

which proves that a is lf-recognizable. □

4. Characterization of rational and algebraic power series

Throughout this section M is always assumed to be a free monoid generated by a finite set X. As a conclusion of the well-known Schützenberger Theorem (cf. [2]) we obtain

Theorem 5 [2]. A power series s of R<<X*>> is rational if and only if s belongs to an invariant finitely generated R-subsemimodule of R<<X*>>. □

Algebraic power series are defined as solutions of systems of equations. Let $Z = \{z_1,\ldots,z_n\}$ be a finite set of variables disjoint from X. An algebraic system S is a set of equations

$$z_i = \varphi_i \qquad i = 1,\ldots,n,$$

where $\varphi_i \in R\langle V^*\rangle$ with $V = X \cup Z$. S is called proper whenever $(\varphi_i,1)=0$ and $(\varphi_i,z_j) = 0$ for each i and j. It is known (cf. [2]) that each proper algebraic system S has a unique solution $|S| = (\sigma_1,\ldots,\sigma_n)$ with $\sigma_i \in R\langle\langle X^*\rangle\rangle$ for $i = 1,\ldots,n$.

Theorem 6. A power series s of R<<X*>> is algebraic if and only if s belongs to an invariant finitely generated R-subsemialgebra of R<<X*>>.

Proof. (1) Let s be an algebraic power series of R<<X*>> determined by a proper algebraic system S. Without loss of generality we may assume that S has the following form

$$z_i = \sum_{x \in X}(r_i(x)x + \sum_{j=1}^{n} r_{ij}(x)xz_j + \sum_{j,k=1}^{n} r_{ijk}(x)xz_jz_k)$$

for $i = 1,\ldots,n$, where $r_i(x)$, $r_{ij}(x)$ and $r_{ijk}(x)$ are elements of R (cf. Theorem 2.3 [2], p. 128). Let $|S| = (\sigma_1,\ldots,\sigma_n)$ be the solution of S. Define A to be the R-subsemialgebra of R<<X*>> generated by $\{\sigma_1,\ldots,\sigma_n\}$. Evidently, s belongs to A because of $s = (s,1) + \sigma_i$ for some σ_i of the generator set.

It remains to prove that A is invariant. Since each σ_i obeys the equation

$$\sigma_i = \sum_{x \in X}(r_i(x)x + \sum_{j=1}^{n} r_{ij}(x)x\sigma_j + \sum_{j,k=1}^{n} r_{ijk}(x)x\sigma_j\sigma_k) \quad ,$$

we derive

$$\varrho_x\sigma_i = r_i(x) + \sum_{j=1}^{n} r_{ij}(x)\sigma_j + \sum_{j,k=1}^{n} r_{ijk}(x)\sigma_j\sigma_k \quad ,$$

which implies

$$\varrho_x \sigma_i \in A \qquad \text{for } x \in X \text{ and } i = 1,\ldots,n.$$

Notice that an arbitrary element a of A is a finite sum $a = \sum r_\pi \pi$, where $r_\pi \in R$ and $\pi \in \{\sigma_1,\ldots,\sigma_n\}^*$. If we prove that $\varrho_x \pi$ belongs to A for each non-empty word π, then $\varrho_x a = \sum r_\pi \varrho_x \pi$ belongs to A too, which implies that A is invariant. Now, assume that $\pi = \sigma_i \pi'$, where π' is an arbitrary word over $\{\sigma_1,\ldots,\sigma_n\}$. Observe that $(\sigma_i,1) = 0$ for $i = 1,\ldots,n$. Because of

$$\varrho_x(\sigma_i \pi') = (\varrho_x \sigma_i)\pi' + (\sigma_i,1)\varrho_x \pi'$$

we thus derive

$$\varrho_x(\sigma_i \pi') = (\varrho_x \sigma_i)\pi'.$$

Since $\varrho_x \sigma_i \in A$ and $\pi' \in A$, the required condition $\varrho_x \pi \in A$ follows.

(2) Conversely, let A be an invariant R-subsemialgebra of $R\langle\langle X^*\rangle\rangle$ generated by the set $\{s_1,\ldots,s_n\}$. It has to be shown that each element a of A is an algebraic power series. For that reason a proper algebraic system S must be constructed. Without loss of generality we may assume $(s_i,1) = 0$ for $i = 1,\ldots,n$. Since A is invariant we derive

$$\varrho_x s_i = \sum_{v \in Z^*} r_{i,v}(x) v[\underline{z}/\underline{\sigma}],$$

where $Z = \{z_1,\ldots,z_n\}$ and $v[\underline{z}/\underline{\sigma}]$ denotes the substitution of each z_i by σ_i in v. Define S as follows

$$z_i = \sum_{x \in X} \sum_{v \in Z^*} r_{i,v}(x) x v \qquad i = 1,\ldots,n.$$

Take into consideration that $(s_i,1) = 0$. Then we conclude

$$s_i = \sum_{x \in X} x \varrho_x s_i = \sum_{x \in X} x\Big(\sum_{v \in Z^*} r_{i,v}(x) v[\underline{z}/\underline{\sigma}]\Big)$$

$$= \sum_{x \in X} \sum_{v \in Z^*} r_{i,v}(x)(xv)[\underline{z}/\underline{\sigma}] = z_i[\underline{z}/\underline{\sigma}]$$

$$= \sigma_i.$$

Therefore, each s_i is an algebraic power series. Since the set of all algebraic power series is closed under scalar product, sum and product, each a of A is algebraic. □

References

1. Nivat, M., Séries formelles algébriques. In "Séries Formelles" (Ed. Berstel, J.). Laboratoire d'Informatique Théorique et Programmation, Paris 1978.

2. Salomaa, A. and Soittola, M., Automata-Theoretic Aspects of Formal Power Series. Springer-Verlag, New York-Heidelberg-Berlin 1978.

3. Wechler, W., The Concept of Fuzziness in Automata and Language Theory. Akademie-Verlag, Berlin 1978.

A CROSSING MEASURE FOR 2-TAPE TURING MACHINES

Gerd Wechsung
Sektion Mathematik der Friedrich-Schiller-Universität Jena
Jena, DDR

We define a complexity measure for 2-tape Turing machines (2TM) that generalizes the usual crossing measure for 1-tape Turing machines (1TM). We prove that the constant resource bound functions yield an infinite hierarchy of complexity classes of the new measure. This shows a completely different behaviour of the new measure compared to the crossing measure for 1TM. In a similar way as in the 1-tape case the new measure allows to prove lower time bounds on 2TM computations.

1. Crossing sequences for 2TM

One could try to define a crossing measure C for 2TM as the maximal number of crossings of the boundaries between adjacent squares performed by one of the two heads. This seems to be the most natural generalization of the crossing measure introduced by B. A. Trachtenbrot [1] and F. C. Hennie [2] for 1TM. We start with

Fact 1: C is not a measure in the sense of Blum [3].

To prove this statement we show that every computable function can be computed with C-complexity not greater than 3. Let M_1 be a 1TM. We contruct a 2TM M_2 that works as follows (having as initial tape content on tape 1 the input of M_1 whereas tape 2 is initially empty). Step 1: Using the input on tape 1 M_2 generates on tape 2 the second configuration K_2 of M_1 and returns the head of tape 2 to the first square of K_2. Step 2: Using K_2 M_1 generates K_3 on tape 1 on the right hand side of the input word and returns the head of tape 1 to the first square of K_3. And so on.

It is obvious that no boundary is crossed more than three times. This shows that C fails to be a Blum measure. □

However, also for 2TM a crossing measure can be introduced in a reasonable way. We consider a tessellation of the Euclidian plane in congruent squares provided with a coordinate system such that every square has a pair of integers as coordinates. If we take the positions of the heads on tape 1 and tape 2 of a 2TM M as square coordinates with respect to our tessellation, every computation of M gives rise to a sequence of pairwise neighbouring squares. The curve connecting the centres of consecutive squares of this sequence will be taken as representation of the considered computation. An illustration is given by figures 1 and 2.

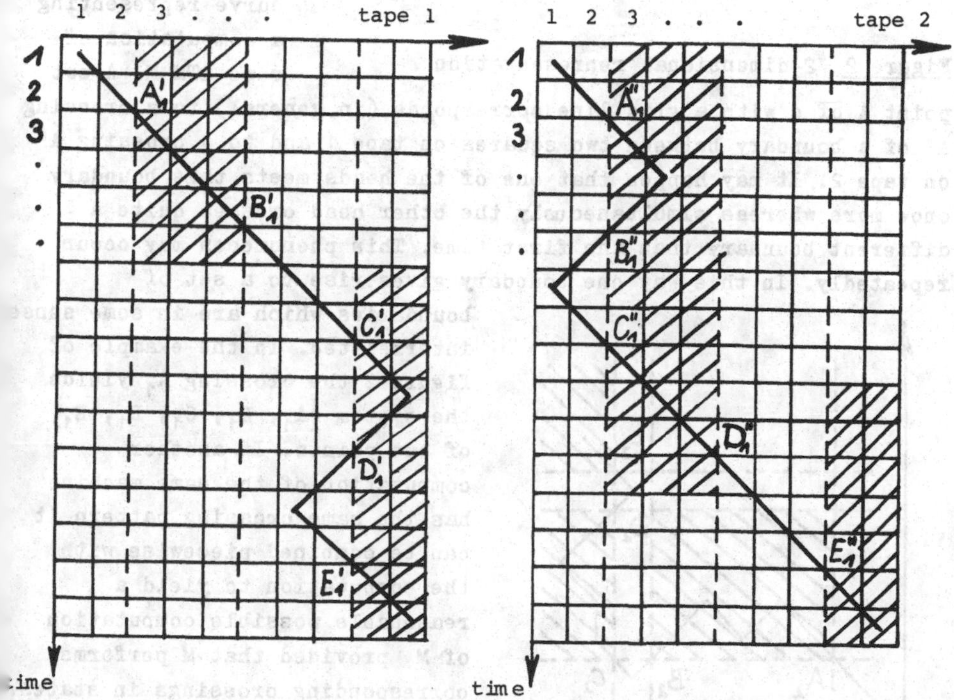

Figure 1 Records of the work of the heads of a 2TM computation

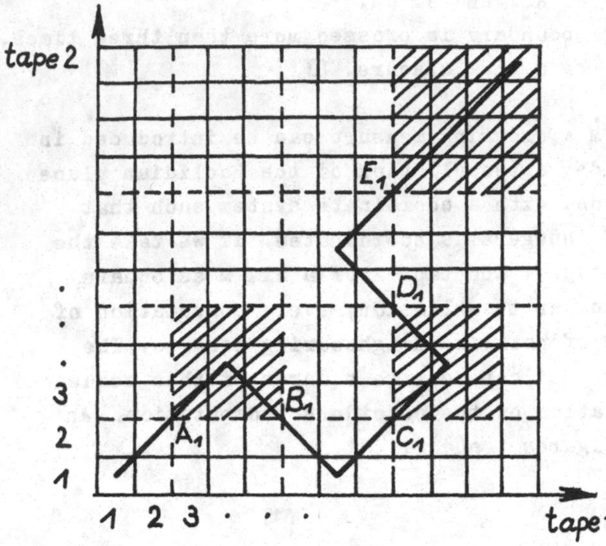

Figure 2 2-dimensional representation

Next we introduce the notion of crossing sequence. This will be done in such a way that we can prove a lemma similar to that of Trachtenbrot [1] and Hennie [2] about the possibility of mixing two different computations having a common crossing sequence.

Let c be a curve representing a computation of some 2TM M. A cut point A of c with a grid line corresponds (in general) to a crossing A' of a boundary between two squares on tape 1 and to a crossing A" on tape 2. It may happen that one of the heads meets this boundary once more whereas simultaneously the other head crosses quite a different boundary than the first time. This phenomenon may occur repeatedly. In this way one boundary gives rise to a set of boundaries which are in some sense interrelated. In the example of figure 2 the crossing A_1 yields the system $\{A_1, B_1, C_1, D_1, E_1\}$ of cut points. If another computation of the same machine has the same crossing pattern it can be combined piecewise with the computation to yield a reasonable possible computation of M provided that M performs corresponding crossings in states.

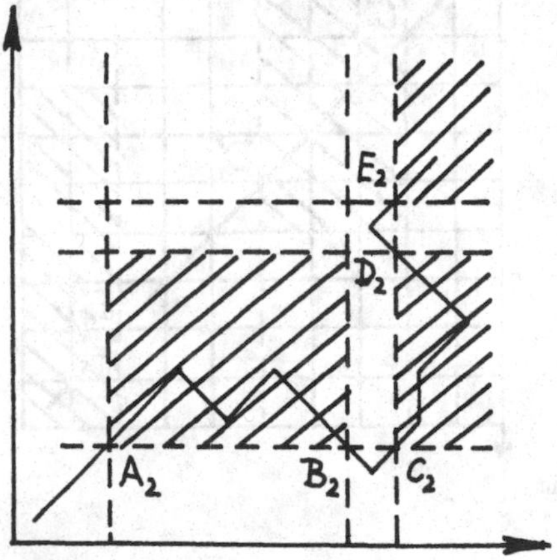

Figure 3

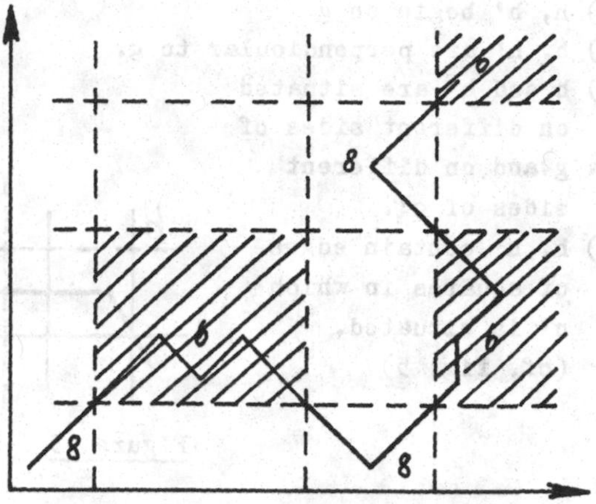

Figure 4

Example: Figure 3 shows a crossing system $\{A_2, B_2, C_2, D_2, E_2\}$ of the same structure as that of figure 2. Assume that the crossing of X_1 is performed in the same state as the crossing of X_2 (X = A,B,C,D,E). Then, replacing the hatched parts of the first computation (fig. 2) with the corresponding parts of the second one (fig. 3) the computation of figure 4 can be constructed.

This leads to the idea to take such sequences of cut points together with the corresponding states as crossing sequences for 2TM.

We need some preliminary notions for the definition of crossing sequences. Let c be a curve representing a computation of a 2TM. By f,g,... we denote gridlines or one sided infinite parts of such gridlines (rays).

Definition 1. A system B of gridlines or rays is called closed with respect to c $\overset{}{\underset{df}{\Longleftrightarrow}}$

(1) Whenever f and g have their cut point on c and $f \in S$, then $g \in S$.

(2) Let c' be a maximal horizontal or vertical part of c. Let $S' = g_1, \ldots, g_m$ be the subset of S of those lines having a cut point with c'. Then either $m \equiv o(2)$, or there is a $g_i \in S'$ such that S contains two rays h,h' with the following properties:

(a) h, h' begin on g.
(b) h, h' are perpendicular to g.
(c) h and h' are situated on different sides of g and on different sides of c'.
(d) h, h' contain edges of squares in which c' is situated. (cf. fig. 5)

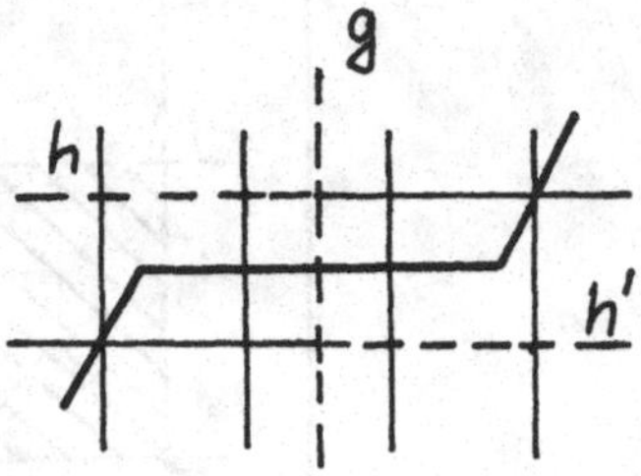

Figure 5

Examples of closed systems are given in figures 2, 3 and 4. A further example (fig. 6) illustrates the case that vertical and horizontal parts occur in c.

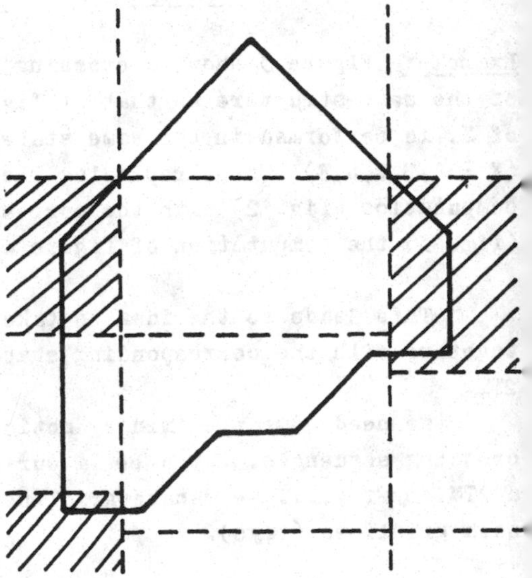

Figure 6

Definition 2. Let c represent a 2TM computation and let S be a system closed with respect to c and minimal with this property. S is called a crossing system of c. Let further $S = \{h_1, \ldots, h_k, p_1, \ldots, p\}$ where h_i (p_j) denote the horizontal (vertical) gridlines. Then

$$\gamma_{S,c} = (Z_1, t_1) \ldots (Z_s, t_s)$$

is called the crossing sequence of c on S $\Leftrightarrow_{df}$ c crosses S exactly s times, and for all $\sigma = 1, \ldots, s$ it holds: If c crosses S for the σth time (in chronological order) then M has in this moment the state Z_σ, and the crossing takes place

(a) across the cut point of h_{i_σ} and p_{j_σ} if $t_\sigma = (i_\sigma, j_\sigma)$,

(b) across h_{i_σ} between p_{j_σ} and $p_{j'_\sigma}$ if $t_\sigma = (i_\sigma; j_\sigma, j'_\sigma)$,

(e) across p_{j_σ} between h_{i_σ} and $h_{i'_\sigma}$ if $t_\sigma = (i_\sigma, i'_\sigma; j_\sigma)$.

s ist called the length of $\gamma_{S,c}$. We write for short $s = |\gamma_{S,c}|$. $(t_1, \ldots, t_s)$ is called the crossing pattern of S.

A difference to the 1-tape case consists in the fact that according to (2) of definition 1 a crossing of c may belong to many different crossing sequences.

2. The crossing measure for 2TM and the combining lemma

In the following we confine ourselves to 2TM M with the property

(1) If $zxy \rightarrow z'x'y'o\ m$ is an instruction of M then $x' = x$ and for every $\bar{x}$

$$z\bar{x}y \rightarrow z'\bar{x}y'\ o\ m \text{ is also an instruction of M.}$$

(2) If $zxy \rightarrow z'x'y'\ m\ o$ is an instruction of M then $y' = y$ and for every $\bar{y}$

$$zx\bar{y} \rightarrow z'x'\bar{y}\ m\ o \text{ is also an instruction of M.}$$

(This means: A nonmoving head does not write, and the behaviour of the machine does not depend on what that head reads.)

Without this additional property which does not restrict the generality and which has no influence on time and tape complexity lemma 3 is not true. We could avoid the restriction of the machine model on the cost of a more complicated definition of crossing sequences, but we find our way preferable.

The measure 2T-CROS for 2TM is defined as follows

<u>Definition 3.</u> Let M_i be the ith 2TM of a fixed Gödel numbering of all 2TM. Let c_w be the 2-dimensional representation of the computation of M_i on input w.

$$\text{2T-CROS}_i(w) =_{df} \begin{cases} \max\{|\gamma_{S,c_w}| : S \text{ is a crossing system of } c_w\} & \text{if } M_i \text{ halts on input } w; \\ \text{undefined otherwise.} \end{cases}$$

<u>Proposition 2:</u> 2T-CROS is a measure in the sense of Blum.

The possible occurrence of horizontal or vertical parts in a computation curve makes the desired combination lemma a little bit more complicated than suggested by the example in section 1 (fig. 2, 3, and 4). The notion of colourability introduced next will allow a precise formulation of our combining lemma.

Let S be a crossing system of a curve c. By S together with the four border lines (that correspond to the borders of the workspace of the two tapes) the minimal rectangle containing c is decomposed into rectangles. These latter ones can be considered as being arranged in rows and columns like the elements of a matrix. They induce a decomposition of c. Let $c_1, \ldots, c_n$ be the order in which the parts of c occur when c is run through. These terminology is used in the next definition.

<u>Definition 4.</u> c is colourable with respect to S $\Longleftrightarrow_{df}$ there exists a mapping
$F : \{c_1, \ldots, c_n\} \longrightarrow \{1,2\}$
($F(c_i)$ is called the colour of c_i)
such that the following conditions are fulfilled

(a) $\bigwedge_{i=1}^{n-1} F(c_i) \neq F(c_{i+1})$.

(b) All c_i belonging to the same row (with possible exception of those which are completely horizontal) have the same colour.

(c) All c_j belonging to the same column (with possible exception of those which are completely vertical) have the same colour.

Figures 2, 3, and 6 show colourable computations. Colouring is indicated by hatching.
This is a general fact. We can casily prove:

Lemma 3: If S is a crossing system of a computation c then c is colourable with respect to S. □

Now we are ready to state our combining lemma.

Lemma 4: Let c and c' be two (not necessarily distinct) computations of a 2TM M, and let S and S' be crossing systems for c resp. c'.

If $\gamma_{S,c} = \gamma_{S',c'}$ and $(c_1,\dots, c_n)$ and $(c_1',\dots, c_n')$ are the decompositions of c resp. c' induced by S resp. S', then the curves $c_1\, c_2'\, c_3\, c_4' \dots$ and $c_1'\, c_2\, c_3'\, c_4 \dots$ represent possible computations of M.

Proof (sketched). The conditions (b) and (c) of definition 4 ensure that whenever a skip is made from one computation to the other one such that at least one of the heads enters a square that it had left previously then on both tapes the heads find exactly those circumstances which they have left by the last skip. The coincidence of the sequence of crossing states guarantees the correct continuation of the computation. As far as horizontal or vertical parts of c and c' are concerned the correctness of the mixing follows from the special properties of the machines discussed at the beginning of this section. □

3. An infinite hierarchy of 2T-CROS classes

Complexity classes with respect to 2T-CROS are defined in the usual way. Let c_A denote the characteristic function of the set A and let φ_i be the function computed by the ith 2TM M_i. For recursive t we define

$$2T\text{-}CROS(t) =_{df} \left\{ A : \bigvee_i \left(\varphi_i = c_A \wedge 2T\text{-}CROS_i \leq_{ae} t \right) \right\}.$$

If k is an integer then the symbol k is also used to denote the function with the constant value k.

Theorem 5: For every natural number $k \geq 1$ it holds

$$2T\text{-}CROS\ (k) \subset 2T\text{-}CROS\ (k+1).$$

Proof (sketched) The set

$$A_{k+1} =_{df} \{ \underbrace{a^n\ b^n\ a^n\ b^n\ \ldots}_{n(k+1)\ \text{symbols}} \quad : \ n \in \mathbb{N} \}$$

belongs to 2T-CROS (k+1) as is easily verified. If it were accepted by a 2TM M within 2T-crossing complexity k then using lemma 4 we could find words accepted by M but not belonging to A_{k+1} □

References

1. Trachtenbrot, B.A., Turing machine computations with logarithmic delay. Algebra i Logika, 3 (1964), 33-48.
2. Hennie, F.C., One-tape off-line Turing machine computations. Information and Control, 8 (1965), 553-573.
3. Blum, M., A machine-independent theory of the complexity of recursive functions. Journal of the Association for Computing Machinery, 14 (1976), 322-336.

THE COMPLEXITY OF LEXICOGRAPHIC SORTING AND SEARCHING

Juraj Wiedermann

Computing Research Centre

Dúbravská 3, 885 31 Bratislava, Czechoslovakia

1. INTRODUCTION

Let $U_1, U_2, \ldots, U_k$ be totally ordered sets and let V be a set of n k-tuples in the Cartesian product $U_1 x U_2 x \ldots x U_k$. For any k-tuple v in V, let $c_i(v)$ denote the i-th component of v.

A lexicographic ordering $\prec$ is defined on V in usual way, that is, for $v, u \in V$, $v \prec u$ if and only if either $c_1(v) < c_1(u)$ or there exists $1 \leq j < k$ such that $c_i(v) = c_i(u)$ for $i = 1, 2, \ldots, j$ and $c_{j+1}(v) < c_{j+1}(u)$, where $<$ is the total ordering on each U_j.

We shall consider the problem of lexicographic sorting k-tuples of V, as well as that of searching for a k-tuple in V.

The computational complexity of both problems will be measured by the number of three-branch component comparisons needed for solving these problems (i.e. two components $c_i(v)$ and $c_i(u)$ will be compared yielding $c_i(v) > c_i(u)$, $c_i(v) = c_i(u)$, or $c_i(v) < c_i(u)$ as an answer). We shall be interested in obtaining the (worst case) upper and lower bounds on the complexity, as a function of both n and k.

Note that the problem of lexicographic sorting can be straightforwardly solved by applying any "onedimensional" sorting algorithm directly to the k-tuples of V which are in this case viewed as "unstructured" elements with respect to the lexicographic ordering $\prec$. However, this approach would require about $\Theta(n \log n)$ "lexicographic" comparisons, which can need as much as $\Theta(kn \log n)$ component comparisons, because in the worst case the lexicographic order of two k-tuples cannot be detected until all k component comparisons have been performed.

In a similar manner the lexicographic search can be done in $\Theta(k \log n)$ steps.

In contrast to these "trivial" upper bounds we shall show that making use of the particular structure of lexicographic ordering the lexicographic sorting and searching can be accomplished using $\Theta(n(\log n + k))$ and $\lceil \log (n+1) \rceil + k+1$ component comparisons, respectively, and that these bounds are asymptotically optimal in the case of sorting and optimal in the case of searching.

In the conclusion of the paper we shall point out some applications of the previous results, mainly in the databases and also in the theory of sorting.

2. LEXICOGRAPHIC SORTING

When it comes to finding a good algorithm for lexicographic sorting the set V of n k-tuples, divide-and-conquer strategy works well:

```
recursive procedure LEXICOSORT (S,i);
comment  S is the set of k-tuples to be lexicographically sorted;
         it is supposed that first i-1 components of all k-tuples of S
         are equal, with  1 ≤ i ≤ k;

begin
    if |S| = 1
       then return(S)
       else let C be the multiset of all i-th components of k-tuples
            of S;
         1. find the median m of the set C;
            let  S1 = { v∈S | ci (v)<m },
                 S2 = { v∈S | ci (v)= m },
                 S3 = { v∈S | ci (v)> m };
         2. if |S1| ≠ 0  then  return(LEXICOSORT ( S1,i)) fi;
         3. if  i = k  then  return(S2)
                       else  return(LEXICOSORT(S2,i+1))
            fi;
         4. if |S3|≠ 0 then  return(LEXICOSORT ( S3,i))  fi
    fi
end
```

The procedure is activated by calling LEXICOSORT(V,1).

The time complexity $T(n,k)$ of LEXICOSORT, applied to the set of n k-tuples, is expressed by a recurrence relation

$$T(n,k) \leqslant cn + T(n_1,k) + T(n_2,k-1) + T(n_3,k),$$
$$T(1,k) = T(n,0) = 0 \qquad \text{for } k \geqslant 1, \quad n \geqslant 1,$$

where the four terms in the right side of the relation correspond to the complexity of steps 1 through 4, respectively, in the algorithm LEXICOSORT, with $n_1 = |S_1|$, $n_2 = |S_2|$, and $n_3 = |S_3|$.

Taking into account that $n_1+n_2+n_3=n$, $n_1 \leqslant n/2$, $n_3 \leqslant n/2$, it is not difficult to verify the solution of the recurrence in the form

$$T(n,k) \leqslant O(n(\log n + k)) .$$

The example of n k-tuples which differ solely in the last component shows that about $\Omega(n(\log n + k))$ comparisons are indeed necessary for lexicographic sorting: we surely need at least $(n-1) \cdot (k-1)$ comparisons to detect the equality of the first k-1 components of all n k-tuples, and it takes $\Omega(n \log n)$ more comparisons to complete the sort with respect to the last components.

Thus we have established the following theorem:

THEOREM 1: The lexicographic sort of n k-tuples can be performed in $\Theta(n(\log n + k))$ three-branch comparisons.

We see that if $k = \Omega(\log n)$, the complexity of the lexicographic sort is linear in the size of the input, i.e. in the number of the components of all k-tuples.

3. LEXICOGRAPHIC SEARCHING

Any lexicographic search algorithm, based on three-branch comparisons, can be viewed as a ternary decision tree. In this tree the components of k-tuples are stored in its vertices.

The search for an unknown k-tuple v starts in the root by comparing the value of $c_1(v)$ with the value $c_1(r)$ stored in the root. If $c_1(v) < c_1(r)$ $(c_1(v) > c_1(r))$, the search proceeds in a similar way in the left (right) subtree by comparing $c_1(v)$ with the root of this subtree; if $c_1(v) = c_1(r)$, then the value of the first component of v has been found and the search for the next component $c_2(v)$ proceeds now in the middle subtree in an analogous manner.

The search is successful if all k components of v are found in the tree; otherwise the search ends unsuccessfully.

Consider now the problem of constructing the appropriate lexicographic search tree for given set V of n k-tuples.

There is an alternative way to see the algorithm LEXICOSORT as an algorithm which constructs recursively a lexicographic search tree for the set S of k-tuples with first i-1 components equal, for $1 \leq i \leq k$.

In the root of this tree the value m found in the step 1 of the algorithm is placed, and the left, middle, and right subtree is the lexicographic search tree for the set S_1, S_2, and S_3 of k-tuples, respectively, in which the first i-1, i, and i-1 components, respectively, are equal (this corresponds to steps 2, 3, and 4, respectively).

We shall call the decision tree, constructed by algorithm LEXICOSORT, an optimal lexicographic search tree (the reason why will become clear later).

When performing the comparisons as dictated by this tree, each unsuccessful comparison $c_i(v) : c_i(r)$ (with an answer '<' or '>') halves the space of the remaining possibilities for v (it always holds $|S_1| \leq \lfloor |S| /2 \rfloor$, $|S_3| \leq \lfloor |S| /2 \rfloor$). On the other hand, the successful comparison need not decrease the cardinality of the space of remaining possibilities for v (if $|S_2| = |S|$), but in any case it makes the step toward the termination of the algorithm by determining the value of one component of v and thus decreasing the "dimensionality" of the remaining search space.

Hence, if $T(n,k)$ denotes the number of three-branch comparisons sufficient to find the k-tuple v in the set of n k-tuples, it obviously holds

$$T(n,k) \leq 1 + \max\{T(\lfloor n/2 \rfloor, k), T(n,k-1)\},$$
$$T(n,0) = 0 \qquad \text{for } n \geq 1,$$
$$T(1,k) = k \qquad \text{for } k \geq 1.$$

The solution of this recurrence is given by

$$T(n,k) \leq \lceil \log(n+1) \rceil + k - 1 ,$$

but again the example of n k-tuples with all but the last components equal shows that in the last expression in fact the equality holds.

This implies the next theorem:

THEOREM 2: The lexicographic search in the set of n k-tuples can be performed in the optimal lexicographic search tree in

$$\lceil \log(n+1) \rceil + k - 1$$

three-branch comparisons, and this number is optimal.

The somewhat weaker form $T(n,k) \leq \log n + 2k$ of the last result was originally obtained by Fredman [1] and van Leeuwen [3]. Our construction of the optimal lexicographic search tree differs from their constructions by more careful selection of the component around which the set of k-tuples is partitioned.

An alternative construction of the optimal lexicographic search tree is given also in [4].

4. APPLICATIONS

It is quite natural to view the set V of n k-tuples as a file F of n records, each record consisting of k attributes (keys). Then the lexicographic search in the set V corresponds to the search for the answer to the exact-match query in the file F [2].

Another interpretation of the set V is to consider it as a k-ary relation on $U_1 x U_2 x \ldots x U_k$. Given two k-ary relations V and W on $U_1 x U_2 x \ldots x U_k$, with $|W| = m$ and $|V| = n$, $m \leq n$, we can compute their intersection $V \cap W$ as follows: first, we lexicographically sort the set W in $O(m(\log m + k))$ comparisons, and then in the resulting optimal lexicographic search tree we perform n succesive searches for elements of V; this consumes another $O(n(\log m + k))$ comparisons.

Thus the intersection of two k-ary relations V and W can be found in $O((m+n)(\log m + k))$ three-branch comparisons.

Theorem 2 can also be used to improve on Fredman´s result about the complexity of sorting $x_1, x_2, \ldots, x_n$, provided we know the subset

G of all the n! orderings on $x_1,x_2,\ldots,x_n$ to which the resulting ordering belongs [1]. Then, following Fredman (and using theorem 2) it can be shown that $\lceil \log (|G|+1)\rceil + n - 1$ comparisons suffice to determine the resulting ordering of $x_1,x_2,\ldots,x_n$ (the original result was $\log|G| + 2n$).

As a further application of this result we can show that if X and Y are n element sets of real numbers, then $n^2 + O(n \log n)$ comparisons suffice to sort the n^2 element set $X+Y$, thereby saving a factor of two over the original Fredman's bound $2n^2 + O(n \log n)$ [1]. Moreover, in the same paper Fredman has shown that $(n-1)^2$ comparisons are in fact necessary, so the complexity of sorting $X+Y$ is known to within the lower order terms.

REFERENCES:

1. Fredman, M.L.: How good is the information theory bound in sorting?, Theoretical Computer Science 1, 1976, pp. 355-361.

2. Rivest, R.L.: Partial-match retrieval algorithms, SIAM J. Computing 5, 1976, pp. 115-174.

3. van Leeuwen, J.: The complexity of data organisation. In: Foundations of computer science II, Part 1. Mathematical centre tracts 81, Mathematisch centrum, Amsterdam 1976.

4. Wiedermann, J.: Search trees for associative retrieval (in Slovak), Informačné systémy 1, 1979.

AN ALGEBRAIC APPROACH TO CONCURRENCE

Józef Winkowski
Instytut Podstaw Informatyki PAN
00-901 Warszawa, PKiN, Skr.p.22, Poland

Introduction

One of the ways to describe a system is to specify its actions and how they can be composed. This leads to some algebraic structures.

If only states of the whole system and their changes are considered then there is essentially one way to compose actions that corresponds to executing them one after another. Thus the suitable algebraic structure is a category, a monoid, or something similar.

If also concurrence (independence) of the states of some parts of the system and of actions has to be reflected, then some more information is necessary. The idea to apply monoidal categories (X-categories) for that purpose has been suggested by Hotz [1]. Another idea is due to Mazurkiewicz [2] who suggested to consider an independence relation. Our approach (Winkowski [5]) is to apply categories with an additional operation reflecting concurrence (partially monoidal categories).

We exploit the fact that some special partially monoidal categories (algebras of partial sequences) can be constructed which are closely related to Petri nets and play the role of free objects. The algebras of partial sequences will be introduced generalizing the concept of string to that of partial sequence, and replacing the string concatenation by two binary partial operations on partial sequences, called the sequential composition and the parallel one. Then we shall show that mappings defined on the generators of an algebra of partial sequences can, under certain conditions, uniquely be extended to homomorphisms of partially monoidal categories. The obtained result will be applied to define interpretations of concurrent schemata corresponding to safe condition-event Petri nets of a class. This will be done assigning an algebra of partial sequences to a scheme and defining the interpretations as homomorphisms defined on such an algebra.

A part of the material has been the subject of the paper [5] which contains however some errors and thus can be consulted for some proofs only and with a certain criticism.

1. Partially monoidal categories

1.1. Notions

A <u>partially monoidal category</u> (p.m. category) is $A=(U,\underline{dom},\underline{cod},\cdot,+,0)$ such that:

(A1) $\underline{cat}(A):=(U,\underline{dom},\underline{cod},\cdot)$ is a (morphisms-only) category ($x.y$ denotes the composition of x and y, $\underline{dom}(x)$ and $\underline{cod}(x)$ denote the domain and codomain identities of x),

(A2) + is an associative and commutative partial binary operation in U ($x+(y+z)=(x+y)+z$ and $x+y=y+x$ whenever either side is defined),

(A3) 0 is a constant which is a neutral element of + ($0+x=x+0=x$ for every $x\in U$),

(A4) if $x+y$ is defined then $\underline{dom}(x)+y$ and $\underline{cod}(x)+y$ are defined,

(A5) if $x+y$ is defined then $\underline{dom}(x+y)=\underline{dom}(x)+\underline{dom}(y)$ and $\underline{cod}(x+y)=\underline{cod}(x)+\underline{cod}(y)$,

(A6) if $x\cdot y$, $z\cdot t$, $x+z$, $x+t$, $y+z$, $y+t$ are defined then $(x\cdot y)+(z\cdot t)$ and $(x+z)\cdot(y+t)$ are also defined and identical.

Given such a p.m. category A, we say that a set $V\subseteq U$ is admissible iff $x\in V$ implies $\mathrm{dom}(x)\in V$ and $\mathrm{cod}(x)\in V$. By closure(V) (resp.: +closure(V)) we denote the closure of V with respect to the operations $\cdot$,+, and 0 (resp.: the operation +). An admissible set $V\subseteq U$ such that closure(V)=U is called a set of generators of A. If closure(W)$\neq$U for every proper part W of V then V is said to be a minimal set of generators of A.

A mapping $h:V\longrightarrow U'$ from an admissible set V of A into the carrier U' of a p.m. category $A'=(U',\mathrm{dom}',\mathrm{cod}',\cdot',+',0')$ is said to be admissible iff $h(\mathrm{dom}(x))=\mathrm{dom}'(h(x))$, $h(\mathrm{cod}(x))=\mathrm{cod}'(h(x))$ for every $x\in V$ and $h(x)+'h(y)$ is defined and identical with $h(x+y)$ for every $x,y\in V$ such that $x+y$ is defined and belongs to V. A homomorphism $h:A\longrightarrow A'$ is an admissible mapping $h:U\longrightarrow U'$ which is a functor from cat(A) into cat(A') and transforms 0 onto 0'.

1.2. Examples

Let us consider a memory Mem=(Loc,Cont) consisting of a set Loc of locations and of a set Cont of contents of locations. Let Fields be the set of fields of the memory, i.e. of the subsets of Loc. To every field $f\in$ Fields the set States(f) of states $s:f\longrightarrow$ Cont corresponds (States($\emptyset$)=$\emptyset$).

Actions on the memory Mem are triples (f,g,R), where $f,g\in$ Fields and R is a binary relation between the states of f (data) and those of g (results).

By U we denote the set of actions on Mem. Next, we define:

$$\mathrm{dom}(f,g,R)=(f,f,\text{the identity in States}(f)),$$

$$\mathrm{cod}(f,g,R)=(g,g,\text{the identity in States}(g)),$$

$$(f,g,R)\cdot(h,i,S)=(f,i,R\circ S)\ \text{whenever}\ g=h,$$

where $s(R\circ S)t$ iff sRu and uSt for some $u\in$ States(g),

$$(f,g,R)+(h,i,S)=(f\cup h,g\cup i,R\|S)\ \text{whenever}\ (f\cup g)\cap(h\cup i)=\emptyset,$$

where $s(R\|S)t$ iff $(s|f)R(t|g)$ and $(s|h)S(t|i)$ ($s|f$, $t|g$, $s|h$, $t|i$ denote the corresponding restrictions of s and t),

$$0=(\emptyset,\emptyset,\emptyset).$$

Then A=(U,dom,cod,$\cdot$,+,0) is a p.m. category (the p.m. category of actions on Mem).

Usually, only a set V of elementary actions is given (including necessary actions of the form (f,f,the identity in States(f))) and we can execute them sequentially or in parallel obtaining thus a set U' of available actions. Restricting A to such a set U' we obtain a p.m. category A' of available actions with V as a set of generators. The inclusion $U'\subseteq U$ is a homomorphism from A' into A.

In what follows we shall give a method of constructing a sort of free p. m. categories. This will be done introducing the concept of partial sequence and suitable operations on partial sequences.

2. Partial sequences

The concept of partial sequence is a generalization of the string concept. We shall define partial sequences as isomorphism classes of labelled partially ordered sets satisfying certain conditions.

2.1. Notions

Given a partially ordered set $(X,\leq)$, by an antichain we mean a set of mutually incomparable elements of X. Any maximal antichain $Y\subseteq X$ determines two subsets of X:

$Y^- := \{x \in X: x \leq y \text{ for some } y \in Y\}$ and $Y^+ := \{x \in X: y \leq x \text{ for some } y \in Y\}$.

We deal with labelled partially ordered sets (l.p.o. sets) $H=(X,\leq,l)$ such that:

(H1) $(X,\leq)$ is a partially ordered set,

(H2) every chain $Z \subseteq X$ is finite,

(H3) $l:X \longrightarrow L$ is a mapping (a labelling) assigning elements of a certain fixed set L (of labels) to the elements of X,

(H4) $l(x)=l(y)$ implies $x \leq y$ or $y \leq x$,

(H5) given a maximal antichain Y and a maximal chain Z, for every $x \in Z \cap Y^-$, $y \in Z \cap Y^+$ there are $p \in Z \cap Y^-$, $q \in Z \cap Y^+$ such that $x \leq p \leq q \leq y$ and $l(p)=l(q)$.

An occurrence $f:H \longrightarrow H'$ of H in another l.p.o. set $H'=(X',\leq',l')$ is an injection $f:X \longrightarrow X'$ such that:

(O1) $x \leq y$ iff $f(x) \leq' f(y)$,

(O2) $l(x)=l'(f(x))$ for every $x \in X$,

(O3) $f(x) \leq' z' \leq' f(y)$ implies $z'=f(z)$ for some $z \in X$,

(O4) there is an antichain $Y' \subseteq X'$ such that for every maximal antichain $Y \subseteq X$ the sets $f(Y)$ and Y' are disjoint and $f(Y) \cup Y'$ is a maximal antichain.

If $f:X \longrightarrow X'$ is a bijection (in this case (O1) implies (O3) and (O4)) then $f:H \longrightarrow H'$ is said to be an isomorphism and H,H' are said to be isomorphic.

Every class of all the l.p.o. sets which are isomorphic with an l.p.o. set H is called a partial sequence (p. sequence) and is denoted by [H].

Given l.p.o. sets H,I,J,K, we say that occurrences $f:H \longrightarrow I$, $g:J \longrightarrow K$ are equivalent iff there are isomorphisms $h:H \longrightarrow J$, $i:I \longrightarrow K$ such that $f \circ i = h \circ g$. Every class of all the occurrences which are equivalent with an occurrence $f:H \longrightarrow I$ of an l.p.o. set H in an l.p.o. set I is called an occurrence of the p. sequence [H] in the p. sequence [I] and is denoted by $[f]:[H] \longrightarrow [I]$.

Given p. sequences P,Q,R, we define the composition $W:P \longrightarrow R$ of two occurrences $U:P \longrightarrow Q$, $V:Q \longrightarrow R$ choosing l.p.o. sets $H \in P$, $I \in Q$, $J \in Q$, $K \in R$, some occurrences $f:H \longrightarrow I$, $g:J \longrightarrow K$ such that $f \in U$ and $g \in V$, an isomorphism $i:I \longrightarrow J$, and taking $W=[f \circ i \circ g]$ (such a composition depends on P,Q,R,U,V only so that our definition is correct).

2.2. Examples

Let us consider the l.p.o. sets in Fig.2.2.1 and in Fig.2.2.2 (the arrows resulting from the transitivity of the orderings are omitted).

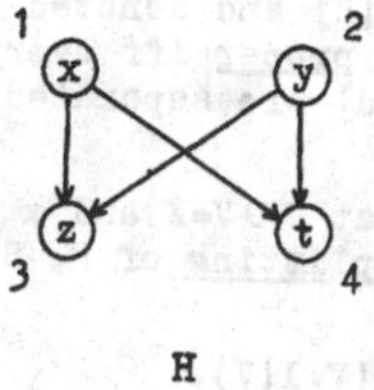

Fig.2.2.1

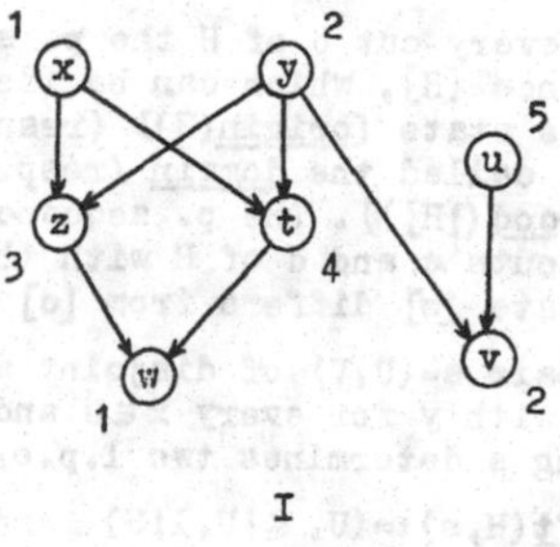

Fig.2.2.2

One may interpret the l.p.o. set I in Fig.2.2.2 as a history of a process of activating certain resources denoted 1,2,3,4,5. The history consists of particular activations: x,w of 1, y,v of 2, z of 3, t of 4, and u of 5. The ordering of the activations reflects a causal relation between them. The activations z and t are direct consequences of x and y. The activation w is a direct consequence of z and t. The activation v is a direct consequence of u and of the (hidden) passivation of 2 separating the activations y and v of 2. The l.p.o. set H in Fig.2.2.1 may be interpreted as a part of such a history. There is a unique occurrence (inclusion) of H in I.

The l.p.o. sets H and I determine two p. sequences (l.p.o. sets with "unnamed elements") which are shown in Fig.2.2.3 and in Fig.2.2.4, respectively.

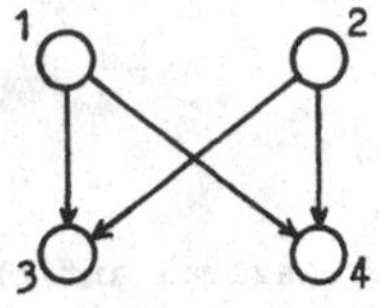

Fig.2.2.3

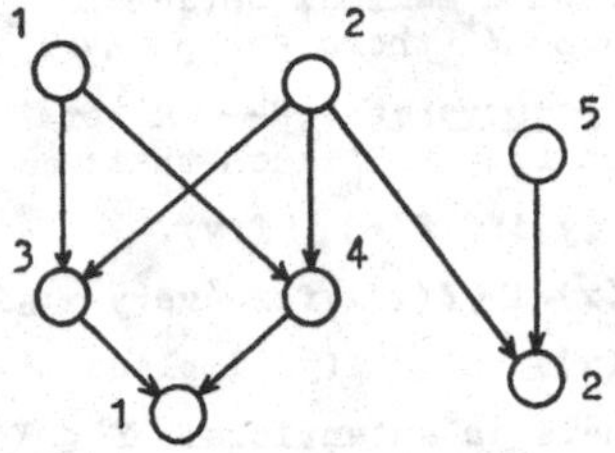

Fig.2.2.4

3. Operations on partial sequences

The needed operations on p. sequences can be introduced considering two internal structures of l.p.o. sets. One structure is determined by maximal antichains. Another is determined by suitable partitions.

3.1. Notions

The restriction of an l.p.o. set $H=(X,\leq,l)$ to a maximal antichain $Y\subseteq X$ is an l.p.o. set called a cut of H. Such a cut c determines two l.p.o. sets:

$\underline{head}(H,c):=(Y^-,\leq|Y^-,l|Y^-)$ and $\underline{tail}(H,c):=(Y^+,\leq|Y^+,l|Y^+)$.

The set of cuts of H with the ordering:

$c\sqsubseteq d$ iff c is a cut of $\underline{head}(H,d)$

is called the cut structure of H. To the set of minimal (resp.: maximal) elements of X the least cut denoted by origin(H) (resp.: the greatest cut denoted by end(H)) corresponds.

To every cut c of H the p. sequence [c] corresponds, called a state of the p. sequence [H], which can be identified with the set of labels of the elements of c. The state [origin(H)] (resp.: [end(H)]), depending on the p. sequence [H] only, is called the domain (resp.: the codomain) of [H] and denoted by dom([H]) (resp.: cod([H])). The p. sequence [H] is said to be proper iff every two comparable cuts c and d of H with the same states [c]=[d] are separated by a cut e whose state [e] differs from [c] and [d].

A pair $s=(U,V)$ of disjoint subsets of X such that $U\cup V=X$ and x is incomparable with y for every $x\in U$ and $y\in V$ is called a splitting of H. Such a splitting s determines two l.p.o. sets:

$\underline{left}(H,s):=(U,\leq|U,l|U)$ and $\underline{right}(H,s):=(V,\leq|V,l|V)$.

The set of splittings of H with the ordering:

$(U,V) \subseteq (U',V')$ iff $U \subseteq U'$

is called the <u>splitting structure</u> of H.

Given two p. sequences P=[H] and Q=[I], due to (H5) there may be at most one p. sequence R=[J] such that there is a cut c of J with <u>head</u>(J,c) $\in$ P and <u>tail</u>(J,c) $\in$ Q. Such a p. sequence is called the <u>sequential composition</u> of P and Q and denoted by P·Q. The occurrences corresponding to the inclusions <u>head</u>(J,c) $\subseteq$ J and <u>tail</u>(J,c) $\subseteq$ J are called the <u>canonical occurrences</u> of P and Q in P·Q, respectively.

Given two p. sequences P=[H] and Q=[I], there may be at most one p. sequence R=[J] such that there is a splitting s of J with <u>left</u>(J,s) $\in$ P and <u>right</u>(J,s) $\in$ Q. Such a p. sequence is called the <u>parallel composition</u> of P and Q and denoted by P+Q. The occurrences corresponding to the inclusions <u>left</u>(J,s) $\subseteq$ J and <u>right</u>(J,s) $\subseteq$ J are called the <u>canonical occurrences</u> of P and Q in P+Q, respectively.

Let P=[(X, $\leq$,l)] be a p. sequence. We say that P is a <u>one-element</u> p. sequence iff X contains exactly one element (such a p. sequence can be identified with the label l(x) of x $\in$ X). We say that P is a <u>prime</u> p. sequence iff all the elements of X are minimal or maximal, every minimal element is comparable with every maximal element and vice-versa, and at least two elements have different labels. One-element and prime p. sequences are said to be <u>elementary</u>.

3.2. Properties

1. The cut structure of an l.p.o. set is a lattice.
2. The splitting structure of an l.p.o. set is a Boolean algebra.
3. Given two p. sequences P and Q, the sequential composition P·Q exists iff <u>cod</u>(P)=<u>dom</u>(Q).
4. Given two p. sequences P and Q, the parallel composition P+Q exists iff the sets of labels occurring in P and Q are disjoint.
5. Given two p. sequences P and Q, the parallel composition P+Q exists iff for every state s of P and every state t of Q there exists the parallel composition s+t.
6. The class of p. sequences, when endowed with the operations of taking domains, codomains, sequential compositions, parallel compositions, and the constant $0:=[(\emptyset,\emptyset,\emptyset)]$, is a p.m. category.
7. Given p. sequences P,Q,R,S, if (P·Q)+(R·S) exists then P+R, P+S, Q+R, Q+S, (P+R)·(Q+S) also exist and (P·Q)+(R·S)=(P+R)·(Q+S).
8. If P·Q=R+S then there are T,U,V,W such that P=T+U, Q=V+W, R=T·V, S=U·W.
9. If P+Q=R+S then there are T,U,V,W such that P=T+U, Q=V+W, R=T+V, S=U+W.
10. If P·Q=R·S then there are T,U,V,W such that P=T·(<u>dom</u>(U)+V), Q=(U+<u>cod</u>(V))·W, R=T·(U+<u>dom</u>(V)), S=(<u>cod</u>(U)+V)·W.
11. Every P $\in$ <u>closure</u>($\{P_1,\dots,P_m\}$) can be represented in the following "sequential" form:

$$P = (P_{11}+\dots+P_{1n_1}+P_{i_1})\cdot \;\dots\; \cdot(P_{m1}+\dots+P_{mn_m}+P_{i_m}),$$

where $P_{i_1},\dots,P_{i_m}$ is a permutation of $P_1,\dots,P_m$ and $P_{11},\dots,P_{1n_1},\dots,P_{m1},\dots,P_{mn_m}$ are of the form <u>dom</u>(P_k) or <u>cod</u>(P_k) for some $k \in \{1,\dots,m\}$.
12. Every proper p. sequence which is finite (corresponds to a finite l.p.o. set) can be decomposed into elementary p. sequences.
13. If there is an occurrence F:P$\longrightarrow$Q of a p. sequence P in a p. sequence Q then there are p. sequences R,S,T such that Q=R·(P+S)·T.

14. Every occurrence $U:S \longrightarrow R$ of a prime p. sequence S in a p. sequence R which is the sequential or the parallel composition of two p. sequences P and Q with the canonical occurrences $F:P \longrightarrow R$ and $G:Q \longrightarrow R$ can uniquely be represented either as $S \xrightarrow{V} P \xrightarrow{F} R$ with an occurrence $V:S \longrightarrow P$ or as $S \xrightarrow{W} Q \xrightarrow{G} R$ with an occurrence $W:S \longrightarrow Q$.

Property 1 is due to (H2) and (H5). Property 2 is simple to verify. The sequential composition P•Q in 3 can be obtained identifying every final element of P with the initial element of Q that has the same label, and extending the orderings of P and Q to the weakest ordering such that the result is a p. sequence. The parallel composition P+Q in 4 can be obtained taking the p. sequence which consists of two independent parts P and Q. 5 is a reformulation of 4. Proving the axiom (A6) of p.m. categories in 6 and proving properties 7,8,9,10 needs only manipulating with cuts and splittings of an l.p.o. set. Property 11 can be proved transforming (with the help of 6 and 7) the tree which describes the way of obtaining P from $P_1, \ldots, P_m$. To prove 12 it suffices to consider a maximal chain of cuts. Proving 13 and 14 is just a verification.

3.3. Examples

The p. sequence Q in Fig.2.2.4 can be obtained composing the p. sequences: P in Fig.2.2.3, and R,S,dom(S) in Fig.3.3.1. Namely, it can be represented as (dom(S)+P)•(R+S). Applying properties 6 and 7 we can reduce such an expression to the sequential form (dom(S)+P)•(dom(R)+S)•(cod(S)+R) or to another sequential form (dom(S)+P)•(dom(S)+R)•(cod(R)+S). P,R,S are prime p. sequences and have one occurrence each in the resulting p. sequence Q. These occurrences determine uniquely the corresponding occurrences in one of the components: dom(S)+P or R+S.

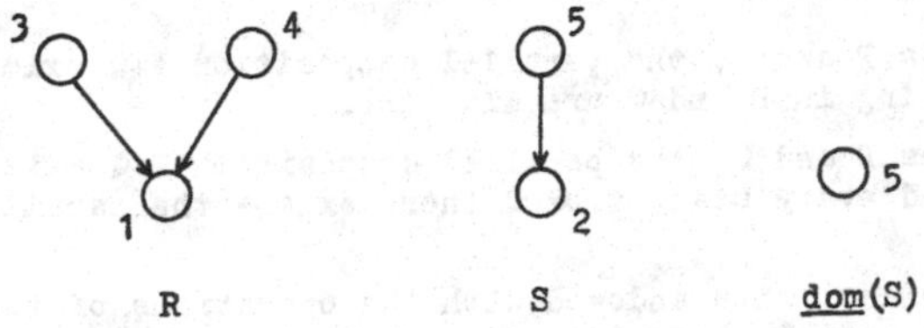

Fig.3.3.1

4. Algebras of partial sequences

Now we shall concentrate on restrictions of the p.m. category of p. sequences to some sets of p. sequences. We are interested in those restrictions which are p.m. categories.

4.1. Notions

Given a set U of p. sequences, by the restriction of the p.m. category of p. sequences to the set U we mean the partial algebra A=(U,dom′,cod′,•′,+′,0′), where dom′,cod′,•′,+′,0′ are the intersections $U^2 \cap$ dom, $U^2 \cap$ cod, $U^3 \cap •$, $U^3 \cap +$, $U \cap 0$, respectively. Such a restriction is called an *algebra of p. sequences* iff it is a p.m. category. To simplify denotations we shall use the symbols dom, cod,•,+,0 also for the restricted operations dom′,cod′,•′,+′,0′, respectively.

A set U of p. sequences is said to be *complete* iff $P \in U$ and P=Q•R or P=Q+R implies $Q \in U$ and $R \in U$.

An algebra A=(U,dom,cod,•,+,0) of p. sequences is said to be *complete* iff its carrier U is complete.

4.2. Properties

1. If V is a set of elementary p. sequences such that dom(x) and cod(x) are in +closure(V) for $x \in V$ then U=closure(V) is complete.
2. Let A=(U,dom,cod,·,+,0) be a complete algebra of p. sequences which has a minimal set V of generators such that every $P \in V$ is an elementary p. sequence or is a parallel composition of one-element p. sequences belonging to V. Let $h: V \longrightarrow U'$ be an admissible mapping from V into the carrier U' of a p.m. category A' =(U',dom',cod',·',+',0'). Then $h: V \longrightarrow U'$ can be extended to a (unique) homomorphism $H: A \longrightarrow A'$ iff it has an admissible extension h' : +closure(V)$\longrightarrow$U'.

Property 1 can be proved by induction on the number of occurrences of elementary p. sequences from V in a p. sequence from U. Property 2, which is one of our main results, can be proved reducing every $P \in U$ to a sequential form:

$$P = (P_{11}+\ldots+P_{1n_1}+P_1) \cdot \quad \ldots \quad \cdot (P_{m1}+\ldots+P_{mn_m}+P_m),$$

and defining:

$$H(P) = h'(P_{11}+\ldots+P_{1n_1}+P_1) \cdot' \quad \ldots \quad \cdot' h'(P_{m1}+\ldots+P_{mn_m}+P_m).$$

By induction on the number of occurrences of prime p. sequences in P it can be shown (due to properties 7 - 10,3.2) that H(P) does not depend on the particular representation of P. That the obtained mapping is a homomorphism is almost immediate.

4.3. Examples

Let A=(U,dom,cod,·,+,0) be a complete algebra of p. sequences with a minimal set V of generators as in 2,4.2, and let A' =(U',dom',cod',·',+',0') be a p. m. category of actions on a memory Mem as in 1.2. Let C be the set of those parallel compositions of one-element p. sequences from V which belong to V.

Suppose that we are given an allocation Field:C$\longrightarrow$Fields such that Field(c) ∩ Field(d)=∅ and Field(c+d)=Field(c) ∪ Field(d) whenever $c+d \in C$. Suppose that we are also given binary relations:

$$R(P) \subseteq \text{States}(\text{Field}(\text{dom}(P))) \times \text{States}(\text{Field}(\text{cod}(P)))$$

for prime $P \in V$. Then defining h(P):=(Field(dom(P)),Field(cod(P)),R(P)) we obtain an admissible mapping $h: V \longrightarrow U'$ which can uniquely be extended to an admissible mapping h' : +closure(V)$\longrightarrow$U', and thus to a homomorphism $H: A \longrightarrow A'$.

5. Concurrent schemata and their algebras

A class of algebras of p. sequences with minimal sets of generators and admitting homomorphic extensions of admissible mappings defined on such sets can be defined by means of concurrent schemata (cf. Mazurkiewicz [2]) which correspond to finite condition-event Petri nets with certain classes of safe initial markings (cf. Petri [4], Peterson [3]).

5.1. Notions

A Petri net is a triple N=(B,E,F) such that:

(N1) $B \cap E = \emptyset$,

(N2) $F \subseteq B \times E \cup E \times B$,

(N3) domain(F) ∪ range(F)=B ∪ E≠∅.

Each $b \in B$ (resp.: $e \in E$) is called a state element (resp.: a transition element) of N, and F is called the flow relation of N. Every subset of state elements is called a constellation.

Let FU:={$b \in B$: bFe for some $e \in U$}, UF:={$b \in B$: eFb for some $e \in U$}, Fe:=F{e}, and $eF:=\{e\}F$. We say that a constellation l is reachable in one step from a constellation k iff there is $U \subseteq E$ such that: $(Fu \cup uF) \cap (Fv \cup vF) = \emptyset$ for distinct $u,v \in U$, $FU \subseteq k$, $UF \subseteq l$, and $k-FU=l-UF$. Then we write $k \overset{U}{\Longrightarrow} l$ and call this triple a reachability step. We say that a constellation l is reachable from a constellation k iff there is a finite sequence of reachability steps:

$$k = k_0 \overset{U_1}{\Longrightarrow} k_1 \overset{U_2}{\Longrightarrow} \cdots \overset{U_n}{\Longrightarrow} k_n = l.$$

A concurrent scheme is a quadruple $S=(B,E,F,C)$ such that:

(S1) $N=(B,E,F)$ is a finite Petri net,

(S2) $Fe \neq \emptyset$ and $eF \neq \emptyset$ for every $e \in E$,

(S3) there is no $e \in E$ such that $eF=Fe$,

(S4) $Fu=Fv$ and $uF=vF$ implies $u=v$ for every $u,v \in E$,

(S5) C is a set of constellations of N,

(S6) $c \in C$ and $d \subseteq c$ implies $d \in C$,

(S7) $Fe \in C$ and $eF \in C$ for every $e \in E$,

(S8) for every constellation k and every $e \in E$:
$(Fe \cap k=\emptyset$ and $Fe \cup k \in C)$ iff $(eF \cap k=\emptyset$ and $eF \cup k \in C)$.

The constellations belonging to C are called configurations. Maximal configurations are called cases (thus C is the set of parts of cases).

A process P(b) corresponding to a state element b of S (i.e. of N) is the one-element p. sequence {b} with the label b. A process P(e) corresponding to a transition element e of S is the prime p. sequence whose domain is Fe and whose codomain is eF.

Given a subset $A \subseteq B \cup E$, by P(A) we denote the set of processes corresponding to the elements of A. By a process of S we mean every p. sequence $P \in$ closure$(P(B \cup E))$ satisfying the condition:

(P1) for every states u,v of P we have: $u,v \in C$ and for every constellation k:
$(u \cap k=\emptyset$ and $u \cup k \in C)$ iff $(v \cap k=\emptyset$ and $v \cup k \in C)$.

The set of processes of S is called the behaviour of S and is denoted by processes(S). The corresponding algebra of p. sequences (observe property 8, 5.2) is called the algebra of the scheme S.

A set U of p. sequences is said to be regular iff:

(R1) all the p. sequences belonging to U are finite and proper,

(R2) the labels of all $P \in U$ are from a finite set L,

(R3) U is complete (if $P \in U$ and $P=Q \cdot R$ or $P=Q+R$ then $Q \in U$ and $R \in U$),

(R4) if $P,Q \in U$ and $P \cdot Q$ exists then $P \cdot Q \in U$,

(R5) given $P,Q \in U$, the parallel composition $P+Q$ exists and belongs to U iff $c+d \in U$ for a state c of P and a state d of Q.

An algebra of p. sequences is said to be regular iff its carrier is regular.

5.2. Properties

1. If a constellation of a scheme S is reachable from a constellation k and one of the constellations is a configuration then the other is also a configuration.
2. $C \cup P(E) \subseteq$ processes(S).
3. If $P \in$ closure$(P(B \cup E))$ and dom$(P) \in C$ or cod$(P) \in C$ then $P \in$ processes(S).

4. If P·Q exists then P·Q $\in$ processes(S) iff P $\in$ processes(S) and Q $\in$ processes(S).
5. The parallel composition of P,Q $\in$ processes(S) exists and is in processes(S) iff there is a state c of P and a state d of Q such that c+d exists and belongs to C.
6. If P=[H] with H=(X, $\leq$,1) is a process of S then P is proper and for every maximal antichain Y $\subseteq$ X and every maximal chain Z $\subseteq$ X the intersection Y $\cap$ Z is non-empty.
7. A configuration d $\in$ C is reachable from a configuration c $\in$ C iff there is a process P of S such that dom(P)=c and cod(P)=d.
8. A set of p. sequences is the behaviour of a scheme iff it is regular.
9. An algebra of p. sequences is the algebra of a scheme iff it is regular.

Properties 1,2,4 are easy to prove. Properties 3,5,6 can be proved by induction on the number of occurrences of prime processes in suitable p. sequences. Property 7 can also be proved by induction. From 1 - 6 it follows that the behaviours of concurrent schemes are regular. The converse can be proved constructing a scheme S for a given regular set U. Due to 12,3.2, one can find one-element and prime p. sequences generating U and take them as the state elements and the transition elements of the constructed scheme. Configurations can be defined as the states of p. sequences from U.

5.3. Examples

Let us consider an iterative procedure of solving the following set of numerical equations:

$$X = F(X,Y),$$
$$Y = G(X,Y).$$

Such a procedure starts with some initial values X_0,Y_0 of X,Y and computes successively:

$$X_{n+1} = F(X_n,Y_n),$$
$$Y_{n+1} = G(X_n,Y_n).$$

Suppose that there are independent processors F and G capable to execute the operations (x,y):=(F(x,y),y) and (z,t):=(z,G(z,t)), respectively, in disjoint fields fF={x,y} and fG={z,t} of a memory Mem, respectively. Suppose that there is a processor D which takes care of the data and the results of the computation, and distributes the work among the processors F and G. Then the procedure can be described by the scheme shown in Fig.5.3.1 with a suitable interpretation corresponding to the exhibited intended meaning of the transition elements, where the configurations are $\emptyset$,{1},{2},{3},{4},{1,2},{1,4},{2,3},{3,4}. To such a scheme its algebra corresponds with the following minimal set of generators:

$$V = \{1,2,3,4,1+2,3+4,P(F),P(G),P(D)\},$$

where P(F),P(G),P(D) are as in Fig.5.3.2.

The algebra of the scheme contains those elements of closure(V) which have not the states {1,3} or {2,4}.

Now we can take the allocation:

$$\underline{\text{Field}}(1)=\underline{\text{Field}}(3)=\{x,y\},$$
$$\underline{\text{Field}}(2)=\underline{\text{Field}}(4)=\{z,t\},$$
$$\underline{\text{Field}}(1+2)=\underline{\text{Field}}(3+4)=\underline{\text{Field}}(1+4)=\underline{\text{Field}}(2+3)=\{x,y,z,t\},$$

the relations:

$$R(F)=\{((x,y),(x',y')) : x'=F(x,y),\ y'=y\},$$
$$R(G)=\{((z,t),(z',t')) : t'=G(z,t),\ z'=z\},$$
$$R(D)=\{((x,y,z,t),(x',y',z',t')) : x'=x,\ y'=t,\ z'=x,\ t'=t\},$$

and define an admissible mapping h from V into the set U' of actions on Mem as in 4.3. Such a mapping can uniquely be extended to a homomorphism H from the algebra of the scheme into the p.m. category of actions on Mem. This homomorphism is the interpretation corresponding to the intended meaning of the transition elements.

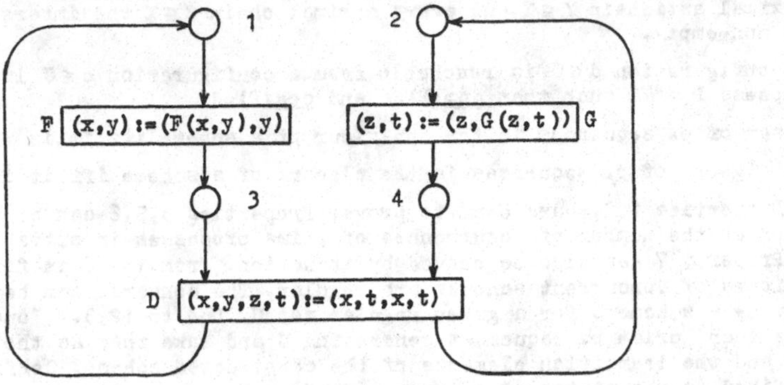

Fig.5.3.1

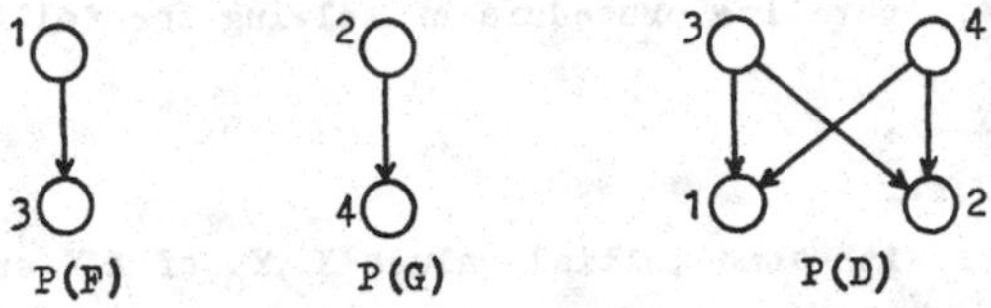

Fig.5.3.2

References

1. Hotz, G.: Grundlagen einer Theorie der Programmiersprachen, Mathematisches Institut und Institut für Angewandte Mathematik, Universität des Saarlandes, A 71/01, A 71/02, 1971.

2. Mazurkiewicz, A.: Concurrent Program Schemes and their Interpretations, DAIMI PB - 78 Aarhus Univ. Publ. , July 1977.

3. Peterson, J. L.: Petri Nets, Computing Surveys, Vol.9, No.3, September 1977.

4. Petri, C. A.: Interpretations of Net Theory, ISF GMD, Bonn, 1975.

5. Winkowski, J.: An Algebraic Approach to Non-Sequential Computations, ICS PAS Report 312, Warsaw, 1978.

ON MULTITAPE AUTOMATA

Hideki Yamasaki
Department of Information Science
Tokyo Institute of Technology
Oh-Okayama, Meguroku, Tokyo 152, Japan

1. Introduction

The notion of an n-tape deterministic (one-way) automaton was introduced by Rabin and Scott [1] and it has been shown by Luckham, Park and Paterson [2] that the equivalence problem for n-tape deterministic automata is reducible to the strong equivalence problem for independent location schemata with n locations and vice versa. Bird [3] has shown that the equivalence problem for 2-tape deterministic automata is decidable. As for the cases of more than two tapes, Lewis [4] has shown that the equivalence problem is decidable for a restricted class of n-tape automata, i.e., for those with each state being in at most one loop in the state diagram.

But it is still open and seems to be hard whether the equivalence problem for arbitrary n-tape deterministic automata ($n \geq 3$) is decidable or not. In passing we note that the inclusion problem for n-tape deterministic automata is undecidable and so is the equivalence problem for nondeterministic ones. (These are direct consequences of the undecidability of the Post correspondence problem.) As a way to approach the problem we derive certain reducibility results for n-tape automata, including those from the open question stated above.

In this paper we consider n-tape automata which move nondeterministically in the following sense; namely those which may have several available tapes to read in each state, but after selecting a tape the move is uniquely determined by the symbol on that tape and the current state. We mostly concentrate our study on n-tape automata with an additional property which guarantees deterministic behavior in the weakest sense. We call these automata determinate (after Karp and Miller [5]). Informally speaking, a determinate automaton may reach many states after it has processed an input n-tape word, but they must be all contained either in the set of final states or in its complement.

Then among others we can show that two deterministic automata are equivalent iff they can be mapped into a determinate automaton in a reasonable manner. Our main result is that the following three decision problems are reducible each other: (1) Whether two n-tape deterministic automata are equivalent or not. (2) Whether two n-tape determinate automata are equivalent or not. (3) Whether an n-tape automaton is determinate or not.

We give basic notions such as determinate automata in section 2, and introduce some auxiliary notions being useful to prove our main result in section 3. Then we establish the reducibility result in secti 4. Concluding remarks are in section 5.

2. Definitions and Properties of Determinate Automata

We will define an n-tape automaton as a special type of automaton over $\Sigma_1 \cup \Sigma_2 \cup \ldots \cup \Sigma_n$, where each Σ_i is a finite alphabet for the i-th tape.

First we will define n-tape words and n-tape languages. Let $\Sigma = \Sigma_1 \cup \Sigma_2 \cup \ldots \cup \Sigma_n$, where $\Sigma_i \cap \Sigma_j = \phi$ if $i \neq j$. Let us denote the direct produ $\Sigma_1^* \times \Sigma_2^* \times \ldots \times \Sigma_n^*$ by $\Pi\Sigma_i^*$. Each element of $\Pi\Sigma_i^*$ is called an n-tape word over Σ, and a subset of $\Pi\Sigma_i^*$ is called an n-tape language over Σ. The concate nation of $\alpha = \langle \alpha_1, \alpha_2, \ldots, \alpha_n \rangle$ and $\beta = \langle \beta_1, \beta_2, \ldots, \beta_n \rangle$ in $\Pi\Sigma_i^*$ is defined by $\alpha\beta = \langle \alpha_1\beta_1, \alpha_2\beta_2, \ldots, \alpha_n\beta_n \rangle$. $(\alpha)_i$ denotes the i-th component of α. We use λ to denote the unit element of Σ_i^*, for $i = 1,2,\ldots,n$. We also use λ to denote the unit element $\langle \lambda, \lambda, \ldots, \lambda \rangle$ of $\Pi\Sigma_i^*$.

Next we define a mapping $\rho : \Sigma \to \{1,2,\ldots,n\}$ to assign to each symbol the name of the alphabet it belongs to, by $\sigma\rho = i$ iff $\sigma \in \Sigma_i$. (We write xf or sometimes f(x) for the value assigned to x by function f.) We also define homomorphisms $h_i : \Sigma^* \to \Sigma_i^*$, for $i = 1,2,\ldots,n$, by $\sigma h_i = \sigma$ if $\sigma \in \Sigma_i$ and $\sigma h_i = \lambda$ otherwise. Then for each $w \in \Sigma^*$ we write $\langle w \rangle$ for the n-tape word $\langle wh_1, wh_2, \ldots, wh_n \rangle$. The derivative of an n-tape language L with respect to $\alpha \in \Pi\Sigma_i^*$ is defined similarly to the 1-tape case. That is, $\alpha \setminus L = \{\beta \mid \alpha\beta \in L\}$.

DEFINITION 1. An n-tape automaton A is a quintuple (Q,Σ,δ,s,F), where (1) Q is a (finite or infinite) set of states, (2) $s \in Q$ is called the initial state, (3) $F \subset Q$ is the set of final states, (4) $\delta : Q \times \Sigma \to Q$ is a partial function called the transition function. When $\delta(p,\sigma) = q$ with $\sigma \in \Sigma_i$, we say that q is an i-successor of p. The partial function δ is either totally defined or totally undefined on each $\{p\} \times \Sigma_i$. That is, if p has an i-successor then $\delta(p,\sigma)$ is defined for all $\sigma \in \Sigma_i$.

The transition function $\delta : Q \times \Sigma \to Q$ can be naturally extended to the partial function $\delta : Q \times \Sigma^* \to Q$. When $\delta(p,w) = q$, we write $p \xrightarrow{w} q$. If Q is finite, we say that A is a finite n-tape automaton. In the sequel we sometimes omit the modifier "n-tape" for n-tape automata and n-tape languages.

DEFINITION 2. Let $A = (Q,\Sigma,\delta,s,F)$ be an automaton. For each $q \in Q$ we define $T_A(q) \subset \Sigma^*$, and $\tau_A(q) \subset \Pi\Sigma_i^*$, by $T_A(q) = \{w \mid \delta(q,w) \in F\}$, and $\tau_A(q) = \{<w> \mid w \in T_A(q)\}$. Then we define $\tau(A) = \tau_A(s)$. The set $\tau(A)$ is called the language accepted by n-tape automaton A.

DEFINITION 3. Two automata A and B over the same alphabet are equivalent iff they accept the same language, i.e., $\tau(A) = \tau(B)$.

DEFINITION 4. Let $A = (Q,\Sigma,\delta,s,F)$ be an automaton, and define $I_p = \{i \mid$ p has i-successors$\}$, for each $p \in Q$. Then a state $p \in Q$ is (1) deterministic, (2) closed, or (3) determinate, respectively, if (1) I_p is either a singleton or the empty set, (2) $p \xrightarrow{\sigma} q$ and $p \xrightarrow{\sigma'} q'$ with $\sigma\rho \neq \sigma'\rho$ imply the existence of $r \in Q$ such that $q \xrightarrow{\sigma'} r$ and $q' \xrightarrow{\sigma} r$, or (3) $p \xrightarrow{\sigma} q$ implies $\tau_A(q) = <\sigma> \setminus \tau_A(p)$. An automaton A is deterministic, closed, or determinate, if all states of A are deterministic, closed, or determinate, respectively.

EXAMPLE. Let A be the automaton in Fig. 1, where the initial state is 1, the final state is 2, $\Sigma_1 = \{a\}$, $\Sigma_2 = \{b\}$, and $\Sigma_3 = \{c\}$. Then $\tau(A) = \{<a^n,b^m,c^{n+m-1}> \mid n+m \geq 1\}$. The automaton A is determinate, but not closed.

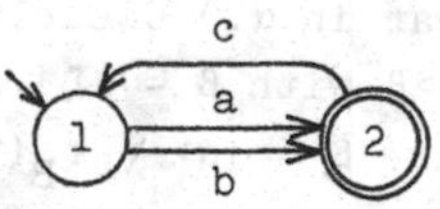

Fig. 1. The 3-tape automaton A

Clearly, if an automaton A is deterministic then it is closed. If A is closed then it is determinate. Indeed, if A is closed, $p \xrightarrow{\sigma} q$ in A, and $\sigma \in \Sigma_i$, then $\tau_A(q) = <\sigma> \setminus \tau_A(p)$ since $u\sigma v \in T_A(p)$ and $uh_i = \lambda$ imply $uv \in T_A(q)$ for all $u,v \in \Sigma^*$.

DEFINITION 5. For two automata $A = (Q_A,\Sigma,\delta_A,s_A,F_A)$ and $B = (Q_B,\Sigma,\delta_B,s_B,F_B)$, a mapping $f : Q_A \to Q_B$ is called a morphism from A to B and written as $f : A \to B$ if $s_A f = s_B$, $F_B f^{-1} = F_A$ and $pf \xrightarrow{\sigma} qf$ in B for each $p \xrightarrow{\sigma} q$ in A. If a morphism $f : A \to B$ is a mapping onto Q_B, and for each $p' \xrightarrow{\sigma} q'$ in B there exists $p \xrightarrow{\sigma} q$ in A such that $pf = p'$ and $qf = q'$, then we say that B is the <u>morphic image</u> of A by the morphism f.

Note that if $f : A \to B$ and $g : B \to C$ are morphisms, then the composition $fg : A \to C$ is a morphism. Let A and B be automata and $f : A \to B$ be a morphism. Then $\tau_A(p) \subset \tau_B(pf)$ for each state p of A. We will use a notation $\tau_A \subset f\tau_B$ to mean that $\tau_A(p) \subset \tau_B(pf)$ for each state p of A. Likewise we write $\tau_A = f\tau_B$ if $\tau_A(p) = \tau_B(pf)$ for each state p of A.

<u>THEOREM 1.</u> An automaton A is determinate iff there exist a closed automaton C and a morphism $f : A \to C$ such that $\tau_A = f\tau_C$.

Proof. If: Clear. Only if: If A is determinate, we can construct a "universal" automaton U such that $\tau_A : A \to U$ is a required morphism, as follows: Let $U = (\{L \mid L \subset \Pi\Sigma_i^*\}, \Sigma, \delta, \tau(A), \{L \mid \lambda \in L \subset \Pi\Sigma_i^*\})$, where $\delta(L,\sigma) = \langle\sigma\rangle \setminus L$. Clearly, U is closed and τ_A is a morphism from A to U. Since τ_U is the identity mapping, $\tau_A = \tau_A\tau_U$.

<u>THEOREM 2.</u> Let $f : A \to B$ be a morphism from an automaton A to a determinate automaton B. A necessary and sufficient condition for $\tau_A = f\tau_B$ is that for each state p of A, if $\alpha \in \tau_B(pf) - \{\lambda\}$ then $(\alpha)_i \neq \lambda$ for some $i \in I_p = \{i \mid$ p has i-successors in A$\}$.

Proof. It is enough to show that $\tau_A \subsetneq f\tau_B$ iff there exist a state p of A and $\alpha \in \tau_B(pf) - \{\lambda\}$ such that $(\alpha)_i = \lambda$ for all $i \in I_p$. "If" part is clear. To prove "only if" part, let α be a shortest word in $\tau_B(pf) - \tau_A(p$ for some state p of A. (The length of $\alpha \in \Pi\Sigma_i^*$ is defined to be the tota number of symbols which appear in α.) Clearly $\alpha \neq \lambda$. Assume $(\alpha)_i \neq \lambda$ for some $i \in I_p$, and $\alpha = \langle\sigma\rangle\beta$ with $\beta \in \Pi\Sigma_i^*$ and $\sigma \in \Sigma_i$. Let $p \xrightarrow{\sigma} q$ in A. Then $pf \xrightarrow{\sigma} qf$ in B, and we have $\beta \in \langle\sigma\rangle \setminus \tau_B(pf) = \tau_B(qf)$, since B is determinate. As β is shorter than α, it follows that $\beta \in \tau_A(q)$, whic however conflicts with our assumption that $\alpha \notin \tau_A(p)$. Thus $(\alpha)_i = \lambda$ for all $i \in I_p$.

<u>COROLLARY.</u> In theorem 2, if both A and B are finite, then it is decidab whether $\tau_A = f\tau_B$ or not.

Proof. Since $T_B(pf) - \{\lambda\}$ is a regular set over Σ, it is decidable whether for each $w \in T_B(pf) - \{\lambda\}$ there exists an $i \in I_p$ with $wh_i \neq \lambda$ or not.

3. Schemes

Theorem 2 gives a condition for a morphism f from A to a determinate automaton B to satisfy $\tau(A) = \tau(B)$. In this section we consider a certain type of automata, called schemes, which guarantees the equality $\tau(A) = \tau(B)$ for any morphism f from a scheme A to a determinate scheme B. The notion appears to be a useful tool for the proof of our main theorem.

DEFINITION 6. An (n-tape) scheme A is an n-tape automaton such that:
(1) For each state p of A, if $\alpha \in \tau_A(p)$ then $(\alpha)_i \neq \lambda$ for all $i \in I_p$.
(2) Any state except the final states has successors.

Note that the final states of schemes cannot have successors.

COROLLARY. It is decidable whether a finite automaton is a scheme or not.

THEOREM 3. Let $f : A \to B$ be a morphism from a scheme A to a determinate scheme B. Then $\tau_A = f\tau_B$ and A is determinate.
Proof. If p is a final state of A, then $\tau_A(p) = \tau_B(pf) = \{\lambda\}$. If p is not a final state of A, then p and pf have i-successors for some i. Since B is a scheme, $(\alpha)_i \neq \lambda$ for all $\alpha \in \tau_B(pf)$. Therefore $\tau_A = f\tau_B$. It follows that A is a determinate scheme.

THEOREM 4. Let B be the morphic image of a determinate automaton A by a morphism $f : A \to B$. Assume that $pf = qf$ implies $\tau_A(p) = \tau_A(q)$ for all states p and q of A. Then $\tau_A = f\tau_B$ and B is determinate.
Proof. Let U be a "universal" automaton for A which is defined in the proof of theorem 1. We define a mapping g from the states of B to the states of U, by $(pf)g = \tau_A(p)$. By the assumptions, g is well defined. Furthermore, g is a morphism from B to U. Since $\tau_A \subset f\tau_B \subset fg\tau_U = \tau_A$, $\tau_A = f\tau_B$ and B is determinate.

COROLLARY. Let $A_k = (Q_k, \Sigma, \delta_k, s_k, F_k)$, $k \in K$, and $B = (Q, \Sigma, \delta, s, F)$ be schemes where K is a (finite or infinite) index set. Let $f_k : A_k \to B$ be morphisms. Assume that for each $p \xrightarrow{\sigma} q$ in B, there exists $p_k \xrightarrow{\sigma} q_k$ in A_k with $p_k f_k = p$ and $q_k f_k = q$ for some $k \in K$. Then the following two statements are equivalent.
(1) B is determinate.
(2) A_k is determinate for all k in K, and if $p_i f_i = p_j f_j$ then $\tau_{A_i}(p_i) = \tau_{A_j}(p_j)$.

Proof. (1) $\Rightarrow$ (2): Immediate from theorem 3. (2) $\Rightarrow$ (1): We assume Q_k ($k \in K$) are pairwise disjoint and let $A = (\bigcup_{k \in K} Q_k, \Sigma, \delta', s', \bigcup_{k \in K} F_k)$, where s' is one of the intial states of A_k's, and $\delta'(p,\sigma) = \delta_k(p,\sigma)$ if $p \in Q_k$. Define a morphism $f : A \to B$, by $pf = pf_k$ if $p \in Q_k$. By the assumptions these are well defined, and A is a determinate scheme. We may assume that B is the morphic image of A by the morphism f. Thus the scheme B is determinate by theorem 4.

This corollary asserts that a "union" of a class of morphic images of schemes is determinate, providing the schemes are all determinate and "mutually consistent"; conversely if the "union" is determinate then the component schemes must be determinate and "mutually consistent".

We close this section by showing that for any finite automaton A, we can construct a scheme that accepts precisely the words in $\tau(A)$ followed by an endmarker on each tape. Let $\$_i$ be an endmarker for the i-th tape which does not belong to Σ. Let $\Sigma_i^\$ = \Sigma_i \cup \{\$_i\}$ for $i = 1,2,\ldots,n$ and $\Sigma_\$ = \Sigma_1^\$ \cup \Sigma_2^\$ \cup \ldots \cup \Sigma_n^\$$. For each subset $I = \{i_1, i_2, \ldots, i_k\}$ of $N = \{1,2,\ldots,n\}$, let $\$_I = \langle \$_{i_1} \$_{i_2} \ldots \$_{i_k} \rangle$. We will write \$ for $\$_N$.

Now for any automaton $A = (Q,\Sigma,\delta,s,F)$, we construct a scheme $A\$ = (\{q_I \mid q \in Q \text{ and } I \subset N\} \cup \{d\}, \Sigma_\$, \delta_\$, s_N, F_\$)$, where $F_\$ = \{q_\phi \mid q \in F\}$ and $\delta_\$$ is defined as follows:

(1) If q has i-successors in A and $i \in I$, then
$\delta_\$(q_I,\sigma) = \delta(q,\sigma)_I$ for each $\sigma \in \Sigma_i$, and $\delta_\$(q_I,\$_i) = q_{I-\{i\}}$.

(2) If $q \in F$ has no i-successors in A for any $i \in I$, then
$\delta_\$(q_I,\sigma) = d$ for each $\sigma \in \bigcup_{i \in I} \Sigma_i$, and $\delta_\$(q_I,\$_i) = q_{I-\{i\}}$ for each $i \in I$.

(3) If $q \notin F$ has no i-successors in A for any $i \in I$, then
$\delta_\$(q_I,\sigma) = d$ for each $\sigma \in \bigcup_{i \in I} \Sigma_i^\$$, and $\delta_\$(q_\phi,\sigma) = d$ for each $\sigma \in \Sigma_\$$.

(4) $\delta_\$(d,\sigma) = d$ for each $\sigma \in \Sigma_\$$.

<u>EXAMPLE.</u> Let A be the automaton in Fig. 1. Then the scheme A\$ is shown in Fig. 2. The states not reachable from the initial state $1_{\{1,2,3\}}$ are deleted for simplicity.

By the construction of A\$, $\delta_\$(q_I, u_1\$_{i_1} u_2\$_{i_2} \ldots u_k\$_{i_k}) = p_\phi \in F_\$$ implies $\delta(q, u_1 u_2 \ldots u_k) = p \in F$, $\langle \$_{i_1} \$_{i_2} \ldots \$_{i_k} \rangle = \$_I$, and $(u_1 u_2 \ldots u_k)h_i = \lambda$ for all $i \in N - I$, which however imply $\delta_\$(q_I, u_1 u_2 \ldots u_k) = p_I$ and $\delta_\$(p_I, \$_{i_1} \$_{i_2} \ldots \$_{i_k}) = p_\phi \in F_\$$.
Hence $\tau_{A\$}(q_I) = (\tau_A(q) \cap \{\alpha \mid (\alpha)_i = \lambda \text{ for all } i \in N - I\})\{\$_I\}$.
From the above equation it is easy to see that A is determinate iff

A$ is determinate. Thus we have the following.

THEOREM 5. For any automaton A there exists a scheme A$ such that $\tau(A\$) = \tau(A)\{\$\}$ and A is determinate iff A$ is determinate.

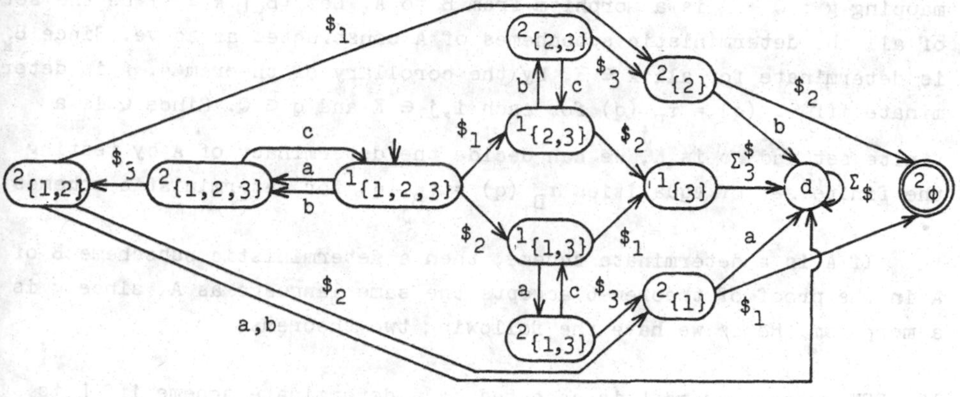

Fig. 2. A$ where $\Sigma_1^\$ = \{a,\$_1\}$, $\Sigma_2^\$ = \{b,\$_2\}$ and $\Sigma_3^\$ = \{c,\$_3\}$.

4. Main Theorem

THEOREM 6. The following three decision problems are reducible each other.
(1) Whether two finite deterministic n-tape automata are equivalent or not.
(2) Whether two finite determinate n-tape automata are equivalent or not.
(3) Whether a finite n-tape automaton is determinate or not.

Proof. (1) is reducible to (2): Clear. (2) is reducible to (3): We may assume $n \geq 2$. Let $A = (Q_A,\Sigma,\delta_A,s_A,F_A)$ and $B = (Q_B,\Sigma,\delta_B,s_B,F_B)$ be determinate automata. Choose arbitrary $\sigma_1 \in \Sigma_1$ and $\sigma_2 \in \Sigma_2$, and construct an automaton $C = (Q_A \cup Q_B \cup \{s,t,u,d\},\Sigma,\delta,s,F_A \cup F_B)$ as described in Fig. 3. Then C is determinate iff s is determinate in C iff $\tau(A) = \tau(B)$.

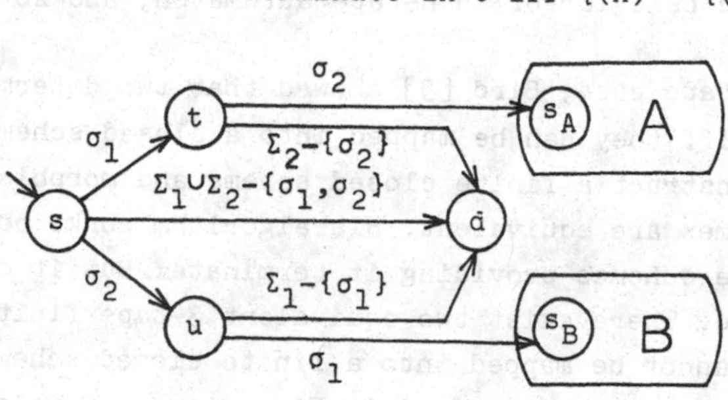

Fig. 3. The automaton C.

(3) is reducible to (1): Let A be a finite automaton. By theorem 5 we may assume that $A = (Q,\Sigma,\delta,s,F)$ is a scheme. Then we can obtain a deterministic subscheme $B = (Q,\Sigma,\delta',s,F)$ of A, where δ' is defined as follows. Choose an arbitrary $i \in I_q$ for each $q \in Q$, and let $\delta'(q,\sigma) = \delta(q,\sigma)$ for each $\sigma \in \Sigma_i$, and $\delta'(q,\sigma)$ be undefined for all $\sigma \in \Sigma - \Sigma_i$. Then the identity mapping $g : Q \to Q$ is a morphism from B to A. Let $\{B_k | k \in K\}$ be the set of all the deterministic subschemes of A constructed as above. Since B_k is determinate for all $k \in K$, by the corollary of theorem 4, A is determinate iff $\tau_{B_i}(q) = \tau_{B_j}(q)$ for each $i,j \in K$ and $q \in Q$. Since Q is a finite set and so is K, we can decide the determinacy of A by testing the finite set of equalities $\tau_{B_i}(q) = \tau_{B_j}(q)$ for deterministic schemes.

If A is a determinate scheme, then a deterministic subscheme B of A in the proof of theorem 6 accepts the same language as A, since g is a morphism. Hence we have the following two theorems.

THEOREM 7. A language L is accepted by a determinate scheme iff L is accepted by a deterministic scheme.

THEOREM 8. If a language L is accepted by a determinate automaton then L{\$} is accepted by a deterministic scheme.

5. Concluding Remarks

We have shown: whether two deterministic automata are equivalent or not is decidable iff whether an automaton is determinate or not is decidable. In the proof of the main theorem we have shown that two deterministic automata are equivalent iff they can be mapped in a reasonable manner into a determinate automaton. The determinacy guarantees a kind of deterministic behavior of an n-tape automaton, and so does the closedness.

For the 2-tape case, Bird [3] showed that two deterministic schemes are equivalent iff they can be mapped into a closed scheme, and gave an algorithm to construct a finite closed scheme and morphisms when the given two finite schemes are equivalent. His algorithm works correctly for arbitrary n-tape schemes providing it terminates. But it does not necessarily terminate. There exist two equivalent 3-tape finite deterministic schemes which cannot be mapped into a finite closed scheme. (Indeed, the deterministic subschemes of A\$ in Fig. 2 are equivalent. But they cannot be mapped into a finite closed scheme.)

Acknowledgements

I wish to thank H. Tanaka for helpfull discussions about this work. Also I wish to thank Prof. M. Takahashi for several suggestions which aided in the clarification of this paper.

References

1. Rabin, M.O. and Scott, D., Finite Automata and Their Decision Problems. IBM J. Res. Develop., 3(1959), 114-125.

2. Luckham, D.C., Park, D.M.R. and Paterson, M.F., On Formalized Computer Programs. JCSS, 4(1970), 220-249.

3. Bird, M., The Equivalence Problem for Deterministic Two-Tape Automata. JCSS, 7(1973), 218-236.

4. Lewis, H.R., A New Decidable Problem, With Applications. Proc. 18th Annual Symp. on Foundations of Computer Science (1977).

5. Karp, R.M. and Miller, R.E., Parallel Program Schemata. JCSS, 3(1969), 147-195.

A TURING MACHINE ORACLE HIERARCHY

Stanislav Žák
Mathematical Institute
Czechoslovak Academy of Sciences
Žitná 25, 115 67 Praha 1
Czechoslovakia

INTRODUCTION

The paper introduces three complexity measures of computations on Turing machines with oracles. The complexity of a computation on a Turing machine with an oracle is given either by the number of interactions with the oracle during the computation, or by the sum of the lengths of the questions asked by the machine of its oracle during the computation, or by the maximum of lengths of these questions. The complexity depends neither on the amount of the tape required by the computation nor on the number of its steps. For the case of these measures we try to find the least enlargement of the complexity bound which increases the computing power.

For oracles of a minimal level (they can decide the acceptance of words on Turing machines without oracles) we construct five complexity hierarchies. The first three hierarchies are constructed on the set of languages accepted by deterministic Turing machines with a fixed oracle according to the three measures mentioned above. Another hierarchy with respect to the first measure is constructed on the set of languages accepted by nondeterministic Turing machines of a special type with an oracle, and the last hierarchy is constructed on the set of languages accepted by nondeterministic Turing machines with an oracle according to the second measure. The hierarchy on the same set according to the third measure does not differ from the deterministic case.

The paper is a by-product of the work (Žák [3]) on the same problem for the case of space complexity of computations on Turing machines and is based on the same principle of diagonalization. It consists of two chapters. The first chapter is concerned with diagonalization and second contains all complexity results.

CHAPTER 1

The aim of this chapter is to introduce a principle of diagonalization. The first part of this chapter contains a basic definition of a mapping called the result of the testing process (rtp) and a theorem which exhibits the logical structure of the diagonalization principle without taking care of existence and complexity aspects. The second part of the chapter contains a lemma which ensures the existence of the rtp-mappings and introduces first complexity aspects. The proofs of the theorem and of two lemmas of the chapter can be found in Žák [3] .

Let us first recall some usual definitions and conventions. An alphabet is a nonempty finite set of symbols, all alphabets are subsets of a fixed infinite set containing, among others, the symbols b,0,1,2,S. A string or a word over an alphabet is a finite sequence of its symbols, Λ denotes the empty word, $|u|$ is the length of the word u. A language over an alphabet is a set of strings over this alphabet. If X is an alphabet then X^* (X^+,X^n) is the language of all words (of positive length, of the length n, respectively) over X. Two words may be concatenated which yields a similar operation for languages. N denotes the set of natural numbers. If a is a symbol and $i \in N$ then a^i is a string of a's of the length i. By a function or by a bound we always mean a mapping of N into itself. The identity function will be denoted id, id(n) = n, and the integer part of the binary logarithm will be denoted log. For two functions f,g we shall write $f \leq g$ iff $(\forall n)(f(n) \leq g(n))$, and $f \lessapprox g$ iff $(\exists n_0 \in N)(\forall n \geq n_0)(f(n) \leq g(n))$. From time to time in the following text we shall use the if .. then .. else construction well-known from the programing languages.

We shall call two languages L_1,L_2 equivalent ($L_1 \sim L_2$) iff they differ only in a finite number of strings. If W is a class of languages then EW will be the class of all languages for which there are equivalent languages in W.

By a program system we mean a pair (P,F), where P is a language and F is a mapping of P into a set of languages over an alphabet. In this context, P is called the set of programs and its elements are called programs. In what follows if we use the phrase "Let P be a set of programs", we implicitly understand that P is the first item of a program system. Its second item will have the general denotation L and L_p will mean the language which corresponds to the program $p \in P$. The set of all such L_p for all $p \in P$ will be denoted $\mathcal{L}(P)$.

For a program p and a word u, we say that p accepts u (p!u) iff $u \in L_p$.

Let p be a program and Q a set of programs. We say that p diagonalizes Q iff there is a finite set F such that $(\forall q \in Q-F)(p!q \longleftrightarrow \neg q!q)$.

Lemma 1. Let p be a program and Q a set of programs. If there are infinitely many programs from Q with the same language as the program p then p does not diagonalize Q.

Definition. Let Q,R be sets of programs, RTP a mapping of N without some initial segment into the set Q and e a function. If for all $q \in Q$ the sets

$$R_q = \{ r \in R \mid RTP(e(|r|)) = q \; \& \; \neg(q!r \longleftrightarrow \neg r!r) \}$$

are infinite then RTP is called the result of a testing process with the function e on the sets Q,R in short, rtp with e on Q,R.

Theorem 1. Let Q,R be sets of programs, RTP an rtp with e on Q,R X a program and z a mapping from R into N. If for all $q \in Q$ there are infinitely many $r \in R_q$ such that

(1) $X!rl^{z(r)} \longleftrightarrow \neg r!r$, and

(2) $(\forall j,\ 0 \leq j < z(r))(X!rl^{j} \longleftrightarrow RTP(e(|r|))!rl^{j+1})$,

then $L_X \notin E\mathcal{L}(Q)$.

We say that a Turing machine (TM) T accepts a word u if there is a computation of T on u which stops in a final (accepting) state. If T is a deterministic single-tape TM and accepts a word u, then T(u) denotes the word written on the tape after the computation of T on u has been finished.

A Turing machine with oracle A ($A \subseteq N$) is a Turing machine which has among its tapes a fixed one, on which a special symbol S may be written. The set of states of the machine includes three special states q,YES,NO . If it enters the state q, then in the next step if the number of occurrences of S on its fixed tape belongs to A, it must enter the state YES, otherwise the state NO.

A function e will be called (A-)recursive if there is a deterministic Turing machine T (with oracle A) such that for all $n \in N$ $T(1^n) = 1^{e(n)}$.

A language over an alphabet X is called (A-)recursively enumerable if it is accepted by a Turing machine (with oracle A) and it is called (A-)recursive if moreover its complement in X^+ is also (A-)recursively enumerable.

If P is a set of programs then $!_P$ is the binary relation $\{(p,u) \mid p \in P, u \in L_p\}$. - The graph of a binary relation H on a set of strings is the set $\{u2v \mid (u,v) \in H\}$.

Let A be an oracle. We say that a sequence $\{a_i\}$ of words over an alphabet is (A-)effective iff there is a deterministic Turing machine (with oracle A) rewriting the unary code of any $i \in N$ to a_i.

Lemma 2 (rtp-lemma). (a) Let Q,R be nonempty sets of programs and e a function. If no program from Q diagonalizes R and if $e \leq id$ and $\lim e = \infty$ then there is an rtp with e on Q,R .
(b) Let A be an oracle. If, in addition, the sets Q,R, $Q \subseteq \{b,1\}^*$, are (A-)recursively enumerable languages and the graphs of the relations $!_Q, !_R$ are (A-)recursive and also the function e is (A-)recursive then there is a deterministic Turing machine T (with oracle A) with one tape and with one head such that
(1) during the computation on the input word 1^k, T uses only the input cells and two adjacent cells,
(2) T writes only the symbols 1,b (1,b,S),
(3) there is a constant c such that the mapping RTP = $= \{(k,T(1^k)) \mid k \in N, k \geq c\}$ is an rtp with e on Q,R.

In fact, we have two lemmas - the version without an oracle and the relativized version.

CHAPTER 2

In this chapter, by an oracle machine we shall mean a deterministic or nondeterministic single-tape single-head Turing machine with an oracle such that its tape is infinite in both directions and with input words over the alphabet $\{0,1\}$ only.

We shall say that a machine M asks its oracle if M is in the state q. By the length of such a question we shall mean the number of symbols S currently written on the tape.

For an oracle machine M and for an input word u, we define
$oracle_M^1(u) = oracle_M^2(u) = oracle_M^3(u) = \infty$ if M does not accept u, otherwise
$oracle_M^1(u)$ = the minimal number of questions asked by M of its oracle during an accepting computation of M on u,
$oracle_M^2(u) = \min \{s \mid s =$ sum of lengths of all questions asked by M of its oracle during an accepting computation of M on u$\}$, and
$oracle_M^3(u) = \min \{s \mid s =$ maximum of lengths of questions asked by M during an accepting computation of M on u$\}$.
(In short, the complexity of acceptance of a word by a machine is given by the complexity of the most modest accepting computation.)

In what follows, by a machine we shall mean a nondeterministic machine with a fixed oracle A.

Lemma 3 (Universal machine). There is a recursive set S, $S \subseteq 1^+\{b,1\}^*$, in a one-one correspondence with the set of all machines, and a machine U such that for all i = 1,2,3, for all $s \in S$ and fo all input words u the equality $oracle_U^i(su) = oracle^i_{M_s}(u)$ holds (where M_s stands for the machine corresponding to s).

A similar lemma holds for the case of deterministic machines. Then U_D is a deterministic universal machine and S_D a set of codes of deterministic machines.

Let us fix the set S from Lemma 3. In what follows the language accepted by the machine M_s will be denoted $L(M_s)$ or $L(s)$. For i = = 1,2,3, we shall also write $\mathrm{oracle}_s^i(u)$ instead of $\mathrm{oracle}_{M_s}^i(u)$, where $s \in S$ and $u \in \{0,1\}^*$.

Definition (for i = 1,2,3). (a) If t is a bound and M_s a machine then by i-t-cut off of the language L(s) we mean the set $L_t^i(s) = L_t^i(M_s) = \{u \mid \mathrm{oracle}_s^i(u) \le t(|u|)\}$.
(b) We say that a machine M_s accepts its language within i-bound t if $L(s) = L_t^i(s)$.
(c) For a bound t we define

$\mathrm{ORACLE}^i(t) = \{L \mid (\exists s \in S)(L = L(s) = L_t^i(s))\}$,

$\mathrm{CORACLE}^i(t) = \{L_t^i(s) \mid s \in S\}$,

$\text{D-ORACLE}^i(t) = \{L \mid (\exists s \in S_D)(L = L(s) = L_t^i(s))\}$,

$\text{D-CORACLE}^i(t) = \{L_t^i(s) \mid s \in S_D\}$.

Let us repeat some standard definitions. For $A,B \subseteq N$ we shall write $A \le_m B$ iff there is a recursive function f such that for all x, $x \in N$, $x \in A \longleftrightarrow f(x) \in B$.

K will be a set of natural numbers,

$K = \{\langle T,u\rangle \mid T$ is a TM without oracle, $u \in \{0,1\}^*$, T accepts u$\}$,

where $\langle\,\rangle$ is a standard coding.

Definition (for i = 1,2,3). Let A be an oracle and f,g functions. We say that f is (i,g,A)-recursive iff there is a deterministic machine T with oracle A such that $L(T) = 1^*$, $T(1^n) = 1^{f(n)}$ for all $n \in N$, and $L(T) = L_g^i(T)$. We say that f is (i,rec,A)-recursive iff it is (i,h,A)-recursive for a recursive function h.

Lemma 4 (for i = 2,3). If $K \le_m A$ and if t is an A-recursive bound then the language $L = \{su \mid u \in L_t^i(s),\ s \in S\}$ is A-recursive. If moreover t is (i,rec,A)-recursive then there is a recursive function f such that the language L can be decided by a machine D for which the equality $L(D) = L_f^i(D)$ holds.

Remark. Lemma 4 yields trivial separation results such as: for i = 2,3 $\mathrm{ORACLE}^i(f) - \mathrm{CORACLE}^i(t) \ne \emptyset$. It seems that f is not a small function (with respect to t).

For a word u, a language L, a family of languages $\mathcal{L}$ we define Shadow $u = 1^{|u|}$, Shadow $L = \{\text{Shadow } u \mid u \in L\}$ and Shadow $\mathcal{L} =$ $= \{\text{Shadow } L \mid L \in \mathcal{L}\}$.

Theorem 2. Let t be a recursive bound, $\lim t = \infty$. If $K \leq_m A$ then there is a language L such that
(1) $L \subseteq 1^*$,
(2) $L \in \text{ORACLE}^3(t)$,
(3) $L \notin \text{Shadow ORACLE}^3(t')$, where $t'(0) = 0$, $t'(n) = t(n-1)$ for n >

Sketch of the proof. First, we shall choose a set Q of programs such that $\mathcal{L}(Q) = \text{Shadow CORACLE}^3(t')$ and a set R of programs both satisfying the conditions of the rtp-lemma. Secondly, we shall construct a machine X such that X accepts its language L(X) within 3-oracle bound t and this language has the properties (1),(2) from Theorem 1. By application of this theorem we shall get $L(X) \notin E\mathcal{L}(Q) =$ $= E \text{ Shadow CORACLE}^3(t') = \text{Shadow ORACLE}^3(t')$.

Let us write Q = S and, for $q \in Q$, $L_q = \text{Shadow } L^3_{t'}(q)$.

Let $\{s_i\}$ be an effective sequence of programs from S in which each s, $s \in S$, occurs infinitely many times.

Now we define a recursive sequence $r_1, z(r_1), r_2, z(r_2), \ldots, r_i$, $z(r_i), \ldots$, where $r_i \in 1^+$, $z(r_i) \in N$, $r_1 = 1$, $r_{i+1} = r_i + z(r_i) +$ and
$z(r_i) = \min\{n \mid$ the tape of the length $t(|r_i| + n)$ is sufficient for deciding whether $r_i \in L_{r_i} = \text{Shadow } L^3_{t'}(s_i)\}$

- see Lemma 4. We put $R = \{r_i \mid i \in N\}$.

Let us define, for $i \in N$,

$$e(|r_i|) = \min(\{|r_i|\} \cup \{t(|r_i| + j) \mid 0 \leq j < z(r_i)\}),$$

and $e(n) = n$ for $n \in N$ such that $n \neq |r_i|$ for each $i \in N$.

There is an rtp RTP with e on Q,R constructive in sense of the rtp-lemma since Q,R,e satisfy the conditions of this lemma.

Now, we are going to construct the machine X. X accepts only strings of 1's and during the computation on the input word 1^n X never asks of A a question of the length greater than $t(n)$ – see conditions (1),(2) of the theorem. After computing $t(n)$, X recursively constructs the sequence $r_1, z(r_1), \ldots, r_i, z(r_i), \ldots$.

If $(\exists r \in R)(1^n = r1^{z(r)})$ then X decides whether $r \in L_r$ and accepts iff $\neg r!r$.

If $(\exists r \in R)(1^n = r1^j)$, where $0 \leq j < z(r)$, then X recursively constructs the number $e(|r|)$ and then the program q, $q = RTP(e(|r|))$. Further X nondeterministically rewrites the input word 1^n to any word from $\{0,1\}^{n+1}$ and on this word computes in the same way as the universal machine U (Lemma 3) according to the program q. If there is an accepting computation of U on a word from $\{0,1\}^{n+1}$ according to the program q and this computation does not require to ask of A a question of a length greater than $t(n)$ then X accepts. Formally:

(1) $1^n \in L(X) \longleftrightarrow (\exists u \in \{0,1\}^{n+1})(\mathrm{oracle}_U^3(RTP(e(|r_i|))u) \leq t(n))$.

Now, we shall apply Theorem 1. We have $r1^{z(r)} \in L(X) \leftrightarrow \neg r!r$ for all $r \in R$ and therefore it suffices to prove for all sufficiently large r, $r \in R$, that

$$(\forall j,\ 0 \leq j < z(r))(r1^j \in L(X) \longleftrightarrow RTP(e(|r|))!r1^{j+1})$$

– condition (2) of Theorem 1. Let us fix r,j , $r \in R$, $0 \leq j < z(r)$, and write $n = |r| + j$. We know that the following statements are equivalent:

i) $r1^j \in L(X)$,

ii) $(\exists u \in \{0,1\}^{n+1})(\mathrm{oracle}_U^3(RTP(e(|r|))u) \leq t(n))$ – see (1) ,

iii) $(\exists u \in \{0,1\}^{n+1})(\mathrm{oracle}_q^3(u) \leq t(n) = t'(n+1))$
– see Lemma 3, $q = RTP(e(|r|))$,

iv) $(\exists u \in \{0,1\}^{n+1})(u \in L_{t'}^3(q))$,

v) $1^{n+1} \in \text{Shadow } L_{t'}^3(q) = L_q$,

vi) $q!1^{n+1}$,

vii) $RTP(e(|r|))!r1^{j+1}$.

The language L(X) satisfies the conditions of Theorem 1 and therefore $L(X) \notin E\mathcal{L}(Q) = E \text{ Shadow } CORACLE^3(t') = \text{Shadow } ORACLE^3(t')$.

Q.E.D.

From the fact D-ORACLE^3(t) = ORACLE^3(t) follows that the same theorem holds for the deterministic case.

Remark. Condition (3) in Theorem 2 may be changed as follows:
$L \notin \text{Shadow} \cup \{\text{ORACLE}^3(t') \mid \liminf (t(n) - t'(n+1)) \geq 0\}$.

Example. ORACLE^3(n + 1) - ORACLE^3(n) $\neq \emptyset$.

Now, we turn our attention to the complexity measures oracle^1 and oracle^2 .

Lemma 5 (for i = 1,2). If $K \leq_m A$ then for each k, $k \in N$, $k \geq 1$, there is an (i,e'/k,A)-recursive function e' such that:

(1) e' is nondecreasing and unbounded, $e' \leq id$,

(2) Val e' = $\{e'(n) \mid n \in N\}$ is a recusive set,

(3) for each nondecreasing and unbounded recursive function d the inequality $e' \preccurlyeq d$ holds.

Definition. We say that a machine with an oracle is an r-machine if each its infinite computation contains infinitely many questions to its oracle.

Lemma 6. If $K \leq_m A$ then there is a mapping F, $F: S \rightarrow S$, such that:

(a) For each $s \in S$, $M_{F(s)}$ is an r-machine.

(b) If M_s is an r-machine, then for each $u \in \{0,1\}^+$ the equality $\text{oracle}^1_{F(s)}(u) = 1 + 2.\text{oracle}^1_s(u)$ holds.

(c) The set $F(S) = \{F(s) \mid s \in S\}$ is recursive .

(d) F is realizable on a TM.

Let us fix the mapping F from the lemma and write $F(M_s)$, $F(M_s) =_{df} M_{F(s)}$.

Lemma 7. Let A be an oracle, $K \leq_m A$. If t is an A-recursive bound then the languages $\{su \mid u \in L^1_t(s) , s \in F(S)\}$ and $\{su \mid u \in L^1_t(s) , s \in S_D\}$ are A-recursive.

Definition. For a bound t we define
$\text{F-ORACLE}^1(t) = \{L \mid (\exists s \in F(S))(L = L(s) = L^1_t(s))\}$, and

$\text{F-CORACLE}^1(t) = \{ L_t^1(s) \mid s \in F(S) \}$.

By application of Theorem 1 we can prove the following theorem.

Theorem 3. Let A be an oracle, $K \leq_m A$, t a recursive bound. The following sets contain a language over the alphabet $\{1\}$:

$$\text{F-ORACLE}^1(t) - \text{Shadow F-CORACLE}^1(t') ,$$

$$\text{ORACLE}^2(t) - \text{Shadow CORACLE}^2(t') ,$$

$$\text{D-ORACLE}^1(t) - \text{D-CORACLE}^1(t') ,$$

$$\text{D-ORACLE}^2(t) - \text{D-CORACLE}^2(t') ,$$

where $t'(n+1) = t(n) - e'(n)$ for all $n \in N$ and e' is $(1,e'/8,A)$- or $(2,e'/8,A)$-recursive function from Lemma 5, providing $e' \leq t$.

Example. Languages over $\{1\}$ are also contained in

$$\text{ORACLE}^2(n + \log^k n) - \text{Shadow ORACLE}^2(n) \quad \text{for } k > 0 ,$$

$$\text{D-ORACLE}^i(n + \log^k n) - \text{D-ORACLE}^i(n) \quad \text{for } i = 1,2 \text{ and } k > 0.$$

A trivial diagonalization yields results such as

$$\text{D-ORACLE}^1(2 + 2n) - \text{D-ORACLE}^1(n) \neq \emptyset \quad .$$

Remark after Lemma 4 gives trivial results for $i = 2$.

REFERENCES

1. Rogers, H.,Jr. , Theory of Recursive Functions and Effective Computability, McGraw-Hill, New York, 1967 .
2. Simon, I. , On some subrecursive reducibilities, Tech. Rep. STAN-CS-77-608, June 1977 .
3. Žák, S. , A Turing machine space hierarchy, to appear in Kybernetika.

A SURVEY OF SOME SYNTACTIC RESULTS IN THE λ-CALCULUS

Gérard Berry
Ecole des Mines
Sophia-Antipolis
06560 - VALBONNE

Jean-Jacques Lévy
IRIA-LABORIA
Domaine de Voluceau
78150 - ROCQUENCOURT

Introduction

In this paper, we consider the λ-calculus as a formal setting for studying procedure calls in programming languages. These languages (as in LISP) allow procedures to take procedure as arguments and also to deliver procedures as results. The λ-calculus could be typed or untyped. In the first case, it is more realistic to add a recursion operator as in LCF[Mi1, Pl2, Mi2, Be2]. In the untyped case, the recursion is already in the language. (Take Y= λf·(λx·f(xx))(λx·f(xx))). In both cases, the λ-calculus has constants which correspond to the basic functions symbols of terms in recursive program schemes [Ni,Vu]. Therefore, one can say that the λ-calculus framework is a generalisation of recursive programs schemes to any functional order.

Here, we shall look at syntactic results in [Be3, Le1]. We only consider the untyped λ-calculus, because definitions are simpler in it. Of course, we do not claim that programming languages should be untyped. But the pure λ-calculus has already enough properties which exist too in other formalisms [Be4, Be3, Hu1, Hu2]. As usual, the survey will be very partial. For a full treatment of the syntax and the semantics of the λ-calculus, it is strongly recommended to look at the future book [Ba2]. Furthermore, a lot of our results could not exist without the λ-calculus models [Sc1, Sc2, Pl1, Pl3] and the study of their connections with the syntax [Ba1, Wa1, Hy1, Na1, Bo1, Mo1, Ba3, Pl2, Mi2, Be3, Cu2].

1. Basic definitions

Let $V = \{x_1, x_2, \cdots\}$ be an infinite enumerable set of *variables* also written $x,y,z\cdots f,g,h\cdots$ and $C = \{c_1, c_2, \cdots\}$ a set of *constant* symbols. Then the set $\Lambda = \Lambda(V,C)$ of λ-*expressions* on V and C is the minimal set containing V and C such that :

1) If $M,N\in\Lambda$, then $(MN)\in\Lambda$ (*application*)
2) If $x\in V$ and $M\in\Lambda$, then $(\lambda x M)\in\Lambda$ (*abstraction* w.r.t. x)

Parenthesis can be suppressed in λ-expressions with the usual following abbreviations [Ch1]. For applications, parenthesis are implicitly to the left. Thus $(MN_1N_2\cdots N_n)$ stands for $(\cdots((MN_1)N_2)\cdots N_n)$ when $n\geq 1$. A new symbol "·" is added and permits to gather abstractions. Thus $(\lambda x_1 x_2 \cdots x_m \cdot M)$ is for $(\lambda x_1(\lambda x_2\cdots(\lambda x_m M)\cdots))$ when $m\geq 1$. Finally, the outermost parenthesis and also the ones corresponding to the

the expression following a dot symbol can be suppressed. For instance, one has $(\lambda xy\cdot cxy)z(\lambda x\cdot x)$ for $(((\lambda x(\lambda y((cx)y)))z)(\lambda xx))$.

Intuitively, the expression $(\lambda x\cdot M)$ corresponds to a unary procedure with x as a formal parameter and M as result. Procedures with several arguments are represented by a sequence of unary procedures (*curryfication*). For instance $(\lambda xy\cdot M)$ can be considered as the procedure with parameters x and y, and M as result.

In any expression, variables can be or not under the scope of a corresponding abstraction operator. They are respectively called *bound* or *free*. For instance, in $(\lambda x\cdot xy)$ x is a bound variable and y is free. The binders act exactly as universal or existential quantifiers in logical formulae. The *substitution* of the free variable x by an expression N in M, written $M[x\backslash N]$, is defined as follows :

$c[x\backslash N] = c$

$x[x\backslash N] = N$

$y[x\backslash N] = y \qquad (x\neq y)$

$(MM')[x\backslash N] = (M[x\backslash N]\ M'[x\backslash N])$

$(\lambda x\cdot M)[x\backslash N] = (\lambda x\cdot M)$

$(\lambda y\cdot M)[x\backslash N] = (\lambda z\cdot M[y\backslash z][x\backslash N])$ if $x\neq y$ and z is the variable such that

1) if x is not free in M or y not free in N, then z=y

2) otherwise, z is the first variable in V not free in M and N. (Thus, one forbids to any occurrence of a free variable in N to become bound in $M[x\backslash N]$).

The usual *rules of conversion* in the λ-calculus are the following :

(α-rule) $(\lambda x\cdot M)\rightarrow(\lambda y\cdot M[x\backslash y])$ if y is not free in M

(β-rule) $(\lambda x\cdot M)N\rightarrow M[x\backslash N]$

(η-rule) $\lambda x\cdot Mx\rightarrow M$ if x is not free in M

(δ-rule) $cM_1M_2\cdots M_n\rightarrow N$

The first rule means renaming of bound variables. The second one is the usual ALGOL copy-rule which substitutes the actual parameter N to the formal parameter x of $(\lambda x\cdot M)$. The η-rule reflects the extensional aspects of functions. The δ-rules permits to describe the behaviour of constants. There can be a large number of them. For instance, if we suppose that the truth values constants $\{tt,ff\}$ and the conditional *if* are in C, we can have the δ-rules :

if tt M N → M

if ff M N → N

Now, an expression M can be *immediatly reduced* to N, also written M→N, if one can get N from M by converting a subexpression of M. This subexpression is called a *redex*. And M can be *reduced* to N, written $M\overset{*}{\rightarrow}N$, if $M\rightarrow M_1\rightarrow M_2\cdots\rightarrow M_n=N$ for some $n\geq 0$ and expressions $M_1,M_2,\cdots M_n$. Between two given expressions M and N, there may be several reductions. In order to specify one of them, we should have to precise the occurrences of the redexes converted at each elementary step. This will not be done here. We shall make no difference between redexes and their occurrences, as long as it will be clear from the context. Thus $M \xrightarrow{R} N$ means that R is the redex converted from M to N. Furthermore, we use letters ρ,σ,τ for giving names to reductions and write $\rho:M \xrightarrow{*} N$ to indicate that M and N are the initial and final expressions of the reduction ρ. If σ starts at the final expression of ρ, the *concatenation* of ρ and σ is written $\rho\sigma$. The empty reduction starting at M is denoted by 0_M or simply by 0, when M is clear from the context.

An expression without redexes is called a *normal form*. And if $M \xrightarrow{*} N$ where N is in normal form, then M has N as a normal form.

Thus, reductions map computations in the λ-calculus considered as a programming language. Normal forms correspond to final results. We shall restrict attention here to the only case of the α and β-rules of conversion. The η-rule is more a logical rule than a rule existing in programming languages. The δ-rules, although very important from a computer scientist point of view, are more complicated and there is not yet a full treatment of them (see for instance [Hu1, Hu2] for first order terms). Furthermore, the α-rule and all the corresponding tedious problems will be ignored. We shall just remember that bound variables names are not important, the set of λ-expressions being understood as the quotient set by α-interconvertibility and equality of λ-expressions as the α-interconvertibility relation. Hence, we just consider the λ-calculus with the β-rule, i.e. a programming language with just procedure calls as rules of computations. By reductions, redexes and normal forms, we shall mean β-reductions, β-redexes and β-normal forms. Finally, as constants and free variables have the same behaviour for the β-rule, we can simplify by assuming $C=\emptyset$, i.e. a λ-calculus without constants.

2. Confluency. The lattice structure of reductions

The following result is well-known in the λ-calculus.

Theorem 2.1 (The Church-Rosser theorem). If $M \xrightarrow{*} N$ and $M \xrightarrow{*} P$, there is an expression Q such that $N \xrightarrow{*} Q$ and $P \xrightarrow{*} Q$.

As a corollary, if M has a normal form, then this normal form is unique. The proof of the Church-Rosser theorem is not too simple. One can prove easily that, if $M \to N$ and $M \to P$, then there is a Q such that $N \xrightarrow{*} Q$ and $P \xrightarrow{*} Q$. But this local confluency property is not enough for deriving the theorem. The solution is to define *parallel reductions*. Let $\mathcal{F}$ be a given set of redexes (occurrences) in an expression M. The simultaneous contraction of redexes in $\mathcal{F}$, written $M \xrightarrow{\mathcal{F}} N$, can be defined as the successive contractions of the redexes of $\mathcal{F}$ in an inside-out way. This is the Martin-Löf method (see [Ba1]). An alternative solution is used in [Ch1, Cu1], gives more information, but is more complicated. Roughly speaking, it says that the order in which the redexes in $\mathcal{F}$ are contracted is not relevant (finite developments theorem [Ba2, Hi1, Cu1, Le1]. Now, the Church-Rosser theorem can be proved by showing the following lemma.

Lemma 2.2. Let F and G be two sets of redexes in M, and $M \xrightarrow{F} N$ and $M \xrightarrow{G} P$. Then there are two sets G' and F' of redexes in N and P, and an expression Q such that $N \xrightarrow{G'} Q$ and $P \xrightarrow{F'} Q$.

The previous lemma can be strengthened, because the sets F' and G' are connected to F and G. In order to state the stronger property, one needs to trace redexes along a given reduction. Let R and S be two redexes (occurrences) in M, and $\rho : M \xrightarrow{S} N$. Then the set R/ρ of *residuals* of R by ρ is the set of redexes in N corresponding to R. The exact definition is done by tedious cases on the relative positions of R and S (see [Ch1]). It is simpler to look at some examples. Take $I = \lambda x \cdot x$, $\Delta = \lambda x \cdot xx$ and underline redex R and their residuals in the following elementary reductions :

$\rho : M = \Delta(\underline{Ix}) \rightarrow (\underline{Ix})(\underline{Ix}) = N$

$\rho : M = (\underline{Ix})(\Delta(Ix)) \rightarrow (\underline{Ix})(Ix)(Ix)) = N$

$\rho : M = \underline{I(\Delta(Ix))} \rightarrow \underline{I((Ix)(Ix))} = N$

$\rho : M = \underline{\Delta(Ix)} \rightarrow (Ix)(Ix) = N$

$\rho : M = (Ix)(\Delta(\underline{Ix})) \rightarrow (Ix)(\underline{Ix})(\underline{Ix}) = N$

$\rho : M = \underline{\Delta\Delta} \rightarrow \Delta\Delta = N$

Now, if $\rho : M \xrightarrow{*} N$ is a non elementary reduction and R is a redex in M, then the residuals R/ρ of R by ρ are defined by transitivity. Similarly, if F is a set of redexes in M, then F/ρ is the natural set extension. Finally, the residuals definition works too with parallel reductions. (The finite development theorem even tells that the order in which redexes are contracted at each elementary step is not relevant for the residual relation).

In order to state the strenghtened lemma, it will be convenient to call *coinitial* two reductions starting at the same expression and *cofinal* two reductions ending at the same expression. Furthermore, if G is a set of redexes in M and ρ and σ are two parallel reductions of the form $\rho : M \xrightarrow{*} N$ and $\sigma : M \xrightarrow{G} P$, we denote by σ/ρ the reduction $\sigma/\rho : N \xrightarrow{G'} Q$ where $G' = G/\rho$ is the set of residuals of G by ρ. Finally, let also write $\rho \sqcup \sigma = \rho(\sigma/\rho)$. Now, the following lemma is fundamental for the rest of the paragraph.

Lemma 2.3 (The parallel moves lemma). Let ρ and σ be two coinitial elementary parallel reductions steps starting at M. Then :

1) $\rho \sqcup \sigma$ and $\sigma \sqcup \rho$ are cofinal,

2) $R/(\rho \sqcup \sigma) = R/(\sigma \sqcup \rho)$ for any redex R in M.

(One proves it usually as a corollary of the already mentioned finite developments theorem). Now, we shall consider the structure of coinitial parallel reductions and generalise to the λ-calculus the computation lattice defined in [Vu] for recursive programs schemes. The trouble with the λ-calculus is the total absence of structure for expressions which can be obtained by reductions from a given expression. (This is true too in recursive programs schemes without the restrictions used in [Vu]). This difficulty will be overcomed by defining a permutation equivalence on reductions as the smallest congruence for concatenation generated from the parallel moves lemma and from the suppression of the empty steps.

Formally, there are some slight difficulties with the empty steps. Let denote by $\emptyset_M$ the elementary parallel reduction $M \xrightarrow{\emptyset} M$ and by $\emptyset_M^n$ the reduction $\emptyset_M \emptyset_M \cdots \emptyset_M$ (n steps of elementary empty parallel reduction). When M is clear from the context, we just write $\emptyset$ and $\emptyset^n$.

Definition 2.4. The *equivalence* $\equiv$ of two (parallel) reductions *by permutations*, written $\rho\equiv\sigma$, is the smallest equivalence relation such that :

1) $\rho\sqcup\sigma\equiv\sigma\sqcup\rho$ if ρ and σ are two coinitial elementary parallel reductions,
2) $\rho\emptyset\equiv\emptyset\rho\equiv\rho$,
3) $\rho\tau_1\sigma\equiv\rho\tau_2\sigma$ if $\tau_1\equiv\tau_2$.

Clearly, if $\rho\equiv\sigma$, then ρ and σ are coinitial and cofinal. But the converse could not be true. Take for instance $I=\lambda x\cdot x$ and $M=I(Ix)$. Let R and S be the two redexes in M and $\rho:M\xrightarrow{R}N$, $\sigma:M\xrightarrow{S}N$. Then $\rho\not\equiv\sigma$. The permutation equivalence will be shown to be easily decidable. This is achieved by extending the / and $\sqcup$ operators to any coinitial parallel reductions. We have already defined σ/ρ and $\rho\sqcup\sigma$ when σ is an elementary parallel reduction step and ρ is an arbitrary reduction. Now suppose that ρ and σ are two coinitial (parallel) reductions. Then let

$$\sigma/\rho = (\sigma_1/\rho)(\sigma_2/(\rho/\sigma_1))$$

if $\sigma=\sigma_1\sigma_2$ and σ_2 is an elementary reduction step. Furthermore, we always write $\rho\sqcup\sigma=\rho(\sigma/\rho)$. This is better described by figure 1, in which each elementary square is an application of the parallel moves lemma.

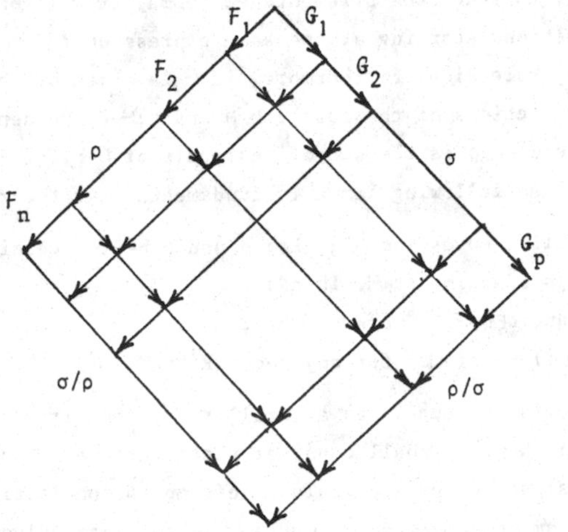

Figure 1

Lemma 2.5 (The generalised parallel moves lemma). Let ρ and σ be two coinitial parallel reductions. Then :

1) $\rho\sqcup\sigma$ and $\sigma\sqcup\rho$ are cofinal,
2) $\tau/(\rho\sqcup\sigma) = \tau/(\sigma\sqcup\rho)$ for any reduction τ coinitial with ρ and σ.

This lemma can be proved by algebraic manipulations from the previous definitions, which also give the below list of properties (forgetting the appropriate conditions on initial and final expressions of the considered reductions)

$\rho/(\sigma\tau) = (\rho/\sigma)/\tau$,

$\rho\sqcup\sigma \equiv \sigma\sqcup\rho$,

$\rho\equiv\sigma$ iff $\rho/\sigma=\emptyset^n$ and $\sigma/\rho=\emptyset^p$ for some $n,p\geq 0$, ($\equiv$ is decidable)

$\rho\equiv\sigma$ iff $\forall\tau\cdot\tau/\rho=\tau/\sigma$,

$\rho\sqcup\rho\equiv\rho$,

$\rho\sqcup(\sigma\sqcup\tau)=(\rho\sqcup\sigma)\sqcup\tau$,

$\rho\sigma\equiv\rho\tau$ implies $\sigma\equiv\tau$,

$\rho\equiv\sigma$ implies $\rho/\tau\equiv\sigma/\tau$.

We can now consider the associate prefix ordering on coinitial (parallel) reductions, by defining :

$\rho\sqsubseteq\sigma$ iff $\exists\tau\cdot\rho\tau\equiv\sigma$

Again, a list of properties of $\sqsubseteq$ can be easily done :

$\rho\sqsubseteq\rho$,

$\rho\sqsubseteq\sigma\sqsubseteq\tau$ implies $\rho\sqsubseteq\tau$,

$\rho\sqsubseteq\sigma\sqsubseteq\rho$ iff $\rho\equiv\sigma$,

$\rho\sqsubseteq\sigma$ iff $\rho/\sigma=\emptyset^n$ for some $n\geq 0$, ($\sqsubseteq$ is decidable)

$\rho\sqsubseteq\sigma$ implies $\rho/\tau\sqsubseteq\sigma/\tau$,

$\rho\sqsubseteq\sigma$ implies $\tau\rho\sqsubseteq\tau\sigma$.

Theorem 2.6 (The semi-upper lattice property). The set of reductions starting at a given expression forms a semi-upper lattice. Formally, if ρ and σ are coinitial, then

1) $\rho\sqsubseteq\rho\sqcup\sigma$ and $\sigma\sqsubseteq\rho\sqcup\sigma$

2) For any τ such that $\rho\sqsubseteq\tau$ and $\sigma\sqsubseteq\tau$, then $\rho\sqcup\sigma\sqsubseteq\tau$.

The proof works again by easy algebraic manipulations. For the category theory people, the category whose objects are expressions and morphisms are equivalence classes of reductions is a pushout category where every arrow is an epimorphism. Remark too that this structure of reductions depends only on the parallel moves lemma and thus exists in a lot of confluent rewriting systems (see for instance [Be2, Be4, Hu2]). The case of non-confluent systems seems more complicated (see[Bo]), but not impossible.

3. Correct strategies for the normal form

Leftmost outermost redexes have nice properties in the λ-calculus. For instance, if R is the leftmost outermost redex in M and $\rho:M\overset{*}{\to}N$ is a reduction such that $R/\rho\neq\emptyset$, then $R/\rho=\{S\}$ and S is the leftmost outermost redex in N (Remark : this may not be true of any outermost redex). Furthermore, let the *normal reduction* starting at M be the reduction

$$M \xrightarrow{R_1} M_1 \xrightarrow{R_2} M_2 \xrightarrow{R_3} \cdots$$

contracting successively the leftmost outermost redexes R_i of M_{i-1}.

Then the following theorem is shown in [Cu1].

Theorem 3.1. If M has a normal form, then the normal reduction of M reaches the normal form.

In fact, it is a corollary of a stronger theorem. Let the reduction

$$M \overset{R_1}{\rightarrow} M_1 \overset{R_2}{\rightarrow} M_2 \cdots \overset{R_n}{\rightarrow} M_n = N$$

be a *standard reduction* iff, for all i,j such that $1 \leq i < j \leq n$, the redex R_j is not a residual (along ρ) of a redex containing R_i or to the left of R_i. Then, the standardisation theorem of [Cu1] can be strengthened as follows.

Theorem 3.2 (The standardisation theorem). For any reduction ρ, there is a unique standard reduction ρ_s such that $\rho \equiv \rho_s$.

Again this theorem exists in other formalisms, for instance [Be2, Be4, Hu2] although in this last case standard reductions are more difficult to define. In terms of programming languages, we just rediscover that a call-by-name evaluation strategy can produce more result than a call-by-value one. But, further properties of normal forms can be studied.

Let us introduce a constant Ω and consider the set $\Lambda_\Omega = \Lambda(V,\{\Omega\})$ of corresponding λ-Ω-expressions. Let N_Ω and N be the set of (β-) normal forms in Λ_Ω and Λ. Consider now the *prefix ordering* on λ-Ω-expressions, written $M \leq N$, defined by :

1) $\Omega \leq M$ for all $M \in \Lambda_\Omega$,

2) $\lambda x \cdot M \leq \lambda x \cdot N$ if $M \leq N$,

3) $MN \leq PQ$ if $M \leq P$ and $N \leq Q$.

Let say that M and N are *compatible*, written $M \uparrow N$, if there is a P such that $M \leq P$ and $N \leq P$. The greatest lower bound (glb) of two λ-Ω-expressions M and N, written $M \sqcap N$, and the least upper bound (lub) of two compatible λ-Ω-expressions M and N, written $M \sqcup N$, are defined by :

$$\begin{cases} M \sqcap N = \Omega \text{ if } M = \Omega \text{ or } N = \Omega \text{ or } M \text{ and } N \text{ are incompatible} \\ (\lambda x \cdot M) \sqcap (\lambda x \cdot N) = \lambda x \cdot (M \sqcap N) \\ (MN) \sqcap (PQ) = (M \sqcap P)(N \sqcap Q) \end{cases}$$

$$\begin{cases} \Omega \sqcup M = M \sqcup \Omega = M \\ (\lambda x \cdot M) \sqcup (\lambda x \cdot N) = \lambda x \cdot (M \sqcup N) \\ (MN) \sqcup (PQ) = (M \sqcup P)(N \sqcup Q) \end{cases}$$

Finally, let us order the set $\mathbb{T} = \{tt, ff\}$ of truth values by $ff \sqsubseteq tt$ and denote by ∧ the glb on $\mathbb{T}$ (i.e. the usual "and" on boolean values). We give a name to the property of having a normal form by defining :

$$nf(M)=tt \text{ iff } M\overset{*}{\rightarrow}N \text{ for some } N\in\mathcal{N}$$

Then, it is straightforward to check that nf is monotonic, i.e. $M\leq N$ implies $nf(M)\sqsubseteq nf(N)$. Now, the following theorems can be proved.

<u>Theorem 3.3</u> (The stability theorem). If $M\uparrow N$, then $nf(M\sqcap N) = nf(M)\wedge nf(N)$.

<u>Corollary</u>. For any λ-Ω-expression M, there is a (unique) minimum prefix N of M such that $nf(M) = nf(N)$.

Thus, if an expression M has a normal form, the redexes of M which are also in this minimum prefix are really *needed* in order to get the normal form. It can be proved that a redex is needed iff it has a residual contracted in the normal reduction. Call-by-need reductions strategies (see [Vu, Le1]) can be defined as contracting at each step one needed redex. And they can be shown to reach the normal form. The stability theorem is proved in [Be1] in a more general statement. It exists too in other frameworks [Hu2].

But, needed redexes can be found without look-ahead with the use of the following theorem (see [Be1, Hu2]), which corresponds to the activity lemma in [Pl2].

<u>Theorem 3.4</u> (The sequentiality theorem). Let M be any given λ-Ω-expression such that $nf(M)=ff$ and there exists a λ-Ω-expression N verifying $nf(N)=tt$ and $M\leq N$. Then there is an occurrence of an Ω in M such that, for every $N\geq M$ such that $nf(N)=tt$, there is no more Ω at that occurrence in N.

Thus, in any λ-Ω-expression M such that $nf(M)=ff$, there is always one uniformly critical Ω to increase in order to get a normal form. Take for instance $M=(\Omega\Omega)$. Then the first Ω is critical, but not the second one since $(\lambda x\cdot y)\Omega\rightarrow y$. But in $M = (\lambda x\cdot x\Omega\Omega)\Omega$, the last Ω is the only critical one.

<u>4. Reductions costs</u>

In order to compute a normal form, useless redexes can be avoided by contracting only the needed redexes (for instance the leftmost outermost one). But needed redexes can be duplicated. For instance, take $M=\Delta((\lambda x\cdot xy)I)$ with $\Delta=\lambda x\cdot xx$ and $I=\lambda x\cdot x$. Then all the redexes in M are needed and the normal reduction takes five steps. However, with a *sharing* mechanism as in [Wa1, Vu], this can be reduced to three steps :

shared λ-calculus

$M = \Delta((\lambda x\cdot xy)I)$

$M'_1 =$ 

usual λ-calculus

$M = \Delta((\lambda x\cdot xy)I)$

$M_1 = ((\lambda x\cdot xy)I)((\lambda x\cdot xy)I)$

M'_2 =  Iy $\qquad$ M_2 = (Iy)(Iy)

M'_3 = 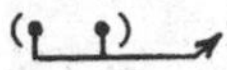 y $\qquad$ M_3 = yy

From this example, we can guess that the (parallel) reductions, which correspond to shared reductions, contract at each step residuals of a single redex. Remark that this redex may not be on the considered reduction. In the third step of the example, the two (Iy) redexes are only residuals of a single (Iy) redex when the two first steps are permuted. This means that we shall consider duplications (or residuals) modulo the permutation equivalence on reductions.

For this, we relativize redexes to their history, i.e. to the reductions which permit to get them. Formally, we write the pair (ρ,R) for meaning that R is a redex in the final expression N of $\rho:M\overset{*}{\to}N$. Now, for two coinitial reduction ρ and σ and two corresponding redexes (ρ,R) and (σ,S), the *duplication* relation (*modulo permutations*) can be defined by

$(\rho,R)\leq(\sigma,S)$ iff $\exists\tau\cdot\ \rho\tau\equiv\sigma$ and $S\in R/\tau$

By remembering that $\rho\equiv\sigma$ implies $R/\rho=R/\sigma$ and by using the laws of $\equiv$ and $\underline{[}$, a list of properties of $\leq$ can again be done by easy algebraic manipulations.

$(\rho,R)\leq(\sigma,S)$ iff $(\rho',R)\leq(\sigma',S)$, when $\rho\equiv\rho'$ and $\sigma\equiv\sigma'$,

$(\rho,R)\leq(\sigma,S)$ iff $\rho\underline{[}\sigma$ and $S\in R/(\sigma/\rho)$, ($\leq$ is decidable)

$(\rho,R)\leq(\rho,R)$,

$(\rho,R)\leq(\sigma,S)\leq(\tau,T)$ implies $(\rho,R)\leq(\sigma,T)$,

$(\rho,R)\leq(\sigma,S)\leq(\rho,R)$ iff $\rho\equiv\sigma$ and $R=S$.

The duplication relation can be closed by symmetry and transitivity. Thus redexes R and S with histories ρ and σ are in a same *redex family*, written $(\rho,R)\sim(\sigma,S)$, iff

1) either $(\rho,R)\leq(\sigma,S)$,
2) or $\quad(\sigma,S)\leq(\rho,R)$,
3) or $\quad\exists(\tau,T)$ such that $(\rho,R)\sim(\tau,T)\sim(\sigma,S)$.

Again by easy manipulations, it can be proved that :

$(\rho,R)\sim(\sigma,S)$ iff $(\rho',R)\sim(\sigma',S)$, when $\rho\equiv\rho'$ and $\sigma\equiv\sigma'$,

$(0,R)\sim(\sigma,S)$ iff $S\in R/\sigma$ (remember 0 is the empty reduction).

But, further properties of redexes families are more complicated, mainly because one needs to look inside expressions. For instance, (ρ,R) and (σ,S) seem to be in a same family iff R and S are "created in a same way" along ρ and σ. But creation of redexes has not been yet expressed in simple terms in the λ-calculus. However, this can be done [Vu, Be4, Le1, Le2] by extracting along any reduction ρ the steps necessary for getting a redex R in the final expression of ρ, or by giving names to redexes and a corresponding law for generating new names. These tedious methods show that $(\rho,R)\sim(\sigma,S)$ is decidable.

An easier property is to prove the existence of a *canonical* element in each family class. Let ρ and σ be two coinitial reductions. Then say that σ is to the right of ρ iff the standard reductions ρ_s and σ_s of ρ and σ are of the form $\rho_s=\tau\rho'$ and $\sigma_s=\tau\sigma'$ with the first redex contracted in σ' being internal or to the right of the first redex contracted in ρ'.

<u>Lemma 4.1</u>. In each redex family class, there is an element (ρ,R), unique up to $\equiv$, such that ρ is maximum to the right and minimum for $\underline{\sqsubset}$.

This canonical element (ρ,R) can also be characterised as the only one in the class such that the length of the associated standard reduction ρ_s is minimum. Furthermore, when (ρ,R) is canonical and $\rho\underline{\sqsubset}\sigma$, one has $(\rho,R)\sim(\sigma,S)$ iff $(\rho,R)\leq(\sigma,S)$, (thus generalising the previously mentioned case when ρ=0). These families properties can be summarised on the example of figure 2.

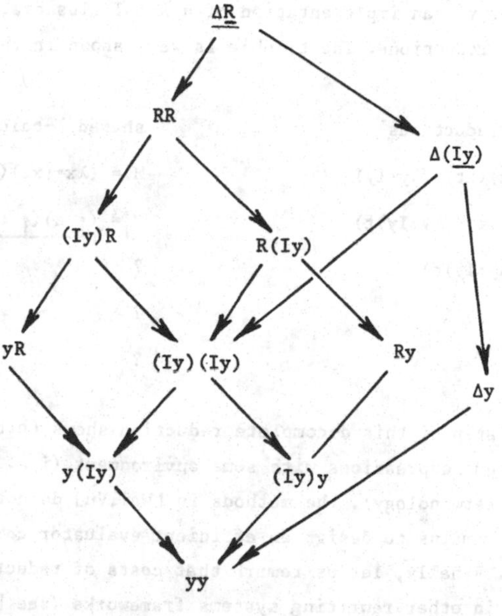

<u>Figure 2</u> ($\Delta=\lambda x\cdot xx, I = \lambda x\cdot x, R = (\lambda x\cdot xy)I$ and the three canonical redexes corresponding to the three redexes families are underlined).

Now, let us also use the notation (ρ,F) for meaning that F is a set of redexes (occurrences) in the final expression N of $\rho:M\overset{*}{\to}N$. Then, the set F with history ρ is a *set of duplications of a single redex*, if there exists a redex S with history σ such that $(\sigma,S)\leq(\rho,R)$ for all $R\in F$. We say too that F is complete for duplications, in short *d-complete* , iff F is a maximum set of duplications of a single redex. Similarly, an *f-complete* set F of redexes with history ρ is a maximum set such that (ρ,R)~(σ,S) for every R,S∈F. Now, d-complete (parallel) reductions can be defined as contracting at each step a d-complete set of redexes. Similarly for f-complete reductions. It is not hard to prove that f-complete and d-complete reductions coincide. Thus, as the family relation is decidable, the d-complete reductions are effective in some sense. Furthermore, the d-complete reduction form a sublattice of the reductions lattice. Finally, by associating needed redexes and d-complete reductions, the following optimality result can be stated.

<u>Theorem 4.2.</u> Any d-complete reduction, which contracts a needed redex at each step, reaches the normal form in a minimum number of steps.

Thus, the d-complete reduction which contracts the leftmost outermost redex is optimal. However, remark that our cost measure unit is the simultaneous contraction of a set of duplications of one single redex. One may not trust in it, mainly because there is not yet an implementation of a λ-calculus evaluator corresponding to the d-complete reductions. The trouble is well shown in the following example :

d-complete reductions	shared λ-calculus
$M = (\lambda x\cdot(xz)(xt))(\lambda y\cdot Iy)$	$M = (\lambda x\cdot(xz)(xt))(\lambda y\cdot Iy)$
$M_1 = ((\lambda y\cdot Iy)z)((\lambda y\cdot Iy)t)$	$M_1' = (\;\;z)(\;\;t) \rightarrow \lambda y\cdot Iy$
$M_2 = (Iz)((\lambda y\cdot Iy)t)$	?
$M_3 = z((\lambda y\cdot y)t)$	?
$M_4 = zt$	?

The second step of this d-complete reduction shows that we can no more share subexpressions, but expressions with some environment (i.e. closures in the programming languages terminology). The methods in [Wa1,Vu] do not treat the previous example, and it remains to design an efficient evaluator corresponding to d-complete reductions. Finally, let us remark that costs of reductions can be studied in a similar way in other rewriting systems frameworks (see [Vu, Be4, Hu2]).

<u>5. Approximations Bohm's trees</u>

We now consider more extensional properties. Instead of looking at reductions, one can study the behaviour of λ-expressions on the outside world. For this, let a *context* C[ ] be a λ-expression with one missing subexpression, and C[M] be the expression obtained when this subexpression is M. (Remark : free variables of M may become bound in C[M]). Now, we can say that M is *totally undefined* iff, for all contexts C[ ] and λ-expressions N, when C[M] has a normal form, then C[N] has also a normal form. Hence, if M is totally undefined, when there is a non termination possibility for C[N], then C[M]does not terminate too. Fortunately, there is an internal characterization of totally undefined λ-expressions [Wal, Bal]. Let *a head normal form* (in short *hnf*) be any expression of the form $\lambda x_1 x_2 \cdots x_m \cdot x M_1 M_2 \cdots M_n$, where $m,n \geq 0$ and $x_1, x_2, \cdots, {}_m, x \in V$ and $M_1, M_2, \cdots M_n \in \Lambda$. Furthermore, let us say that M *has a hnf* iff there is a hnf N such that $M \overset{*}{\rightarrow} N$. Then M is totally undefined iff M has not a hnf. (Remark : totally undefined is also called unsolvable in [Bal]).

Thus, we have new objects to consider : the head normal forms. We can remark that hnfs have a behaviour similar to normal forms. For instance, if M has a hnf, the normal reduction (see §3) always reaches a (minimal) hnf. Moreover, although an expression can have several hnfs, they have all the same "first level structure". This suggests that we can decompose any λ-expression M in its hnf structure. In the literature,this is called the *Bohm's tree* of M.We shall not define it properly here, because one needs to define infinite λ-expressions. But the following (unproper) definition will be enough speaking. The Bohm's tree of M, written BT(M), is such that :

$BT(M) = \Omega$ if M has no hnf,

$BT(M) = \lambda x_1 x_2 \cdots x_m \cdot x(BT(M_1))(BT(M_2)) \cdots (BT(M_n))$ if $M \overset{*}{\rightarrow} \lambda x_1 x_2 \cdots x_m \cdot x M_1 M_2 \cdots M_n$.

Hence, a Bohm's tree can be infinite. For instance $BT(Y) = \lambda f \cdot f(f(f(\cdots)))$ (see introduction for Y). The prefix ordering ≤ and the glb operator ⊓ for λ-Ω-expressions can be extended to Bohm's trees in an obvious way. For instance, $BT(M) \leq BT(N)$ iff BT(M) matches BT(N) except in some Ω's. We use letters a,b,c··· for finite Bohm's trees, i.e. a λ-Ω-expression whose all subexpressions (except Ω's) are in hnf. Finally, when $a \leq BT(M)$, the finite Bohm's tree a will be called a *finite approximation* of M.

<u>Theorem 5.1</u> (The stability theorem). If $M \uparrow N$, then $BT(M \sqcap N) = BT(M) \sqcap BT(N)$.

Again, as a corollary, for any λ-Ω-expression M and for any finite approximation a, there is a (unique) minimum prefix N of M such that $a \leq BT(N)$. This theorem is proved in [Bel]. Sequentiality can also be expressed for Bohm's trees. Let M be a λ-Ω-expression and u an occurrence of some Ω in M. Write $M <_u N$ for meaning $M \leq N$ and that the subexpression of occurrence u in N is not any longer Ω. Thus $M <_u N$ means that N increases at least at occurrence u. This notation can also be extended to infinite Bohm's trees.

<u>Theorem 5.2</u> (The sequentiality theorem). Let M be a λ-Ω-expression and v an occurrence of an Ω in BT(M). If there is an N such that $M \leq N$ and $BT(M) <_v BT(N)$, then there is an occurrence u of an Ω in M such that, for all $N \geq M$, $BT(M) <_v BT(N)$ implies $M <_u N$.

Thus, BT is a sequential function in the sense of [KP]. This means that we can evaluate BT(M) in a coroutine-like way (see also [KP]). Finally, application and λ-abstraction may be shown continuous with respect to Bohm's trees.

<u>Theorem 5.3</u> (The continuity theorem). Let C[ ] be any context, M any expression. Then, for any b≤BT(C[M]), there is an a≤BT(M) such that b≤BT(C[a]).

As a corollary, Bohm's trees can be considered as a semantics for λ-expressions, since $M \overset{*}{\to} N$ implies BT(M)=BT(N), and that BT(M)=BT(N) implies BT(C[M])=BT(C[N]) for any context C[ ]. One may wonder whether these three theorems are true in the usual models of the λ-calculus. In fact, the only continuity property always holds. Stability and sequentiality are for instance not true in the D_∞ model [Sc1]. A model, in which every function is stable, is built in [Be3]. Finally, a proof of these last theorems can be found respectively in [Be1, Le1, Wa2].

Finally, by writing $\overset{*}{\underset{\eta}{\to}}$ for η-reduction and by composing the $\overset{*}{\underset{\eta}{\to}}$ and ≤ binary relations, the following table establishes the correspondance between Bohm's trees and models of the λ-calculus :

$M \sqsubseteq_{T^\omega} N$ iff $\forall a \le BT(M)\ \exists b \le BT(N)$ s.t. $a \le b$, (i.e. BT(M)≤BT(N)) [Ba3]

$M \sqsubseteq_{P\omega} N$ iff $\forall a \le BT(M)\ \exists b \le BT(N)$ s.t. $a \overset{*}{\underset{\eta}{\leftarrow}} \le b$ [Hy1,Ba3]

$M \sqsubseteq_{Mo} N$ iff $\forall a \le BT(M)\ \exists b \le BT(N)$ s.t. $a \overset{*}{\underset{\eta}{\leftarrow}} \le \overset{*}{\underset{\eta}{\to}} b$ [Hy1]

$M \sqsubseteq_{D_\infty} N$ iff $\forall k \ge 0\ \forall a \le BT(M)\ \exists b \le BT(N)$ s.t. $a \overset{*}{\underset{\eta}{\leftarrow}} \le^k \overset{*}{\underset{\eta}{\to}} b$. [Hy1,Wa2]

By $a \le^k b$, we means that a matches b until the k-level except in some Ω's, i.e. :

$a \le^0 b$ for all a and b,

$\Omega \le^k b$ for all b and k≥0,

$\lambda x_1 x_2 \cdots x_m \cdot x a_1 a_2 \cdots a_n \le^{k+1} \lambda x_1 x_2 \cdots x_m \cdot x b_1 b_2 \cdots b_n$ if $a_i \le^k b_i$ for $1 \le i \le n$.

The models T^ω, Pω, D_∞ are described in [P13,P11,Sc2,Sc1]. By $\sqsubseteq_{Mo}$, we mean the extensional relation in [Mo1] defined by:

$M \sqsubseteq_{Mo} N$ iff $\forall$C[ ] nf(C[M])=*tt* implies nf(C[N])=*tt*.

Furthermore, one corresponding Bohm's theorem [Hy1,Wa2] can also be proved for D_∞:

$M \sqsubseteq_{D_\infty} N$ iff $\forall$C[ ] hnf(C[M])=*tt* implies hnf(C[N])=*tt*

where hnf(M)=*tt* means that M has a hnf. (This kind of theorem should also be true in T^ω and Pω if one adds δ-rules, because otherwise it is impossible to distinguish η-interconvertible expressions with the only β-rule.)

<u>Acknowledgments</u>: to G.Huet for permitting to use notions and notations of [Hu2].

Bibliography

[Ba1] Barendregt H.,Some extensional terms models for combinatory logics and lambda-calculi,PhD Thesis,Utrecht,1971.

[Ba2] Barendregt H.,The lambda-calculus,Its syntax and semantics,North-Holland,Amsterdam,to appear.

[Ba3] Barendregt H.,Longo G.,Equality of lambda terms and recursion theoretic reducibility in the model T^{ω},University of Utrecht,Math.report n°107,1979.

[Be1] Berry G.,Séquentialité de l'évaluation formelle des lambda-expressions,3ème Colloque international sur la programmation,Paris,1978,Dunod éditeur.Rapport IRIA-LABORIA n°304.

[Be2] Berry G.,Stable models of typed lambda-calculi,5th ICALP Conf.,Udine,1978.Springer Verlag LNCS n°62,p.72-90.

[Be3] Berry G.,Modèles complètement adéquats et stables des lambda-calculs typés, University of Paris 7,Thèse d'état,1979.

[Be4] Berry G.,Lévy J-J.,Minimal and optimal computations of recursive programs, JACM,vol 26,n°1,Jan 1979,p.148-175.

[Bo1] Bohm C.,Alcune proprieta della forme $\beta\eta$-normali del $\lambda\beta\kappa$-calcolo,Publicazioni dell' istituto per le applicazioni del calcolo,n°696,Rome,1968.

[Bo] Boudol G.,Sur la sémantique des shémas de programmes récursifs non-déterministes,1st colloque AFCET-SMF de mathématiques appliquées,Ecole polytechnique,Palaiseau, 1978.

[Ch1] Church A.,The calculi of lambda-conversions,Annals of Math.Studies n°6,Princeton University Press,1941.

[Cu1] Curry H.B.,Feys R.,Combinatory logic,Vol 1,North-Holland,1958.

[Cu2] Curien P-L.,Algorithmes séquentiels sur structures de données concrètes,Thèse de 3ème cycle,University of Paris 7,1979.

[Hi1] Hindley R.,Reductions of residuals are finite,Transactions of american math. soc.,vol 240,Jun 1978.

[Hu1] Huet G.,Confluent reductions:Abstract properties and applications to term rewriting systems,18th FOCS Conf.,1977.

[Hu2] Huet G.,Lévy J-J.,Computations in non-ambiguous linear term rewriting systems, to appear as IRIA-LABORIA report,1979.

[Hy1] Hyland M.,A syntactic characterisation of equality in some models of the lambda calculus,Journal of London math.soc. 12,2,1976,p.361-370.

[KP] Kahn G.,Plotkin G.,Structures de données concrètes,IRIA-LABORIA report n°336, 1978.

[Le1] Lévy J-J.,Réductions correctes et optimales dans le lambda-calcul,Thèse d'état, University of Paris 7,1978.

[Le2] Lévy J-J.,Le problème du partage dans l'évaluation des lambda-expressions,1st Colloque AFCET-SMF de math.appliquées,Ecole polytechnique,Palaiseau,1978.IRIA-LABORIA report n°325.

[Le3] Lévy J-J.,An algebraic interpretation of the lambda calculus and an application of a labelled lambda calculus,TCS,vol 2,n°1,1976.IRIA-LABORIA report N°103.

[Mi1]Milner R.,Implementation and applications of Scott's logic for computable functions,ACM-SIGACT Conf.on proving assertions about programs,Las Cruces,1972.

[Mi2]Milner R.,Fully abstract models of typed lambda-calculi,TCS,vol 4,n°1,1977.

[Na] Nakajima R.,Infinite normal forms in the lambda-calculus,Lambda calculus and Computer science theory,Springer Verlag,LNCS n°37,1975.

[Ni] Nivat M.,On the interpretation of recursive program schemes,Symposia mathematica vol 15,Istituto nazionale di alta matematica,1975,p.225-281.

[Pl1]Plotkin G.,A set theoritical definition of application,University of Edinburgh, AI memo. MIP-R-95,1972.

[Pl2]Plotkin G.,LCF considered as a programming language,TCS,vol 5,n°3,1977,p.223-256

[Pl3]Plotkin G.,T^{ω} as a universal domain,JCSS,vol 17,n°2,Oct 1978.

[Sc1]Scott D.S.,Continuous lattices,Oxford PRG 7,1971.Springer Verlag,LNM,n°274,1971.

[Sc2]Scott D.S.,Data types as lattices,SIAM Journal on computing,vol 5,n°3,Sept 1976.

[Vu] Vuillemin J.,Proof techniques for recursive programs,PhD thesis,Stanford,1973.

[Wa1]Wadsworth C.,Semantics and pragmatics of the lambda-calculus,PhD thesis,Oxford, 1971.

[Wa2]Wadsworth C.,The relation between computational and denotational properties for Scott's D_{∞} model of the lambda-calculus,SIAM Journal on computing,vol 5,n°3,Sept 1976

[We] Welch P.,Continuous semantics and inside-out reductions,Lambda-calculus and computer science theory,Springer Verlag,LNCS n°37,1975.

ON RATIONAL EXPRESSIONS REPRESENTING INFINITE RATIONAL TREES : APPLICATION TO THE STRUCTURE OF FLOW CHARTS

Guy Cousineau and Maurice Nivat

Laboratoire d'Informatique Théorique et Programmation, Paris

0. INTRODUCTION

The idea to associate to a flow-chart or non recursive program scheme an infinite tree obtained by development of the flow-chart is an ancient one : it is suggested if not formalized in [14, 16, 21, 25]. The earliest neat mathematical formulation is [29]. This tree is also the solution of a system of tree equations which can easily be written using the labels in the flow-chart represented by an ordinary Algol-like program with go to statements [7, 23, 25] it happens that this tree is a rational infinite tree where we define such a tree by the fact that its set of subtrees is finite : an alternative characterization is that a tree is rational if and only if it is the limit of a sequence of finite trees $\{t_n \mid n \in N\}$ whose elements form a rational subset of the set of finite trees (in the sense of Donner [10], the limit can be taken either in the natural metric topology on the set of all finite and infinite trees [22] or as a least upper bound in the structure of complete partial order defined on the same set [29]).

Clearly two flow-charts with the same infinite tree are equivalent in the most obvious sense already introduced in [13, 28]. Also two expressions representing the same infinite tree will be two equivalent ways of expressing the same flow-chart : one standard way of expressing a flow chart is the set of labeled instructions and conditional go to statements which is immediately translated into a system of tree equations, but long ago and ever since many people have been interested in using to express the same flow-chart other language constructs such as while repeat and exit statements, do loops etc... [1, 3, 4, 11, 17, 18, 19, 20]. What is expected is to get more compact, or more easily readable programs to express the same flow-chart : and indeed, it should be possible to define the same tree from this different expression.

The analogy of the while construct with the Kleene's star which is one of the basic operations in the theory of rational languages has lead to several studies [3, 4, 16, 17, 18, 19, 20] which have mainly proved that the while statement is not powerful enough to build all infinite rational trees. Several authors introduced then the "iteration with multiple exits" [1, 18, 20, 24, 27] which is indeed powerful enough in that sense that every infinite rational tree can be represented by a finite expression containing the basic operations and predicates, the composition, and the iteration (repeat statement).

In order to prove algebraically this result the first author of the present paper introduced in [8] the set of trees with indexed leaves which we describe below. The indices which appear as labels of the leaves are to be interpreted as exit statements, and if one adds in the set of basic operations the star (which corresponds to iteration) and the product (which corresponds to composition) the resulting set of trees $M(F \cup \mathbb{N} \cup \{*, \cdot\})$ is no longer a free-magma : on the contrary a set of reduction rules can be given which allows to transform any finite tree $t \in M(F \cup \mathbb{N} \cup \{*, \cdot\})$ into an infinite tree $T = \varphi(t)$ which does not contain $*$ or $\cdot$, that is an element of $M^{\infty}(F \cup \mathbb{N})$. The main theorem is that, given a system of tree equations of a special type which we call regular, there exists an algorithm to build a finite tree $t \in M(F \cup \mathbb{N} \cup \{*, \cdot\})$ such that $\varphi(t)$ is the solution of the system (this theorem is also the main one in [9]).

This is immediately interpreted as an algorithm to transform an Algol-like program with go to statements into an iterative program containing no go to's but instead iterations and answers the question raised by Arsac [1, 2]. The interesting fact about the set of trees with indexed leaves is that one can define a natural product and a natural star operation, this being not possible in the ordinary set of trees. The formal analogy with the product and star defined on the subsets of a free monoid is however very formal as proved by the consideration of the set of branches of the trees and the operations on these sets of branches corresponding to the tree product and tree star [23 bis, see also 6].

A number of problems arise if one considers non regular systems of equations, whose solutions have been proved by C. Henry to be algebraic trees (as defined and studied by B. Courcelle [6]).

This is a survey paper mainly based on previous work by the authors [8, 9, 23, 23 bis] students of them [5, 12], and by several people with whom they have been working in close connection [1, 2, 7, 24, 26, 27].

1. REGULAR TREES

We use the same notations as Courcelle-Nivat [6 bis]. Given an alphabet X with arity, $M_{\Omega}^{\infty}(X)$ [resp. $M_{\Omega}(X)$] denotes the set of trees [resp. finite trees] that can be built with the symbols in X and a special 0-ary symbol Ω in accordance with the arities. The infinite trees will be denotes by capital letters $T_1, T_2, \ldots$ and finite trees by small letters $t_1, t_2, \ldots$ For various purposes, it is very convenient to present these trees as mappings : $\mathbb{N}_+^* \to X \cup \{\Omega\}$ [10].

The order $<$ is defined by :

$$T_1 < T_2 \text{ iff } \operatorname{dom}(T_1) \subseteq \operatorname{dom}(T_2) \text{ and}$$

$$\forall m \in \operatorname{dom}(T_1) \quad T_1(m) \neq \Omega \Rightarrow T_2(m) = T_1(m)$$

It can be shown that $M_\Omega^\infty(X)$ with the free operations corresponding to elements of X and the order $<$ is the free complete ordered X-magma (or X-algebra).

A central notion in this paper is the notion of sub-tree : the subtree T/m of root m in the tree T.

$$\operatorname{dom}(T/m) = \{m' \in N_+^* \,/\, m\,m' \in \operatorname{dom}(T)\} \text{ and } (T/m)\,(m') = T(m\,m')$$

A tree T is <u>regular</u> if the set of its subtrees is finite. Regular trees are exactly the solutions of regular systems of equations. Such a system is of the form $\alpha_1 = \tau_1, \ldots, \alpha_n = \tau_n$ where $\alpha_1, \ldots, \alpha_n$ are 0-ary variables and $\forall i \in [n]$ $\tau_i \in M_\Omega(X \cup \{\alpha_1, \ldots, \alpha_n\})$.

At that this point, a comparison with languages can be useful. To the alphabet X is associated $\bar{X}$ which contains letters $x_1, \ldots, x_n$ for each n-ary symbol x in X. So every tree $T \in M_\Omega^\infty(X)$ has a branch language $Br(T) \subseteq \bar{X}^*$ which can be defined in the following way : to each element $m = i_1 \ldots i_k$ in dom (T) is associated the word $br(m) = x_{i_1}^{(1)} \ldots x_{i_k}^{(k)}$ where $T(\varepsilon) = x^{(1)}$ and $\forall j \;\; 1 \leq j \leq k-1$ $T(i_1 \ldots i_j) = x^{(j+1)}$ and $Br(T) = \bigsqcup_{m \in \operatorname{dom}(T)} br(m)$.

We have the following facts :

- $\forall m \in \operatorname{dom}(T) \quad Br(T/m) = Br(T) \,/\, br(m)$ where for any language L and word $u \in L$, L/u denotes the quotient of L by u i.e. $\{v \,/\, u v \in L\}$.

- $(T_1, \ldots, T_n)$ is the solution of the system $\alpha_i = \tau_i$, $i = 1, \ldots, n$ iff $(Br(T_1), \ldots, Br(T_n))$ is the solution in $(2^{\bar{X}^*})^n$ of the system $\alpha_i = Br(\tau_i)$, $i = 1, \ldots, n$ which is a right linear algebraic grammar.

It follows immediatly that the branch languages of regular trees are exactly those regular languages L which are prefix-closed and complete in the sense that whenever $u x_i v \in L \subseteq \bar{X}^*$ where x is an n-ary symbol of X, for each $j \in [n]$ there exists some v' such that $u x_j v' \in L$ [6, 23 bis].

Now we would like to extend this comparison to regular expressions and find a way to solve regular systems of equations on $M_\Omega^\infty(X)$. For that purpose, we shall have to particularize our tree domains and introduce the notion of trees with indexed leaves.

2. <u>TREES WITH INDEXED LEAVES</u>

This paragraph is mainly taken from [9] where all proofs are to be found.

We assume now that our alphabet X is the disjoint union of F and N where N is the set of integers which will be considered as 0-ary symbols.

On $M_\Omega^\infty(F \cup N)$, we define the following operations :

$$\forall k \in N \quad \uparrow_k : M_\Omega^\infty(F \cup N) \rightarrow M_\Omega^\infty(F \cup N) \qquad \uparrow_k(T) = T[i+1/i,\ i \geq k]$$

$$C_k : M_\Omega^\infty(F \cup N)^2 \rightarrow M_\Omega^\infty(F \cup N) \qquad C_k(T_1, T_2) = T_1\ [T_2 / k]$$

These operations are continuous in the sense of Scott. We shall write $\uparrow$ instead of $\uparrow_0$ and call this operation the positive shift. Similarly, we define a negative shift $\downarrow$ by $\downarrow(T) = T\,[\Omega/0\ ;\ i/i+1]$. We shall also write $T_1.T_2$ instead of $C_0(T_1, T_2)$ and call this operation product. With this product $M_\Omega^\infty(F \cup N)$ is a monoid : indeed the product is associative and has a neutral element 0. We shall define $T^0 = 0$ and $T^{n+1} = T^n.T$.

For any tree T, $\{\downarrow(T^n)/n \in N\}$ is an increasing sequence for the order $<$ and we shall denote $*(T) = \text{lub}\{\downarrow(T^n)/n \in N\}$. The least upper bound of the sequence $\{\downarrow(T^n)\}$ exists since $M_\Omega^\infty(F \cup N)$ is a complete partial order.

<u>Exemple</u> : T = 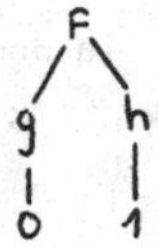

$\downarrow(T^0) = \downarrow(0)$

$\downarrow(T^1) =$ 

$*(T) =$ 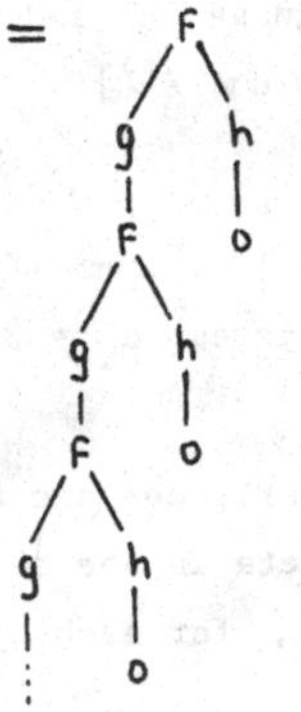

$\downarrow(T^2) =$ 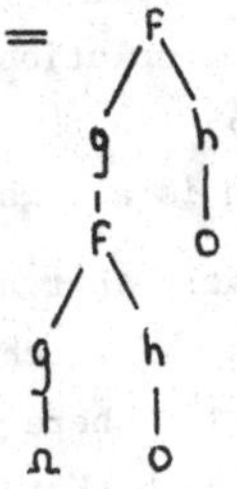

The operations $\uparrow_k$, C_k and $*$ do preserve regular trees. Thus every expression E in $M_\Omega(F \cup N \cup \{*, c_k, \uparrow_k\}$ represents a regular tree T in $M_\Omega^\infty(F \cup N)$ which we denote $\varphi(E)$. Moreover, $\uparrow_k$, c_k, $*$ are related by the following properties :

<u>Property</u> : $\forall T \quad \forall k \quad \uparrow_k(*(T)) = *(\uparrow_{k+1}(T))$

<u>Property</u> : $\forall\, T_1, T_2 \quad \forall k \quad C_k\,(*(T_1), T_2) = *(C_{k+1}(T_1, \uparrow(T_2)))$.

The set formed of the equations which express the fact that c_k and $\uparrow_k$ are morphisms on $M^{\infty}_{\Omega}(F \circ \mathbb{N})$, the intuitive simplifications rules $C_k(k, T) = T$, $C_k(k', T) = k'$ for $k' \neq k$, $\uparrow_k(k') = k'+1$ for $k' \geq k$, $\uparrow_k(k') = k'$ for $k' < k$, is a (infinite) confluent reduction system as defined in [5] : any expression built with $\uparrow_k$, c_k, $*$ as operations and trees as variables can be transformed by a sequence of reduction into a normal form containing only the $*$ operation which is equivalent to the first one in the sense that it denotes the same tree. The operations $\uparrow_k$ and c_k are thus redundant.

An example of reduction is :

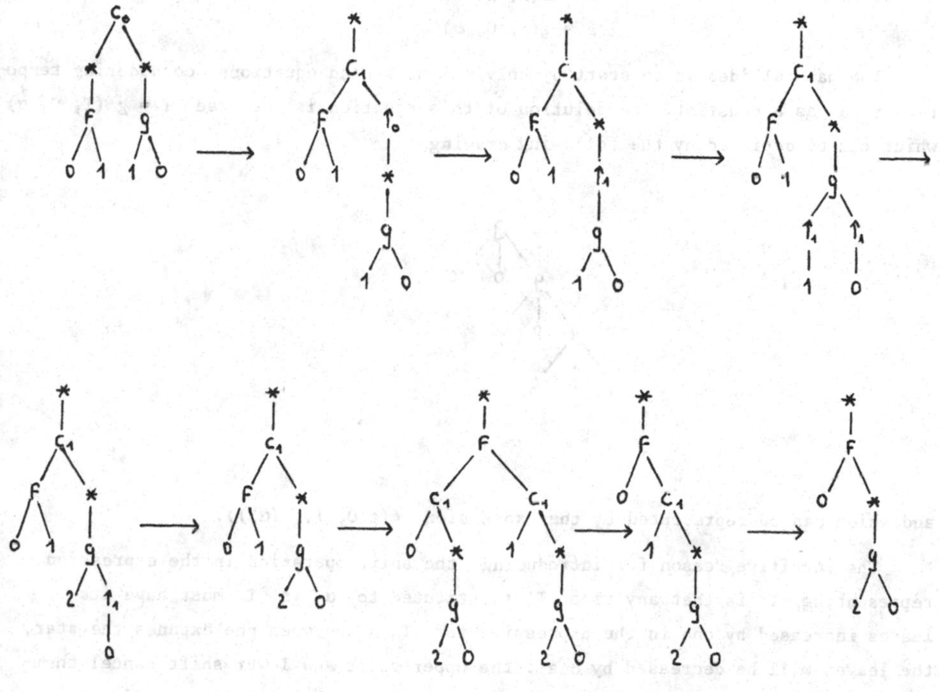

3. <u>SOLVING REGULAR EQUATIONS</u>

The main property established in [8] which enables us to solve regular equations is the following : Let M be a subset of $\mathrm{dom}(T)\setminus\{\varepsilon\}$. Then

<u>Theorem</u> : $\forall m \in M \quad T/m = T$ implies $T = *((\uparrow(T))\,[M \leftarrow 0])$.

where we denote by $T[M \leftarrow 0]$ the tree obtained by substituting 0 to every subtree of root $m \in M$.

Concretely, let us be given an equation $\alpha = \tau$ with $\tau \in M^{\infty}(F \cup N \cup \{\alpha\})\setminus\{\alpha\}$. It has a unique solution $T = \tau[T/\alpha]$ and this solution is equal to the tree denoted by $*(\tau')$ where τ' is obtained from τ by increasing every integer and replacing all occurrences of α by 0.

<u>Example</u> : $\alpha = f(\alpha, g(0, \alpha))$
implies $\alpha = *(f(0, g(1, 0)))$.

It is not entirely trivial to extend this way of solving an equation to the problem of solving a system of equations. Let us consider the following system :

$$\alpha = f(\alpha, \beta, 0)$$
$$\beta = g(\beta, 0, \alpha)$$

The natural idea is to start by solving the second equations, considering temporarily α as a constant. The solution of this equation is the tree $T = g(T, 0, \alpha)$ which can be depicted by the following drawing

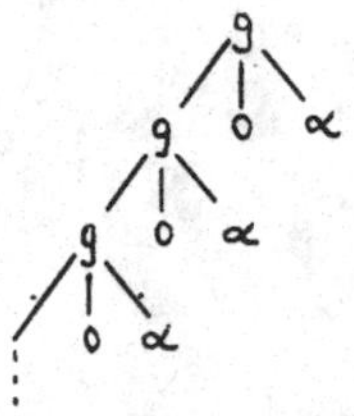

and which can be represented by the expression $*(g(0, 1, \uparrow(\alpha)))$.

The intuitive reason for introducing the shift operation in the expression representing T is that any tree T' substituted to α in T must have its leaves increased by one in the expression for T since when one expands the star, the leaves will be decreased by one : the upper shift and lower shift cancel themselves so that eventually it is really T' which is substituted to α in T.

Now we substitute the result T in the first equation and obtain :

$$\alpha = f(\alpha, T, 0)$$

Pictorially this gives

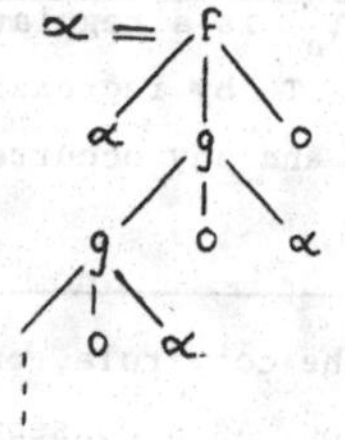

This equation in α has a unique solution which is the tree T' represented by the (infinite) expression :

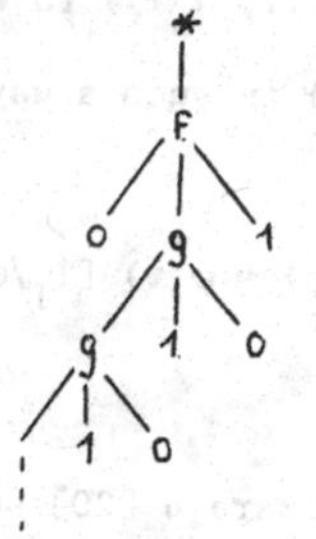

It is easy to see that the equation $\alpha = f(\alpha, T, 0)$ can also be written

$$\alpha = f(\alpha, *(g(0, 1, \uparrow(\alpha))), 0)$$

since T is represented by $*(g(0, 1, \uparrow(\alpha)))$ and its solution T' represented by $*(f(0, *(g(0, 2, 1)), 1)$ (it suffices to compute this expression).

The way we obtained this expression from equations will be now precisely described. We shall call <u>regular expression</u> any element of $M(F \cup \{*\} \cup N)$ and <u>regular τ-expression</u> any element of $M(F \cup \{*\} \cup N \cup M(\{\uparrow\} \cup \{\alpha_1, \ldots, \alpha_n\}))$ (where elements of $M(\{\uparrow\}, \{\alpha_1, \ldots, \alpha_n\})$ have arity 0) which respects some restrictions on the places where shift operations occur (these restrictions will be given in the sequel). These expressions can be considered as trees and described by mappings with domain in N_+^*.

Given an expression E and $m \in \mathrm{dom}(E)$, the <u>depth</u> $\delta(m, E)$ of point m in E is the number of left-factors m' of E such that $E(m) = *$. An integer leaf m of a regular expression or regular τ-expression E is said to be a <u>terminal sign</u> of E iff $E(m) \geq \delta(m, E)$.

The restriction we put on regular τ-expressions E is that whenever $E(m) = \uparrow^k(\alpha_i)$, k is equal to $\delta(m, E)$.

Now the resolution rule can be fully stated :

Given an equation $\alpha_n = \tau_n$ where τ_n is a regular τ-expression the solution is $*(\tau')$ where τ' is obtained from τ by increasing every terminal sign, replacing all occurrences of α_n by 0 and any occurrence of $\uparrow^k(\alpha_i)$ with $i \neq n$ by $\uparrow^{k+1}(\alpha_i)$.

Using this rule together with the copy rule, one can solve any regular system of equations into a regular expression. As a consequence we have the following.

Theorem : Regular expressions denote exactly the regular trees.

The deux ex machina is that regular τ-expressions are exactly those expressions τ in $M_\Omega(F \cup \{*\} \cup N \cup M(\{\uparrow\} \cup \{\alpha_1,\ldots, \alpha_n\}))$ to which we can associate a regular term $\varphi(\tau) \in M^\infty_\Omega(F \cup N \cup \{\alpha_1,\ldots, \alpha_n\})$ in such a way that φ commutes with substitution to the variables $\alpha_1,\ldots, \alpha_n$:

$$\varphi(\tau[T_1/\alpha_1,\ldots, T_n/\alpha_n]) = \varphi(\tau)\ [T_1/\alpha_1,\ldots, T_n/\alpha_n]$$

4. HIERARCHY RESULTS (adapted from Kosaraju [20] see also [15, 27]

We now show that the classes of regular trees C_n that can be described by an expression τ in $M(F \cup \{*\} \cup \{0, 1,\ldots, n\})$ form a strictly increasing hierarchy. We take $F = \{f_{ij} / 0 \le i,j \le n+1\}$ all symbols in F being binary and the tree T defined by

- $T(\varepsilon) = f_{oo}$
- if $T(m) = f_{ij}$ $T(m1) = f_{i'j}$ where $i' = i+1 \mod n+2$

  $T(m2) = f_{ij'}$ where $j' = j+1 \mod n+2$

We shall use projections π_1 and π_2 : $\pi_1(f_{ij}) = i$ $\pi_2(f_{ij}) = j$

It should be clear that T is the first component of the solution of the system of equations $\alpha_{ij} = f_{ij}(\alpha_{i'j}, \alpha_{ij'})$ and that this system can be solved with only $n+2$ applications of the resolution rule. So, $T \in Cn+1$. (The first application creates a zero 0 and the following can only increase it by one).

We now show that $T \notin C_n$. First observe that T has the two following properties

(1) $\forall m, m' \in N^*_+$ $T/m = T/mm' \Rightarrow |m'| \ge n+1$
and $\exists m'_1,\ldots, m'_{n+2}$ prefixes of m' such that
either $\forall i,j \in [n+2]$ $\pi_1(T(mm'_i)) \neq \pi_1(T(mm'_j))$
or $\forall i,j$ $[n+2]$ $\pi_2(T(mm'_i)) \neq \pi_2(T(mm'_j))$

(2) $\forall m \in dom(T)$ $\forall p \in N$ $\pi_2(T(m1^p)) = \pi_2(T(m))$
$\pi_1(T(m2^p)) = \pi_1(T(m))$

Let E be a regular expression denoting T and $m_o \in dom(E)$ such that $E(m_o) = *$, $E/m_o 1 \in M(F, \mathbb{N})$. We can assume that there exists some m such that $E(m_o 1 m) = 0$. If not, E can be simplified using the rule $*(\uparrow(T)) = T$.

Let $\bar{m}_o$ be $\sigma_E(m_o 1)$ where σ_E is defined as follows :

$$\sigma_E(\varepsilon) = \varepsilon \qquad \sigma_E(i\,m) = \sigma_{E/i}(m) \text{ if } E(\varepsilon) = *$$
$$= i.\sigma_{E/i}(m) \text{ if } E(\varepsilon) \in F$$

We have $\forall m \in dom(E/m_o 1)$

$$\text{if } E(m_o 1 m) \in F \text{ then } T(\bar{m}_o m) = E(m_o 1 m)$$
$$\text{if } E(m_o 1 m) = 0 \text{ then } T/m_o m = T/m_o$$

$\forall m_1, m_2 \in dom(E/m_o 1)$

$$\text{if } E(m_o 1 m_1) = E(m_o 1 m_2) \in N \text{ then } T(\bar{m}_o m_1) = T(\bar{m}_o m_2)$$

Now if $E(m_o 1 m) = 0$, we have $T/\bar{m}_o = T/\bar{m}_o m$ and by property (1) there exists $m_1, \ldots, m_{n+2}$, prefixes of m such that either

$$\forall i,j \in [n+2] \quad \pi_1(T(\bar{m}_o m_i)) \neq \pi_1(T(\bar{m}_o m_j))$$
$$\text{or} \quad \forall i,j \in [n+2] \quad \pi_2(T(\bar{m}_o m_i)) \neq \pi_2(T(\bar{m}_o m_j))$$

Let us consider the first case : let p_i be for each $i \in [n+2]$ the greatest integer such that $m_o 1 m_i 2^{p_i} \in dom(E)$.

We have $\forall i \in [n+2] \quad \pi_1(T(\bar{m}_o m_i 2^{p_i})) = \pi_1(T(\bar{m}_o m_i))$

Therefore $\forall i,j \in [n+2] \quad \pi_1(T(\bar{m}_o m_i 2^{p_i})) \neq \pi_1(T(\bar{m}_o m_j 2^{p_j}))$

hence $T(\bar{m}_o m_i 2^{p_i}) \neq T(\bar{m}_o m_j 2^{p_j})$

hence $E(m_o m_i 2^{p_i}) \neq E(m_o 1 m_j 2^{p_j})$

Then, E must contain $(n+2)$ distinct integers and consequently cannot belong to $M(F \cup \{*\} \cup \{0,\ldots, n\})$.

The proof can be easily extended to show that T cannot be denoted either by an expression $\tau \in M(F \cup \{*, \bullet\} \cup \{0,\ldots, n\})$ in which we allow concatenation. We would have only to change the definition of σ_E and use the fact that if $E/m_o 1 \in M(F \cup \{\bullet\} \cup \{0, 1,\ldots, n\})$, it can be transformed into an equivalent expression in $M(F \cup \{0, 1,\ldots, n\})$ using the rules of section 2.

5. NON-REGULAR EQUATIONS

We have seen two kinds of systems of equations having regular solutions : the regular ones and the systems in which the right member of equations are regular τ-expressions. One can obtain systems having non-regular solutions in many different ways : for example by allowing the use of products in the right members or by deleting the condition we put in the definition of regular τ-expressions. The two following equations have non-regular solutions :

In fact, they both have for solution the tree T defined by :

- dom(T) is the Lukasiewicz language on the alphabet {1, 2}
- T(m) is equal to 0 if m is maximal in dom(T) and f otherwise T can be depicted by the following drawing where single arrows point out subtrees equal to T, double arrows substrees equal to T^2, triple arrows, trees equal to T^3

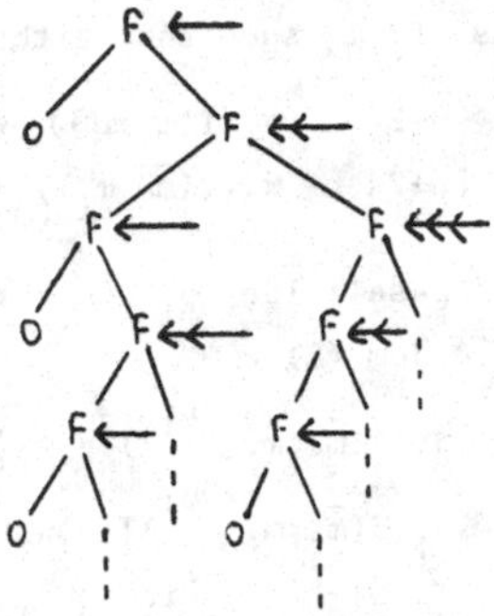

The following result is due to Henry [12].

<u>Theorem</u> : The solution of the systel $\alpha_i = \tau_i$ where

$\tau_i \in M(F \cup \{*, \cdot\} \cup N \cup \{\alpha_1, \ldots, \alpha_n\})$ are exactly the algebraic trees [6].

<u>Note</u> : Algebraic trees are usually defined on some magma $M^{\infty}(F, V)$ where V is a denumerable set of variables that can be put in one to one correspondance with N.

6. APPLICATION TO PROGRAMS

In this paragraph we show how the rule given above solves the problem of translating go to statements into repeat and exit statements.

It has already been mentioned in the introduction that star and product correspond to iteration and concatenation of programs. So any rule used to obtain and modify our rational expressions can be used as program transformations. We shall use a tree representation for programs writing repeat
E for <u>repeat</u> E <u>end</u> ,

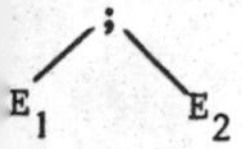

for E_1 ; E_2 if E_1 is a compound statement , 

for a ; E if a is a basic statement and making exit 0 (skip) statements explicit.

For example the program

```
repeat
    if  p  then  a
           else  b ; exit 1
    fi
end
repeat
    if  q  then  c ; exit 1
           else  d
    fi
end
```

is represented by :

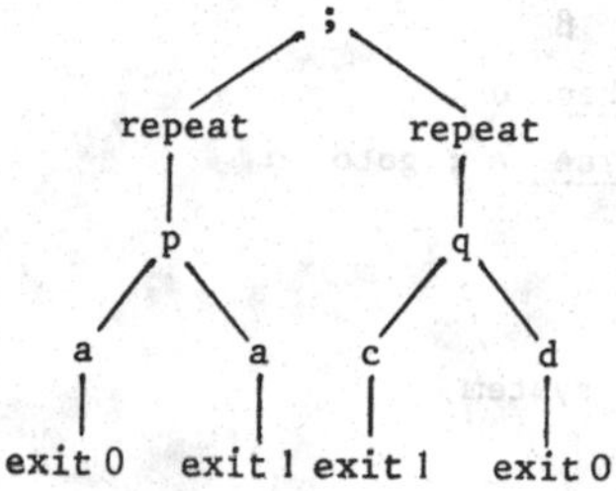

This tree is very similar to the example in section 2 . Applying the same rules we obtain :

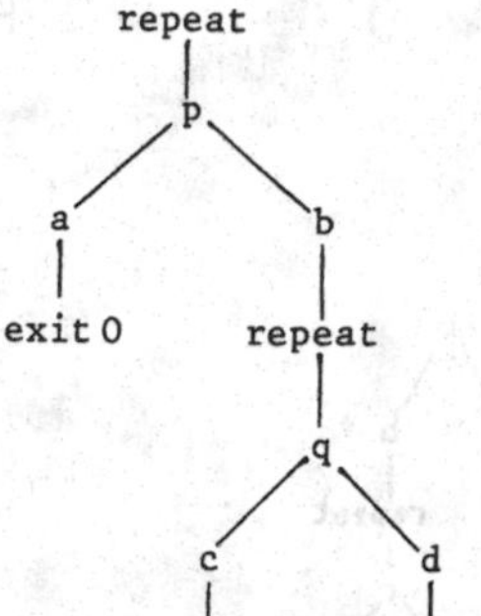

or in more usual notations :

```
repeat
    if  p  then  a
           else  b ;
                 repeat
                     if  q  then  c ; exit 2
                            else  d
                     fi
                 end
    fi
end
```

Similarly, the way we solved equations in section 3 can be used to put programs into iterative form. The program

```
α :  if  p  then  a ; goto  α
             else  b ; goto  β
     fi
β :  if  r  then  c ; goto  β
             else  if  s  then  d
                              else  e ; goto  α
                   fi
     fi
```

which can be represented by the system

and solved in the following way :

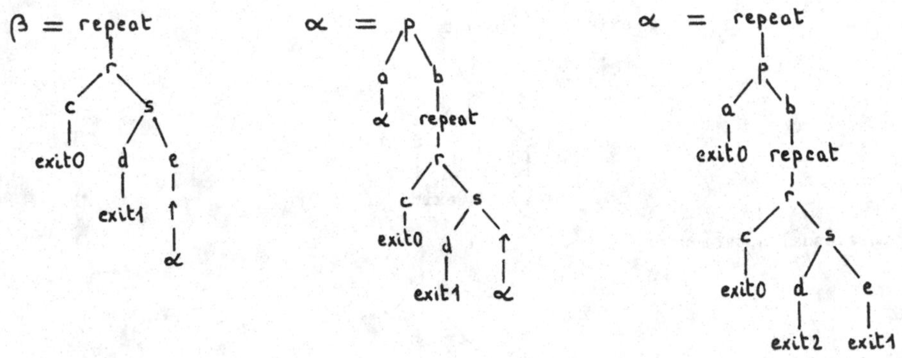

We obtain therefore the following program :

```
repeat  if  p  then  a
|                else  b ;
|                      repeat  if  r  then  c
|                      |              else  if  s  then  d ; exit 2
|                      |                             else  e ; exit 1
|                      |                        fi
|                      |           fi
|                      end
|         fi
end
```

BIBLIOGRAPHY

1. Arsac J., Nolin L., Vasseur J.P. and G. Ruggiu : Le système de programmation structurée EXEL. Revue Thomson-CSF, Vol. 6 (1974).

2. Arsac J. : La construction de programmes structurés. Dunod, Paris (1977).

3. Ashcroft E. and Z. Manna : The translation of goto programs to while programs, in Proc. IFIP Cong, Amsterdam (1971).

4. Böhm C. and G. Jacopini : Flow diagrams, turing machines and languages with only two formation rules, Proc. IFIP Cong. Amsterdam (1971).

5. Casteran P. : Structures de contrôle : définitions opérationnelles et algébriques. Thèse de 3ème cycle, Université Paris VII (1979).

6. Courcelle B. : Recursive schemes, algebraic trees and deterministic languages, in Proc. 15th Symp. Found. Comp. Sci. (1974).

7. Courcelle B., G. Kahn and J. Vuillemin : Algorithmes d'équivalence et de réduction à des expressions minimales dans une classe d'équations récursives simples, in Automata, Languages and Programming (J. Loeckx ed.) LNCS n° 14, Berlin (1974).

8. Cousineau G. : Arbres à feuilles indicées et transformations de programmes. Thèse d'Etat, Université Paris VII (1977).

9 Cousineau G. : An algebraic definition for control structures, to appear in Theoretical Computer Science (1979).

0. Doner J.E. : Tree acceptors and some of their applications, Jour. Comp. Syst. Sci. Vol 4 (1970), pp. 406-451.

1. Ershov A.P. : Theory of program schemata, in Proc. IFIP Cong., Amsterdam (1971).

2. Henry C. : Resolution d'équations algébriques sur le magma libre : application aux transformations de programmes, Thèse de 3ème cycle, Université Paris VII, (1978).

3. Ianov I.I. : The logical scheme of algorithms, english translation in Problems of Cybernetics, Vol. 1 (1960), pp. 82-140.

4. Igarashi S. : An axiomatic approach to the equivalence problem of algorithms, Report of the Computer Center, Univ. of Tokyo, vol. 1 (1969) pp. 1-101.

5. Jacob G. : Structural invariants for some classes of structured programs in Mathematical Foundations of Computer Science, LNCS n° 64, Springer Verlag, Berlin (1978).

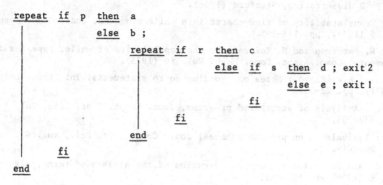

16. Kaplan D.M. : The formal theoretic analysis of strong equivalence for elemental programs, Ph D dissertation, Stanford (1968).

17. Kasai T. : Translatability of flow-charts into while programs. Jour. Comp. Syst. Sci., vol. 9 (1974), pp. 177-195.

18. Kasami T., W. Peterson and N. Tokura : On the capabilities of while, repeat and exit statements. Comm. Assoc. Comp. Mach. Vol. 16 (1973).

19. Knuth D.E. and R.W. Floyd : Notes on avoiding go to statements. Inf. Proc. Letter Vol. 1 (1971), pp. 23-31.

20. Kosaraju R. : Analysis of structured programs, Jour. Compt. Syst. Sci., Vol. 9 (1974) pp. 232-255.

21. Milner R. : Equivalence on program schemes, Jour. Comp. Syst. Sci., vol. 4 (1970) pp. 205-219.

22. Mycielski J. and W. Taylor : A compactification of the algebra of terms, Alg. Univ. Vol. 6 (1976) pp. 159-163.

23. Nivat M.: Langages algébriques sur le magma libre et sémantique des schémas de programmes in Automata, languages and programming (M. Nivat ed.) Amsterdam (1973)

24. Nolin L. and G. Ruggiu : Formalization of EXEL, in Proc. ACM Symp. Princ. Prog. Lang., Boston (1973).

25. Paterson M.S. : Equivalence problems in a model of computation Ph. D dissertation Trinity College, Cambridge (1967).

26. Robinet B. : Un modèle fonctionnel des structures de contrôle, RAIRO Inf. The, Vol. 11 (1977) pp. 213-236.

27. Ruggiu G. : De l'organigramme à la formule, Thèse d'Etat, Université Paris VII, (1973).

28. Rutledge J.D. : On Ianov's program schemata. Journal Assoc. Comp. Mach. Vol. 11 (1964) pp. 1-9.

29. Scott D. : The lattice of flow diagrams, in Symposium on semantics of algorithmic languages (E. Engeler ed.), in Lecture Notes Math. n° 188, Springer Verlag, Berlin (1971).

6 bis. Courcelle B. and M. Nivat : Algebraic families of interpretations, in Proc. 17th Symp. Found. Comp. Sci. (1976).

23 bis. Nivat M. : Chartes, arbres, programmes itératifs. Revue Thomson-CSF, Vol. 10 (1978) pp. 705-731.

Vol. 49: Interactive Systems. Proceedings 1976. Edited by A. Blaser and C. Hackl. VI, 380 pages. 1976.

Vol. 50: A. C. Hartmann, A Concurrent Pascal Compiler for Minicomputers. VI, 119 pages. 1977.

Vol. 51: B. S. Garbow, Matrix Eigensystem Routines – Eispack Guide Extension. VIII, 343 pages. 1977.

Vol. 52: Automata, Languages and Programming. Fourth Colloquium, University of Turku, July 1977. Edited by A. Salomaa and M. Steinby. X, 569 pages. 1977.

Vol. 53: Mathematical Foundations of Computer Science. Proceedings 1977. Edited by J. Gruska. XII, 608 pages. 1977.

Vol. 54: Design and Implementation of Programming Languages. Proceedings 1976. Edited by J. H. Williams and D. A. Fisher. X, 496 pages. 1977.

Vol. 55: A. Gerbier, Mes premières constructions de programmes. XII, 256 pages. 1977.

Vol. 56: Fundamentals of Computation Theory. Proceedings 1977. Edited by M. Karpiński. XII, 542 pages. 1977.

Vol. 57: Portability of Numerical Software. Proceedings 1976. Edited by W. Cowell. VIII, 539 pages. 1977.

Vol. 58: M. J. O'Donnell, Computing in Systems Described by Equations. XIV, 111 pages. 1977.

Vol. 59: E. Hill, Jr., A Comparative Study of Very Large Data Bases. X, 140 pages. 1978.

Vol. 60: Operating Systems, An Advanced Course. Edited by R. Bayer, R. M. Graham, and G. Seegmüller. X, 593 pages. 1978.

Vol. 61: The Vienna Development Method: The Meta-Language. Edited by D. Bjørner and C. B. Jones. XVIII, 382 pages. 1978.

Vol. 62: Automata, Languages and Programming. Proceedings 1978. Edited by G. Ausiello and C. Böhm. VIII, 508 pages. 1978.

Vol. 63: Natural Language Communication with Computers. Edited by Leonard Bolc. VI, 292 pages. 1978.

Vol. 64: Mathematical Foundations of Computer Science. Proceedings 1978. Edited by J. Winkowski. X, 551 pages. 1978.

Vol. 65: Information Systems Methodology, Proceedings, 1978. Edited by G. Bracchi and P. C. Lockemann. XII, 696 pages. 1978.

Vol. 66: N. D. Jones and S. S. Muchnick, TEMPO: A Unified Treatment of Binding Time and Parameter Passing Concepts in Programming Languages. IX, 118 pages. 1978.

Vol. 67: Theoretical Computer Science, 4th GI Conference, Aachen, March 1979. Edited by K. Weihrauch. VII, 324 pages. 1979.

Vol. 68: D. Harel, First-Order Dynamic Logic. X, 133 pages. 1979.

Vol. 69: Program Construction. International Summer School. Edited by F. L. Bauer and M. Broy. VII, 651 pages. 1979.

Vol. 70: Semantics of Concurrent Computation. Proceedings 1979. Edited by G. Kahn. VI, 368 pages. 1979.

Vol. 71: Automata, Languages and Programming. Proceedings 1979. Edited by H. A. Maurer. IX, 684 pages. 1979.

Vol. 72: Symbolic and Algebraic Computation. Proceedings 1979. Edited by E. W. Ng. XV, 557 pages. 1979.

Vol. 73: Graph-Grammars and Their Application to Computer Science and Biology. Proceedings 1978. Edited by V. Claus, H. Ehrig and G. Rozenberg. VII, 477 pages. 1979.

Vol. 74: Mathematical Foundations of Computer Science. Proceedings 1979. Edited by J. Bečvář. IX, 580 pages. 1979.